AutoCAD Electrical 2027 Black Book

By
Gaurav Verma
Matt Weber
(CADCAMCAE Works)

Edited by
Kristen

ISBN # 978-1-77459-210-6

DEDICATION

To teachers, who make it possible to disseminate knowledge
to enlighten the young and curious minds
of our future generations

To students, who are the future of the world

THANKS

To my friends and colleagues

To my family for their love and support

To Jim Pytel, Columbia Gorge Community College for allowing us to reference his YouTube channel on basic concepts of electrical technology in this book

Training and Consultant Services

At CADCAMCAE Works, we provide effective and affordable one to one online training on various software packages in Computer Aided Design(CAD), Computer Aided Manufacturing(CAM), Computer Aided Engineering (CAE), Computer programming languages(C/C++, Java, .NET, Android, Javascript, HTML and so on). The training is delivered through remote access to your system and voice chat via Internet at any time, any place, and at any pace to individuals, groups, students of colleges/universities, and CAD/CAM/CAE training centers. The main features of this program are:

Training as per your need

Highly experienced Engineers and Technician conduct the classes on the software applications used in the industries. The methodology adopted to teach the software is totally practical based, so that the learner can adapt to the design and development industries in almost no time. The efforts are to make the training process cost effective and time saving while you have the comfort of your time and place, thereby relieving you from the hassles of traveling to training centers or rearranging your time table.

Software Packages on which we provide basic and advanced training are:

CAD/CAM/CAE: CATIA, Creo Parametric, Creo Direct, SolidWorks, Autodesk Inventor, Solid Edge, UG NX, AutoCAD, AutoCAD LT, EdgeCAM, MasterCAM, SolidCAM, DelCAM, BOBCAM, UG NX Manufacturing, UG Mold Wizard, UG Progressive Die, UG Die Design, SolidWorks Mold, Creo Manufacturing, Creo Expert Machinist, NX Nastran, Hypermesh, SolidWorks Simulation, Autodesk Simulation Mechanical, Creo Simulate, Gambit, ANSYS and many others.

Computer Programming Languages: C++, VB.NET, HTML, Android, Javascript and so on.

Game Designing: Unity.

Civil Engineering: AutoCAD MEP, Revit Structure, Revit Architecture, AutoCAD Map 3D and so on.

We also provide consultant services for Design and development on the above mentioned software packages

For more information you can mail us at:
cadcamcaeworks@gmail.com

Table of Contents

Chapter 4 : Inserting Components

Chapter 5 : Wires, Circuits, and Ladders

Chapter 6 : Editing Wires, Components, and Circuits

Chapter 7 : PLCs and Components

Chapter 8 : Practical and Practice

Chapter 9 : Panel Layout

Chapter 10 : Reports

Chapter 11 : Project

Chapter 12 : AutoCAD Electrical with Autodesk Inventor

Annexure I : Basics of Electrical System

Annexure II : Basics Components of Electrical Control System

Preface

AutoCAD Electrical 2027 is an extension to AutoCAD package. Easy-to-use CAD-embedded electrical schematic and panel designing enable all designers and engineers to design most complex electrical schematics and panels. You can quickly and easily employ engineering techniques to optimize performance while you design, to cut down on costly prototypes, eliminate rework and delays, and save you time and development costs.

The **AutoCAD Electrical 2027 Black Book**, the 12th edition of AutoCAD Electrical Black book, has been updated as per the enhancements in the AutoCAD Electrical 2027. Following the same strategy as for the previous edition, the book follows a step by step methodology. It covers almost all the information required by a learner to master the AutoCAD Electrical. The book starts with basics of Electrical Designing, goes through all the Electrical controls related tools and discusses practical examples of electrical schematic and panel designing. Chapter on Reports makes you able to create and edit electrical component reports. We have also discussed the interoperability between Autodesk Inventor and AutoCAD Electrical which is need of industry these days. Two annexures have been added to explain basic concepts of control panel designing. More than 550 MCQs make it easy to evaluate your progress. Some of the salient features of this book are :

In-Depth explanation of concepts

Every new topic of this book starts with the explanation of the basic concepts. In this way, the user becomes capable of relating the things with real world.

Topics Covered

Every chapter starts with a list of topics being covered in that chapter. In this way, the user can easy find the topic of his/her interest easily.

Instruction through illustration

The instructions to perform any action are provided by maximum number of illustrations so that the user can perform the actions discussed in the book easily and effectively. There are about 900 small and large illustrations that make the learning process effective.

Tutorial point of view

At the end of concept's explanation, the tutorial make the understanding of users firm and long lasting. Almost each chapter of the book has tutorials that are real world projects. Moreover, most of the tools in this book are discussed in the form of tutorials.

Project

Projects and exercises are provided to students for practicing.

For Faculty

If you are a faculty member, then you can ask for video tutorials on any of the topic, exercise, tutorial, or concept. As faculty, you can register on our website to get electronic desk copies of our latest books, self-assessment, and solution of practical. Faculty resources are available in the **Faculty Member** page of our website (**www.cadcamcaeworks.com**) once you login. Note that faculty registration approval is manual and it may take two days for approval before you can access the faculty website.

Formatting Conventions Used in the Text

All the key terms like name of button, tool, drop-down etc. are kept bold.

Free Resources

Link to the resources used in this book are provided to the users via email. To get the resources, mail us at ***cadcamcaeworks@gmail.com*** with your contact information. With your contact record with us, you will be provided latest updates and informations regarding various technologies. The format to write us mail for resources is as follows:

Subject of E-mail as ***Application for resources of .... book***.
Also, given your information like
Name:
Course pursuing/Profession:
Contact Address:
E-mail ID:

Note: We respect your privacy and value it. If you do not want to give your personal informations then you can ask for resources without giving your information.

About Authors

The author of this book, Matt Weber, has authored many books on CAD/CAM/CAE available already in market. **SolidWorks Electrical Black Books** are one of the most selling books in SolidWorks Electrical field. The author has hands on experience on almost all the CAD/CAM/CAE packages. If you have any query/doubt in any CAD/CAM/CAE package, then you can contact the author by writing at cadcamcaeworks@gmail.com

The author of this book, Gaurav Verma, has written and assisted in more than 19 titles in CAD/CAM/CAE which are already available in market. He has authored AutoCAD Electrical Black Books and SolidWorks Electrical Black Books. He has also authored books on various modules of Creo Parametric and SolidWorks. He has provided consultant services to many industries in US, Greece, Canada, and UK. He has assisted in preparing many Government aided skill development programs. He has been speaker for Autodesk University. He has assisted in preparing AutoCAD Electrical course for Autodesk Design Academy. He has worked on Sheetmetal, Forging, Machining, and Casting designs in Design and Development departments of various manufacturing firms.

For Any query or suggestion

If you have any query or suggestion, please let us know by mailing us on ***cadcamcaeworks@gmail.com***. Your valuable constructive suggestions will be incorporated in our books and your name will be addressed in special thanks area of our books on your confirmation.

Page left blank intentionally

Chapter 1

Basics of Electrical Drawings

Topics Covered

The major topics covered in this chapter are:

- ***Need of Drawings***
- ***Electrical Drawings***
- ***Common Symbols in Electrical Drawings***
- ***Wire and its Types***
- ***Labeling***

NEED OF DRAWINGS

In this book, you will learn about electrical wiring and schematics created by using tools in AutoCAD Electrical. Most of the readers of this book will be having prior experience with electrical drawings but there are a few topics that should be revised before we move on to practical on software.

When you work in an electrical industry, you need to have a lot of information ready like the wire type, location of switches, load of every machine, and so on. It is very difficult to remember all these details if you are working on electrical system of a big plant since there might be thousands of wires connecting hundreds of switches and machines. To maintain accuracy in wiring of such big systems, you need electrical drawings. Earlier, electrical drawings were handmade but now, we use printed documentation for these informations. Figure-1 shows an electrical drawing (electronic).

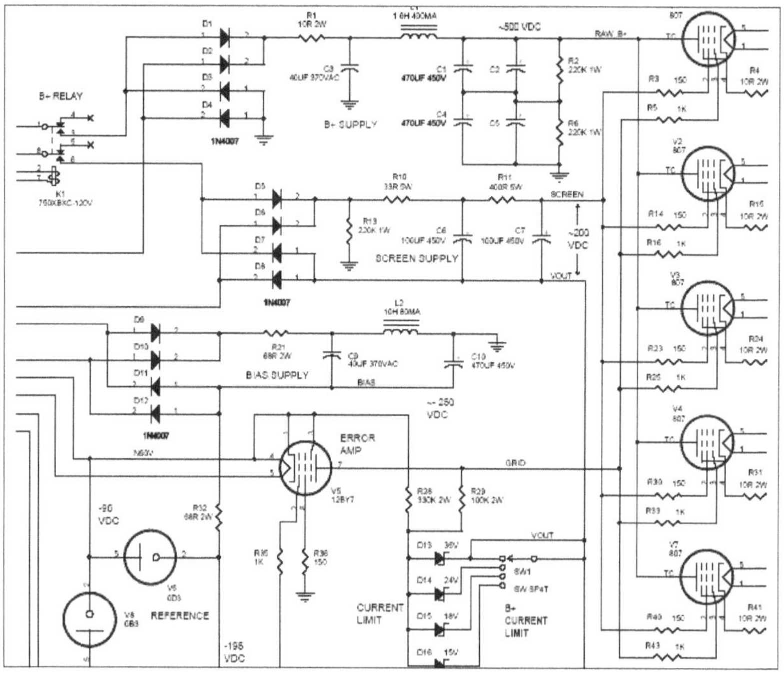

Figure-1. Circuit diagram

ELECTRICAL DRAWINGS

Electrical drawings are the representation of electrical components connected with wiring to perform specific tasks. An electrical drawing can be of a house, industry, or an electrical panel. Electrical drawings can be divided into following categories:

- Circuit diagram
- Wiring diagram
- Wiring schedule
- Block diagram
- Parts list

Circuit Diagram

A circuit diagram shows how the electrical components are connected together. A circuit diagram consists:

- Symbols to represent the components;
- Lines to represent the functional conductors or wires which connect the components together.

A circuit diagram is derived from a block or functional diagram (see Figure-2). It does not generally bear any relationship to the physical shape, size or layout of the parts. Although, you could wire up an assembly from the information given in it, they are usually intended to show the detail of how an electrical circuit works.

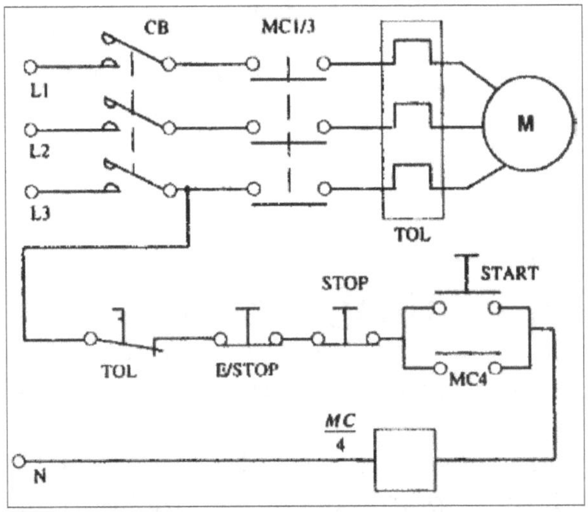

Figure-2. Circuit diagram

Wiring Diagram

A wiring diagram is the drawing which shows complete wiring between the components. We use wiring diagrams when we need to represent:

- Control or signal functions;
- Power supplies and earth connections;
- Termination of unused leads, contacts;
- Interconnection via terminal posts, blocks, plugs, sockets, and lead-throughs.

The wiring diagrams have details, such as the terminal identification numbers which enable us to wire the unit together. Note that internal wiring of components is generally not displayed in the wiring diagrams. Figure-3 shows a wiring diagram.

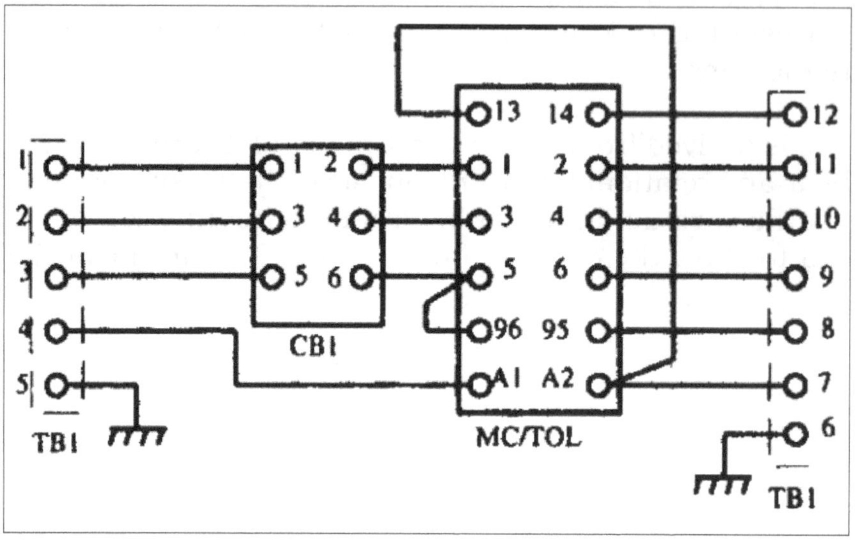

Figure-3. Wiring diagram

Wiring Schedule

A wiring schedule defines the wire reference number, type (size and number of conductors), length, and the amount of insulation stripping required for soldering.

In complex equipment, you may also find a table of interconnections which will give the starting and finishing reference points of each connection as well as other important information such as wire color, identification marking, and so on. Refer to Figure-4.

Schedule: Motor Control					206-A
Wire No	**From**	**To**	**Type**	**Length**	**Strip Length**
01	TB1/1	CB1/1	16/0.2	600 mm	12 mm
02	TB1/2	CB1/3	16/0.2	650 mm	12 mm
03	TB1/3	CB1/5	16/0.2	600 mm	12 mm
04	TB1/4	MC/A1	16/0.2	800 mm	12 mm
05	TB1/5	Ch/1	16/0.2	500 mm	12 mm

Figure-4. Wiring Schedule

Block Diagram

The block diagram is a drawing which is used to show and describe the main operating principles of the equipment. The block diagram is usually drawn before the circuit diagram is started.

It will not give the detail of the actual wiring connections or even the smaller components. Figure-5 shows a block diagram.

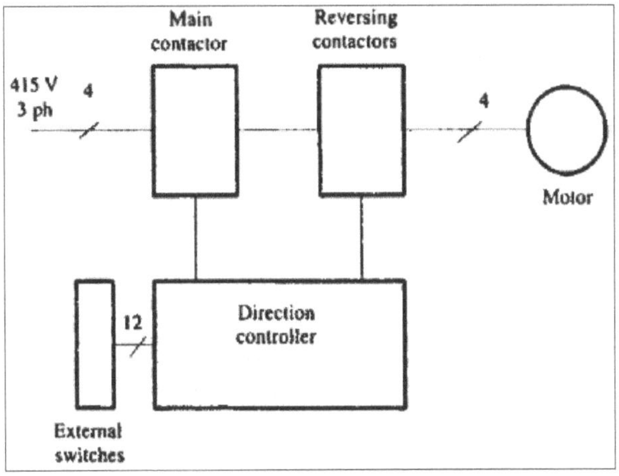

Figure-5. Block diagram

Parts list

Although, Part list (Bill Of Materials) is not a drawing in itself but most of the time it is a part of the electrical drawing project; refer to Figure-6. The parts list gives vital information:

- It relates component to circuit drawing reference numbers.
- It is used to locate and cross refer actual component code numbers to ensure you have the correct parts to commence a wiring job.

PARTS LIST			
REF	BIN	DESCRIPTION	CODE
CB1	A3	KM Circuit Breaker	PKZ 2/ZM-40-8
MC	A4	KM Contactor	DIL 2AM 415/50
TOL	A4	KM Overload Relay	Z 1-63

Figure-6. Parts list

We have discussed various types of electrical drawings and you may have noticed that there are various symbols to represent components in these drawings. Following section will explains some common electrical symbols.

SYMBOLS IN ELECTRICAL DRAWINGS

Symbols used in electrical drawings can be divided into various categories discussed next.

Conductors

There are 12 types of symbols for conductors; refer to Figure-7 and Figure-8. These symbols are discussed next.

1. General symbol, conductor or group of conductors.
2. Temporary connection or jumper.
3. Two conductors, single-line representation.
4. Two conductors, multi-line representation.

5. Single-line representation of n conductors.
6. Twisted conductors. (Twisted pair in this example.)

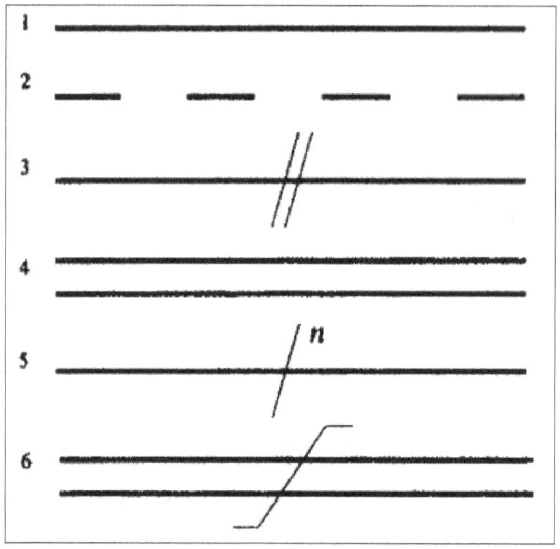

Figure-7. Symbols for conductors

7. General symbol denoting a cable.
8. Example: eight conductor (four pair) cable.
9. Crossing conductors – no connection.

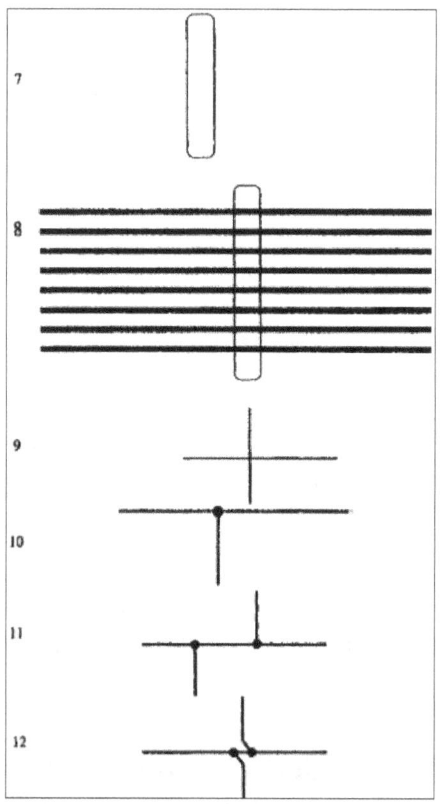

Figure-8. Symbols for conductors

10. Junction of conductors (connected).
11. Double junction of conductors.
12. Alternatively used double junction.

Connectors and terminals

Refer to Figure-9.
13. General symbol, terminal or tag.
14. Link with two easily separable contacts.
15. Link with two bolted contacts.
16. Hinged link, normally open.
17. Plug (male contact).
18. Socket (female contact).
19. Coaxial plug.
20. Coaxial socket.

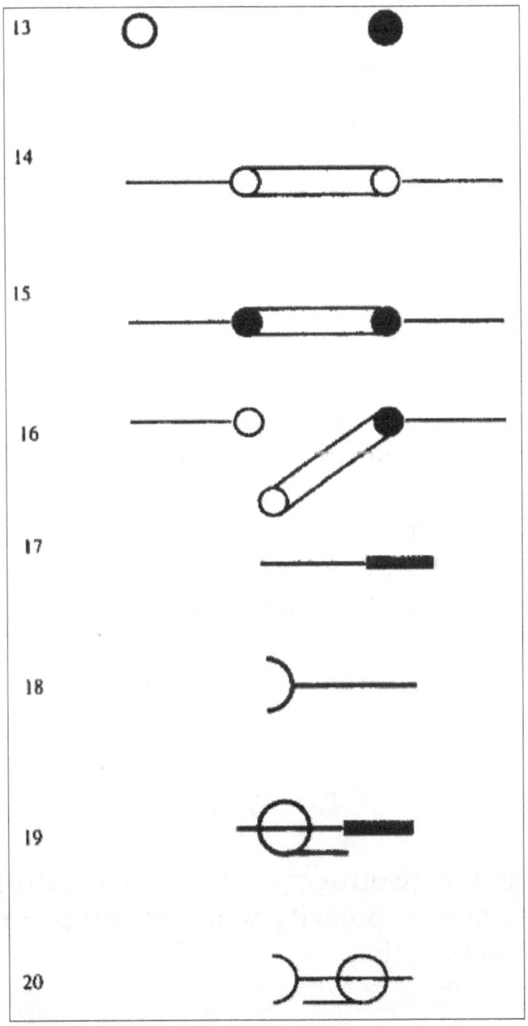

Figure-9. Connectors symbols

These symbols are used for contacts with moveable links. The open circle is used to represent easily separable contacts and a solid circle is used for bolted contacts.

Inductors and transformers

Refer to Figure-10.
21. General symbol, coil or winding.
22. Coil with a ferromagnetic core.
23. Transformer symbols.

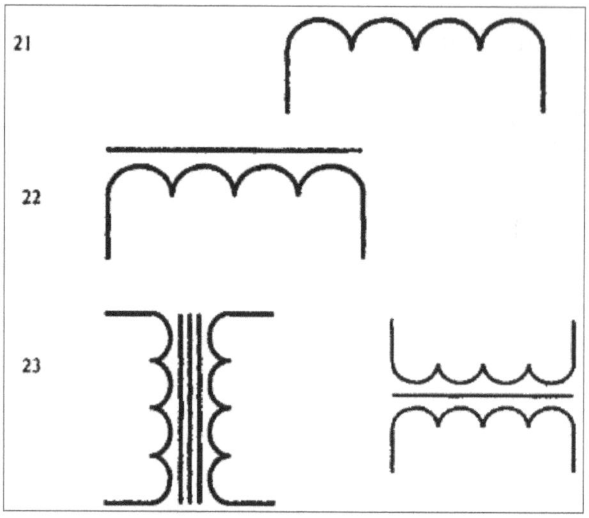

Figure-10. Inductors symbols

Resistors

Refer to Figure-11.
24. General symbol.
25. Old symbol sometimes used.
26. Fixed resistor with a fixed tapping.
27. General symbol, variable resistance (potentiometer).
28. Alternative (old).
29. Variable resistor with preset adjustment.
30. Two terminal variable resistance (rheostat).
31. Resistor with positive temperature coefficient (PTC thermistor).
32. Resistor with negative temperature coefficient (NTC thermistor).

Capacitors

Refer to Figure-12.
33. General symbol, capacitor. (Connect either way round.)
34. Polarised capacitor. (Observe polarity when making connection.)
35. Polarized capacitor, electrolytic.
36. Variable capacitor.
37. Preset variable.

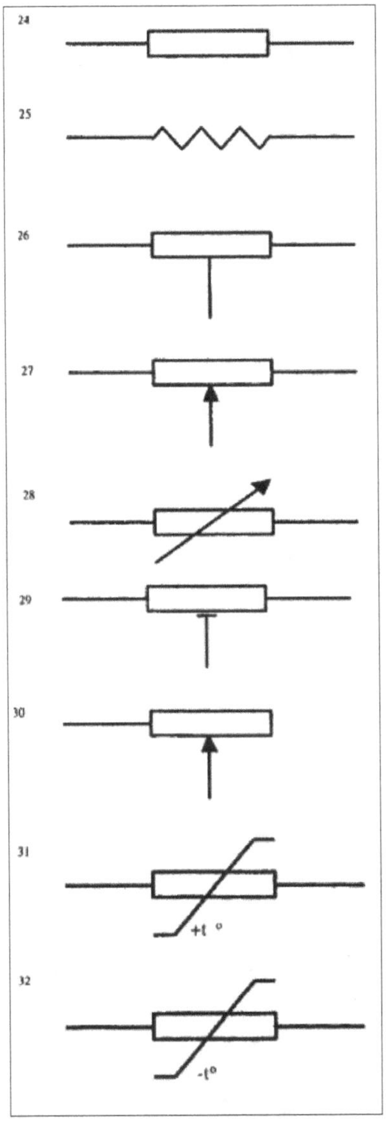

Figure-11. Resistors symbol

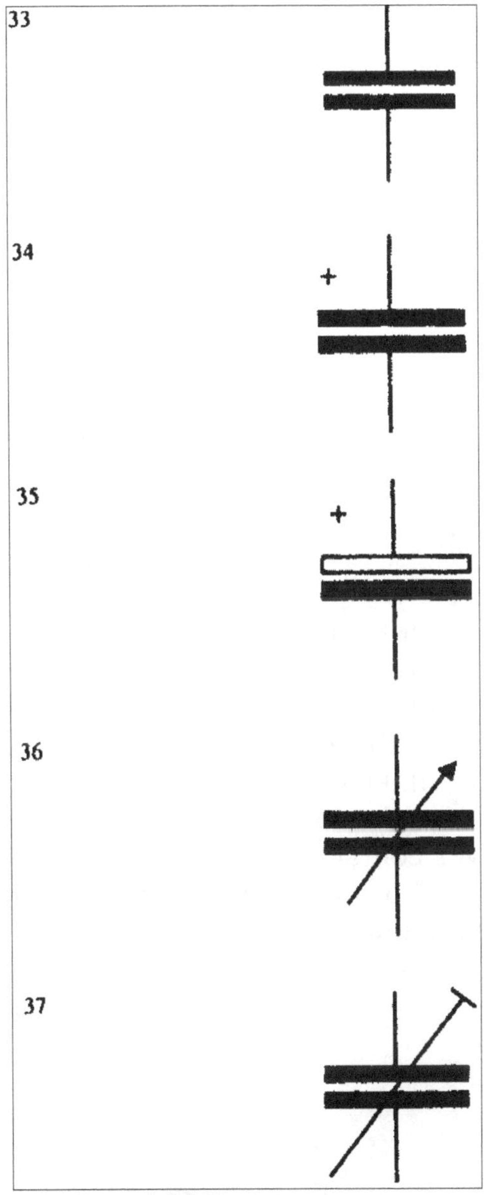

Figure-12. Capacitors symbols

Fuses

Refer to Figure-13.
38. General symbol, fuse.
39. Supply side may be indicated by thick line: observe orientation.
40. Alternative symbol (older).

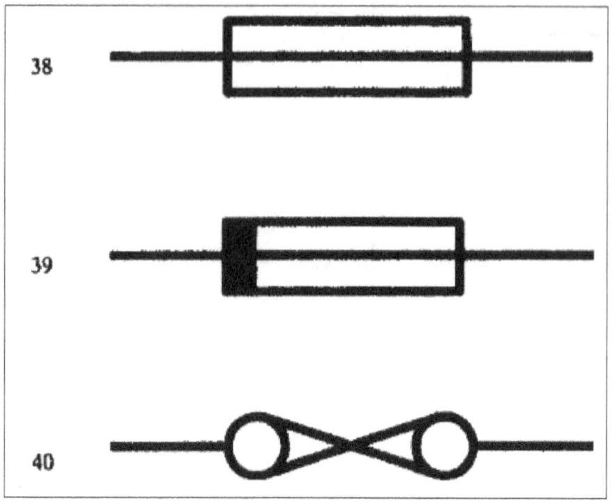

Figure-13. Fuses symbols

Switch contacts

Refer to Figure-14.

41. Break contact (BSI).
42. Alternative break contact version 1 (older).
43. Alternative break contact version 2.
44. Make contact (BSI).
45. Alternative make contact version 1.
46. Alternative make contact version 2.
47. Changeover contacts (BSI).
48. Alternative showing make-before-break.
49. Alternative showing break-before-make.

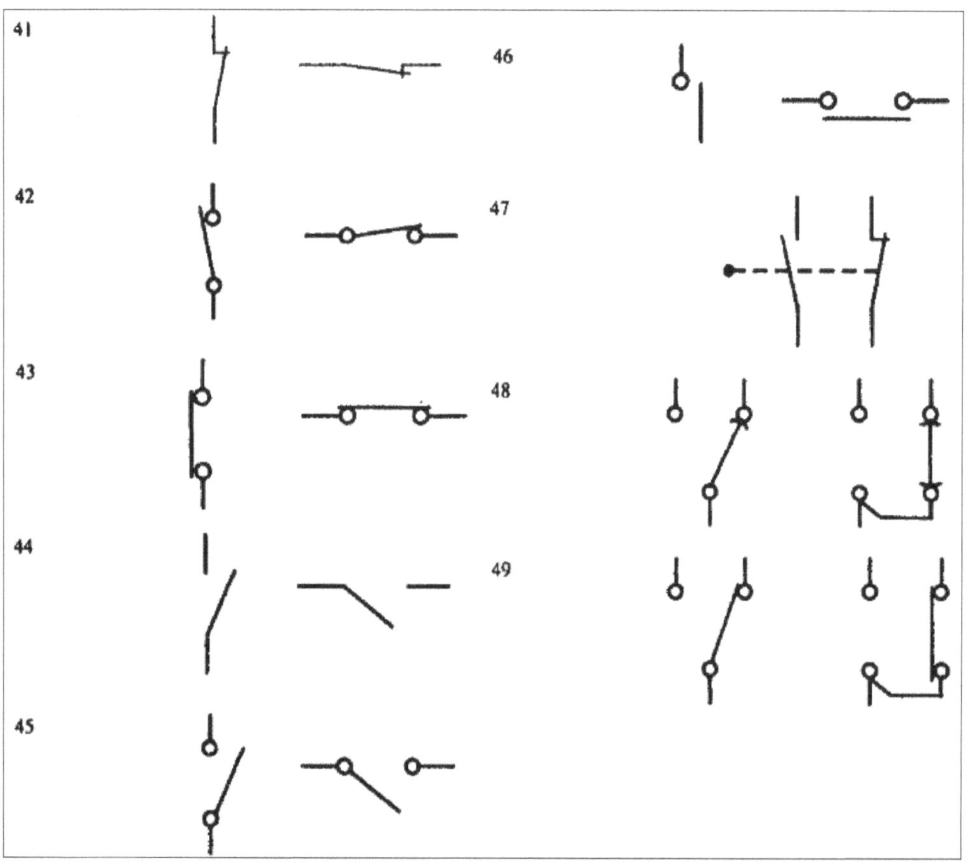

Figure-14. Switch Contact symbols

Switch types

Refer to Figure-15.
50. Push button switch momentary.
51. Push button, push on/push off (latching).
52. Lever switch, two position (on/off).
53. Key-operated switch.
54. Limit (position) switch.

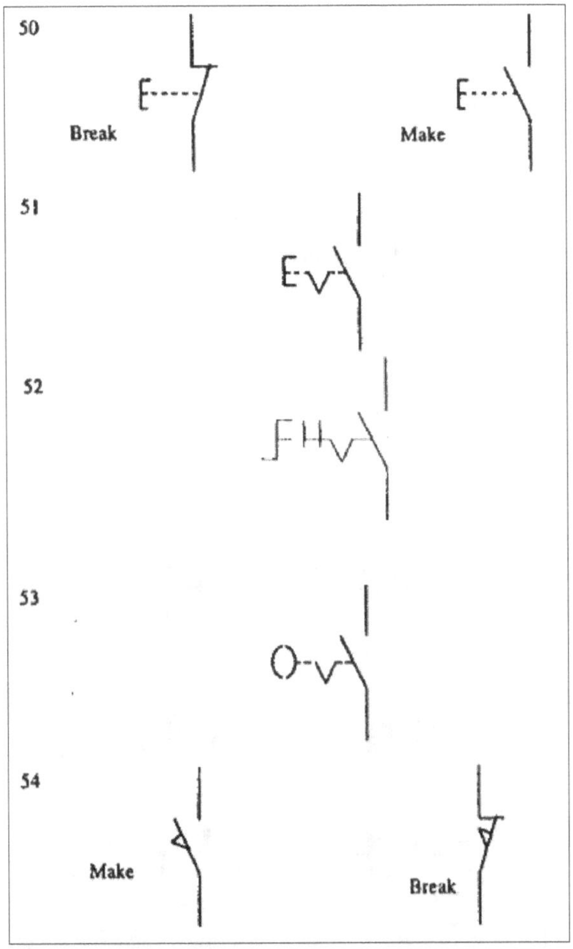

Figure-15. Switch symbols

Diodes and rectifiers

Refer to Figure-16.
55. Single diode. (Observe polarity.)
56. Single phase bridge rectifier.
57. Three-phase bridge rectifier arrangement.
58. Thyristor or silicon controlled rectifier (SCR) general symbol.
59. Thyristor – common usage.
60. Triac – a two-way thyristor.

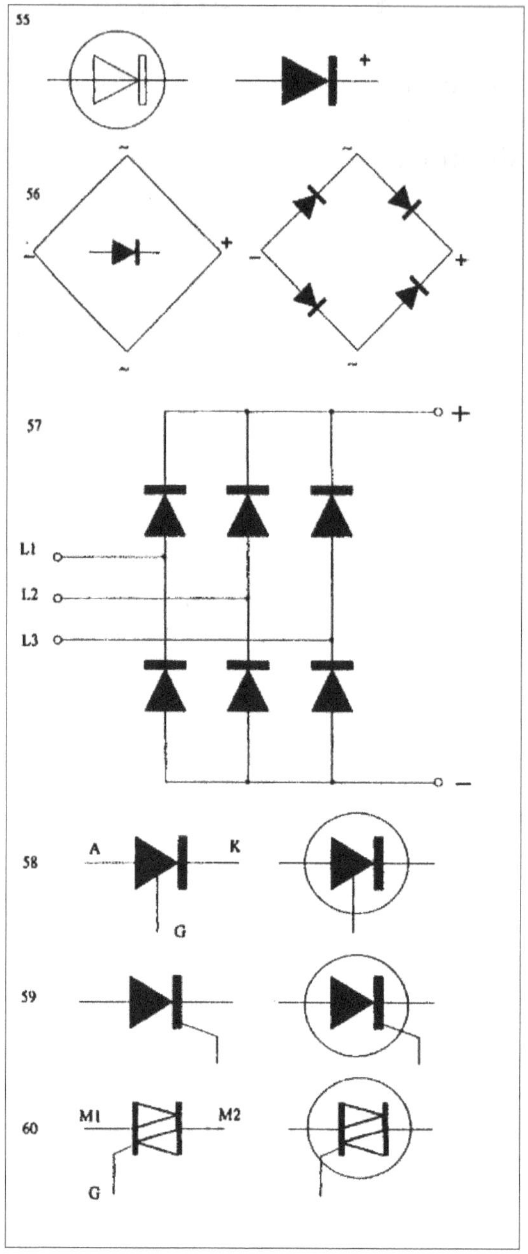

Figure-16. Diode Symbols

Earthing

Refer to Figure-17.

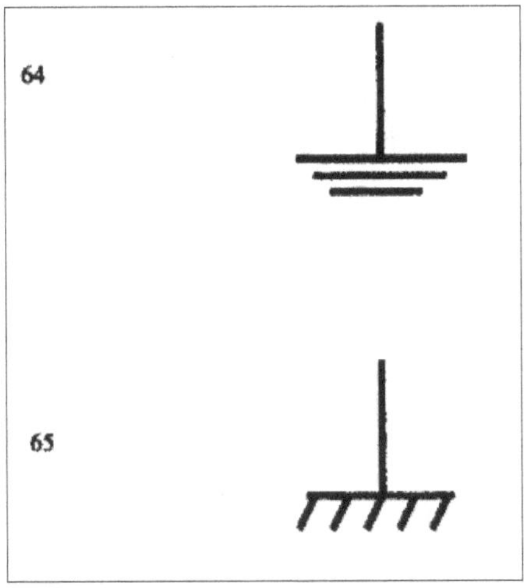

Figure-17. Earthing

Although there are lots of standard symbols in the electrical library, you might still need some user defined symbols for your drawing. You can create these symbols and give map keys/legends for them in the drawing. You will learn about creating symbols later in the book

After learning about various symbols, the next important topic is wire and its specifications.

WIRE AND SPECIFICATIONS

Electrical equipment use a wide variety of wires and cables. It is the responsibility of designer to correctly identify and create the wires in the drawing which are suitable for the current application. The wrong wire types can cause operational problems and could render the unit unsafe. Some factors to be considered for wire selection are:

- The insulation material;
- The size of the conductor;
- Conductor material;
- Solid or stranded and flexible.

Types of Wires

- **Solid or single-stranded wires** are not very flexible and are used where rigid connections are acceptable or preferred usually in high current applications like in power switching contractors. This type of wire can be un-insulated.
- **Stranded wire** is flexible and most interconnections between components are made with it.
- **Braided wire**, also called Screened wire, is an ordinary insulated conductor surrounded by a conductive braiding. In this type of wire, the metal outer is not used to carry current but is normally connected to earth to provide an electrical shield to the internal conductors from outside electromagnetic interference.

Wire specifications

There are several ways to represent wire specifications in electrical drawings. The most common method is to specify the number of strands in the conductor, the diameter of the strands, the cross sectional area of the conductor, and then the insulation type. Refer to Figure-18:

- The 1 means that it is single conductor wire.
- The conductor is 0.6 mm in diameter and is insulated with PVC.
- The conductor has a cross-sectional area nominally of 0.28 mm.

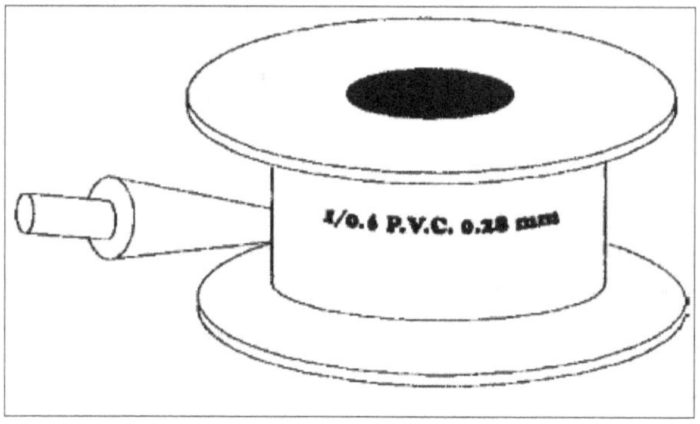

Figure-18. Example of wire specification

Standard Wire Gauge

If you are using solid wire in your drawings then it can be represented in drawing by using the Standard Wire Gauge or SWG system. The SWG number is equivalent to a specific diameter of conductor; refer to Figure-19.

For example; 30 SWG is 0.25 mm diameter.
14 SWG is 2 mm in diameter.

The larger the number – the smaller the size of the conductor.

There is also an American Wire Gauge (AWG) which uses the same principle, but the numbers and sizes do not correspond to those of SWG.

SWG table

SWG No.	Diameter
14 swg	2 mm
16 swg	1.63 mm
18 swg	1.22 mm
20 swg	0.91 mm
22 swg	0.75 mm
24 swg	0.56 mm
25 swg	0.5 mm
30 swg	0.25 mm

Figure-19. SWG table

Till this point, we have learned about various schematic symbols and wires. Now, we will learn about labeling of components.

LABELING

Labeling is the marking of components for identifying incoming and outgoing supply; refer to Figure-20.

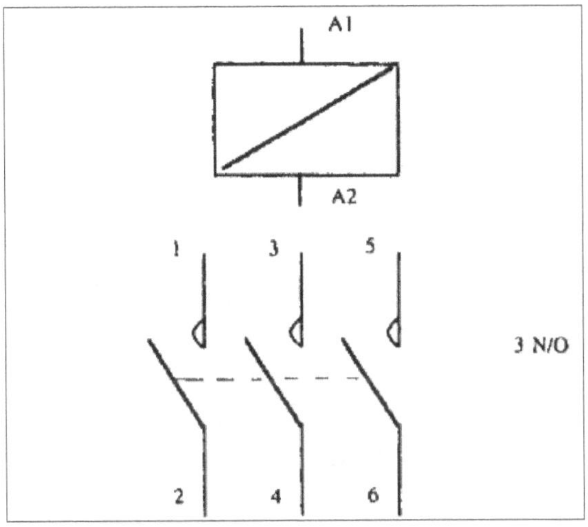

Figure-20. Contacts

Coils are marked alphanumerically, e.g. A1, A2.
Odd numbers – incoming supply terminal.
Even numbers – outgoing terminal.

Main contacts are marked with single numbers:
Odd numbers – incoming supply terminal.
Next even number – outgoing terminal.

We will find different type of markings while working on the electrical drawings.

Cable Markers

Cable or wire markers are used to identify wires, especially in multi way cables or wiring harnesses. Both ends of the wires are marked with the same numbers to identify start and end of the wires.

Often the cable/wire numbers are same as those of connectors to which they are connected. In any case, the wiring drawing or run-out sheet will give the wire numbers to be used. The markers are placed so that the number is read from the joint as illustrated. This example shows wire number 27; refer to Figure-21.

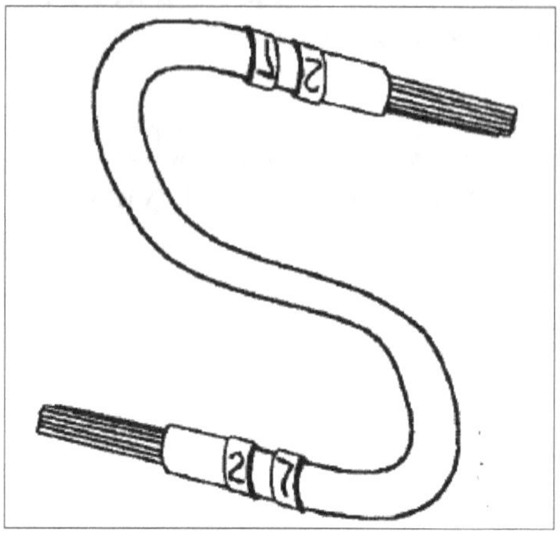

Figure-21. Wire marking

Most of the cables have numbers printed as well as being colored, although you may find some wires/cables colored only.

Some cable markers are wrapped round the wire and are adhesive, while others are like small sleeves which slip over the insulation.

SELF-ASSESSMENT

Q1. Define Electrical Drawings and its types.

Q2. What is the difference between circuit diagram and wiring diagram?

Q3. .... is a type of drawing which defines the wire reference number, type (size and number of conductors), length, and the amount of insulation stripping required for soldering.

Q4. The .... is a drawing which is used to show and describe the main operating principles of the equipment and is usually drawn before the circuit diagram is started.

Q5. Which of the following is symbol for twisted pair conductor?

A.
B.
C.
D.

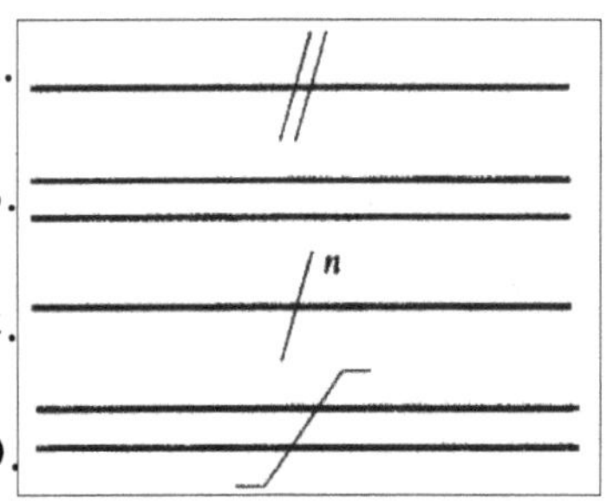

Q6. Which symbol is used to represent easily separable contact?

A.
B.

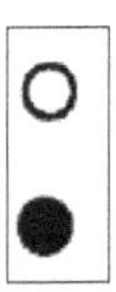

Q7. Which of the following wires are not very flexible and are used where rigid connections are accept able or preferred usually in high current applications ?

A. Single-stranded wire
B. Stranded Wire
C. Braided Wire
D. Coaxial Wire

Q8. The larger the SWG number of wire – the smaller the size of the conductor (T/F)

REVIEW QUESTIONS

Q1. Why are electrical drawings necessary in large industrial setups?
A. To enhance physical appearance of equipment
B. To reduce cost of installation
C. To maintain accuracy in wiring complex systems
D. To reduce the need for skilled labor

Q2. What is typically not included in a circuit diagram?
A. Component symbols
B. Physical layout of parts
C. Functional conductors
D. Electrical connections

Q3. Which type of diagram shows control or signal functions and terminal identification numbers?
A. Circuit diagram
B. Block diagram
C. Wiring diagram
D. Parts list

Q4. What does a wiring schedule primarily define?
A. Operating principles
B. Symbol meanings
C. Wire reference number and specifications
D. Electrical current values

Q5. When is a block diagram usually created?
A. After making the wiring schedule
B. After completing the wiring diagram
C. Before starting the circuit diagram
D. After the physical layout is done

Q6. What information does a parts list provide?
A. Instructions for soldering
B. Circuit behavior analysis
C. Cross reference for component code numbers
D. Wiring termination details

Q7. What symbol is used for a male contact in connectors and terminals?
A. Coaxial socket
B. Plug
C. Socket
D. Terminal or tag

Q8. Which symbol represents a resistor with a negative temperature coefficient?
A. PTC thermistor
B. Fixed resistor
C. NTC thermistor
D. Variable resistor

Q9. Which fuse symbol requires observing orientation due to supply side indication?
A. General symbol
B. Old alternative
C. Symbol with thick line
D. Circular symbol

Q10. What is the function of the metal outer in braided (screened) wire?
A. Carry electric current
B. Increase flexibility
C. Provide shielding from interference
D. Strengthen wire

Q11. How is wire specification typically represented in electrical drawings?
A. Using circuit diagram reference numbers
B. Only by wire color
C. By conductor strands, diameter, cross-section, and insulation
D. With schematic symbols

Q12. What does 14 SWG represent?
A. 14 mm wire diameter
B. 0.25 mm diameter wire
C. 2 mm diameter wire
D. A 14-strand conductor

Q13. How are coils labeled for identification in drawings?
A. With colors
B. With symbols
C. Using A1, A2 alphanumeric codes
D. Using numeric-only codes

Q14. What do odd numbers on main contact markings typically indicate?
A. Fuse ratings
B. Outgoing terminals
C. Earthing points
D. Incoming supply terminals

Q15. Why are cable markers used?
A. To identify voltage levels

B. To mark insulation types
C. To identify wires in multi-way cables
D. To measure wire length

Q16. What is a characteristic of stranded wire?
A. Very rigid
B. Used in power switching contractors
C. Flexible, used for interconnections
D. Uninsulated by default

Q17. What distinguishes the American Wire Gauge (AWG) from SWG?
A. It uses color coding
B. It has same numbers as SWG
C. It uses different number-size relations
D. It is only used for stranded wires

FOR STUDENT NOTES

Answer for Self-Assessment:

Ans3. Wiring Schedule, Ans4. Block Diagram, Ans5. D, Ans6.A, Ans7. A, Ans8. T

Chapter 2

Introduction to AutoCAD Electrical and Interface

Topics Covered

The major topics covered in this chapter are:

- ***Introduction***
- ***System Requirement***
- ***Starting AutoCAD Electrical***
- ***Components of AutoCAD Electrical Interface***
- ***Starting Drawing***
- ***3D Printing***

INTRODUCTION TO AUTOCAD ELECTRICAL

In today's world, AutoCAD Electrical is a well-known name in the Electrical CAD industry. If we look back into the history, then first version of AutoCAD desktop application came out around 1982 with the name AutoCAD Version 1.0. From 1982 to till today AutoCAD has gone through continuous enhancements and modifications. The AutoCAD Electrical is built on this well known platform: AutoCAD. The latest version AutoCAD Electrical 2027 is the most advanced model of AutoCAD Electrical available for us. The software has expanded more into the user interface of AutoCAD Electrical and has become the most user friendly one. The 2027 version of software has rich capabilities to operate with Autodesk Inventor for 3D electrical CAD and electromechanical models. It features dark theme, cloud storage connectivity, AI assistant, and many performance improvements. For every tool/command it has more that one ways to invoke. This software also gives you the access to customize it as per your requirements.

Although the software is capable to perform 3D operations but in this book, we will concentrate on 2D drawing creation. Now, we will begin with starting AutoCAD Electrical and then we will discuss the interface of AutoCAD Electrical. But before we discuss about starting AutoCAD Electrical, Please check the system requirements to run AutoCAD Electrical 2027 properly. The system requirements are given next.

System requirements for AutoCAD Electrical 2027

Operating System
64-bit Windows 11

CPU Type
2.5 GHz (3+ GHz recommended)

Memory

Basic : 16 GB (32 GB recommended)

Display Resolution

1920 x 1080 with True Color.
Resolutions up to 3840 x 2160 supported (with capable display card).

Display Card
Basic: 2 GB GPU with 29 GB/s Bandwidth and DirectX 11 compliant
Recommended: 8 GB GPU with 106 GB/s Bandwidth and DirectX 12 compliant

Disk Space
20.0 GB (suggested SSD)

Pointing Device
MS-Mouse compliant device

Media (DVD)
Download and installation from DVD

Browser
Latest version of 64-bit Google Chrome, Mozilla Firefox, Microsoft Edge (Chromium-based), or Apple Safari

.NET Framework
.NET Framework Version 10 or later
*DirectX12 recommended by supported OS

For Mac system, you can check the requirement by scanning the QR code given below. Make sure that you fulfill all the requirements for the software before running it.

Note that there are various ways to perform the same operation in AutoCAD Electrical. Sometimes it may not be necessary to discuss all the ways so, we will be using the best practice to perform an operation.

INSTALLING AUTOCAD ELECTRICAL

The commercial version of software can be purchased from www.autodesk.com website. Here, we will discuss the procedure to download and install educational version of AutoCAD Electrical.

- Open your default web browser application and go to the web page https://www.autodesk.com/education/home.
- Log in using your student/educator credentials. Click on the **Select product** link button in the web page. Scroll down on the displayed page and click on the **Select** link button for AutoCAD; refer to Figure-1. The web page to get access to the software will be displayed; refer to Figure-2.

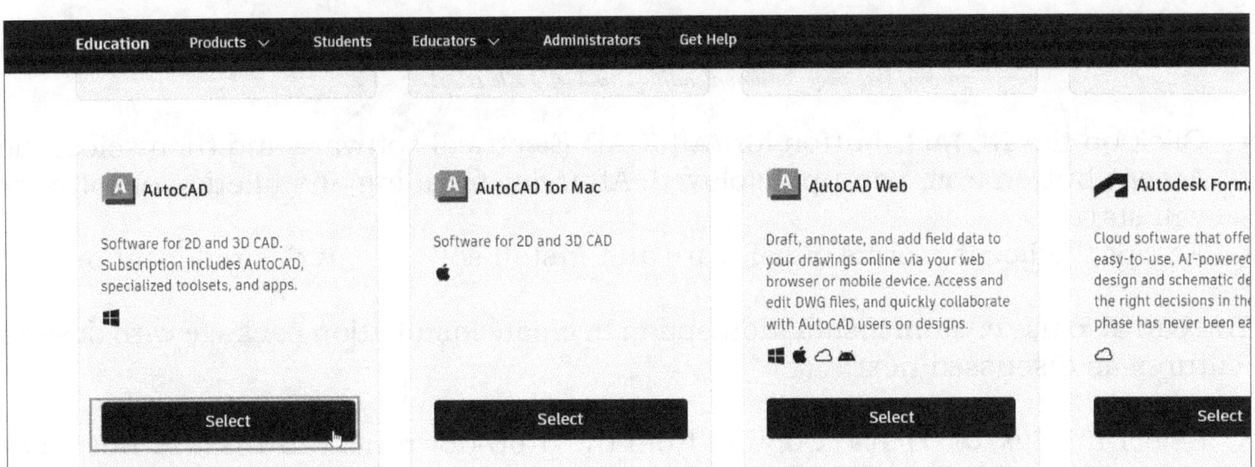

Figure-1. Select button for AutoCAD

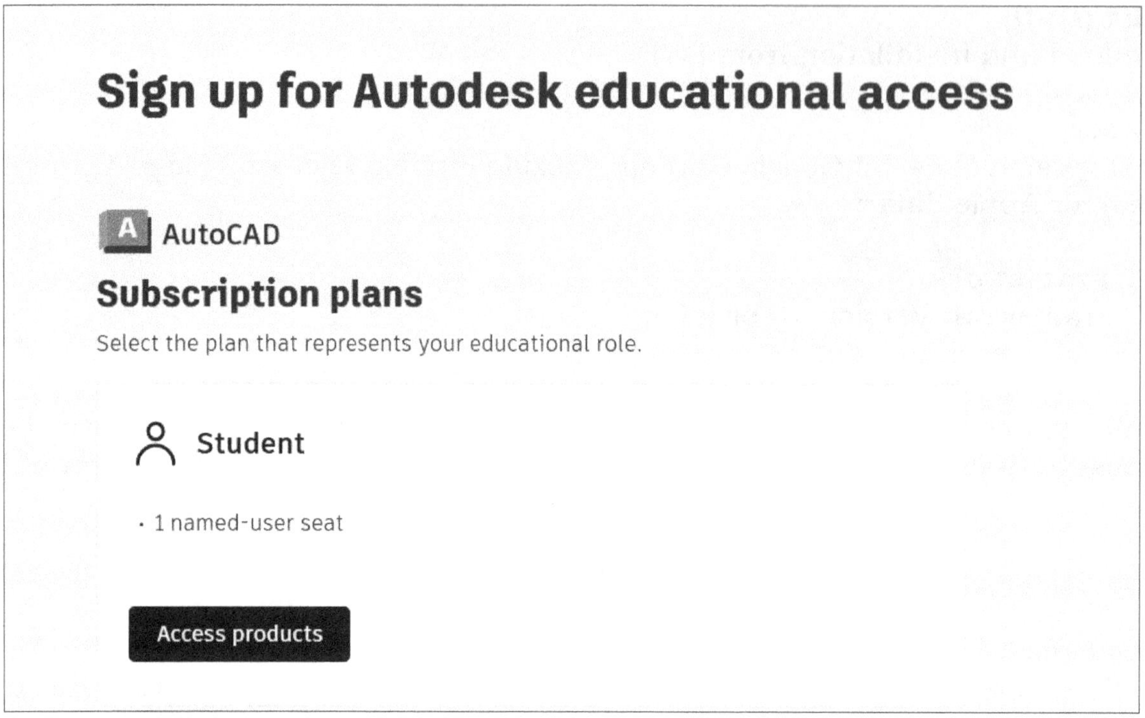

Figure-2. Webpage for accessing subscription

- Click on the **Access products** link button from the web page. The web page will be displayed as shown in Figure-3.

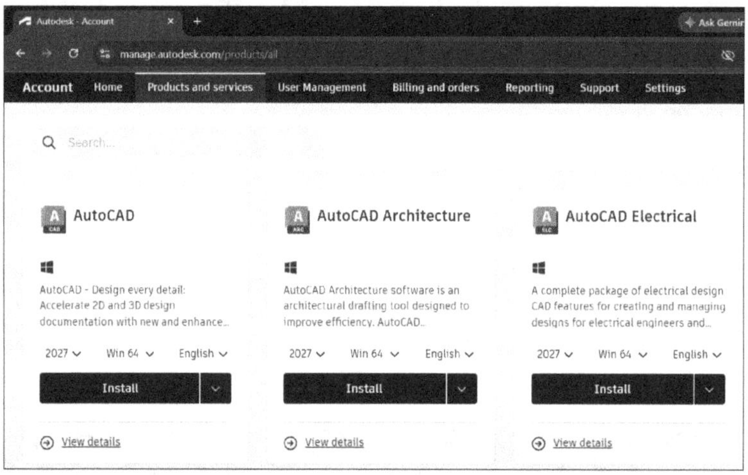

Figure-3. Web page for downloading AutoCAD Electrical

- Click on the **INSTALL** button for AutoCAD Electrical software and then select the **Accept** button from pop-up displayed. After downloading, installation of software will start.
- Accept the license terms of software and install software at desired location.

You can also use custom installation option to create installation package with desired features as discussed next.

- Select the **Custom Install** option from the drop-down next to **Install** button at the web page for downloading the software; refer to Figure-4 or click on the **Custom install** tab at the top in the web page and select the **Create package** button. The **New package** web page will be displayed; refer to Figure-5. Select check box for AutoCAD Electrical to customize the software.

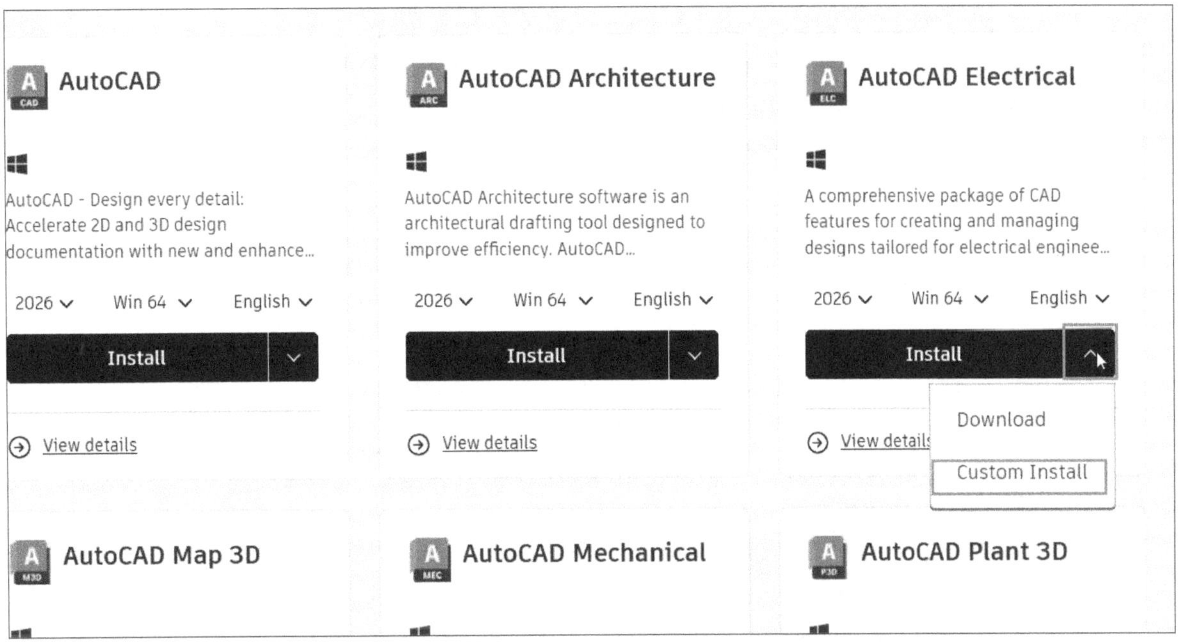

Figure-4. Custom Install option

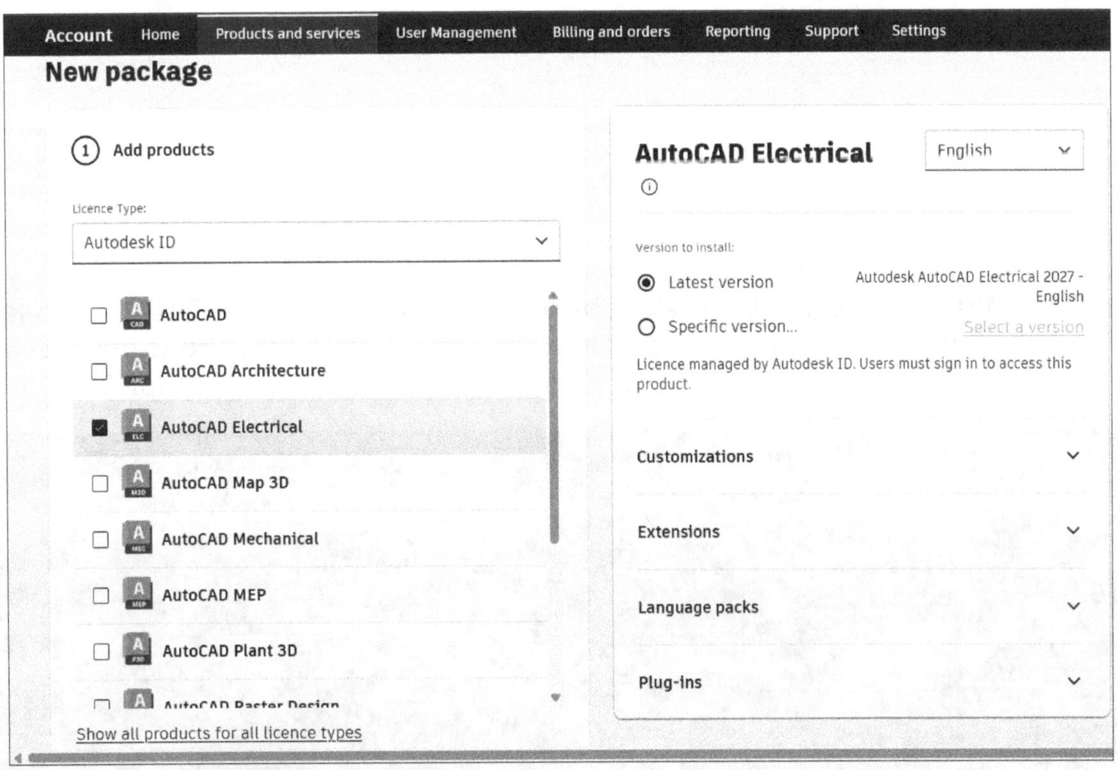

Figure-5. Custom Install page

- Expand the **Customizations**, **Extensions**, **Language packs**, and **Plug-ins** sections and select desired check boxes to add more functions to the software.
- After setting desired parameters, click on the **Next** button from the web page. You will be asked to specify installation location for the software. Set desired path location in **Installation Path** field and select the I agree to the Terms of Use check box. The **Download** button will become active on page.
- Save the file at desired location and run it. Once the files have been downloaded, install the software.

STARTING AUTOCAD ELECTRICAL

- Click on the **Start** button in the Taskbar and type AutoCAD Electrical in the search box. The list of options related to AutoCAD Electrical will be displayed; refer to Figure-6 (in Microsoft Windows 11).

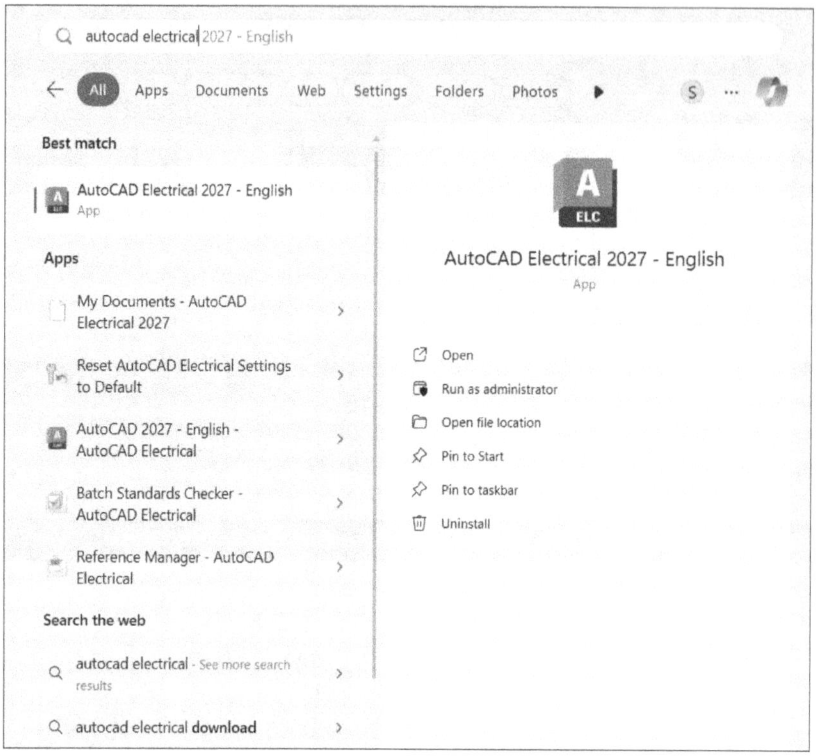

Figure-6. Start menu

- Click on **AutoCAD Electrical 2027** button from the list. AutoCAD Electrical will initialize and once the background processing is complete, the interface will be displayed as shown in Figure-7. (You may need to sign in again for licensing).

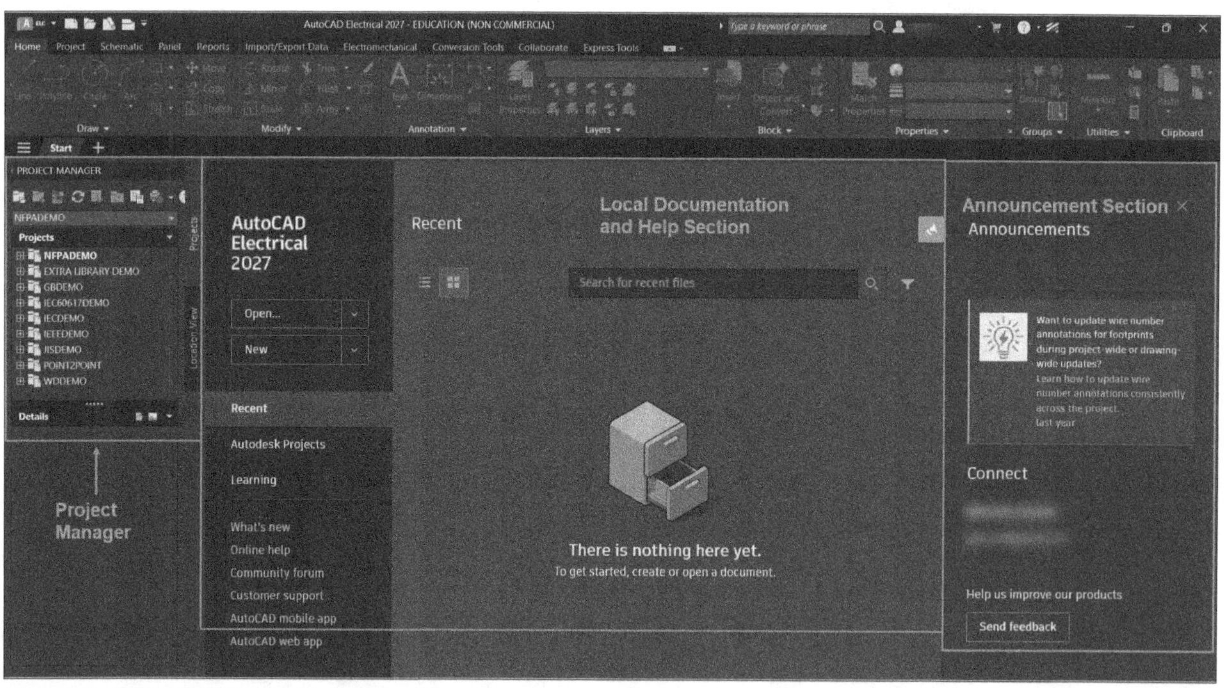

Figure-7. AutoCAD Electrical interface

The first page of interface of AutoCAD Electrical is divided into various sections like PROJECT MANAGER, Local Documents and help, and announcements. You will learn about the options in these sections later.

Now, our first task is to create a drawing document. There are two major unit systems to create drawings: Metric (SI) and Imperial. You will learn to create drawings in both the unit systems.

CREATING A NEW DRAWING DOCUMENT

- Click on the down arrow next to **New** button in the first page of interface; refer to Figure-8. Select the **Browse templates** option from the drop-down list. The **Select Template** dialog box will be displayed; refer to Figure-9.

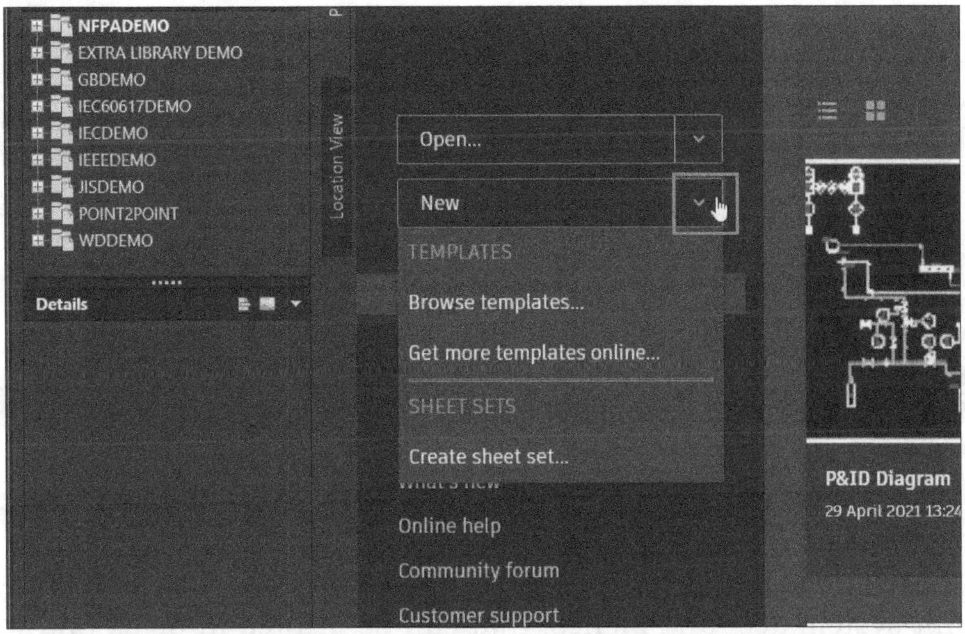

Figure-8. New document options

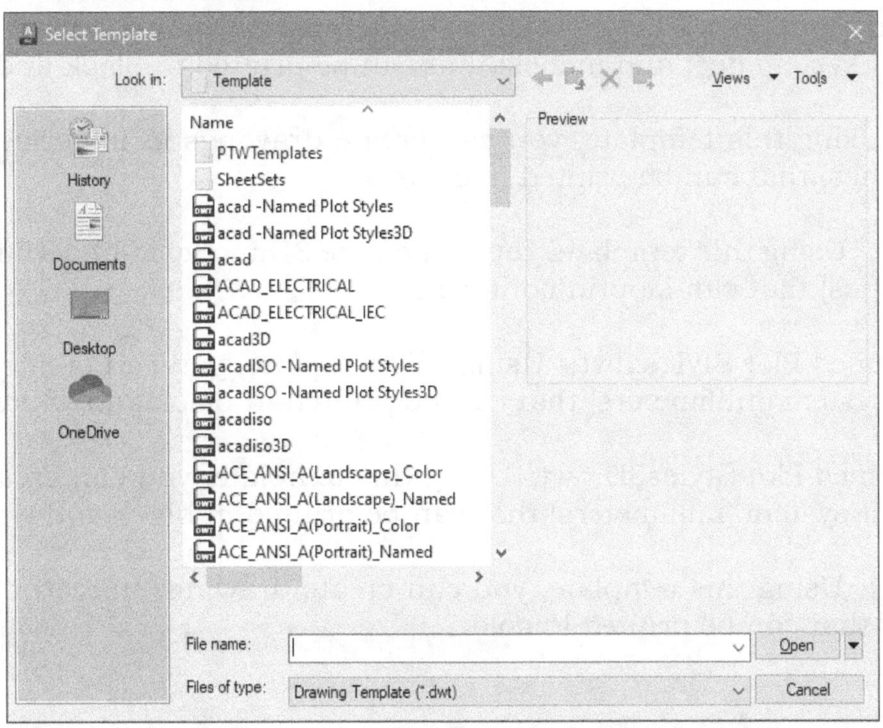

Figure-9. List of templates

- Select desired template from the list (ACAD ELECTRICAL.dwt in our case). The drawing environment will open according to the selected template; refer to Figure-10.

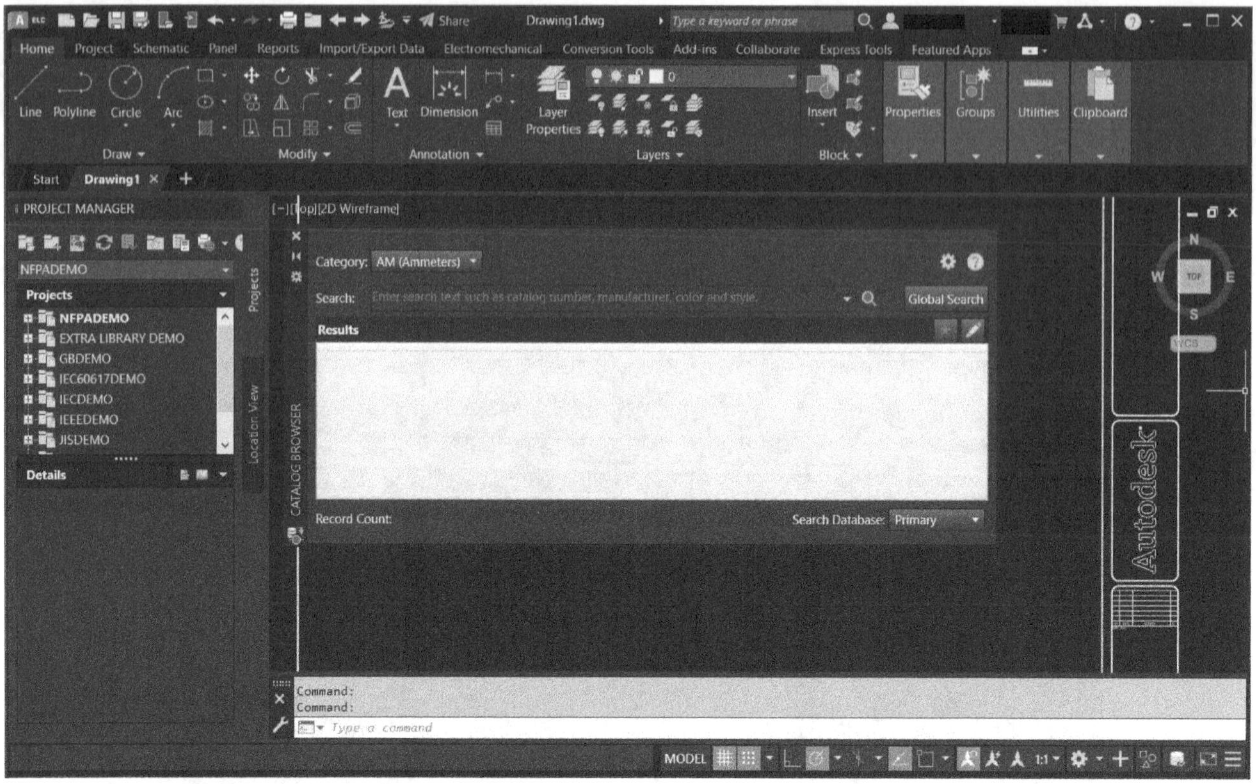

Figure-10. AutoCAD Electrical with acadiso template

Meaning of Default templates

acad -Named Plot Styles.dwt :- Using this template, you can create drawings in imperial unit system (feet and inches) that can be printed in black and white.

acad -Named Plot Styles3D.dwt :- Using this template, you can create 3D model in imperial unit system (feet and inches) that can be printed in black and white.

acad.dwt :- Using this template, you can create drawings in imperial unit system (feet and inches) that can be printed in color.

acad3D.dwt :- Using this template, you can create 3D model in imperial unit system (feet and inches) that can be printed in color.

acadISO -Named Plot Styles.dwt:- Using this template, you can create drawings in metric unit system (millimeters) that can be printed in black and white.

acadISO -Named Plot Styles3D.dwt:- Using this template, you can create 3D model in metric unit system (millimeters) that can be printed in black and white.

acadiso.dwt :- Using this template, you can create drawings in metric unit system (millimeters) that can be printed in color.

acadiso3D.dwt :- Using this template, you can create 3D model in metric unit system (millimeters) that can be printed in color.

Tutorial-iArch.dwt :- Using this template, you can create architectural drawing that are compatible with the tutorial in default library. Note that the unit for this template is imperial (Feet and Inches).

Tutorial-iMfg.dwt :- Using this template, you can create mechanical manufacturing drawing that are compatible with the tutorial in default library. Note that the unit for this template is imperial (Feet and Inches).

Tutorial-mArch.dwt :- Using this template, you can create architectural drawing that are compatible with the tutorial in default library. Note that the unit for this template is metric (Millimeters).

Tutorial-mMfg.dwt :- Using this template, you can create mechanical manufacturing drawing that are compatible with the tutorial in default library. Note that the unit for this template is metric (Millimeters).

Note that you can download more templates from the Autodesk server by using **Get more templates online** option from the **New** drop-down of first page of interface.

Electrical Templates

The template files that start with ACE are meant for electrical drawings. Also, you can use ACAD Electrical.dwt and ACAD Electrical IEC.dwt for creating AutoCAD Electrical drawings. You will learn about these templates later in the book.

Now, you know about the default templates. You have learned to create a new document with selected template. Now, you will learn about interface. We have divided the AutoCAD Electrical interface into following sections for easy understanding:

- Title Bar
- Application Menu
- Ribbon
- File Tab bar
- Drawing Area
- Command Window
- Bottom Bar

Now, we will discuss each of the sections one by one.

TITLE BAR

Title bar is the top strip containing quick access tools, name of the document, and connectivity options; refer to Figure-11.

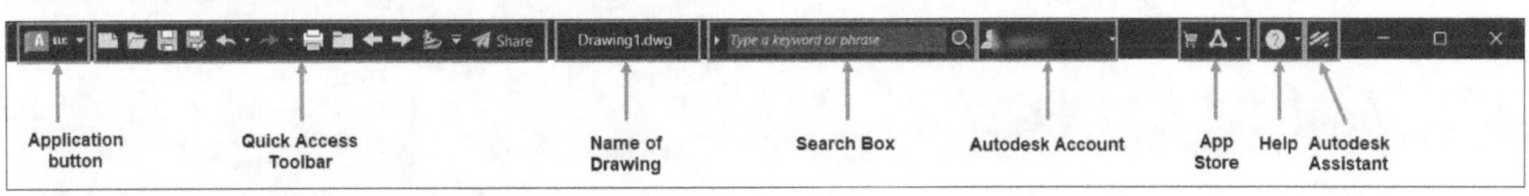

Figure-11. Title bar

- **Application** button is used to display **Application** menu, which we will discuss later.
- **Quick Access Toolbar** contains tools for common file handling. The **Quick Access Toolbar** contains tools for creating new file, saving current file, printing the file, and so on. You can add desired tools in the **Quick Access Toolbar** by using customizing options.
- The center of the title bar shows the name of the drawing.
- The **Search Box** is used to search desired topic in the **AutoCAD Electrical Help**. Type the keyword for the topic whose information is to be found in the text box adjacent to search button and press **ENTER** from the keyboard.
- **Sign In** button is used to sign into the Autodesk account. If you are not having an Autodesk account then you can create one from Autodesk website. If you have an Autodesk account then you can save and share your files through Autodesk cloud and you can render on cloud which are very fast services. To sign into your Autodesk account, click on the **Sign In** button. As a result, a drop-down will display; refer to Figure-12. Click on **Sign In to Autodesk account** option from the drop-down. The **Autodesk - Sign In** dialog box will be displayed; refer to Figure-13. Enter your ID and click **NEXT** button. You will be asked to specify password. Type your password and click on the **Sign In** button to login. Note that a sign-in is needed to even start the software as license of software is now linked to the account.

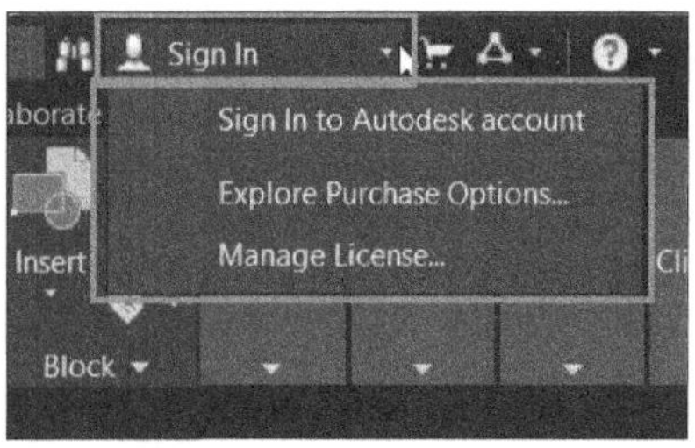

Figure-12. Sign In option

Figure-13. Autodesk sign in dialog box

- **Autodesk App Store** button in the Title Bar is used to install or share apps for Autodesk products. Click on the **Autodesk App Store** button, exchange apps web page will open in the browser where you can buy or try various apps as per your requirement; refer to Figure-14.

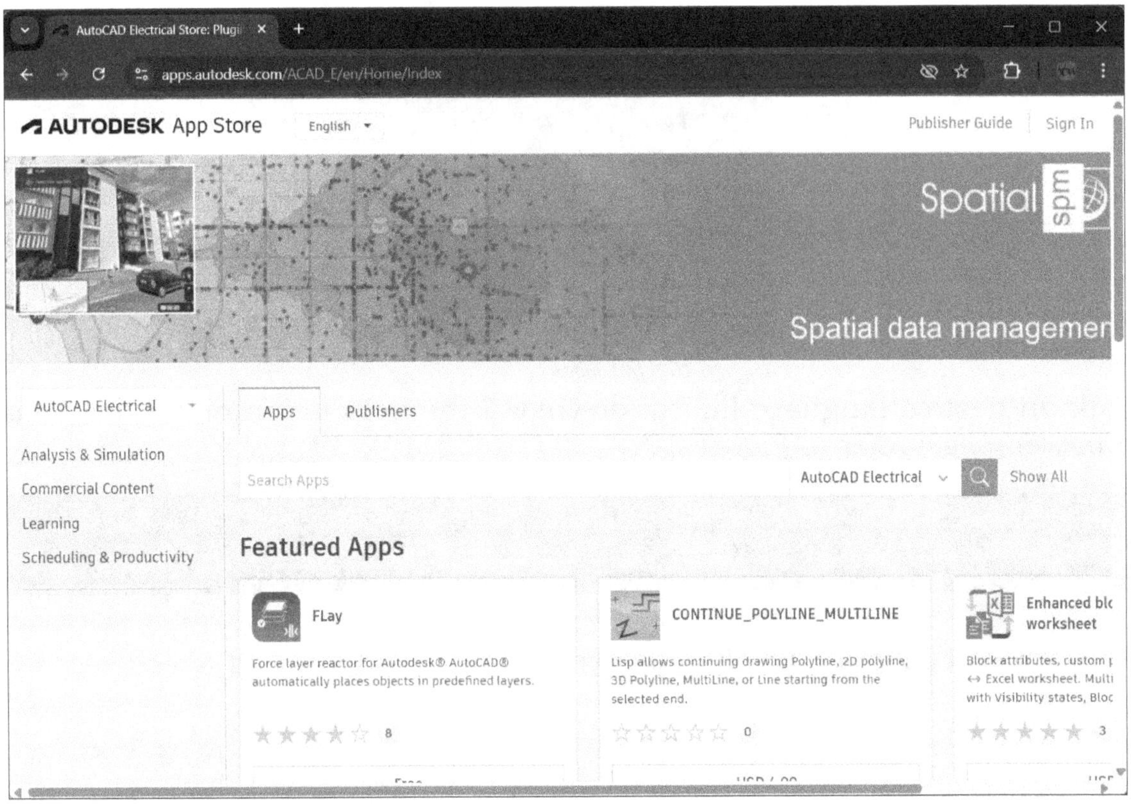

Figure-14. Autodesk App Store

- Options in the **Stay Connected** drop-down are used to manage autodesk account and access the resources of software available online on Autodesk website; refer to Figure-15.

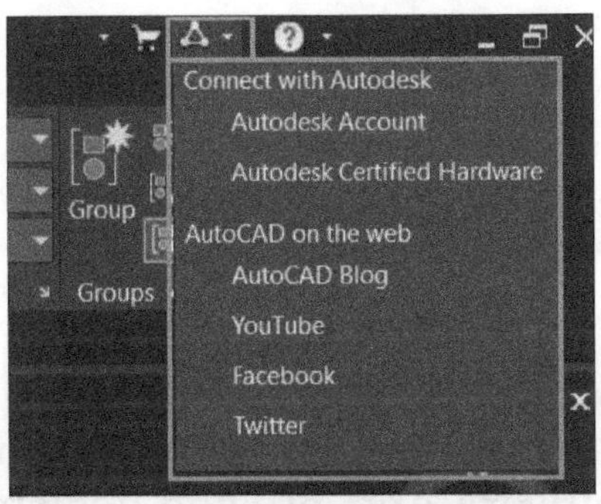

Figure-15. Stay Connected drop-down

- **Help** button is used to display online help of AutoCAD Electrical on the Autodesk server. If you click on the down arrow next to help button then a list of options is displayed; refer to Figure-16. If you want to use offline help then click on the **Download Offline Help** option and select the version of software from web page.

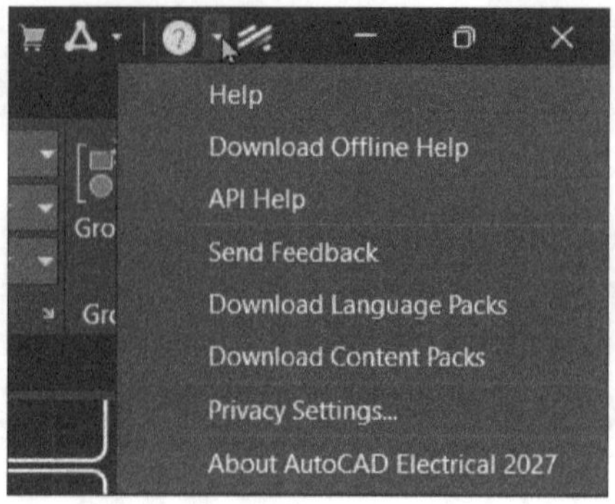

Figure-16. Help menu

- Click on desired language link to download the help file in respective language from web page; refer to Figure-17.

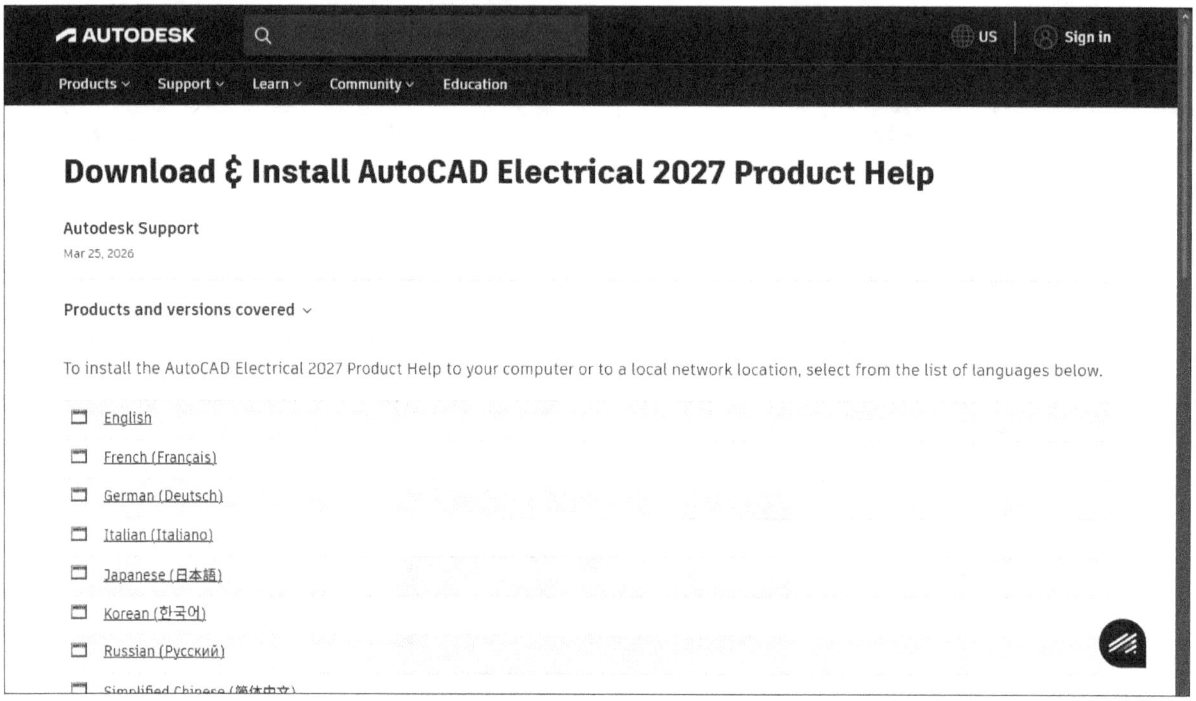

Figure-17. Offline help link

- **Autodesk Assistant** is AI powered assistant tool that can answer your queries related to tools in the software and provide assistance in further support. On clicking this button for first time, the assistant window will be displayed as shown in Figure-18. Click on the **Continue** button to use preview features of assistant as well. Now, type your query just like you use any other AI assistant app (ChatGPT, Gemini etc.) and you will receive information based on data available on Autodesk servers.

Now, we will change the color scheme of AutoCAD Electrical because of printing compatibility as white color will consume less ink while printing. You can retain the present color scheme if you are comfortable with it. To change the color scheme, steps are given next.

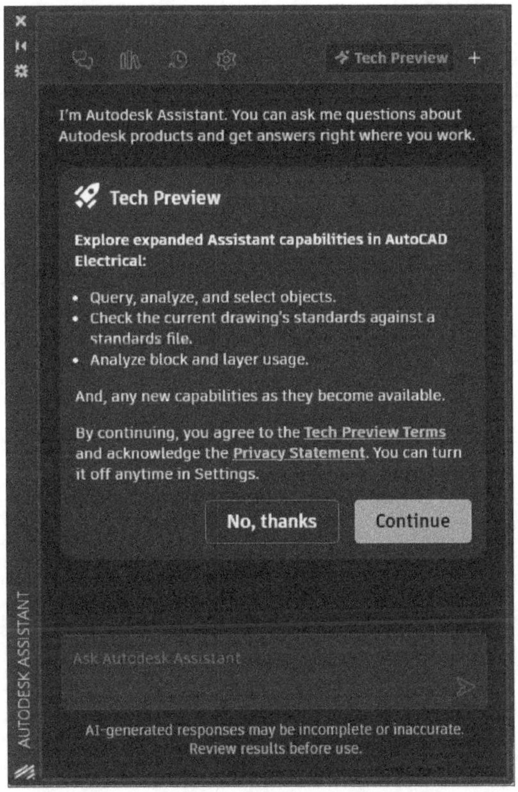

Figure-18. AUTODESK ASSISTANT

Changing Color Theme

- Click on the **Application** button . The **Application Menu** will be displayed; refer to Figure-19.

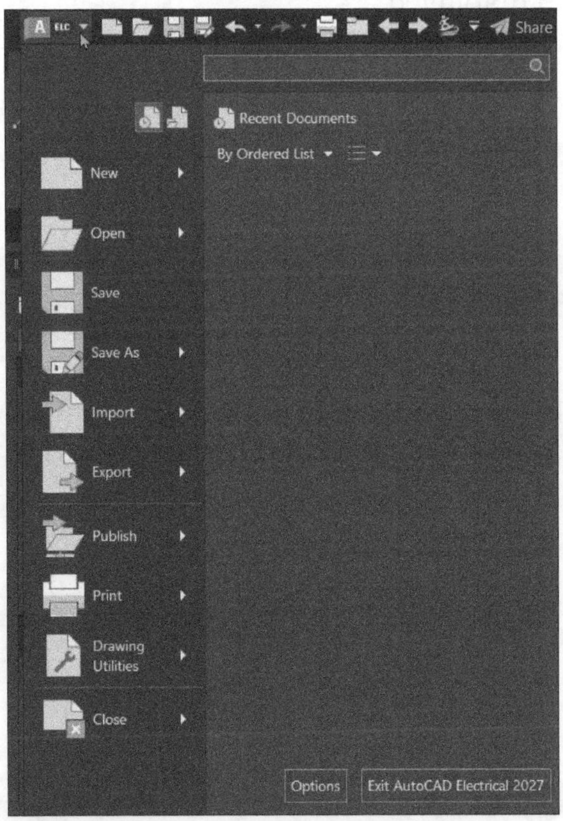

Figure-19. Application menu

- Click on the **Options** button at the bottom right of the **Application Menu**. The **Options** dialog box will be displayed.
- Click on the **Display** tab. The dialog box will be displayed as shown in Figure-20.

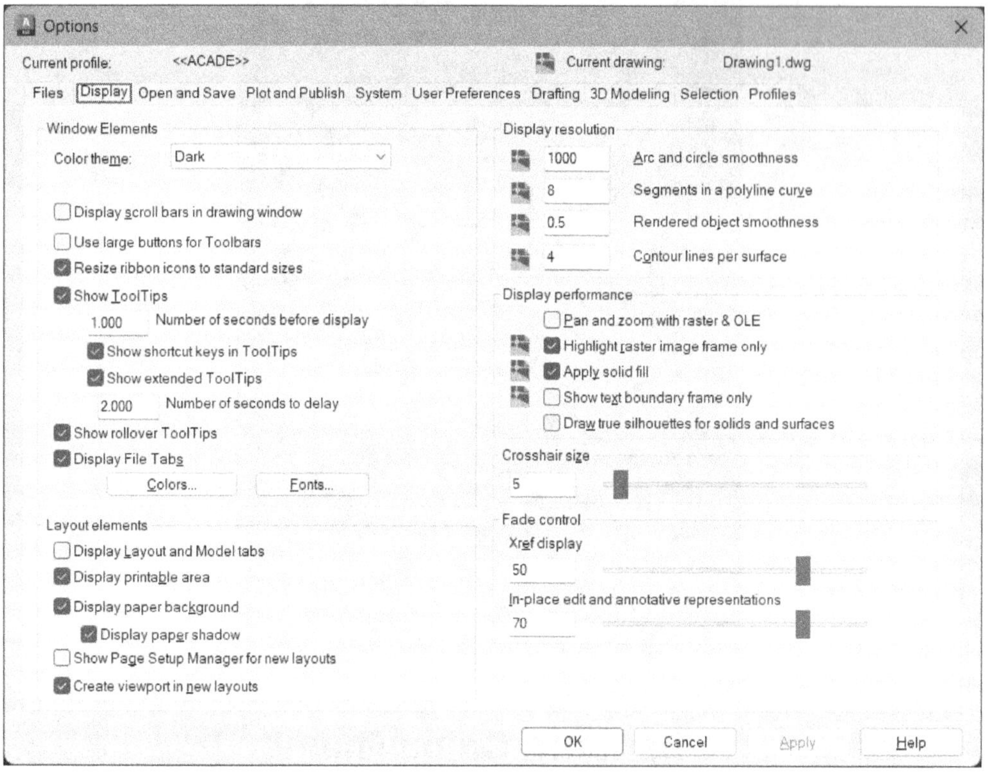

Figure-20. Options dialog box

- Click on the **Dark** option of **Color theme** drop-down in the **Window Elements** area. List of options will be displayed.
- Select the **Light** option from the drop-down; refer to Figure-21.

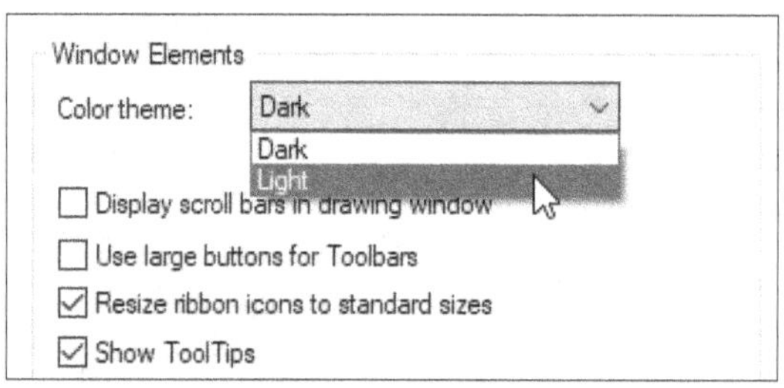

Figure-21. Light option in Color theme drop-down

- Click on the **Colors** button from same area in the dialog box. The **Drawing Window Colors** dialog box will be displayed as shown in Figure-22.

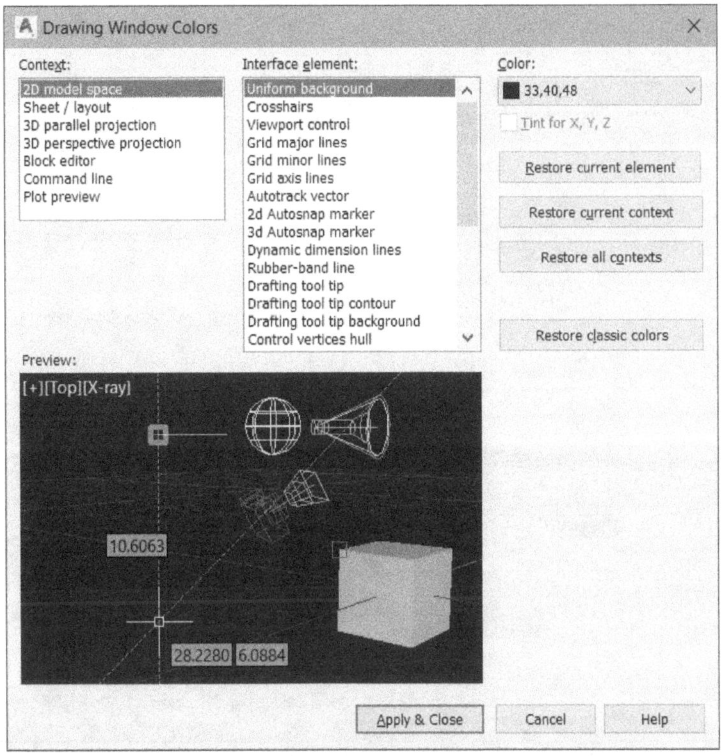

Figure-22. Drawing Window Colors dialog box

Note that using the options in this dialog box, you can change the color of any element of interface.

- Select the **Uniform background** option from **Interface element** list box, click on the drop-down for colors and select the **White** option; refer to Figure-23.

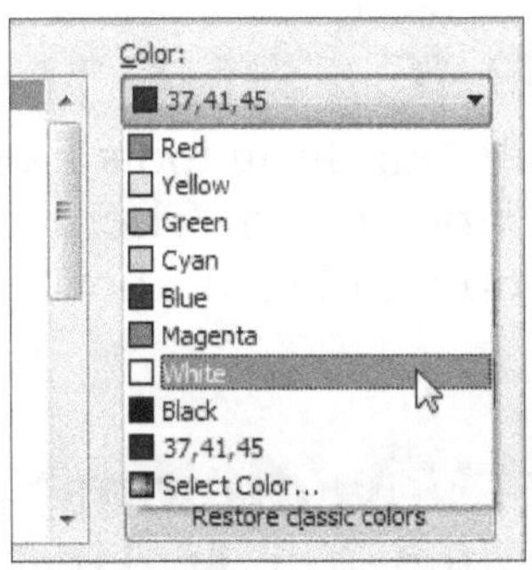

Figure-23. Color drop-down

- Similarly, one by one select the other background options from the **Interface element** box and set their colors to white for options in **Context** box; refer to Figure-24.

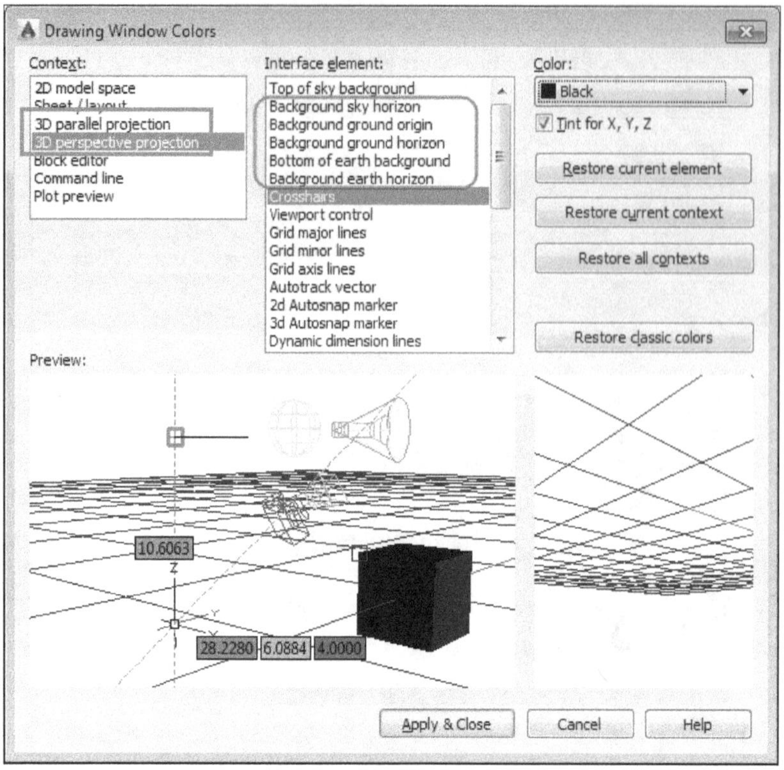

Figure-24. Background options

- Select the **Crosshairs** option from the **Interface element** box and select the **Black** option from the **Color** drop-down.
- Click on the **Apply & Close** button from the **Drawing Window Colors** dialog box and click on the **OK** button from the **Options** dialog box. The drawing area and interface will be displayed in colors as defined.

APPLICATION MENU

Application menu contains the file handling tools like creating new document, saving document, printing document, and so on. There are also options to define basic settings of the software. Now, you will learn about various tools and options of the **Application Menu**.

New options

- Click on the **Application** button. The **Application** menu will be displayed as shown in Figure-25 when no file is open in AutoCAD Electrical and as shown in Figure-26 when you have opened a file in AutoCAD Electrical. Click on the arrow next to **New** button in **Application** menu. More options for creating new documents will be displayed.

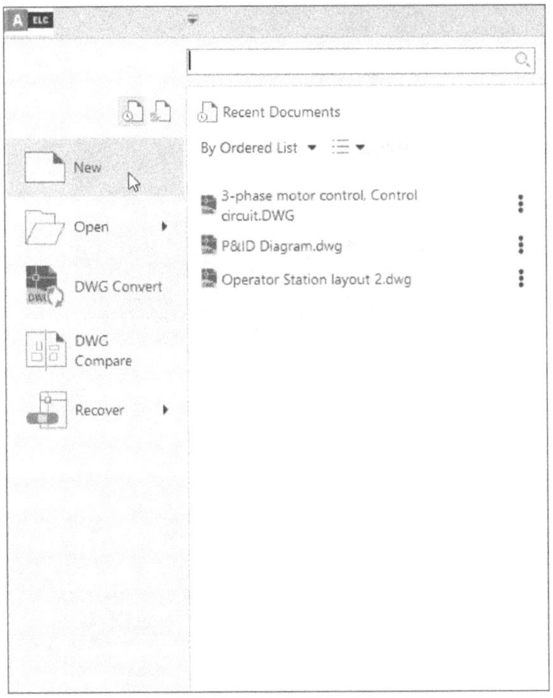

Figure-25. New option when no file is open

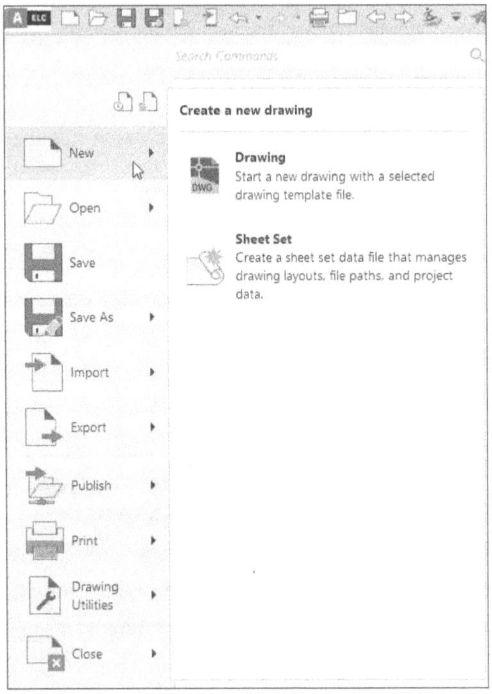

Figure-26. Options for new documents

- In AutoCAD Electrical, there are two options to create new documents; **Drawing** and **Sheet Set**.

Using the **Drawing** option, you can create an individual drawing file which contains model and various views of the model in orthographic projection. Using the **Sheet Set** option, you can create a group of inter-related drawing files in the form of sheets.

Creating Drawings

- AutoCAD Electrical also uses command prompt to perform various operations. In some cases, if you try to create a new drawing using **New** option of **File** menu then dialog box may not be displayed by default. Enter **FILEDIA** in the command prompt and then enter **1** in the command prompt; refer to Figure-27.

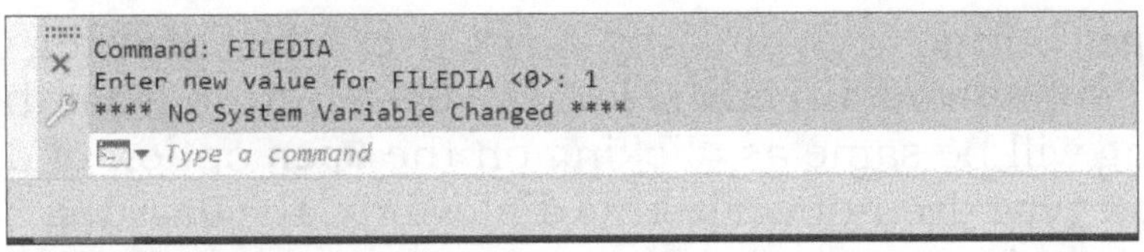

Figure-27. FILEDIA variable

- Click on the **Drawing** option from the **New** options. The **Select template** dialog box will be displayed; refer to Figure-28.
- Select desired template from the dialog box. Use of each template has already been discussed.
- If you want to create a document without using any of the template but with desired units then click on the down arrow next to **Open** button in the dialog box. A drop-down will be displayed; refer to Figure-29.

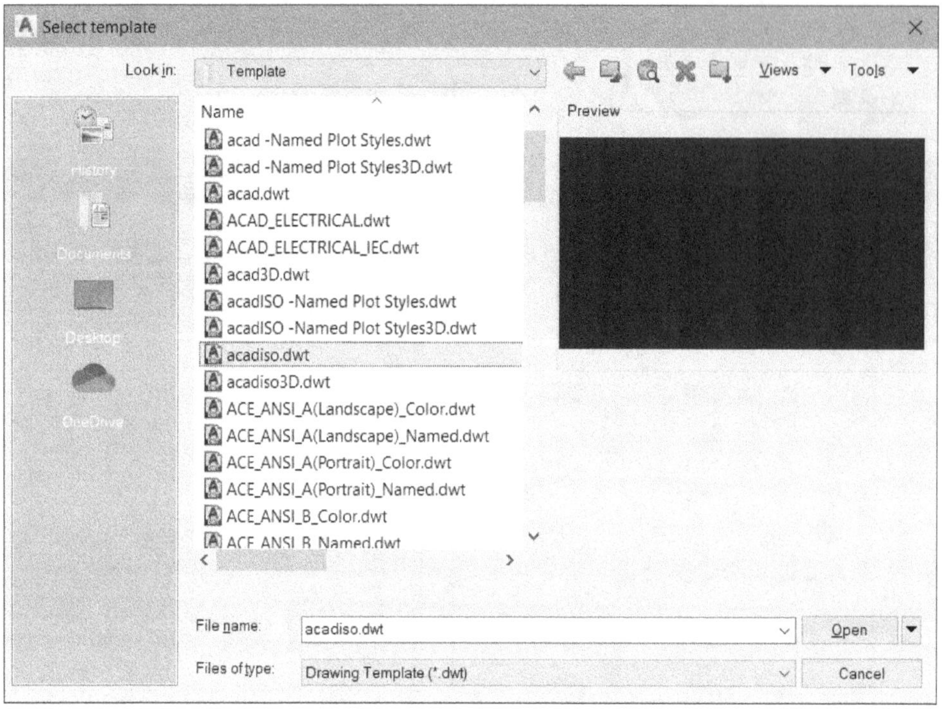

Figure-28. Select template dialog box

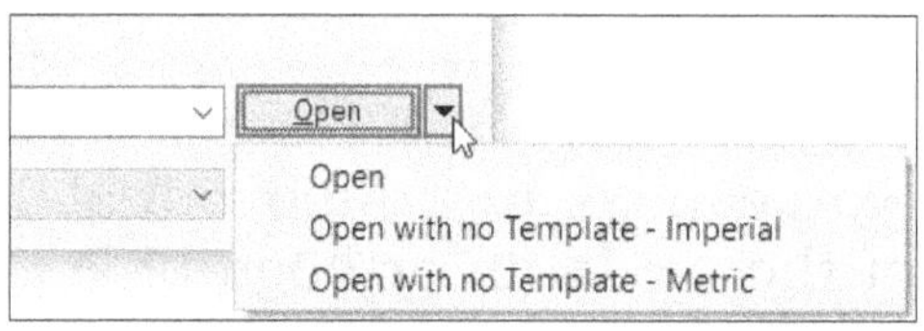

Figure-29. Open drop-down

- Click on the **Open with no Template - Imperial** button to work in Feet and Inches without loading any template.

Or

- Click on the **Open with no Template - Metric** button to work in millimeters without loading any template.

Or

- Click on the **Open** button from the drop-down to create a drawing using the selected template. Note that if you directly select the **Open** button from the dialog box then the functioning will be same as clicking on the **Open** option of drop-down.
- In the **Files of type** drop-down in this dialog box, the **Drawing Template (*.dwt)** option is selected which means you can select the templates. You can also use any previous drawing as a template. To do so, click on the **Drawing Template (*.dwt)** option. The **Files of type** drop-down will expand as shown in Figure-30.

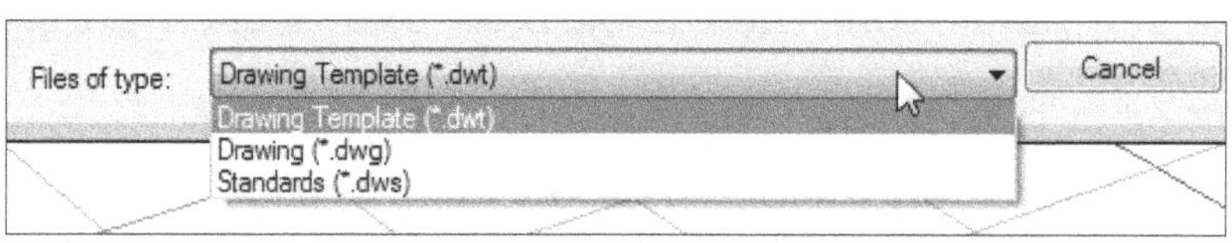

Figure-30. Files of type drop down

- Select the **Drawing (*.dwg)** option from the list and you will be able to select any drawing in place of template. Select the **Standards (*.dws)** option from the drop-down if you want to use a template saved as standard file in AutoCAD. Rest of the procedure is same.

Creating Sheet Sets

- Click on the **Sheet Set** option from the **New** options in the **Application Menu**. The **Begin** page of **Create Sheet Set** dialog box will be displayed; refer to Figure-31.

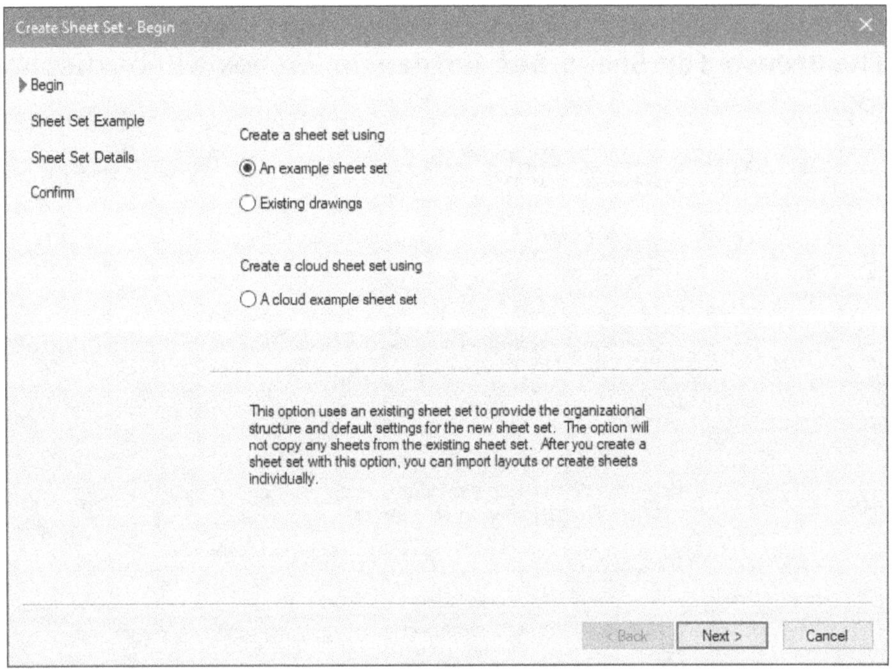

Figure-31. Begin page of Create Sheet Set dialog box

- If you want to use the example sheet set (template) then select the **An example sheet set** radio button from this page. You can club more than one drawing files in the form of a sheet set by using the **Existing drawings** radio button. If you want to create a cloud sheet then select the **A cloud example sheet set** radio button. We are using the **An example sheet set** radio button in our case.
- Click on the **Next** button from the dialog box. The **Sheet Set Example** page of **Create Sheet Set** dialog box will be displayed; refer to Figure-32.

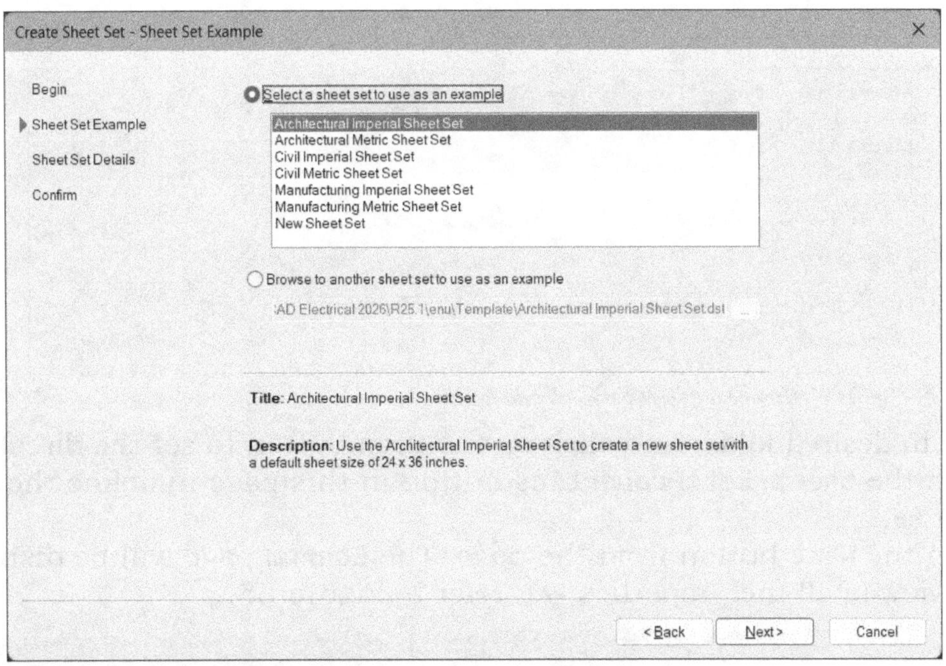

Figure-32. Sheet Set Example page of Create Sheet Set dialog box

- Select desired sheet set template from the box and click on the **Next** button from the dialog box. The **Sheet Set Details** page will be displayed as shown in Figure-33.
- Click in the **Name of new sheet set** edit box and specify the name of the sheet set as desired.
- Click on the **Browse** button next to **Store sheet set data file (.dst) here** edit box. The **Browse for Sheet Set Folder** dialog box will be displayed as shown in Figure-34.

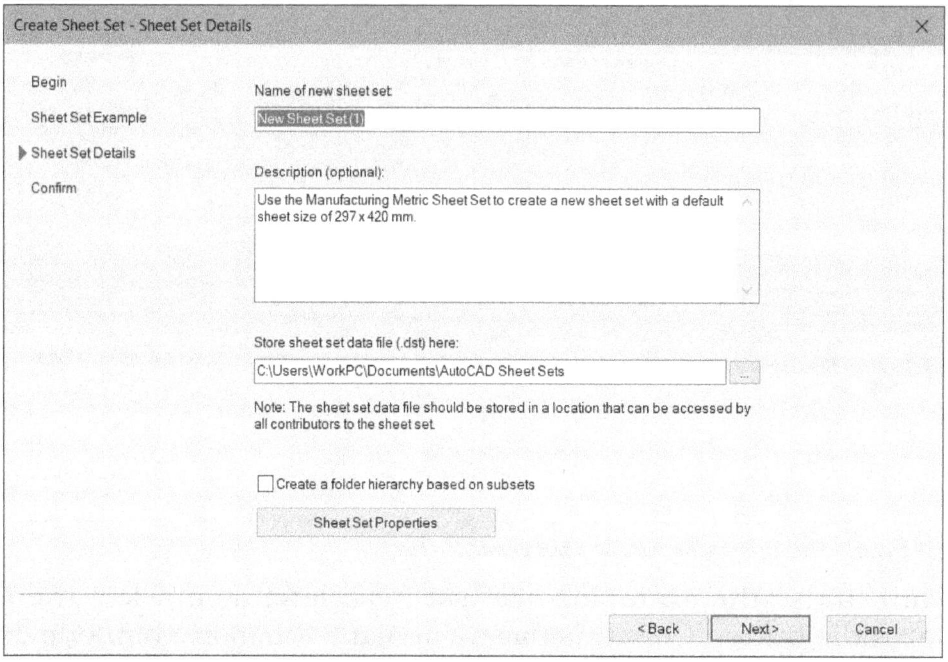

Figure-33. Sheet Set Details page of Create Sheet Set dialog box

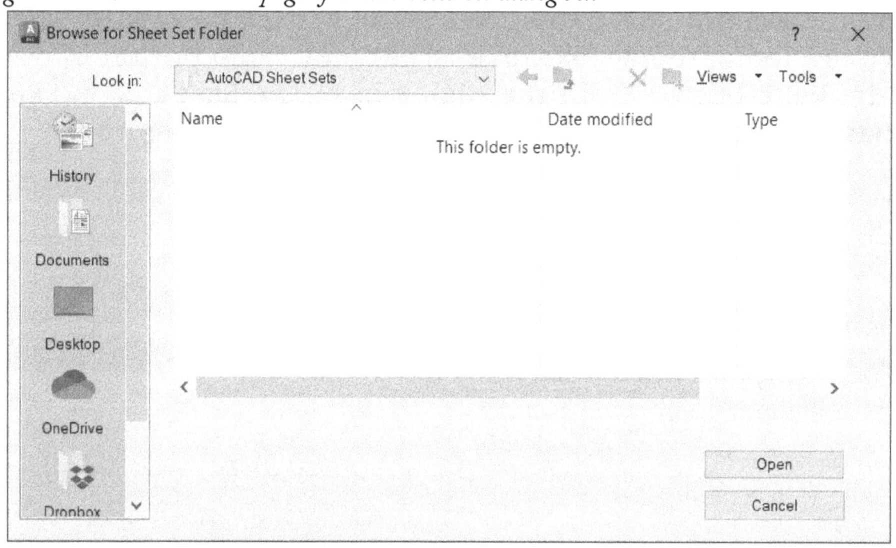

Figure-34. Browse for Sheet Set Folder dialog box

- Browse to desired folder and click on the **Open** button to set the directory.
- Click on the **Sheet Set Properties** button in this page to define the properties of sheet set.
- Click on the **Next** button from the page. The **Confirm** page will be displayed with the preview of all the properties set; refer to Figure-35.

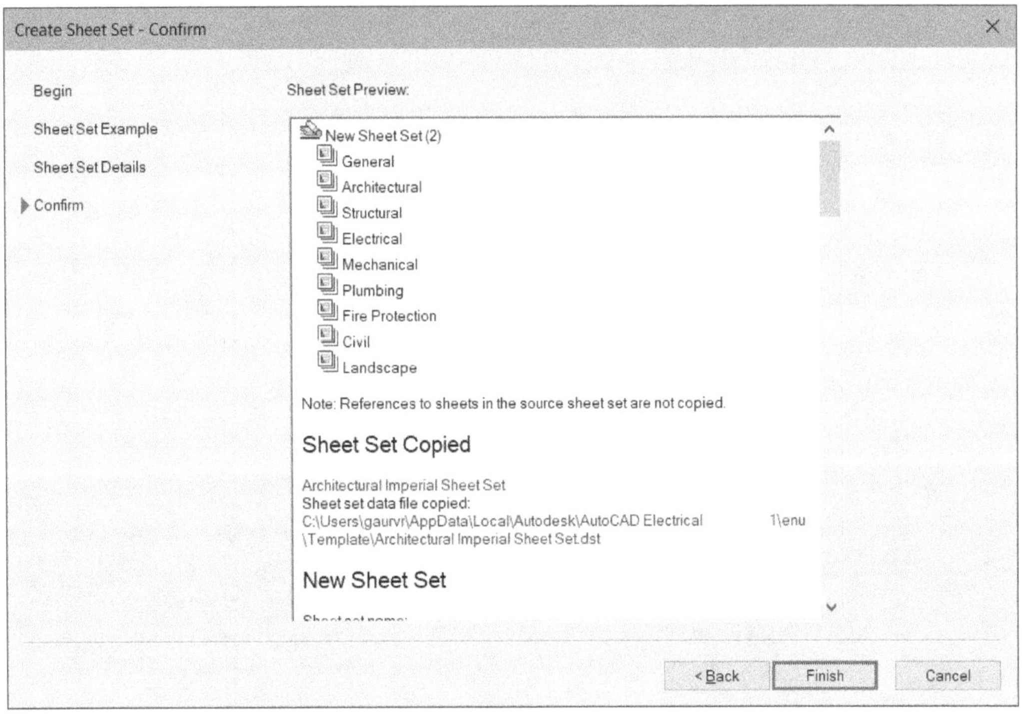

Figure-35. Confirm page of Create Sheet Set dialog box

- Click on the **Finish** button from the page. The **SHEET SET MANAGER** will be displayed with the current sheet set; refer to Figure-36.

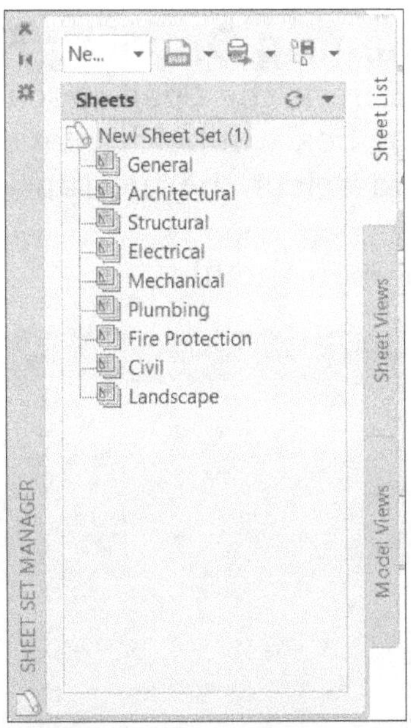

Figure-36. Sheet Set Manager

Open Options

The options to open different type of files are available in the **Open** cascading menu of **Application** menu; refer to Figure-37. Using these options, you can open drawing file from local drive, you can open sheet sets from local drive, you can import DGN format files, and you can open sample files from local drive as well as online. Note that the procedure to open all the type of files is similar, so here we will discuss the procedures to open a drawing file from local drive and from AutoCAD Web.

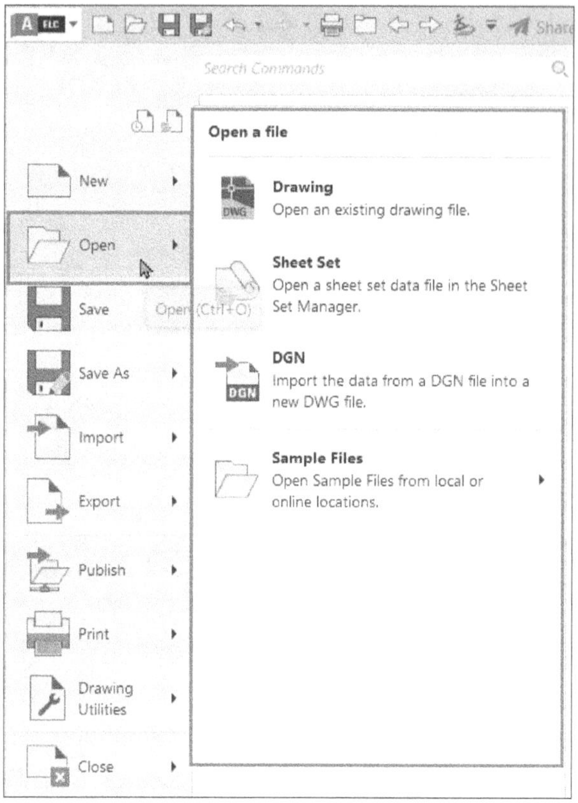

Figure-37. Open cascading menu

Opening Drawing File

- Click on the **Drawing** option from the **Open** cascading menu in the **Application Menu**. The **Select File** dialog box will be displayed as shown in Figure-38.
- Browse to desired folder and select the file that you want to open in AutoCAD Electrical.
- Click on the **Open** button to open the file.

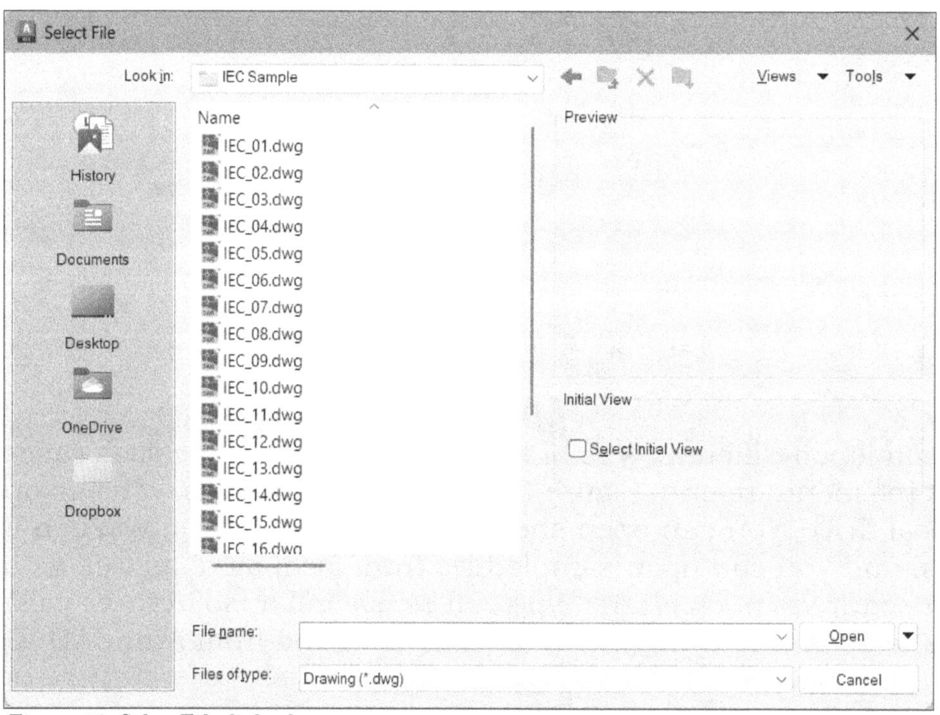

Figure-38. Select File dialog box

Note that there are three more methods to open a file; opening as read only, opening partially, and opening as read only with partial content. These options are displayed on clicking at the down arrow next to **Open** button once you have selected a drawing with multiple layers; refer to Figure-39. Partially opening a drawing file mean skipping some of the layers that are of no use to us. A layer can be assumed as transparent sheet on which we draw something in AutoCAD Electrical. In AutoCAD Electrical, each drawing is stacking of multiple layers one over the other.

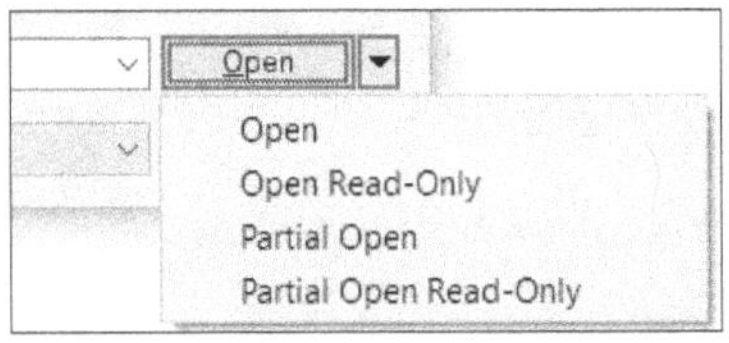

Figure-39. Open options

- If you want to open the file as read only and don't want to make the changes in it then select the **Open Read-Only** option from the drop-down.
- If you want to open a drawing file partially, then select the **Partial Open** option from the drop-down. The **Partial Open** dialog box will be displayed as shown in Figure-40.

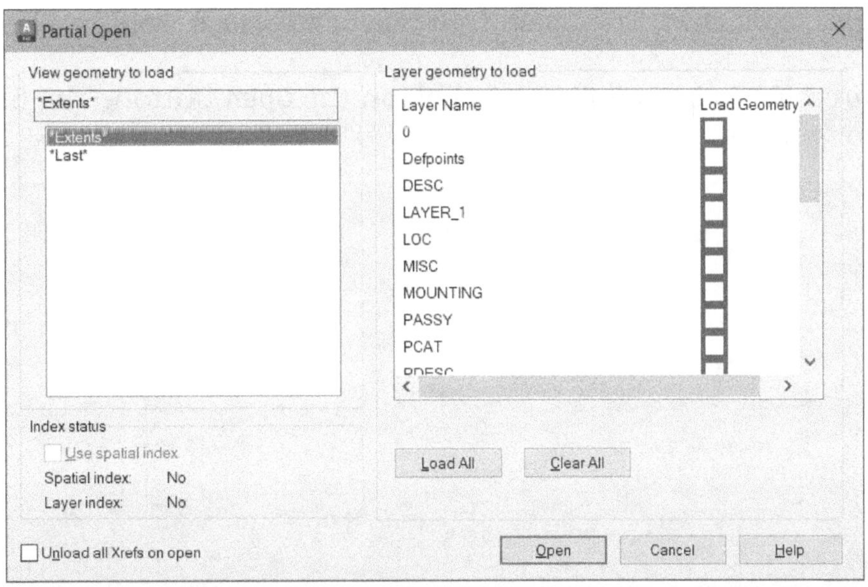

Figure-40. Partial Open dialog box

- Select the check boxes corresponding to the layers that you want to include while opening the drawing.
- Click on the **Open** button from the dialog box. The file will open with only the selected layers.

Similarly, you can open a file partially and read only by using the **Partial Open Read-Only** option.

Opening Drawing from AutoCAD Web & Mobile (Commercial version)

- Click on **Drawing from AutoCAD Web & Mobile** option from the **Open** cascading menu of the **Application** menu. An update dialog box will be displayed; refer to Figure-41.

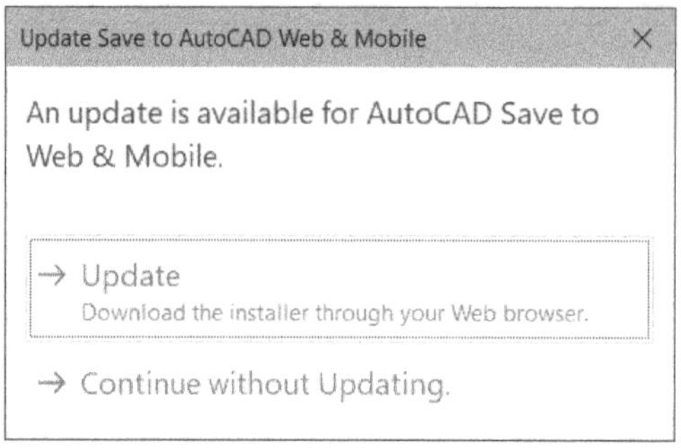

Figure-41. Update Save to AutoCAD Web & Mobile dialog box

- Click on the **Update** button from the dialog box, save the setup file at desired location and install it. Note that you will be required to close the AutoCAD applications for installation. So, run the AutoCAD Electrical application again and start the tool again. The **Open from AutoCAD Web & Mobile** dialog box will be displayed; refer to Figure-42.
- Select desired file to be opened and click on the **Open** button. The file will open in application window.

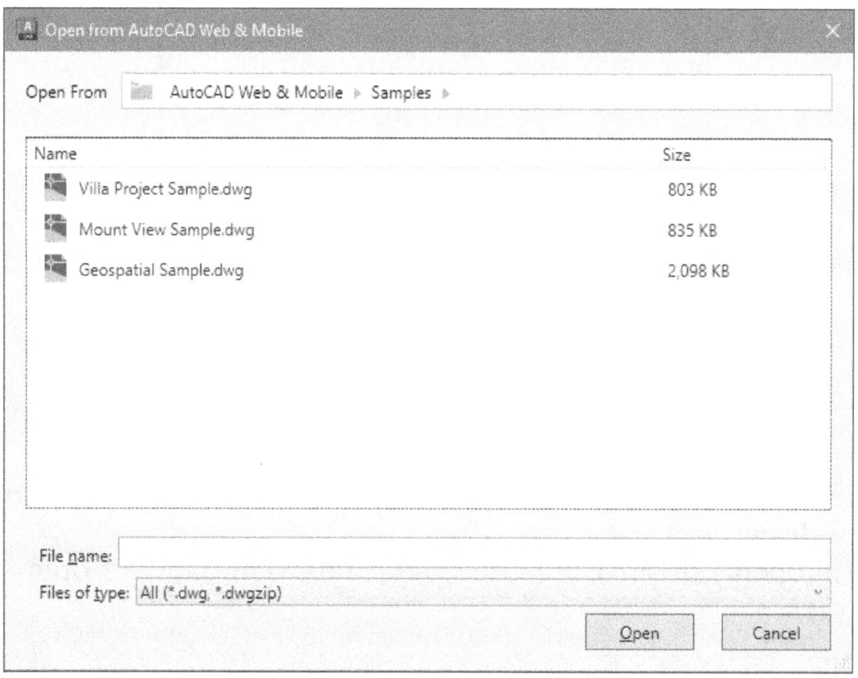

Figure-42. Open from AutoCAD Web & Mobile dialog box

Similarly, you can use **DGN** option in **Open** cascading menu to import DGN file in current drawing. You can open sample files from the **Sample Files** cascading menu.

Save

This option is used to save current opened file in the local drive for its later use. The steps to save the file are given next.

- Click on the **Save** option from the **Application** menu. The **Save Drawing As** dialog box will be displayed if you are saving the file for the first time, refer to Figure-43.

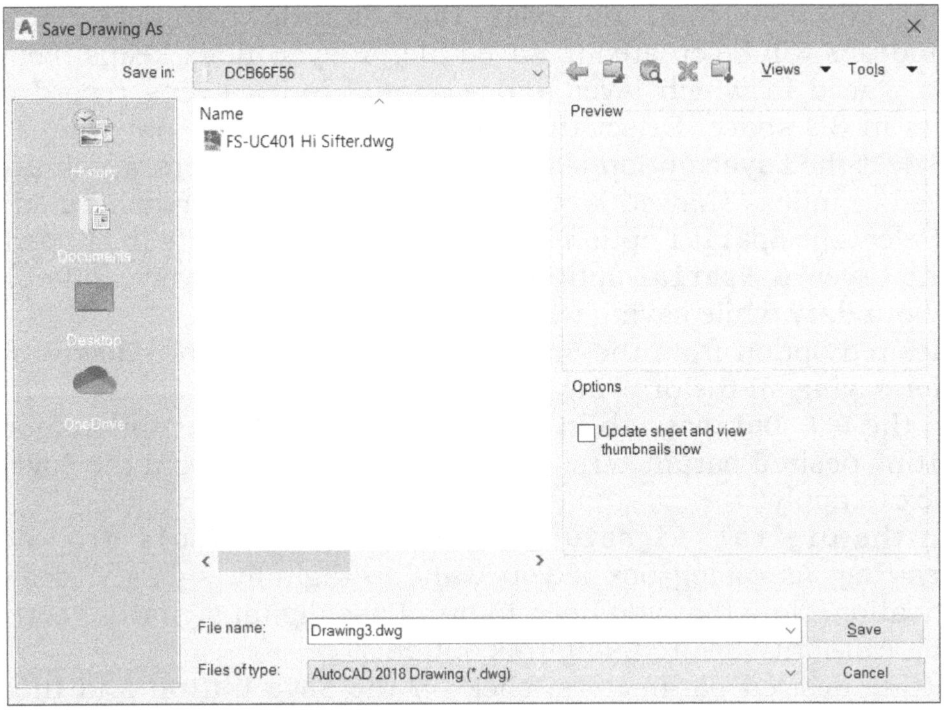

Figure-43. Save Drawing As dialog box

- Change the file name as desired and click in the **Files of type** drop-down to change the file format. On clicking at the **Files of type** drop-down, the list of formats is displayed as shown in Figure-44.

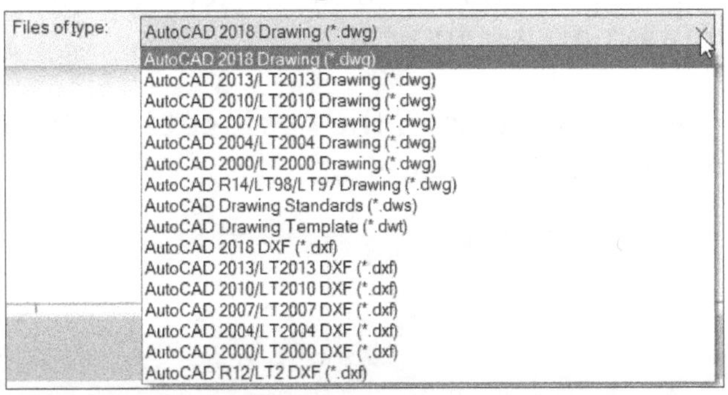

Figure-44. AutoCAD Electrical file formats

- Browse to desired location where you want to save the file. Note that the buttons in the left area of the dialog box are used to access some common locations like Documents, Desktop, FTP, and so on.
- Click on the **Options** option from the **Tools** drop-down at the top right in the dialog box. The **Saveas Options** dialog box will be displayed; refer to Figure-45.

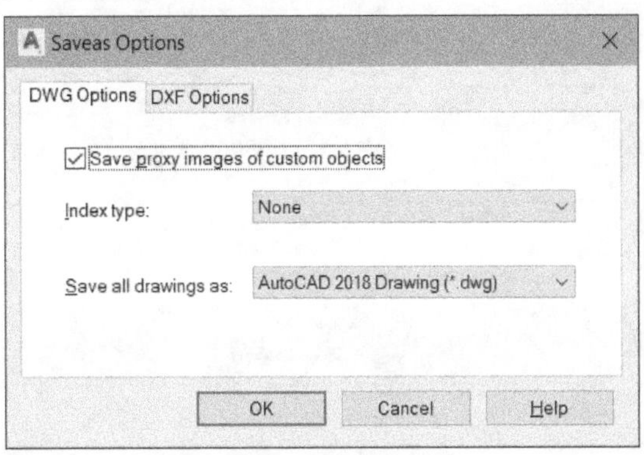

Figure-45. Saveas Options dialog box

- Select the **Save proxy image of custom objects** check box to display the image of custom objects in preview while opening the current file.
- Select desired option from the **Index type** drop-down to define how layer and spatial indexes will be created in saved file. A layer index keeps record of which object is placed in which layer and a spatial index keeps record of locations of objects in 3D space. Select the **None** option if you do not want to create an index. Select the **Layer** option if you want to include layers which are either ON or Thawed in index. Thawed layers are those which are regenerated after being frozen. Select the **Spatial** option if you want to load objects in clipped boundary. Select the **Layer & Spatial** option if you want to load layers and objects within clipped boundary while saving the file.
- Select desired option from the **Save all drawings as** drop-down to set default format for saving all the drawing files.
- Click on the **DXF Options** tab in the dialog box to define options for DXF files. After setting desired parameters, click on the **OK** button from the **Saveas Options** dialog box to apply.
- Click on the **Digital Signatures** option from the **Tools** drop-down in the **Save Drawing As** dialog box if you want to digitally sign the drawing for its authentication. Note that you need to purchase digital signature certificate from Certifying Authority before using this option.
- After setting desired parameters, click on the **Save** button and the file will be saved with specified settings.

Save As

This option is used to save the file in formats other than the default format (**.dwg**). The steps to use this option are given next.

- Click on the arrow next to **Save As** in the **Application Menu**. The **Save As** options will be displayed as shown in Figure-46.

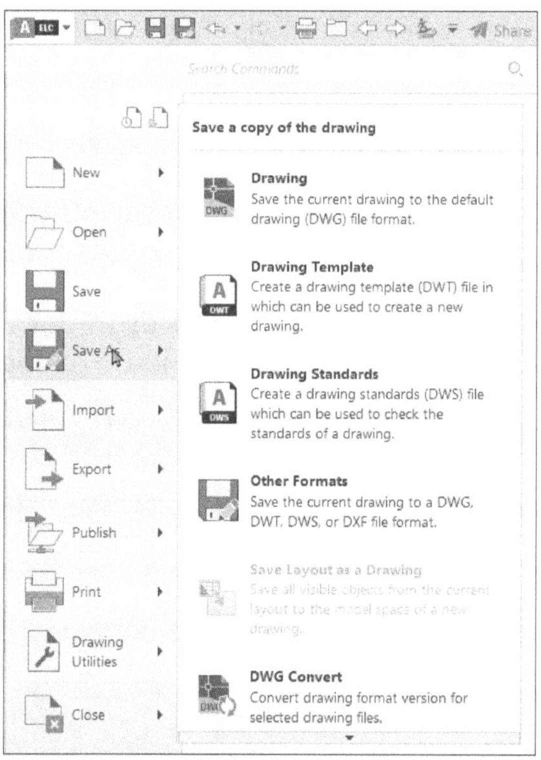

Figure-46. Save As options

- If you want to save the file with a format different from the AutoCAD Electrical format then click on the **Other Formats** button from the list of options. The **Save Drawing As** dialog box will be displayed as discussed earlier. Note that some additional formats will be displayed in the **Files of type** drop-down.
- Select desired file format from the list and click on the **Save** button from the dialog box to save the file. Note that later when we will work on 3D models and some more options will be available in this drop-down.

Import

The options in **Import** cascading menu are used to insert model from files of non-native (non-AutoCAD) formats. The procedure to use **Other Formats** tool of this cascading menu is given next. You can apply the same procedure to other tools in this cascading menu.

- Click on the **Other Formats** tool from the **Import** cascading menu of the **Application** menu. The **Import File** dialog box will be displayed; refer to Figure-47.

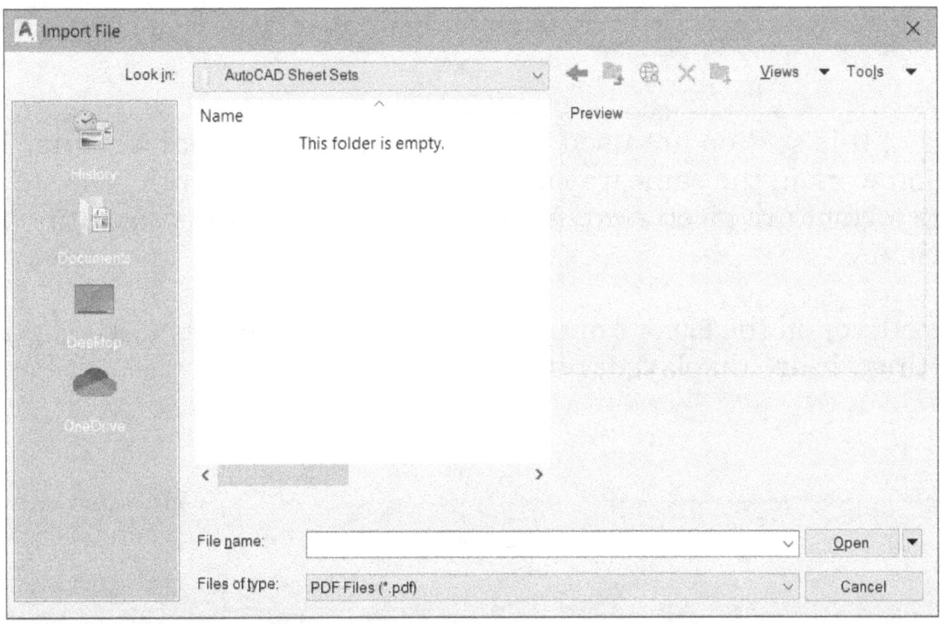

Figure-47. Import File dialog box

- Select desired option from the **Files of type** drop-down to define format of the file to be imported.
- Browse to the directory of file and double-click on the file to open it. A message box will be displayed notifying that import process is running in the background.
- Click on the **Close** button from the message box. Depending on the format of file selected for import, you may get a dialog box like in our case; refer to Figure-48 (for importing PDF).
- Set desired parameters in the dialog box and click on the **OK** button. The file will be imported.

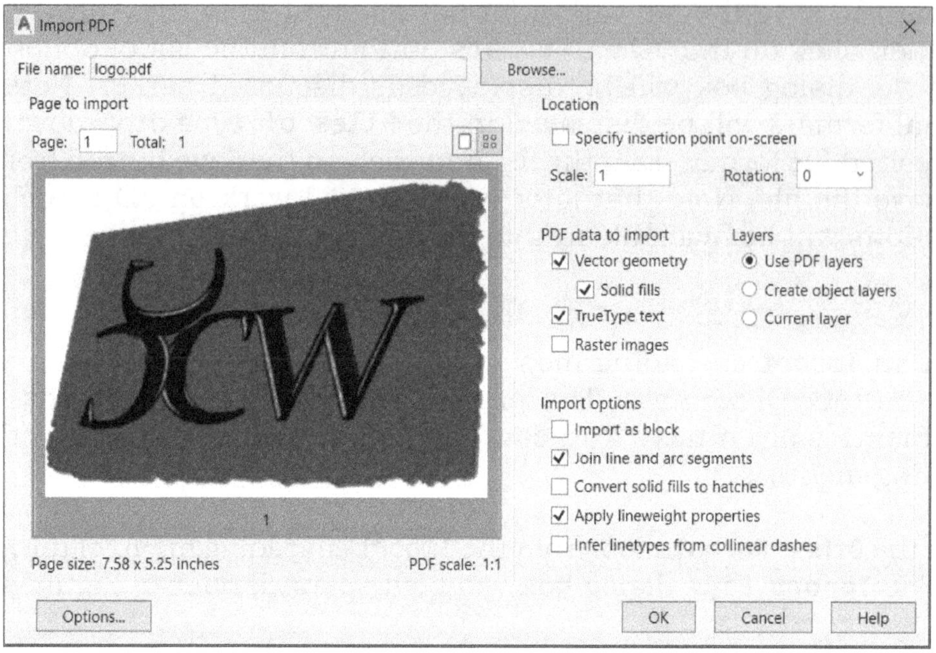

Figure-48. Import PDF dialog box

Export

The options in this section are used to save the file in external formats. The options in this section work in the same way as the **Save As** options work. The steps to export a file in iges format are given next. You can apply the same procedure to save files in other formats.

- Hold the cursor on the **Export** option in the **Application Menu**. The options related to exporting file are displayed; refer to Figure-49.

Figure-49. Export options

- Click on the **Other Formats** button from the list. The **Export Data** dialog box will be displayed as shown in Figure-50.

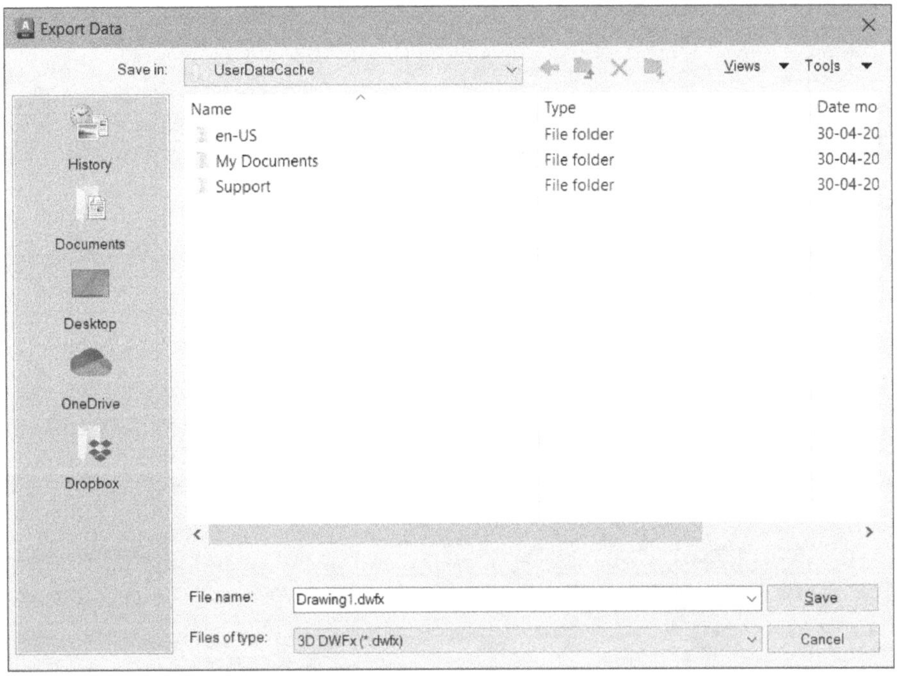

Figure-50. Export Data dialog box

- Specify desired name in the **File name** edit box and click on the **Files of type** drop-down. The list of formats will be displayed as shown in Figure-51.

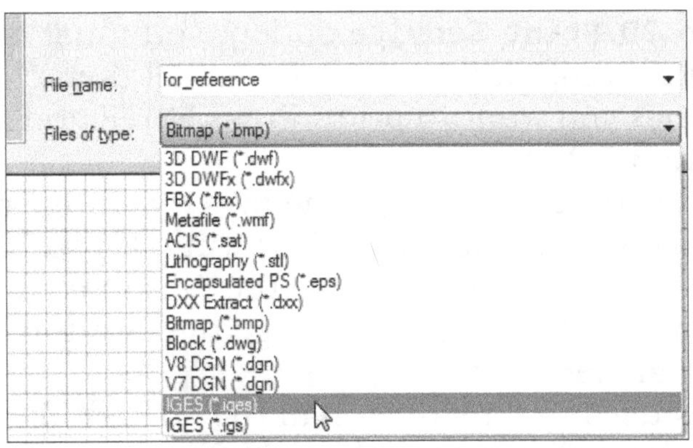

Figure-51. Formats for export

- Select the **IGES (*.iges)** option from the list to save file in IGES format. Click on the **Save** button from the dialog box to save the file with the selected format.

Publish

The options for this section are used to package and transfer the file to external sources like 3D printer, web services, and e-mail transfer; refer to Figure-52.

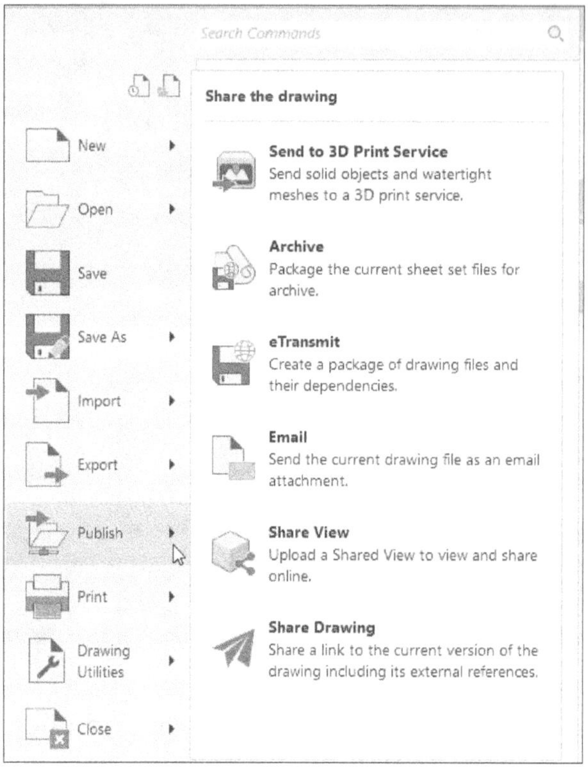

Figure-52. Publish options

- To use the **Send to 3D Print Service** option, you must have a 3D model which can be sent for 3D printing. If you have the model then click on this button and select the models that you want to print. Press **ENTER** and follow the instructions given by software to 3D Print.
- The **Archive** option is used to package the files for transfer. Note that to use this option, you need to create a sheet set of drawing. If you have the sheet set then click on this button and you can package the files in a zip format.
- The **eTransmit** option is also used to package the files but it compresses the files for fast electronic transfer.
- The **Email** option is used to directly e-mail the current file to your client.
- The **Share View** option (in commercial version) is used to share the current view or model & layout views of the opened file online. Click on this option from the **Publish** cascading menu. The **Share View** dialog box will be displayed; refer to Figure-53. Set desired options in the dialog box and click on the **Share** button. A message box will be displaying informing you about running background process of sharing. Click on the **Proceed** button. The indication of running process will be displayed at the bottom right corner of application window; refer to Figure-54. Once the process is complete, click on the **View in Browser** link button from message box display at the bottom right corner of application window. The views will open in default browser. You can copy the link from browser and share it with your colleagues or clients.

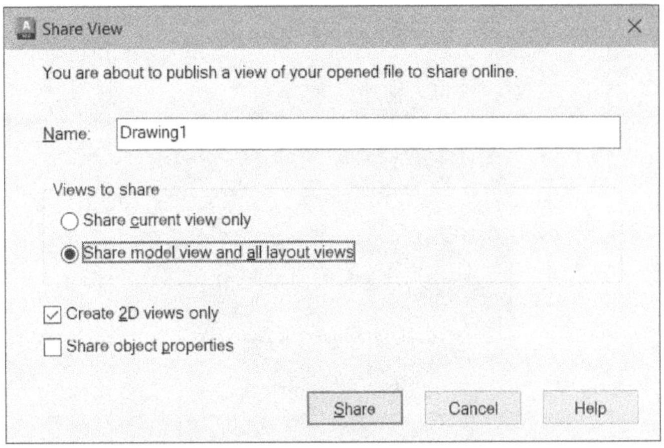

Figure-53. Share View dialog box

Figure-54. Indication of running process

- The **Share Drawing** option is used to share current drawing with others using a link. On clicking this option, the **Share a link to this drawing** dialog box will be displayed; refer to Figure-55. You can copy the link for sharing via E-mail and other methods. Select the **View only** button if you want to share the file only for viewing. Select the **Edit and save a copy** button to share the file as an edit-able copy. Click on the **Close** button at the top right corner of dialog box to exit.

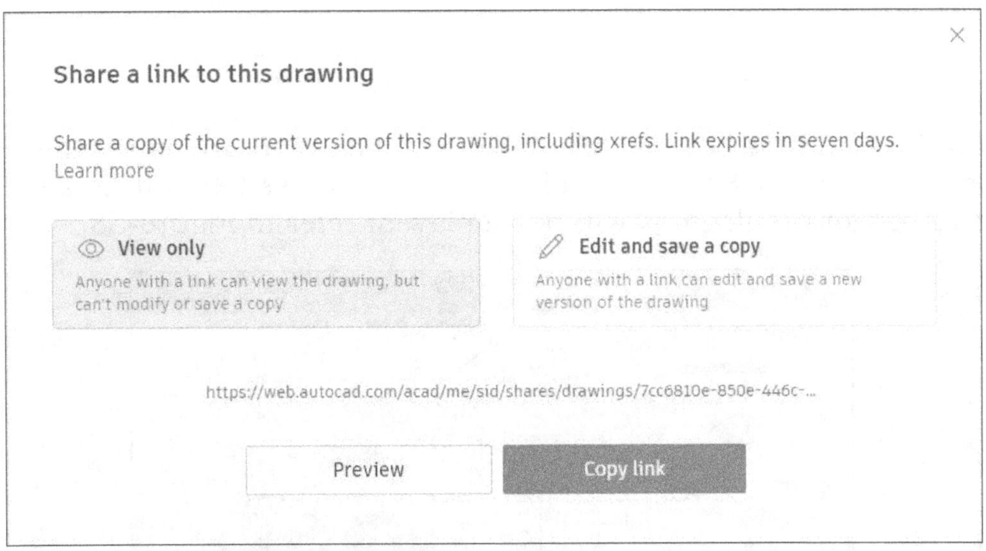

Figure-55. Share a link to this drawing dialog box

Sending Part for 3D Printing

Although, this topic is not of much interest for electrical control designers but since the technology is latest in CAD, so we are discussing it here. The steps for 3D printing through AutoCAD based software are given next.

- Create the 3D model as required using the modeling tools; refer to Figure-56.

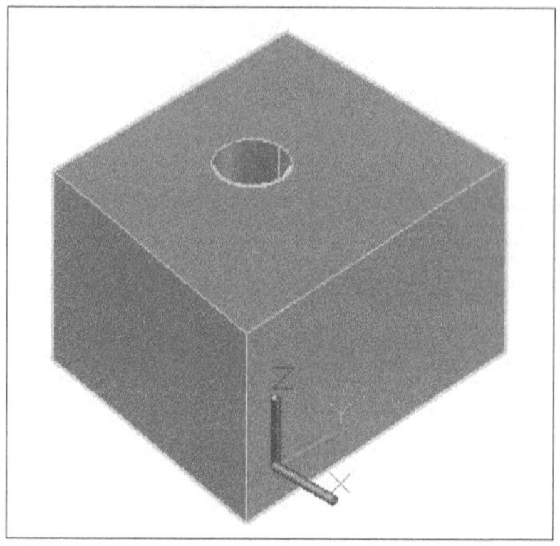

Figure-56. Model for 3D printing

- Click on the **Send to 3D Print Service** tool from the **Publish** cascading menu in the **Application** menu. The **3D Printing** dialog box will be displayed as shown in Figure-57.

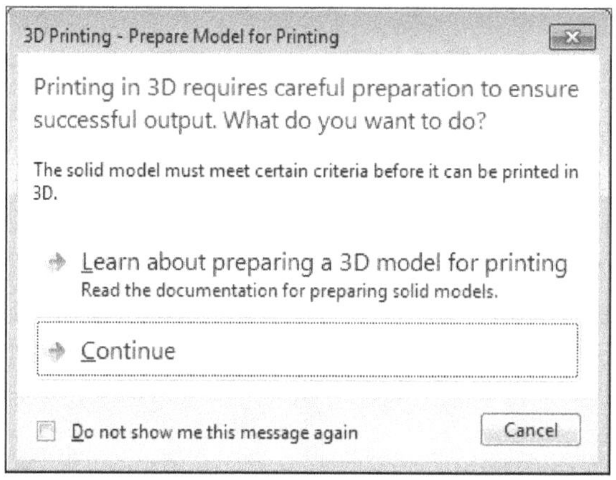

Figure-57. 3D Printing dialog box

- Click on the **Continue** button from the dialog box. You are asked to select solids or watertight meshes. Select the solid model and press **ENTER** from keyboard. The **3D Print Options** dialog box will be displayed; refer to Figure-58.

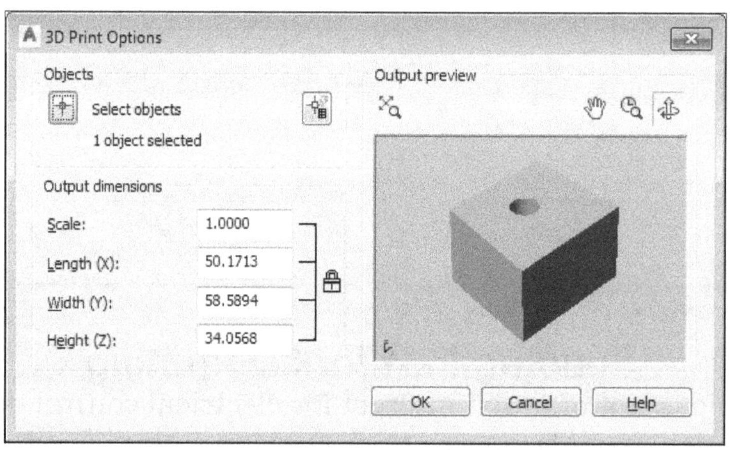

Figure-58. 3D Print Options dialog box

- Check the parameters in the dialog box and make the modifications if required.
- Click on the **OK** button from the dialog box. The **Create STL File** dialog box will be displayed; refer to Figure-59.

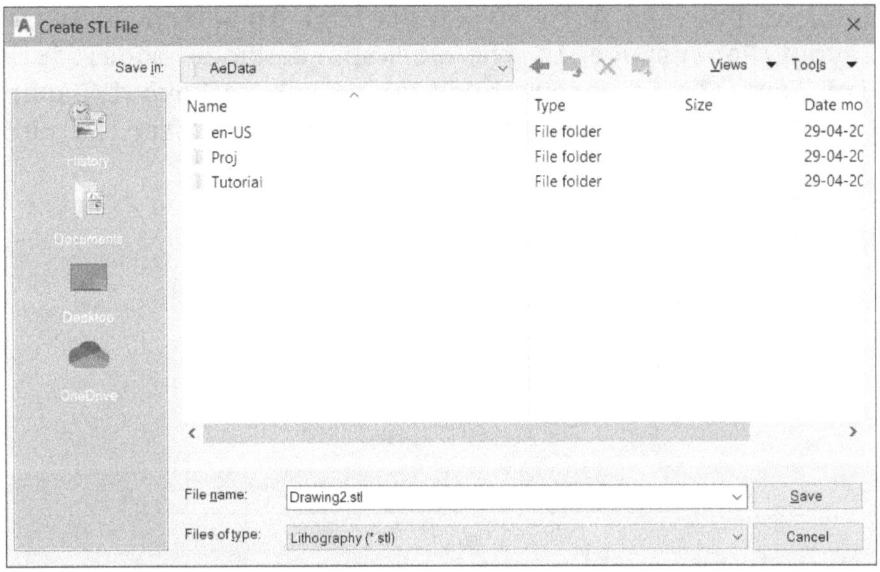

Figure-59. Create STL File dialog box

- Specify desired name of file in the **File name** edit box and click on the **Save** button from the dialog box. You can send this file to your 3D printing service provider to get 3D printed model.

Print

Printing is an important requirement of the CAD industry. All the designs are manufactured and controlled by printed copy of the drawing. To get the print out follow the steps given next.

- Click on the **Plot** option from the **Print** cascading menu in the **Application** menu. The **Plot** dialog box will be displayed for **Model** option; refer to Figure-60.

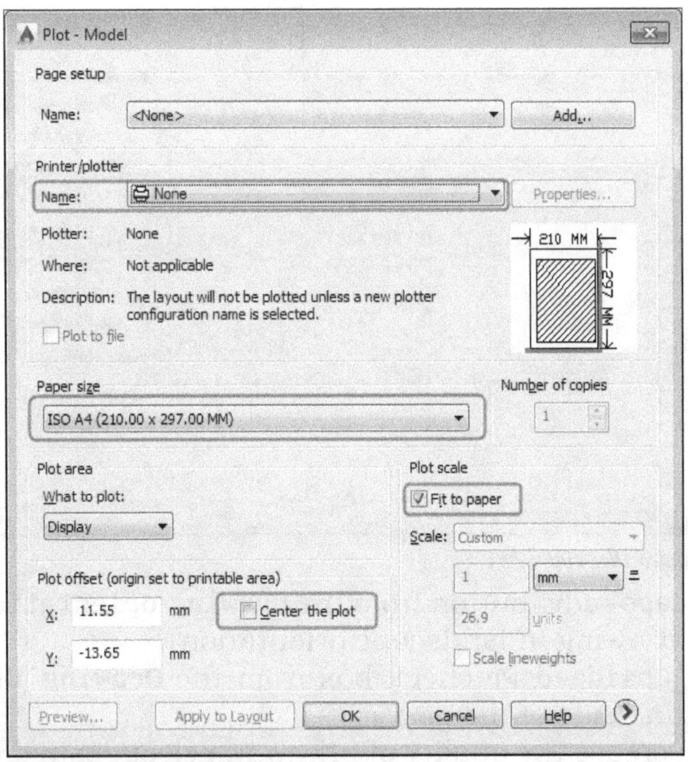

Figure-60. Plot dialog box

(Model is the workspace in AutoCAD Electrical in which 3D view of model is displayed. Layout/Paperspace is the workspace where model is displayed in its paper print form. It is recommended to switch to Layout/Paperspace before printing on the paper. The procedure is discussed later in next topic.)

- Select the Printer/plotter from the **Name** drop-down in the **Printer/plotter** area of the dialog box; refer to Figure-61.

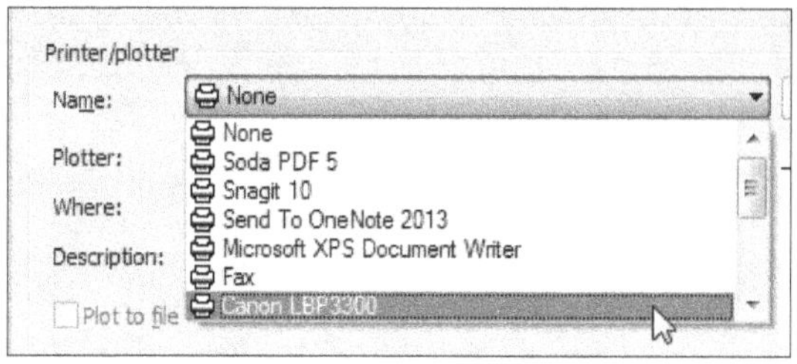

Figure-61. Printer selection

- Select the paper size from the drop-down in the **Paper size** area of the dialog box.
- Select the **Center the plot** check box to print the drawing at the center of the paper.
- Select the **Fit to paper** check box to fit it in the current paper size.
- Click on **More options** button or press **ALT + >** to expand the dialog box. The dialog box will be displayed as shown in Figure-62.

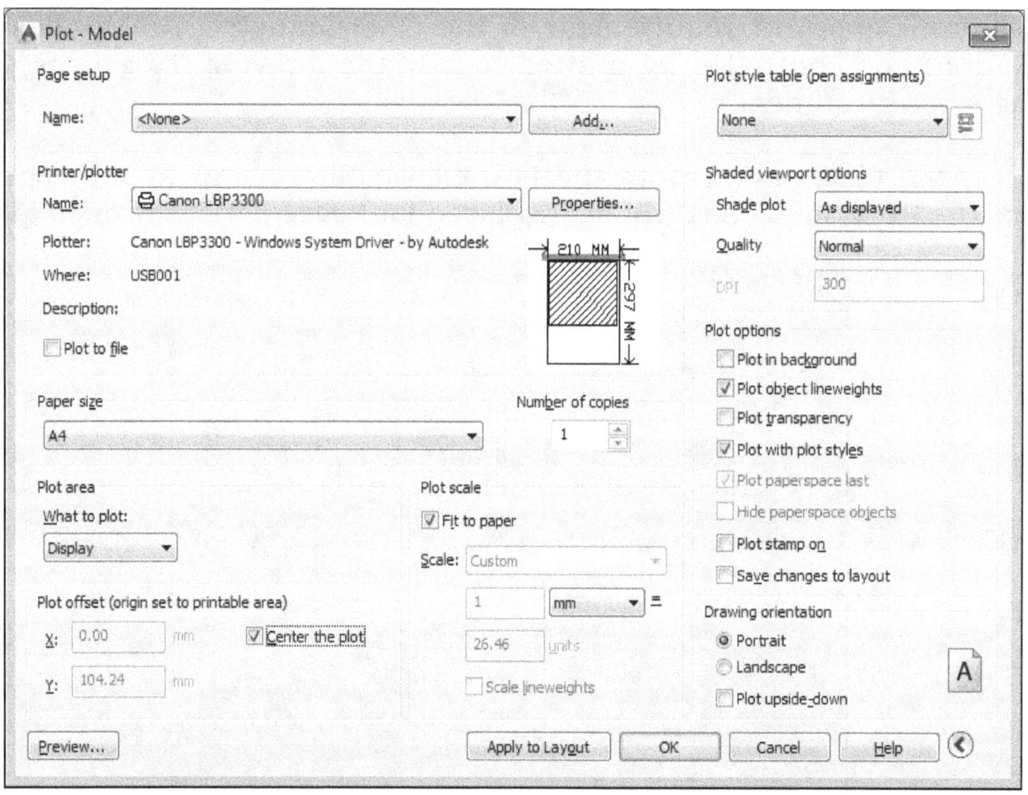

Figure-62. Expanded Plot dialog box

- Select the **Landscape** radio button from the **Drawing orientation** area of the dialog box to print the drawing in landscape orientation.
- Select the **Plot upside-down** check box from the **Drawing Orientation** area to reverse the printing side of the paper.
- To increase or decrease the quality of printout, select desired quality option from the **Quality** drop-down in the **Shaded viewport options** area.

- Set the other parameters as desired and click on the **OK** button from the dialog box to print the drawing.

FILE TAB BAR

The **File Tab bar** contains tabs for each drawing that is opened; refer to Figure-63.

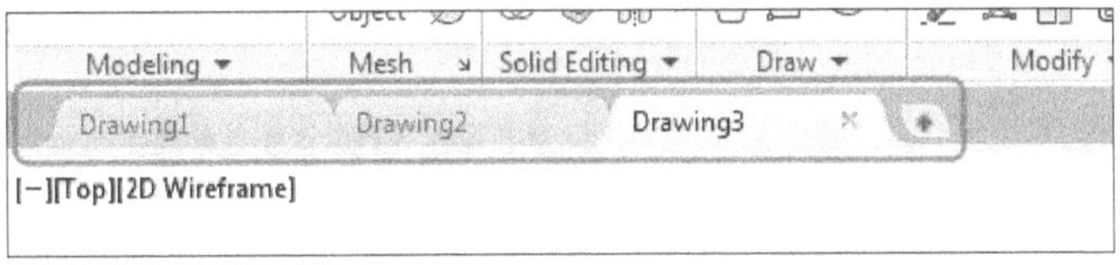

Figure-63. File Tab bar

There are various functions that can be performed by using the File Tab bar, which are discussed next.

- Hold the cursor over the tile of any drawing file in the File Tab bar. An interactive box will be displayed allowing you to switch between model and their paper spaces/ layouts; refer to Figure-64.

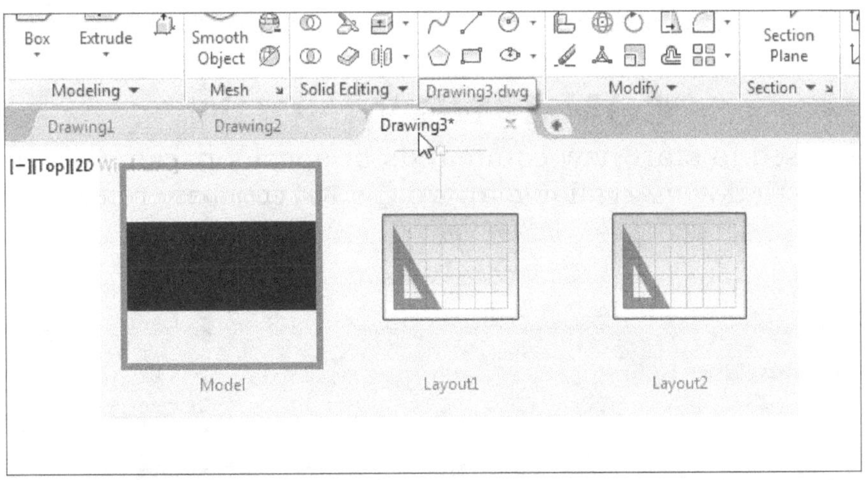

Figure-64. Interactive box for layout switching

- Click on the **Layout1** or **Layout2** to switch to paper space mode. You can switch back to Model space by using the same procedure.
- You can create a new drawing file by clicking on the plus sign next to the drawing file tab. On doing so, the Start page of AutoCAD Electrical will be displayed.
- Using the options in this page, you can start a new drawing as discussed earlier.

DRAWING AREA

The blank area in the application window below **File Tab bar** is called **Drawing Area**. This area is used to create sketches and models. Figure-65 shows the annotated drawing area. You will learn about more options related to drawing area later in this book.

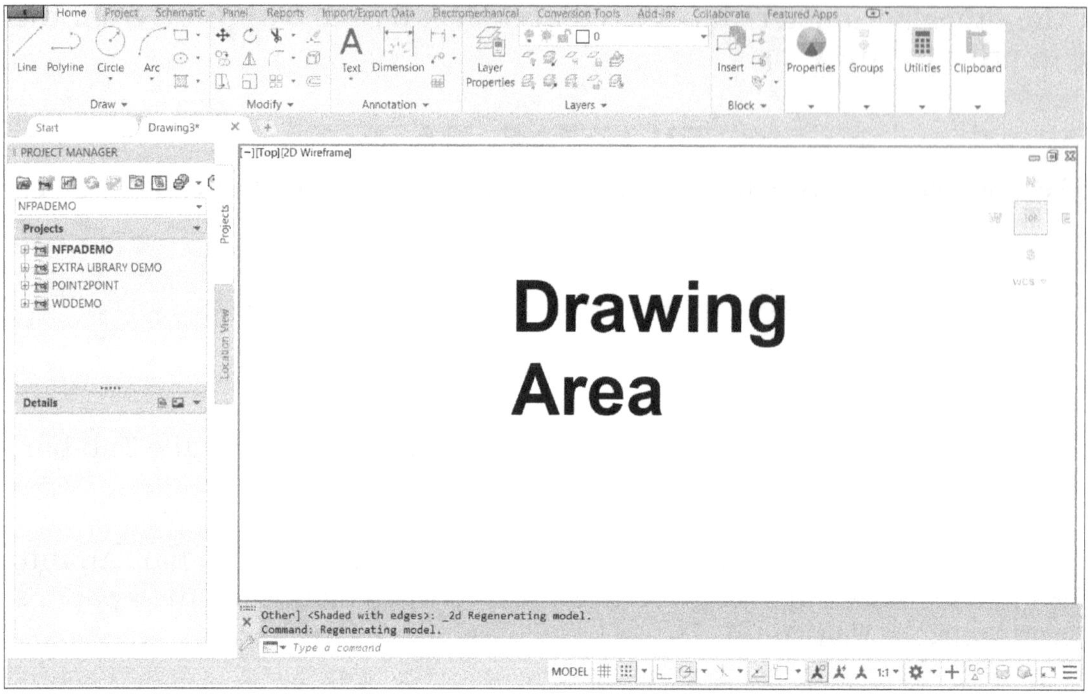

Figure-65. Drawing area

COMMAND WINDOW

This window is used to start new commands or specify parameters for the running commands. This window is available above the **Bottom bar**; refer to Figure-66.

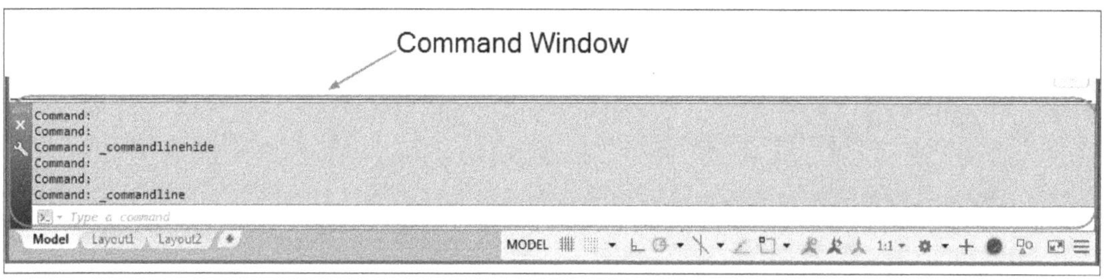

Figure-66. Command Window

You can press **CTRL+9** to display/hide the command window. The methods to use the command window are given next.

- Click in the **Type a Command** box in the command window. You are prompted to specify the command.
- Type a few alphabets of your desired command, a list of commands relevant to your specified alphabets will be displayed; refer to Figure-67.

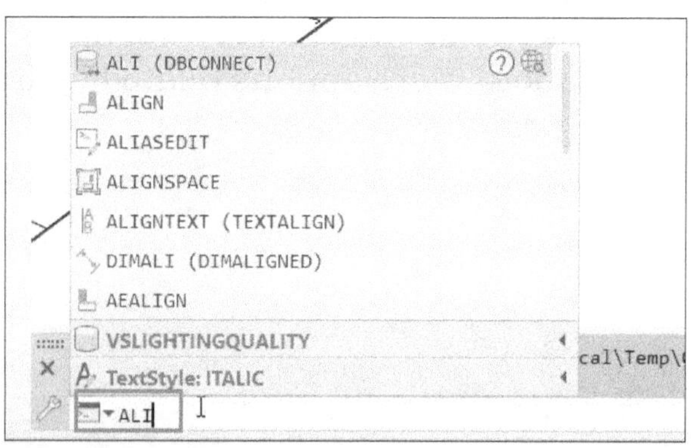

Figure-67. Command list

- Hold the cursor on the name of a command, the description of the command will be displayed; refer to Figure-68.

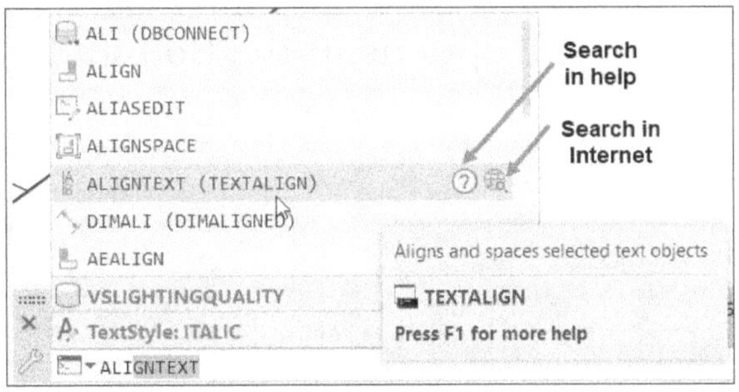

Figure-68. Tool description

- To know more about the highlighted command, click on the **Search in Help** or **Search on Internet** button next to the command name.
- You can scroll down in the box to browse the command not visible currently in the box. You can also specify the values of various variables.
- Click on the **<** sign next to the variables tile in the box; refer to Figure-69. The list of variables related to the specified alphabets will be displayed; refer to Figure-70.

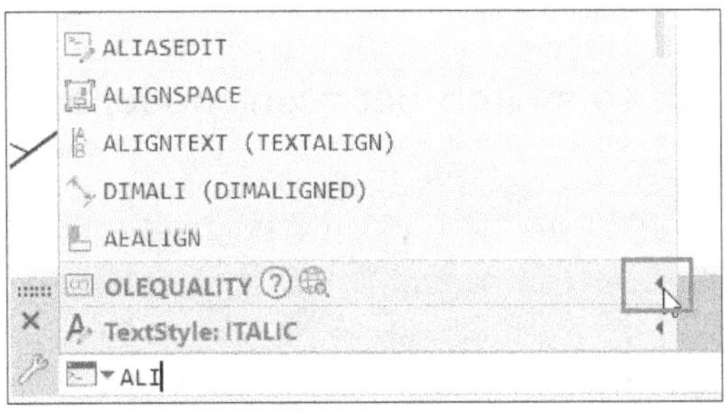

Figure-69. Variables

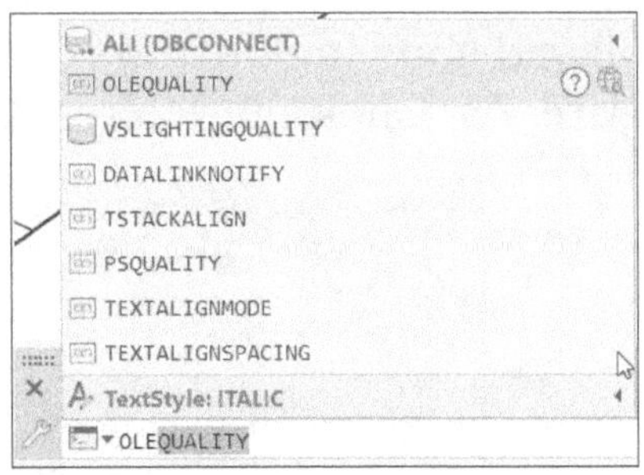

Figure-70. List of variables

- To start any command or specify value of any variable, click on it in the box. The respective options will be displayed in the command window and you will be prompted to specify desired values. For example, type **L** at the command prompt in the **Command Window** and click on the **Line** tool at the top in the command box. You will be prompted to specify the position of the starting point of the line; refer to Figure-71. Enter the coordinate values for first point in command prompt. Press **ESC** to exit the tool.

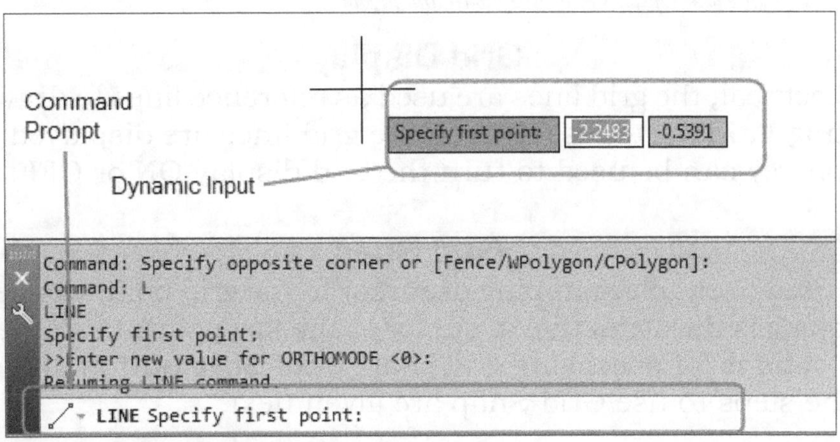

Figure-71. Command prompt

Bottom Bar

The tools in the **Bottom Bar** are used to perform various functions like creating a new layout or switching between existing ones, enabling orthogonal movement of cursor and so on; refer to Figure-72. The options in the Bottom Bar are discussed next.

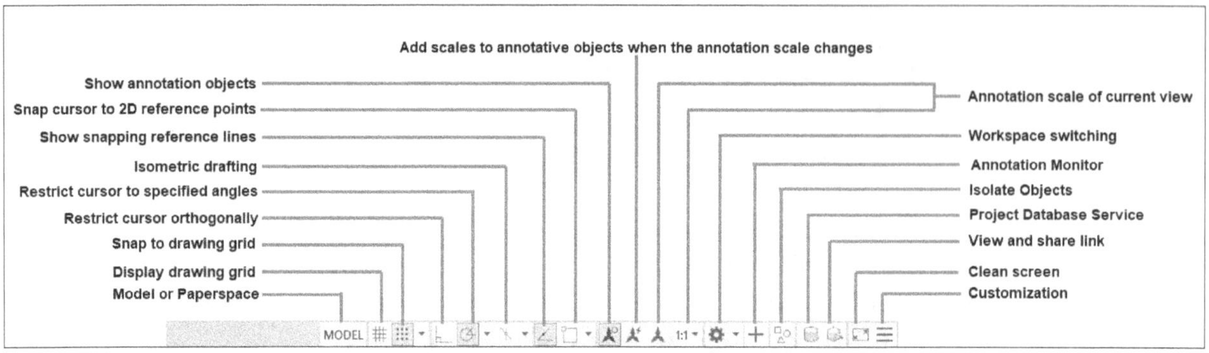

Figure-72. Bottom bar

Modelspace and Paperspace

In a layout, you can switch between paper space and model space by using the **MODEL** button in the Bottom Bar. To display layout or switch back to Model environment, hover the cursor on the File name tile in the Tile bar below **Ribbon**. An interactive box will be displayed showing buttons to switch between Model and Layout environment; refer to Figure-73. Click on desired button to switch between model and layouts.

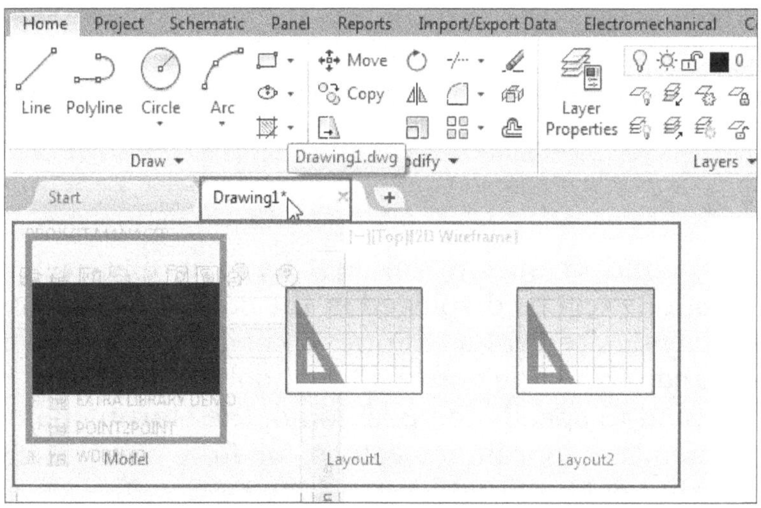

Figure-73. Options to switch model and layout

Grid Display

In AutoCAD Electrical, the grid lines are used as reference lines to draw objects. If the **Display Drawing Grid** button is ON then the grid lines are displayed on the screen. The **F7** function key can be used to turn the grid display ON or OFF.

Grid Snap

This option is used to enable snapping of cursor to the grid points as per the specified settings. Snapping is the attraction of cursor to the key points of AutoCAD Electrical. There are two options for specifying grid snap: **Grid Snap** button and corresponding drop-down. The steps to use Grid Snap are given next.

- Click on the **Snap to drawing grid** button to enable/disable snapping of cursor to the key points on grid.

- Click on the down arrow next to **Snap to drawing grid** button; a flyout will be displayed as shown in Figure-74.

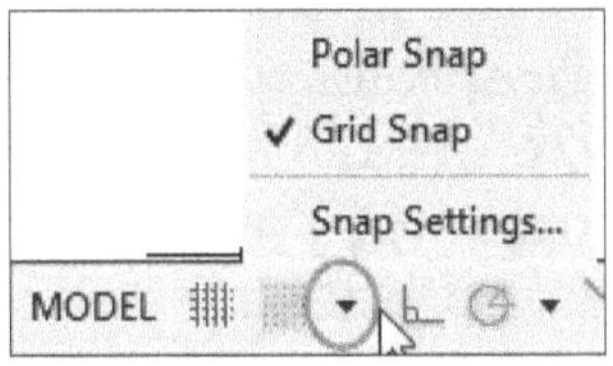

Figure-74. Snap flyout

- Select the **Grid Snap** option if you want to snap the cursor to the rectangular grid intersections. Select the **Polar Snap** option from the flyout if you want to snap the cursor to major angle lines.
- Select the **Snap Settings** option from the flyout to specify the setting related to selected type of Snap. The **Drafting Setting** dialog box will be displayed with the **Snap and Grid** tab selected; refer to Figure-75.

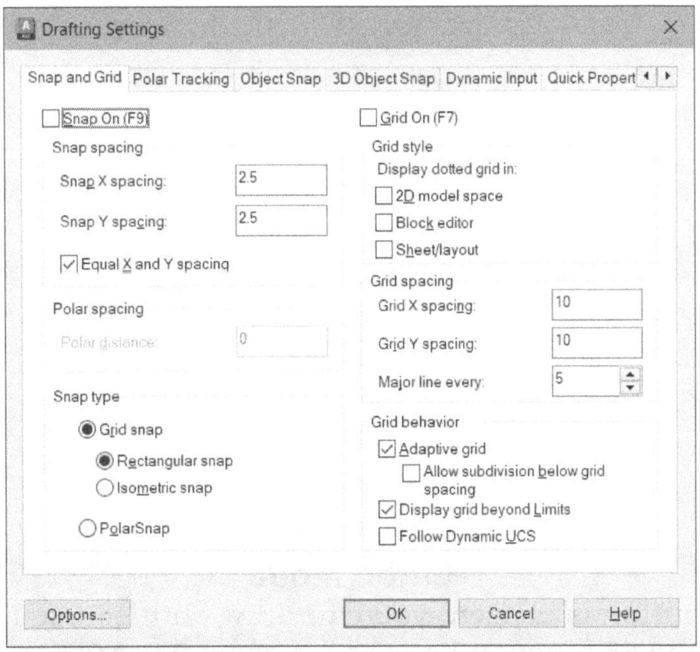

Figure-75. Drafting Settings dialog box

- After specifying desired settings, click on the **OK** button from the dialog box. The cursor will automatically start snapping to the key points as specified by the settings. Various options in the **Drafting Setting** dialog box are discussed next.

Drafting Settings dialog box

You can use the **Drafting Settings** dialog box to set the drawing modes such as grid, snap, Object Snap, Polar, Object snap tracking, and Dynamic Input. All these modes help you draw accurately and also increase the drawing speed. You can right-click on the **Snap Mode**, **Grid Display**, **Ortho Mode**, **Polar Tracking**, **Object Snap**, **3D Object Snap**, **Object Snap Tracking**, **Dynamic Input**, **Quick Properties**, or **Selection Cycling** button in the **Status Bar** to display a shortcut menu. In this shortcut menu, choose **Settings** option to display the **Drafting Settings** dialog box as shown in Figure-75. This dialog box has seven tabs: **Snap and Grid**, **Polar Tracking**, **Object Snap**, **3D Object Snap**, **Dynamic Input**, **Quick Properties**, and **Selection Cycling**. On starting AutoCAD Electrical, these tabs have default settings. You can change them according to your requirements.

Snap and Grid

Grid lines are the checked lines on the screen at predefined spacing, see Figure-76. In AutoCAD Electrical, by default, the grids are displayed as checked lines. You can also display the grids as dotted lines, refer to Figure-77. To do so, select the **2D model space** check box in the **Grid style** area. These dotted lines act as a graph that can be used as reference lines in a drawing. You can change the distance between grid lines as per your requirement. If grid lines are displayed within the drawing limits, it helps to define the working area. The grid also gives you an idea about the size of the drawing objects. To display the grid up to the limits, clear the **Display grid beyond Limits** check box. Now, the grids will be displayed only up to the limits set.

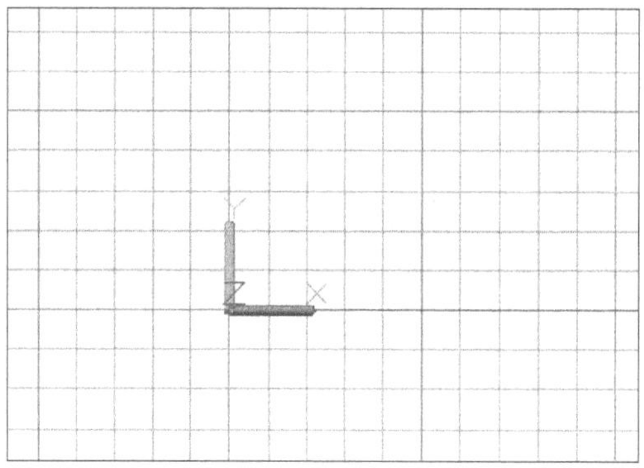

Figure-76. Grid

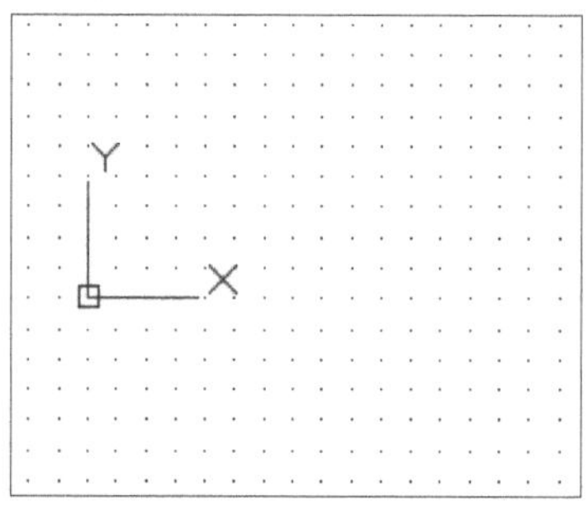

Figure-77. Dotted grid

Ortho Mode

If the **Ortho Mode** button is chosen, you can draw entities like lines at right angles only. You can use the **F8** function key to turn ortho **ON** or **OFF**.

Polar Tracking

- Click on the **Polar Tracking** button  to restrict the cursor movements along the angle specified for polar tracking.
- To set the polar tracking angle, click on the down arrow next to the **Polar Tracking** button. A list of options will be displayed; refer to Figure-78.

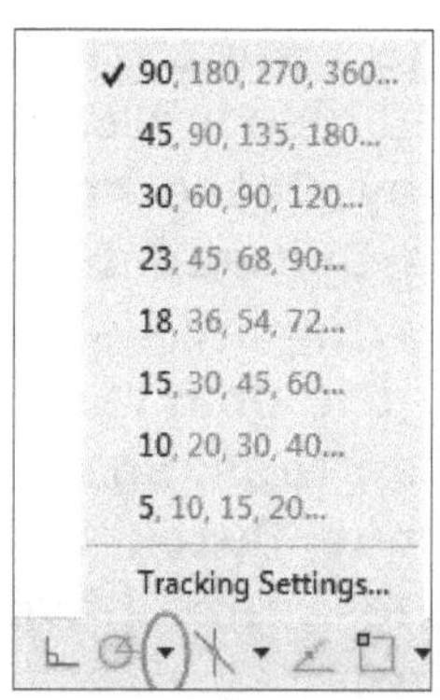

Figure-78. Polar tracking angles

- Click on desired angle option from the list. The cursor will start tracking as per the selected angle option.
- To specify the settings related to **Polar Tracking**, click on the **Tracking Settings** button from the list of options. The **Drafting Settings** dialog box will be displayed with the **Polar Tracking** tab selected; refer to Figure-79.

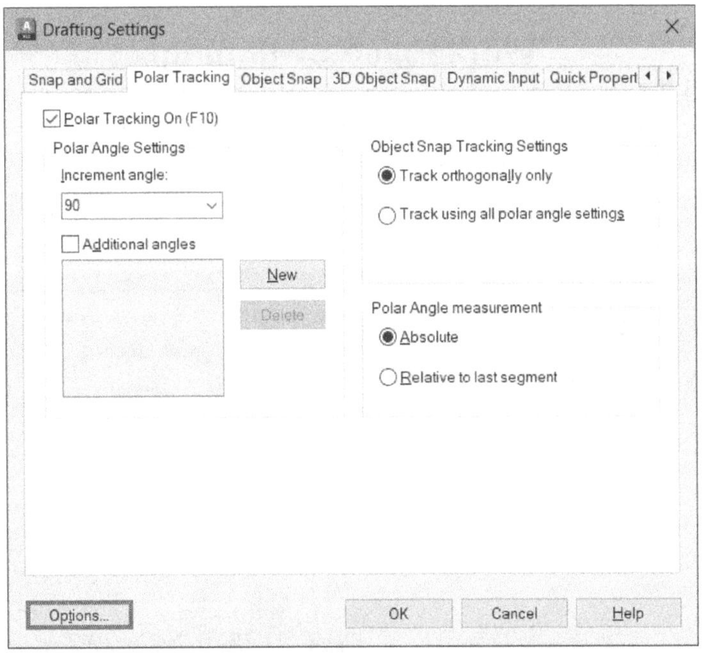

Figure-79. Drafting Settings dialog box with Polar Tracking tab

- Click in the **Increment angle** drop-down and select desired angle.
- You can specify more than one angle for tracking. For that, click on the **Additional angles** check box and click on the **New** button. You will be asked to specify the additional angle.
- Enter desired angle in the edit box. You can follow the same procedure to specify up to 10 additional angles.
- Click on the **OK** button to apply the specified settings.

Isometric Drafting

This button is used to enable isometric drafting. The procedure to use this option is given next.

- Click on the **Isometric Drafting** button  from the Bottom bar. The orientation of the cursor will change as per the current isometric plane selected; refer to Figure-80.
- To select another isometric plane, click on the down arrow next to the **Isometric Drafting** button. A list of isometric planes will be displayed; refer to Figure-81.

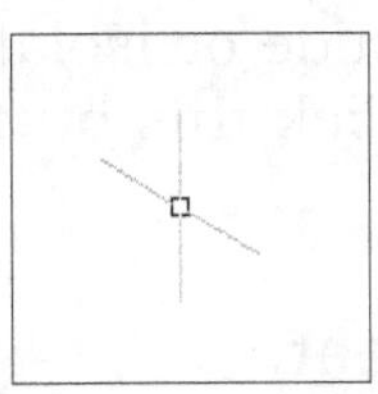

Figure-80. Cursor orientation

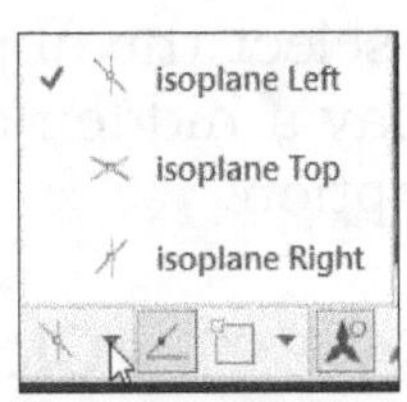

Figure-81. Isometric planes

- Select desired isometric plane and the cursor will be oriented accordingly.

Annotation buttons

There is a group of four buttons that you call annotation buttons in AutoCAD Electrical; refer to Figure-82. These buttons are used to manage annotation and their display in the drawing area. When you create an annotation in drawing, you can change its scale by using these buttons.

Workspace Switching

This flyout is used to switch between various workspaces available in AutoCAD Electrical. A workspace is defined as a customized arrangement of **Ribbon**, toolbars, menus, and window palettes in the AutoCAD Electrical environment. Click on the down arrow next to gear symbol in the Bottom bar. A flyout will be displayed; refer to Figure-83.

Figure-82. Annotation buttons

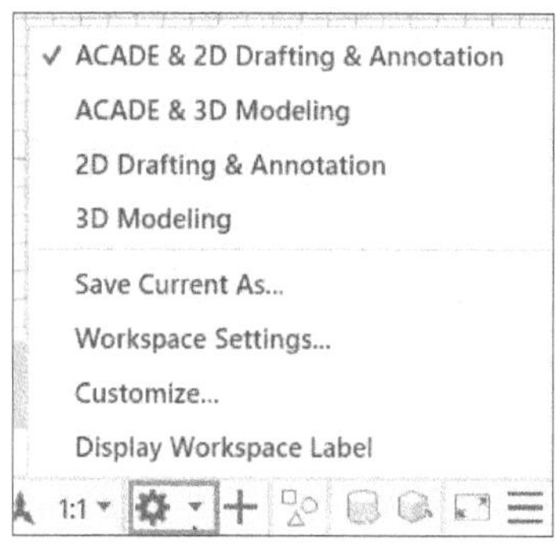

Figure-83. Workspace switching flyout

There are four default workspaces; **ACADE & 2D Drafting & Annotation**, **ACADE & 3D Modeling**, **2D Drafting & Annotation**, and **3D Modeling**. **ACADE & 2D Drafting & Annotation** workspace displays the tools and options related to 2D electrical drafting. **ACADE & 3D Modeling** workspace displays the tools and options related to basic 3D electrical modeling. In this environment, you can generate isometric drawings. The **2D Drafting & Annotation** workspace displays the tools to perform 2D drafting in general. The **3D Modeling** workspace displays almost all the tools required to perform advanced modeling.

Hardware Acceleration On

This button is used to enable use of graphic cards if installed for graphical operations. You should keep this option **ON** unless it causes performance drop while working. If this button is not visible in status bar then select the **Graphics Performance** option from the **Customization** flyout which is discussed later.

Isolate Objects

This button is used to hide or isolate objects from the drawing area. On choosing this button, a flyout will be displayed with two options. Choose the **Isolate Objects** option from this flyout, select the objects to hide or isolate, and then press **ENTER**. To end isolation or display a hidden object, click this button again and choose the **End Object Isolation** option.

Clean Screen

The **Clean Screen** button is at the lower right corner of the screen. This button, when chosen, displays an expanded view of the drawing area by hiding all the toolbars except the command window, Status Bar, and menu bar. The expanded view of the drawing area can also be displayed by choosing **View > Clean Screen** from the menu bar or by using the CTRL+0 keys. Choose the **Clean Screen** button again to restore the previous display state.

Customization

This option is used to customize the **Bottom** bar. You can add or remove the buttons from the **Bottom** bar as per the requirement. Click on the **Customization** button at the right corner in the **Bottom** bar. A flyout will display as shown in Figure-84. The options that are displayed with tick mark are available in the **Bottom** bar and other options are not. To display the other options or hide the current displaying options click on them in the flyout. For example, click on the **Units** option from the flyout. It will be added in the **Bottom** bar; refer to Figure-85. Now, you can change the unit style by clicking on this option in the Bottom bar any time while working.

Figure-84. Customization flyout

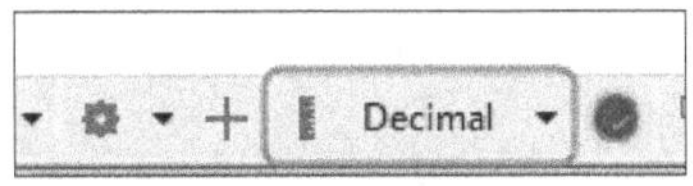

Figure-85. Unit option

SELF-ASSESSMENT

Q1. The first version of AutoCAD was named as ....

Q2. To install AutoCAD Electrical on Windows 10, we must have installed .NET Framework version 4.8 or later on the system. (T/F)

Q3. What is the unit of length in Imperial Unit System?

Q4. What is the unit of length in Metric Unit System?

Q5. Which of the following is a template for electrical drawing?

A. acad.dwt
B. acadiso.dwt
C. acad3D.dwt
D. ACE IEC a0 Color.dwt

Q6. The option to download offline help is available in the .... menu.

Q7. The .... option is used to save the file in formats other than the default format (.dwg).

Q8. The **Send to 3D Print Service** option is available in the .... cascading menu of Application menu.

Q9. The .... command in AutoCAD Electrical is used to print the documents on paper.

Q10. If the .... button is chosen, you can draw entities like lines at right angles only.

REVIEW QUESTIONS

Q1. Which year was the first version of AutoCAD desktop application released?
A. 1975
B. 1982
C. 1995
D. 2000

Q2. AutoCAD Electrical is built on the platform of which software?
A. Autodesk Revit
B. AutoCAD
C. Autodesk Inventor
D. AutoCAD LT

Q3. What feature allows AutoCAD Electrical 2026 to connect to files stored remotely?
A. External Hard Drive Sync
B. USB Sync
C. Cloud Storage Connectivity
D. LAN Drive Sharing

Q4. From where can the educational version of AutoCAD Electrical be downloaded?
A. www.autodesk.com/products
B. www.autodesk.com/education/home
C. www.autodesk.edu
D. www.autocadelectrical.com

Q5. What is the first step to start AutoCAD Electrical in Windows 11?
A. Click File > Start
B. Open Windows Explorer
C. Click Start and search for AutoCAD Electrical
D. Use the Run dialog box

Q6. What are the two unit systems used in creating drawings in AutoCAD Electrical?
A. British and European
B. Metric and Imperial
C. US and UK
D. Decimal and Binary

Q7. What does the acadISO - Named Plot Styles.dwt template allow you to do?
A. Create 3D imperial drawings in color
B. Create metric drawings in black and white
C. Create imperial drawings in color
D. Create architectural metric drawings

Q8. Which of the following templates is used for architectural drawing in metric units?
A. Tutorial-iArch.dwt
B. acadISO.dwt
C. Tutorial-mArch.dwt
D. acad3D.dwt

Q9. What are ACE templates meant for?
A. Mechanical drawings
B. Electrical drawings
C. Plumbing layouts
D. Civil engineering

Q10. What is the function of the Quick Access Toolbar in AutoCAD Electrical?
A. View active layers
B. Edit command scripts
C. Perform common file handling tasks
D. Change drawing scales

Q11. Which element shows the name of the drawing file in AutoCAD Electrical's interface?
A. Ribbon
B. Title Bar
C. Drawing Area
D. Application Menu

Q12. What is the purpose of the Search Box in the Title Bar?
A. Search file names
B. Search system folders
C. Search help topics
D. Search cloud documents

Q13. How can you sign in to your Autodesk account within AutoCAD Electrical?
A. From Help tab
B. From Application Menu
C. By clicking Sign In on the Title Bar
D. From File > Options

Q14. What does the Autodesk App Store button allow you to do?
A. Buy new versions of AutoCAD
B. Access system settings
C. Install or share Autodesk apps
D. Upgrade hardware

Q15. How can you access offline help for AutoCAD Electrical?
A. By going to File > Help
B. Clicking on Help > Download Offline Help
C. Typing 'offline help' in the command window
D. Selecting "Offline Mode" in settings

Q16. To change the color theme to light, what dialog box do you open after clicking the Application button?
A. Display Settings
B. Theme Manager
C. Options
D. Preferences

Q17. What is the primary function of the Application Menu in AutoCAD Electrical?
A. Display tool palettes
B. Handle file operations and basic settings
C. Configure user interface themes
D. Launch command prompt

Q18. What are the two main options available under the New option in the Application Menu?
A. Project and Page
B. Template and Layout
C. Drawing and Sheet Set
D. Model and View

Q19. Which command must be entered in the command prompt to ensure dialog boxes appear when creating new drawings?
A. FILESAVE
B. FILEOPEN
C. FILEDIA
D. FILEPROMPT

Q20. What does the "Open with no Template - Metric" option allow you to do?
A. Work in centimeters
B. Work in millimeters without a template
C. Import metric files
D. Set metric viewports

Q21. What file type must be selected to use a previous drawing as a template?
A. *.dst
B. *.pdf
C. *.dwg
D. *.bak

Q22. Which radio button should you select to create a sample sheet set using a template?
A. Existing drawings
B. An example sheet set
C. Cloud template
D. Use external file

Q23. What is the final step in creating a sheet set in AutoCAD Electrical?
A. Selecting the title block
B. Clicking the Finish button
C. Saving a backup
D. Enabling layers

Q24. What does the "Partial Open" option allow you to do?
A. Open a drawing with just one layout
B. Skip some layers while opening
C. Open read-only copy
D. Open multiple files together

Q25. What does the "Open Read-Only" option do?
A. Prevents editing of the drawing
B. Locks only text layers
C. Allows editing after approval
D. Encrypts the drawing

Q26. Which feature allows you to access files stored on the cloud in AutoCAD Electrical?
A. Cloud Connector
B. WebSync
C. AutoCAD Web & Mobile
D. FileStream

Q27. What must you do before installing the update to use AutoCAD Web & Mobile?
A. Log in to Autodesk account
B. Save the update to USB
C. Close AutoCAD applications
D. Enable macros

Q28. What is the purpose of the Save As option?
A. Automatically saves in .dwg format
B. Saves backups to the cloud
C. Saves the file in different formats
D. Uploads files to FTP

Q29. What type of index keeps track of which object is placed on which layer?
A. Spatial index
B. Layer index
C. Object index
D. Template index

Q30. Which export format is used to save the file as a universal model exchange format?
A. DGN
B. DWG
C. IGES
D. PDF

Q31. What is required before using the Digital Signatures option?
A. Subscription plan
B. Certificate from Certifying Authority
C. Administrator access
D. Password

Q32. What does the eTransmit option do?
A. Encrypts a drawing
B. Compresses files for electronic transfer
C. Sends drawing to a 3D printer
D. Converts drawing to PDF

Q33. What does the Share View option allow you to do?
A. Print drawing
B. Share real-time edits
C. Share model/layout views online
D. Enable password protection

Q34. What does the Share Drawing option allow users to do?
A. Transfer via Bluetooth
B. Share a link for view or edit access
C. Create multiple copies
D. Archive files

Q35. What is the first step in the process of sending a model for 3D printing using AutoCAD-based software?
A. Select the printer from the drop-down list
B. Click the Continue button
C. Create the 3D model using modeling tools
D. Save the file in STL format

Q36. In the 3D printing workflow, after selecting the solid model, which dialog box appears next?
A. Plot Dialog Box
B. Drafting Settings Dialog Box
C. 3D Print Options Dialog Box
D. Create STL File Dialog Box

Q37. What file format is used to send a model to a 3D printing service?
A. DWG
B. PDF
C. STL
D. DXF

Q38. Before printing a drawing, which workspace is recommended for use?
A. Model
B. Annotation
C. Layout/Paperspace
D. Command

Q39. Which option ensures that the drawing is printed at the center of the paper?
A. Fit to paper
B. Center the plot
C. Landscape
D. Print Preview

Q40. What is the function of the File Tab bar in AutoCAD Electrical?
A. Enables grid snapping
B. Displays drawing limits
C. Switches between opened drawings and layouts
D. Expands the command window

Q41. How can you access the command window if it is hidden?
A. Press F7
B. Use CTRL+0
C. Press CTRL+9
D. Click on the File Tab

Q42. What appears when you type a few alphabets of a command in the command window?
A. A drawing window
B. A help file
C. A list of matching commands
D. A variable settings box

Q43. How can you enable grid display in AutoCAD Electrical?
A. Press F8
B. Press F7
C. Use CTRL+9
D. Use Snap to Grid button

Q44. What is the purpose of Grid Snap?
A. Turn grid display ON
B. Enable snapping to rectangular grid intersections
C. Snap to object edges
D. Restrict cursor movement to angles

Q45. What is the default style of grid lines in AutoCAD Electrical?
A. Solid
B. Dashed
C. Checked
D. Dotted

Q46. What key is used to toggle Ortho Mode ON or OFF?
A. F5
B. F7
C. F8
D. F9

Q47. What happens when Polar Tracking is turned ON?
A. Cursor moves only in straight lines
B. Cursor snaps to grid points
C. Cursor movement is restricted to specified angles
D. Drawing is displayed in isometric view

Q48. What is the function of the Isometric Drafting button?
A. Adjusts drawing orientation to landscape
B. Activates 3D modeling tools
C. Changes cursor orientation based on selected isometric plane
D. Hides the grid lines

Q49. What does the Workspace Switching flyout allow the user to do?
A. Switch between open drawings
B. Change printing layout
C. Switch between AutoCAD Electrical workspaces
D. Adjust drawing scale

Q50. What should you do if graphics performance drops due to hardware acceleration?
A. Turn off the command window
B. Turn off grid display
C. Disable hardware acceleration
D. Enable Isometric Drafting

Q51. What is the use of the Isolate Objects button?
A. Zoom in on objects
B. Lock the object layer
C. Hide or isolate specific objects
D. Convert to STL

Q52. What does the Clean Screen button do?
A. Deletes all objects
B. Opens a new file
C. Expands drawing area by hiding toolbars
D. Changes workspace to model

Q53. Where can you customize the Bottom Bar settings?
A. Command window
B. File Tab bar
C. Customization flyout in the Bottom Bar
D. Ribbon menu

FOR STUDENT NOTES

Answer for Self-Assessment:

Ans1. AutoCAD Version 1.0, Ans2. T, Ans3. Feet & Inches, Ans4. Millimeters, Ans5. D, Ans6. Help, Ans7. SaveAs, Ans8. Publish, Ans9. Plot, Ans10. Ortho Mode

Chapter 3
Project Management

Topics Covered

The major topics covered in this chapter are:

- ***Workflow in AutoCAD Electrical***
- ***Starting a New Project***
- ***Changing Properties of a project***
- ***Adding drawings in the project***
- ***Retagging and renumbering ladders in the drawings of project***
- ***Plotting/publishing project files***

WORKFLOW IN AUTOCAD ELECTRICAL

AutoCAD Electrical is a software that can manage a complete project of electrical circuits. But, there is a step by step process to work with AutoCAD Electrical. Before, you start to learn various tools in AutoCAD Electrical, it is necessary to understand the workflow in AutoCAD Electrical. The workflow in AutoCAD Electrical can be described as shown in Figure-1.

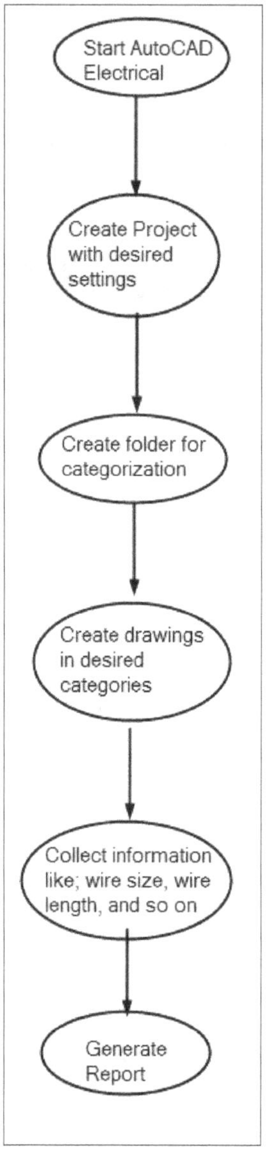

Figure-1. Workflow

You have learned about starting AutoCAD Electrical and now it is time to understand the procedures to manage projects in AutoCAD Electrical.

INITIALIZING PROJECT

In AutoCAD Electrical, project is a well structured group of various inter-related electrical drawings. There are two major categories of drawings in AutoCAD Electrical-Schematic diagrams and Panel drawings. It is mandatory to create a project file before creating electrical drawings in AutoCAD Electrical. The procedure to create a project is given next.

- Start AutoCAD Electrical. If you are starting it for the first time then you will get the screen as shown in Figure-2. The marked left area displays the **Project Manager** which is used to manage projects.

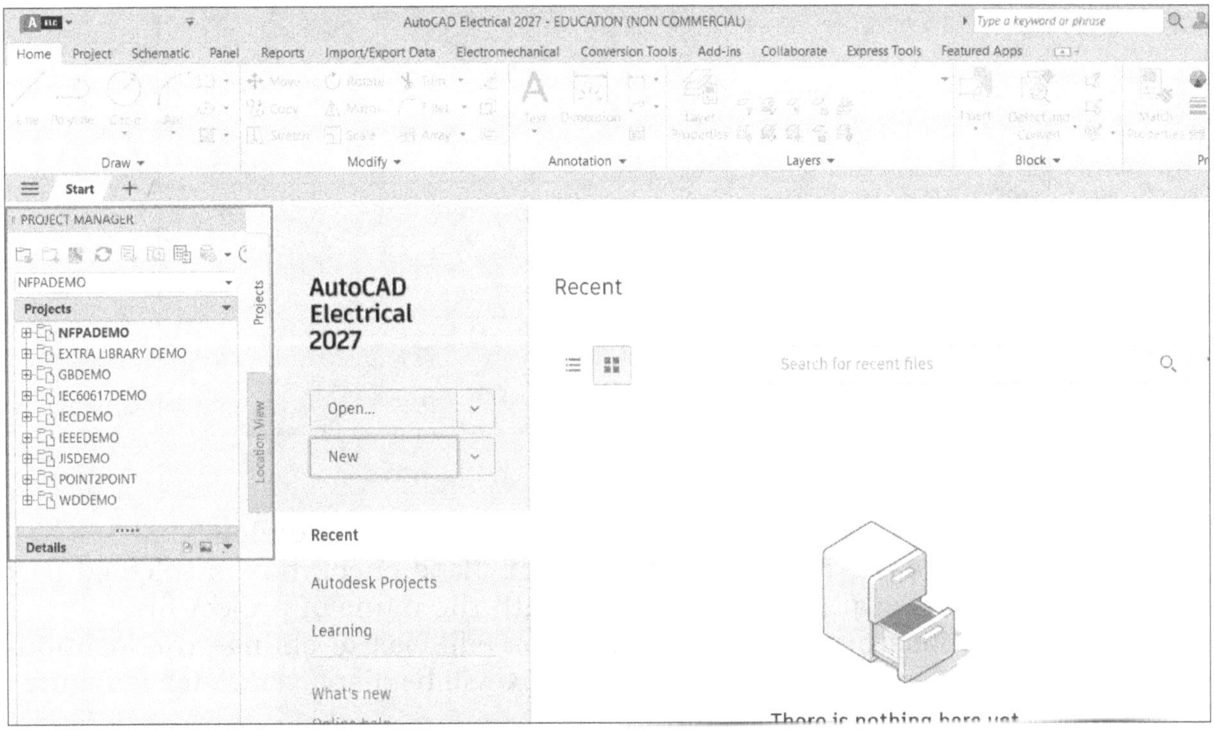

Figure-2. AutoCAD Electrical initial screen

- Click on the **Project Manager** button in the **Quick Access Toolbar** if the **Project Manager** is not displayed; refer to Figure-3.

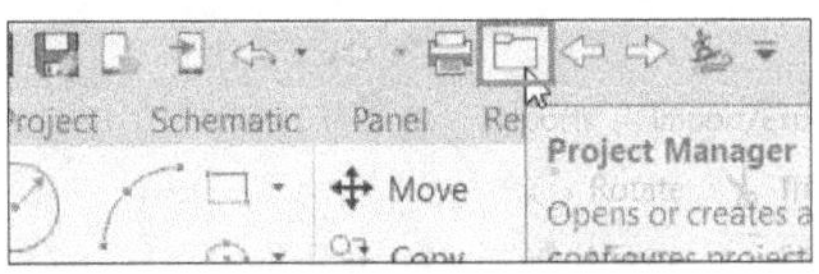

Figure-3. Project Manager button

To be able to use all the tools in the **Project Manager**, open a drawing file or create a new one as discussed earlier. The deactivated tools in the **Project Manager** will become active; refer to Figure-4.

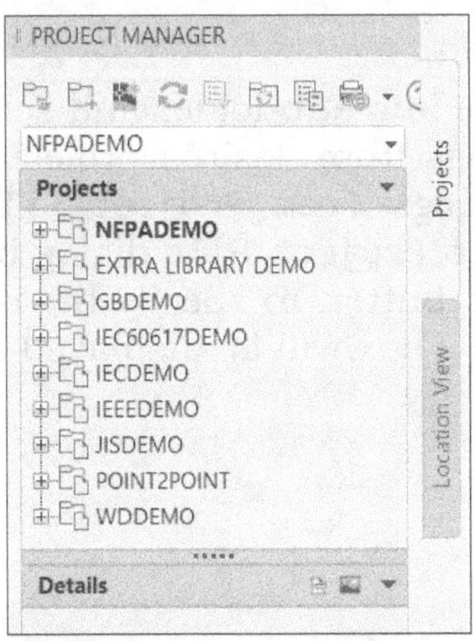

Figure-4. Project Manager

- Click on the **New Project** button in the **Project Manager**. The **Create New Project** dialog box will be displayed; refer to Figure-5.

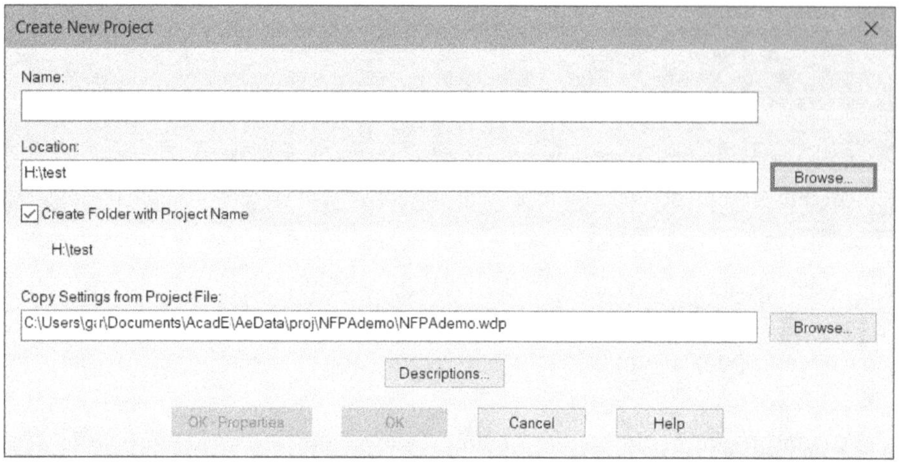

Figure-5. Create New Project dialog box

- Specify the name of the project in the **Name** edit box of the dialog box.
- Make sure the **Create Folder with Project Name** check box is selected if you want to create folder at specified location with the name of project file.
- Click on the **Browse** button next to **Location** edit box to change the location of the project. The **Browse For Folder** dialog box will be displayed; refer to Figure-6.

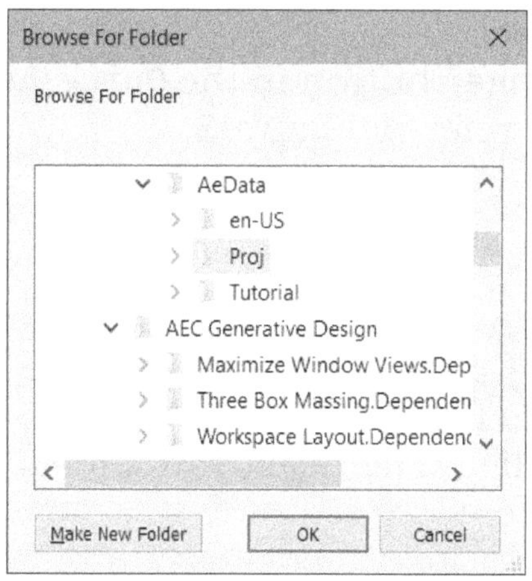

Figure-6. Browse For Folder dialog box

- Click on the nodes to move into sub-folder and select desired location.
- If you want to copy settings from desired project file then click on the **Browse** button next to **Copy Settings from Project File** edit box. Select desired file using options in the **Select Project File** dialog box displayed.
- Click on the **Descriptions** button to specify the description of the Project. The **Project Description** dialog box will be displayed as shown in Figure-7.

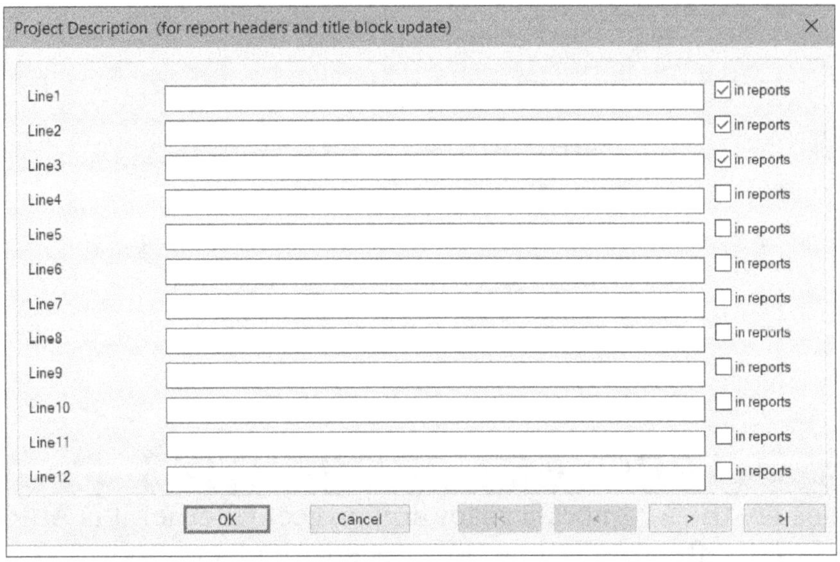

Figure-7. Project Description dialog box

- Specify the descriptions in the lines one by one. If you want to include any of the description line in your report then click on the **in reports** check box next to that line. You can write what this project is about in description.
- Click on the **OK** button to save the descriptions.
- Click on the **OK** button to create the project. If you want to specify the properties of the project while creating then click on the **OK - Properties** button. The **Project Properties** dialog box will be displayed; refer to Figure-8.

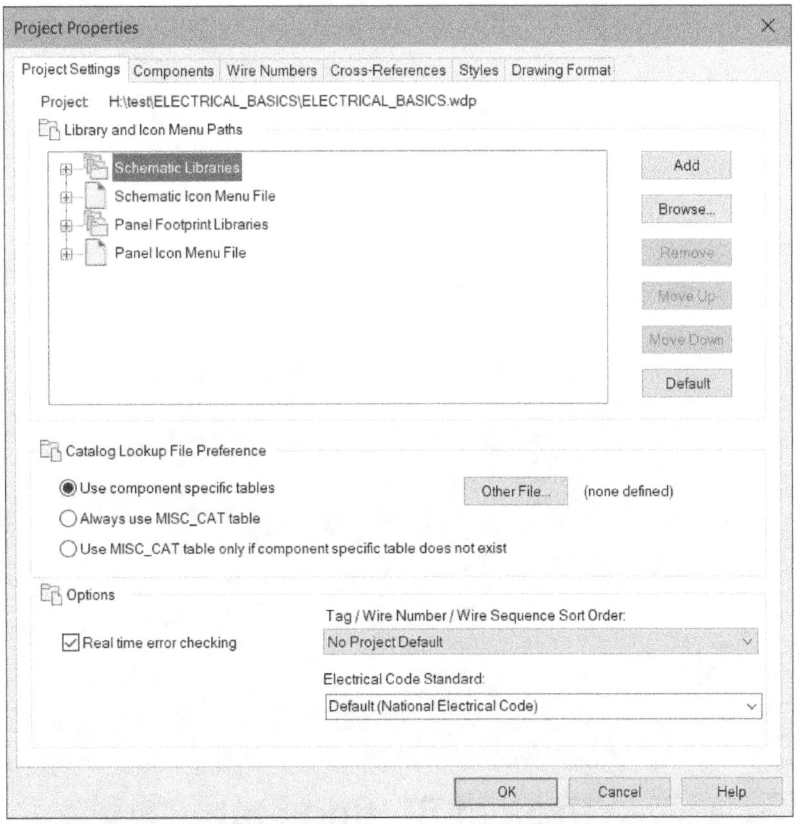

Figure-8. Project Properties dialog box

- Specify desired settings and click on the **OK** button from the dialog box to apply settings. Note that you will learn about project properties in the next section.
- As you click on the **OK** button, the project name is added in the Project list of the **Project Manager**; refer to Figure-9.

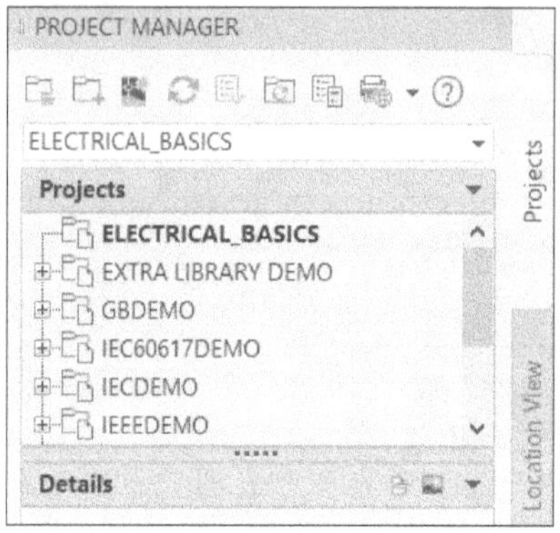

Figure-9. Project name added in Project list

PROJECT PROPERTIES

Project Properties are the key modifiers for any project whether it is AutoCAD Electrical or any other related software. Follow the procedure given next to change properties of a project.

- Right-click on the name of the project in the **Project Manager** whose properties are to be changed. A shortcut menu will be displayed; refer to Figure-10.
- Click on the **Properties** button in shortcut menu. The **Project Properties** dialog box will be displayed as shown in Figure-8.

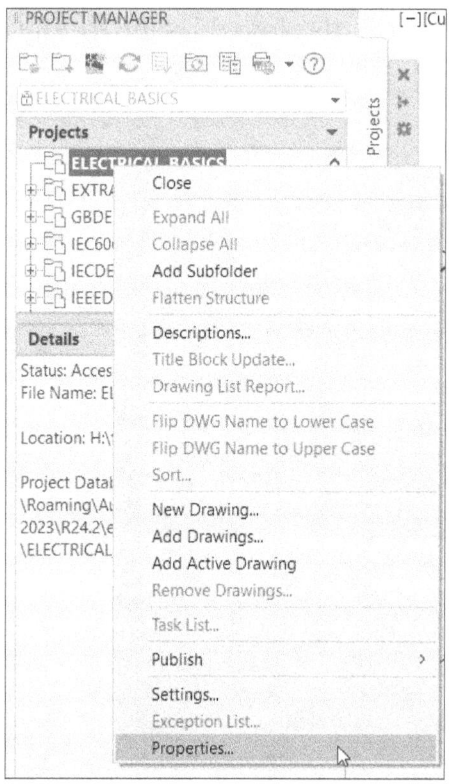

Figure-10. Shortcut menu

Project Settings tab

The options in this tab are used to specify settings for the current project file. Various options in this tab are discussed next.

Library Icon Menu Paths area

- The first area in the dialog box is **Library Icon Menu Paths**. The options in this area are used to specify the directories containing symbols and icons required in Electrical drawings.
- To add any new library, click on the **Add** button from the dialog box. A new link will be added under the **Schematic Libraries** node. Specify the location of the directory that contains the symbols by using **Browse** button.
- Similarly, you can add more files for Icons and panels. Note that for every new symbol library, you need to specify path of library as well as create entry in menu file for it.

Catalog Lookup File Preference area

- Options in the **Catalog Lookup File Preference** area are used to set the database of files that are to be included in the library. If you add any new symbol or icon then make sure that you have entered it in the database file (lookup file) used in this section.
- To set a user defined database, click on the **Other File** button in this area. The **Catalog Lookup File** dialog box will be displayed as shown in Figure-11.

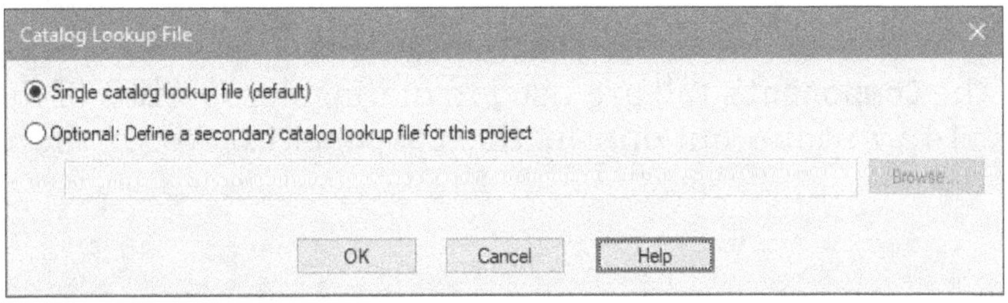

Figure-11. Catalog Lookup File dialog box

- By default, system uses the default Lookup file available with the current library. But if you need an additional file for look up then click on the **Optional : Define a secondary catalog lookup file for this project** radio button and then click on the **Browse** button. The **Secondary Catalog Lookup File** dialog box will be displayed; refer to Figure-12.

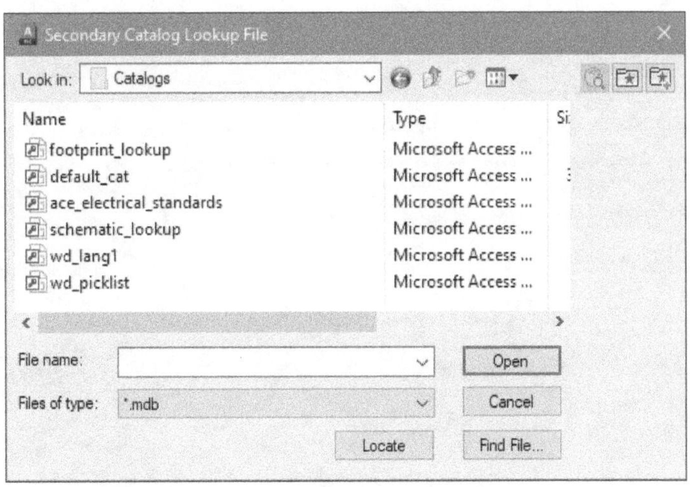

Figure-12. Secondary Catalog Lookup File dialog box

- Select desired database file and click on the **Open** button. The location of file will be added in the edit box.
- Click on the **OK** button from the dialog box.

Options area

- Select the **Real time error checking** check box to allow the system to check for some basic errors in the drawing like duplication of wire numbers and component tags.
- The **Tag/Wire Number/Wire Sequence Sort Order** drop-down is used to specify the order in which the component tags/wire numbers will be arranged in the Project. There are many options in this drop-down to modify the order; refer to Figure-13.

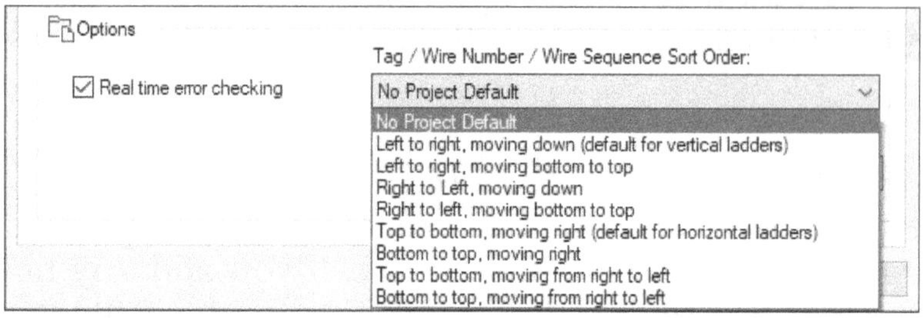

Figure-13. Sort Order drop-down

- The **Electrical Code Standard** drop-down is used to set the standard of electrical codes to be followed in project drawings.

Components tab

The options in the **Components** tab are used to manage properties of the components; refer to Figure-14. Various options in the **Components** tab of the dialog box are discussed next.

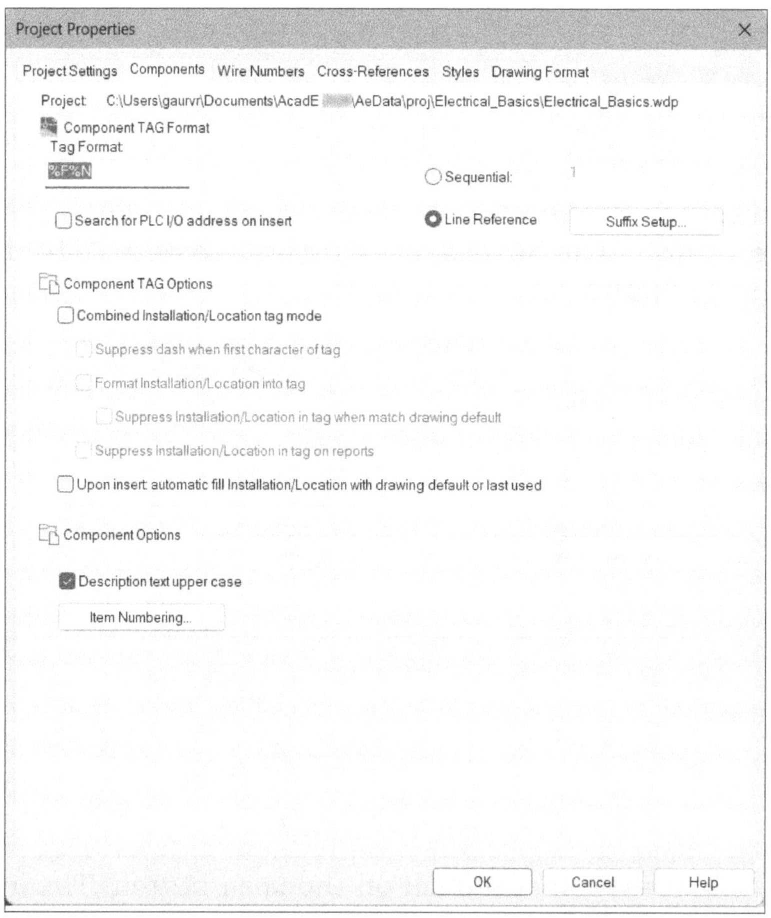

Figure-14. Components tab of Project Properties dialog box

Component TAG Format area

- The **Tag Format** edit box is used to set the format for tag assigned to various components. By default, **%F%N** is specified in the **Tag Format** edit box. **%F** is used to assign Family number to tag. **%N** is used to assign serial number to the tag. Similarly, **%S** is used to assign Sheet number and **%D** is used to assign Drawing number in the tag.
- The **Suffix Setup** button is used to manage suffix for component tag. On clicking this button, the **Suffix List for Reference-Based Component Tags** dialog box will be displayed; refer to Figure-15. These suffix are automatically applied in case of duplicates in component tags.

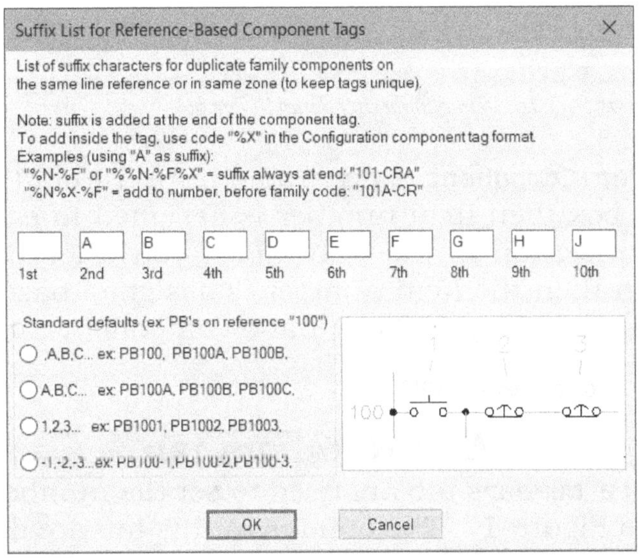

Figure-15. Suffix List for Reference-Based Component Tags dialog box

- Select the **Sequential** radio button if you do not want to use the suffix. In this case, successive numbers will be added in the duplicate component tags.
- If you are working with PLC as in the later chapter, then select the **Search for PLC I/O address on insert** check box to force system for checking the assigned PLC I/O addresses while applying tags.

Component TAG Options area

- Select the **Combined Installation/Location tag mode** check box to use tag created by combining installation and location code. Once you select this check box, the check boxes related to this option will become active. You can use these activated check boxes to further refine the Installation/Location tag mode. You will learn about Installation code and Location code later in the book.
- Select the **Upon insert automatic fill installation/Location with drawing default or last used** check box to insert the default installation/location code in component tag details when placing a component in drawing.

Component Options area

- Select the **Description text upper case** check box to make the description text uppercase in drawing.

- Click on the **Item Numbering** button to setup the item numbering mode for the current project. On clicking this button, the **Item Numbering Setup** dialog box is displayed; refer to Figure-16. Note that in earlier versions of AutoCAD Electrical, it was possible to number items in electrical drawings individually in a project but now, item numbers are always processed across the entire project.

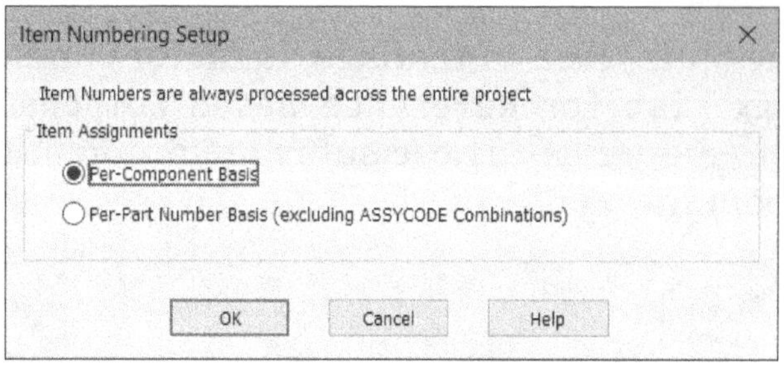

Figure-16. Item Numbering Setup dialog box

- If you select the **Per-Component Basis** radio button from the **Item Assignments** area of the dialog box then Item number is assigned to each component in the catalogue during insertion in the drawing. If you select the **Per-Part Number Basis** radio button then the Item number is assigned based on part number of the component in the catalogue. Note that a component can have more than one part number in the catalogue.

Wire Numbers tab

The options in the **Wire Numbers** tab are used to set the numbering system for wires in the project; refer to Figure-17. The options in this tab are discussed next.

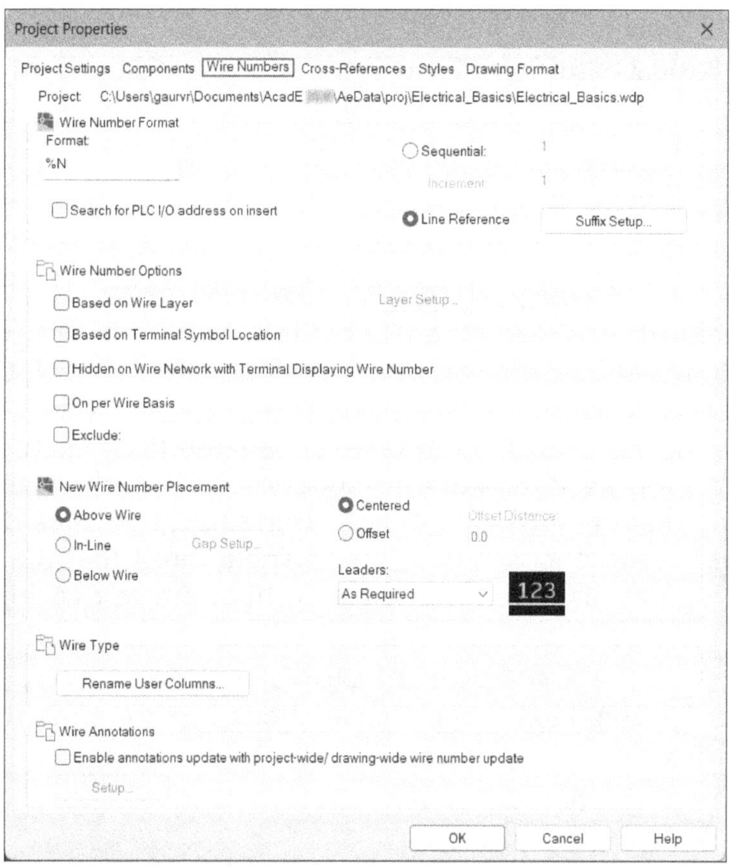

Figure-17. Wire Numbers tab in Project Properties dialog box

Wire Number Format area

- The **Format** edit box in the **Wire Number Format** area is used to set the format for wire numbering in project. You can add any suffix or prefix to **%N** for making wire numbering unique for your project. The other options in this area work in the same way as they do for components in the **Components** tab of the dialog box.

Wire Number Options area

- Select the **Based on Wire Layer** check box to make the wire numbering based on the wire layers. On selecting this check box, the **Layer Setup** button becomes active. Click on this button to display the **Assign Wire Numbering Formats by Wire Layer** dialog box; refer to Figure-18. In this dialog box, you can add, remove, or update the wire layer for wire numbering.

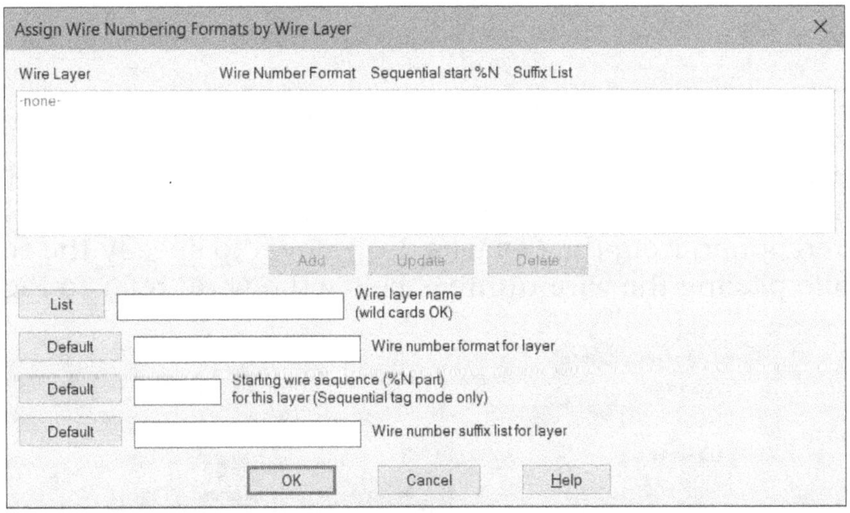

Figure-18. Assign Wire Numbering Formats by Wire Layer dialog box

- Select the **Based on Terminal Symbol Location** check box from **Project Properties** dialog box to make the wire numbering dependent on the terminal symbols. For example, a wire network starts at line reference 100 and drops down and over on line reference 103. If a schematic terminal symbol carries the WIRENO attribute located on line reference 103, and this option is enabled, AutoCAD Electrical calculates a reference-based wire number using 103 instead of 100. If multiple wire number terminals exist on this network, the line reference value of the upper left-most terminal is used.
- Select the **Hidden on Wire Network with Terminal Displaying Wire Number** check box to make the wire numbering hidden in the Wire network when the wire number is linked with terminal location.
- Select the **On per Wire Basis** check box to make the wire numbering on per wire basis rather then the wire network.
- The **Exclude** check box is used to exclude the wire numbering of the specified range from the wire numbering of the network. On selecting this check box, the edit box next to it becomes active and you can specify the range of the wire numbering like, 200-500 or 2;4;5;23.

New Wire Number Placement area

The options in this area are used to define the placement position for wire numbering with respect to wire. These options are discussed next.

- Select the **Above Wire** radio button to place the wire numbering on the wire. This is the default option for the wire numbering placement.
- Select the **In-Line** radio button to place the wire numbering in line with the wire. On selecting this radio button the **Gap Setup** button becomes active. Click on this button to specify the gap between the wire number and wire ends; refer to Figure-19. After specifying the values, click on the **OK** button from the dialog box displayed to apply the gap.

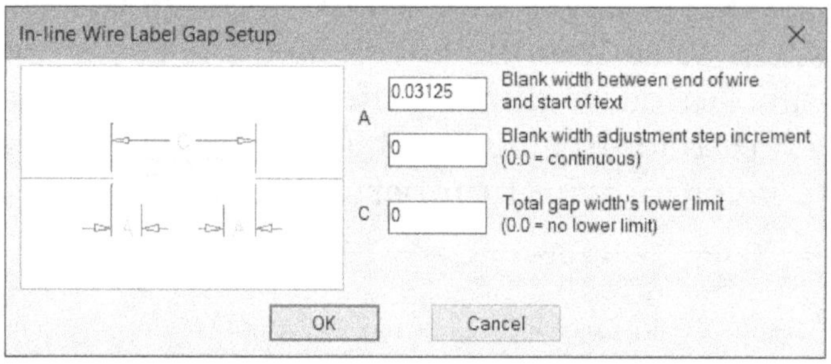

Figure-19. In-Line Wire Label Gap Setup dialog box

- Click on the **Below wire** radio button to place the wire numbering below the wire. On selecting this radio button, the options related to leaders will become active. Select desired option from the **Leaders** drop-down to specify the type of leader to be used while placing the wire number below the wire; refer to Figure-20.

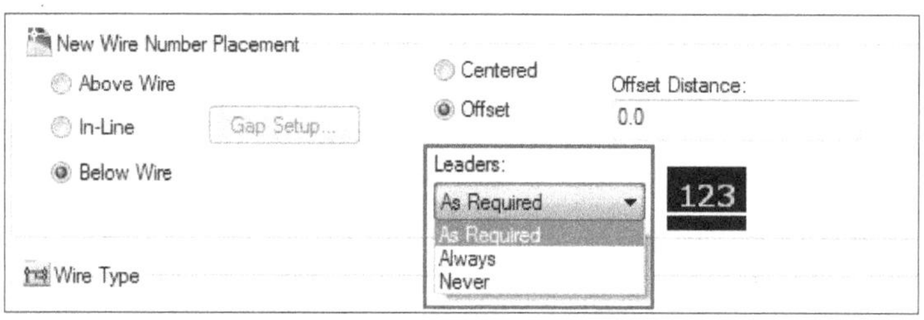

Figure-20. Leaders drop-down

- Click on the **Centered** radio button to make the wiring horizontally centered in the wire ends.
- Select the **Offset** radio button to specify the offset distance of the wire number from one of the end point of the wire. Set the other parameters as desired in this area.

Wire Type Area

There is only one option in this area, **Rename User Columns** button. This button is used to rename the user columns. Note that the user columns are used to specify user defined parameters. For example, you can make a user defined column with the name of manufacturer.

Wire Annotations Area

Select the **Enable annotations update with project-wide/drawing-wide wire number update** check box to automatically update annotations applied to wires when you are updating wire numbers. On selecting this check box, the **Setup** button will become active. Click on this button to define format for annotation of different blocks; refer to Figure-21.

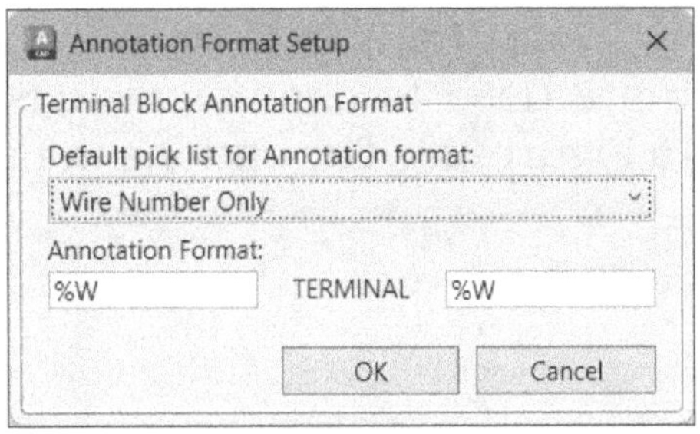

Figure-21. Annotation Format Setup dialog box

Cross-References tab

Options in this tab are used to manage cross-referencing of the project files. These options are discussed next.

Cross-Reference Format area

There are two edit boxes in this area: **Same Drawing** and **Between Drawings** edit boxes. Specify desired format for both the edit boxes.

Cross-Reference Options area

- Select the **Real time signal and contact cross-referencing between drawings** check box to make the real-time changes in the drawing. Means, if there is some change in an object (in native drawing) being cross-referenced then the same changes will be reflected in the current drawing.
- Select the **Peer to Peer** check box to make cross-referencing between components of different categories in the same drawing.
- The **Suppress Installation/Location codes when matching the drawing default** check box does the same as the name suggests. Selecting this check box suppresses the installation codes and locations codes when matching the drawing defaults.

Similarly, options in the **Component Cross-reference Display** area are used to modify the display of component cross-references. Note that some components need contact mapping like coils. You can set their graphical contact display by selecting the **Graphical Format** radio button from the **Component Cross-reference Display** area and then clicking on the **Setup** button. The **Graphical Cross-Reference Format Setup** dialog box will be displayed; refer to Figure-22. Set the parameters as desired in the dialog box and click on the **OK** button.

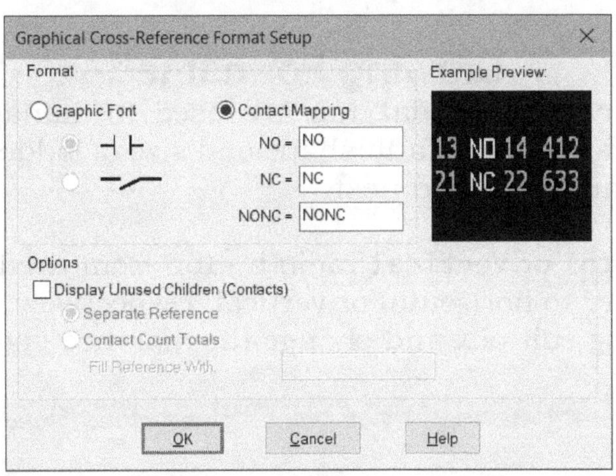

Figure-22. Graphical Cross-Reference Format Setup dialog box

Styles tab

The options in the **Styles** tab are used to change various styles related to Wiring, Arrow, PLCs, and so on; refer to Figure-23. The options in this tab are discussed next.

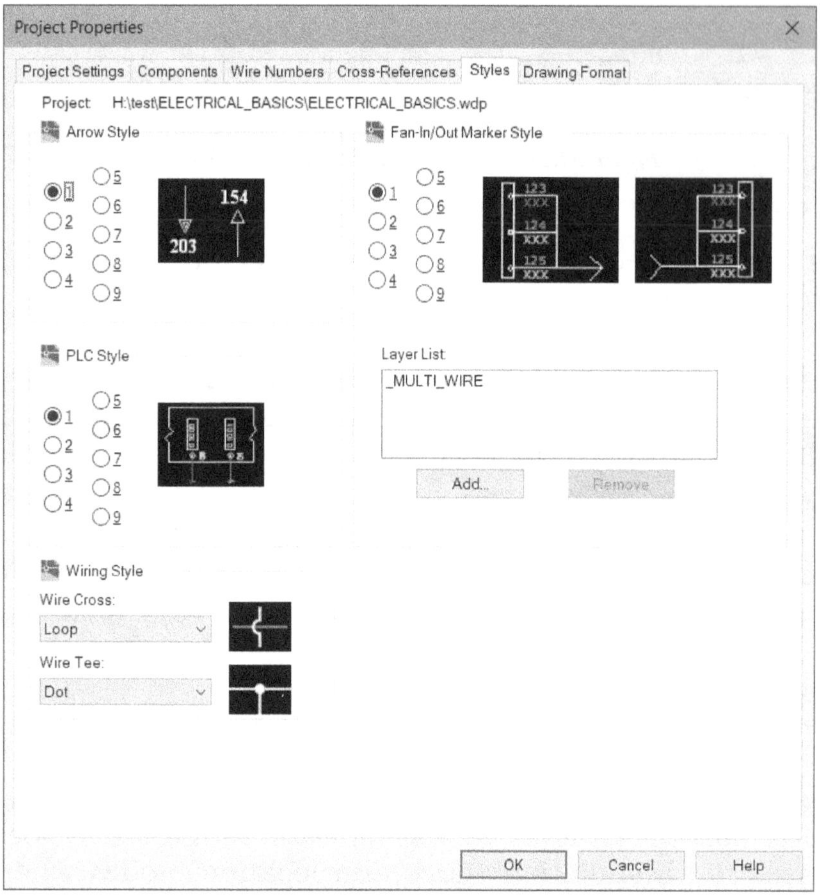

Figure-23. Styles tab of Project Properties dialog box

- Select desired radio button from the **Arrow style** area of the dialog box to change the arrow style.
- Similarly, you can change the PLC style and Fan In/Out Marker style.
- Select desired option from the **Wire Cross** drop-down to change the way wire crossing should be displayed in drawing. Note that preview is also displayed on selecting an option from the drop-down.
- Similarly, you can select desired option from the **Wire Tee** drop-down to change the Wire tee formation in drawing.

Drawing Format tab

The options in the **Drawing Format** tab are used to manage some of the basic parameters of the drawing like default shape and size of ladders, Feature scale, and so on. These options are discussed next.

- Select the **Horizontal** or **Vertical** radio button from the **Ladder Defaults** area to change the ladder to horizontal or vertical, respectively.
- Click in the **Spacing** edit box and change the distance between two consecutive wires in the ladder.

- Similarly, you can setup the other parameters in this tab.
- Click on the **OK** button from the **Project Properties** dialog box to apply the changes.

OPENING A PROJECT FILE

The procedure to open a project file is similar to opening a drawing file. The procedure is given next.

- Click on the **Open Project** button  from the **Project Manager**. The **Select Project File** dialog box will be displayed as shown in Figure-24.

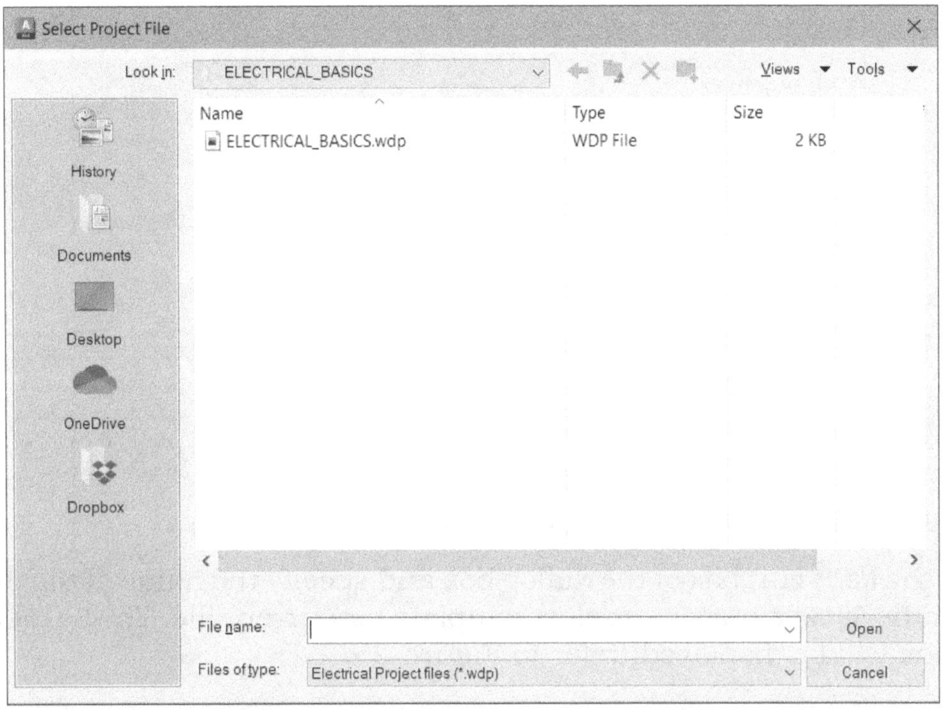

Figure-24. Select Project File dialog box

- Browse to the location of the project file and double-click on the file. The project file will open and display in the **Project Manager**.

NEW DRAWING IN A PROJECT

Creation of a new drawing has been discussed earlier. At this stage, you will learn to create drawings in a project (Although, earlier also we were creating drawing in a project unknowingly). The procedure to create a drawing file in a project is given next.

- Click on the **New Drawing** button from the **Project Manager**. The **Create New Drawing** dialog box will be displayed as shown in Figure-25.

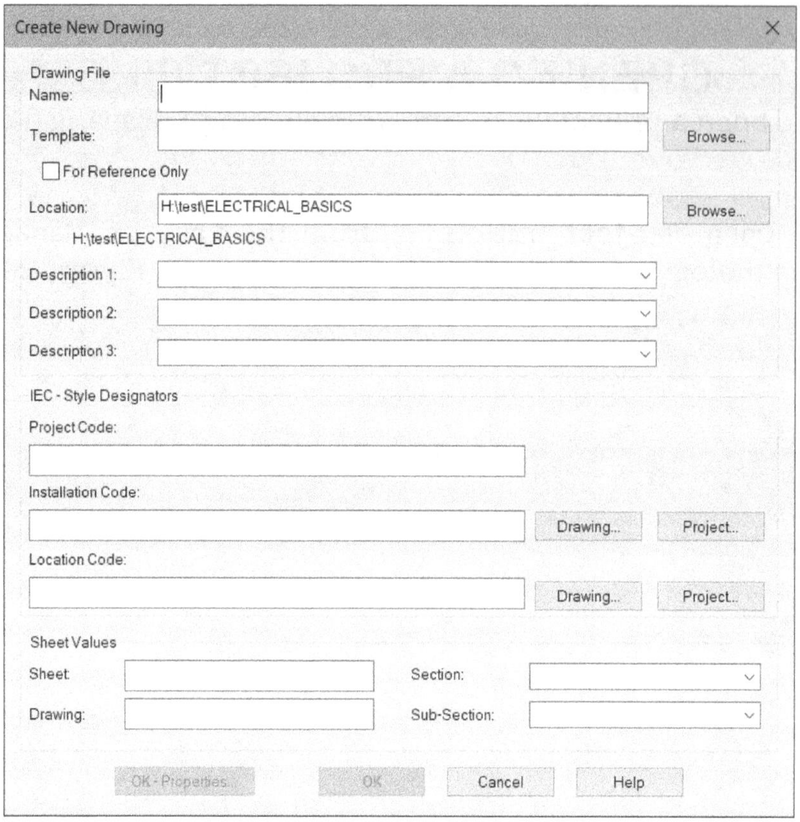

Figure-25. Create New Drawing dialog box

- Click in the **Name** edit box of the dialog box and specify the name of the drawing file.
- Click on the **Browse** button to select template for current file. The **Select template** dialog box will be displayed; refer to Figure-26.

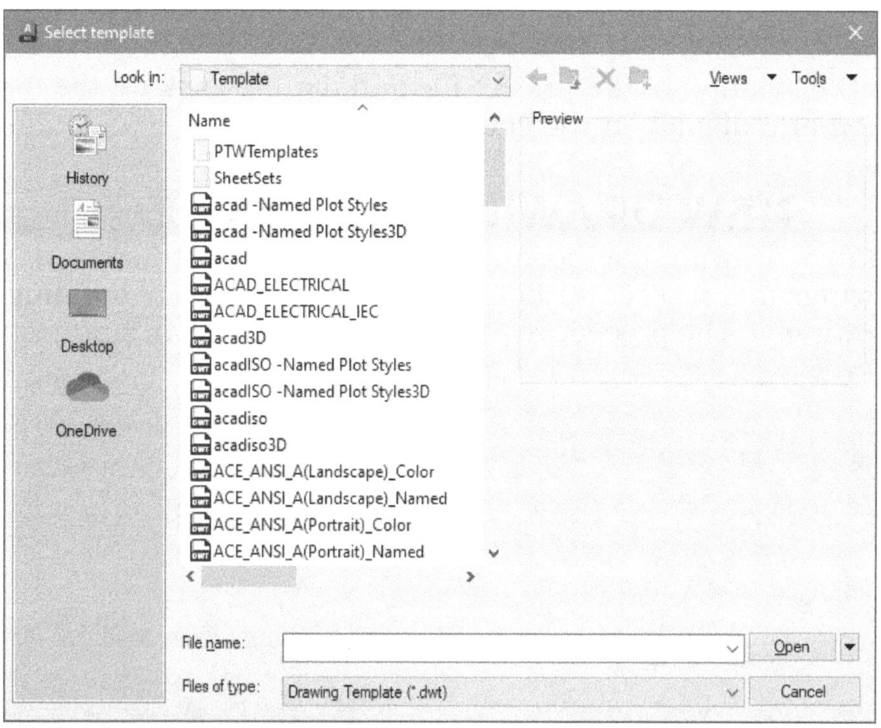

Figure-26. Select Template dialog box

- Select desired template for your drawing.
- After selecting desired template, click on the **Open** button from the dialog box. Path of the template will be added in the **Template** edit box. Note that you can skip selecting template if you want to use default template of project.
- Click in the **Description 1**, **Description 2**, and **Description 3** edit boxes one by one and specify the description as required.
- The options in the **IEC - Style Designators** area are used to specify the identifiers for the project. Click in the **Project Code** edit box and specify the identifier code for the project.
- Similarly, you can specify **Installation Code** and **Location Code**. Note that you can use the codes of the earlier created drawings or projects by selecting the buttons corresponding to them.
- Also, you can specify the **Sheet**, **Drawing**, **Section**, and **Sub-Section** numbers for the current drawing.
- All the parameters starting from **Project Code** to **Sub-Section** number are used as meta-data for the drawing and later help to identify the drawing.
- Click on the **OK** button. You are asked whether to set the default settings of the project or not. Click on the **Yes** or **No** button as desired. The drawing will be created.

REFRESH

The **Refresh** button is used to refresh the files in the Project. If you have performed changes in any drawing of the project then click on this button from the **Project Manager**, the information in the **Project Manager** will get updated.

PROJECT TASK LIST

The **Project Task List** button is used to display the tasks that are to be performed in the current project. Whenever there are multiple users working on a project or an external drawing changes the parameter of project then this button becomes active notifying you the modifications made in project. Click on this button, the list of tasks will be displayed which you can accept to modify drawings.

PROJECT WIDE UPDATE OR RETAG

The **Project Wide Update/Retag** button is used to update or retag component tags, wire numbers, cross-references, and so on. The procedure to update or retag these parameters is given next.

- Click on the **Project Wide Update/Retag** button from the **Project Manager**. The **Project-Wide Update or Retag** dialog box will be displayed as shown in Figure-27.

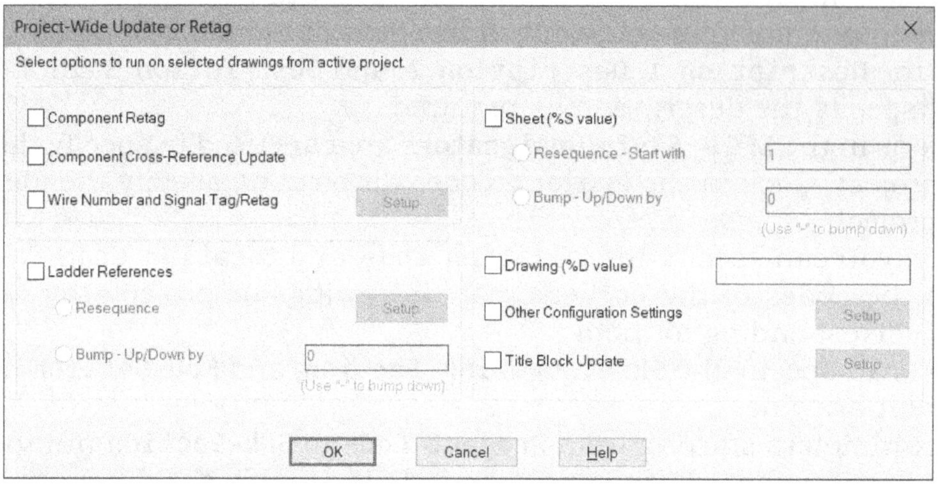

Figure-27. Project-Wide Update or Retag

- Click on the **Component Retag** check box to specify tags of the non-fixed components while adding drawings in the project.
- Select the **Component Cross-Reference Update** check box to update all the component cross-references of the drawing while adding them in the project.
- Select the **Wire Number and Signal Tag/Retag** check box, the **Setup** button next to the selected check box will become active.
- Click on the **Setup** button to change the settings for wire numbers and signal tags. The **Wire Tagging (Project-Wide)** dialog box will be displayed; refer to Figure-28.

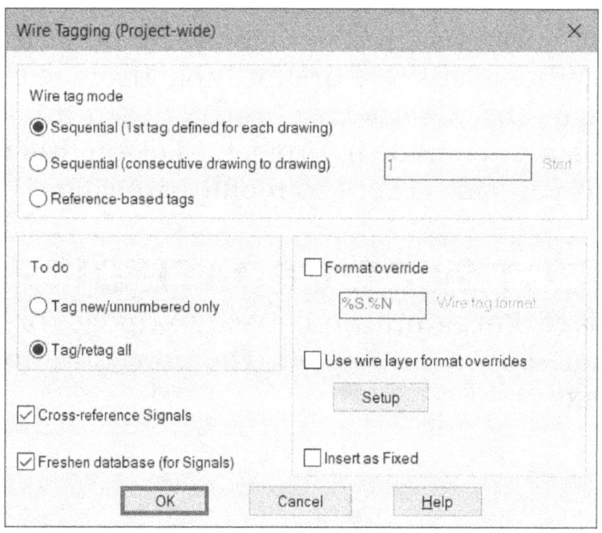

Figure-28. Wire Tagging dialog box

- Using the options in the dialog box, you can update the tags of wires and signals. You can restart wire tag for each drawing or continue wire tags through drawings in project by using the radio buttons in this dialog box.
- Select the **Ladder References** check box in the **Project-Wide Update or Retag** dialog box to renumber the ladders in the circuits. On selecting this check box, the two radio buttons below it will become active.

- Select the **Resequence** radio button and click on the **Setup** button. The **Renumber Ladders** dialog box will be displayed.
- Specify the starting number of the ladders in the edit box displayed at the top in the dialog box. This numbering will be applicable to the first drawing in the project and similarly, you can specify the starting number of ladder for second onwards drawings by selecting desired radio button in the dialog box.
- Click on the **OK** button from the dialog box to renumber the ladders in the current project.
- Select the **Bump - Up/Down by** radio button for ladder references and specify the value in the edit box next to the radio button to increase the ladder numbers by specified value. Specify the negative value to decrease the ladder numbers.
- Select desired check box from the right area of the dialog box and specify the related values. You can modify sheet numbers, drawing number parameters, title block parameters, and other configurations by using the options.
- Click on the **OK** button from the dialog box. The **Select Drawings to Process** dialog box will be displayed refer to Figure-29.

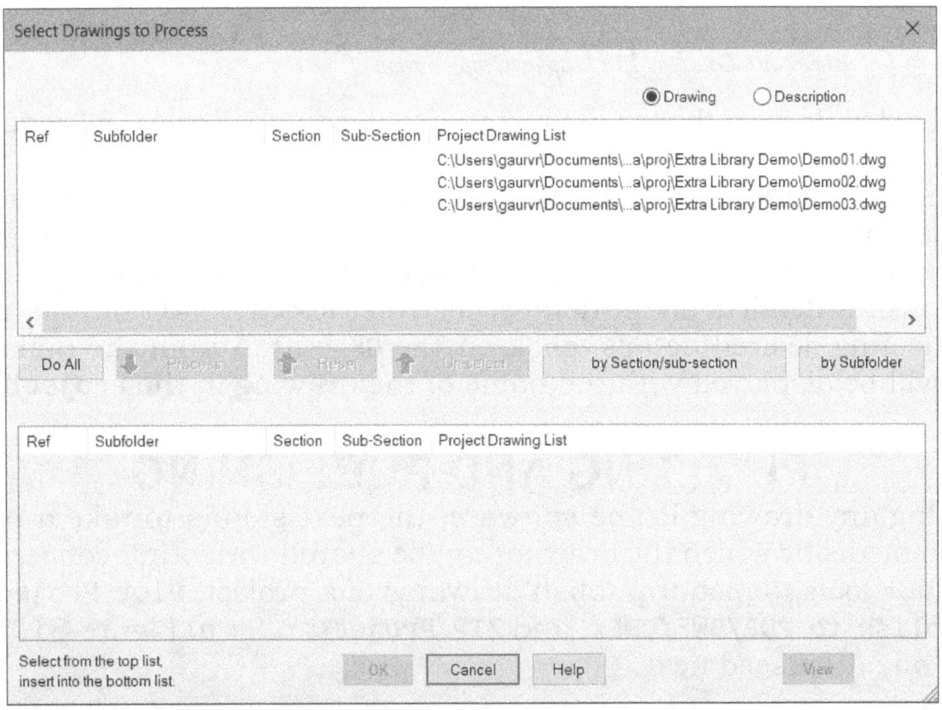

Figure-29. Select Drawings to Process dialog box

Select the drawings from the list by holding the **CTRL** key and then click on the **Process** button to include the drawings for applying the settings specified earlier. The drawings will be added in the bottom list.

- You can include all the drawings by clicking on the **Do All** button from the dialog box.
- Click on the **OK** button from the dialog box. The modified settings will be applied to the selected drawings.

DRAWING LIST DISPLAY CONFIGURATION

This button is used to configure details of the drawings that are to be displayed in the **Project Manager**. The procedure to use this option is given next.

- Click on the **Drawing List Display Configuration** button. The **Drawing List Display Configuration** dialog box will be displayed as shown in Figure-30.

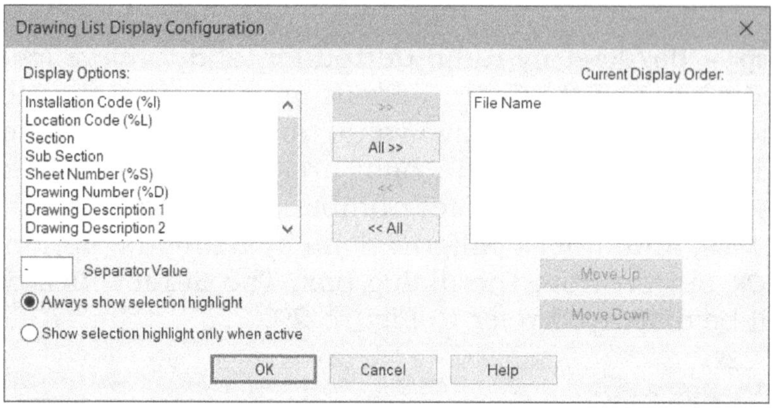

Figure-30. Drawing List Display Configuration

- Select the details from the left list box that you want to display in **Project Manager** one by one.
- Press the **Include** button **>>** from the dialog box. The details will be added in the right list box.
- You can include all the details by clicking on the **All>>** button.
- To exclude any detail or all the details, click on the **<<** or **<<All** button respectively.
- After selecting desired details, click on the **OK** button from the dialog box. The details will be displayed with the name of the drawing in the **Project Manager**.

PLOTTING AND PUBLISHING

After creating any drawing in the software, the next step is to take a hardcopy or distribution media by which the drawing can be shared with other concerned people. There are four tools to plot or publish drawings of a project: **Plot Project**, **Publish to WEB**, **Publish to PDF/DWF/DWFx**, and **ZIP Project**; refer to Figure-31. These tools are one by one discussed next.

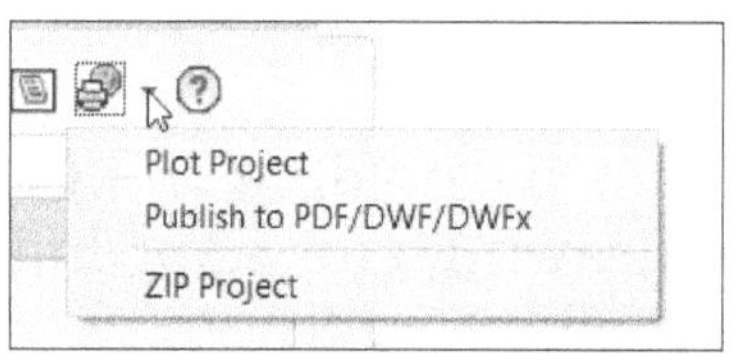

Figure-31. Tools to plot & publish

Plot Project

The **Plot Project** tool is used to plot drawings of the project on the paper by using your printing/plotting device. The steps to use this tool are given next.

- Click on the **Plot Project** tool from the **Project Manager**. The **Select Drawings to Process** dialog box will be displayed; refer to Figure-29.

- Include the drawings that you want to plot by using the **Process** button after selecting them in the list.
- Click on the **OK** button from the **Select Drawings to Process** dialog box. The **Batch Plotting Options and Order** dialog box will be displayed; refer to Figure-32.

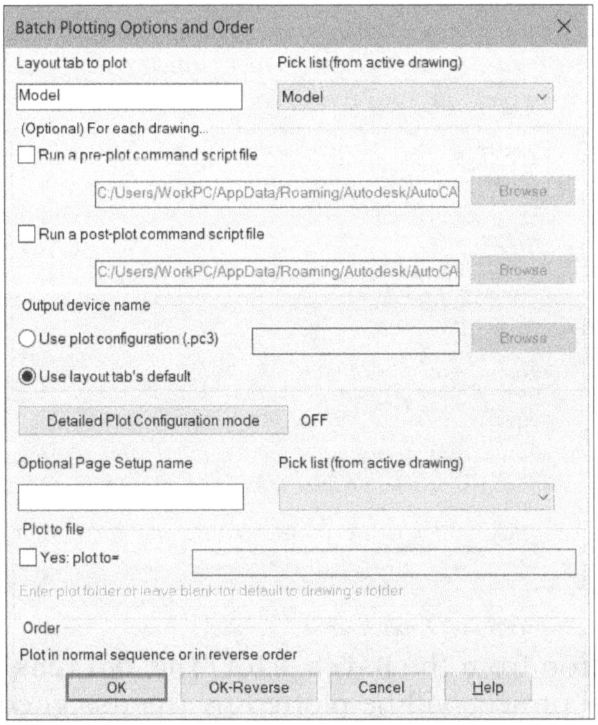

Figure-32. Batch Plotting Options and Order dialog box

- Click on the **Detailed Plot Configuration mode** button to specify settings for the plot. The **Detailed Plot Configuration Option** dialog box will be displayed; refer to Figure-33.

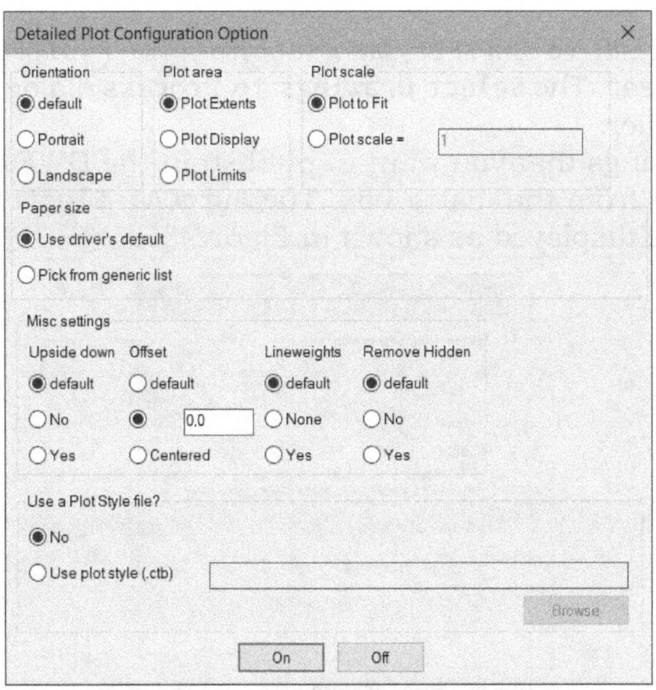

Figure-33. Detailed Plot Configuration Option dialog box

- Modify desired settings of plotting like orientation of page, plot area, paper size, and so on using the options in this dialog box.

- To select desired paper size, click on the **Pick from generic list** radio button in the **Paper size** area. The dialog box with various paper sizes will be displayed; refer to Figure-34.
- Select desired paper size and click on the **OK** button from the dialog box.
- Next, click on the **On** button from the **Detailed Plot Configuration Option** dialog box.

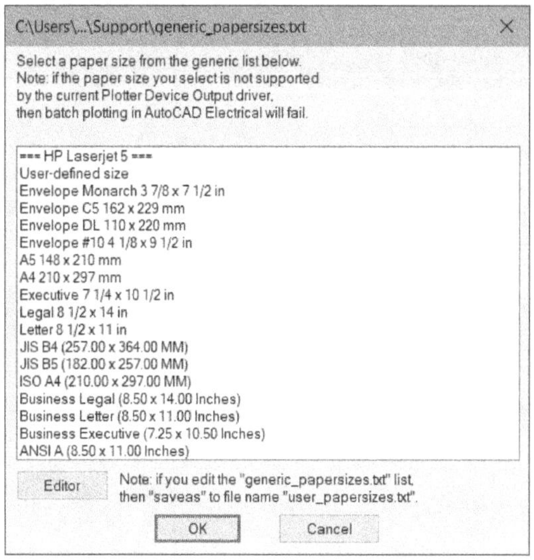

Figure-34. Paper sizes

- Click on the **OK** button from the **Batch Plotting Options and Order** dialog box. The drawings of the project will be plotted by the assigned plotter.

Publish to DWF/PDF/DWFx

This tool is used to publish the drawings in DWF or PDF format. The procedure to use this tool is given next.

- Click on the **Publish to DWF/PDF/DWFx** tool from the **Plot/Publish** drop-down in the **Project Manager**. The **Select Drawings to Process** dialog box will be displayed as discussed earlier.
- Include the drawings that you want to publish in the DWF or PDF format. Click on the **OK** button from the dialog box. The **AutoCAD Electrical-Publish Setup** dialog box will be displayed as shown in Figure-35.

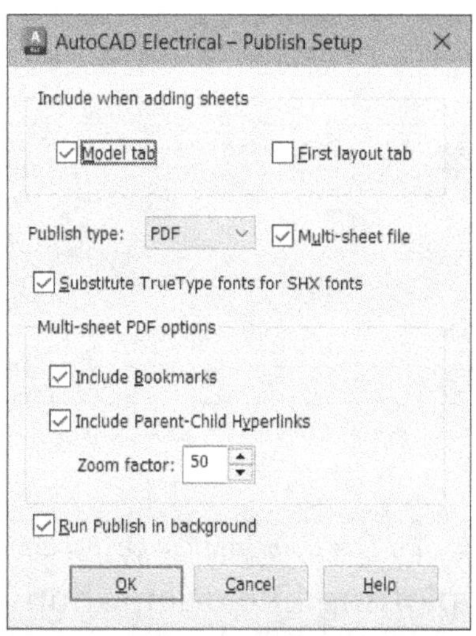

Figure-35. AutoCAD Electrical Publish Setup dialog box

- Select the check boxes for the options that you want to include in your DWFs and PDFs.
- Click in the **Publish type** drop-down to change the type of file for publishing.
- Click on the **OK** button from the dialog box. The **Publish** dialog box will be displayed; refer to Figure-36.

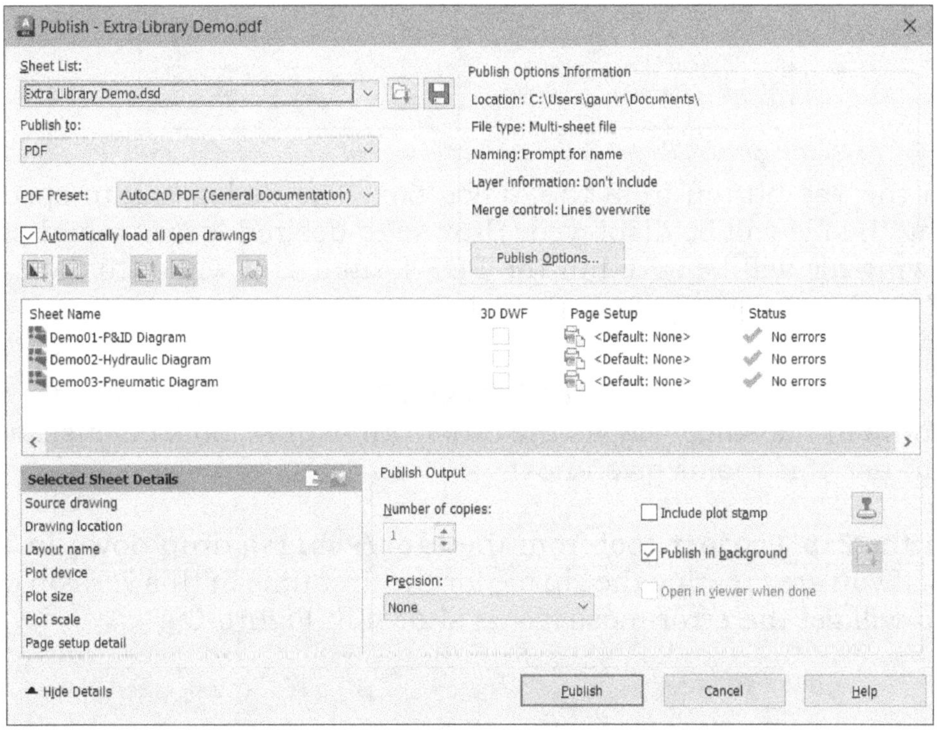

Figure-36. Publish dialog box

- Set desired options and then click on the **Publish** button from the dialog box. The **Specify PDF File** dialog box will be displayed (in case of PDF format selected); refer to Figure-37.

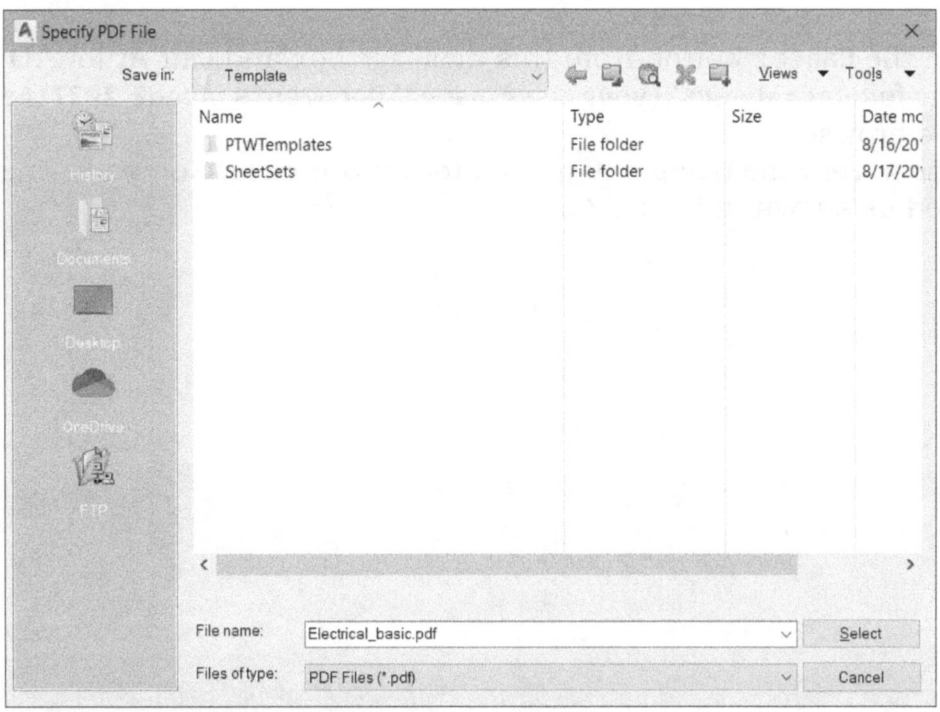

Figure-37. Specify PDF File dialog box

- Browse to desired location and then click on the **Select** button. If you have unsaved sheet list then the **Publish - Save Sheet List** dialog box will be displayed as shown in Figure-38.

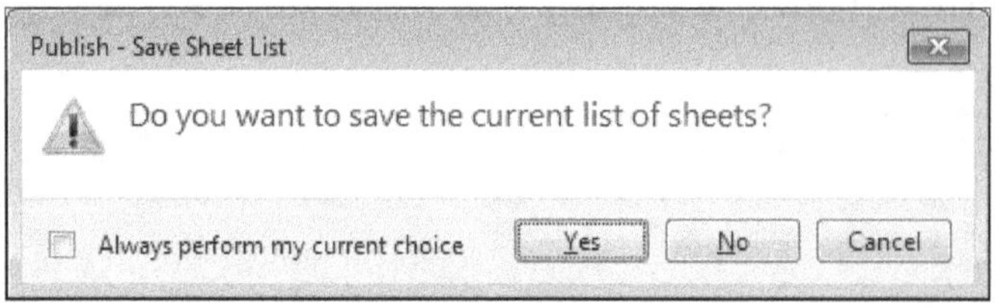

Figure-38. Publish Save Sheet List dialog box

- Click on the **Yes** button from the dialog box. The dialog box to specify location for saving the file will be displayed. Browse to desired location and save the file.
- The drawing file will be saved in the PDF format and will automatically open in the PDF reader of your system.

Zip Project

This tool works in the same way as the other publishing tools discussed above. The procedure to use this tool is given next.

- Click on the **Zip Project** tool from the **Plot/Publish** drop-down in the **Project Manager**. If you are creating the zip file for the first time in this version of AutoCAD, then you will get the error message as shown in Figure-39.

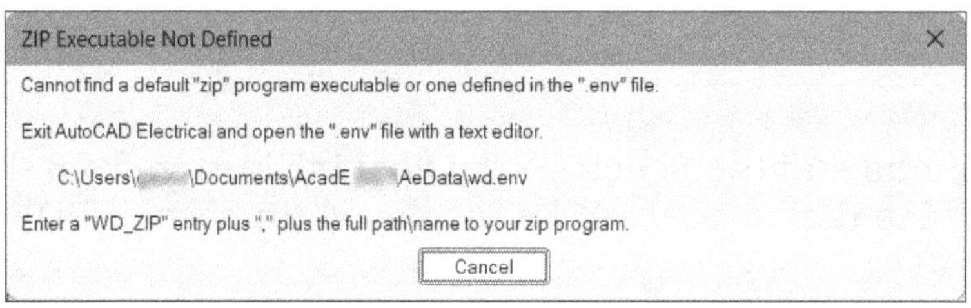

Figure-39. Error message

- Click on the **Cancel** button from this message box and exit AutoCAD Electrical.
- Open the folder **C:\Users\{your user name}\Documents\AcadE 2027\AeData** using windows browser.
- Open the **wd.env** file using any text editor (in our case, Notepad). The file will be displayed as shown in Figure-40.

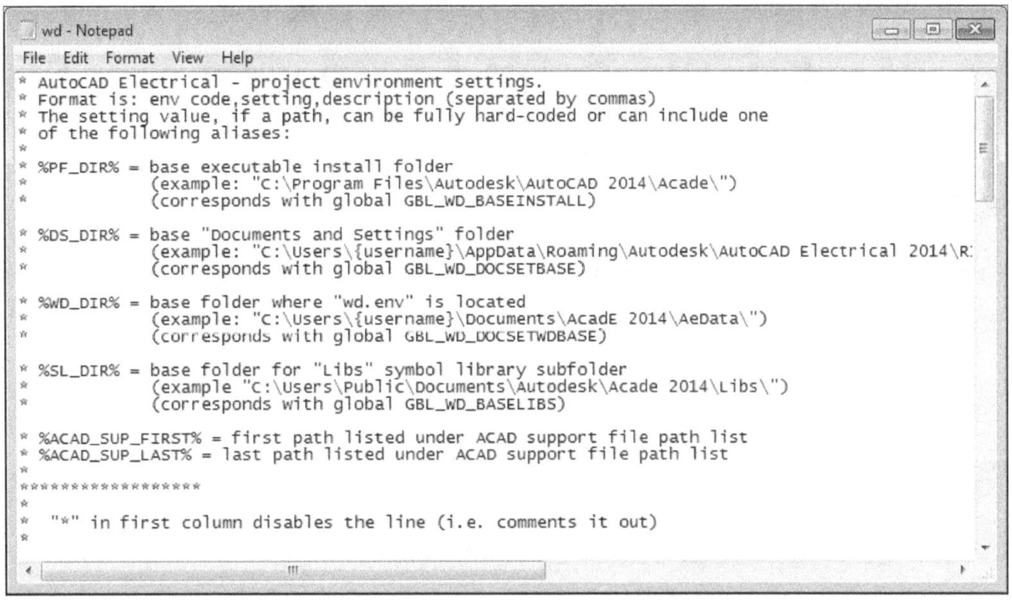

```
wd - Notepad
File Edit Format View Help
* AutoCAD Electrical - project environment settings.
* Format is: env code,setting,description (separated by commas)
* The setting value, if a path, can be fully hard-coded or can include one
* of the following aliases:
*
* %PF_DIR% = base executable install folder
*            (example: "C:\Program Files\Autodesk\AutoCAD 2014\Acade\")
*            (corresponds with global GBL_WD_BASEINSTALL)

* %DS_DIR% = base "Documents and Settings" folder
*            (example: "C:\Users\{username}\AppData\Roaming\Autodesk\AutoCAD Electrical 2014\R:
*            (corresponds with global GBL_WD_DOCSETBASE)

* %WD_DIR% = base folder where "wd.env" is located
*            (example: "C:\Users\{username}\Documents\AcadE 2014\AeData\")
*            (corresponds with global GBL_WD_DOCSETWDBASE)

* %SL_DIR% = base folder for "Libs" symbol library subfolder
*            (example "C:\Users\Public\Documents\Autodesk\Acade 2014\Libs\")
*            (corresponds with global GBL_WD_BASELIBS)

* %ACAD_SUP_FIRST% = first path listed under ACAD support file path list
* %ACAD_SUP_LAST% = last path listed under ACAD support file path list
*
******************
*
*  "*" in first column disables the line (i.e. comments it out)
*
```

Figure-40. wd file in notepad

- Move down in the file at the location as shown in Figure-41.

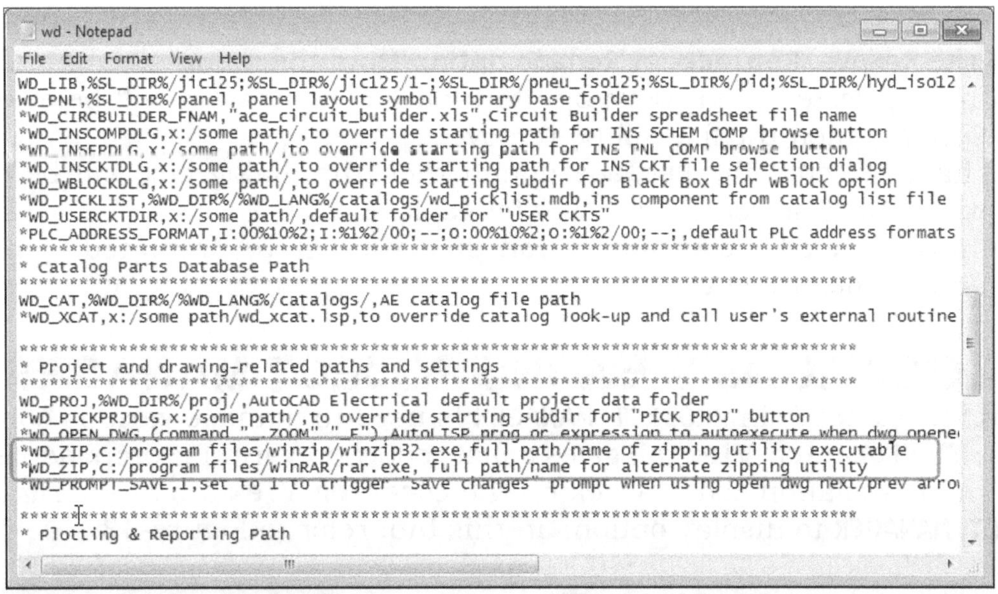

```
wd - Notepad
File Edit Format View Help
WD_LIB,%SL_DIR%/jic125;%SL_DIR%/jic125/1-;%SL_DIR%/pneu_iso125;%SL_DIR%/pid;%SL_DIR%/hyd_iso12
WD_PNL,%SL_DIR%/panel, panel layout symbol library base folder
*WD_CIRCBUILDER_FNAM,"ace_circuit_builder.xls",Circuit Builder spreadsheet file name
*WD_INSCOMPDLG,x:/some path/,to override starting path for INS SCHEM COMP browse button
*WD_INSFPDLG,x:/some path/,to override starting path for INS PNL COMP browse button
*WD_INSCKTDLG,x:/some path/,to override starting path for INS CKT file selection dialog
*WD_WBLOCKDLG,x:/some path/,to override starting subdir for Black Box Bldr WBlock option
*WD_PICKLIST,%WD_DIR%/%WD_LANG%/catalogs/wd_picklist.mdb,ins component from catalog list file
*WD_USERCKTDIR,x:/some path/,default folder for "USER CKTS"
*PLC_ADDRESS_FORMAT,I:00%10%2;I:%1%2/00;--;O:00%10%2;O:%1%2/00;--;,default PLC address formats
*******************************************************************************************
* Catalog Parts Database Path
*******************************************************************************************
WD_CAT,%WD_DIR%/%WD_LANG%/catalogs/,AE catalog file path
*WD_XCAT,x:/some path/wd_xcat.lsp,to override catalog look-up and call user's external routine

*******************************************************************************************
* Project and drawing-related paths and settings
*******************************************************************************************
WD_PROJ,%WD_DIR%/proj/,AutoCAD Electrical default project data folder
*WD_PICKPRJDLG,x:/some path/,to override starting subdir for "PICK PROJ" button
*WD_OPEN_DWG,(command "_.ZOOM" "_E"),AutoLISP prog or expression to autoexecute when dwg opene
*WD_ZIP,c:/program files/winzip/winzip32.exe,full path/name of zipping utility executable
*WD_ZIP,c:/program files/WinRAR/rar.exe, full path/name for alternate zipping utility
*WD_PROMPT_SAVE,1,set to 1 to trigger "save changes" prompt when using open dwg next/prev arro

*******************************************************************************************
* Plotting & Reporting Path
```

Figure-41. Location to identified

- Check carefully the lines marked in the red box. You will notice that these lines link the zip file with two programs **WinZip32** and **WinRAR**.
- Remove the ***** mark from these line. Press **CTRL+S** to save the file.
- Restart the AutoCAD Electrical and reopen the project file and drawing you were working on.
- Click on the **Zip Project** tool from the **Plot/Publish** drop-down in the **Project Manager**. The **Select Drawings to Process** dialog box will be displayed as discussed earlier.
- Rest of the procedure is same as discussed earlier.

REMOVING, REPLACING, AND RENAMING DRAWINGS IN A PROJECT

Till this point, you have learned to add drawings in the Project and you have learned to modify properties of the drawings. Now, you will learn to remove, replace, or rename a drawing file in a project. The steps to perform these actions are given next.

- Select the drawing from the **Project Manager** that you want to remove, replace, or rename. Right-click on it, a shortcut menu will be displayed; refer to Figure-42.

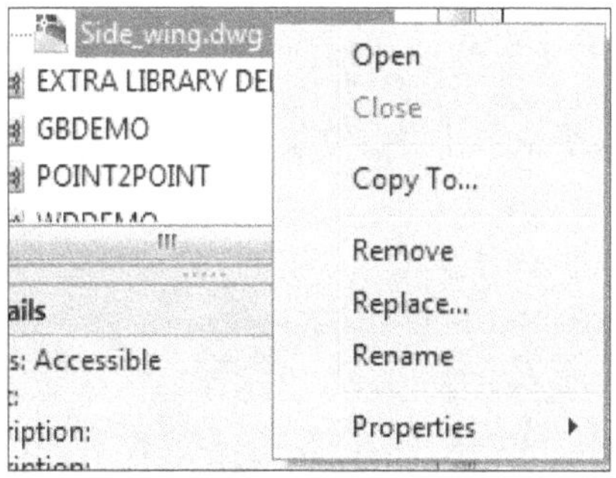

Figure-42. Shortcut menu for drawing

- Select the **Remove**, **Replace**, or **Rename** option from the menu.
- If you have selected the **Remove** option then a confirmation will be asked and if you select **Yes** then the file will be removed.
- If you have selected the **Replace** option then the dialog box will be displayed prompting you to select the drawing for replacement.
- If you have selected the **Rename** option then you will be prompted to specify the new name of the drawing.

LOCATIONS VIEW IN PROJECT MANAGER

The **Location View** tab in **PROJECT MANAGER** is used to manage various components of project. The **Location View** tab gives a fast report on various components by their location and installation codes. Click on the **Location View** tab at the right sidebar of **PROJECT MANAGER** to display options in this tab; refer to Figure-43.

Figure-43. Location View tab in Project Manager

Click on the + sign before (??) in the **Location View** tab to check sub-categories of the category. The (??) is displayed for Installation code. Since, we have not defined the installation code for current project so (??) is displayed there. On expanding the installation code, you will get various components arranged as per their location codes; refer to Figure-44.

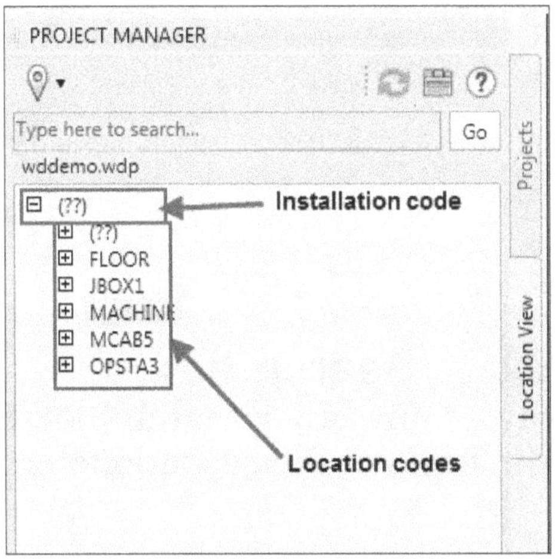

Figure-44. Installation code and location code in Location View tab

Now, expanding the location codes will display components for those locations. The other options in the **Location View** tab are discussed next.

Filter by Installation and Location

Using this option, you can filter out the unwanted installations or locations. The procedure to use this option is given next.

- Click on the **Filter by Installation and Location** button. A drop-box will be displayed; refer to Figure-45.

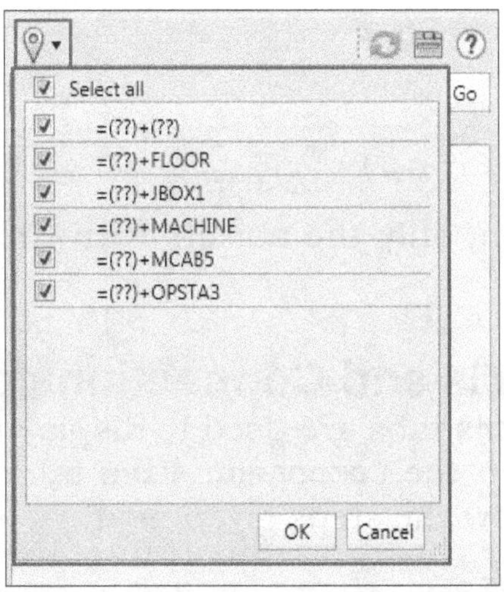

Figure-45. Drop-box for filtering installations and locations

- Clear the check boxes for the locations and installations that you do not want to see in the **Location View** tab and click on the **OK** button from the drop-box. The locations and installations for which the check box is selected will be displayed in the **Location View** tab; refer to Figure-46.

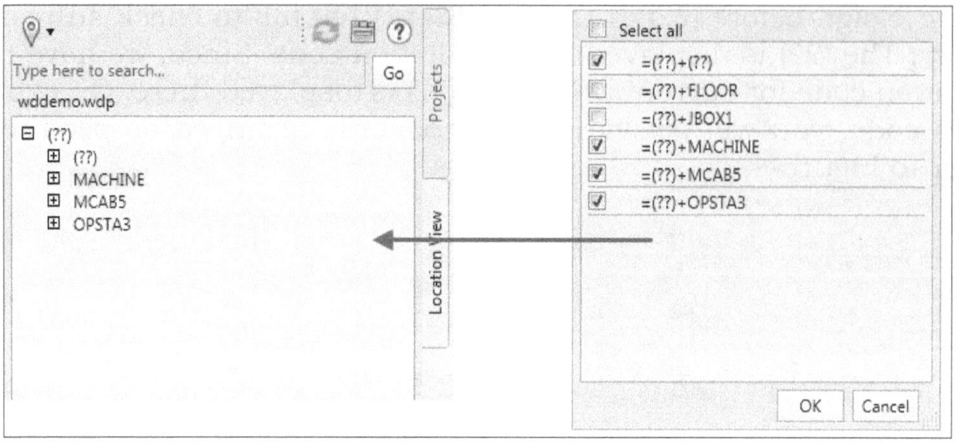

Figure-46. Locations and installations after filtering

Search box

This tool works the same way as any other search tool. Specify your parameters to search for components on the basis of their locations or installations. Procedure to use this tool is given next.

- Type the location code, installation code, or component keyword in the type box for searching and click on the **Go** button next to the search box; refer to Figure-47.

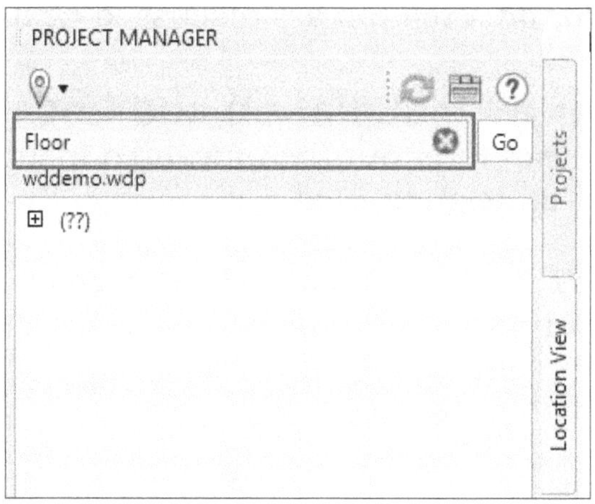

Figure-47. Search box for Location view tab

- On clicking the **Go** button, only the objects with the specified keywords will be displayed.

Details and Connections tabs

The **Details** and **Connections** tabs are used to display the details and connections of the component selected in the **Component View** tab of the **Project Manager**. To learn more about them, follow the steps given next.

- Click on the **Display Details and Connections** button from the **Project Manager**; refer to Figure-48. The **Details** and **Connections** tabs will be added in the **Project Manager**; refer to Figure-49.

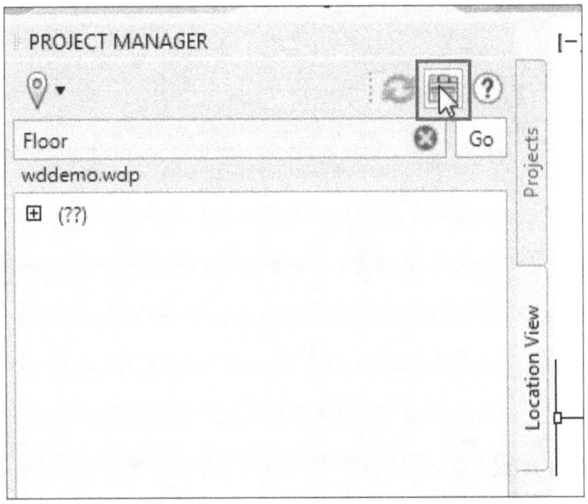

Figure-48. Display Details and Connections button

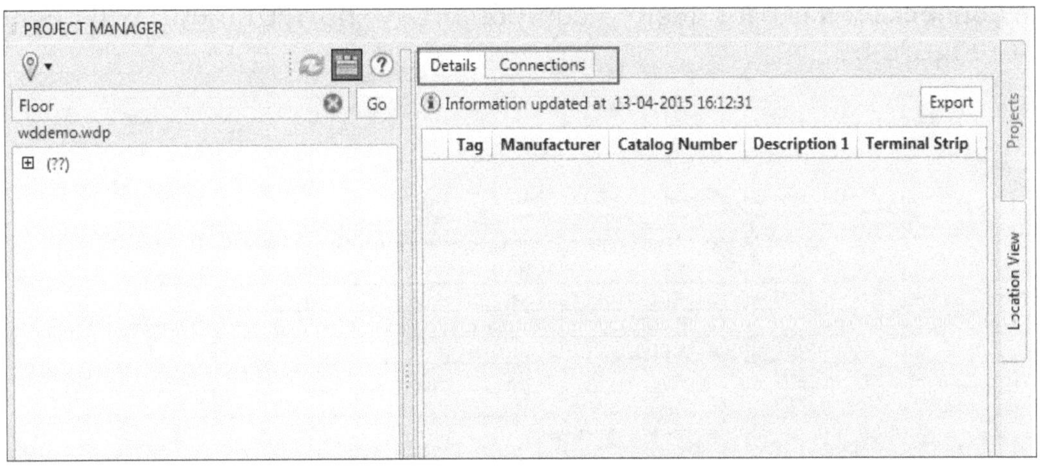

Figure-49. Details and Connections tabs

- Select a component from the left pane of the **Project Manager**. The detail of the component will be displayed in the **Details** tab of the **Project Manager**; refer to Figure-50.

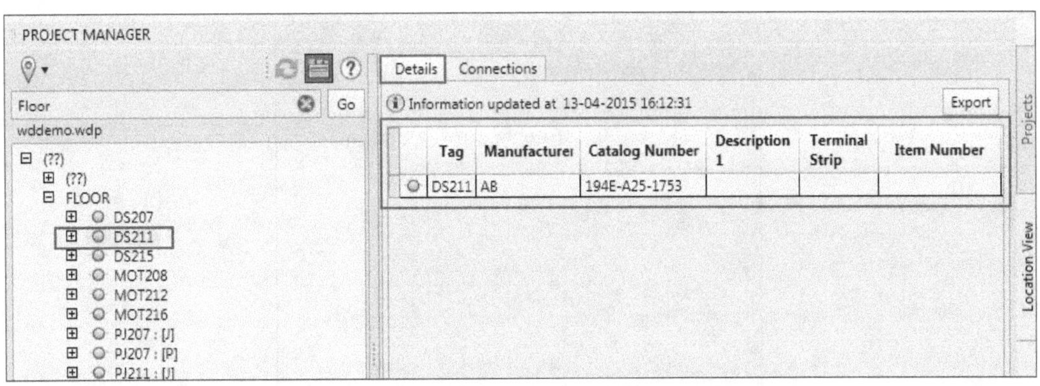

Figure-50. Details of a component

- To check the connections of the component, click on the **Connections** tab. Various connections of the component will be displayed in the **Project Manager**; refer to Figure-51.

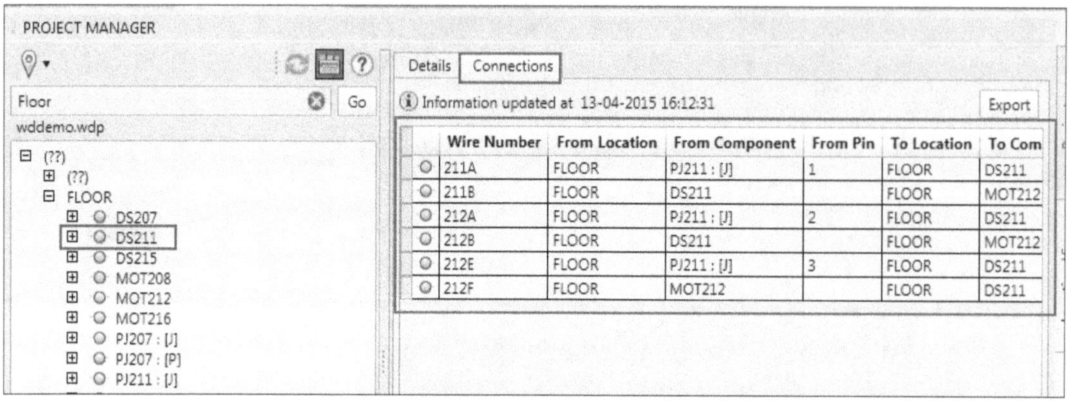

Figure-51. Connections of the component

- If you want to check the component in the drawing then right-click on any field in the **Connections** tab for desired component. A shortcut menu will be displayed; refer to Figure-52.

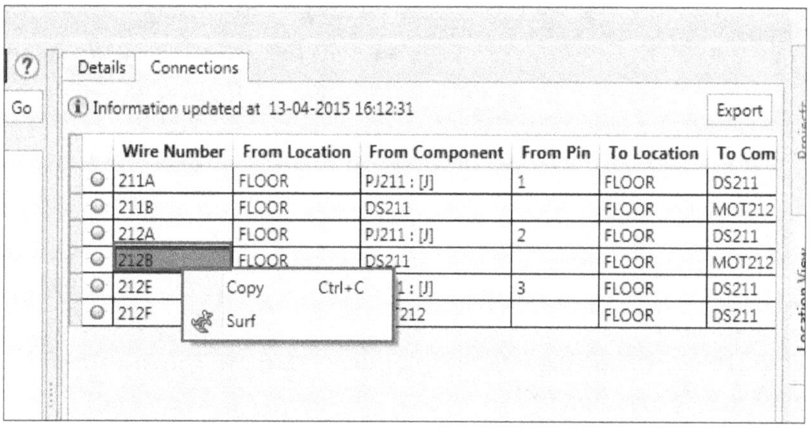

Figure-52. Shortcut menu for component

- Click on the **Surf** button from the shortcut menu. The **Surf** dialog box will be displayed; refer to Figure-53.

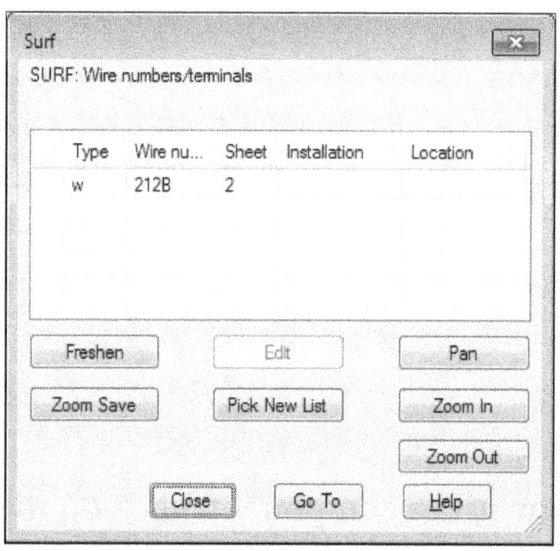

Figure-53. Surf dialog box

- Click on the **Go To** button from the dialog box to check the component in the drawing.

Note that you can export the details of components in csv or xls format by using the **Export** button above the table in the **Project Manager**. Using the exported file, you can share the information with your peers.

UTILITIES

The **Utilities** tool is used to modify multiple wire numbers, component tags, and attribute texts at one place. This tool is also useful when you need to run a script project wide. The procedure to use this tool is given next.

- Click on the **Utilities** tool from the **Project Tools** panel in the **Project** tab of the **Ribbon**. The **Project-Wide Utilities** dialog box will be displayed; refer to Figure-54.

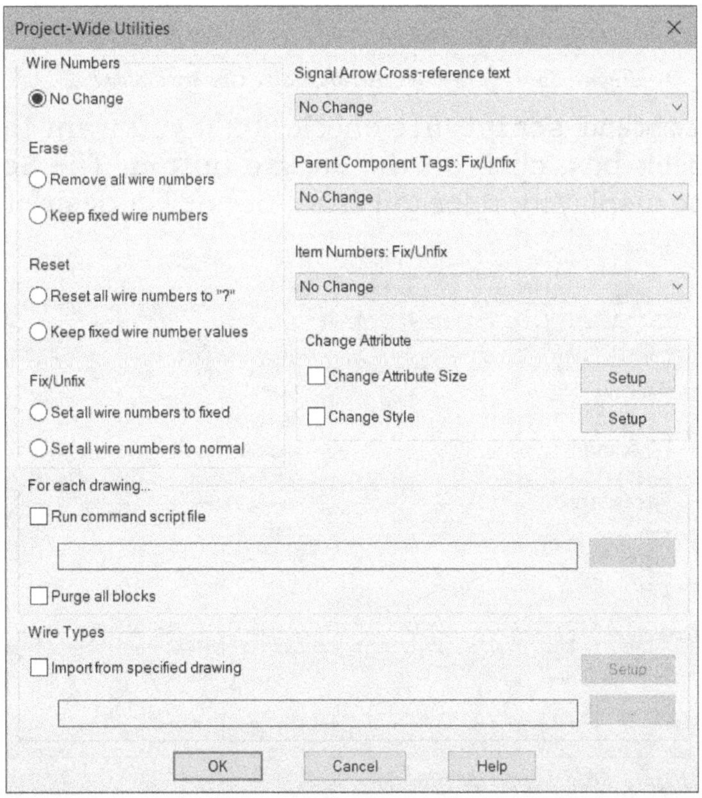

Figure-54. Project-Wide Utilities dialog box

- Select desired radio button from the **Wire Numbers** area of the dialog box to modify wire numbers.
- If you want to remove all the signal arrow cross-reference texts then select the **Remove all** option from the drop-down.
- To fix or unfix component tags and item numbers, select desired option from the **Parent Component Tags** and **Item Numbers** drop-downs, respectively.
- Sometimes, you may get smaller text in the drawing which is not visible on paper. In such cases, select the **Change Attribute Size** check box and click on the **Setup** button. The **Project-wide Attribute Size Change** dialog box will be displayed; refer to Figure-55.
- Select the check boxes of the attributes whose text size is to be changed. Specify desired height and width sizes in the respective edit boxes. Note that if you leave the edit box blank then the respective parameter will be unchanged. Click on the **OK** button to apply attribute.

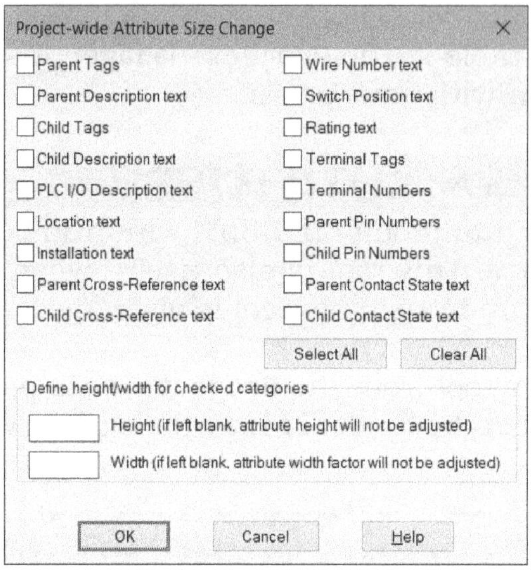

Figure-55. Project-wide Attribute Size Change dialog box

- Select the **Run command script file** check box if you want to run a script. After selecting the check box, click on the **Browse** button. The **Select Script File** dialog box will be displayed; refer to Figure-56. Select desired script and click on the **Open** button.

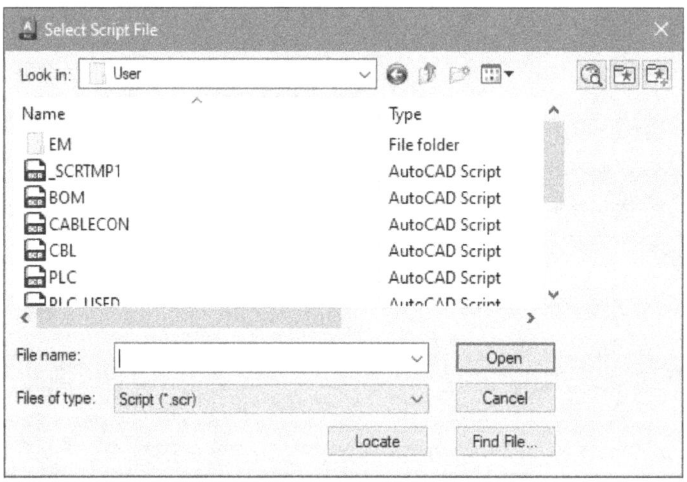

Figure-56. Select Script File dialog box

- Similarly, set the other options as required and click on the **OK** button. The **Batch Process Drawings** dialog box will be displayed; refer to Figure-57.

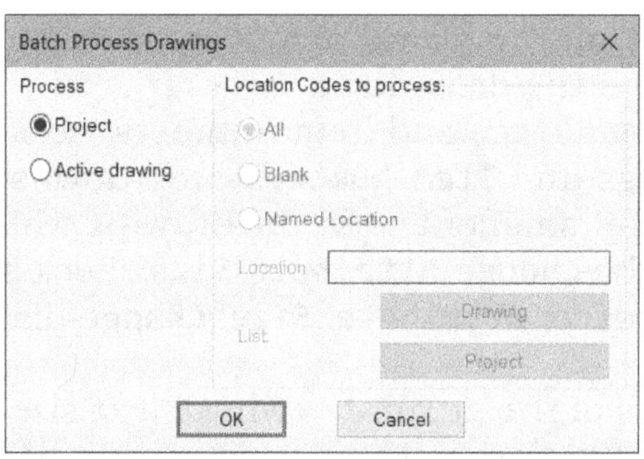

Figure-57. Batch Process Drawings dialog box

- Click on the **OK** button. The **Select Drawings to Process** dialog box will be displayed. Select all the drawings on which you want to perform changes and click on the **Process** button. If you want to include all the drawings then click on the **Do All** button.

- After selecting the drawings, click on the **OK** button from the dialog box. If you have unsaved changes in the current drawing, then save the drawings using the options of **QSAVE** dialog box. The parameters will get changes as specified in the **Project-Wide Utilities** dialog box.

MARKING AND VERIFYING DWGS

The **Mark/Verify DWGs** tool in the **Project Tools** panel is used to mark the drawings for tracking changes and creating a reports of changes made in drawings. The procedure to use this tool is given next.

- Click on the **Mark/Verify DWGs** tool from the **Project Tools** panel in the **Project** tab of **Ribbon**. The **Mark and Verify** dialog box will be displayed; refer to Figure-58.

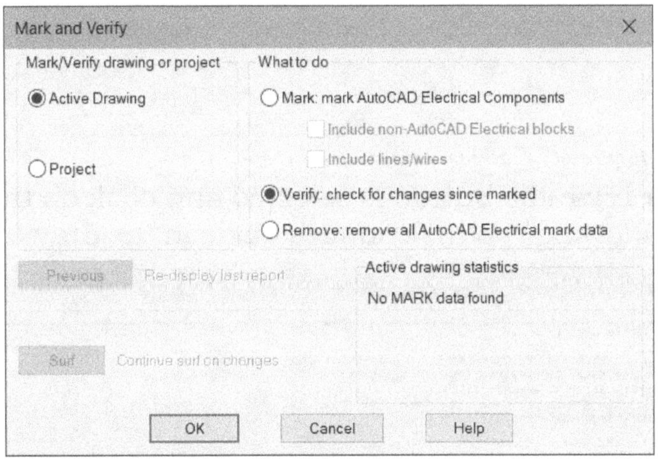

Figure-58. Mark and Verify dialog box

- Select desired option from the **Mark/Verify drawing or project** area. Select the **Active Drawing** option if you want to mark current drawing for tracking changes or select the **Project** radio button if you want to track changes in all the drawings of current project.
- Select desired options from the **What to do** area to define what parameters are to be marked and verified. Select the **Mark: mark AutoCAD Electrical Components** radio button and select the check boxes for components to be marked for checking.
- Click on the **OK** button from the dialog box to create a mark. The **Enter Your Initials** dialog box will be displayed as shown in Figure-59.

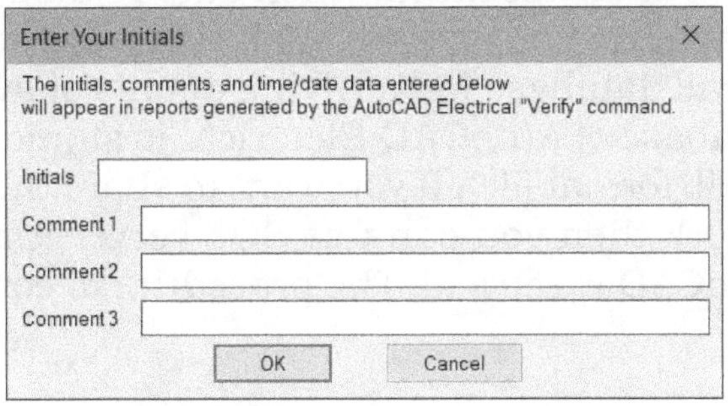

Figure-59. Enter Your Initials dialog box

- Type your initials & comments on the drawing and click on the **OK** button. The **Select Drawings to Process** dialog box will be displayed.

- Add the drawings in process list as discussed earlier and click on the **OK** button from the dialog box.
- Now, make some changes in the drawing like delete a component, add a new one or so on. Click again on the **Mark/Verify DWGs** tool. You will find that a tracking mark is added in the **Active drawing statistics** area of the dialog box; refer to Figure-60.

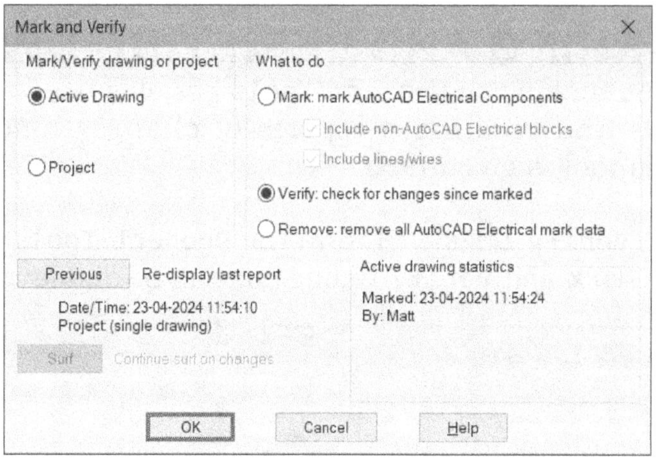

Figure-60. Tracking mark created

- Make sure the **Verify** radio button is selected and click on the **OK** button. A report will be displayed showing you the changes made in the drawing; refer to Figure-61.

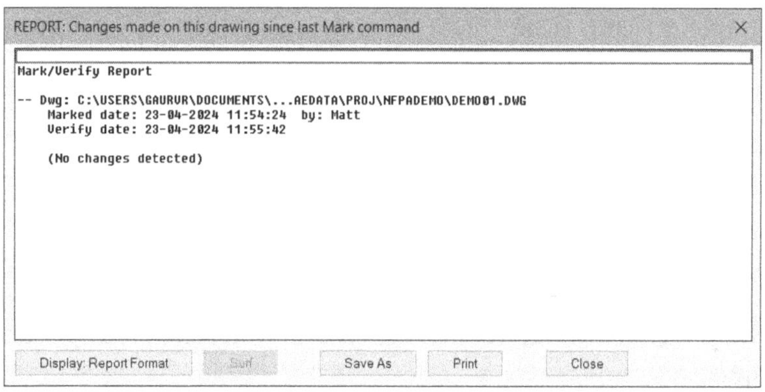

Figure-61. REPORT generated for changes

- Save or print the file if needed and then click on the **Close** button to exit.

CONFIGURING CATALOG DATABASE

The Catalog database is a database file in which data related to components is stored in the form of tables. Whenever you access Catalog Browser in AutoCAD Electrical, in the back-end you also call database for verifying details of available components. By default, the catalog database of AutoCAD Electrical is stored in the form of Microsoft Access database file (.mdb format file). If you want to share the same database file with multiple users on network then you can use SQL server function while configuring catalog database in AutoCAD Electrical. The procedure to configure catalog database is given next.

- Click on the **Config Catalog Database** tool from the **Project Tools** panel in the **Project** tab of **Ribbon**. The **Configure Database** dialog box will be displayed; refer to Figure-62.

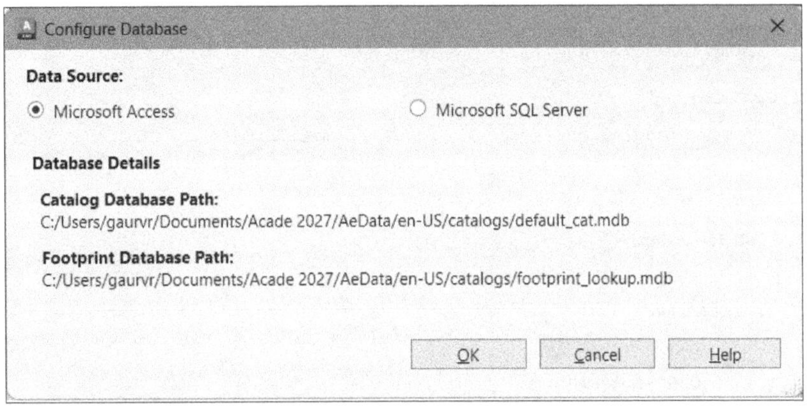

Figure-62. Configure Database dialog box

- By default, the **Microsoft Access** radio button is selected and hence the database file of your local drive is selected.
- Select the **Microsoft SQL Server** radio button from the **Data Source** area to use database saved on SQL server. The options in the dialog box will be displayed as shown in Figure-63.

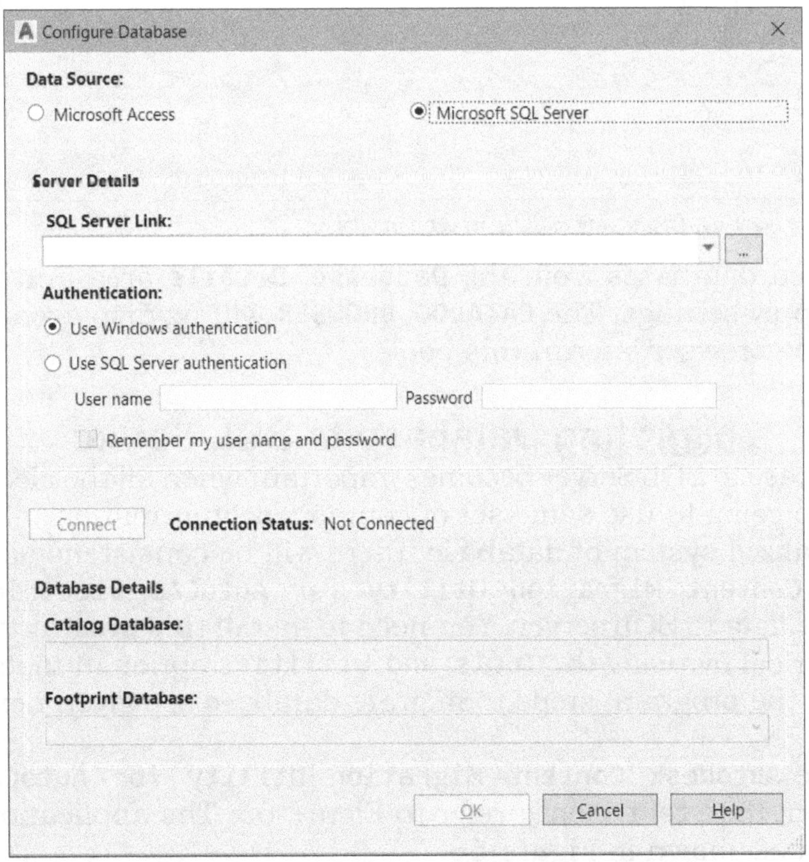

Figure-63. Configure Database dialog box with SQL Server options

- To use this option, you need to migrate all the database on SQL server. The procedure to migrate database is discussed in the next topic.
- Select desired link from **SQL Server Link** field (or use **Browse** button) from the dialog box and connect using desired authentication method. The database details will be displayed in the **Database Details** area; refer to Figure-64.

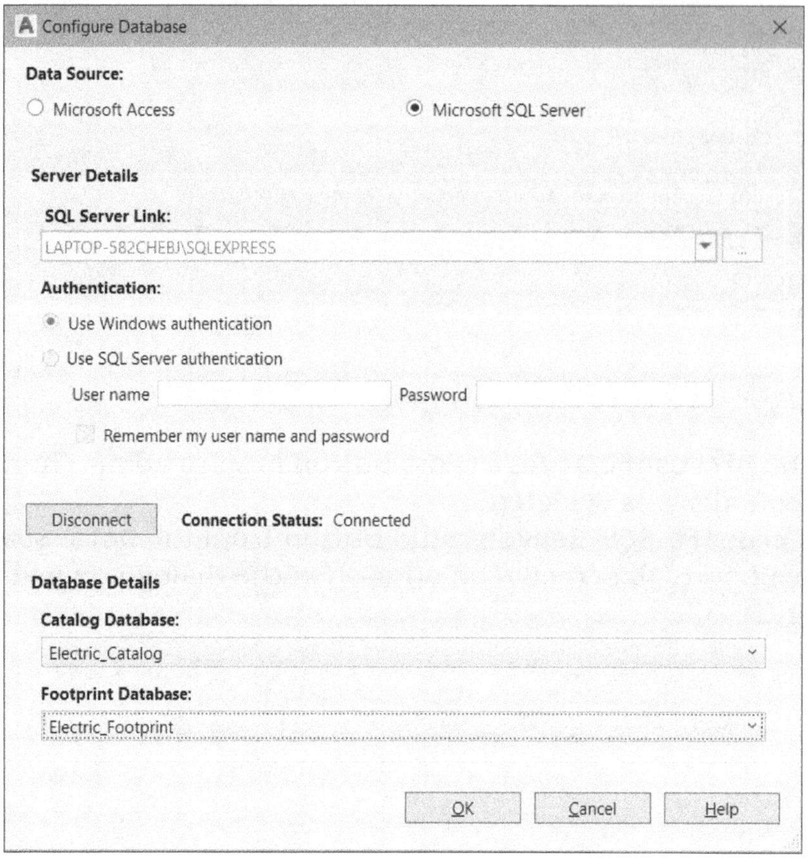

Figure-64. Connecting to SQL Server database

- Select desired databases from the **Database Details** area and click on the **OK** button to apply settings. The **CATALOG BROWSER** will be displayed. (You will learn about catalog browser later in this book.)

Migrating Database to SQL Server

Migrating database to SQL Server becomes important when all the electrical designers in company are going to use same set of component information in their drawings. With the centralized system of database, there will be consistency in their designs. The **Autodesk Content Migration Utility for AutoCAD Electrical** program is used to migrate data to SQL server. You need to install this program while installing AutoCAD Electrical by using the **Tools and Utilities** option at installation screen. After installing the program, steps to migrate database are given next.

- Click on the **Autodesk Content Migration Utility for AutoCAD Electrical** program from the **Start** menu; refer to Figure-65. The application interface will be displayed as shown in Figure-66.

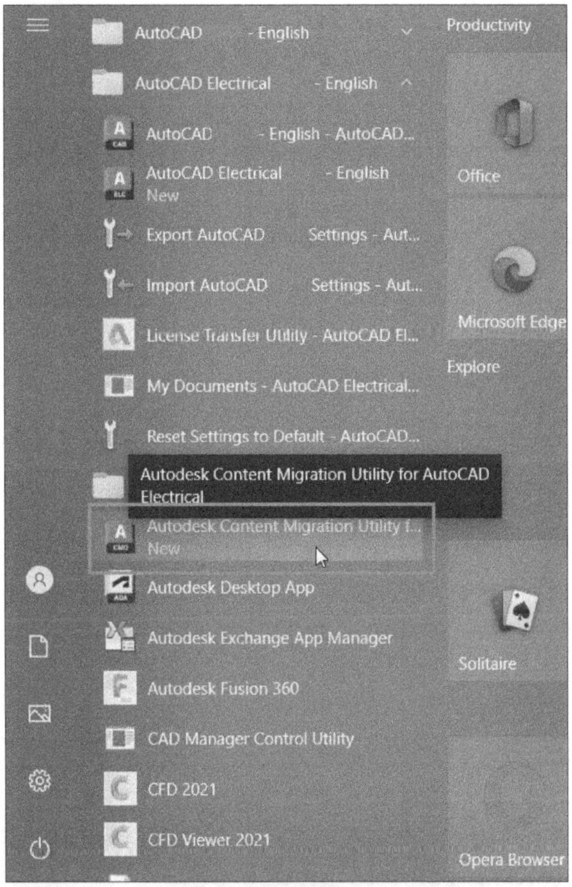

Figure-65. Autodesk Content Migration Utility

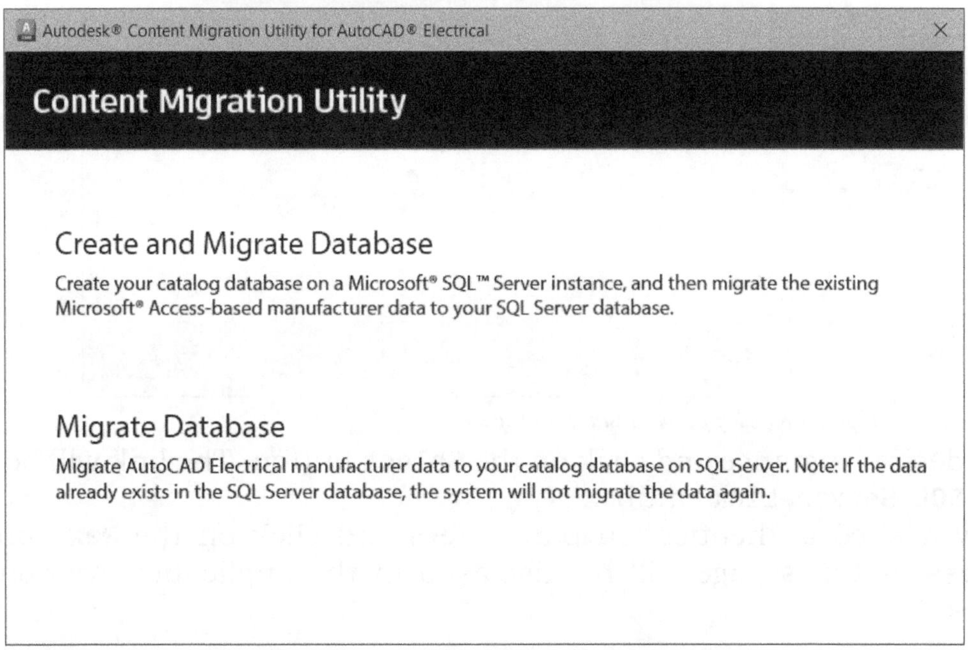

Figure-66. Autodesk Content Migration Utility for AutoCAD Electrical

- Click on the **Create and Migrate Database** tool in the application window. The page will be displayed as shown in Figure-67.

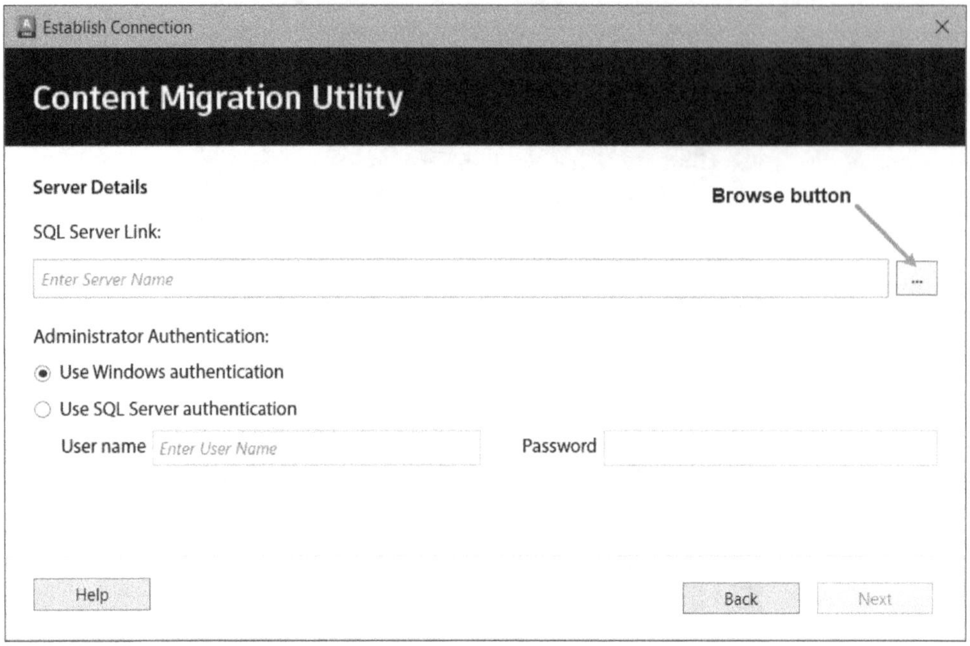

Figure-67. Server Details page

- Click on the **Browse** button next to **SQL Server Link** edit box; refer to Figure-67. The **Select Server Instance** dialog box will be displayed after system has check available server instances; refer to Figure-68.

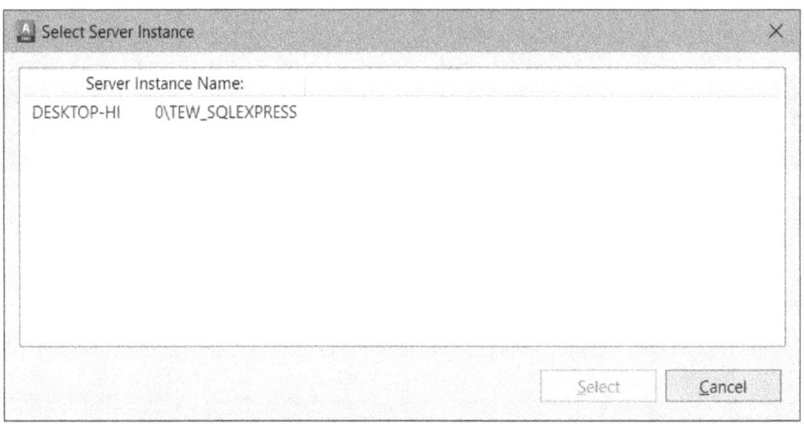

Figure-68. Select Server Instance dialog box

- Select desired instance and click on the **Select** button. The link will be displayed in the **SQL Server Link** edit box.
- Specify desired authentication parameters and click on the **Next** button. The **Database Details** page will be displayed in the application window; refer to Figure-69.

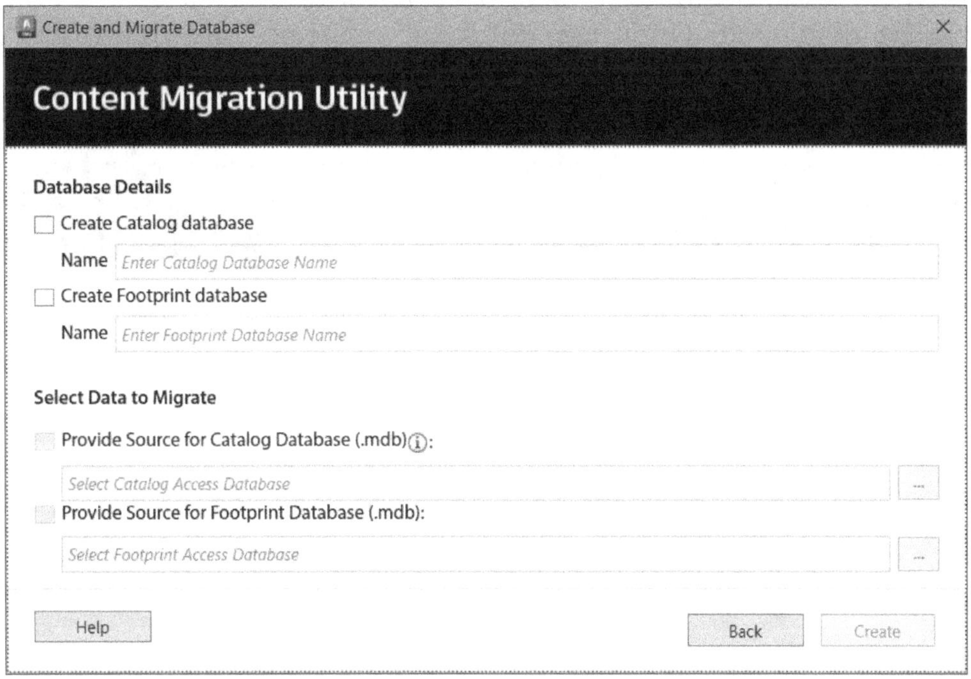

Figure-69. Database Details page

- Select the **Create Catalog database** check box to create a database on SQL server. The **Provide Source for Catalog Database** option will become active. Select the check box for this option and click on the **Browse** button to select desired database file.
- Similarly, set the database for footprints; refer to Figure-70.

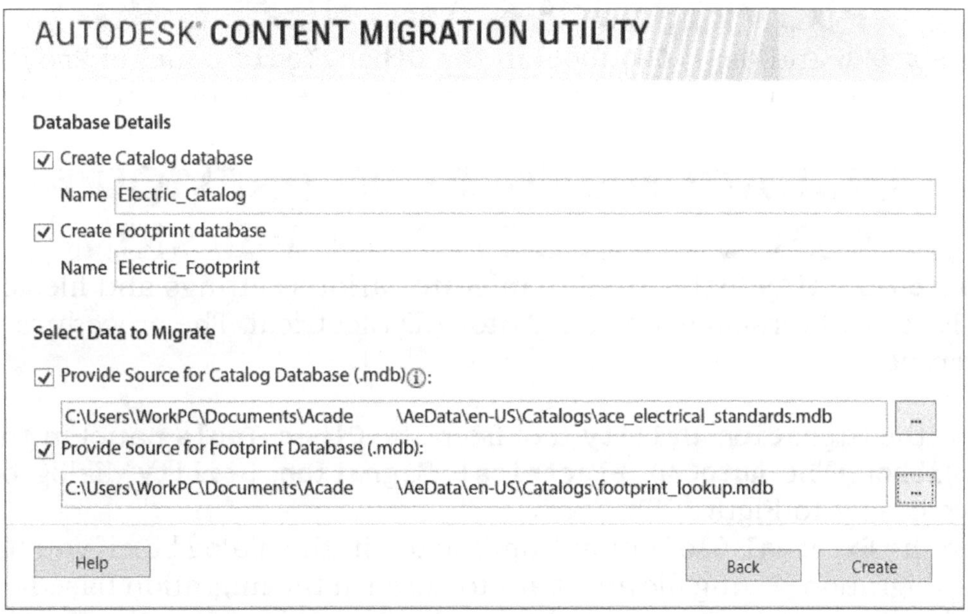

Figure-70. Database parameters specified

- Click on the **Create** button from the application window. The **Completing Creation** page will be displayed. Once the database creation and migration process is complete, the page will be displayed as shown in Figure-71.

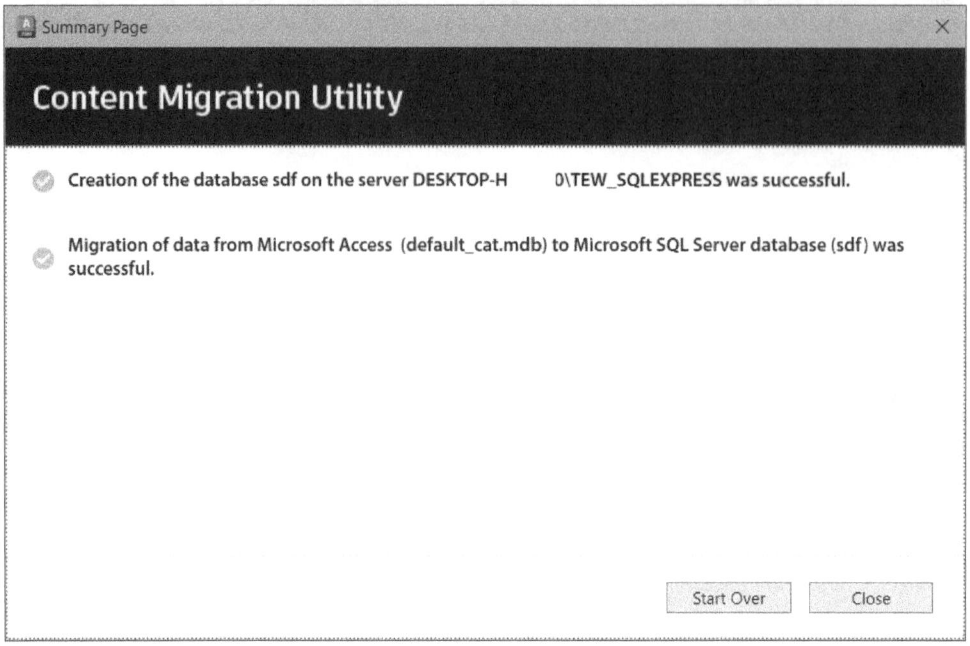

Figure-71. Database migration completion page

- Click on the **Close** button to exit the application. Now, you are ready to use SQL server database in AutoCAD Electrical.

PREVIOUS DWG AND NEXT DWG

The **Previous DWG** and **Next DWG** tools in the **Other Tools** panel of **Project** tab are used to switch to previous and next drawings of current project respectively.

MIGRATION DATA FROM PREVIOUS AUTOCAD ELECTRICAL VERSION

The **Migration Utility** tool is used to migrate various settings and files of previous AutoCAD Electrical version to current AutoCAD Electrical. The procedure to use this tool is given next.

- Click on the **Migration Utility** tool from the **Other Tools** panel in the **Project** tab of **Ribbon**. The **AutoCAD Electrical Migration Utility** dialog box will be displayed; refer to Figure-72.
- Click on the **External file** button from the top in the dialog box if you have earlier saved a migration setting file and want to perform the migration based on that file.
- Select desired item to be migrated from the **Migration items** area at the left in the dialog box and click in the **Copy/Merge Options** field. The option of the field will be activated; refer to Figure-73.

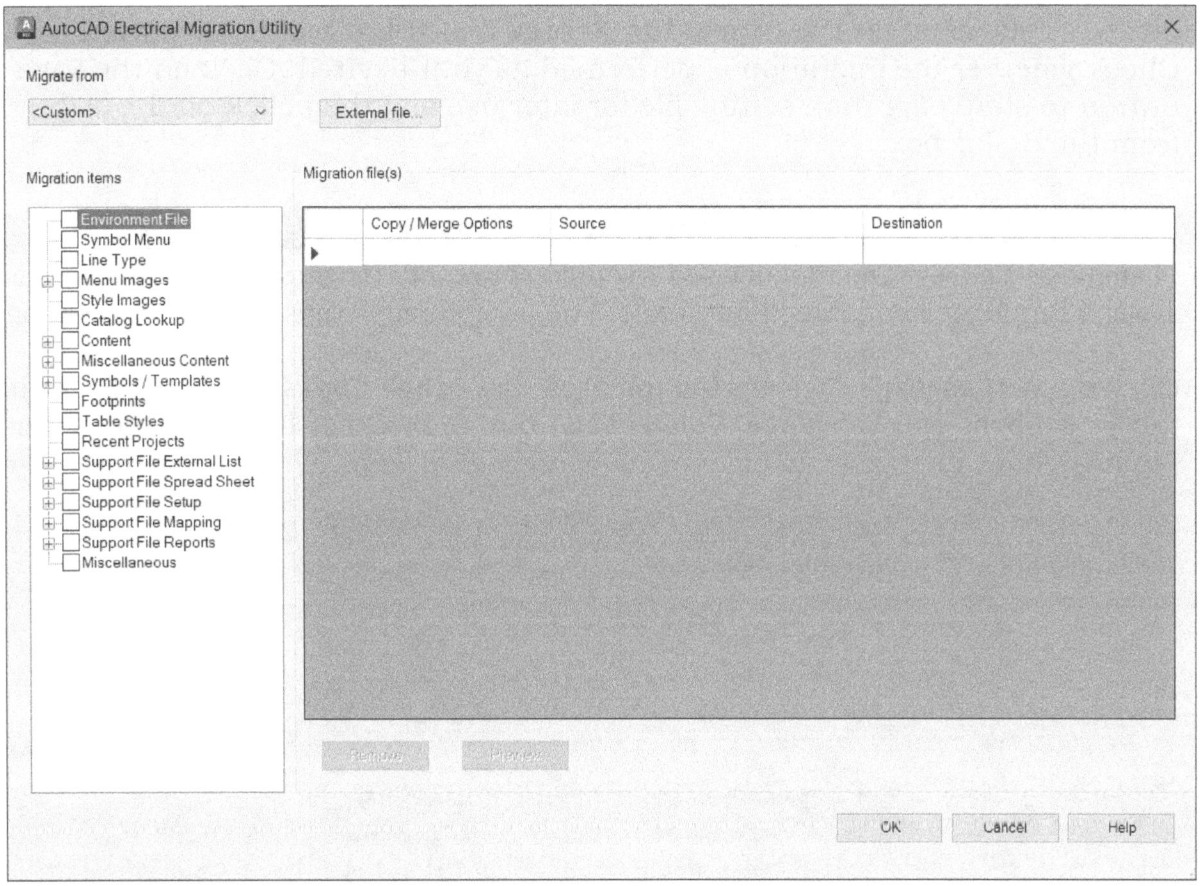

Figure-72. AutoCAD Electrical Migration Utility dialog box

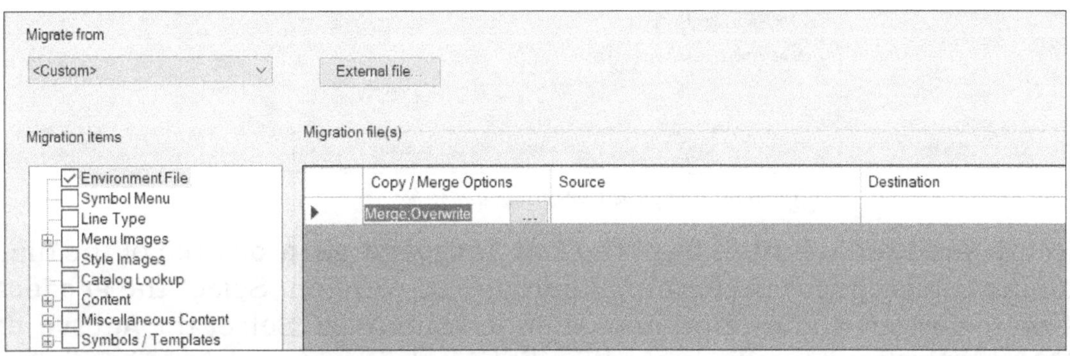

Figure-73. Options activated under Copy/Merge Options field

- Click on the button in the field and set desired options in the **Copy/Merge Options** dialog box. Similarly, click in the **Source** field and select the file to be migrated. Click in the **Destination** field and set the folder location where you want to migrate the file.
- You can migrate the other items in the left area of the dialog box in the same way.

- After setting desired migration parameters, click on the **OK** button from the dialog box to apply settings. The **Migration Review** dialog box will be displayed.
- Check whether the migration is performed as your desired. Click on the **Save As** button to store migration setting file for later use and then click on the **OK** button from the dialog box.

LANGUAGE CONVERSION IN PROJECT

The **Language Conversion** tool is used to convert text of attributes from one language to another language available in database. The procedure to use this tool is given next.

- Click on the **Language Conversion** tool from the **Other Tools** panel in the **Project** tab of **Ribbon**. The **Language Conversion** dialog box will be displayed; refer to Figure-74.

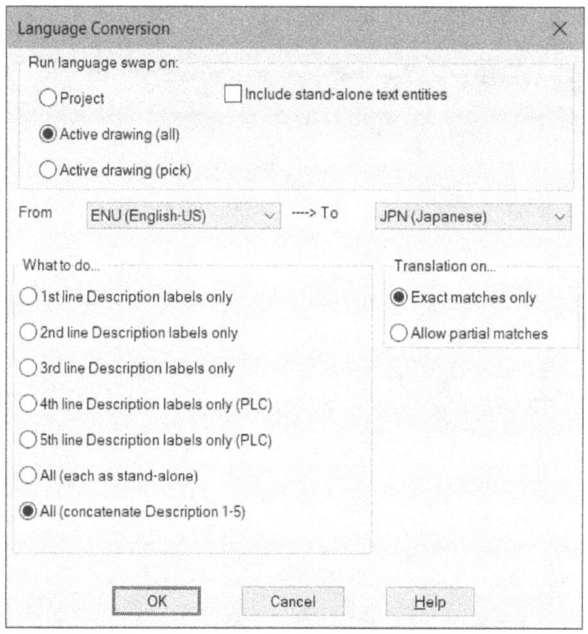

Figure-74. Language Conversion dialog box

- Select desired radio button from the **Run language swap on** area of the dialog box to define the scope for performing language conversion. Select the **Project** radio button to convert all descriptions to defined language. Select the **Active drawing (all)** radio button to perform modifications only in current open drawings. Select the **Active drawing (pick)** radio button to perform modifications only on selected components in the active drawings. Select the **Include stand-alone text entities** check box to include all the text notes which are not part of component attributes.
- Select desired languages in the **From** and **To** drop-downs.
- From the **What to do** area of the dialog box, select desired radio button to define which description(s) is to be changed.
- Select desired radio button from the **Translation on** area of the dialog box to define what kind of translation is desired.
- After setting desired parameters, click on the **OK** button. Note that you will be asked to select the objects to be converted if you have selected the **Active drawing (pick radio)** button and you will be asked to select the drawings to be included in conversion if you have selected the **Project** radio button.

EDITING LANGUAGE DATABASE

The **Edit Language Database** tool is used to modify language conversion database. The procedure to use this tool is given next.

- Click on the **Edit Language Database** tool from the **Other Tools** panel in the **Project** tab of the **Ribbon**. The **Edit: Language Lookup** dialog box will be displayed; refer to Figure-75.

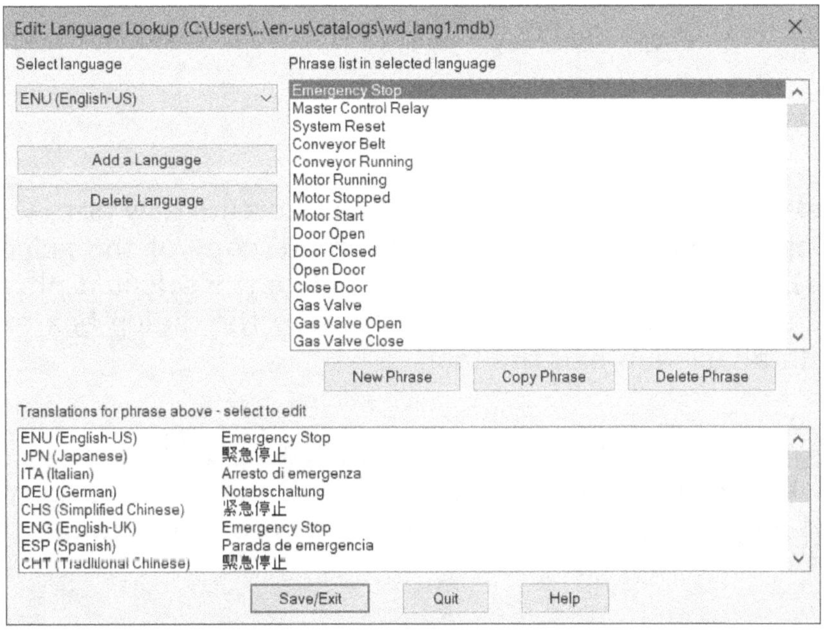

Figure-75. Edit: Language Lookup dialog box

- Click on the **Add a Language** button if you want to add a new language. The **Enter Language Table to Add** dialog box will be displayed; refer to Figure-76.

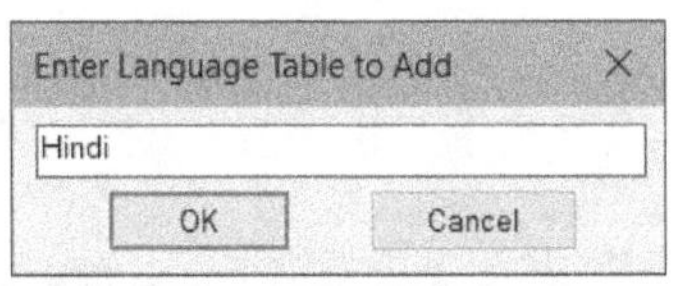

Figure-76. Enter Language Table to Add dialog box

- Specify desired name of language in the edit box and click on the **OK** button. The new language will be added at the bottom in the bottom list of the dialog box; refer to Figure-77.

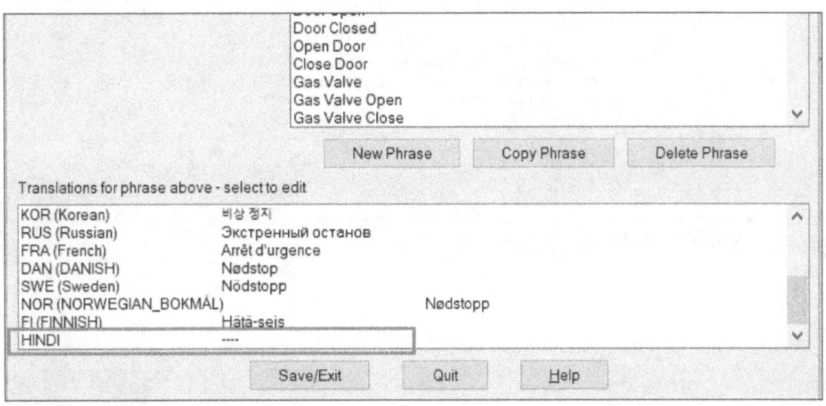

Figure-77. New language added

- Select desired phase whose translation is to be created from the **Phrase list in selected language** list box of the dialog box and then click on the new language added in the bottom list. You will be asked to type translation text for the selected phrase.
- Type desired phrase in respective language (you can use Google translate feature); refer to Figure-78 and click on the **OK** button from the **Edit** dialog box. The new translation will be added next to selected language in bottom list.

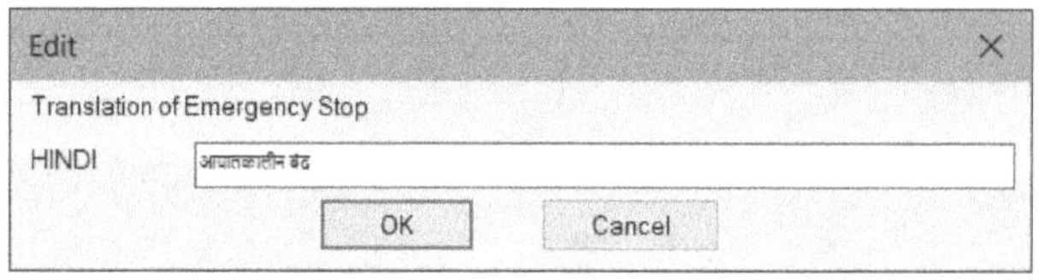

Figure-78. Edit dialog box

- Using the **New Phrase** button, you can similarly create new phrase for translation.
- Using the **Copy Phrase** button, you can create copy of the selected phase and using the **Delete Phrase** button, you can delete the selected phrase.
- Click on the **Save/Exit** button at the bottom in the dialog box to save language conversion database and exit the tool.

TITLE BLOCK SETUP

The **Title Block Setup** tool is used to create and manage link between title block parameters and Project/Drawing parameters. The procedure to use this tool is given next.

- Click on the **Title Block Setup** tool from the **Other Tools** panel in the **Project** tab of **Ribbon**. The **Setup Title Block Update** dialog box will be displayed; refer to Figure-79.

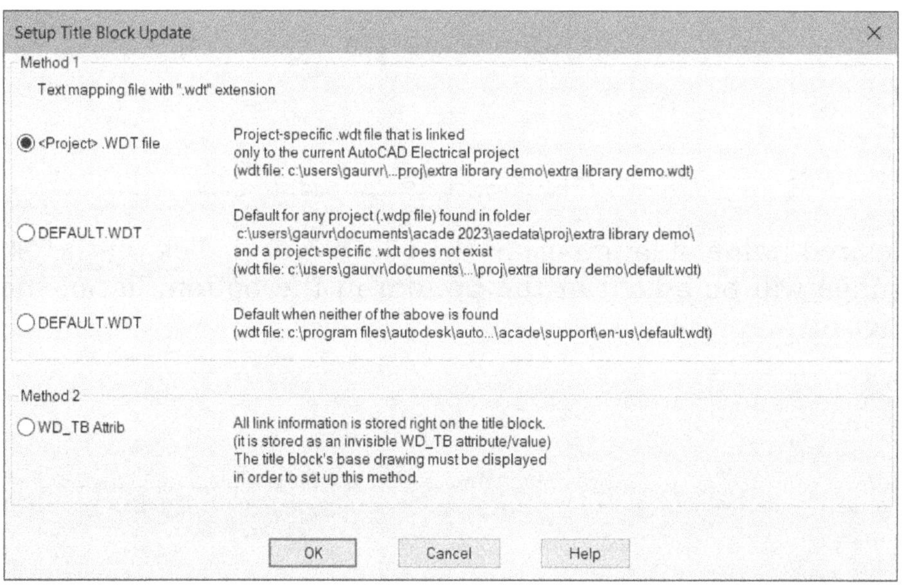

Figure-79. Setup Title Block Update dialog box

- Select desired radio button from the dialog box to define text mapping option. Select the **<Project>.WDT file** or **DEFAULT.WDT** radio button when you want to use project attributes as driving criteria for title block attributes. Select the **WD TB Attrib** radio button when you want to use WD TB attribute for changing parameters in the title block. Note that the WD TB attribute must be placed in the title block while creating it to use this option. In this case, we are selecting the **<Project>.WDT file** radio button. Click on the **OK** button. The **.WDT File Exists** dialog box will be displayed; refer to Figure-80.

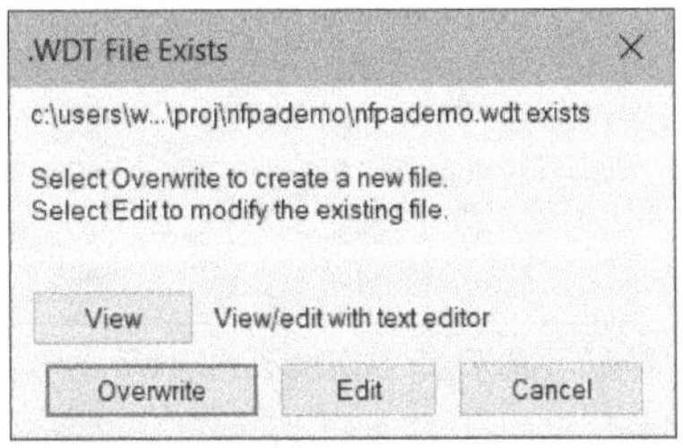

Figure-80. WDT File Exists dialog box

- Click on the **Edit** button from the dialog box to define mapping the attributes. The **Title Block Setup** dialog box will be displayed with default mapping; refer to Figure-81.

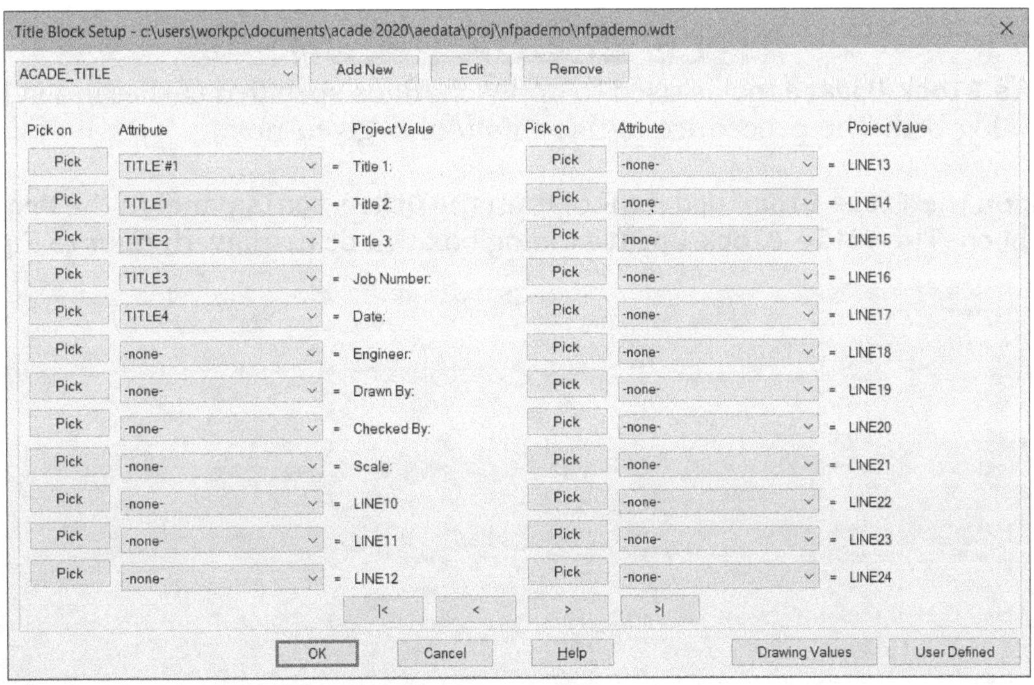

Figure-81. Title Block Setup dialog box

- Set desired options in the **Attribute** drop-downs to map them with respective attributes of blocks.
- If you want to create user defined attribute values for mapping then click on the **User Defined** button at the bottom right in the dialog box. The **Title Block Setup - User Defined** dialog box will be displayed. Define desired values for attributes; refer to Figure-82 and click on the **OK** button.

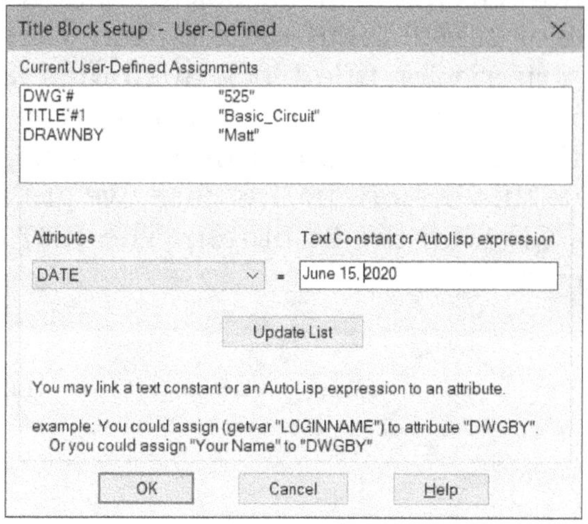

Figure-82. User defined values for attributes

- Now, you can use these newly defined attributes in mapping.
- If you want to use drawing values instead of project values then click on the **Drawing Value** button from the bottom in the dialog box.
- After setting desired mapping parameters, click on the **OK** button from the dialog box to apply changes.

TITLE BLOCK UPDATE

The **Title Block Update** tool is used to apply changes specified earlier in **Title Block Setup** dialog box. The procedure to use this tool is given next.

- Click on the **Title Block Update** tool from the **Other tools** panel in the **Project** tab of **Ribbon**. The **Title Block Update** dialog box will be displayed; refer to Figure-83.

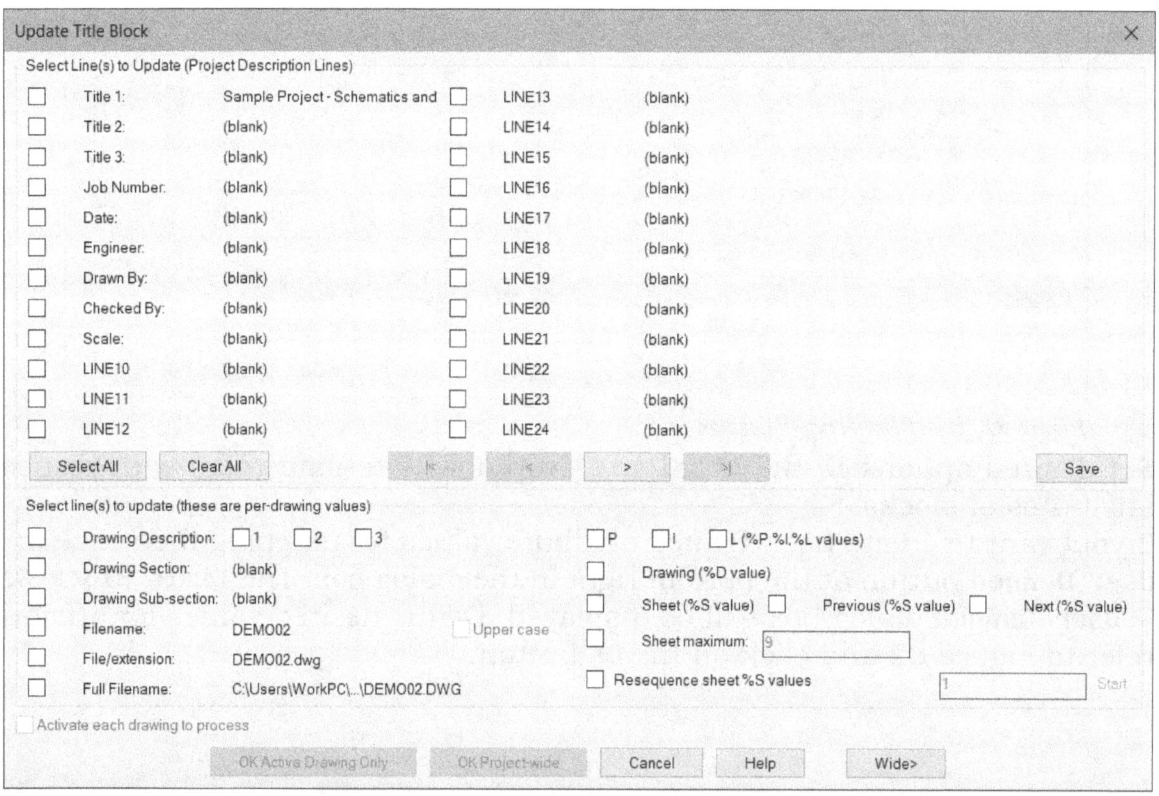

Figure-83. Update Title Block dialog box

- Select the check boxes for lines to be updated in the title block and click on the **OK** button. Click on the **OK Active Drawing Only** button if you want to update title block only in current drawing. Click on the **OK Project-wide** button if you want to update title block in all drawings of the active project.

Note that if you want to edit description lines of project then right-click on the name of Project in the **Project Browser** and select the **Descriptions** option from the shortcut menu; refer to Figure-84. The **Project Descriptions** dialog box will be displayed; refer to Figure-85. Specify desired text in lines and click on the **OK** button.

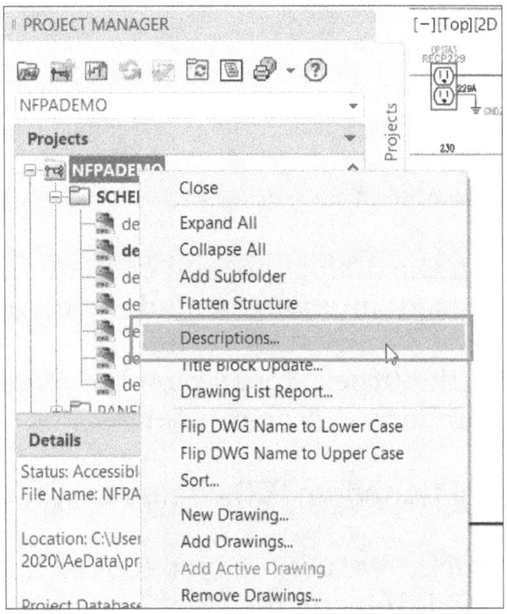

Figure-84. Descriptions option

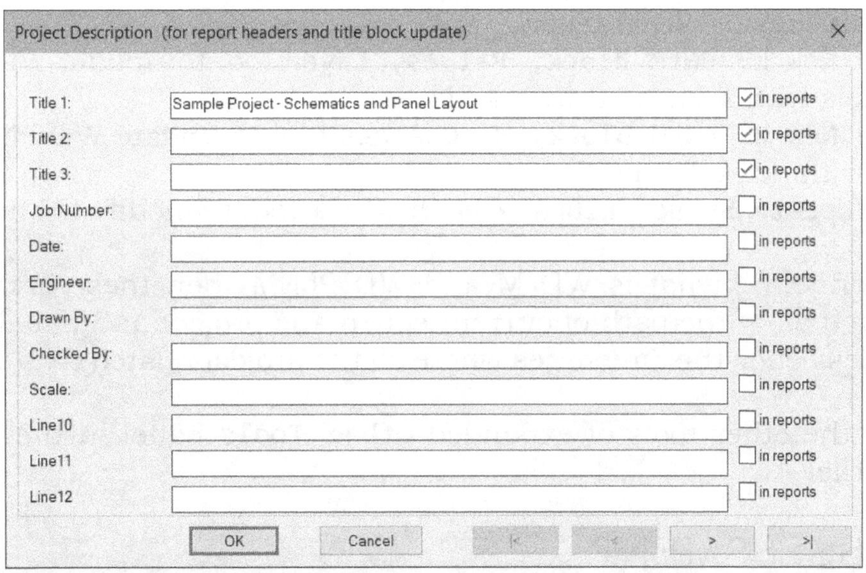

Figure-85. Project Description dialog box

UPDATING COMPONENT BASED ON CHANGES MADE IN CATALOG

The **Component Update From Catalog** tool is used to update the components in current drawing or project-wide if there are changes made to those components in database. The procedure to use this tool is given next.

- Click on the **Component Update From Catalog** tool from the **Other Tools** panel in the **Project** tab of **Ribbon**. The **Component Update From Catalog** dialog box will be displayed; refer to Figure-86.

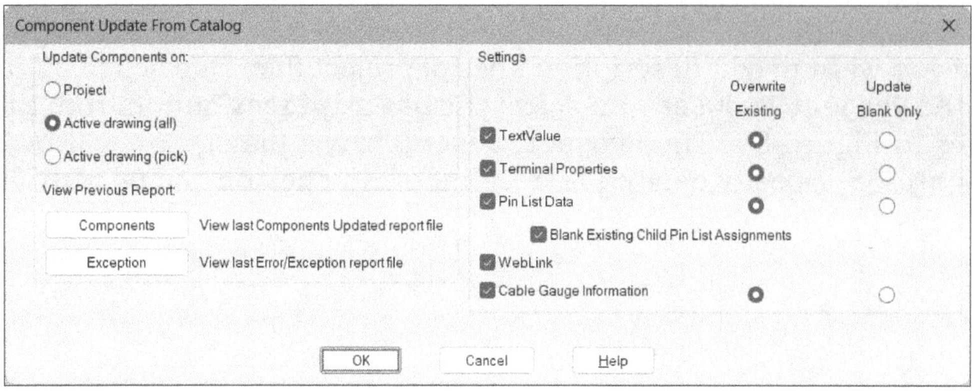

Figure-86. Component Update From Catalog dialog box

- Set desired parameters in the dialog box and click on the **OK** button to update the components in project based on changes made in catalog.

Click on the down arrow in the **Other Tools** panel to extend the panel and click on the pin button to hold it. Similarly, you can use:

- **IEC Tag Mode Update** tool to update IEC component tags in the drawing/project-wide.
- **Update to New WD M Block, Values, Layer** tool to update current WD M block template based on selected drawing file.
- **Update to New WD M Block, No Changes** tool to update the current WD M block with default copy of WD M block.
- **Update to New WD PNLM Block, Values, Layer** tool to update WD PNLM block file of panels with selected drawing file.
- **Update to New WD PNLM Block, No Changes** tool to update WD PNLM block file of panels with default copy.
- Select the **Update Symbol Library WD M Block** tool to modify wd m block file.

If you are wondering what is WD M and WD PLNM then these are environment variables which store the path of current schematic project and panel project files. These variables define the directories where all the project related data will be stored.

You can use the other tools of expanded **Other Tools** panel in the same way as discussed earlier.

SELF-ASSESSMENT

Q1. The .... area of the **Project Properties** dialog box provides options to modify library path for icon menu.

Q2. The **Real time error checking** check box is selected to allow the system to check for some basic errors in the drawing like duplication of wire numbers and component tags. (T/F)

Q3. The .... edit box is used to set the format for tag assigned to various components.

Q4. Write down the steps for changing Wire Tee style for the project.

Q5. Write down steps to change the position of wire number with respect to wire.

Q6. Options to change the default style of ladder creation are available in the **Styles** tab of the **Project Properties** dialog box. (T/F)

Q7. The .... button is used to update or retag component tags, wire numbers, cross-references, and so on.

Q8. Which of the following option is used to print the drawing of current project in image format?

A. Plot Project
B. Publish to Web
C. Publish to PDF/DWF/DWFx
D. ZIP Project

Q9. The .... tool is used to create and manage link between title block parameters and Project/Drawing parameters.

Q10. The .... tool in the **Project Tools** panel is used to mark the drawings for tracking changes and create a reports of changes made in drawings.

REVIEW QUESTIONS

Q1. What are the two major categories of drawings in AutoCAD Electrical?
A. Schematic diagrams and Flowcharts
B. Schematic diagrams and Panel drawings
C. Block diagrams and Circuit drawings
D. Electrical circuits and Architectural layouts

Q2. Which area in AutoCAD Electrical is used to manage projects?
A. Drawing Area
B. Command Line
C. Properties Panel
D. Project Manager

Q3. What is the function of the "Create Folder with Project Name" check box during project creation?
A. Creates a backup folder
B. Assigns a default path
C. Creates a folder with the project's name at the specified location
D. Adds a suffix to the folder name

Q4. Which tab in Project Properties is used to define directories containing symbols and icons?
A. Components tab
B. Styles tab
C. Project Settings tab
D. Drawing Format tab

Q5. What does the "Real time error checking" checkbox enable in Project Properties?
A. Live rendering of drawings
B. Automatic saving
C. Checking for basic errors like duplicate wire numbers
D. Synchronizing files to the cloud

Q6. What does the Tag Format "%F%N" represent in the Components tab?
A. Feature and Name
B. Family and Name
C. Format and Number
D. Family and Serial Number

Q7. What happens when the "Combined Installation/Location tag mode" is selected?
A. Only location is used for tags
B. Tags are generated using both installation and location codes
C. Tags are created manually
D. Tags are sorted alphabetically

Q8. What is the purpose of the Item Numbering button in the Component Options area?
A. Assigns drawing numbers
B. Specifies sheet format
C. Sets the numbering mode for components
D. Defines tag symbols

Q9. Which option allows wire numbering based on wire layers?
A. Based on Terminal Symbol Location
B. Based on Wire Layer
C. Hidden on Wire Network
D. Exclude

Q10. What does the "In-Line" option under New Wire Number Placement do?
A. Places the number above the wire
B. Places the number below the wire
C. Aligns the number horizontally
D. Places the number in line with the wire and activates gap setup

Q11. What is the purpose of the "Enable annotations update with project-wide/ drawing-wide wire number update" checkbox?
A. Saves the project to the cloud
B. Updates only tag names
C. Automatically updates annotations when wire numbers are updated
D. Clears old annotations

Q12. What does selecting the "Real time signal and contact cross-referencing between drawings" checkbox do?
A. Disables cross-referencing
B. Makes changes reflect across drawings in real time
C. Converts drawings into PDFs
D. Adds electrical codes

Q13. Which tab lets you define arrow, PLC, and wire cross styles?
A. Drawing Format
B. Styles
C. Cross-References
D. Components

Q14. What does selecting "Vertical" in the Ladder Defaults area do?
A. Flips the ladder 180 degrees
B. Changes ladder orientation to vertical
C. Locks the ladder position
D. Merges two ladder drawings

Q15. How do you open a project file in AutoCAD Electrical?
A. Double-click the drawing
B. Click the Open Project button in the Project Manager
C. Use Ctrl + O
D. Use the Save As dialog box

Q16. What is the first step to create a new drawing in a project?
A. Open the drawing list
B. Click on the New Drawing button in Project Manager
C. Choose a default template
D. Set the Project Code

Q17. Which box do you use to specify the name of the drawing file?
A. Template edit box
B. Description 1 edit box
C. Name edit box
D. Drawing Number box

Q18. What happens if you skip selecting a template while creating a drawing?
A. The drawing cannot be created
B. You must choose one later
C. The default template of the project is used
D. An error message is displayed

Q19. What does the Refresh button do?
A. Opens the last saved version
B. Refreshes Project Manager with updated drawing information
C. Updates drawing names
D. Restores default settings

Q20. What is the purpose of the Project Task List button?
A. Plot the drawing
B. Display assigned tasks in a project
C. Save the drawing
D. Check spelling errors

Q21. What feature helps update component tags, wire numbers, and cross-references?
A. Drawing Configuration
B. Ladder Setup
C. Project Wide Update/Retag
D. Refresh

Q22. Which option allows you to change the wire number settings during a retag?
A. Ladder Reference Setup
B. Signal Tag/Retag
C. Setup button next to Wire Number and Signal Tag/Retag checkbox
D. Drawing Configuration

Q23. What do you select to renumber the ladders in the project?
A. Drawing List
B. Resequence radio button
C. Project Code
D. Component Setup

Q24. What does the Drawing List Display Configuration button do?
A. Changes drawing file name
B. Removes all drawings
C. Selects details to display in Project Manager
D. Creates new drawings

Q25. What is the purpose of the Plot Project tool?
A. Create zip archive
B. Modify component tags
C. Print project drawings
D. Configure wire settings

Q26. Which format is NOT supported by Publish to DWF/PDF/DWFx tool?
A. DWF
B. PDF
C. DOC
D. DWFx

Q27. What should be edited in the wd.env file to enable Zip Project?
A. Add new drawing path
B. Save as XML
C. Remove * marks from specific lines
D. Change extension to .zip

Q28. What must be done before replacing a drawing?
A. Refresh the page
B. Open Drawing List Configuration
C. Select drawing and use right-click menu
D. Export drawing to CSV

Q29. In the Location View tab, what does (??) indicate?
A. Component error
B. Undeclared installation code
C. Missing drawing file
D. Default drawing name

Q30. How can you filter unwanted installations in Location View?
A. By editing Project Code
B. Using Search Box
C. Using Filter by Installation and Location option
D. By removing drawings

Q31. What does the Surf button in the Connections tab allow you to do?
A. Search for drawings
B. Export drawings to PDF
C. Locate component in the drawing
D. Rename the drawing

Q32. What is the primary function of the Utilities tool in AutoCAD Electrical?
A. Modify component libraries
B. Modify multiple wire numbers, tags, and attribute texts at once
C. Edit schematic layout
D. Design title blocks

Q33. Which option allows you to remove all signal arrow cross-reference texts in the Utilities tool?
A. Fix all
B. Remove signal tags
C. Remove all
D. Clear wire list

Q34. What is displayed when you click the Setup button after selecting the Change Attribute Size checkbox?
A. Layer Manager
B. Drawing Properties dialog box
C. Project-wide Attribute Size Change dialog box
D. Wire Editor

Q35. What should you do to run a command script file in the Utilities tool?
A. Select "Enable Scripts" in the menu
B. Click Run from Tools tab
C. Check "Run command script file" and browse the script
D. Enter the script manually

Q36. What is the purpose of the Mark/Verify DWGs tool?
A. Add wire numbers
B. Mark drawings for change tracking and generate reports
C. Update title blocks
D. Rename layers

Q37. What is prompted after clicking OK in the Mark/Verify DWGs dialog box?
A. Drawing Editor
B. Enter Your Initials dialog box
C. Layer Selection dialog box
D. Block Attribute Manager

Q38. Which file format is used by default for the AutoCAD Electrical catalog database?
A. .xlsx
B. .sql
C. .xml
D. .mdb

Q39. What should be selected in the Configure Catalog Database tool to use a centralized database?
A. Microsoft Access
B. Excel Sheet
C. Microsoft SQL Server
D. Oracle

Q40. What tool is used to migrate data to SQL Server for AutoCAD Electrical?
A. AutoCAD Electrical SQL Manager
B. Autodesk Content Migration Utility
C. SQL Migration Wizard
D. AutoCAD Catalog Converter

Q41. What must you click to create and migrate database in the Autodesk Content Migration Utility interface?
A. Setup Connection
B. Import Database
C. Create and Migrate Database
D. New SQL Instance

Q42. Which tool allows switching to the previous or next drawings in a project?
A. Drawing Navigator
B. Previous DWG and Next DWG tools
C. Drawing Explorer
D. Sheet Selector

Q43. What is the function of the Migration Utility tool?
A. Rename all blocks
B. Convert .dwg to .pdf
C. Migrate files from old to new AutoCAD Electrical version
D. Convert title block format

Q44. What does the Language Conversion tool allow users to do?
A. Change UI language
B. Translate attribute text between languages
C. Import dictionary
D. Add new commands

Q45. Which button should be clicked to add a new language in Edit Language Database?
A. New Language
B. Insert Language
C. Add a Language
D. Create Language

Q46. What is the function of the Title Block Setup tool?
A. Edit wire colors
B. Create link between title block and project parameters
C. Export project data
D. Modify drawing layout

Q47. Which file is selected to use project attributes as driving criteria in title block mapping?
A. .DLG
B. .WDT
C. .WDL
D. .DAT

Q48. What does the Title Block Update tool apply?
A. Page numbering
B. Changes specified in Title Block Setup
C. Component renaming
D. Drawing cleanup

Q49. What allows updating all title blocks in the project at once?
A. OK Active Drawing Only
B. OK Project-wide
C. Update All Blocks
D. Global Refresh

For Student Notes

Answer for Self-Assessment:

Ans1. Library Icon Menu Paths, Ans2. T, Ans3. Tag Format, Ans6. F, Ans7. Project Wide Update/Retag, Ans8.B, Ans9. Title Block Setup, Ans10. Mark/Verify DWGs

Chapter 4

Inserting Components

Topics Covered

The major topics covered in this chapter are:

- ***Inserting Components using Icon menu***
- ***Inserting Components using Catalog Browser***
- ***Inserting Components using User Defined list***
- ***Inserting Components using Equipment list***
- ***Inserting Components using Panel list***
- ***Inserting Components using Terminal (Panel list)***
- ***Pneumatic, Hydraulic, and P&ID components***

ELECTRICAL COMPONENTS

Electrical Components in AutoCAD Electrical are represented by schematic symbols. We have discussed about schematic symbols in Chapter 1 of this book. AutoCAD Electrical can be installed with JIC, IEC, IEEE, JIS, GB, and other standard symbol libraries. Since, the electrical standards adopted by various nations may vary, the markings and symbols used to describe electrical control products vary as well. Whether it is a complex control system on a machine tool or a simple across-the-line motor starter, the need to recognize and understand these symbols becomes more important. It is possible that products from all parts of the world are being used in any one facility. You may be having question that what standard should be followed in your drawings. Answer to this question lies with your country or country of your vendor for which the drawing is being generated. It also depends on the type of equipment you are building because different type of equipment are manufactured under different standards. Without going deeper in the electrical standards, now you will learn to set different type of electrical standard symbol libraries in AutoCAD Electrical before inserting schematic symbols in the drawings.

SETTING SYMBOL LIBRARY FOR PROJECT

You can set desired symbol library in AutoCAD Electrical while starting a new project. Once, you have created an electrical drawing in the new project then it will be very difficult to change the symbol style in that project. The steps to set symbol library for desired electrical standard are given next.

- Once you have started a new project and created a new drawing in it, right-click on the Project name in the **Project Manager**. A shortcut menu will be displayed; refer to Figure-1. Click on the **Properties** option from the menu. The **Project Properties** dialog box will be displayed.

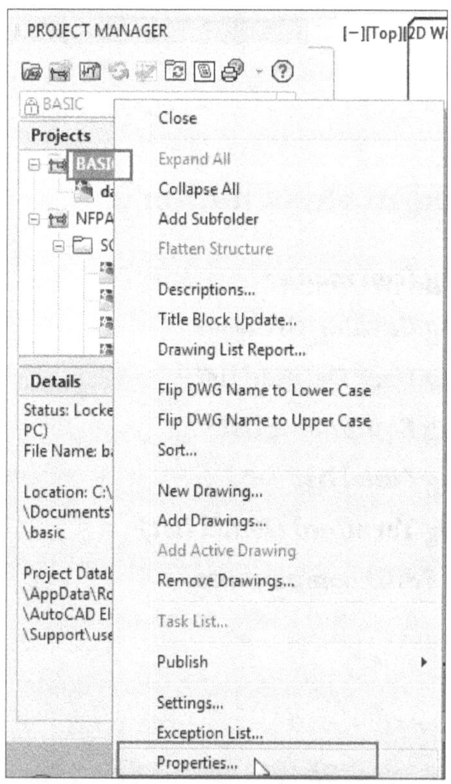

Figure-1. Shortcut menu for project

- Expand **Schematic Libraries** node in the **Library and Icon Menu Paths** area of the **Project Settings** tab in the dialog box. List of libraries will be displayed; refer to Figure-2.

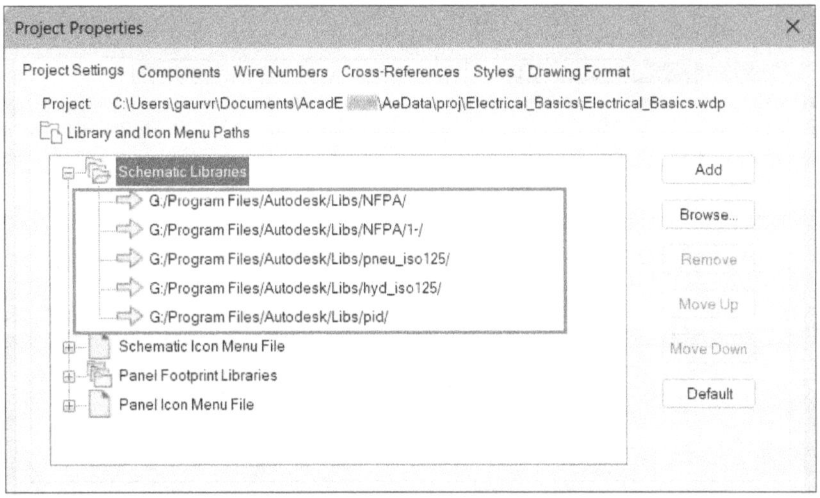

Figure-2. List of libraries

- Select desired library path and click on the **Browse** button. The **Browse For Folder** dialog box will be displayed; refer to Figure-3.

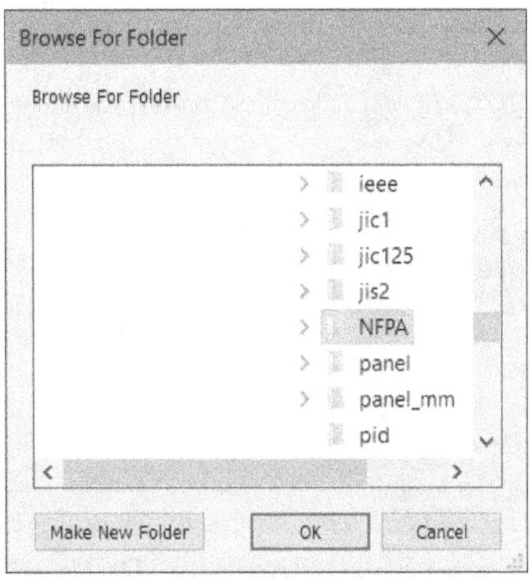

Figure-3. Browse For Folder dialog box

- Select the folder of desired library and click on the **OK** button. The library path in the dialog box will change; refer to Figure-4 (We have changed one library path from NFPA to JIC). Similarly, you can change the other library paths.

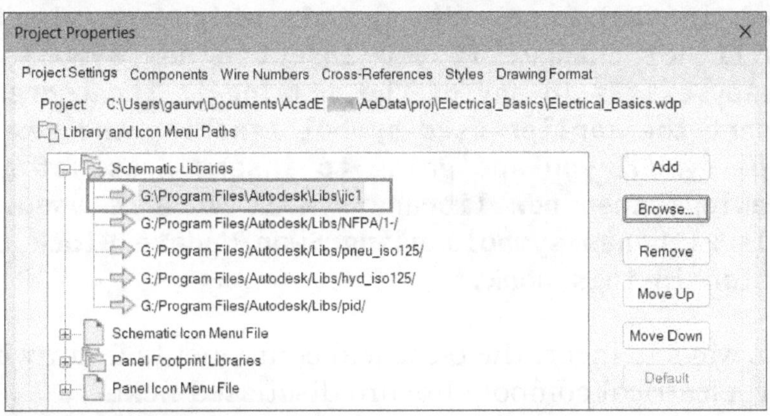

Figure-4. Changed library path

- Expand the **Schematic Icon Menu File** node in the **Library and Icon Manu Paths** area and select the path given under the node; refer to Figure-5.

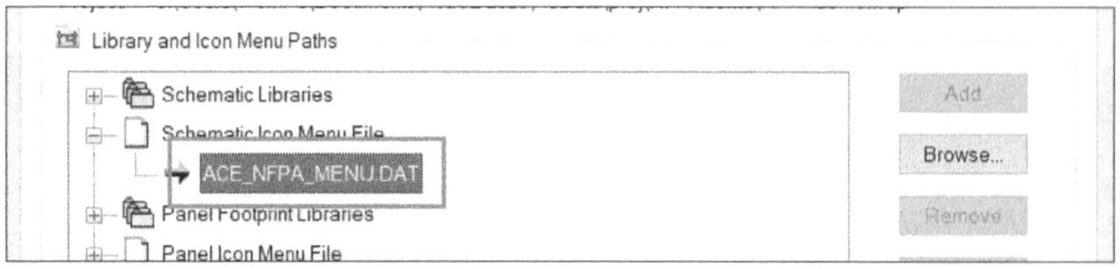

Figure-5. Icon Menu file to be selected

- Click on the **Browse** button and select the data file for selected libraries from the **Select ".dat" AutoCAD Electrical icon menu file** dialog box; refer to Figure-6.

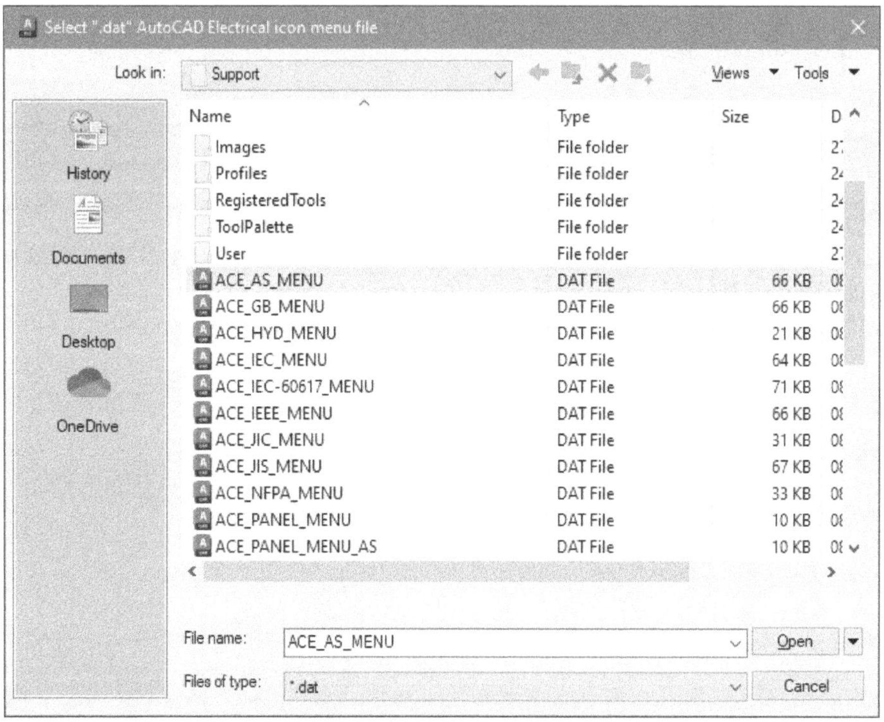

Figure-6. Select "dat" AutoCAD Electrical icon menu file dialog box

- Click on the **Open** button from the dialog box. Path of the data file will be changed.
- Click on the **OK** button from the **Project Properties** dialog box. The symbol library will be changed.

Note that you can change the symbol library later also while working on the project. But in that case, the symbols which are already existing in the drawings will not change. If you insert a new symbol using the Icon Menu which already exists in any of the drawing in the current project then system will insert the earlier used symbol ignoring the new library system selected. If the symbol you are going to insert does not exist in any of the project drawings then new library symbol will be used. You can later swap old symbols with new symbols using Swap/Update Block tool which will be discussed later in this book.

There are various ways to insert the electrical components in drawings. Some of the methods to insert electrical components are discussed next.

INSERTING COMPONENT USING ICON MENU

The **Icon Menu** tool is used to insert schematic symbols of various electrical components by using the **Icon Menu**. The procedure to use this tool is given next.

- Click on the down arrow below **Icon Menu** button in the **Schematic** tab of the **Ribbon**. The list of tools will be displayed; refer to Figure-7.

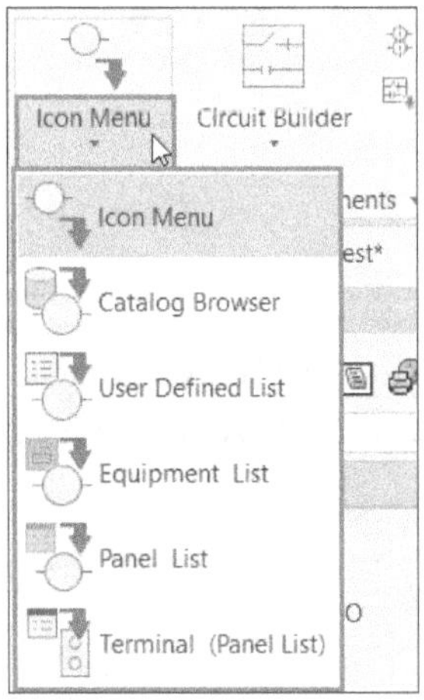

Figure-7. List of tools

- Click on the **Icon Menu** tool from this menu. The **Insert Component** window will be displayed; refer to Figure-8.

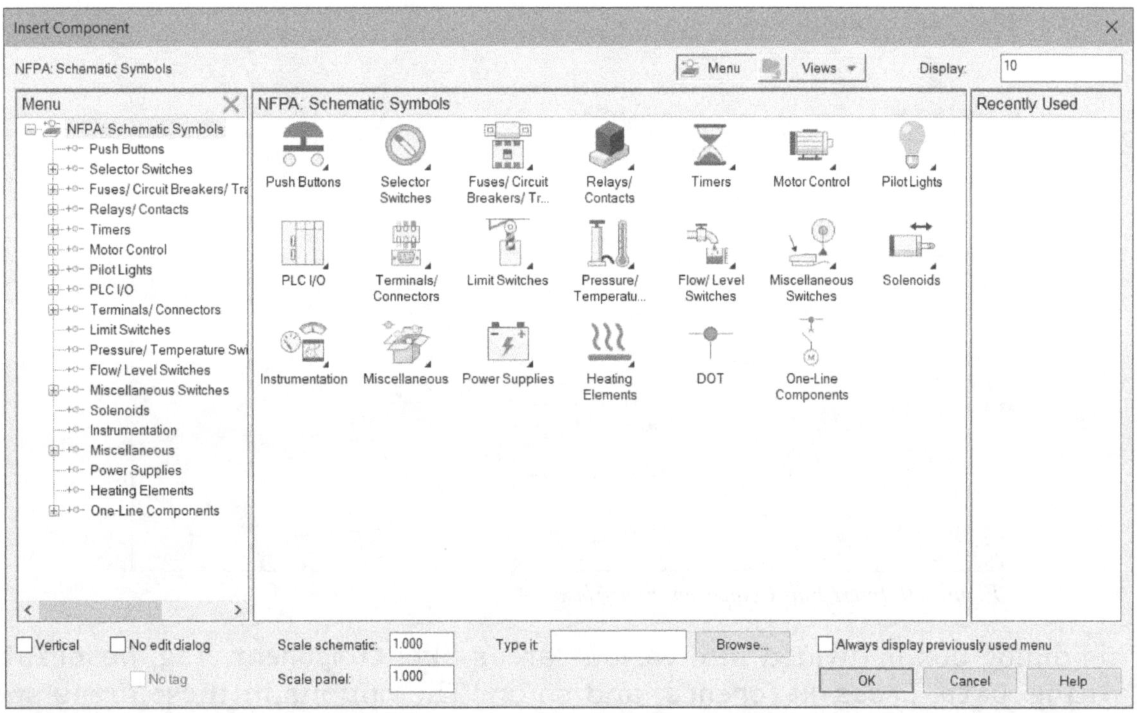

Figure-8. Insert Component window

- Select the category of desired component from the **Menu** box at the left in the dialog box. Note that the categories which have a **+** sign before their name are having sub-categories to browse in.
- You can specify the scale value for schematic symbols by using the **Scale schematic** edit box available at the bottom of the dialog box.

- After selecting the category and subcategory, click on desired symbol. The symbol will be attached to the cursor and you will be prompted to specify location for the symbol; refer to Figure-9.

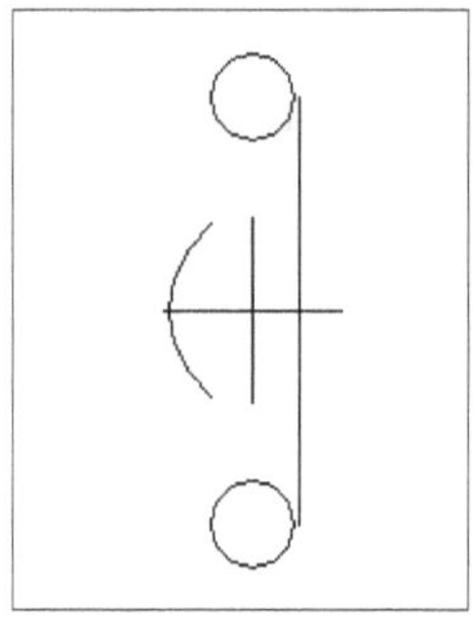

Figure-9. Component attached to cursor

- Click in the drawing area to place the component symbol at desired place. The **Insert/Edit Component** dialog box will be displayed; refer to Figure-10.

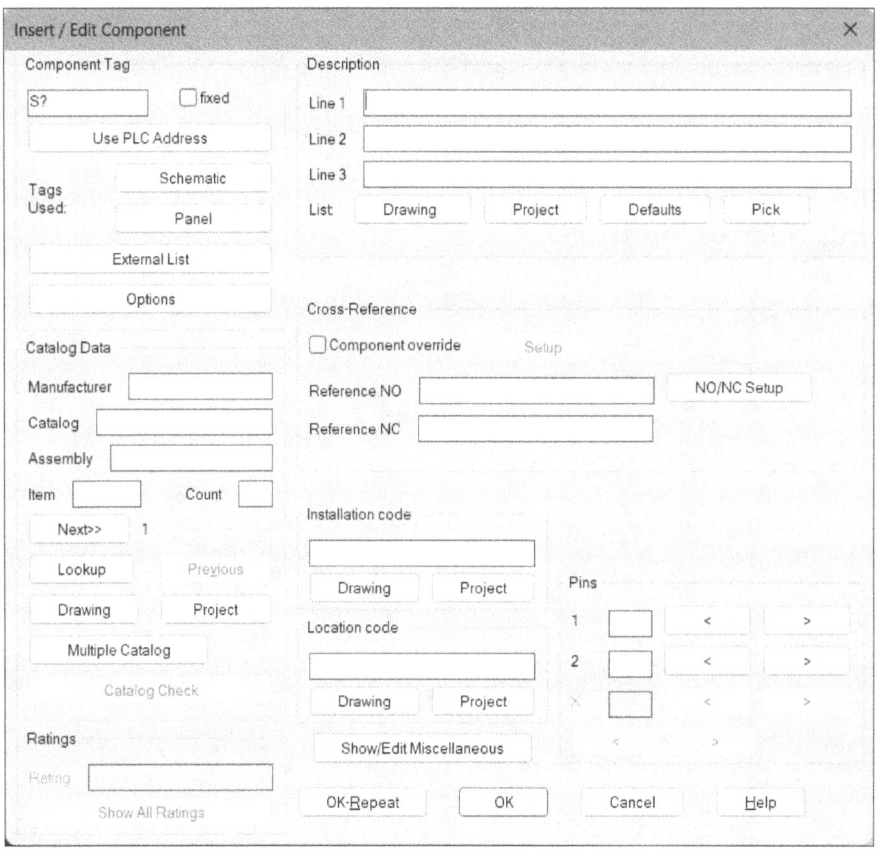

Figure-10. Insert Edit Component dialog box

- This dialog box is divided into various areas like; **Component Tag**, **Description**, **Catalog Data**, **Cross-Reference**, and so on. The options in these areas specify various parameters of the component being inserted. We will start with the **Component Tag** area and then one by one we will use options in other areas.

Component Tag area

- The edit box in this area is used to specify the tag value for your current component. Click in the edit box and specify desired value.

- You can link your component with PLC by using the tag of PLC. To do so, click on the **Use PLC Address** button. If your components are not directly connected to PLC via wires as in our case then the **Connected PLC Address Not Found** dialog box will be displayed; refer to Figure-11.

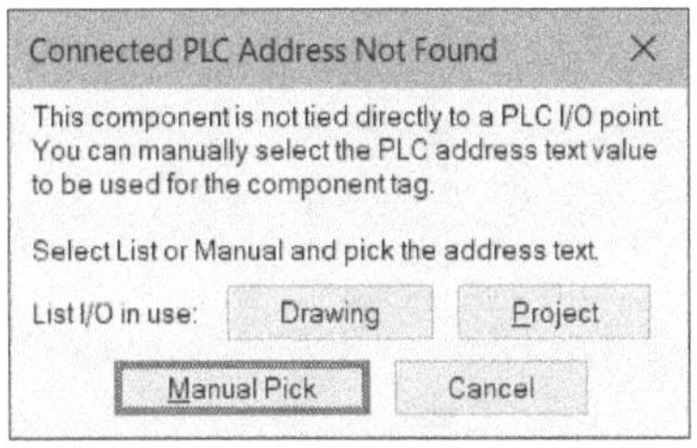

Figure-11. Connected PLC Address Not Found

- Click on the **Drawing** button to display all the PLC connections available in current drawing. The **PLC I/O Point List (this drawing)** dialog box will be displayed; refer to Figure-12 if there is a PLC in drawing. Similarly, you can select the **Project** button to display PLC connections of all drawings in current project or you can use **Manual Pick** button to manually select a PLC address.

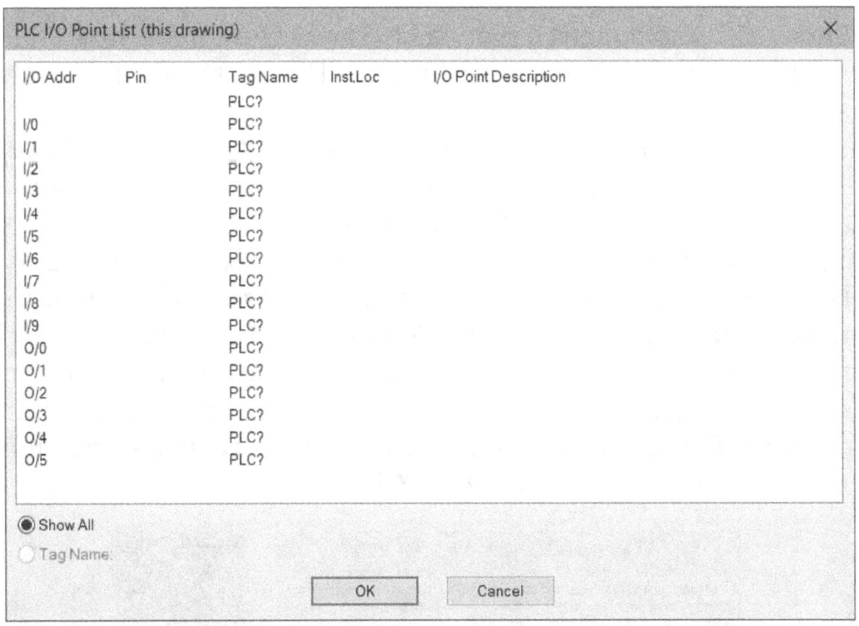

Figure-12. PLC IO Point List dialog box

- Select desired plc pin from the dialog box and click on the **OK** button from the dialog box. The component will be linked to the selected pin of PLC.

Schematic tag

- Using the **Schematic** button in the **Component Tag** area, you can assign schematic tag to the component which has earlier been used in the project or derived from the earlier used tag. The **Tags in Use** dialog box will be displayed as shown in Figure-13.

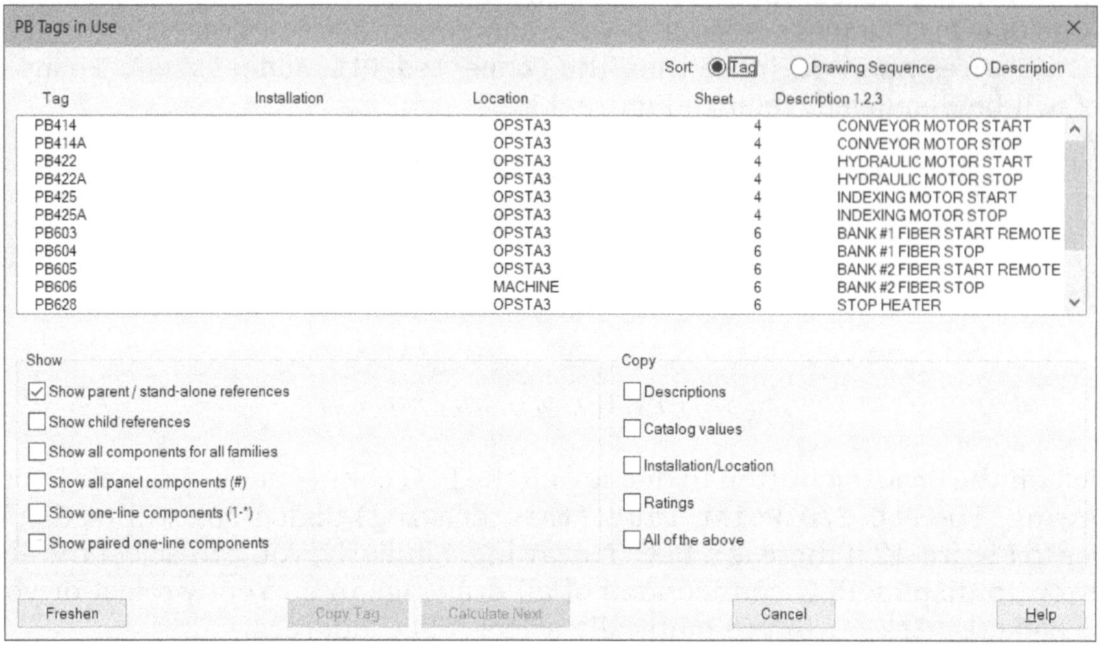

Figure-13. Tags in Use dialog box

- Select the **Show all components for all families** check box. The list of all the tags in the drawing will be displayed. Also, the name of the dialog box will be modified as shown in **All Tags in Use**.
- Select desired tag number and then select **Copy Tag** or **Calculate Next** button. If you select the **Copy Tag** button then same tag will be assigned to the current component. If you select the **Calculate Next** button then the next tag number will be assigned to the component. Note that in both the cases system will ask your permission to overwrite the tag value. Select the options as required.
- Click on the **Options** button from the **Component Tag** area to override naming of the component tags. The **Option** dialog box will be displayed; refer to Figure-14. Specify desired Tag format code like %N for numbers, %F for family code, and so on. Scan the QR code given next to check more tag code definitions. After specifying override codes, click on the **OK** button.

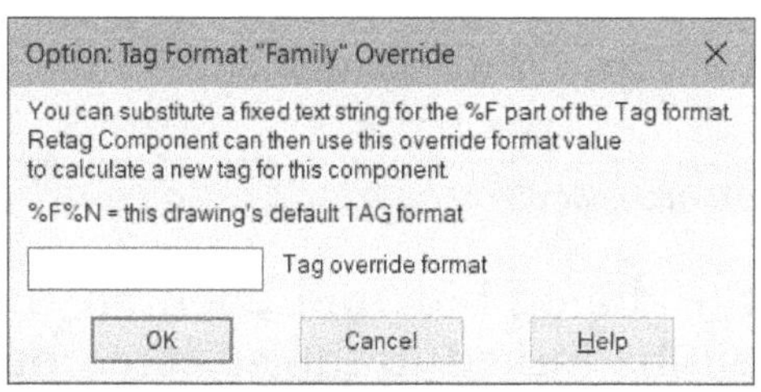

Figure-14. Option dialog box

In the same way, you can use the **Panel** button and the **External List** button from the **Component Tag** area.

Catalog Data area

The options in this area are used to link the inserted component to a catalog. A catalog is a collection of various related components in a categorized form. To specify the details of the Catalog Data follow the steps given next.

- Click in the **Manufacturer** edit box of the **Catalog Data** area and specify the name of the manufacturer. Name of various companies like **AB**, **ABB**, **Siemens**, and so on that manufacture electrical components are available in the AutoCAD Electrical Database. You can also start a new manufacturer by specifying desired name.
- Click in the **Catalog** edit box and specify the value of Catalog number. This number can include alphabets as well as numeric. For example, 300F-1PB.
- Click in the **Assembly** edit box and specify desired value for assembly code.
- Click in the **Item** edit box and specify the item number for the component. Note that each type of component has a unique Item number.
- Click in the **Count** edit box and specify the number of components required in the current drawing. The specified numbers will be automatically added in the Bill of Materials. The area after specifying all the values will be displayed as shown in Figure-15.

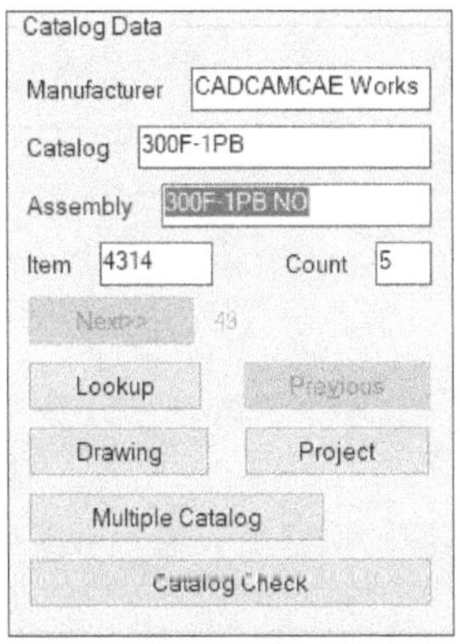

Figure-15. Catalog Data area

- Instead of specifying all the parameters related to Catalog data, you can pick desired component's details from the library of standard components in AutoCAD Electrical. To do so, click on the **Lookup** button in this area. The **Catalog Browser** dialog box will be displayed as shown in Figure-16.

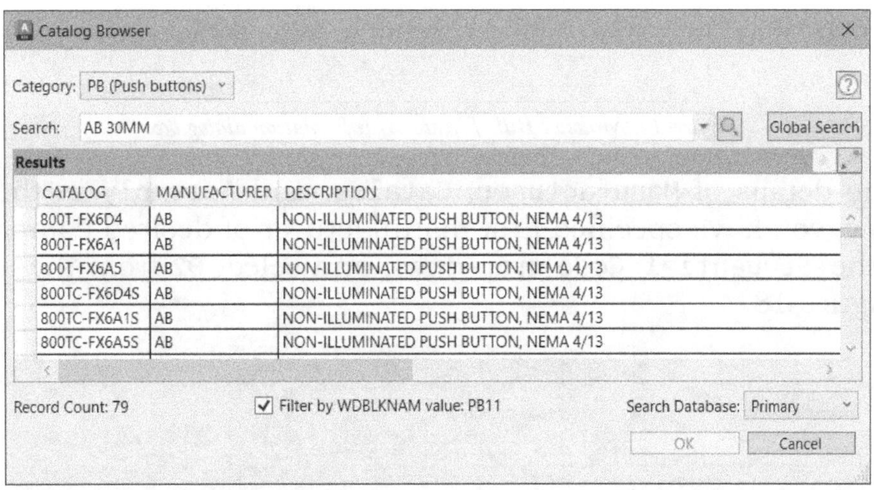

Figure-16. Catalog Browser dialog box

- Select desired option from the **Category** drop-down at the top in the dialog box. If you want to check all the components in this category then click on the search button. The list of all the components in that category will be displayed.
- You can customize user search by specifying desired keywords in the **Search** edit box. After specifying the keywords, click on the **Search** button. The list of components related to the specified keyword will be displayed.

- Double-click on desired component in the list. The related data will be displayed in the **Catalog** area.
- Similarly, you can use the **Drawing** button or the **Project** button to select the catalog data from earlier created components in the drawing or project respectively.

Multiple Catalog Data

- You can use multiple catalog for a component to display alternates for the component. To do so, click on the **Multiple Catalog** button from the **Insert/Edit Component** dialog box. The **Multiple Bill of Material Information** dialog box will be displayed; refer to Figure-17.

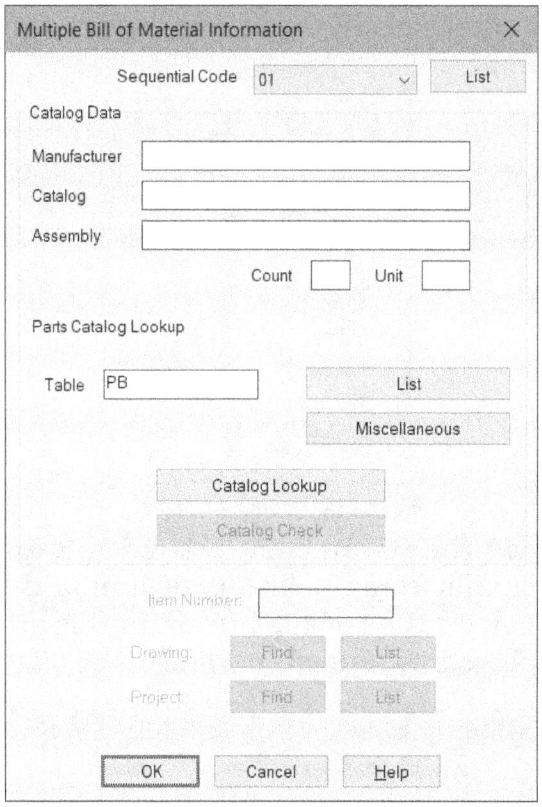

Figure-17. Multiple Bill of Material Information dialog box

- Specify the details of **Manufacturer**, **Catalog**, and **Assembly** in the related edit boxes. Now, you have specified first manufacturer of desired component.
- Click on the **Sequential Code** drop-down and select **02** from the list displayed; refer to Figure-18.

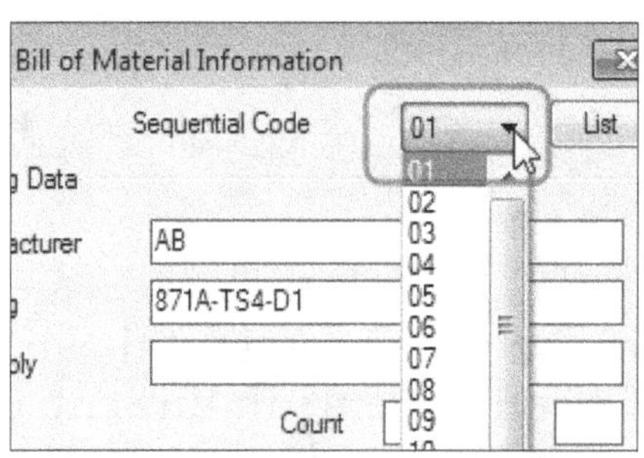

Figure-18. Sequential Code drop down

- Specify the alternate data for **Manufacturer**, **Catalog**, and **Assembly**. Repeat the process until you get desired number of alternates for the selected type of components. You can specify up to 99 alternates at max.
- After specifying the data, click on the **OK** button from the dialog box.

Description Area

The options in this area are used to specify description about the component. To specify the description, follow the steps given next.

- Click in the **Line 1** edit box of this area and specify the first line of description. Similarly, click in the **Line 2** edit box and specify the second line of description.
- If you want to use description of any component in current drawing or project then click on the **Drawing** button or **Project** button respectively.
- The related dialog boxes will be displayed and you will be prompted to select the description of a component from the list. Select desired component description and click on the **OK** button from the dialog box.
- You can also select the description from the default description list. To do so, click on the **Defaults** button in this area. The **Descriptions** dialog box will be displayed; refer to Figure-19.

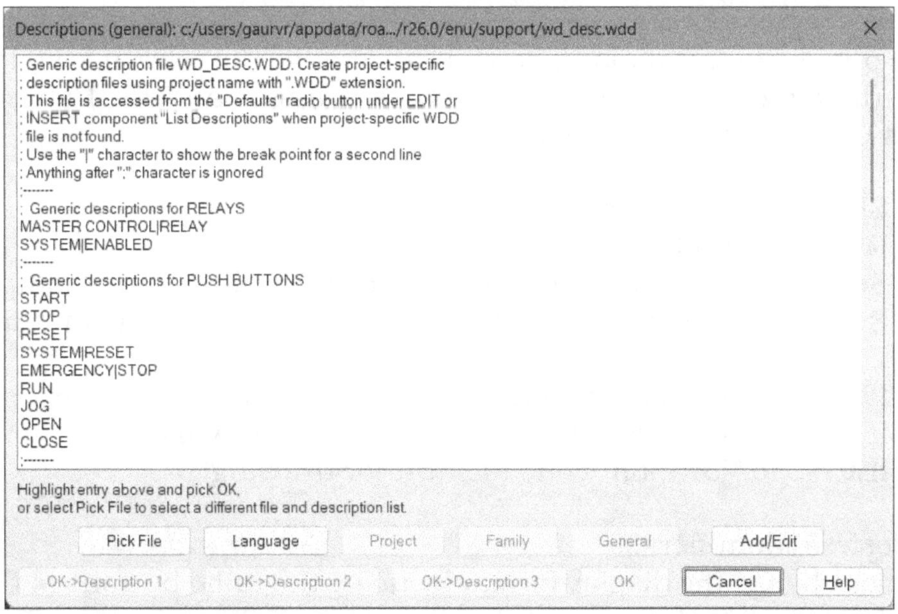

Figure-19. Descriptions dialog box

- Double-click on any of the description in this dialog box. Note that you need to double-click on the lines with all capital characters only.
- If you want to copy description of a component in current open drawing then click on the **Pick** button and select desired component.

Cross-Reference Area

This is a very important area of the dialog box. Using the options in this area, you can link two or more components in same or different drawings of the project. For example, you have a 3 pole switch whose two switch parts are in drawing 1 and one switch part is in drawing 2. In such cases, you need to cross-reference the parts with each other. The procedure to use options in this area is given next.

- When you are inserting a component in the drawing, then based on the parent child relationship components are automatically referenced (Like, relays). This referencing has a specific reference format. To override this format, select the **Component override** check box. The **Setup** button next to it will get activated.
- Click on the **Setup** button. The **Cross-Reference Component Override** dialog box will be displayed; refer to Figure-20.

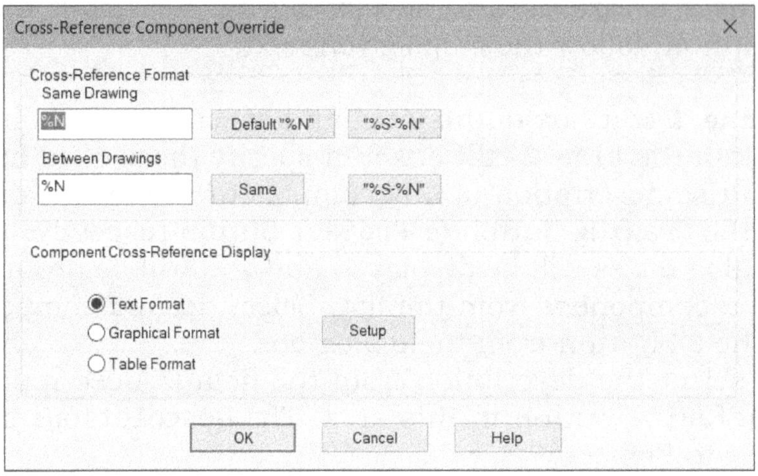

Figure-20. Cross Reference Component Override dialog box

- Using the options in this dialog box, you can change the reference style. By default, for same drawing %N is used as reference number and for Between drawings %S.%N is used. Glossary is given next.

%S Sheet number of the drawing
%D Drawing number
%N Sequential or reference-based number applied to the component
%X Suffix character position for reference-based tagging (not present = end of tag)
%P IEC-style project code (default for drawing)
%I IEC-style "installation" code (default for drawing)
%L IEC-style "location" code (default for drawing)

- Specify desired identifiers.
- You can also change the display format by using the radio buttons in the **Component Cross-Reference Display** area of the dialog box.
- After specifying desired format style, click on the **OK** button from the dialog box to exit.
- Each component need to be specified by NC or NO contact points. NC means Normally Closed and NO means Normally Open. Using the **Reference NO** and **Reference NC** edit boxes, you can specify the references for the contact types.
- A switch generally has two pins, but if you are working on PLC or other integrated circuit components then you need to setup NC/NO for all the pins of components. To do so, click on the **NO/NC Setup** button. The **Maximum NO/NC counts and/or allowed Pin numbers** dialog box will be displayed; refer to Figure-21.

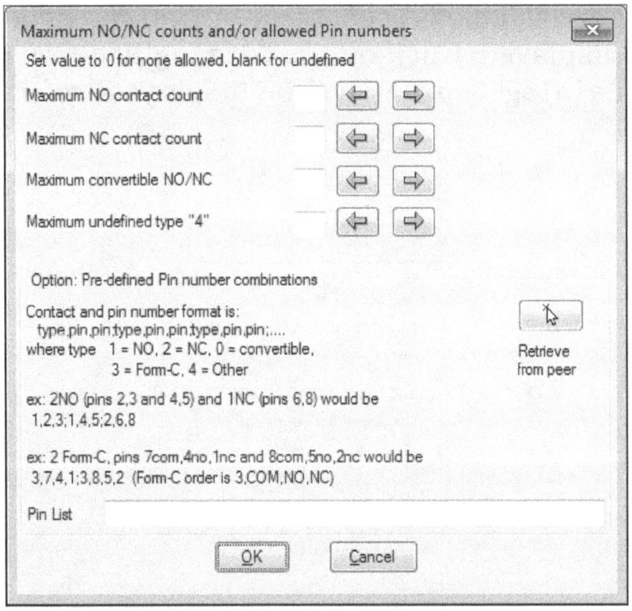

Figure-21. Maximum NO or NC counts dialog box

- Specify the numbers of contact types for different categories in this dialog box.
- In the **Pin List** edit box, you can define the connectivity of one pin with the other for NC, NO, COM, or others. Check the description in **Option** area of the dialog box to know the method of linking pins for switches.
- Click on the **OK** button to apply the changes.
- By default, the reference numbers of drawing are automatically taken in the **Reference NO** and **Reference NC** edit boxes of the **Insert/Edit Component** dialog box if you are inserting child component. You will learn more about cross-references in problems discussed later in the book.

Installation Code and Location Code

- The edit boxes in these areas are used to specify the installation code and location code for the current component. The installation code is used to specify the line or circuit identifier in which you want to add the component. For example, you can specify **Line1 230VAC** as installation code.
- Location code is used to specify the identifier for physical location of the component. For example, you can specify **Sub-Panel 20 Pin** as location code for the component.

Pins area

The options in this area are used to number the pins of the component so that later they can be connected with other components in the line. The procedure to use these options is given next.

- Click on the **<** button or the **>** button to decrease or increase the pin number.
- Click on the **OK** button from the dialog box to exit. The component will be displayed with the specified parameters in the drawing.

CATALOG BROWSER

The **Catalog Browser** is the standard library of various components used in electrical engineering. You can insert any desired component using the options in the **Catalog Browser**. The procedure to use the **Catalog Browser** is given next.

- Click on the down arrow below **Icon Menu** button in the **Ribbon**. The **Component** drop-down will be displayed. Click on the **Catalog Browser** tool from the list; refer to Figure-22. The **Catalog Browser** will be displayed; refer to Figure-23.

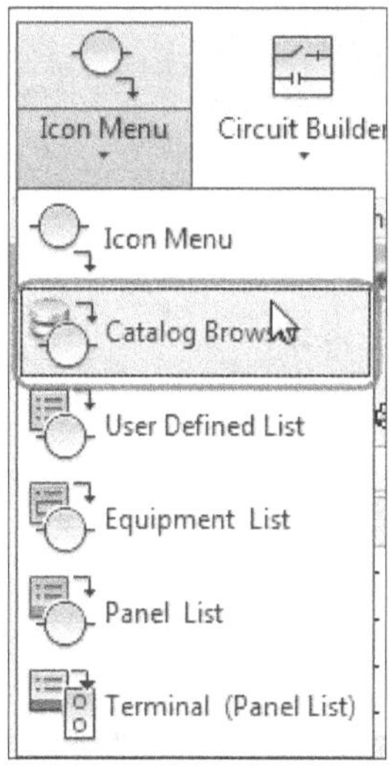

Figure-22. Catalog Browser button

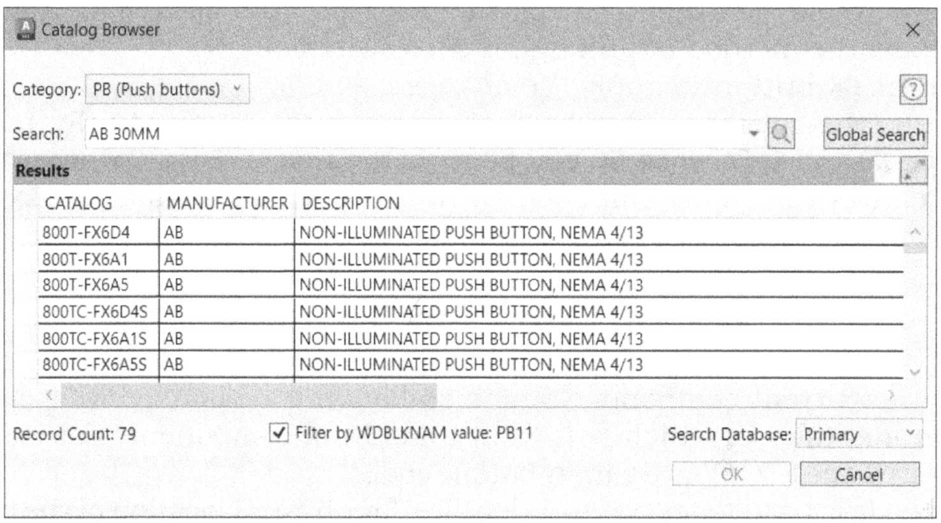

Figure-23. Catalog Browser dialog box

- Click on the **Category** drop-down, a list of various component categories will be displayed; refer to Figure-24.

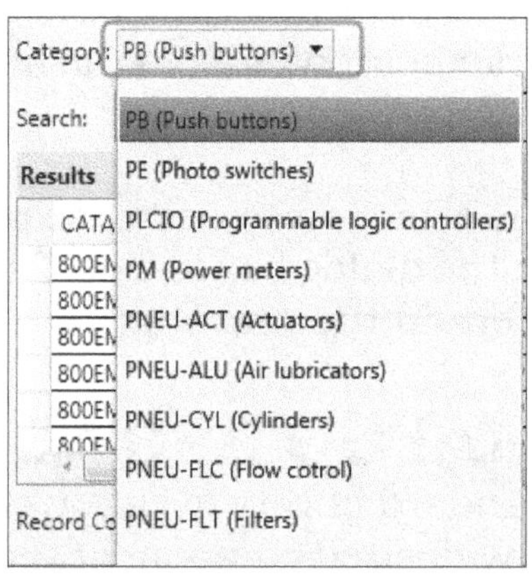

Figure-24. Category drop-down

- Select desired component type from the list.
- Click in the **Search** edit box and specify the requirements in keywords separated by space.
- Click on the **Search** button next to **Search** edit box. The list of components matching the keywords will be displayed; refer to Figure-25.

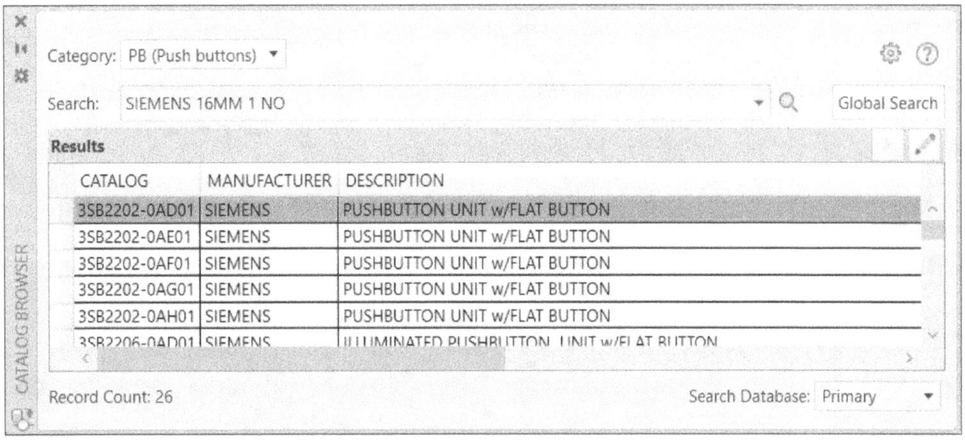

Figure-25. Catalog Browser dialog box with desired components

- Click on the component that you want to include in your drawing, a small toolbar will be displayed; refer to Figure-26.

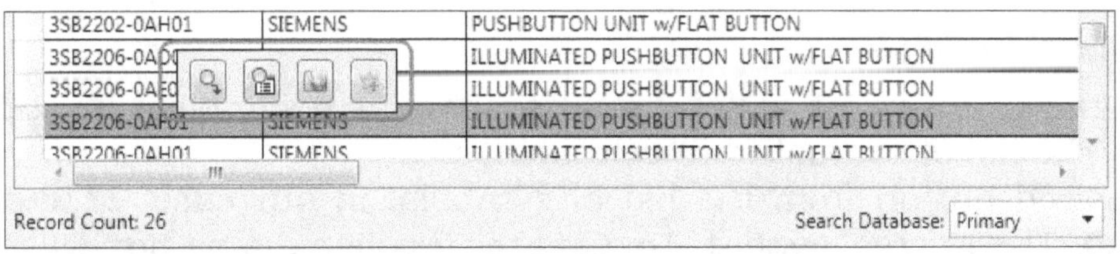

Figure-26. Toolbar in Catalog Browser

- The is used to specify a symbol for the current selected component. The is used to display the assembly details of the selected component. The is used to display the details of the current selected component in the web browser. Using this button, you will reach to the manufacturer's website. The button is used to add the current component in the favorite components list.
- Click on the button to assign a symbol. The **Insert Component** dialog box will be displayed; refer to Figure-27.

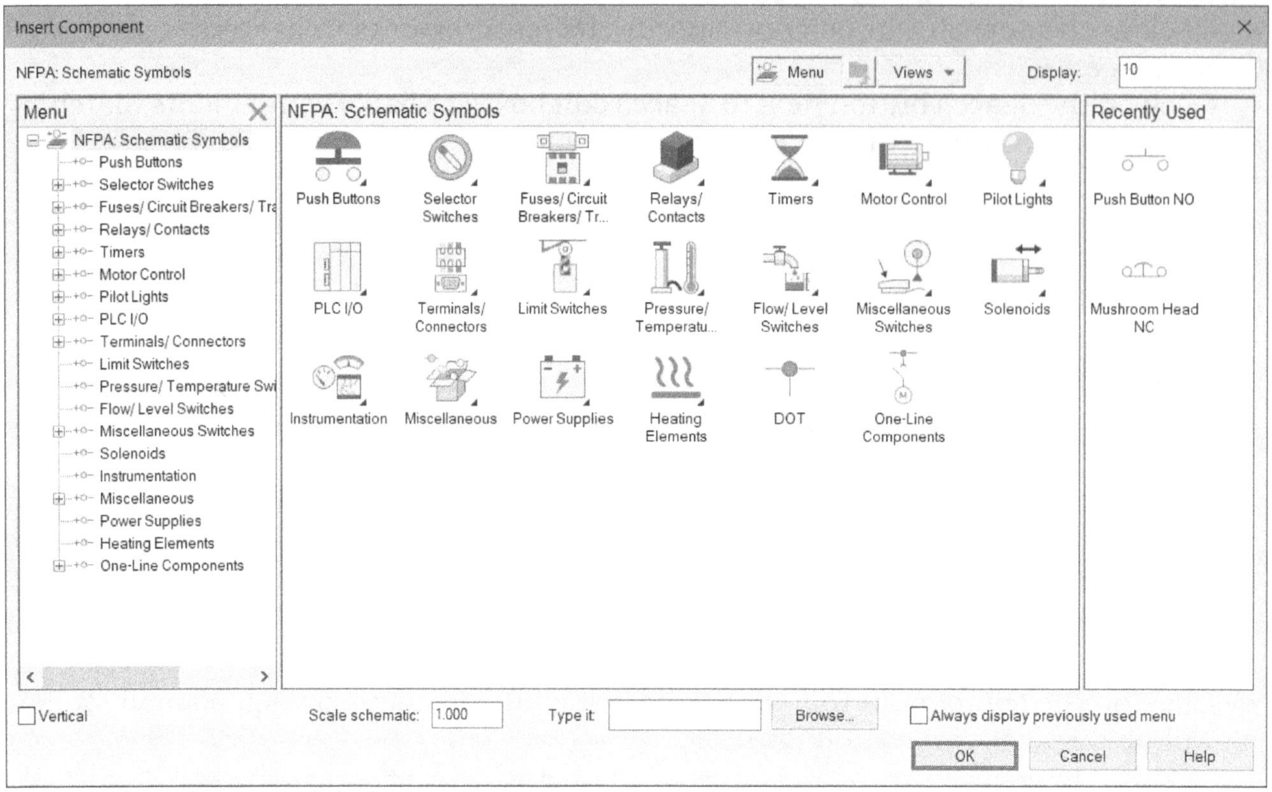

Figure-27. Insert Component dialog box

- Click on the category for which your component resembles. In our case, Push Buttons. The list of symbols will be displayed.
- Select desired button from the list of symbols, in our case, **Illuminated Push Button NO**. The **Assign Symbol To Catalog Number** dialog box will be displayed (if the catalog number is unassigned); refer to Figure-28.

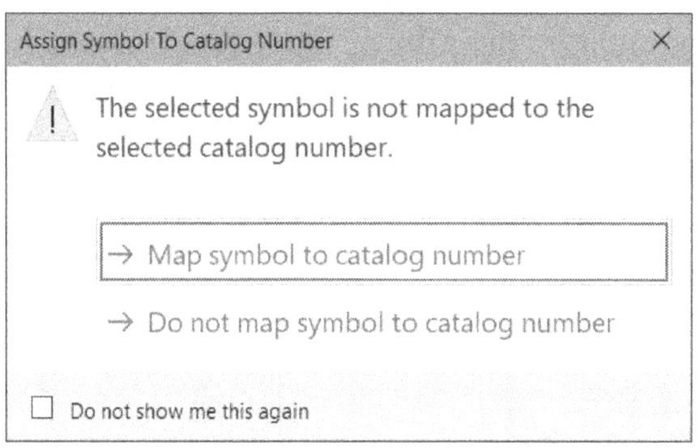

Figure-28. Assign Symbol To Catalog Number dialog box

- Click on the **Map symbol to catalog number** button from the dialog box to map the symbol to catalog. The symbol will get attached to the cursor. Click on the **Do not map symbol to catalog number** button if you want to use a different symbol for selected catalog next time.
- Click on the electrical line to which you want to connect the component or click in the drawing area to place the component; refer to Figure-29. Note that after placing the component, the **Insert/Edit Component** dialog box will be displayed as discussed earlier.

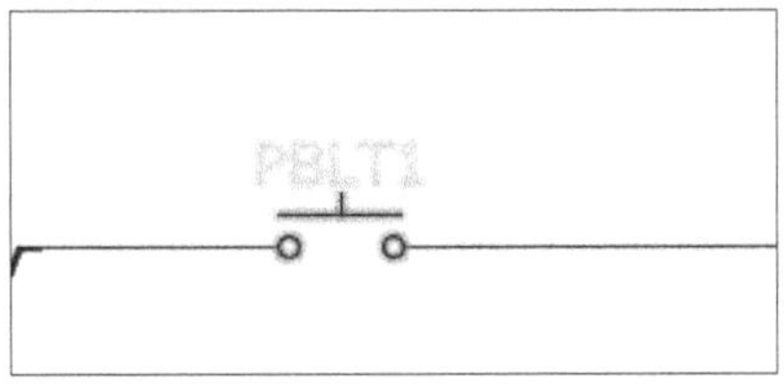

Figure-29. Component after placing on line

- Specify desired parameters and click on the **OK** button from the dialog box.
- If you again go to **Catalog Browser** and click on the same component, then you will get symbols mapped to the component; refer to Figure-30.

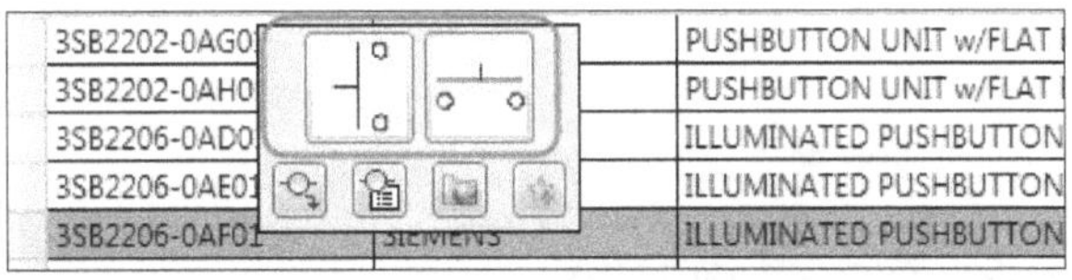

Figure-30. Symbols mapped to component

USER DEFINED LIST

The **User Defined List** tool is used to display the list of components that are collected by user under a common list. The procedure to use this tool is given next.

- Select the **User Defined List** tool from the **Component** drop-down. The **Schematic Component or Circuit** dialog box will be displayed as shown in Figure-31.

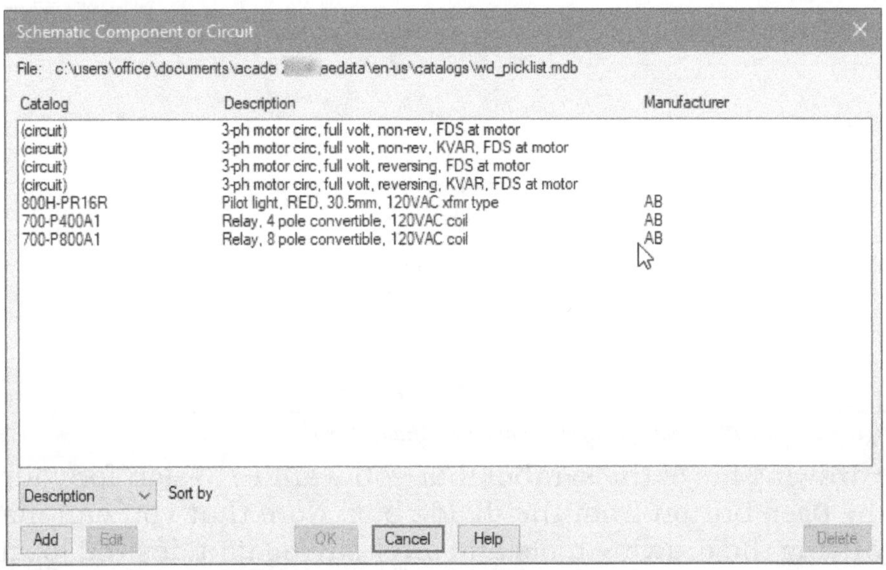

Figure-31. Schematic Component or Circuit dialog box

- Select any component/circuit from the list and click on the **OK** button from the dialog box. The component/circuit will get attached to the cursor. Click to place the component/circuit.
- You can add desired components/circuits in the list by using the **Add** button. Click on the **Add** button. On doing so, the **Add record** dialog box will be displayed; refer to Figure-32.

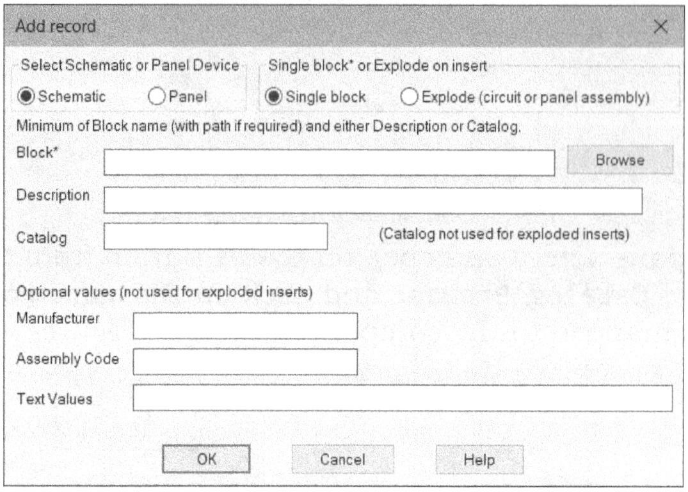

Figure-32. Add record dialog box

- Select the **Schematic** or **Panel** radio button to specify the type of component to be added in the library.
- If it is a single block then select the **Single block** radio button otherwise select the **Explode (Circuit or panel assembly)** radio button.
- Click on the **Browse** button to specify the location of block for component. The **Select Schematic component or circuit** dialog box will be displayed as shown in Figure-33.

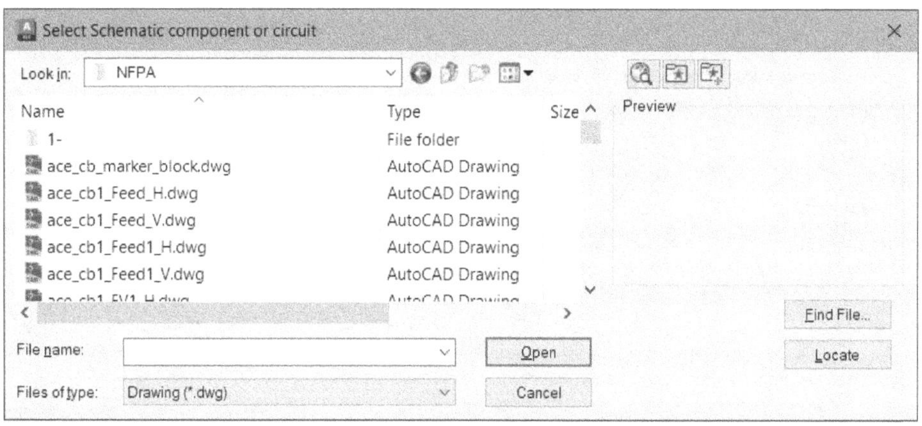

Figure-33. Select Schematic component or circuit dialog box

- Select the drawing file of the symbol that you want to assign for your component. Click on the **Open** button from the dialog box. Note that you can make your own symbol by using the drawing tools and after saving it at desired location, you can add it to the library.
- Click in the **Description** edit box and specify the description from the component.
- Click in the **Catalog** edit box and specify the catalog identifier for the component.
- Click in the **Manufacturer** edit box and specify the name of the manufacturer.
- Click in the **Assembly Code** edit box and specify desired assembly code for the component.
- Click in the **Text Values** edit box to specify the text identifiers for the component.
- Click on the **OK** button from the dialog box to place the newly added symbol.

EQUIPMENT LIST

The **Equipment List** tool is used to display or add the components in the drawing. The procedure to use this tool is given next.

- Click on the **Equipment List** tool from the **Component** drop-down. The **Select Equipment List Spreadsheet File** dialog box will be displayed; refer to Figure-34.

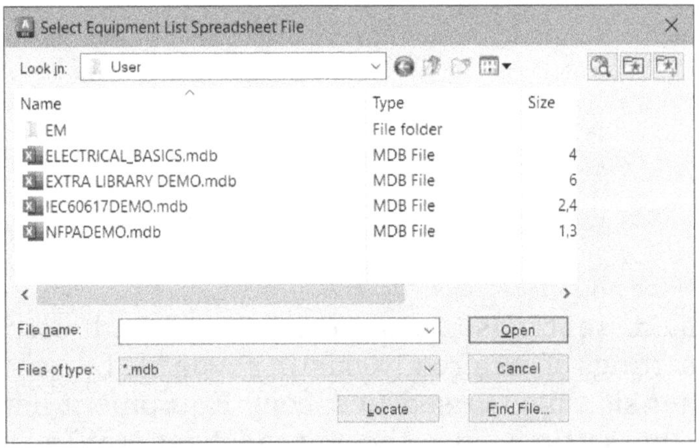

Figure-34. Select Equipment List Spreadsheet File dialog box

- Select desired database file using the options in the dialog box. Click on the **Open** button from the dialog box. The **Table Edit** dialog box will be displayed; refer to Figure 35.

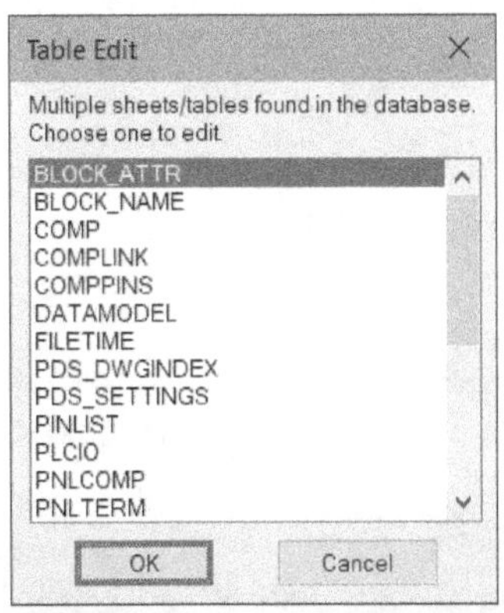

Figure-35. Table Edit dialog box

- Select desired data table from the list and click on the **OK** button from the dialog box. The **Settings** dialog box will be displayed; refer to Figure-36.

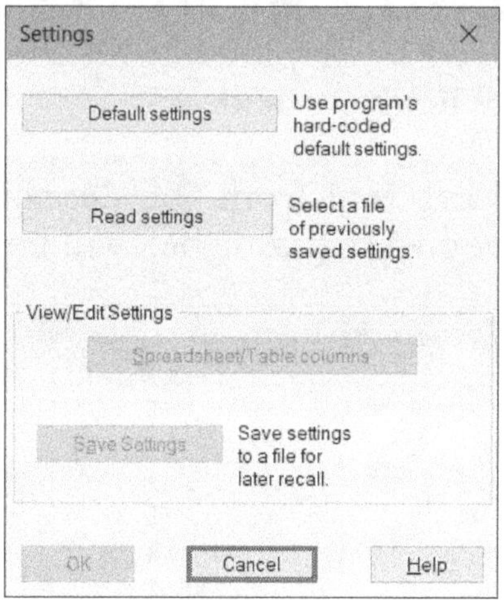

Figure-36. Settings dialog box

- Click on the **Default settings** button to use the default settings of the program. You can use the settings of an already existing **.wde** file by using the **Read settings** button. *.wde extension file is used to specify Equipment list setup file.
- After specifying the settings file, the **Spreadsheet/Table columns** button will become active in the **View/Edit Settings** area of the dialog box. Click on the **Spreadsheet/Table columns** button. The **Equipment List Spreadsheet Settings** dialog box will be displayed; refer to Figure-37.

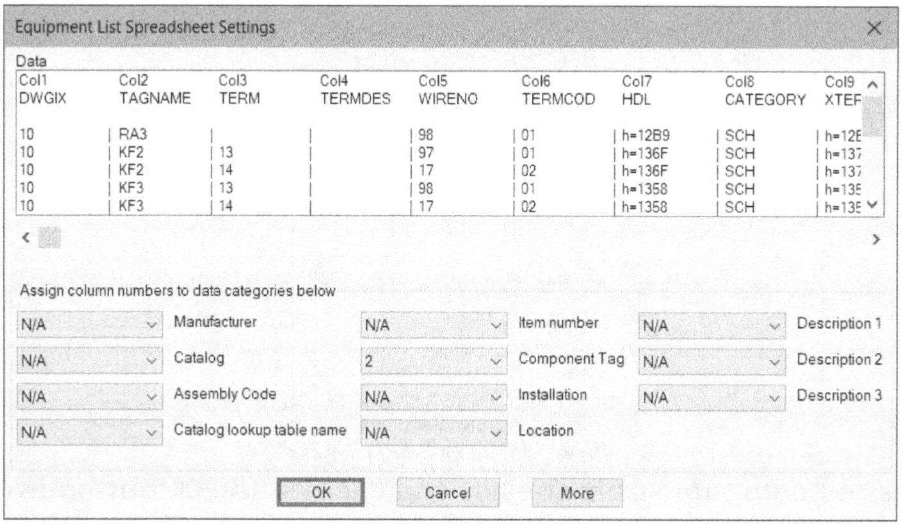

Figure-37. Equipment List Spreadsheet Settings dialog box

- Change the data category of the columns as per the requirement by using the drop-downs in this dialog box. After modifying the category, click on the **OK** button from the dialog box. Click on the **OK** button from the **Settings** dialog box. The **Schematic equipment in** dialog box will be displayed; refer to Figure-38.

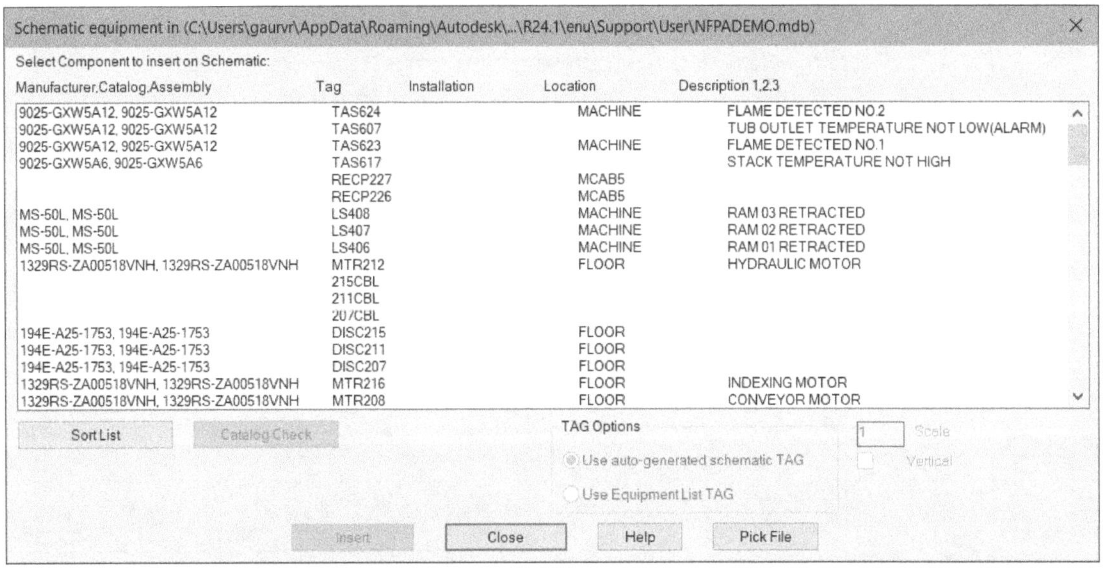

Figure-38. Schematic equipment in dialog box

- Browse through the list of equipment and click on desired equipment from the list.
- Click on the **Insert** button from the dialog box. If there is no symbol assigned to the component then the **Insert** dialog box will be displayed as shown in Figure-39.

Figure-39. Insert dialog box

- Click on the **Icon Menu** button and assign desired symbol to the component. The symbol will get attached to the cursor.
- Click in the drawing area or on the electrical line to place the component/ equipment. The component/equipment will be placed and the **Insert/Edit Component** dialog box will be displayed as discussed earlier.
- Specify desired parameters and click on the **OK** button from the dialog box to exit. Press **ESC** to exit the tool.

PANEL LIST

The **Panel List** tool is used to insert the parts earlier used in panel drawing. The procedure to use this tool is given next.

- Click on the **Panel List** tool from the **Component** drop-down. The **Panel Layout List** dialog box will be displayed as shown in Figure-40.

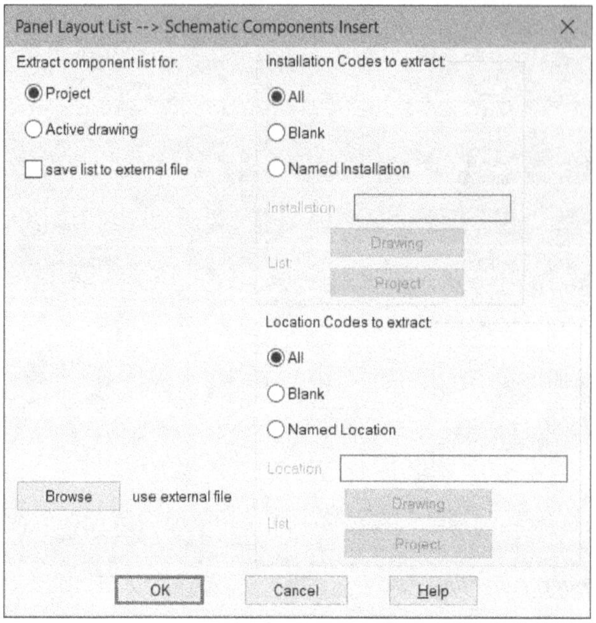

Figure-40. Panel Layout List dialog box

- You can extract the list of components from the current project or active drawing. Select the **Project** radio button or **Active drawing** radio button to get the list of components.
- If you want to save the extracted list of components in an external file, select the **save list to external file** check box.
- You can also use an external file for equipment list by using the **Browse** button.
- After specifying desired parameters, click on the **OK** button from the dialog box. If you have selected the **Project** radio button then the **Select Drawings to Process** dialog box will be displayed as discussed earlier.
- Include the drawings from which you want to extract the list of components. Click on the **OK** button from the dialog box. The list of components will display.

Note that the **Panel List** and **Terminal (Panel List)** tools work only if we have panel parts (Footprints) in our current project or drawing. We will learn more about these tools later in the book.

There are three more categories of components that are available in AutoCAD Electrical; Pneumatic Components, Hydraulic Components, and P&ID Components. The use of these components and their procedure of insertion in drawing is given next.

PNEUMATIC COMPONENTS

Pneumatic components are those components that work because of pressure of the fluid flowing through them. In some electrical systems, you might need to deal with pneumatic components also. So, AutoCAD Electrical provides a special database of these components. The procedure to insert the Pneumatic components is given next.

- Click on the down arrow next to **Insert Component** panel name; refer to Figure-41. The expanded **Insert Component** panel will be displayed; refer to Figure-42.

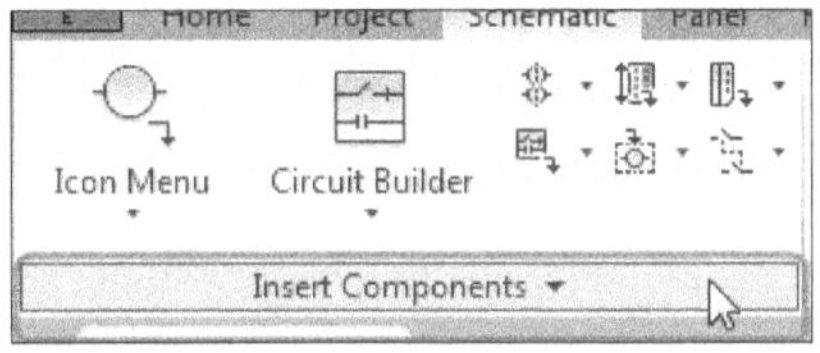

Figure-41. Down arror to be clicked

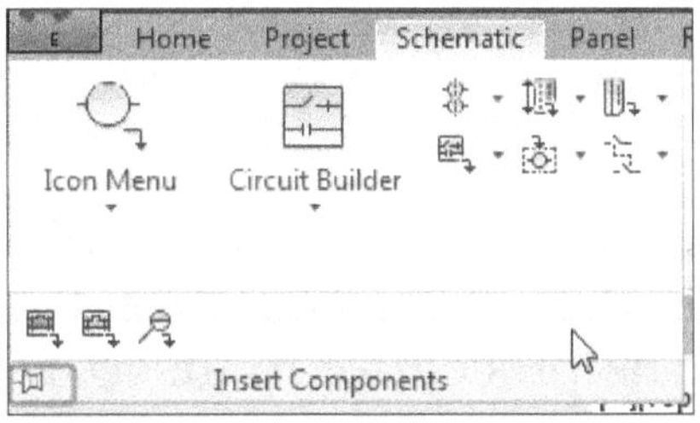

Figure-42. Expanded Insert Components panel

- Click on the pin button (in red box in the above figure) to keep the panel expanded.
- Click on the **Insert Pneumatic Components** button. The **Insert Component** dialog box will be displayed as shown in Figure-43.

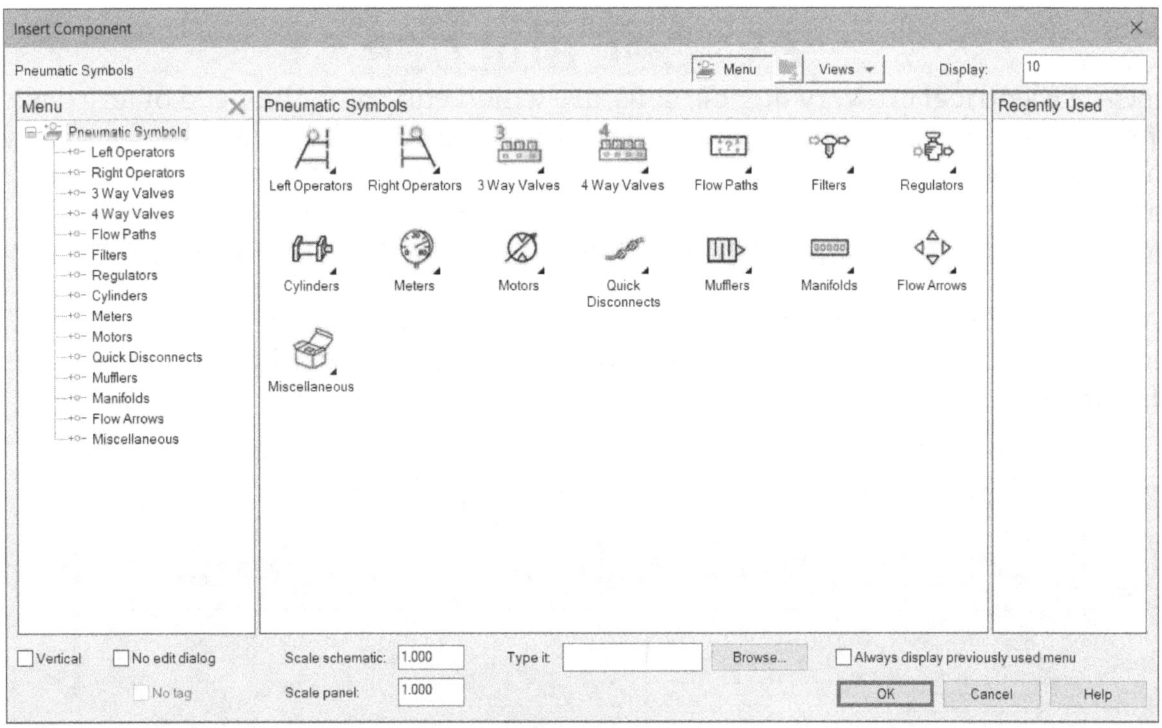

Figure-43. Insert Component dialog box for pneumatic components

- Click on desired category and select the component that you want to insert in the drawing. The component will get attached to the cursor.
- Click in the drawing area to place the component. The **Insert/Edit Component** dialog box will be displayed.
- Specify the parameters as discussed earlier and click on the **OK** button from the dialog box. The component will be placed with the specified parameters.

HYDRAULIC COMPONENTS

The hydraulic components are those components that work due to the pressure of fluids flowing through them. The procedure to insert the hydraulic components is given next.

- Click on the **Insert Hydraulic Components** tool from the expanded **Insert Components** panel. The **Insert Component** dialog box with hydraulic components will be displayed.
- Rest of the procedure is same as for Pneumatic components discussed earlier.

P&ID COMPONENTS

The P&ID components are those components that are used in Piping and Instrumentation Diagrams. The procedure to insert the P&ID components is given next.

- Click on the **Insert P&ID Components** tool from the expanded **Insert Components** panel. The **Insert Component** dialog box with P&ID components will be displayed.
- Rest of the procedure is same as for Pneumatic components discussed earlier.

SYMBOL BUILDER

AutoCAD Electrical is a very flexible software which can fulfill the need of any designer in the field of electrical control panel designing. One of such example is creating your own symbols for use in drawing. It is not possible to add each and every symbol in the Content library of AutoCAD Electrical for Autodesk, so we need to add few symbols based on our needs. The procedure to add new symbols in library is given next.

- After starting a new document with **ACAD Electrical IEC** template, click on the **Home** tab of the **Ribbon**. The basic tools of AutoCAD will be displayed; refer to Figure-44.

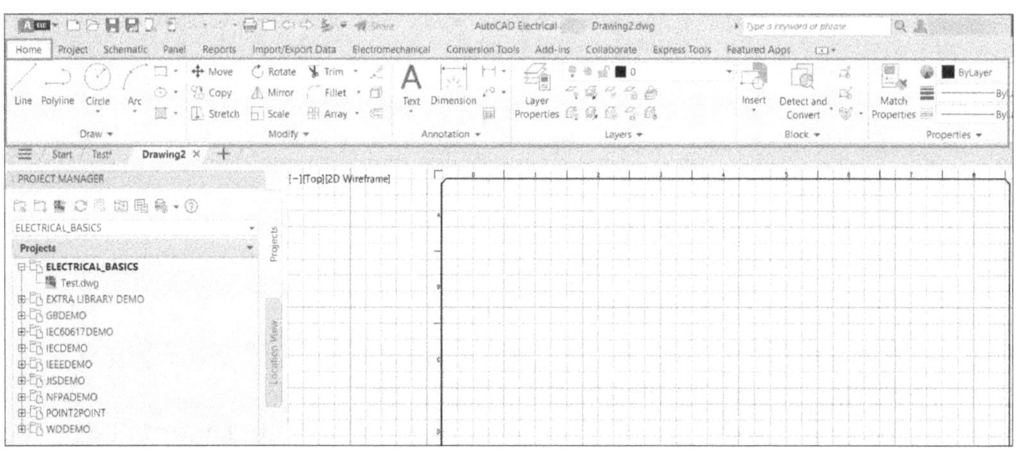

Figure-44. Home tab of Ribbon

- Using the tools in the **Draw** panel, create sketch for desired symbol; refer to Figure-45.

Figure-45. Sketch for symbol

- Click on the **Schematic** tab in the **Ribbon** to display tools related to schematics.
- Click on the **Symbol Builder** tool from the **Other Tools** panel in the **Ribbon**; refer to Figure-46. The **Select Symbol/Objects** dialog box will be displayed; refer to Figure-47.

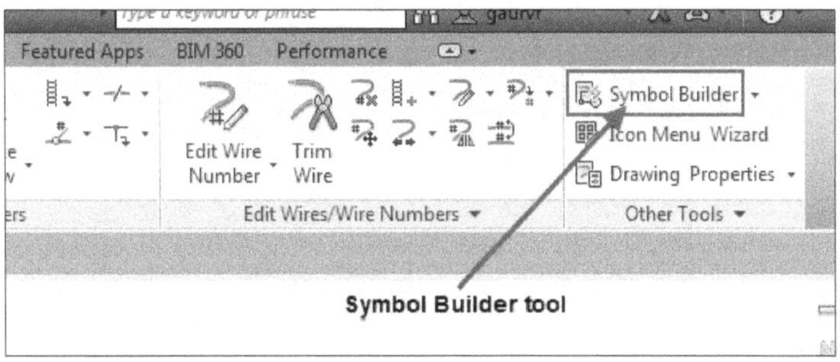

Figure-46. Symbol Builder tool

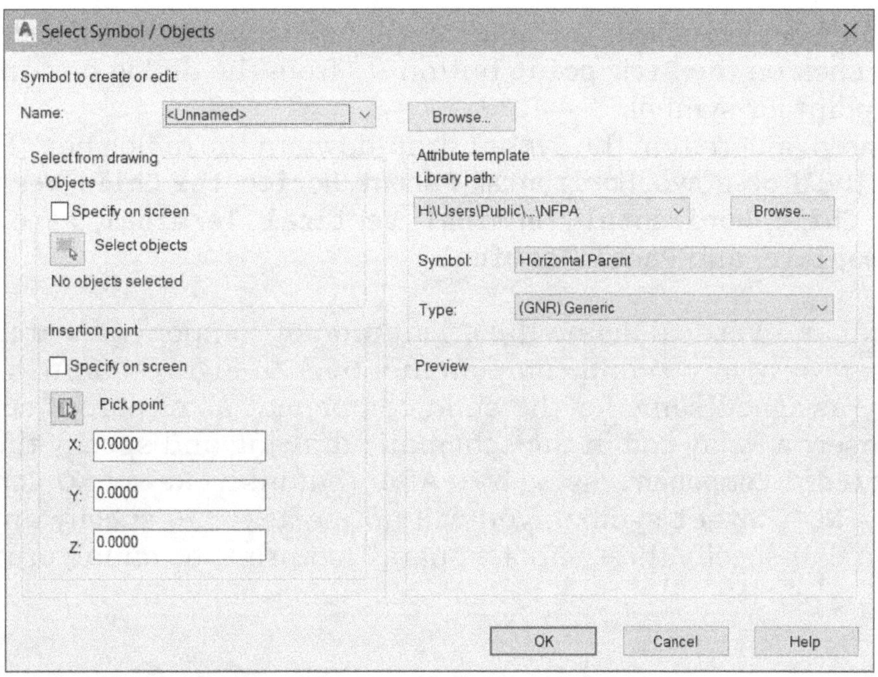

Figure-47. Select Symbol or Objects dialog box

- Click on the **Select objects** button from the dialog box; refer to Figure-48. You will be asked to select the objects.

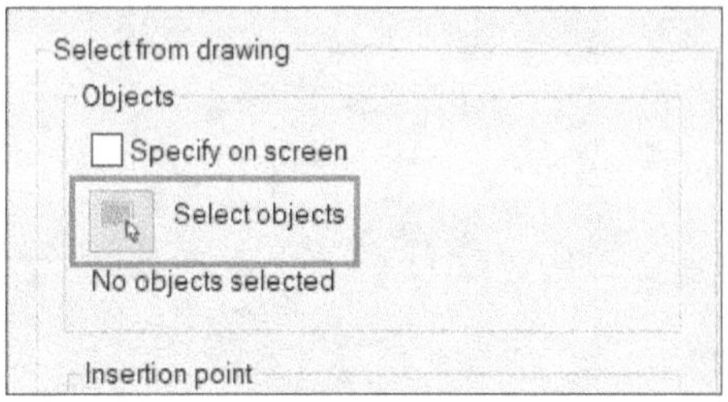

Figure-48. Select Objects button

- Select all the objects from the drawing area; refer to Figure-49.

Figure-49. Objects being selected by window selection

- Press **ENTER** from keyboard, the objects will be displayed as selected.
- Similarly, click on the **Pick point** button from the dialog box and specify the insertion point for symbol.
- Select desired option from the **Symbol** drop-down in the dialog box. There are nine options in the drop-down, **Horizontal Parent**, **Horizontal Child**, **Vertical Parent**, **Vertical Child**, **Horizontal Terminal**, **Vertical Terminal**, **Panel Footprint**, **Panel Nameplate**, and **Panel Terminal**.

The Horizontal and Vertical define the orientation of component. Parent means the component can have more child components; refer to Figure-50. Also, the tags are automatically assigned same for the child component as of parent component. To check this, insert a Relay coil in the schematic drawing and specify the parameters in the **Insert/Edit Component** dialog box. After that insert **Relay NO Contact** symbol and/or **Relay NC Contact** symbol; you will not be asked to specify any parameter. Similarly, you can specify the symbol as panel footprint, terminal, or nameplate.

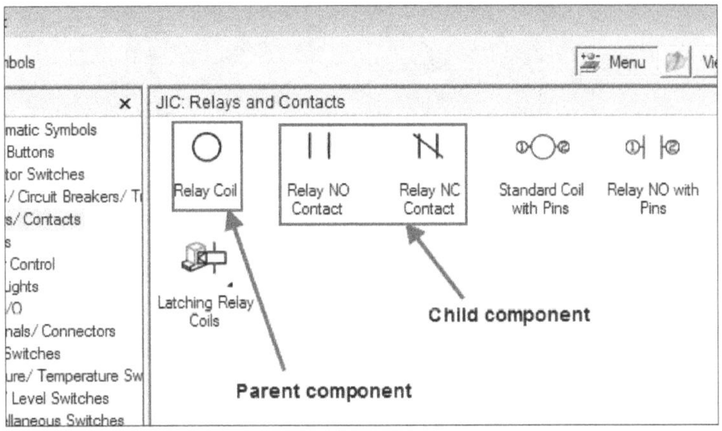

Figure-50. Parent and child components

- Select the type for symbol from the **Type** drop-down in the dialog box.
- Click on the **OK** button from the dialog box. The **Symbol Builder** contextual tab will be selected and the **Symbol Builder** environment will be displayed as shown in Figure-51.

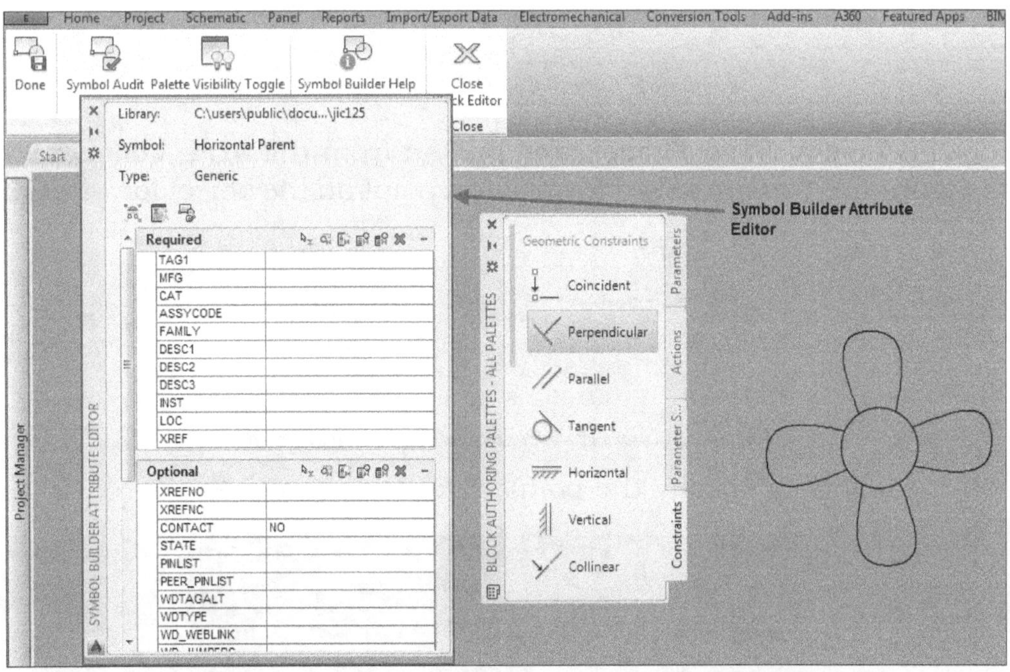

Figure-51. Symbol Builder environment

- Specify the parameters like Tag, Manufacturer, Category, Assembly code etc. in the **Required** rollout of the **Symbol Builder Attribute Editor**; refer to Figure-51.
- Similarly, specify other optional parameters as required.
- Click in the **Direction/Style** drop-down of the **Wire Connection** rollout in the **Symbol Builder Attribute Editor**; refer to Figure-52. Select desired orientation for wire connection from the list displayed. (We will select the **Radial/None** option from the list.)

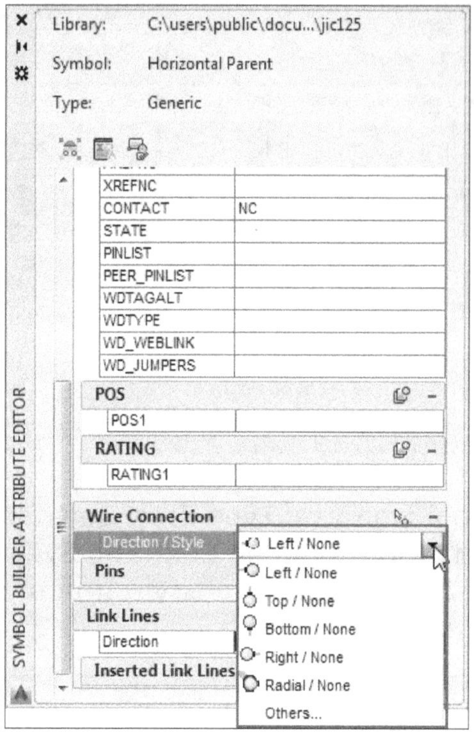

Figure-52. Direction or Style drop-down

- Click on the **Insert Wire Connection** button from the **Wire Connection** rollout; refer to Figure-53. You are asked to specify a point on the object for wire connection.

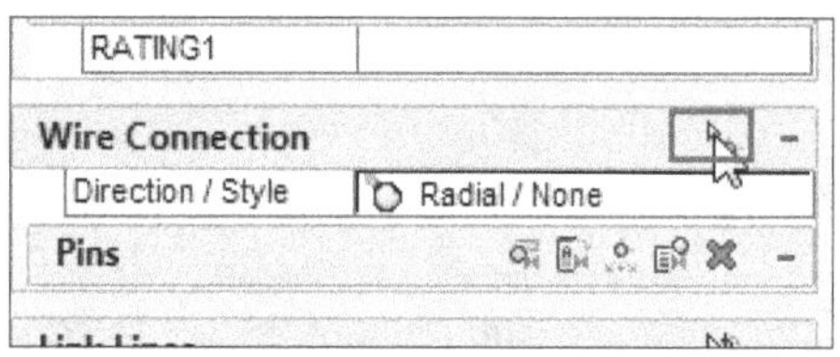

Figure-53. Insert Wire Connection button highlighted

- Click on the object to specify the point; refer to Figure-54.

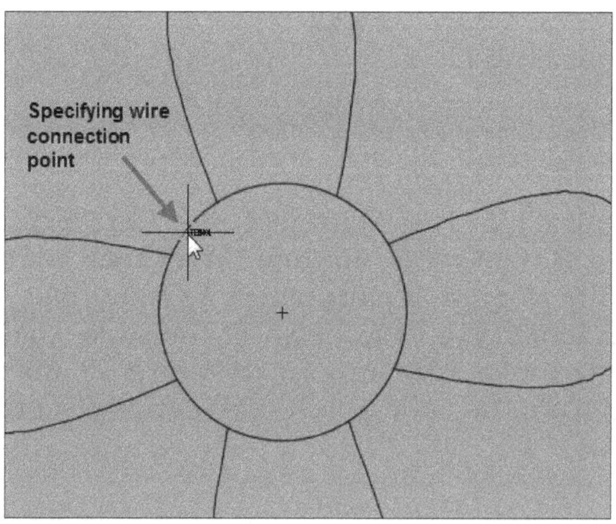

Figure-54. Specifying wire connection point

- Similarly, click to specify the second wire terminal and if the component is 3 Phase then specify third and fourth terminal.

- Press **ENTER** to exit the wire connection insertion mode. The defined terminals will be displayed in the **Pins** rollout; refer to Figure-55.

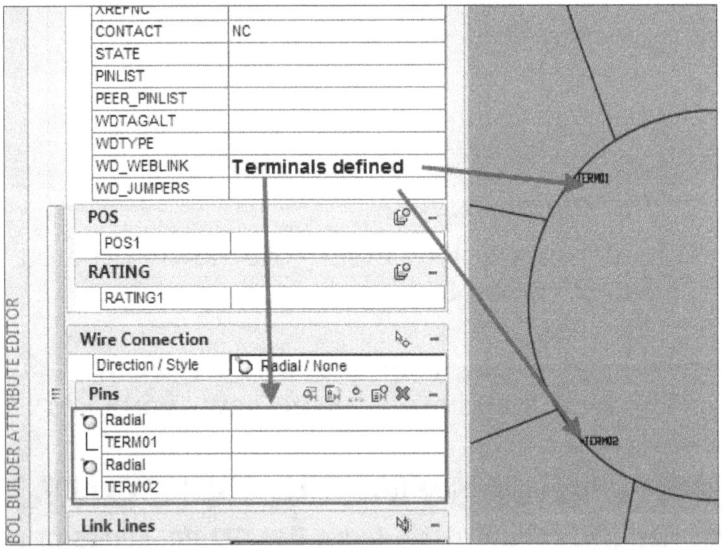

Figure-55. Terminals created

- Specify desired label for TERM01 and TERM02, like you can specify TERM01 as Live and TERM02 as Neutral for connections in their corresponding boxes in the **Pin** rollout.
- Click on the **Done** button from the **Symbol Builder** contextual tab. The **Close Block Editor: Save Symbol** dialog box will be displayed; refer to Figure-56.

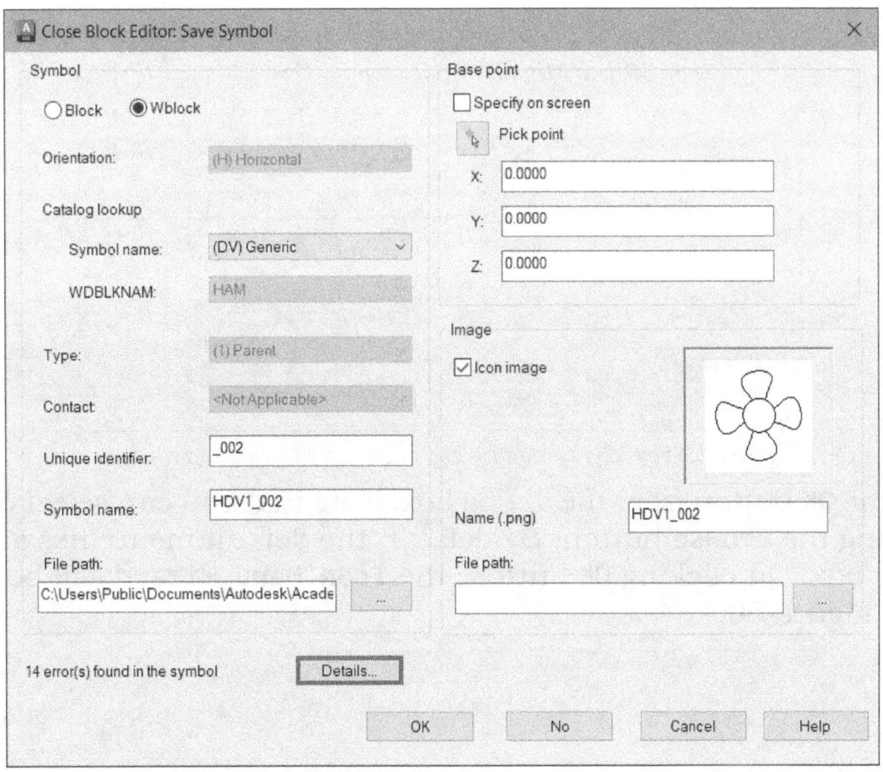

Figure-56. Close Block Editor dialog box

- Click in the **Symbol name** edit box and specify desired name for component. Like, we have specified name as fan.

- Click on the **OK** button from the dialog box. The **Close Block Editor** dialog box will be displayed; refer to Figure-57.

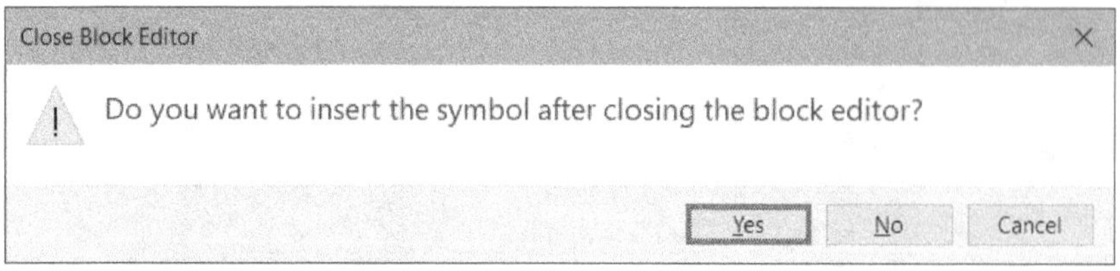

Figure-57. Close Block Editor dialog box

- Click on **Yes** button if you want to insert the newly created symbol in current drawing otherwise select the **No** button.

We have created a new symbol but if we look into the **Icon Menu**, the symbol is not available anywhere. The next step is to add the new symbol in Icon Menu.

ADDING NEW SYMBOL IN ICON MENU

- Click on the **Icon Menu Wizard** button from the **Other Tools** panel in the **Schematic** tab of the **Ribbon**; refer to Figure-58. The **Select Menu file** dialog box will be displayed as shown in Figure-59.

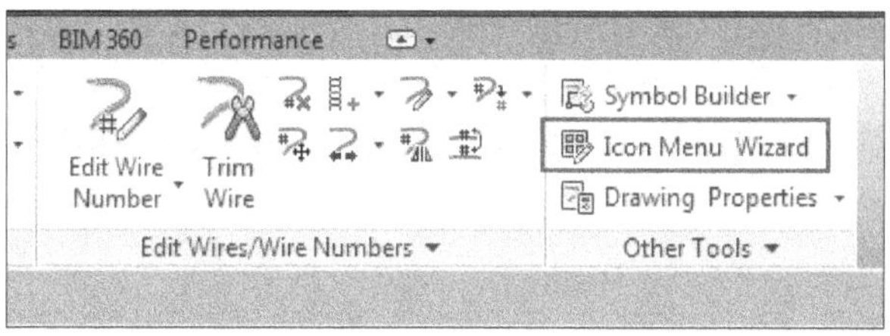

Figure-58. Icon Menu Wizard button

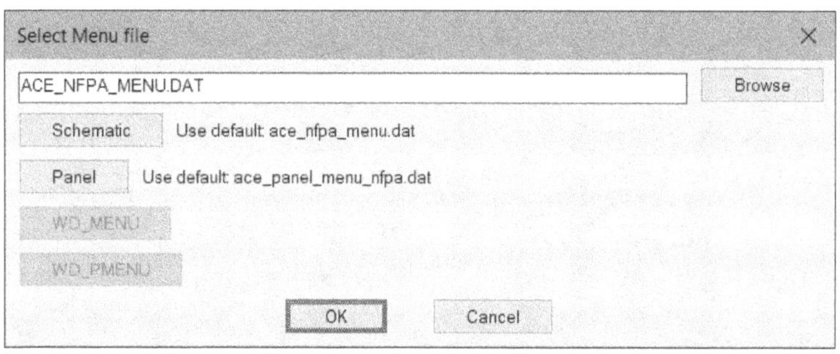

Figure-59. Select Menu file dialog box

- Click on the **OK** button from the dialog box. Note that you can select desired menu file by using the **Browse** button. By default, the default menu files are selected in the dialog box. On clicking **OK** button, the **Icon Menu Wizard** will be displayed as shown in Figure-60.

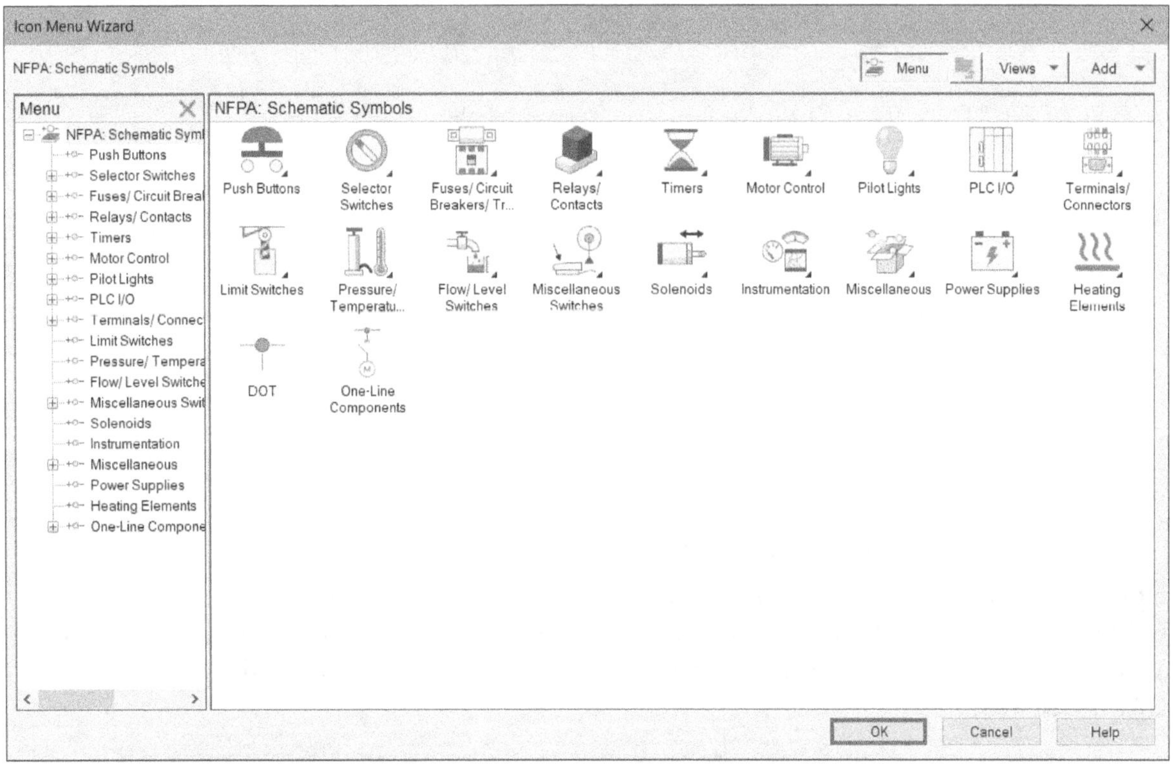

Figure-60. Icon Menu Wizard dialog box

- Click on the **Add** button top-right corner of the dialog box. A list of tools will be displayed; refer to Figure-61.

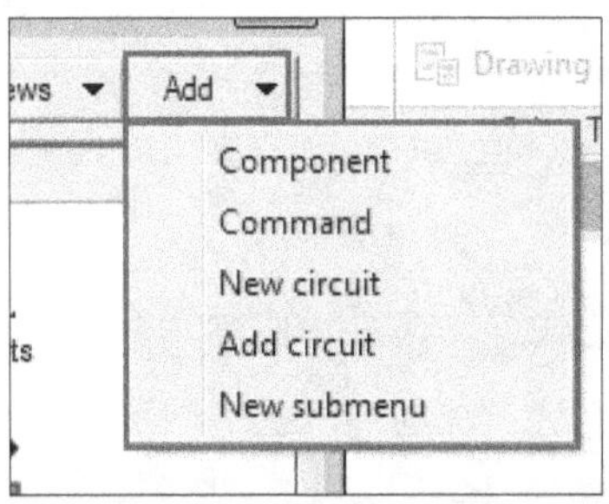

Figure-61. Add menu

- Click on the **Component** option from the menu. The **Add Icon-Component** dialog box will be displayed; refer to Figure-62.

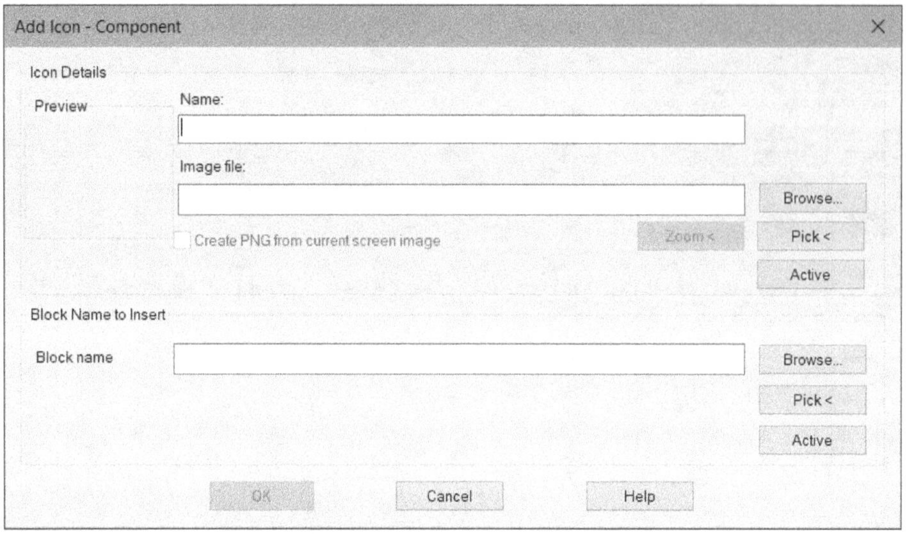

Figure-62. Add Icon-Component dialog box

- Specify desired name for the symbol in the **Name** edit box.
- Click on the **Browse** button next to **Block name** edit box in the **Block Name to Insert** area of the dialog box. The **Select File** dialog box will be displayed; refer to Figure-63.

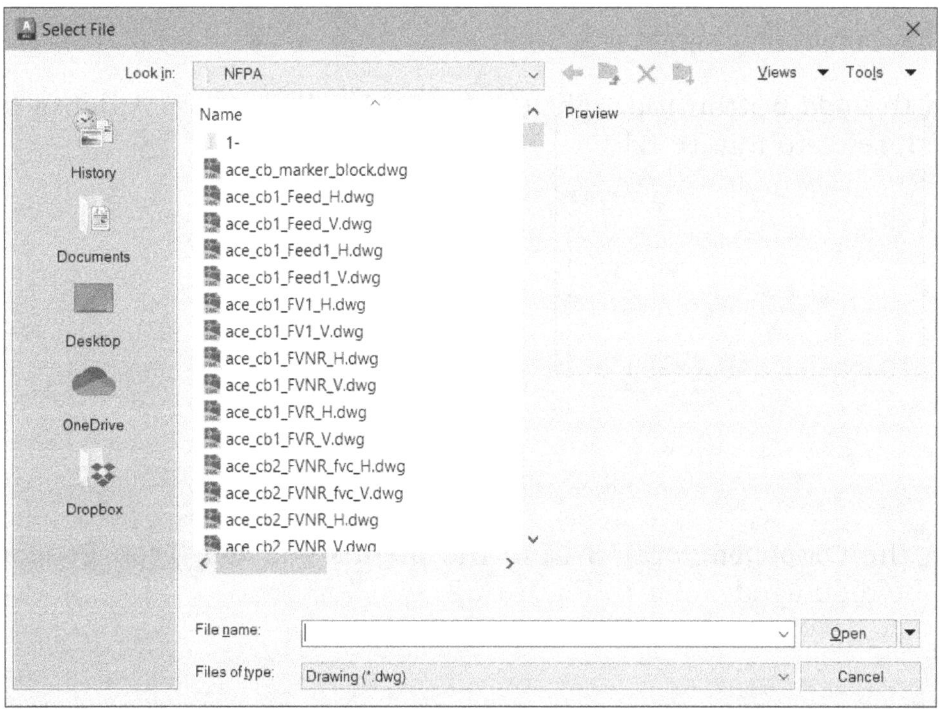

Figure-63. Select File dialog box

- Select the drawing file of the symbol earlier created and click on the **Open** button from the dialog box.
- Click on the **Pick <** button from the **Icon Details** area of the dialog box. You are asked to select the object for creating image.
- Select the sketch of symbol from the drawing area. Note that you must have the symbol drawing open in AutoCAD Electrical to select the sketch of symbol. Preview of symbol will be displayed in the dialog box; refer to Figure-64.

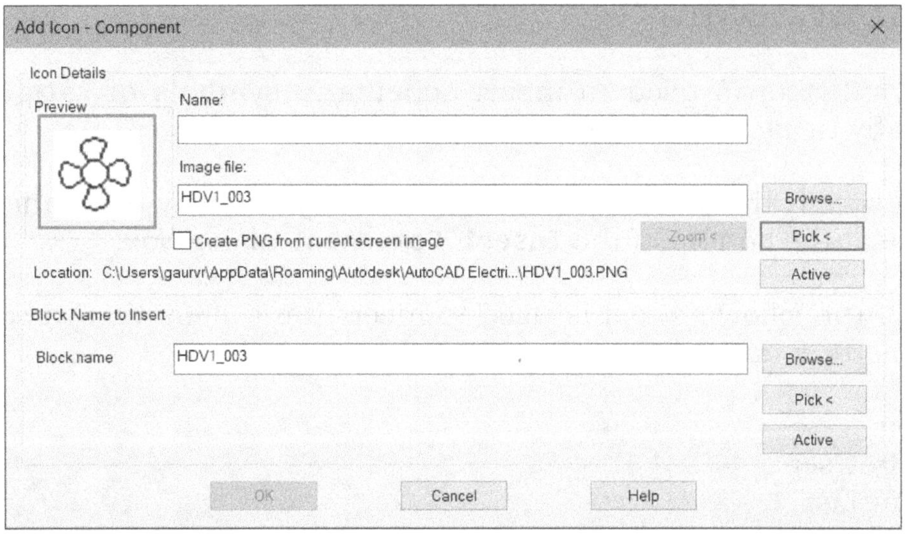

Figure-64. Preview of Icon

- Click on the **OK** button from the **Add Icon - Component** dialog box. The icon will be added in the **Icon Menu Wizard** dialog box; refer to Figure-65.

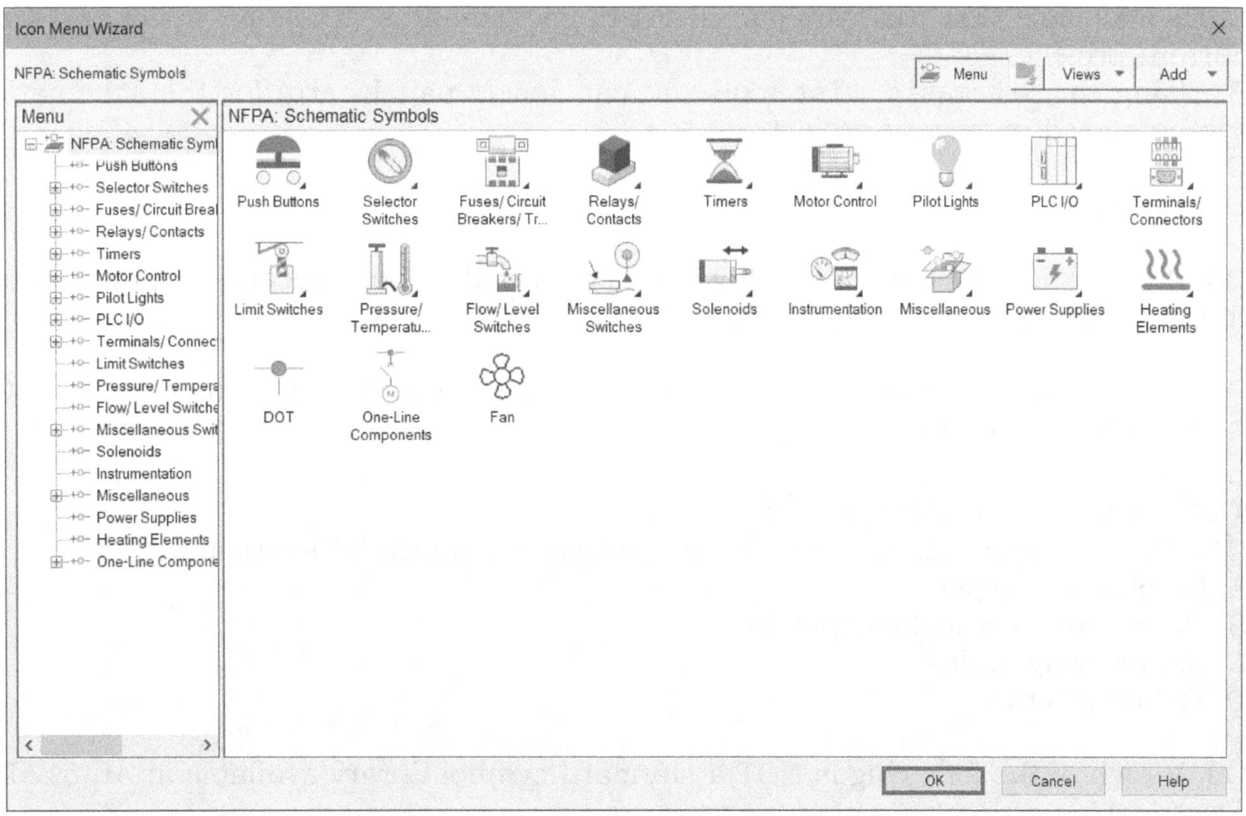

Figure-65. Icon added

- You can drag the new icon in desired folder as per your requirement. Click on the **OK** button from the **Icon Menu Wizard** dialog box to save the icon in **Icon Menu**.

Now, you can use the new symbol in any of the electrical drawing as per your need.

SELF-ASSESSMENT

Q1. The .... tool is used to insert schematic symbols of various electrical components by using the Icon Menu.

Q2. You can specify the scale value for schematic symbols by using the .... edit box available at the bottom of the **Insert Component** dialog box.

Q3. Which of the following tool is used to insert the schematic symbols based on their manufacturer data?

A. Icon Menu
B. Catalog Browser
C. Equipment List
D. User Defined List

Q4. Which of the following is a true statement regarding inserting symbols?

A. Using **Panel List** tool, you can insert the schematic symbols for panel parts in current project.
B. Using the **Schematic List** tool, you can insert panel parts for the schematic symbols used in current project.
C. None of the above
D. Both A and B

Q5. Pneumatic components are those components that work because of pressure of the fluid flowing through them. (T/F)

Q6. You can link your component to PLC with the tag of PLC by using the Use PLC Address button. (T/F)

REVIEW QUESTIONS

Q1. What is the function of schematic symbols in AutoCAD Electrical?
A. Display wire types
B. Represent electrical components
C. Set drawing scales
D. Define layers

Q2. Which of the following is NOT a standard symbol library available in AutoCAD Electrical?
A. JIC
B. IEEE
C. XML
D. IEC

Q3. What should determine the standard library used in an electrical drawing?
A. Manufacturer's brand
B. Drawing file format
C. Country or equipment type
D. Project name

Q4. What must be done before inserting symbols into a drawing?
A. Create a circuit
B. Set the symbol library
C. Print the layout
D. Define wire types

Q5. Where do you go to set the symbol library for a project in AutoCAD Electrical?
A. Ribbon Panel
B. Project Properties
C. Layout Tab
D. Tool Palettes

Q6. Which of these tools is used to insert schematic symbols?
A. Wire Tool
B. Icon Menu
C. Line Command
D. Hatch Tool

Q7. What does the Component Tag area in the Insert/Edit Component dialog box allow you to do?
A. Assign wire numbers
B. Specify description lines
C. Assign a tag to a component
D. Insert terminal blocks

Q8. What happens if you change the symbol library mid-project?
A. All existing symbols are updated
B. Drawings are deleted
C. New symbols use the new library, old ones remain
D. You are prompted to reselect standards

Q9. What is the purpose of the Catalog Data area?
A. Set layer names
B. Manage plot styles
C. Link components to manufacturer data
D. Assign wire numbers

Q10. Which dialog box helps you find and apply catalog data?
A. Insert Symbol
B. Project Manager
C. Catalog Browser
D. Symbol Library

Q11. What is the maximum number of alternates that can be assigned using Multiple Catalog Data?
A. 10
B. 25
C. 50
D. 99

Q12. Which area allows you to describe the physical placement of a component?
A. Catalog Data
B. Location Code
C. Component Tag
D. Cross-Reference

Q13. What is the function of the Cross-Reference area?
A. Define wire types
B. Display PLC address
C. Link related components across drawings
D. Insert title blocks

Q14. Which identifier stands for the sheet number in cross-reference formatting?
A. %D
B. %S
C. %P
D. %X

Q15. What does the NO/NC Setup help define?
A. Project folder structure
B. Location tags
C. Pin connectivity and contact type
D. Manufacturer's code

Q16. How can you specify the number of pins in a component?
A. Using wire types
B. Using the Icon Menu
C. Using the < or > buttons
D. Typing in the layout tab

Q17. What is the Catalog Browser mainly used for?
A. Viewing line types
B. Mapping symbols to wires
C. Searching and inserting catalog components
D. Editing project names

Q18. What happens when you map a symbol to a catalog number?
A. The component is saved for later use
B. A description is automatically applied
C. The symbol gets attached to the cursor
D. Drawing settings are reset

Q19. What allows AutoCAD Electrical to insert the same symbol next time you pick a catalog component?
A. Manufacturer name
B. Catalog number
C. Mapping the symbol
D. Tag override

Q20. Which area allows you to auto-fill details for items like manufacturer, catalog number, and item number?
A. Description area
B. Catalog Browser
C. Component Tag
D. Cross-Reference area

Q21. What is the purpose of the User Defined List tool in AutoCAD Electrical?
A. To define new symbols manually
B. To display a list of components created by AutoCAD
C. To display a list of components selected by the user
D. To modify electrical lines

Q22. What does the Add button in the User Defined List tool allow you to do?
A. Delete components from the list
B. Insert a component in the drawing
C. Add desired components or circuits to the list
D. Connect components with wires

Q23. Which radio button should you select if you want to add a single block to the User Defined List?
A. Explode
B. Circuit
C. Panel
D. Single block

Q24. What is the purpose of the Equipment List tool?
A. To view pneumatic components
B. To draw hydraulic circuits
C. To display or add components in the drawing
D. To modify the title block

Q25. Which file extension is associated with Equipment List setup files?
A. *.dwg
B. *.dxf
C. *.wde
D. *.xml

Q26. What allows you to change the data category of columns in Equipment List?
A. Icon Menu
B. Table Edit dialog box
C. Spreadsheet/Table columns button
D. Component Properties

Q27. What must be done if no symbol is assigned to a component in the Equipment List?
A. Delete the component
B. Re-insert the component
C. Assign a symbol using the Icon Menu
D. Skip the insertion

Q28. What is required for the Panel List tool to function correctly?
A. Electrical lines in drawing
B. Circuit assemblies
C. Panel parts (Footprints)
D. Schematic layouts

Q29. Which type of components operate using the pressure of fluids?
A. Electrical Components
B. Pneumatic Components
C. Hydraulic Components
D. Both B and C

Q30. What tool do you use to insert Pneumatic components in a drawing?
A. Symbol Builder
B. Insert Pneumatic Components
C. Equipment List
D. Component Properties

Q31. How is the procedure to insert Hydraulic components different from Pneumatic components?
A. It uses different drawing tools
B. It requires additional catalog info
C. It is the same as Pneumatic components
D. It does not require insertion

Q32. What kind of diagrams use P&ID components?
A. Electrical Circuit Diagrams
B. Mechanical Layouts
C. Piping and Instrumentation Diagrams
D. Pneumatic Flowcharts

Q33. What is the first step to create a custom symbol in AutoCAD Electrical?
A. Insert the symbol in a drawing
B. Use the Insert Component tool
C. Start a new document using IEC template
D. Create catalog information

Q34. What is the role of the Symbol Builder tool?
A. Create a new project
B. Build custom wire styles
C. Create user-defined circuits
D. Build custom electrical symbols

Q35. In Symbol Builder, what does selecting a Horizontal Parent symbol mean?
A. The component is used in P&ID
B. It connects to a horizontal terminal
C. It can have child components
D. It is used in panel layout only

Q36. What label can you assign to TERM01 and TERM02 in Symbol Builder?
A. Any text label like Live and Neutral
B. Only numbers
C. Only system-generated codes
D. None

Q37. What is the final step to save a newly created symbol using Symbol Builder?
A. Click Insert Component
B. Click Icon Menu
C. Specify Symbol name and click OK
D. Assign Manufacturer name

Q38. What must you do to make a custom symbol appear in the Icon Menu?
A. Save the drawing
B. Add it using the Icon Menu Wizard
C. Assign default catalog code
D. Refresh the panel

Q39. In the Icon Menu Wizard, what must be done to create an image preview of the symbol?
A. Type the symbol name
B. Select the sketch using Pick < button
C. Reload AutoCAD Electrical
D. Run the Symbol Generator

Q40. After adding a symbol to the Icon Menu, how can you organize it?
A. Rebuild the database
B. Drag the icon into a desired folder
C. Assign it a high priority tag
D. Rename the DWG file

FOR STUDENT NOTES

Answer for Self-Assessment:
Ans1. Icon Menu, Ans2. Scale schematic, Ans3. B, Ans4. D, Ans5. F, Ans6. T

Chapter 5

Wires, Circuits, and Ladders

Topics Covered

The major topics covered in this chapter are:

- ***Inserting Wires***
- ***Applying wire numbers***
- ***Inserting user defined circuits***
- ***Inserting ladders***
- ***Cable Markers***
- ***Circuit Builders***

INTRODUCTION

In the previous chapter, you have learned to insert electrical components in the drawings. These components individually can not perform any task. But if we connect them with wires in a desired manner then they can perform desired tasks. During your electrical designing career, you will be find yourself overwhelmed by sheer number of wires to be managed in drawing. In this chapter, you will learn to add wires. Later, you will learn to create circuits, multiple buses, ladders, labeling and other methods to organize the wires.

WIRES

Wires are the life line of any circuit. Wires are used to carry current and make the appliances run. The tools to create wires are available in the **Wire** drop-down in the **Insert Wires/Wire Numbers** panel of the **Schematic** tab in the **Ribbon**; refer to Figure-1.

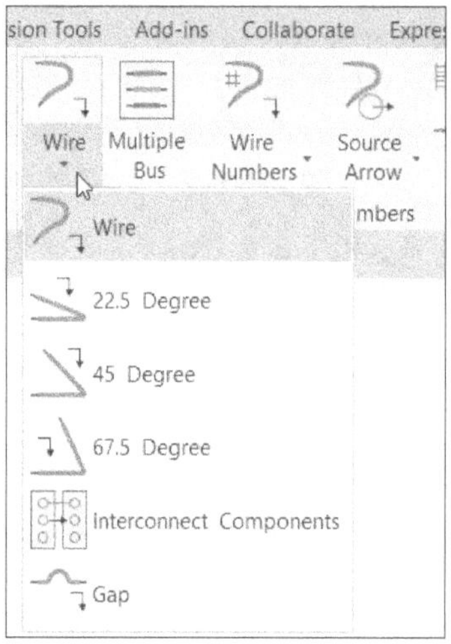

Figure-1. Wire drop-down

The tools in this drop-down are discussed next.

Wire

The **Wire** tool is used to add straight wires in the drawing to connect various components. The procedure to use this tool is given next.

- Click on the **Wire** tool from the **Wire** drop-down. The command prompt will be displayed as shown in Figure-2.

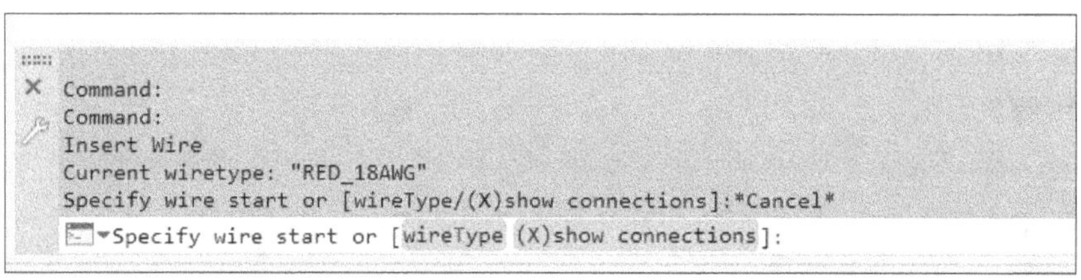

Figure-2. Command prompt for wire insertion

- Click on the **wireType** button in the command prompt or type **T** and press **ENTER** to change the wire type. The **Set Wire Type** dialog box will be displayed as shown in Figure-3.

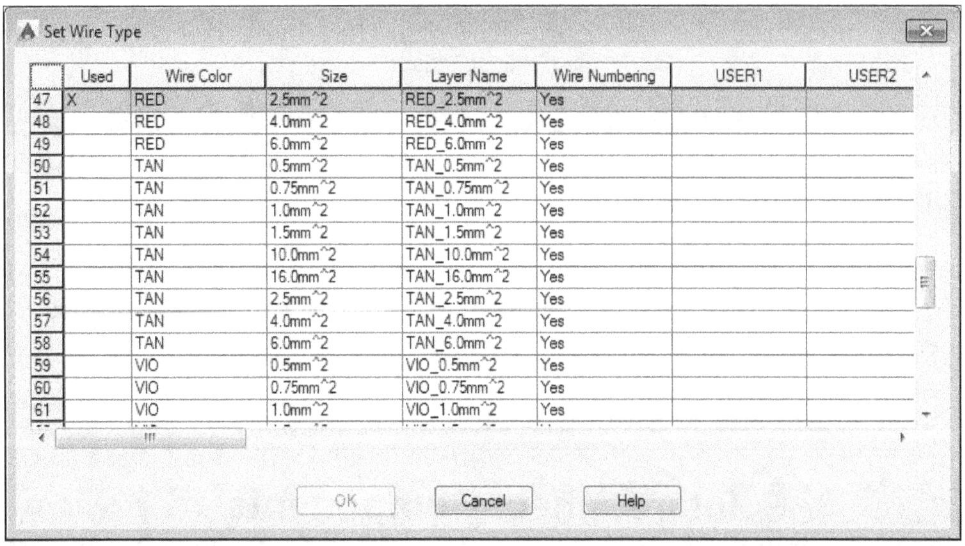

Figure-3. Set Wire Type dialog box

- Select desired wire type from the list and click on the **OK** button from the dialog box. The selected wire will be set for insertion.
- Click in the drawing area to specify the start point of the wire. You will be asked to specify the end point of the wire and the command prompt will be displayed as shown in Figure-4.

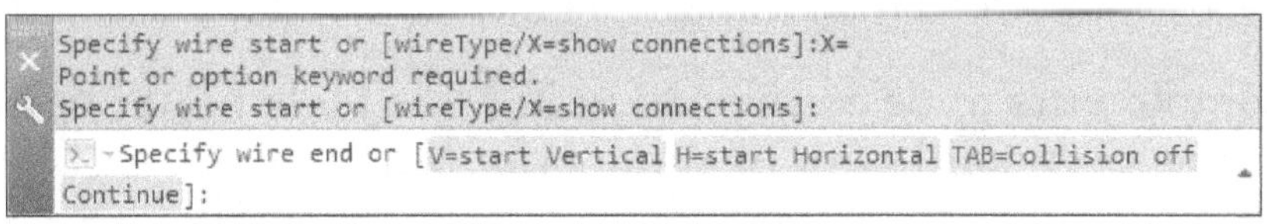

Figure-4. Command prompt after specifying start point

- If you are using the command prompt to specify length of the wire then you can use **V=start Vertical** or **H=start Horizontal** button in the command prompt to force AutoCAD Electrical draw only vertical or horizontal wires respectively.
- Click on the **Tab=Collision off** button to make the wires avoid collision with components. If you click again on this button then the wires will be able to pass through the components (at least in schematic drawings). Figure-5 shows two wirings one with collision OFF and one with collision ON.

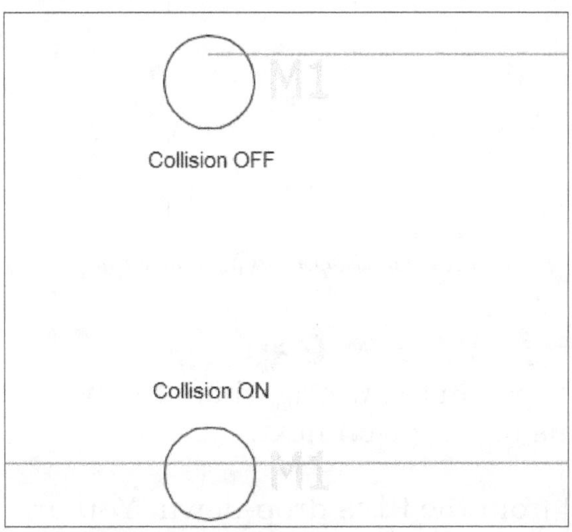

Figure-5. Collision ON or OFF

- Click to specify the end point of the wire. You can go on creating the wires until you end up at a connector of a component. From there you need to start a new wire.

22.5 Degree, 45 Degree, and 67.5 Degree

These tools are used to create wires at predefined angles. The procedure to use these tools is given next.

- Click on the **22.5 Degree**, **45 Degree**, or **67.5 Degree** tool from the **Wire** drop-down. You will be prompted to specify the starting point of the wire.
- Click at desired point to start wire. You are asked to specify the end point of the wire.
- Click to specify the end point of the wire.

Interconnect Components

The **Interconnect Components** tool is used to interconnect two components with each other. The procedure to use this tool is given next.

- Click on the **Interconnect Components** tool from the **Wire** drop-down. You are asked to select the first component.
- Select the first component. You will be prompted to select the second component.
- Select the second component. The wire will be created connecting both the components; refer to Figure-6.

Note that the components to be connected using this tool should be aligned properly so that their connection points are in same orientation.

Figure-6. Components interconnected

Gap

This tool is used to create gap in the wiring where the two wires intersect each other. The procedure to use this tool is given next.

- Click on the **Gap** tool from the **Wire** drop-down. You are asked to select the wire that you want to be remained solid while applying this tool.

- Select desired wire. You are asked to select the crossing wire.
- Select the intersecting wire. Gap in the wires will be created; refer to Figure-7.

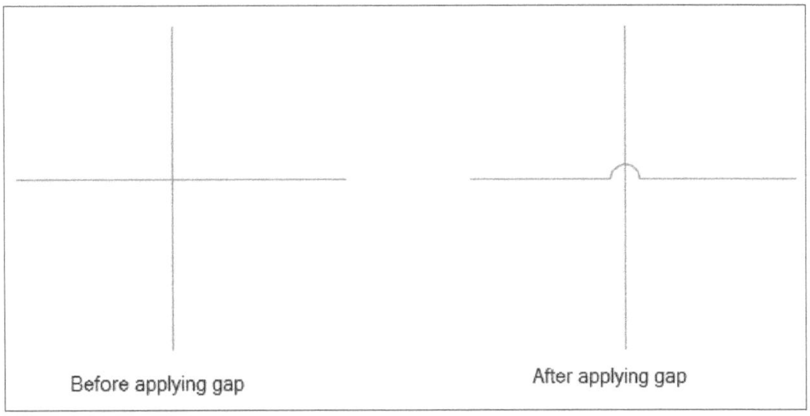

Figure-7. Wires on using Gap tool

MULTIPLE BUS

The **Multiple Bus** tool is used to create multiple lines of wires. To create a multiple bus, follow the steps given next.

- Click on the **Multiple Bus** tool from the **Insert Wires/Wire Numbers** drop-down. The **Multiple Wire Bus** dialog box will be displayed as shown in Figure-8.
- Click in the **Spacing** edit boxes so specify the horizontal and vertical spacing between the wires.
- By default, the **Another Bus (Multiple Wires)** radio button is selected. As a result, you can create the multiple wires from a point on another wire.
- Select the **Empty Space, Go Horizontal** radio button or the **Empty Space, Go Vertical** radio button to create multiple wires in horizontal or vertical direction respectively. Procedures to create the wire bus using these options are discussed next.

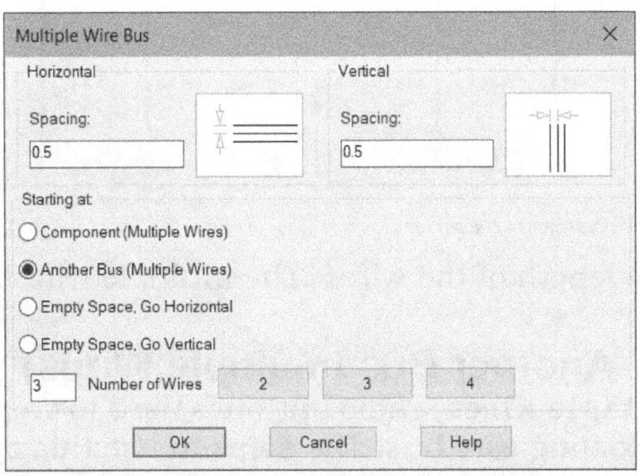

Figure-8. Multiple Wire Bus dialog box

Component (Multiple Wires)

- Select the **Component (Multiple Wires)** radio button from the **Multiple Wire Bus** dialog box.
- Click on the **OK** button from the dialog box. You are prompted to create a selection window to select the connection ports of devices.

- Select the connection ports that you want to use for creating multiple wire bus; refer to Figure-9. The selected ports will be highlighted as shown in Figure-10.

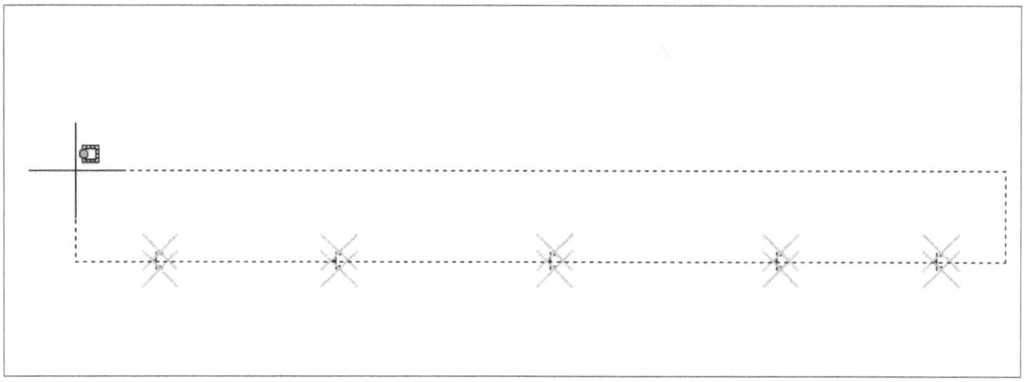

Figure-9. Window selection of ports

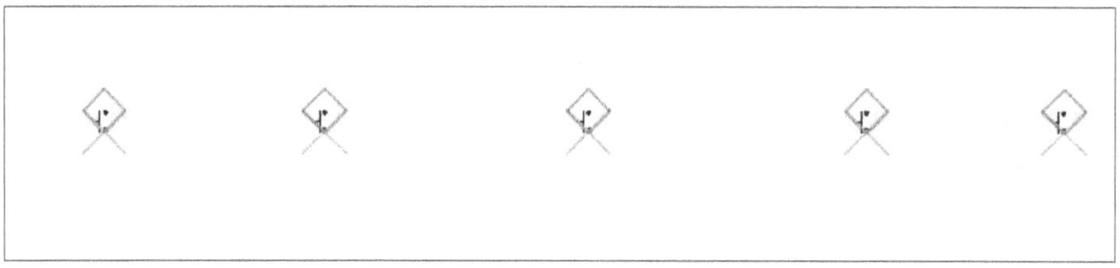

Figure-10. Connection ports highlighted

- Press **ENTER** at the command prompt. Wires will get attached to the cursor; refer to Figure-11.

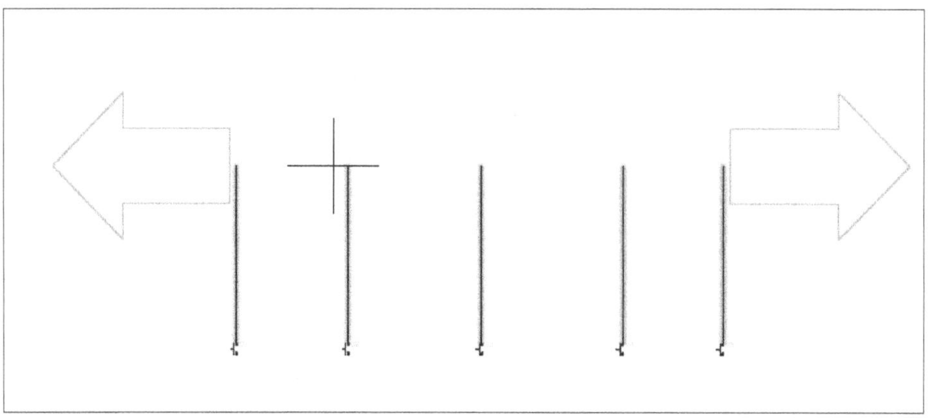

Figure-11. Wires attached to the cursor

- Click to specify the length of the wires. The multiple wire bus will be created.

Another Bus (Multiple Wires)

The **Another Bus (Multiple Wires)** radio button is used to create a multiple wire bus by using an already existing wire bus. The steps to use this option are given next.

- Select the **Another Bus (Multiple Wires)** radio button from the **Multiple Wire Bus** dialog box. Specify the number of wires in the **Number of Wires** edit box. Note that if the number of wires are more than the wires in the already existing bus then extra wires will be connected to the first wire in the bus.
- Click on the **OK** button from the dialog box. You are asked to select a wire.
- Select the middle wire from the already existing wire bus. The ends of the current wire bus will get attached to the cursor; refer to Figure-12.

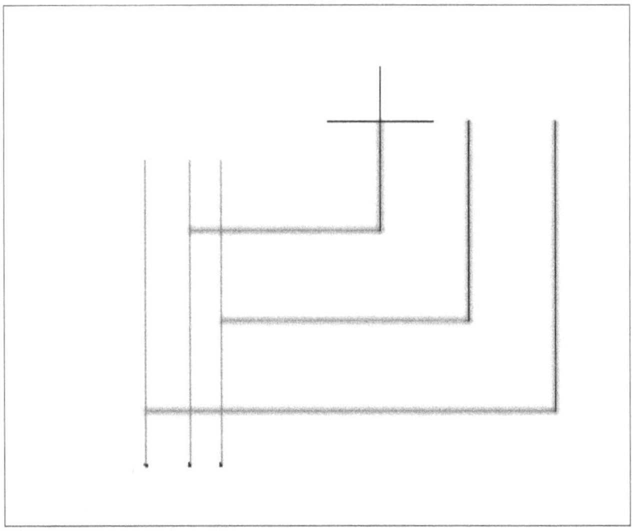

Figure-12. Wire bus from another wire bus

- Click to specify the end point of the wire bus.

Empty Space, Go Horizontal

The **Empty Space, Go Horizontal** radio button is used to create a multiple wire bus in horizontal direction starting from the specified start point. The steps to use this option are given next.

- Select the **Empty Space, Go Horizontal** radio button from the **Multiple Wire Bus** dialog box. Specify the number of wires in the **Number of Wires** edit box.
- Click on the **OK** button from the dialog box. You are asked to specify the starting point for the bus.
- Click in the drawing area. You are asked to specify the end point of the bus.
- Click to specify the end point.

Empty Space, Go Vertical

The **Empty Space, Go Vertical** radio button is used to create a multiple wire bus in vertical direction starting from the specified start point. The steps to use this option are given next.

- Select the **Empty Space, Go Vertical** radio button from the **Multiple Wire Bus** dialog box. Specify the number of wires in the **Number of Wires** edit box.
- Click on the **OK** button from the dialog box. You are asked to specify the starting point for the bus.
- Click in the drawing area. You are asked to specify the end point of the bus.
- Click to specify the end point.

LADDERS

In Electrical systems, multiple circuits that are powered by a common power source can be combined in the form of ladders generally in Automobile electrical circuits. Ladders are arrangement of wires as shown in Figure-13.

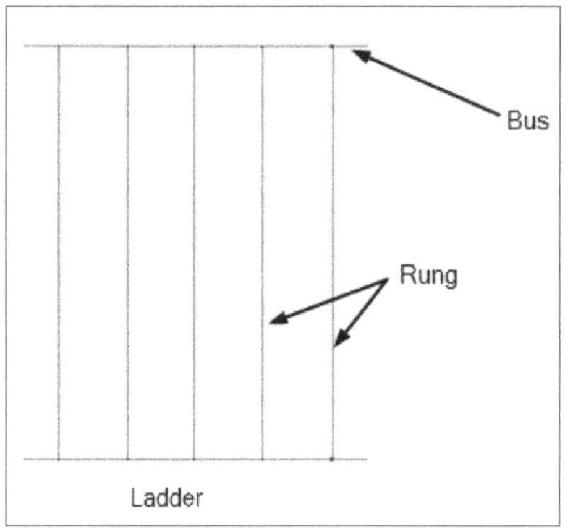

Figure-13. Ladder

The Ladder Diagrams are backbone to PLCs circuits. The tools to create and control ladders are available in the **Ladder** drop-down; refer to Figure-14. The procedures to use the tools of this drop-down are given next.

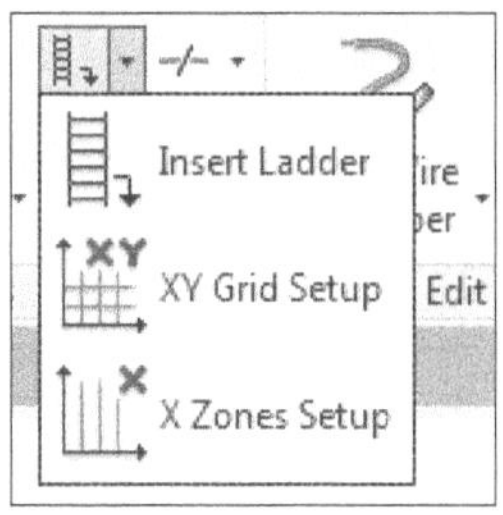

Figure-14. Ladder drop-down

Insert Ladder

The **Insert Ladder** tool is used to insert the ladder of wires in the drawing. This tool is specially useful in creating PLC circuits. The procedure to use this tool is given next.

- Click on the **Insert Ladder** tool from the **Ladder** drop-down. The **Insert Ladder** dialog box will be displayed as shown in Figure-15.

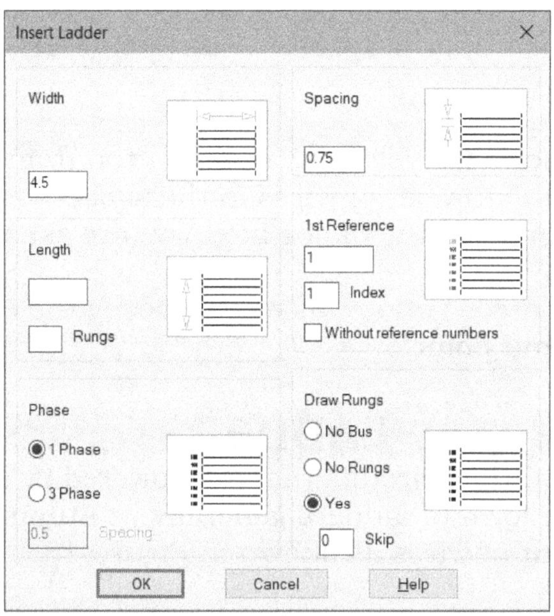

Figure-15. Insert Ladder dialog box

- Click in the edit box in **Width** area and specify the distance between two wire buses.
- Click in the edit box in **Spacing** area and specify the distance between two rungs.
- Click in the **Length** edit box and specify the total length of the ladder or click in the **Rungs** edit box and specify the number of rungs in the ladder. Note that you can specify the value in any one of the two edit boxes. Value in the other edit boxes will be calculated automatically.
- Select desired number of phases for the ladder. If you select the **3 Phase** radio button from the **Phase** area then the ladder will be created with three phase wire lines and a value of distance between phase lines will be required in the **Spacing** edit box.
- You can skip the creation of bus or rung in the ladder by selecting the respective radio button from the **Draw Rungs** area of the dialog box. **Note that ladder without rungs act as reference for cross-references of components**.
- Click on the **OK** button from the dialog box to create the ladder.

XY Grid Setup

The **XY Grid Setup** tool is used to setup a grid for referencing. If you have a parent component and child components in the drawing then cross-references will be generated according to the XY Grid. Before you use this tool, you must enable the X-Y Grid format referencing. To enable this option, follow the steps given next.

- Right-click on the name of current drawing in the **Project Manager**. A shortcut menu will be displayed as shown in Figure-16.

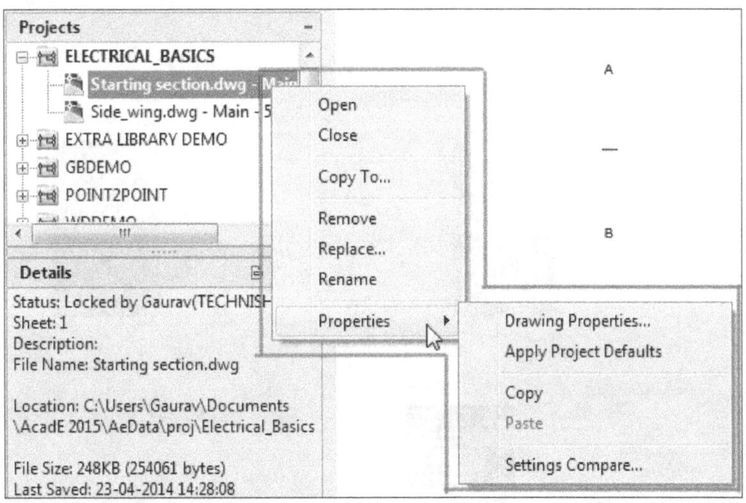

Figure-16. Shortcut menu

- Move the cursor to **Properties** option in the shortcut menu. A cascading menu will be displayed.
- Click on the **Drawing Properties** option from the menu. The **Drawing Properties** dialog box will be displayed; refer to Figure-17.

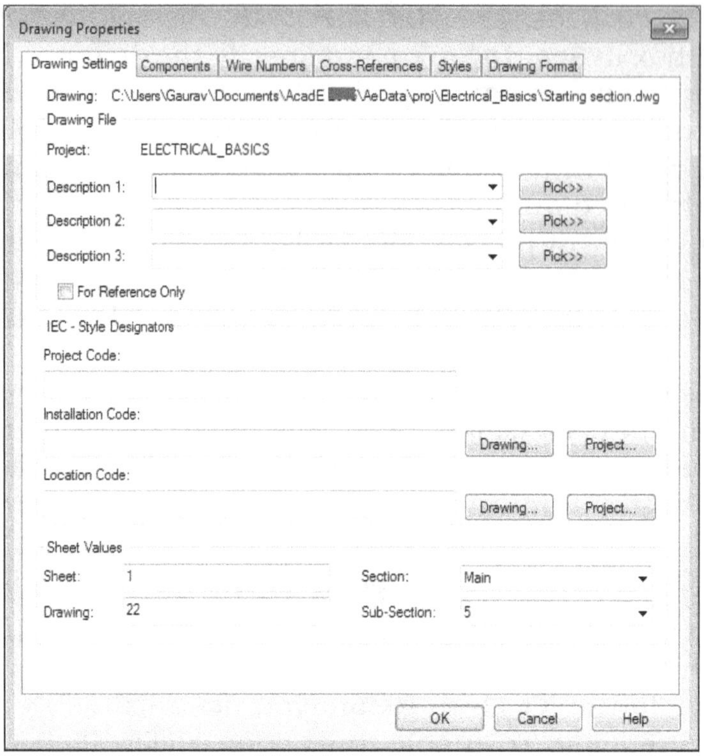

Figure-17. Drawing Properties dialog box

- Click on the **Drawing Format** tab in the dialog box. The options in the dialog box will be displayed as shown in Figure-18.

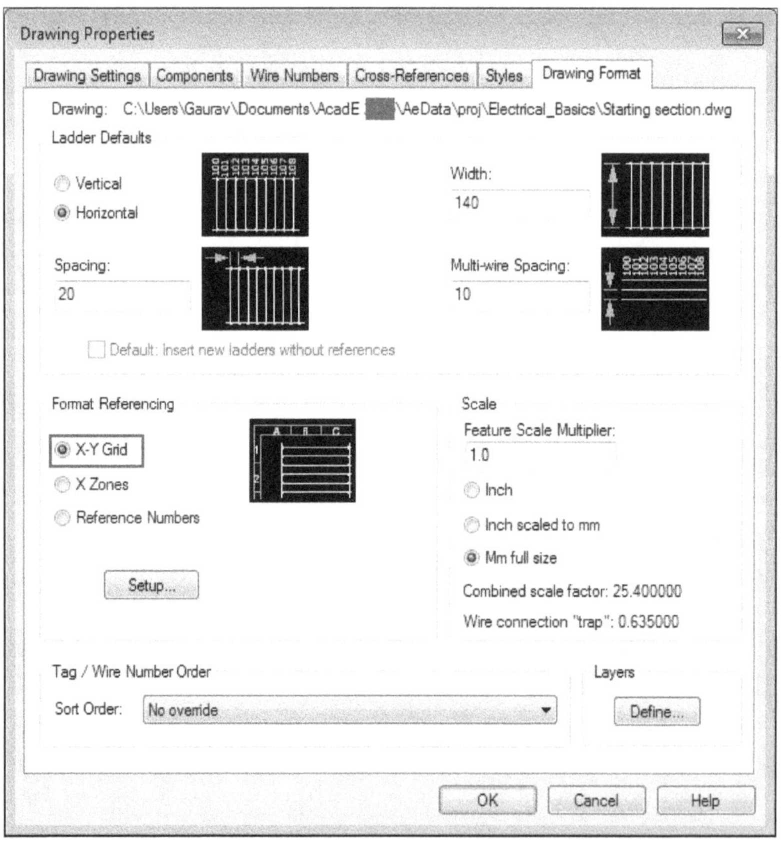

Figure-18. Drawing Properties dialog box with Drawing Format tab

- Select the **X-Y Grid** radio button from the **Format Referencing** area of the dialog box.

- Note that you can also define the default direction of ladder creation by selecting desired radio button from the **Ladder Defaults** area of the dialog box.
- Click on the **OK** button from the dialog box to apply the specified settings.

Now, we have enabled the **X-Y Grid** format and we can use it in our drawing. The procedure to use the **X-Y Grid Setup** tool is given next.

- Click on the **X-Y Grid Setup** tool from the **Ladder** drop-down. The **X-Y Grid Setup** dialog box will be displayed as shown in Figure-19.

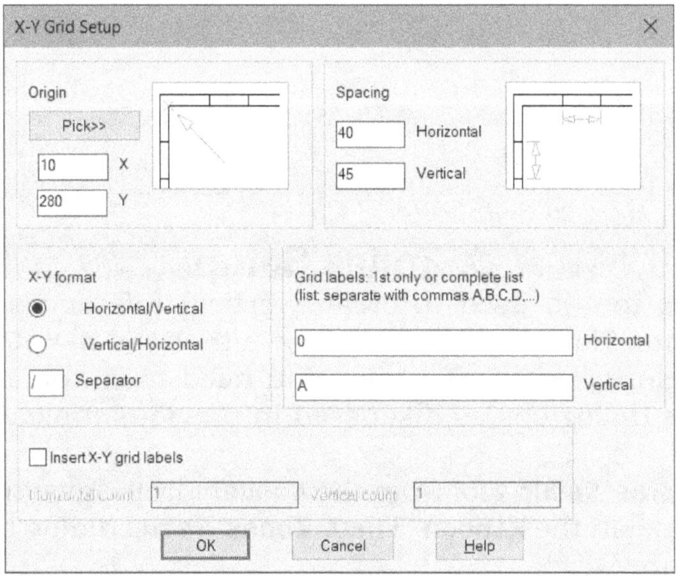

Figure-19. X-Y Grid Setup dialog box

- Click on the **Pick>>** button from the dialog box to specify the origin of the grid. You are prompted to pick a point in the drawing area.
- Click in the drawing area to specify the point.
- Specify the spacing between horizontal and vertical grid lines in the edit boxes available in the **Spacing** area of the dialog box.
- Select desired radio button from the **X-Y format** area of the dialog box.
- Select the **Insert X-Y grid labels** check box to display the labels of the grid.
- After selecting the check box, specify the horizontal and vertical counts in the respective edit boxes below the check box.
- Click on the **OK** button from the dialog box. The grid will be created; refer to Figure-20. Note that the grid created is for reference purpose. Using this grid, you can precisely place the components in the drawing.

Figure-20. Grid created

X Zones Setup

The **X Zones Setup** tool is used to create vertical references in the drawing for inserting components. Note that you can use either the **X-Y Grid Setup** tool or **X Zone Setup** tool. Before creating X zones, you need to enable this option from the **Drawing Properties** dialog box as discussed in previous topic.

- Click on the **X Zones Setup** tool from the **Ladder** drop-down in the **Insert Wires/ Wire Numbers** panel in the **Ribbon**. The **X Zones Setup** dialog box will be displayed as shown in Figure-21.

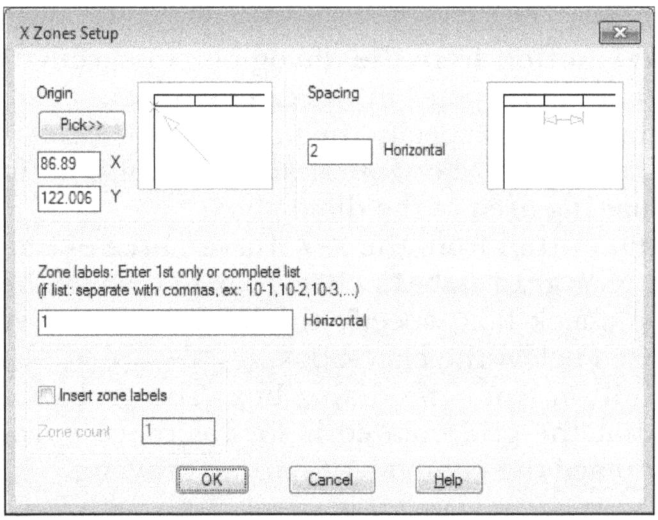

Figure-21. X Zones Setup dialog box

- Click on the **Pick>>** button from the **Origin** area of the dialog box. You are prompted to select a point in the drawing area to specify the origin.
- Click on desired point in the drawing area.
- Click in the **Horizontal** edit box in the **Spacing** area of the dialog box and specify desired distance value.
- Select the **Insert zone labels** check box to include the labels on the zone.
- Click on the **OK** button from the dialog box. The X zone will be created; refer to Figure-22.

Figure-22. X zones

When we work on real world projects of electrical systems, then we generally require references to insert our components. At that time, XY grids and X Zones are required to create references.

WIRE NUMBERING

Wire numbering is an important aspect of wires. Wire number helps to identify various things like circuits related to that wire, components connected to that wire, total length of the wire, and other details. The tools to specify wire numbers are available in the **Wire Numbers** drop-down of the **Insert Wires/Wire Numbers** panel. There are three tools in this drop-down; **Wire Numbers**, **3 Phase**, and **PLC I/O**. Refer to Figure-23. These tools are discussed next.

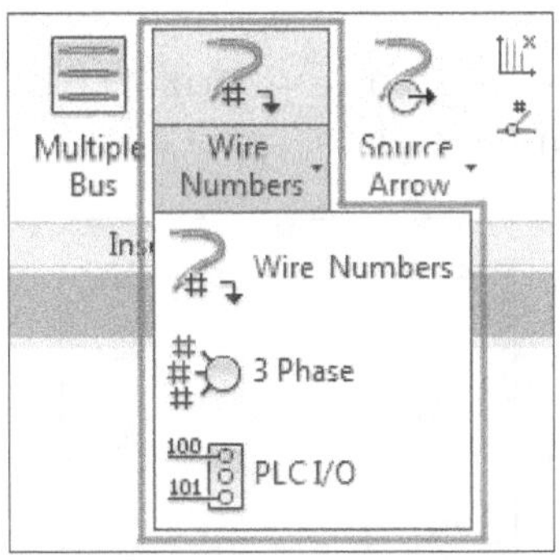

Figure-23. Wire Numbers drop-down

Wire Numbers

The **Wire Numbers** tool is used to specify wire numbers for individual wires or project wide. The procedure to use this tool is given next.

- Click on the **Wire Numbers** tool from the **Wire Numbers** drop-down. The **Wire Tagging** dialog box will be displayed as shown in Figure-24.

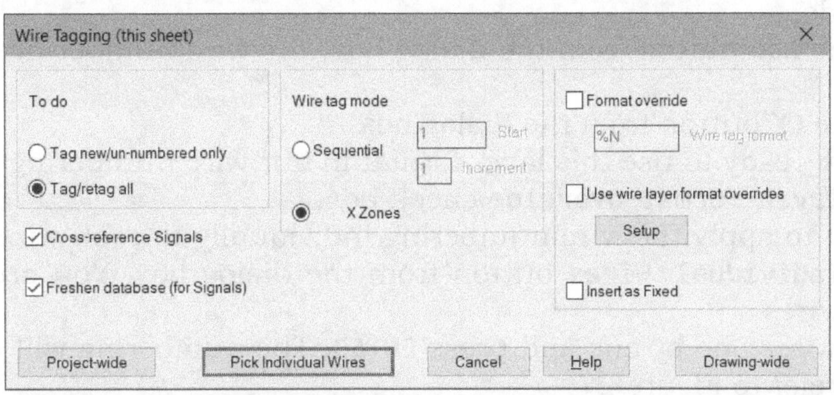

Figure-24. Wire Tagging dialog box

- Click in the **Start** edit box in the **Wire tag mode** area and specify the starting number for wire tagging(numbering).

- Select the **Tag new/un-numbered only** radio button if you want to specify the wire numbers for un-numbered wires only.
- Click on the **Setup** button from the dialog box to specify the format of wire numbering. The **Assign Wire Numbering Formats by Wire Layers** dialog box will be displayed as shown in Figure-25.

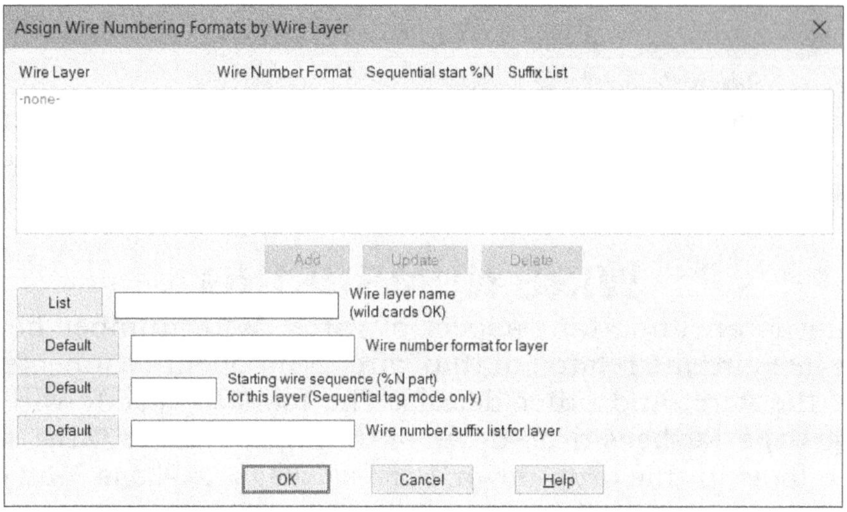

Figure-25. Assign Wire Numbering Formats by Wire Layers dialog box

- Click on the **List** button in this dialog box. The **Select Wire Layer** dialog box will be displayed; refer to Figure-26.

Figure-26. Select Wire Layer dialog box

- Select desired color layer for the wire.
- Click on the **OK** button from the **Select Wire Layer** dialog box.
- Click on the three **Default** buttons one by one to specify the default values in the related fields.
- Click on the **Add** button from the dialog box, the wire numbering format will be added.
- Click on the **OK** button from the dialog box.
- Now, we are ready to use this layer format in our wire numbering. So, select the **Use wire layer format override** check box.
- If you want to apply the wire numbering individually to each wire then click on the **Pick Individual Wires** button from the dialog box. You are prompted to select the wires.
- Select the wires one by one and press **ENTER**. The numbering will be assigned to the wires; refer to Figure-27.

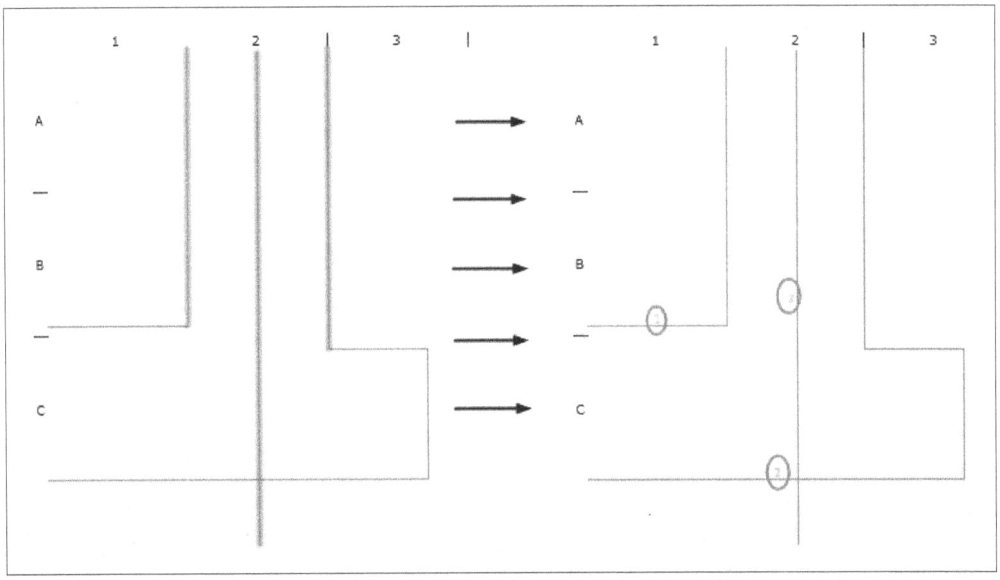

Figure-27. Wire numbering

3 Phase

The **3 Phase** tool is used to number three phase wiring. The procedure to use this tool is given next.

- Click on the **3 Phase** tool from the **Wire Numbers** drop-down in the **Ribbon**. The **3 Phase Wire Numbering** dialog box will be displayed as shown in Figure-28.

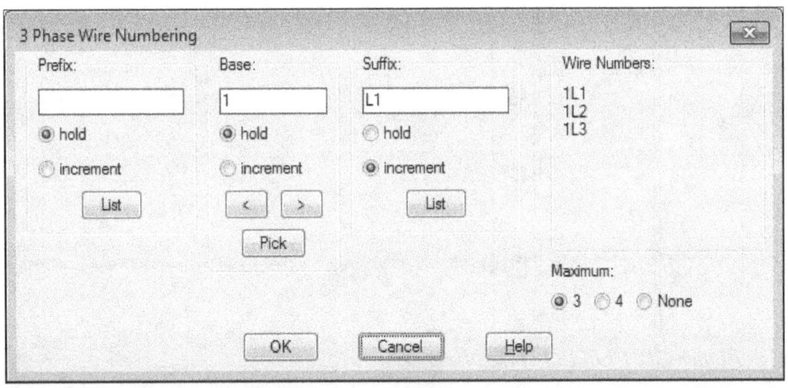

Figure-28. 3 Phase Wire Numbering dialog box

- Specify desired values in the **Prefix**, **Base**, and **Suffix** edit boxes.
- Select desired radio button from the **Maximum** area of the dialog box. If you have 3 wires in the connection then select **3** radio button and if you have 4 wires then select the **4** radio button.
- Click on the **OK** button from the dialog box. You are asked to select first wire of the 3 phase line.
- Select the respective wire. You are asked to select the second wire.
- Select the second wire and similarly, select the third wire.
- After selecting the third wire, the dialog box will be displayed again. Keep on selecting the individual wire that you want to be numbers.
- After numbering, click on the **Cancel** button from the dialog box to exit.

PLC I/O

The **PLC I/O** tool is used to number the input and output ports of the PLC. The procedure to use this tool is given next.

- Click on the **PLC I/O** tool from the **Wire Numbers** drop-down in the **Ribbon**. The **PLC I/O Wire Numbers** dialog box will be displayed as shown in Figure-29.

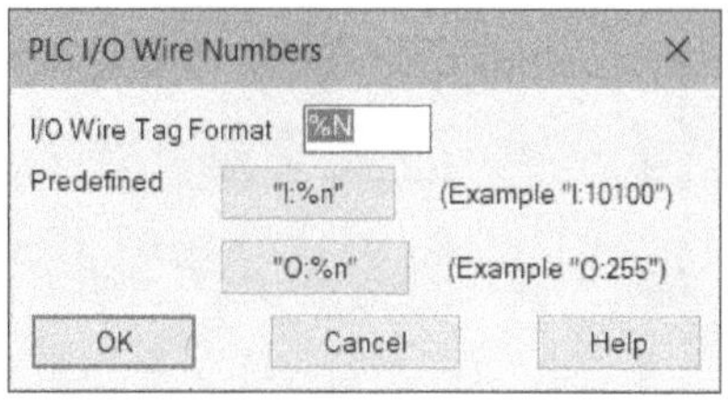

Figure-29. PLC I O Wire Numbers dialog box

- Specify desired tag format for the PLC I/O. The format codes have already been discussed.
- Click on the **OK** button from the dialog box. You are prompted to select the PLC.
- Select a PLC, you are asked to select the wires connected to the PLC.
- One by one select the wires that are connected to the PLC or make the window selection to select multiples and then press **ENTER**.
- The tags defined as per the PLC are extracted and applied to the wires; refer to Figure-30.

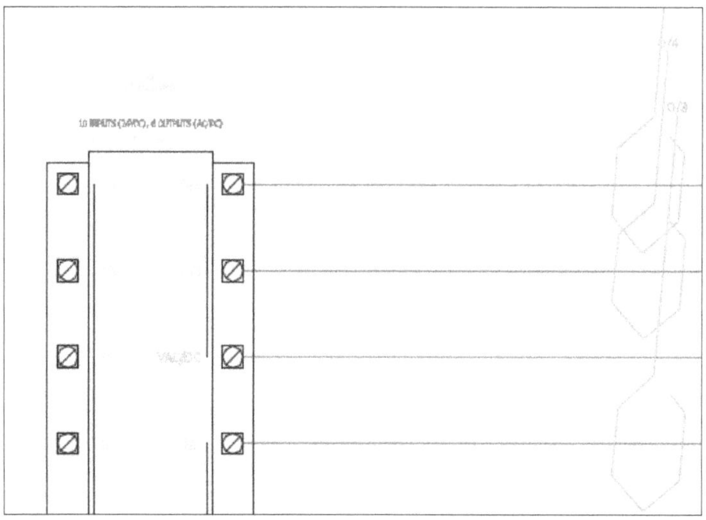

Figure-30. PLC I O Wire Numbers

WIRE NUMBER LEADERS AND LABELS

In the previous figure, the wire number leaders were assigned automatically. We can modify the wire number leaders and assign them desired labels. The tools to do so are available in the **Wire Number Leader** drop-down; refer to Figure-31.

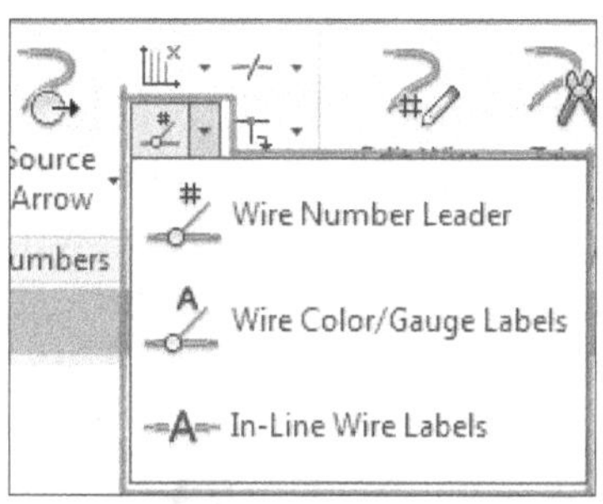

Figure-31. Wire Number Leader drop-down

The procedure to use the tools in this drop-down is discussed next.

Wire Number Leader

The **Wire Number Leader** tool is used to assign leader to a wire number. The procedure to use this tool is given next.

- Click on the **Wire Number Leader** tool from the **Wire Number Leader** drop-down in the **Ribbon**. You will be prompted to select a wire number; refer to Figure-32.

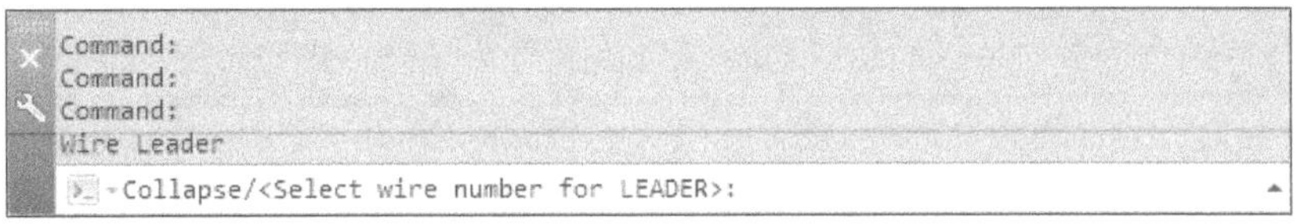

Figure-32. Command prompt for wire numbers

- Select desired wire number from the drawing. You will be prompted to specify the leader end point; refer to Figure-33.

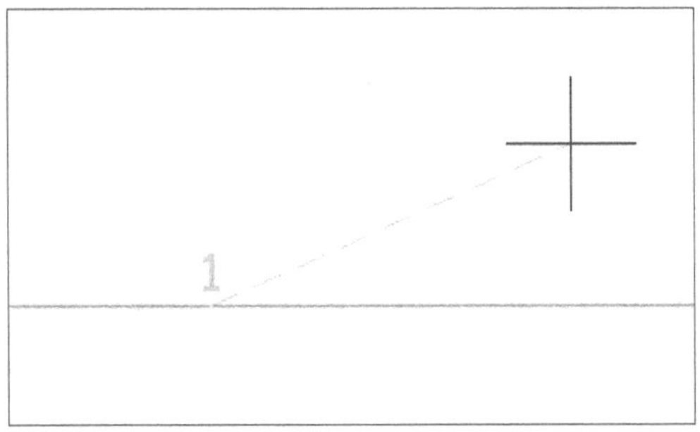

Figure-33. After selecting wire number

- Click on the drawing area to specify the end point of the wire number leader. You will be prompted to specify the next point of the leader landing.
- Click in the drawing area to specify the point or press **ENTER**. The wire number leader will be assigned; refer to Figure-34. Note that your leader style might be different than the one shown in the figure. To change the leader style, type **MLEADERSTYLE** in the command prompt and modify the leader as desired.

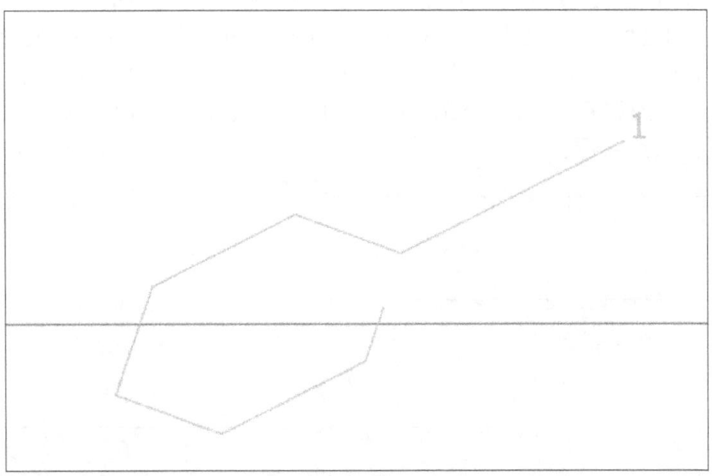

Figure-34. Wire number leader assigned

- Similarly, you can assign leaders to other wire numbers.

Wire Color/Gauge Labels

The **Wire Color/Gauge Labels** tool is used to attach the wire color and/or gauge labels to the wires so that we can easily identify the wires in the circuit. The procedure to use this tool is given next.

- Click on the **Wire Color/Gauge Labels** tool from the **Wire Number Leader** drop-down in the **Ribbon**. The **Insert Wire Color/Gauge Labels** dialog box will be displayed as shown in Figure-35.
- Click on the **Setup** button to specify the settings. The **Wire label color/gauge setup** dialog box will be displayed as shown in Figure-36.

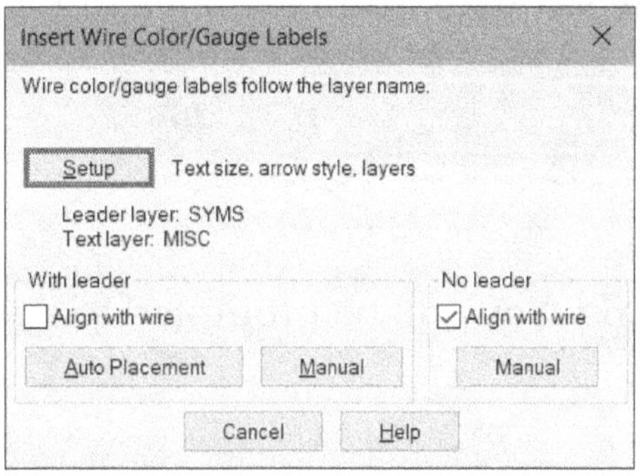

Figure-35. Insert Wire Color or Gauge Labels dialog box

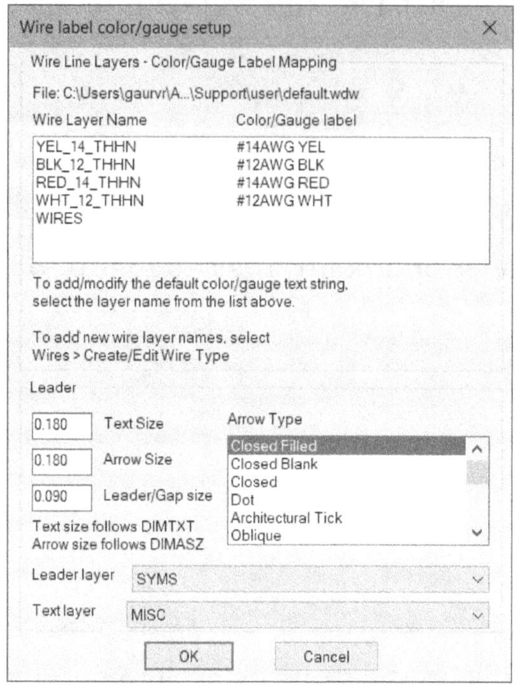

Figure-36. Wire label color or gauge setup dialog box

- Specify desired settings in the dialog box and click on the **OK** button from the dialog box.
- Click on the **Manual** button from the dialog box to manually assign the color/gauge labels. Note that you can assign the labels with leaders or without leaders by selecting the **Manual** button from the respective area.
- Click on the **Auto Placement** button to automatically place the color/gauge labels.
- After selecting desired button, click on the wire and follow the instructions given by system. Press **ENTER** to create the labels. The labels will be attached to the wires; refer to Figure-37.

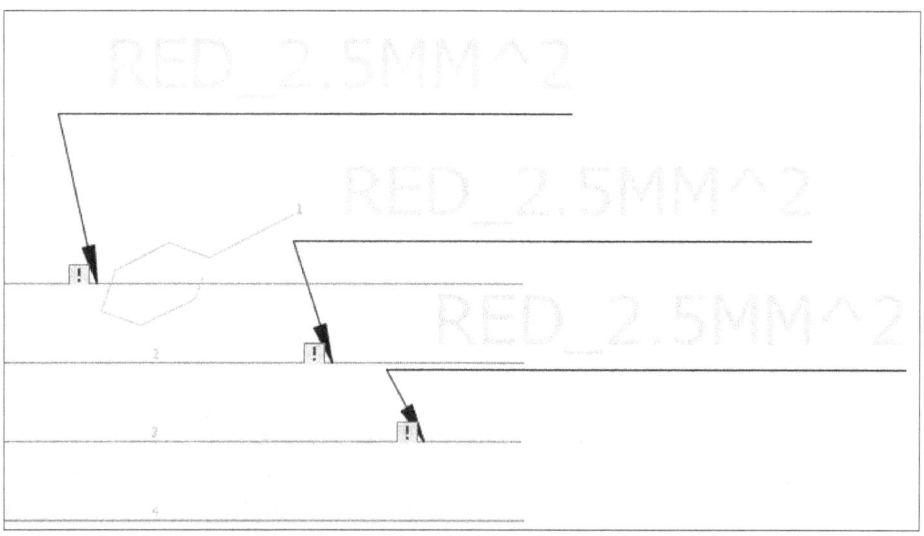

Figure-37. Wire labels

In-Line Wire Labels

The previous tool is used to insert labels at the sides of the wires. The **In-Line Wire Labels** tool is used to insert the in-line wire labels. The labels can include informations like color, gauge, and so on. The procedure to use this tool is given next.

- Click on the **In-Line Wire Labels** tool from the **Wire Number Leader** drop-down. The **Insert Component** dialog box will be displayed; refer to Figure-38.

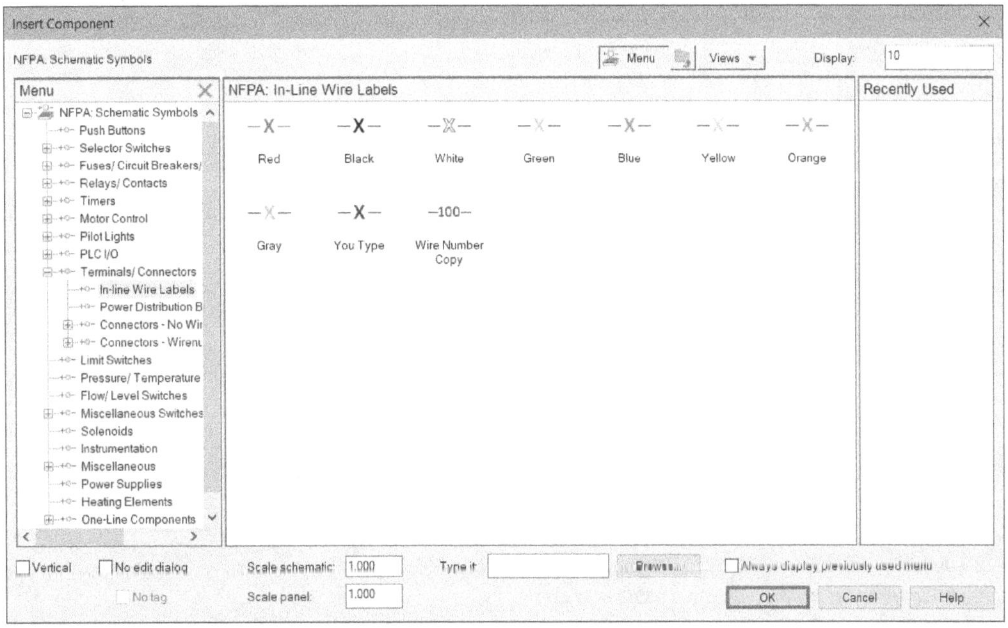

Figure-38. Insert Component dialog box

- Select desired symbol from the dialog box. You will be prompted to specify the insertion point for the symbol.
- Click on the wire to insert the label. The label will be inserted; refer to Figure-39. Press **ESC** to exit the tool. Note that you can press **SPACEBAR** to reactivate the previous tool.

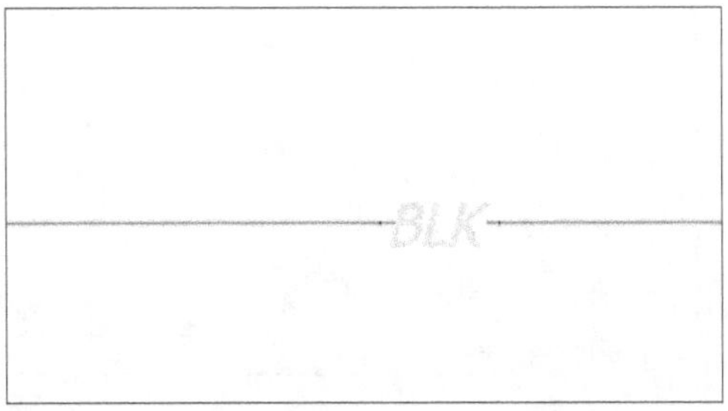

Figure-39. In-Line label

MARKERS

The markers are used to mark the wires for categorizing them. The tools to apply markers are available in the **Cable Markers** and **Insert Dot Tee Markers** drop-downs. The procedures to use these tools are discussed next.

Cable Markers

This tool is used to insert cable markers. The procedure is given next.

- Click on the **Cable Markers** tool from the **Cable Markers** drop-down. The **Insert Component** dialog box will be displayed with the cable markers; refer to Figure-40.

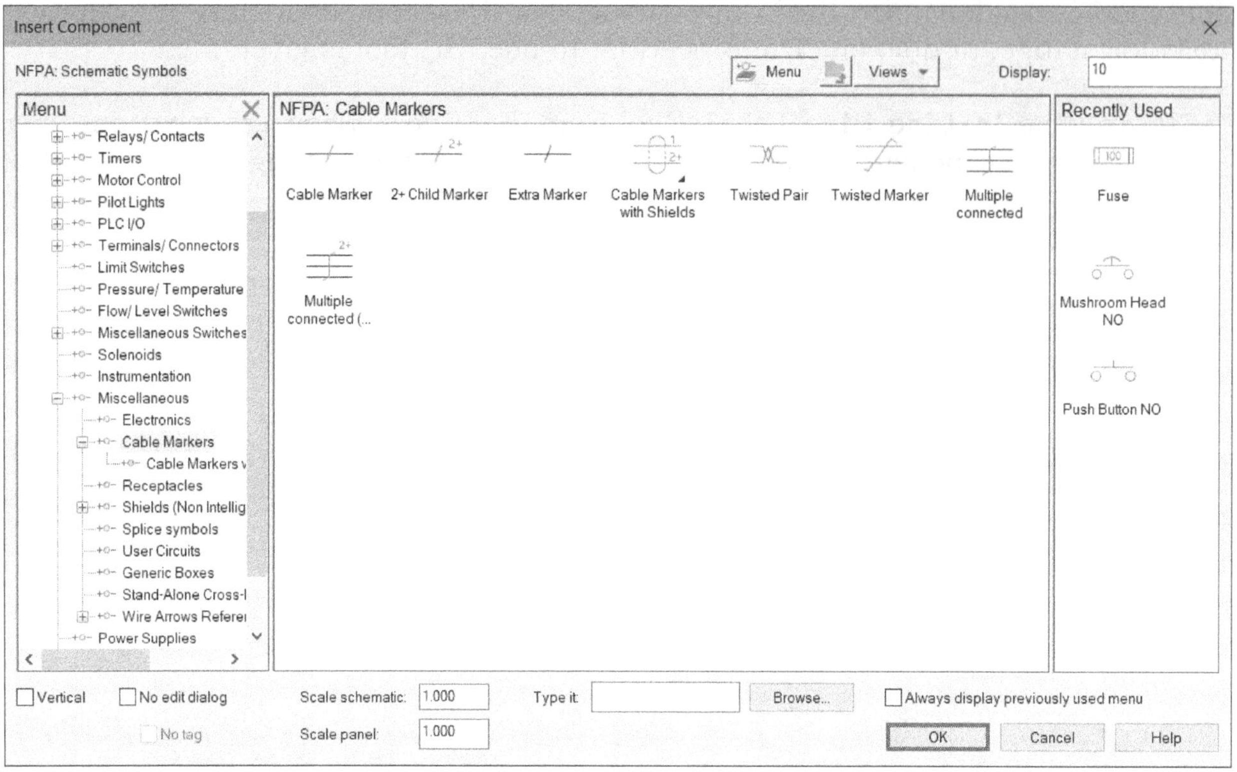

Figure-40. Insert Component dialog box with cable markers

- Click on desired symbol and then click on the wire to place the symbol. The **Insert/Edit Cable Marker (Parent wire)** dialog box will be displayed as shown in Figure-41.

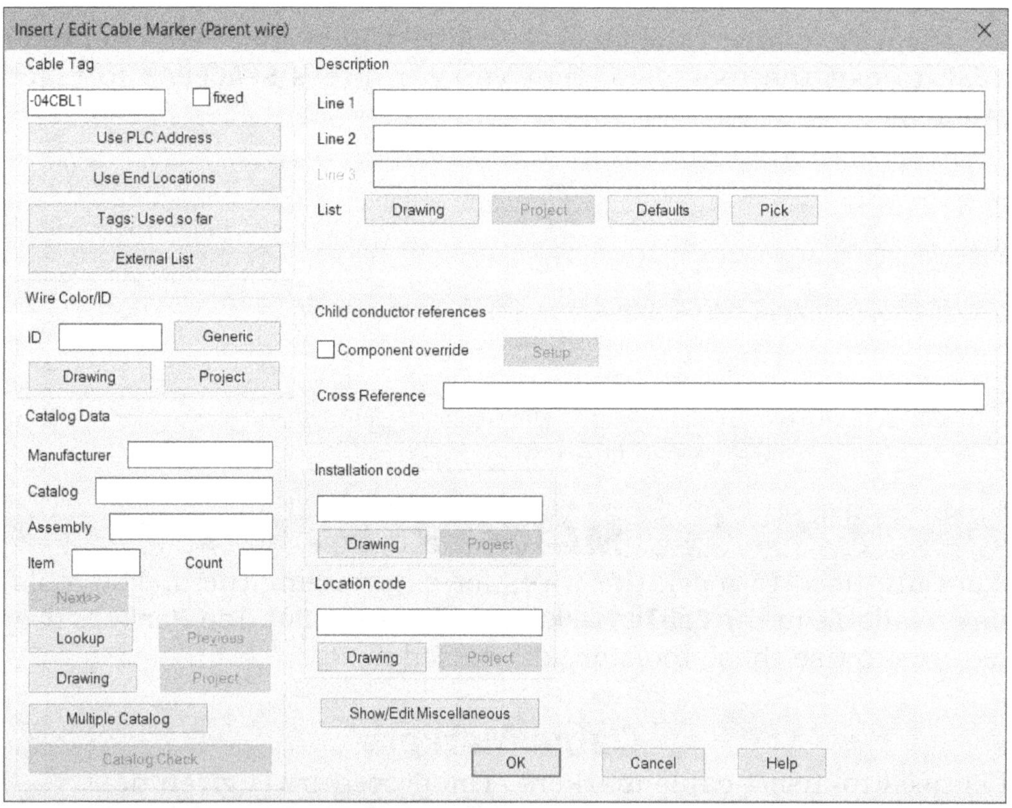

Figure-41. Insert or Edit Cable Marker dialog box

- Specify the parameters as desired and click on the **OK** button from the dialog box. The **Insert Some Child Components** dialog box will be displayed; refer to Figure-42.

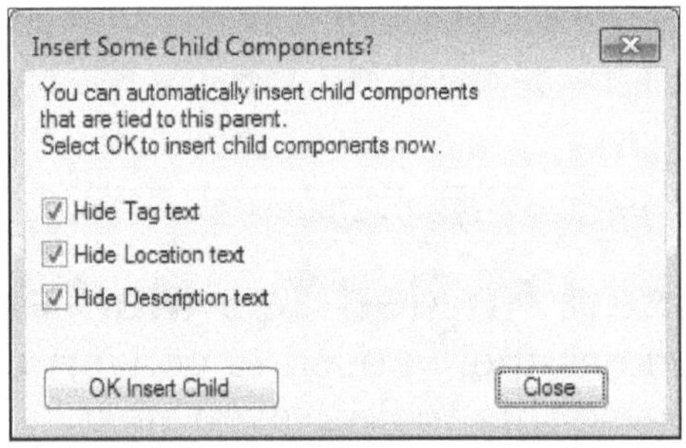

Figure-42. Insert Some Child Components dialog box

- If you want to insert child components then click on the **OK Insert Child** button otherwise click on the **Close** button from the dialog box.
- If you click on the **OK Insert Child** button then you need to click on the child wires and specify the tags as desired.

Multiple Cable Markers

The **Multiple Cable Markers** tool is used to insert multiple cable markers in the drawing. The procedure to use this tool is given next.

- Click on the **Multiple Cable Markers** tool from the **Cable Markers** drop-down. The **Multiple Cable Markers** dialog box will be displayed as shown in Figure-43.

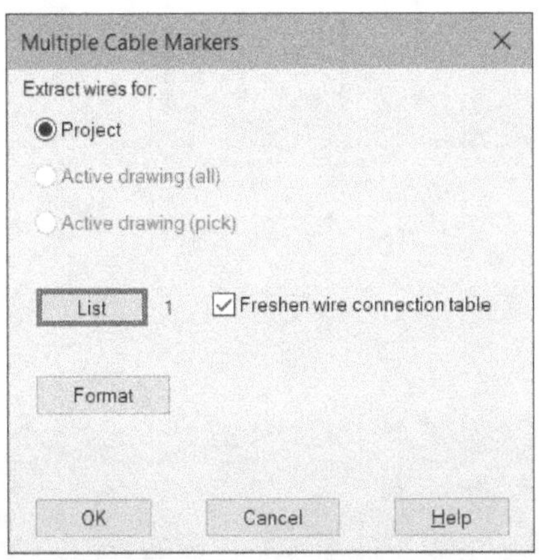

Figure-43. Multiple Cable Markers dialog box

- Select desired radio button from the dialog box and click on the **OK** button from the dialog box. The cable markers will be inserted automatically.

Insert Dot Tee Markers

The **Insert Dot Tee Markers** tool is used to insert dot tee mark on the wire. The procedure to use this tool is given next.

- Click on the **Insert Dot Tee Markers** tool from the **Insert Dot Tee Markers** drop-down in the **Ribbon**. You will be prompted to click on the wire to create the dot tee mark.
- Click on the wire at desired point. The dot tee mark will be created; refer to Figure-44.

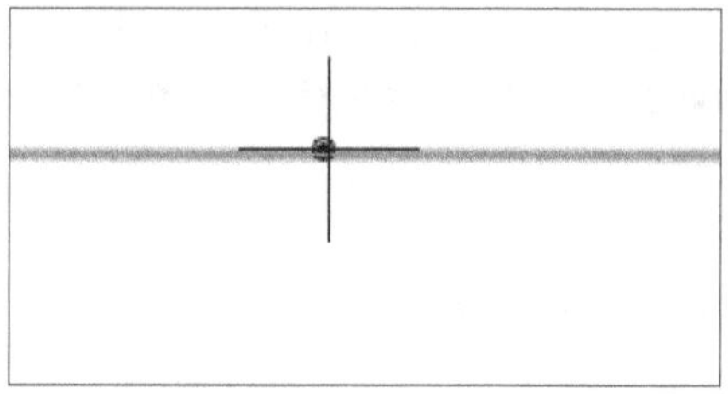

Figure-44. Dot tee mark

Insert Angled Tee Markers

The **Insert Angled Tee Markers** tool is used to convert the tee mark into an angled tee mark. The procedure to use this tool is given next.

- Click on the **Insert Angled Tee Markers** tool from the **Insert Dot Tee Markers** drop-down. You will be prompted to select the tee of wires.
- Click on the tee, it will be converted to angled tee mark; refer to Figure-45.

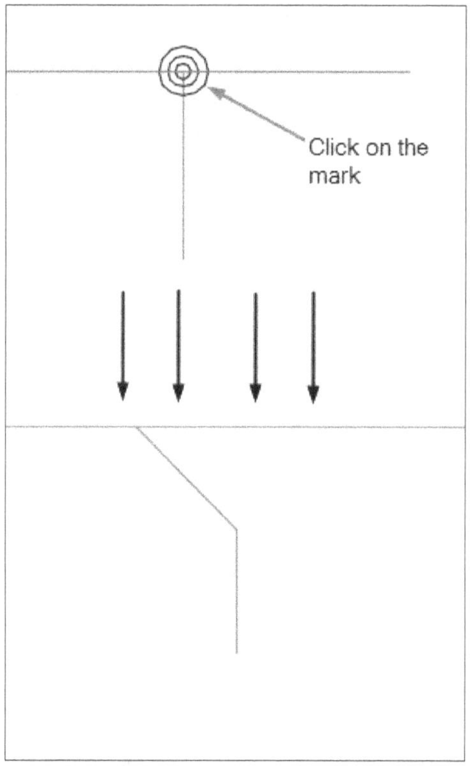

Figure-45. Angled tee

CIRCUIT BUILDER

The **Circuit Builder** tool is used to insert circuits in the drawing. The procedure to use this tool is given next.

- Click on the **Circuit Builder** tool from the **Circuit Builder** drop-down in the **Insert Components** panel in the **Ribbon**; refer to Figure-46. The **Circuit Selection** dialog box will be displayed as shown in Figure-47.
- Select desired circuit from the list.
- Click in the edit boxes of **Scale** area and specify the scale values for circuits and components.
- Specify the spacing between rungs in the **Rung Spacing** area.

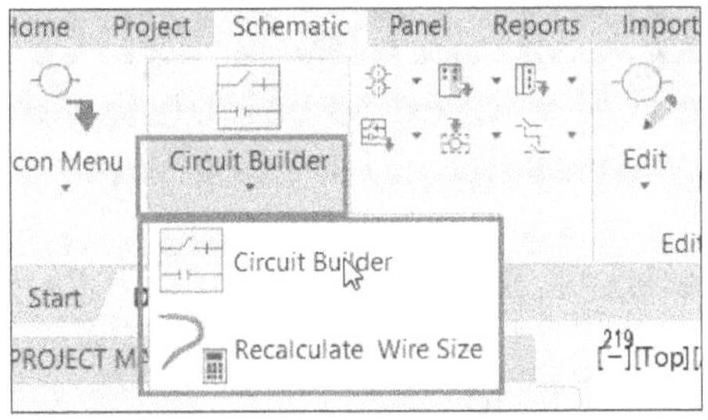

Figure-46. Circuit Builder tool

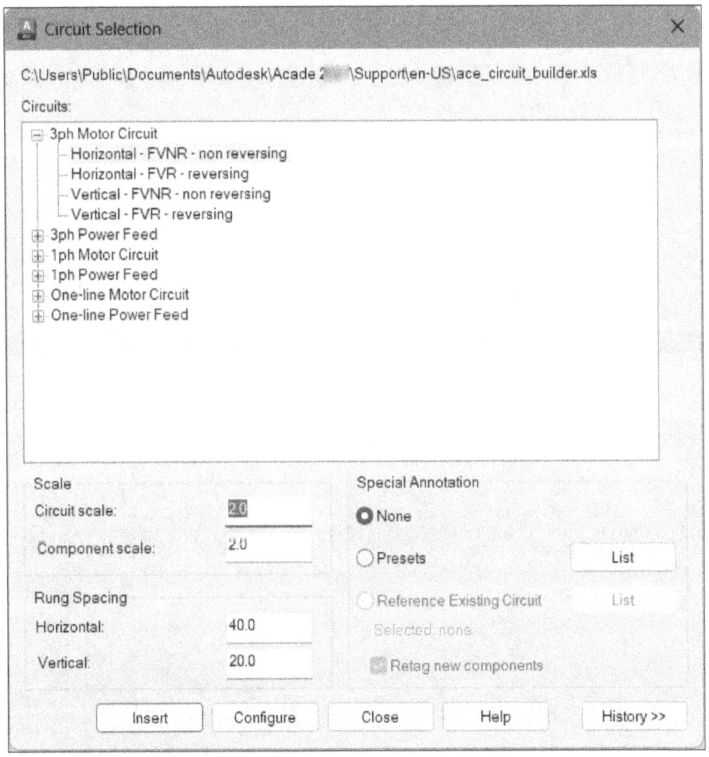

Figure-47. Circuit Selection dialog box

- Select desired radio button from the **Special Annotation** area of the dialog box.
- Click on the **Insert** button from the dialog box. You will be prompted to specify the insertion point of the circuit.
- Click to specify the insertion point, the circuit will be placed; refer to Figure-48.

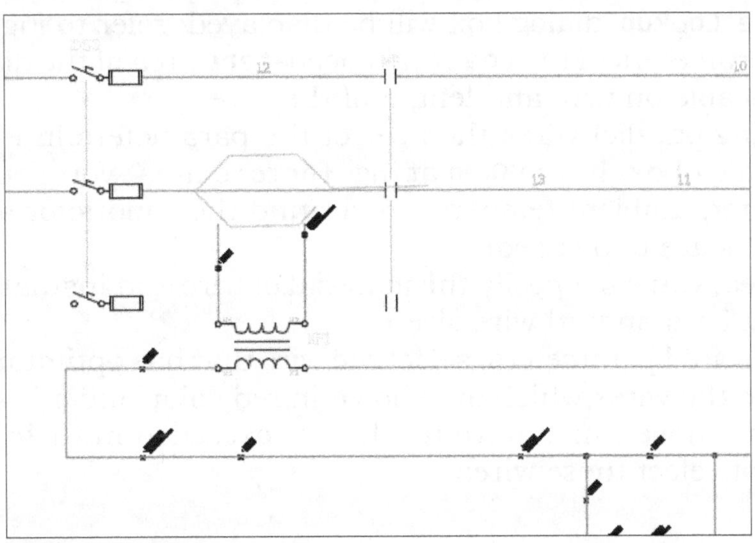

Figure-48. Circuit inserted

RECALCULATING WIRE SIZE

The **Recalculate Wire Size** tool is used to calculate the sizes of wires connected to selected load. The procedure to use this tool is given next.

- Click on the **Recalculate Wire Size** tool from the **Circuit Builder** drop-down in the **Insert Components** panel of **Schematic** tab in the **Ribbon**. You will be asked to select a motor or electrical load.
- Select the motor or electrical load for which you want to calculate the wire sizes. The **Select Motor** dialog box will be displayed; refer to Figure-49.

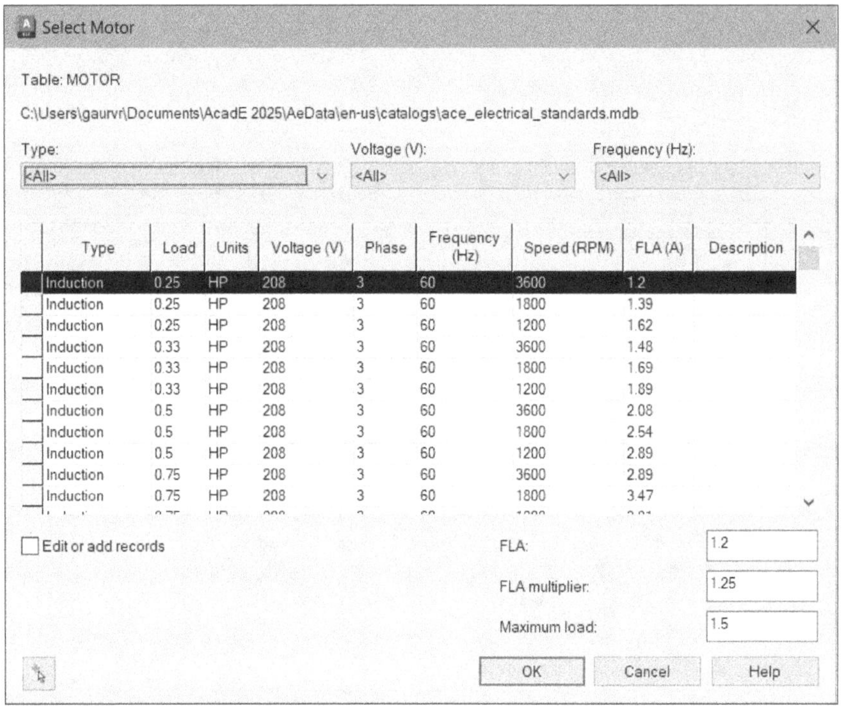

Figure-49. Select Motor dialog box

- Set desired parameters in filter drop-downs and then select the motor load from the table. If you want to create a new entry then select the **Edit or add records** check box.
- Click on the **OK** button from the dialog box after specifying desired parameters. The **Wire Size Lookup** dialog box will be displayed; refer to Figure-50.
- Set desired parameters in the **Load** and **Parameters** area of the dialog box to define the load applicable on wire and length of the wire.
- If you are using parallel wires then select the parameters in **Paralleled wires** area of the dialog box. In the **De-rating factors** area, you need to specify load correction factor, ambient temperature around the conductor and fill correction (if parallel wires are to be used).
- In the **Wire** area, you can specify the material of wire and insulation. You can also select the unit for standard wire size.
- After setting desired parameters, select the wire that has optimum capacity to bear load. Note that the wires which are shown in red color under **% Ampacity** column in the table are those which have reached or exceeded in their load capacity, so you should not select these wires.

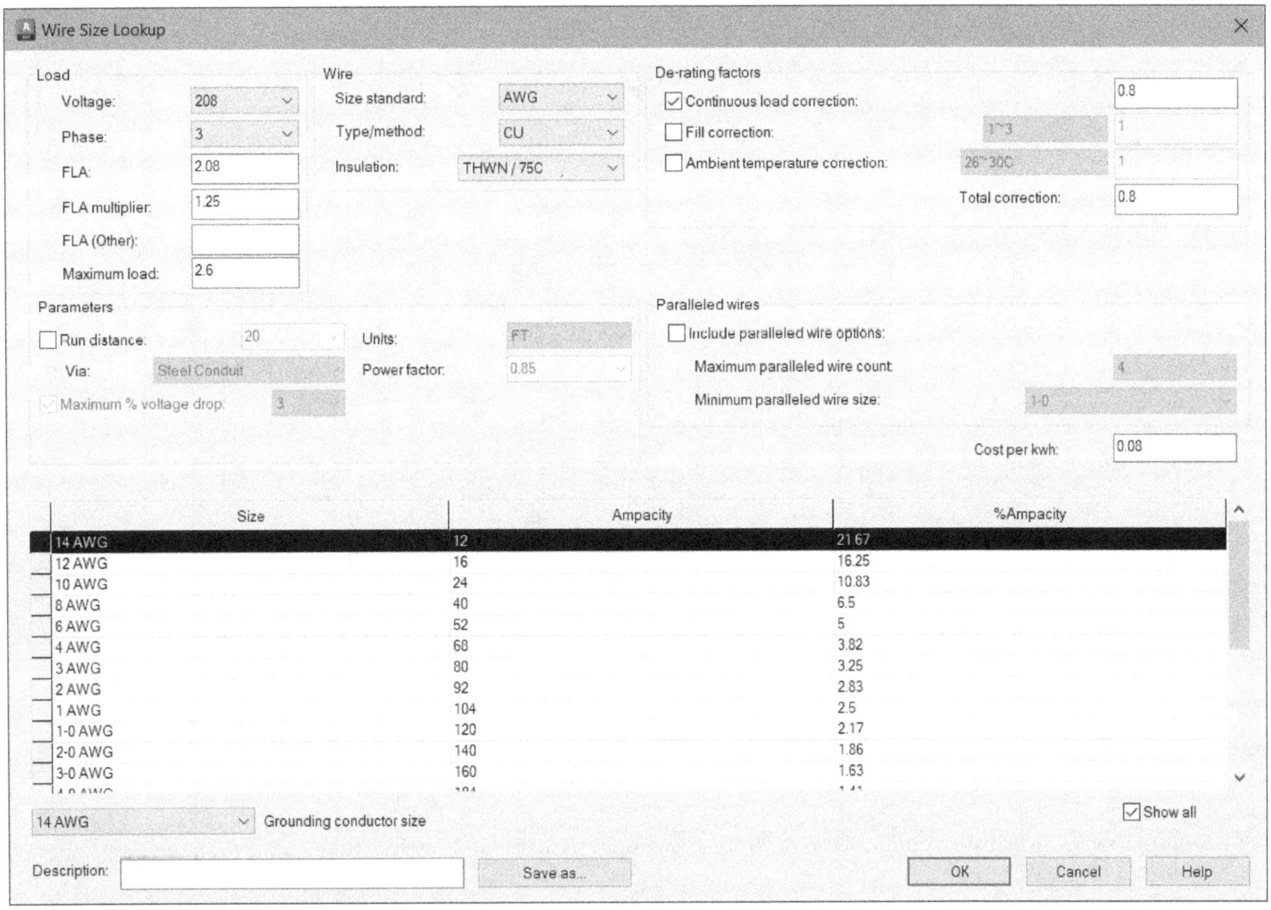

Figure-50. Wire Size Lookup dialog box

- After selecting desired wire, click on the **OK** button from the dialog box. The selected wire properties will be applied to wires connected to selected load/motor.

PRACTICAL

In this practical, you will create the circuit diagram of an electrical system; refer to Figure-51.

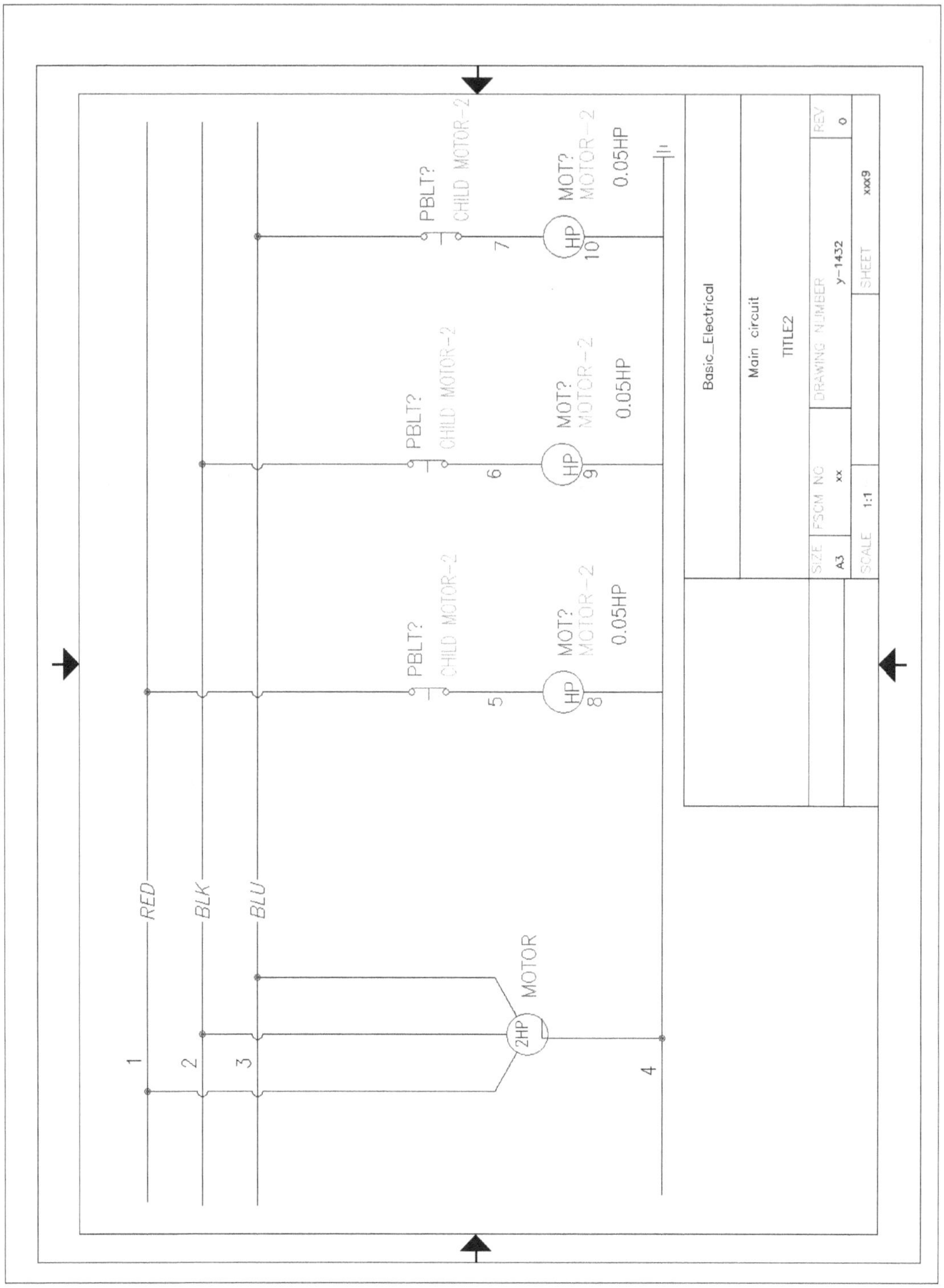

Figure-51. Practical

Starting a New drawing

- Start AutoCAD Electrical and click on the down arrow next to **New** button in the home page and select the **Browse templates** option; refer to Figure-52. The **Select Template** dialog box will be displayed.

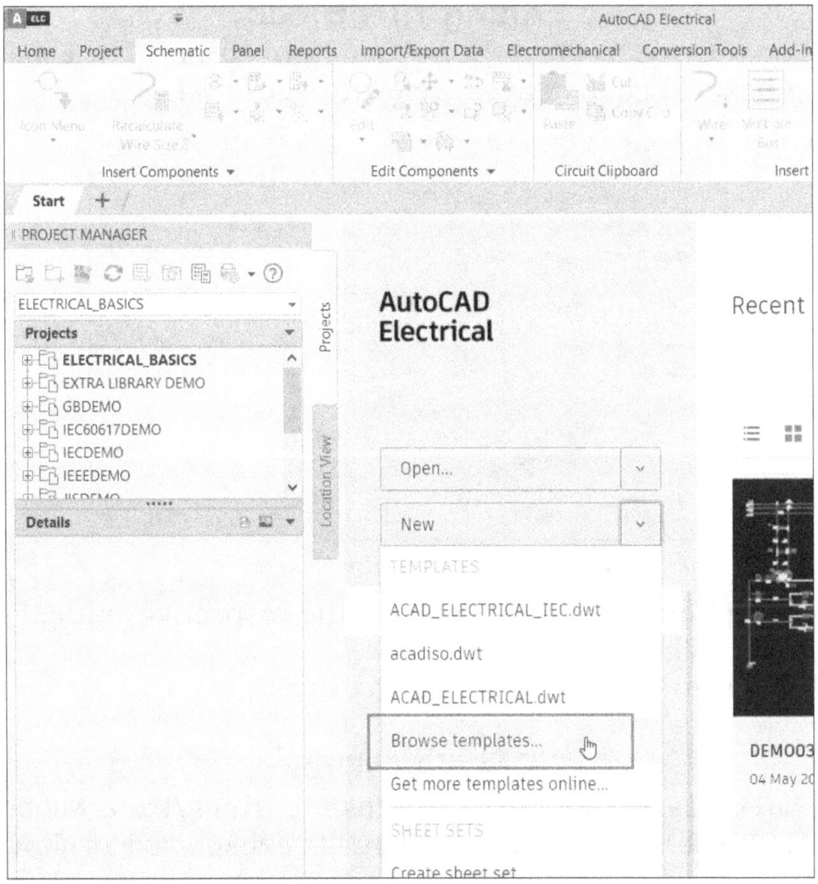

Figure-52. Browse templates option

- Select the **ACE ANSI A(Landscape) Color.dwt** template from the list of templates and click on the **Open** button from the dialog box. A new drawing will open with selected template; refer to Figure-53.

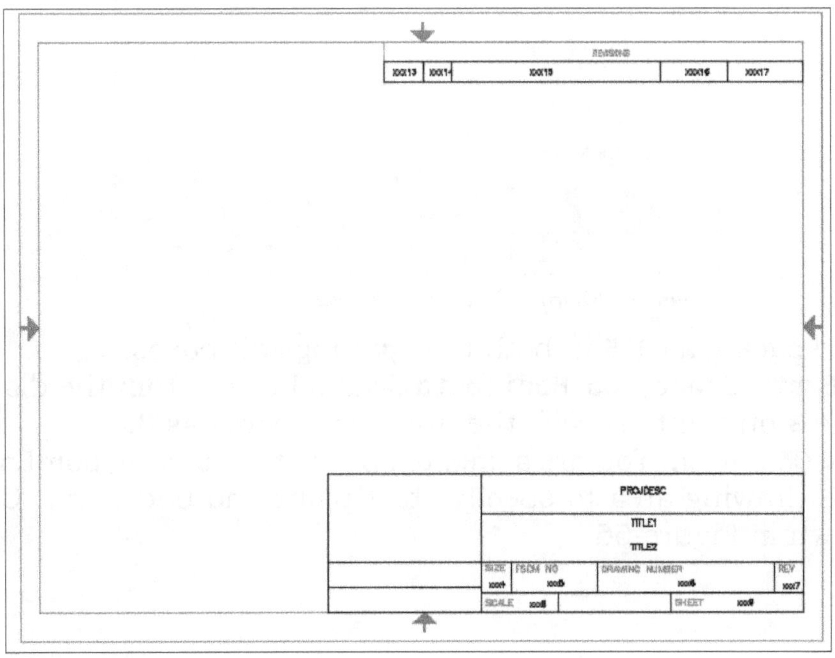

Figure-53. Drawing opened with the template

Editing Title Block

- Double-click on the title block at the bottom right corner. The **Enhanced Attribute Editor** dialog box will be displayed; refer to Figure-54.

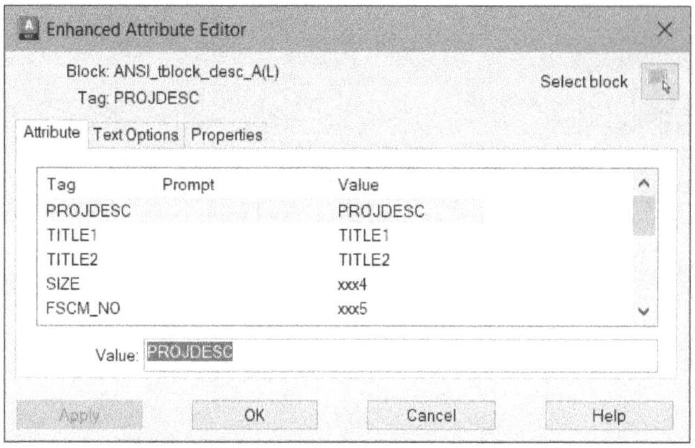

Figure-54. Enhanced Attribute Editor dialog box

- Select desired tag from the list and specify the respective value. For example title of drawing can be WirePractical.
- Click on the **OK** button from the dialog box to exit.

Creating Wires

- Click on the **Multiple Bus** tool from the **Insert Wires/Wire Numbers** panel in the **Ribbon**. The **Multiple Wire Bus** dialog box will be displayed as shown in Figure-55.

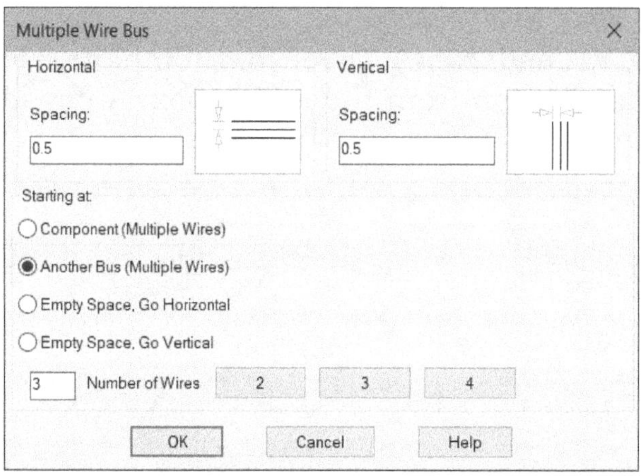

Figure-55. Multiple Wire Bus dialog box

- Specify the spacing as **0.5** in both the **Spacing** edit boxes.
- Select the **Empty Space, Go Horizontal** radio button from the dialog box.
- Click on the **3** button to specify the number of wires as 3.
- Click on the **OK** button. You are asked to specify the starting point of the wire bus.
- Click in the drawing area to specify start point and end point. Create the wire bus as shown in Figure-56.

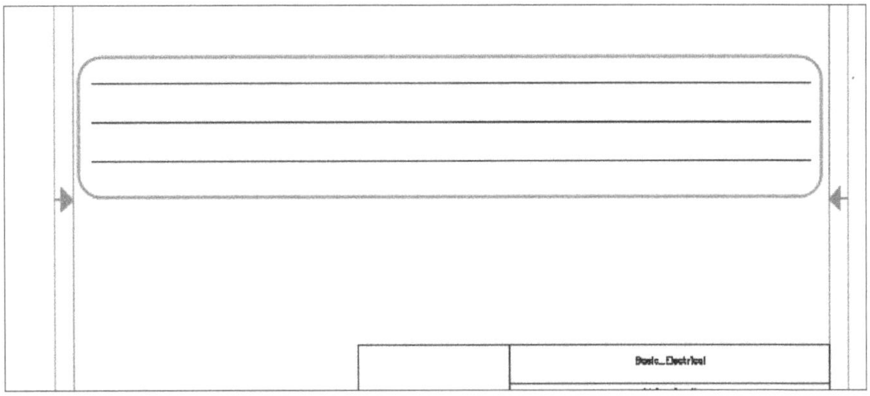

Figure-56. Wire bus to be created

- Similarly, create two more buses each having 3 wires with proper spacing connected to earlier created bus using **Another Bus (Multiple Wires)** method; refer to Figure-57.

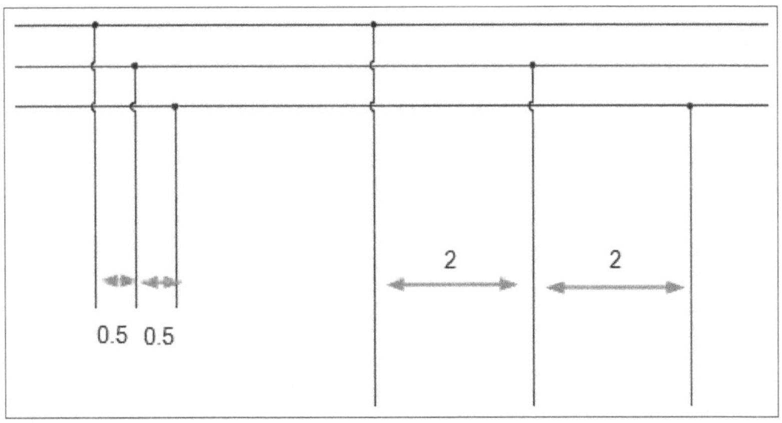

Figure-57. Other two wire buses

- Click on the **Wire** tool from **Insert Wires/Wire Numbers** panel and create a wire for earth; refer to Figure-58.

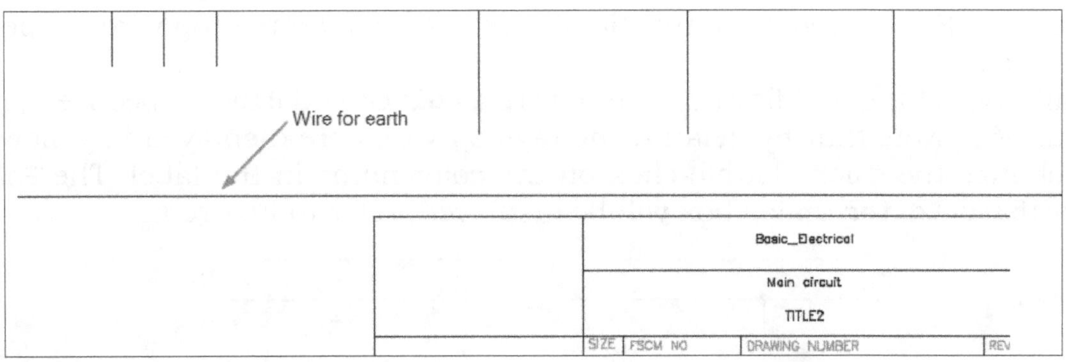

Figure-58. Wire created for earth

Assigning Numbers and Labels to Wires

- Click on the **In-Line Wire Labels** tool from the **Wire Number Leader** drop-down; refer to Figure-59. The **Insert Component** dialog box with wire labels will be displayed; refer to Figure-60.

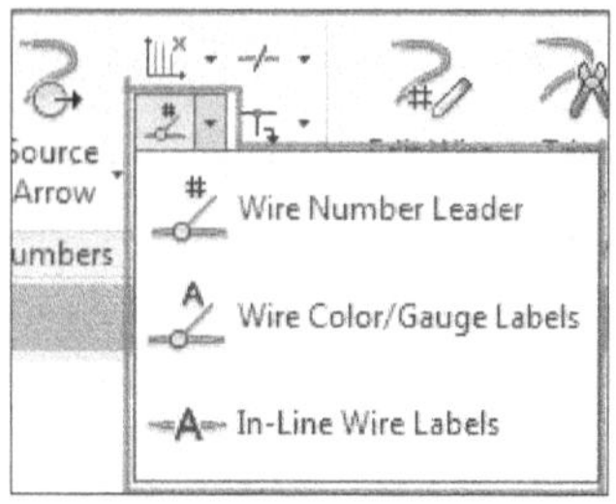

Figure-59. Wire Number Leader drop-down

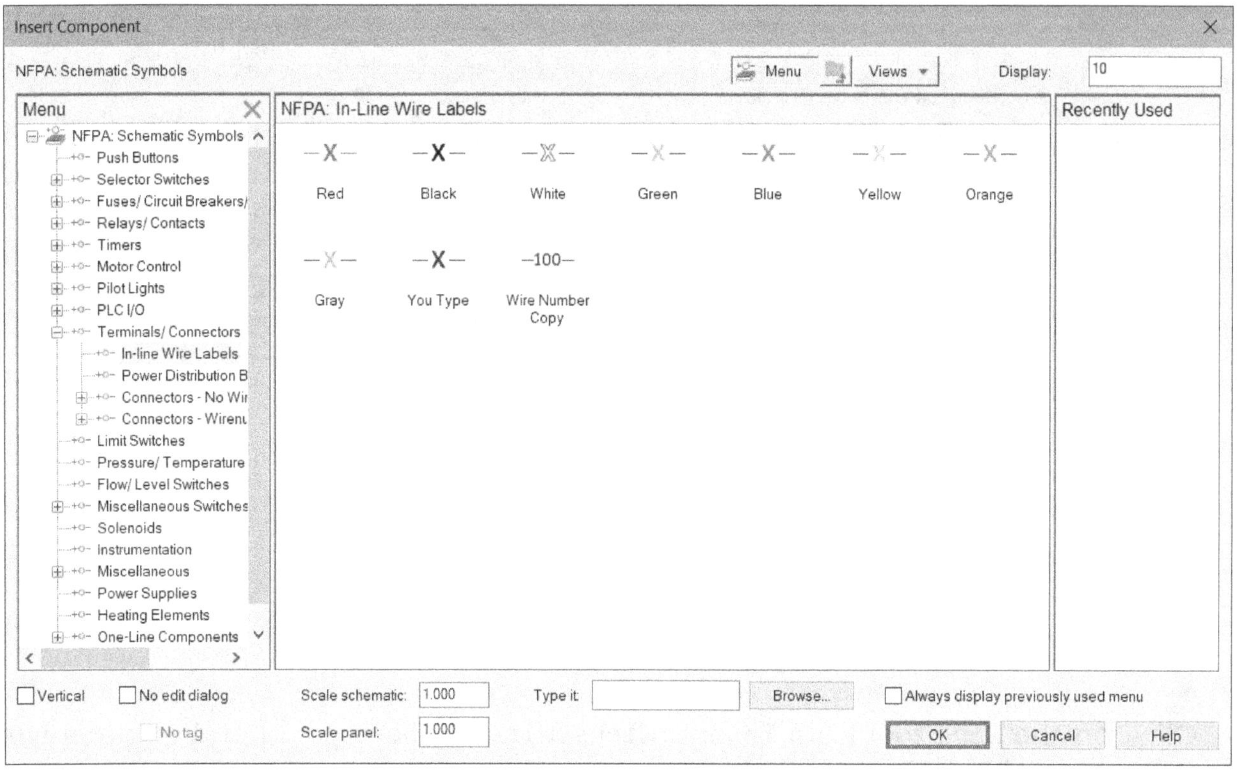

Figure-60. Insert Component dialog box

- Select the Red symbol from the dialog box and click on the top wire to specify its color code.
- Similarly, set the middle and bottom wire as **Black** and **Blue** respectively; refer to Figure-61. Note that by default, the tags on wires are displayed in yellow color. To change the color, double-click on the color name in the label. The **Enhanced Attribute Editor** dialog box will be displayed; refer to Figure-62.

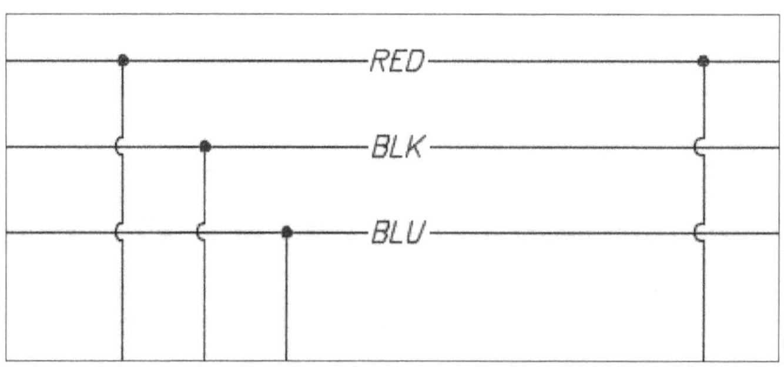

Figure-61. Wires after color coding

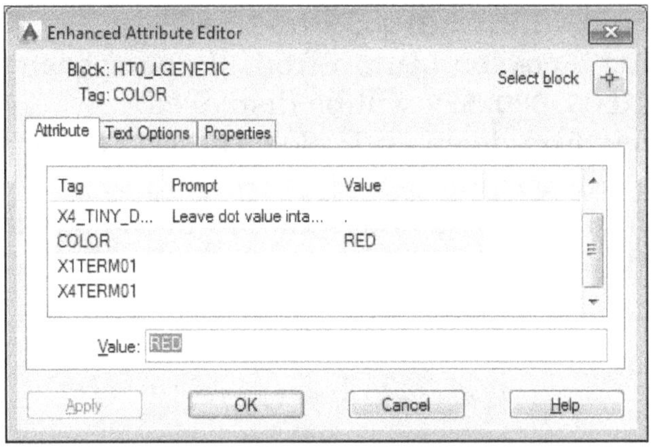

Figure-62. Enhanced Attribute Editor dialog box for color

- Click on the **Properties** tab and change the color of layer as per your requirement.
- Click on the **OK** button from the dialog box.
- Click on the **Wire Numbers** tool from the **Wire Numbers** drop-down. The **Wire Tagging** dialog box will be displayed as shown in Figure-63.

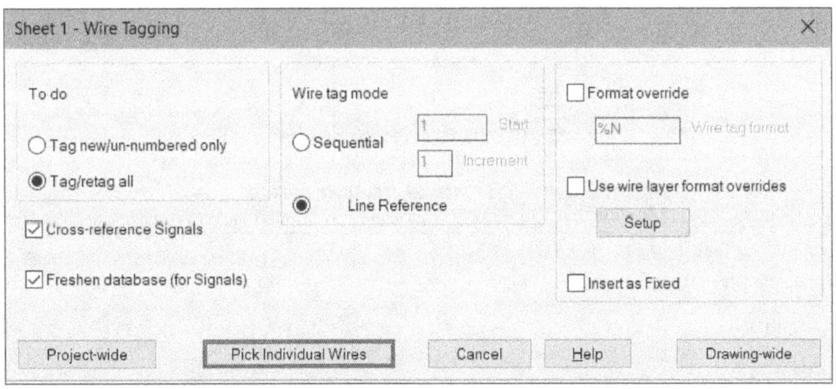

Figure-63. Sheet1 Wire Tagging dialog box

- Select the **Sequential** radio button from the dialog box and click on the **Drawing-wide** button from the dialog box. The wire numbers will be assigned to the wires; refer to Figure-64.

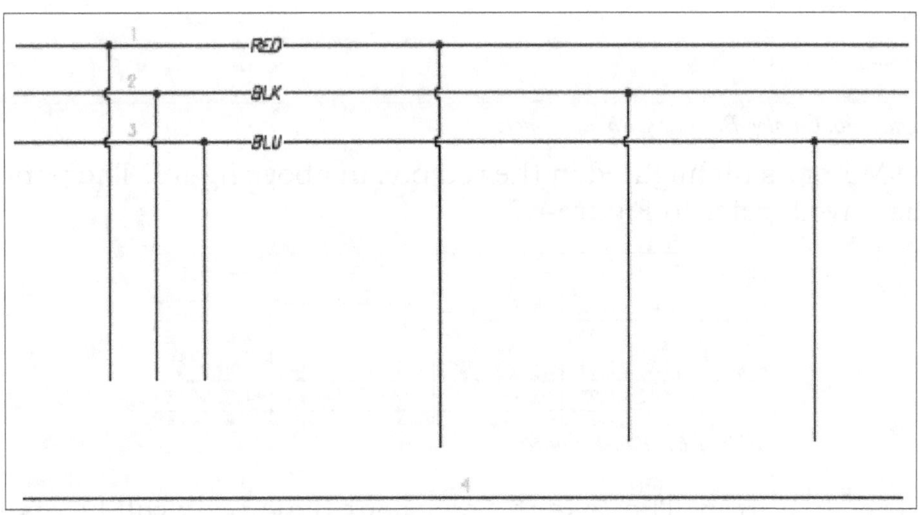

Figure-64. Wire numbers assigned to the wires

- You can change the color of wire numbers as we did for the wire labels.

Inserting 3 Phase Motor

- Click on the **Catalog Browser** button from the **Components** drop-down; refer to Figure-65. The **CATALOG BROWSER** will be displayed.
- Click in the **Category** drop-down and select the **MO (Motors)** option from the list.
- Click in the **Search** edit box and specify **4-Pole** in it.
- Click on the **Search** button. The list of motors in the catalog will be displayed as shown in Figure-66.

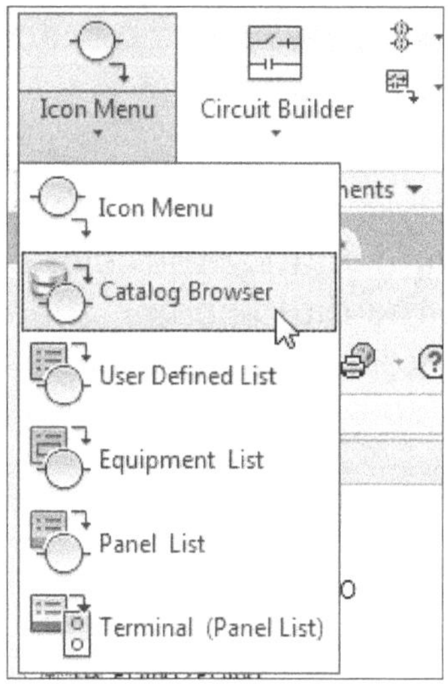

Figure-65. Catalog Browser button

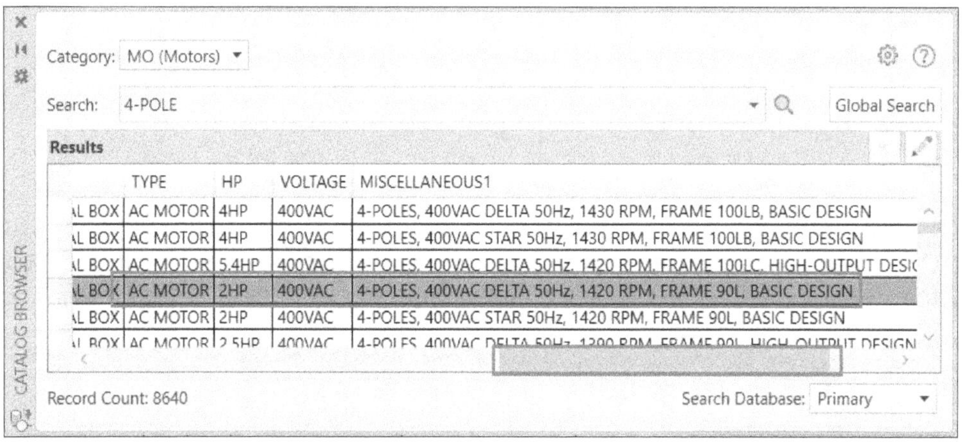

Figure-66. Catalog Browser with list of motors

- Select the Motor as highlighted in the red box in above figure. The panel of buttons will be displayed; refer to Figure-67.

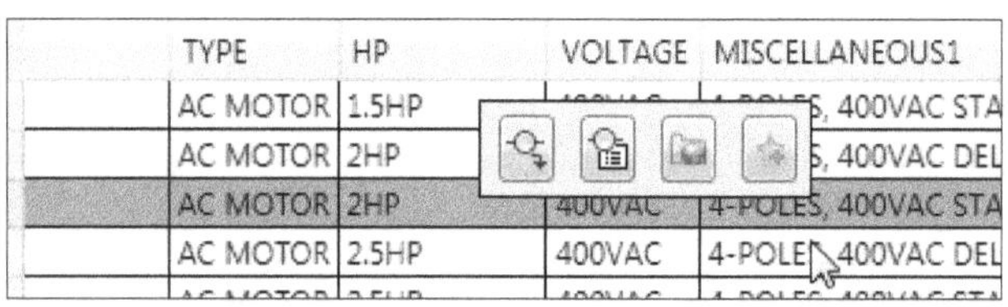

Figure-67. Panel of buttons

- Click on the  button. The **Insert Component** dialog box will be displayed; refer to Figure-68.
- Select the **Motor Control** category from the menu and select the icon highlighted by the cursor; refer to Figure-69.

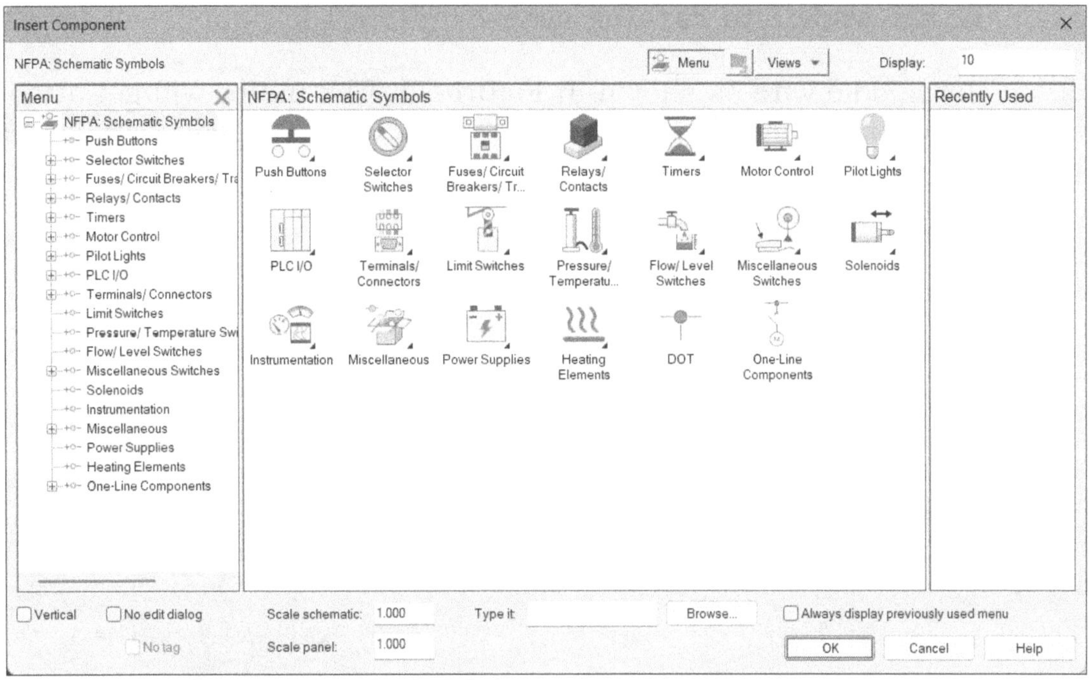

Figure-68. Insert Component dialog box by default

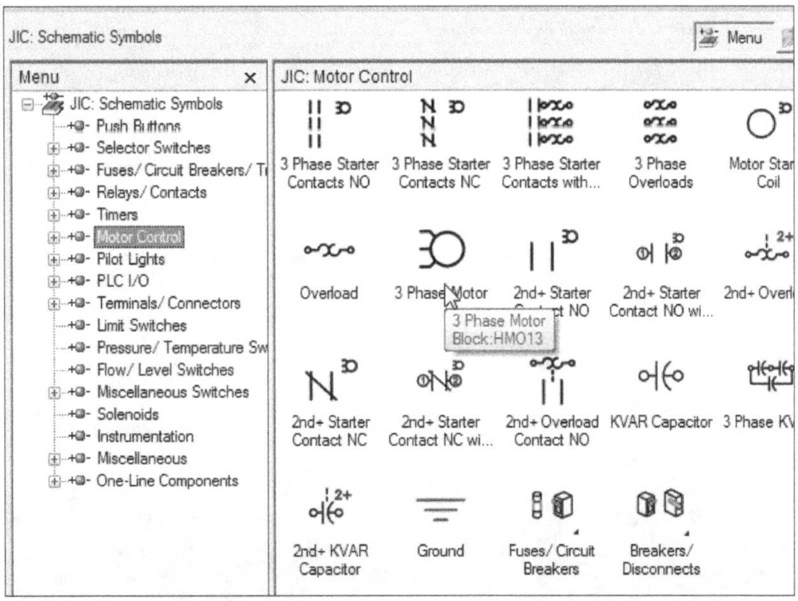

Figure-69. Motor symbol

- On doing so, the **Assign Symbol To Catalog Number** dialog box will be displayed; refer to Figure-70.

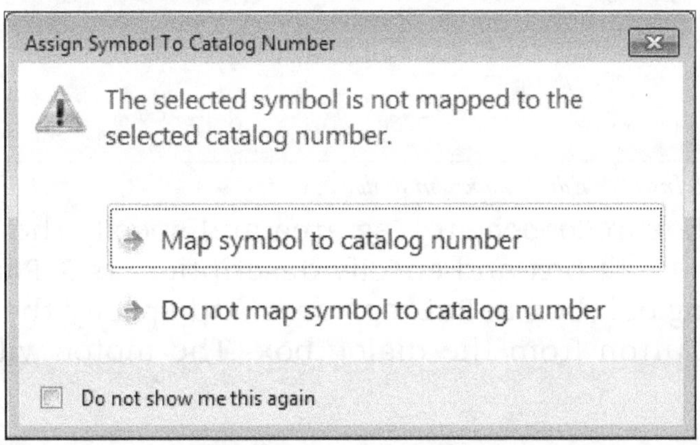

Figure-70. Assign Symbol To Catalog Number dialog box

- Click on the **Map symbol to catalog number** button. The symbol will get attached to cursor.
- Click on the middle wire as shown in Figure-71. The motor will get attached to the three wires automatically and the **Insert / Edit Component** dialog box will be displayed as shown in Figure-72.

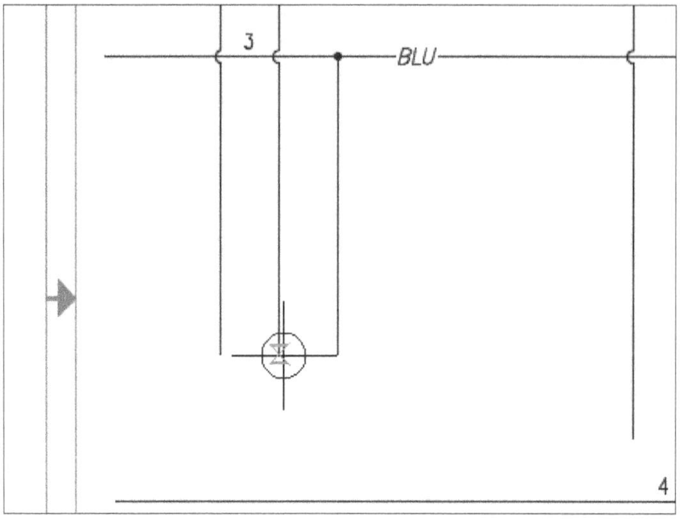

Figure-71. Motor to be placed

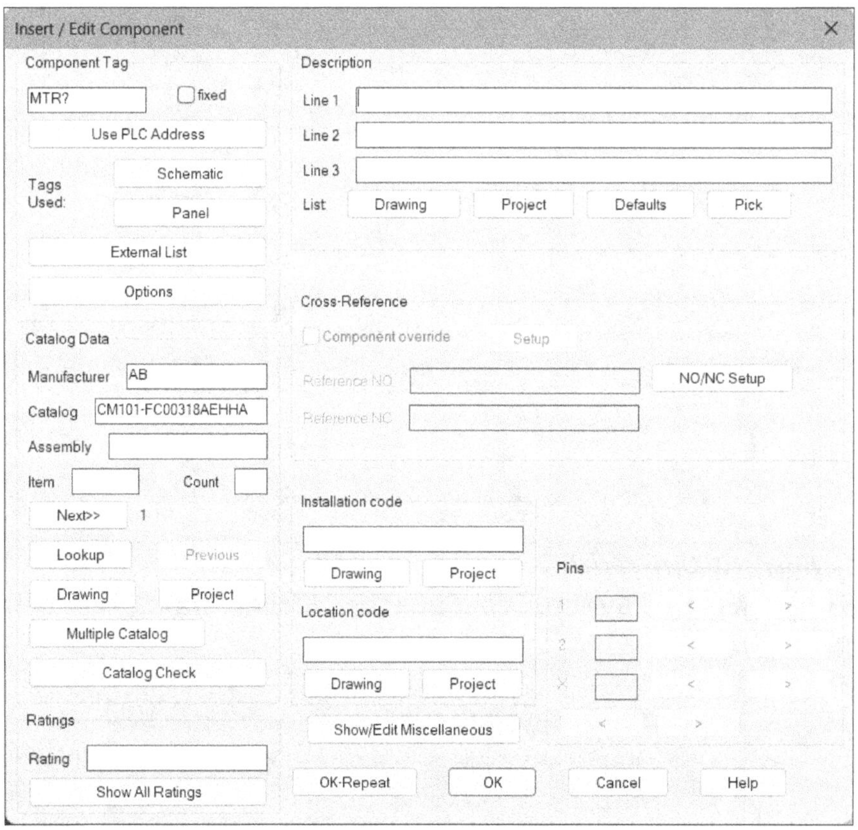

Figure-72. Insert or Edit Component dialog box

- Click in the edit box in **Component Tag** area and specify the tag as **Motor**.
- Click in the **Line 1** edit box and specify description as **3 Phase**.
- Click in the **Rating** edit box in **Ratings** area and specify the rating as **2 HP**.
- Click on the **OK** button from the dialog box. The motor will be placed; refer to Figure-73.

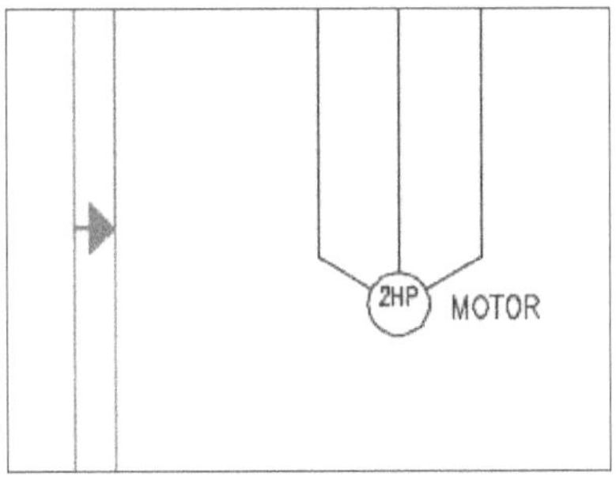

Figure-73. Motor placed

Adding Ground symbol

- Click on the **Icon Menu** button from the **Icon Menu** drop-down. The **Insert Component** dialog box will be displayed.
- Select the **Motor Control** category from the **Menu** area and click on the **Ground** symbol from the right area of the dialog box; refer to Figure-74. You are prompted to specify the location of the ground symbol.

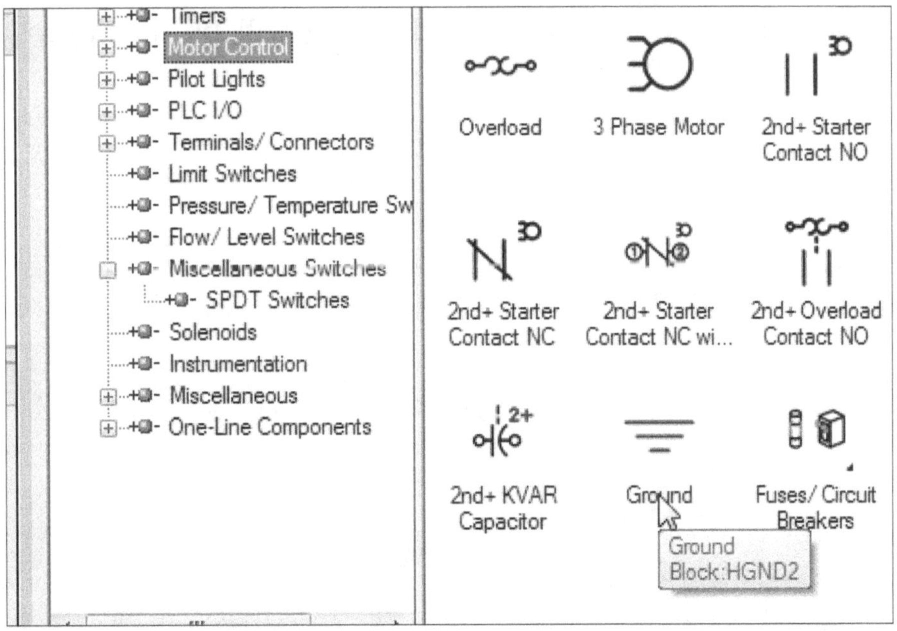

Figure-74. Ground symbol

- Click at the end of wire with 4 wire number; refer to Figure-75.

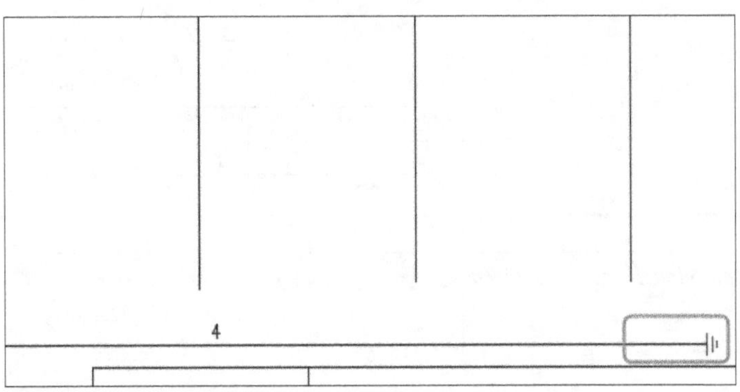

Figure-75. After placing ground symbol

Adding symbols for various components

- Click on the **Catalog Browser** button and insert the components as shown in Figure-76.

- Connect the **2 HP Motor** to the ground wire using the **Wire** tool; refer to Figure-77.

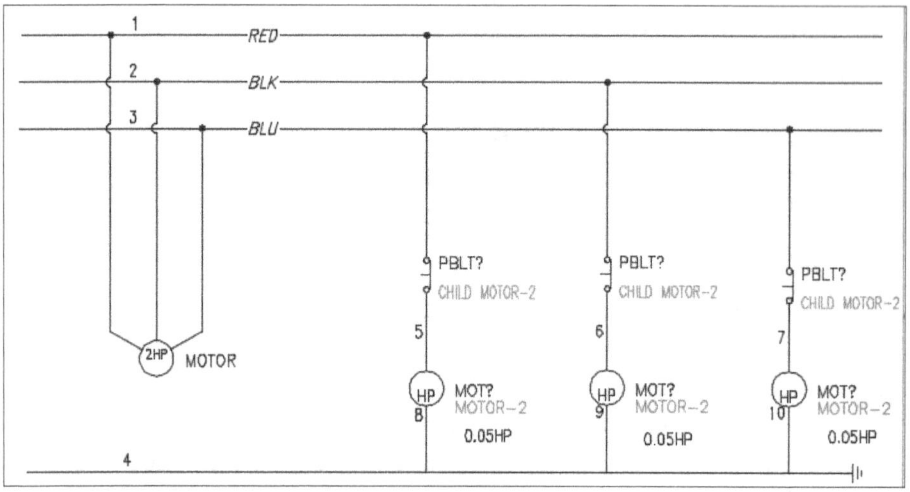

Figure-76. Circuit after connecting components

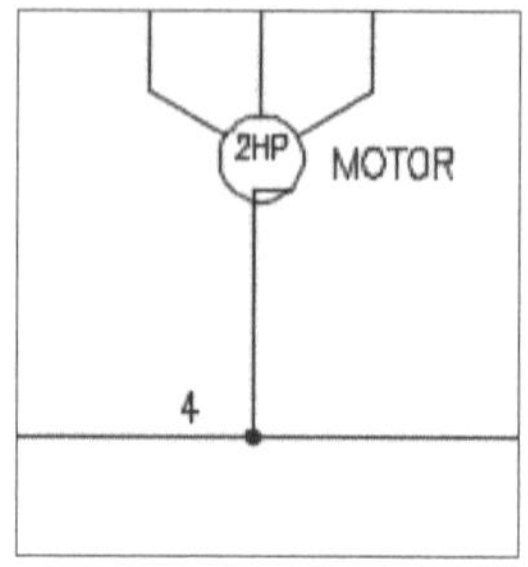

Figure-77. Wire connected to motor

- After adding all the components, the drawing will be displayed as shown in Figure-78.

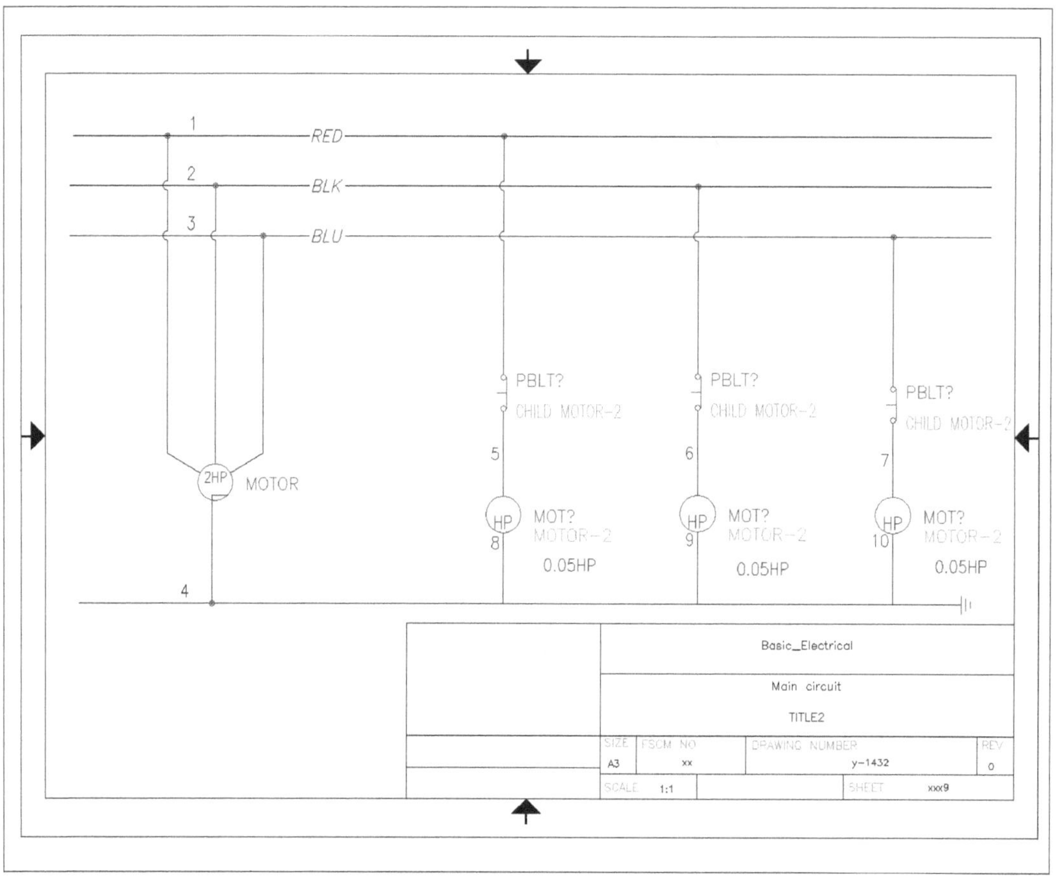

Figure-78. Practical-Model

PRACTICE

Create the circuit diagram as shown in Figure-79.

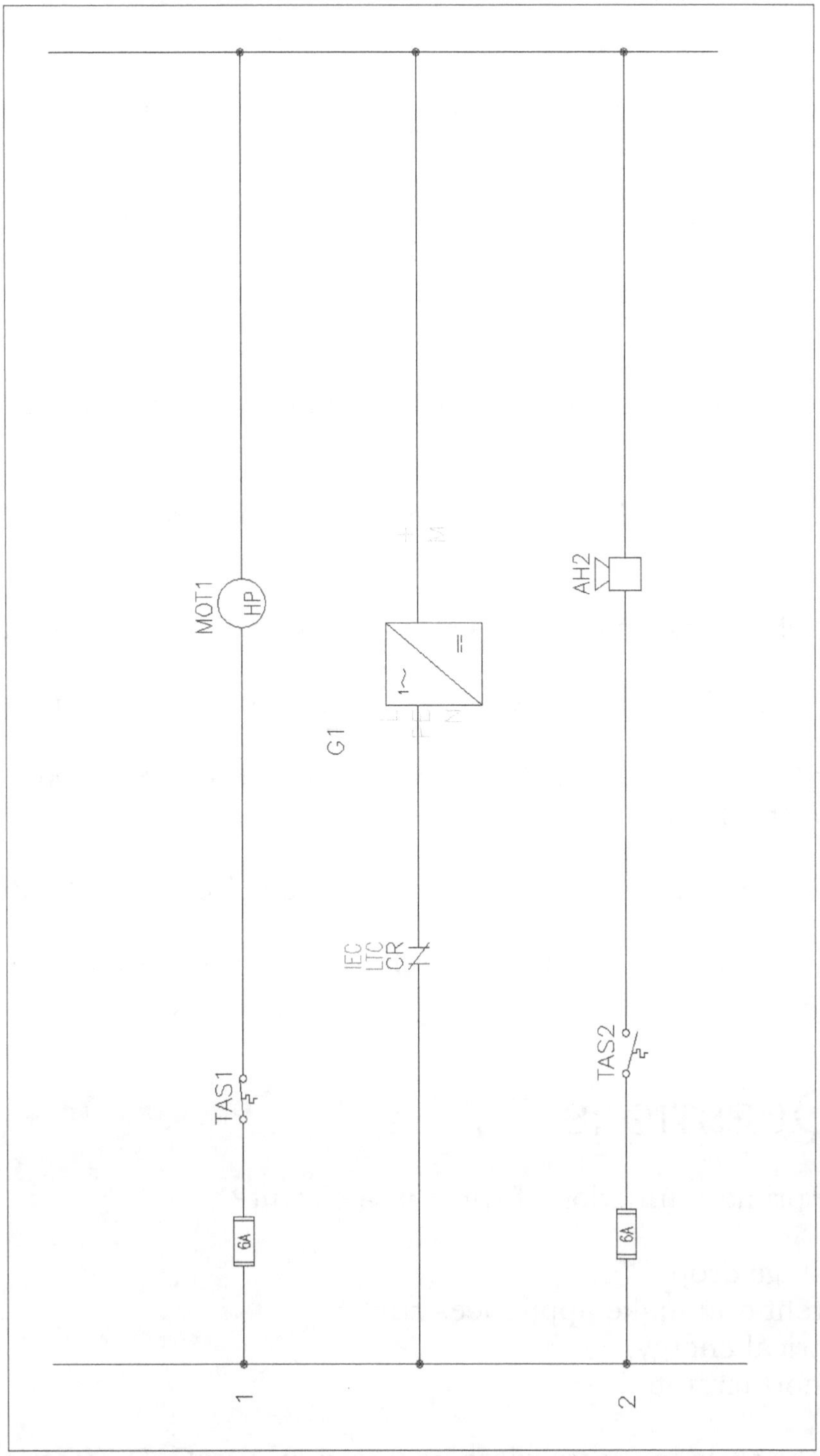

Figure-79. Problem-Model

Self-Assessment

Q1. Once you have selected the **Wire** tool, you can go on creating the wires until you end up at a connector of a component. From there you need to start a new wire. (T/F)

Q2. The components to be connected using **Interconnect Components** tool may or may not be aligned to each other for functioning of the tool. (T/F)

Q3. The **Gap** tool is used to check possible gap in the wiring. (T/F)

Q4. The .... radio button is used to create a multiple wire bus by using an already existing wire bus.

Q5. Multiple circuits that are powered by a common power source can be combined in the form of ....

Q6. Select the .... radio button if you want to specify the wire number to unnumbered wires only.

Q7. The **Wire Number Leader** tool is used to assign leader to a wire number. (T/F)

Q8. To change the leader style, type .... in the command prompt.

Q9. The **Wire Color/Gauge Labels** tool is available in the **Wire Number Leader** drop-down in the **Ribbon**. (T/F)

Q10. The .... tool is used to insert already made circuits in the drawing.

Q11. The .... tool is used to recalculate the size of wire based on connected load.

Review Questions

Q1. What is the primary function of wires in a circuit?

A. To reduce voltage drop
B. To carry current and make appliances run
C. To store electrical energy
D. To prevent short circuits

Q2. Where can the tools to create wires be found in AutoCAD Electrical?

A. Format tab
B. Home tab
C. Wire drop-down in the Insert Wires/Wire Numbers panel of the Schematic tab
D. Drawing Properties dialog box

Q3. What does the 'Wire' tool specifically do?

A. Adds curved wires
B. Connects ladders automatically
C. Adds straight wires in the drawing to connect components
D. Assigns wire numbers

Q4. What is the purpose of the "Tab=Collision off" button?

A. To turn on wire snapping
B. To hide components while wiring
C. To make wires avoid collisions with components
D. To align wires automatically

Q5. Which tools allow creation of wires at predefined angles?

A. Wire Numbers and 3 Phase
B. Interconnect and Gap
C. 22.5 Degree, 45 Degree, and 67.5 Degree tools
D. Multiple Bus and Component Wire

Q6. What is the function of the Interconnect Components tool?

A. It creates wires at custom angles
B. It removes unnecessary components
C. It connects two components that are properly aligned
D. It changes the type of a component

Q7. What does the Gap tool do?

A. Hides intersecting wires
B. Makes wires thicker at intersections
C. Creates a visible gap where two wires intersect
D. Merges wires automatically

Q8. What does the 'Multiple Bus' tool help to create?

A. One long wire
B. A circuit breaker
C. Multiple lines of wires in a bus
D. Only horizontal wires

Q9. What happens when the 'Another Bus (Multiple Wires)' radio button is selected?

A. New wires are added parallel to a component
B. The wires are created from scratch
C. Wires are created from an existing wire bus
D. Wire bus is converted to single line

Q10. What is the purpose of selecting the 'Empty Space, Go Horizontal' radio button?

A. To create a ladder
B. To insert components
C. To create multiple wires horizontally from a point
D. To gap intersecting wires

Q11. What do ladder diagrams typically represent?

A. Electrical insulation
B. Physical wire lengths
C. Multiple circuits powered by a common source
D. Component lists

Q12. In the Insert Ladder dialog box, what does the '3 Phase' radio button do?

A. Sets the wire thickness
B. Changes to single line diagram
C. Creates ladder with three phase wire lines
D. Enables collision detection

Q13. What must be enabled before using the X-Y Grid Setup tool?

A. Insert Zone Labels option
B. Default Ladder Format
C. X-Y Grid format referencing in Drawing Properties
D. Grid Snap setting

Q14. What is the function of the X Zones Setup tool?

A. Creates reference zones horizontally
B. Connects wires to buses
C. Draws ladder rungs
D. Creates vertical reference zones for components

Q15. What is the main benefit of using wire numbering?

A. Increases voltage
B. Helps identify wire details and connected components
C. Decreases resistance
D. Prevents circuit overload

Q16. Which tool is used to assign wire numbers to individual or all wires in a project?

A. Gap
B. Insert Ladder
C. Wire Numbers
D. Interconnect Components

Q17. What does the "Tag new/un-numbered only" option do?

A. Renames existing wire numbers
B. Adds numbers to already numbered wires
C. Assigns numbers only to wires that haven't been numbered yet
D. Deletes wire numbers

Q18. What is the final step to apply a custom wire layer format?

A. Click on the Gap tool
B. Press ESC twice
C. Select the "Use wire layer format override" checkbox
D. Choose Insert Ladder again

Q19. How are components selected for creating a multiple wire bus using the 'Component (Multiple Wires)' option?

A. Click one by one manually
B. Select via their reference names
C. Create a selection window to select ports
D. Type their coordinates

Q20. What happens if extra wires are specified in 'Another Bus' option?

A. They are ignored
B. AutoCAD warns the user
C. They connect to the first wire in the existing bus
D. The command fails

Q21. What is the purpose of the 3 Phase tool in AutoCAD Electrical?
A. To assign color codes to wires
B. To draw three wires simultaneously
C. To number three phase wiring
D. To label PLC components

Q22. In the 3 Phase Wire Numbering dialog box, which field is NOT required to be filled?
A. Prefix
B. Layer
C. Base
D. Suffix

Q23. What happens after selecting all wires for 3 Phase numbering?
A. Labels are printed
B. Dialog box disappears
C. You must click Cancel to exit
D. Wires are removed

Q24. What is the main function of the PLC I/O tool?
A. To assign leader lines
B. To label color codes
C. To number input/output ports of PLC
D. To draw PLC blocks

Q25. What does the Wire Number Leader tool allow you to do?
A. Assign numbering format
B. Assign a leader to a wire number
C. Create rungs for ladder
D. Place markers on wires

Q26. Which command is used to modify the style of wire number leaders?
A. STYLEWIRE
B. MLSTYLE
C. MLEADERSTYLE
D. LEADERFORMAT

Q27. What does the Wire Color/Gauge Labels tool help with?
A. Identifying phase wires
B. Determining length of wire
C. Attaching color/gauge labels to wires
D. Connecting components

Q28. Which button must be clicked for automatic placement of color/gauge labels?
A. Auto Assign
B. Auto Placement
C. Apply Labels
D. Quick Insert

Q29. What does the In-Line Wire Labels tool insert?
A. Angled markers
B. Vertical connectors
C. Labels inside the wire line
D. Dashed lines

Q30. What are cable markers mainly used for?
A. To display grid origin
B. To insert child components
C. To categorize wires
D. To create wire gaps

Q31. What additional step can be taken after inserting a cable marker?
A. Insert a breaker
B. Assign a bus
C. Insert child components
D. Place a resistor

Q32. What is the function of the Insert Dot Tee Markers tool?
A. Create multiple wire buses
B. Insert circular symbol at wire junction
C. Connect ladder rungs
D. Measure wire spacing

Q33. How do you create an angled tee mark?
A. Use Angled Wire tool
B. Select angled junction from Ribbon
C. Use Insert Angled Tee Markers tool
D. Press ALT + T

Q34. What is the function of the Circuit Builder tool?
A. Automatically draw PLCs
B. Label all wires
C. Insert circuits into drawing
D. Create custom components

Q35. What parameter is NOT adjusted in Circuit Builder dialog box?
A. Ladder count
B. Scale
C. Rung Spacing
D. Special Annotations

Q36. Which tool calculates sizes of wires for a given load?
A. Load Calculator
B. Recalculate Wire Size
C. Auto Gauge Estimator
D. Ampacity Checker

Q37. What are red-colored wires in the % Ampacity column?
A. Wires without labels
B. Wires that have optimal capacity
C. Wires nearing or exceeding capacity
D. Wires below threshold

Q38. In Recalculate Wire Size tool, which parameter is NOT user-defined?
A. Ambient temperature
B. Wire insulation
C. Rung count
D. Load correction factor

Q39. What happens after selecting a wire in Recalculate Wire Size tool?
A. Color is changed
B. Dialog closes without change
C. Wire settings are applied
D. Load is reset

Q40. Where is the Recalculate Wire Size tool located?
A. Wire Number Leader drop-down
B. Cable Marker drop-down
C. Insert Components panel
D. Grid Setup dialog box

Answer for Self-Assessment:
Ans1. T, Ans2. F, Ans3. F, Ans4. Another Bus (Multiple Wires), Ans5. ladders, Ans6. Tag new/un-numbered only, Ans7. T, Ans8. MLEADERSTYLE Ans9. T, Ans10. Circuit Builder, Ans 11. Recalculate Wire Size

Chapter 6

Editing Wires, Components, and Circuits

Topics Covered

The major topics covered in this chapter are:

- ***Editing Components***
- ***Internal Jumpers***
- ***Editing Components like fixing tags, copying catalog assignments***
- ***Editing Circuits***
- ***Editing Wires and Wire Numbers***

INTRODUCTION

Till this point, you have learned to insert components, wires, and circuits. Now, we are ready to edit the components, wires and circuits. The tools for editing are available in three panels; **Edit Components**, **Circuit Clipboard**, and **Edit Wires/Wire Numbers** panel; refer to Figure-1, Figure-2, and Figure-3.

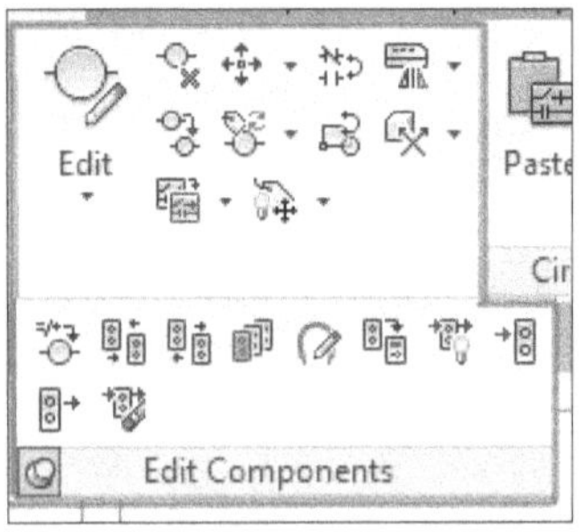

Figure-1. Edit Components panel

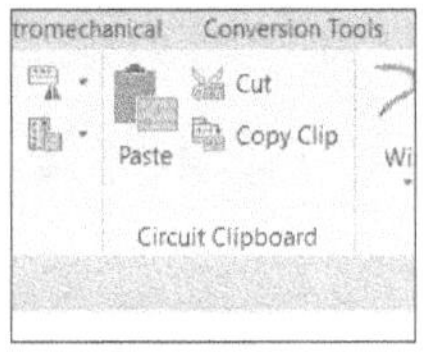

Figure-2. Circuit Clipboard panel

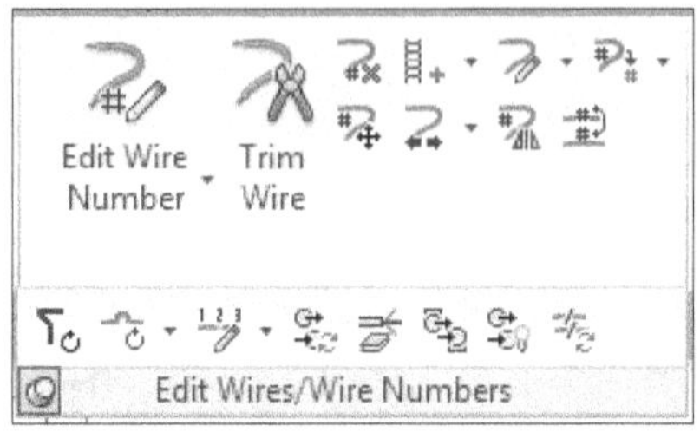

Figure-3. Edit Wires or Wire Numbers panel

EDIT TOOL

 The **Edit** tool is available in the **Edit Components** drop-down of the **Edit Components** panel in the **Ribbon**; refer to Figure-4. This tool is used to edit the details of any component inserted in the drawing area. The procedure to use this tool is given next.

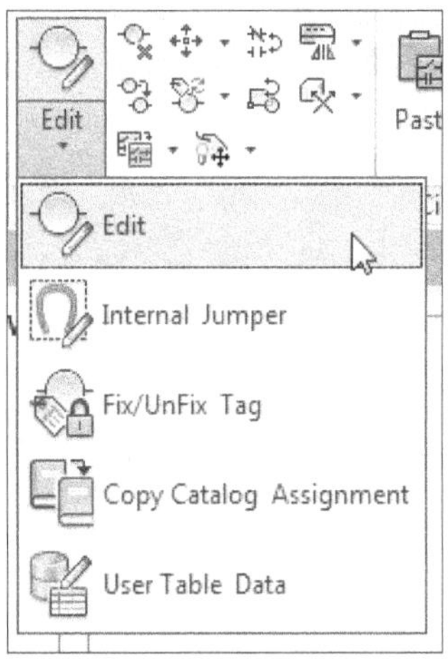

Figure-4. Edit components drop-down

- Click on the **Edit** tool from the **Edit Components** drop-down. You will be asked to select the components/cables/location boxes.
- Select the component. The **Insert/Edit Component** dialog box will be displayed; refer to Figure-5. The options in this dialog box have already been discussed.
- Specify desired settings in the dialog box and click on the **OK** button from the dialog box.
- If you want to use the component with the specified settings again then click on the **OK-Repeat** button from the dialog box.

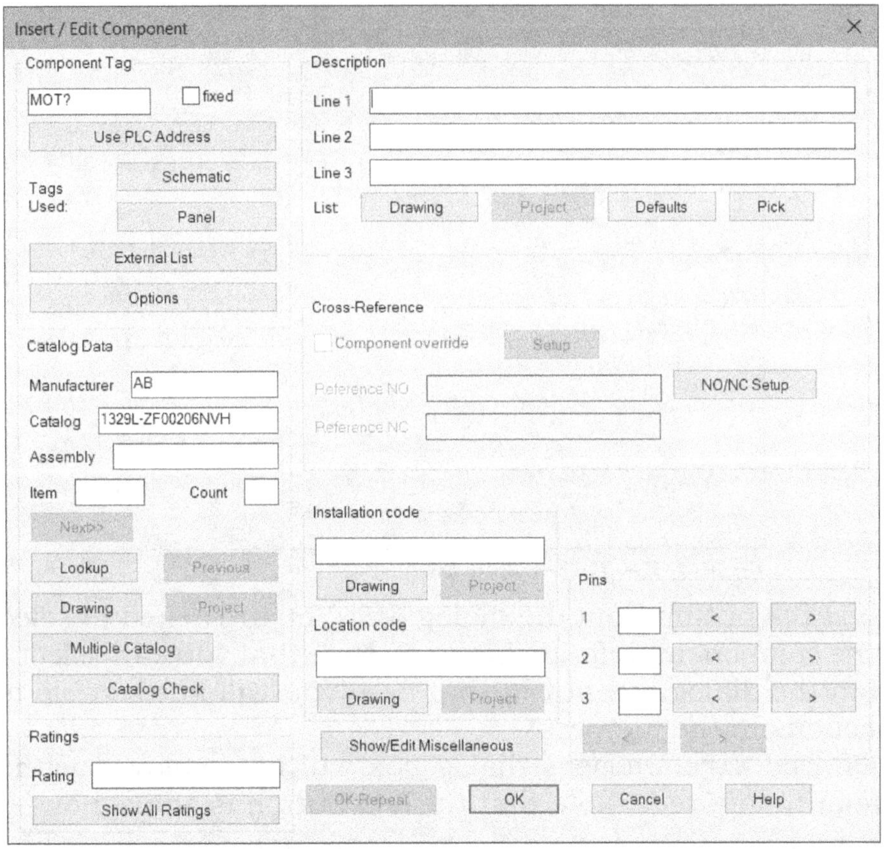

Figure-5. Insert or Edit Component dialog box

INTERNAL JUMPER

An Internal Jumper is used to interconnect terminals of the electrical components. Some components in electrical system may have jumpers to change their functionality. For example, you might have seen jumpers in transformers to change the output voltage between 110V and 230V. The **Internal Jumper** tool in the **Edit Components** drop-down is used to edit the functions of internal jumpers. The procedure to use this tool is given next.

- Click on the **Internal Jumper** tool from the **Edit Components** drop-down. You are asked to select a component for editing jumpers.
- Select the component; refer to Figure-6. The **Wire Jumpers** dialog box will be displayed; refer to Figure-7.

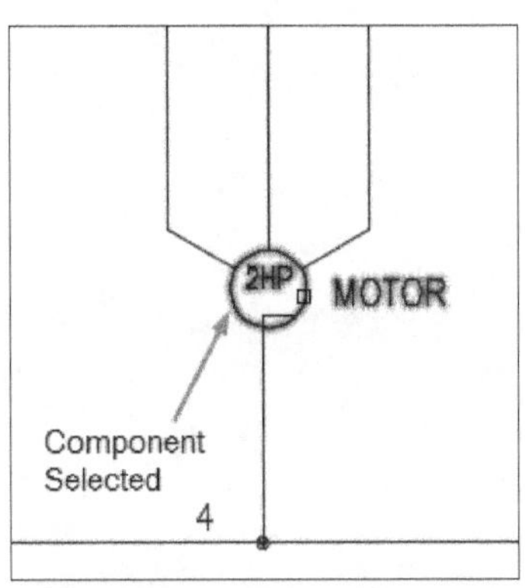

Figure-6. Component selected

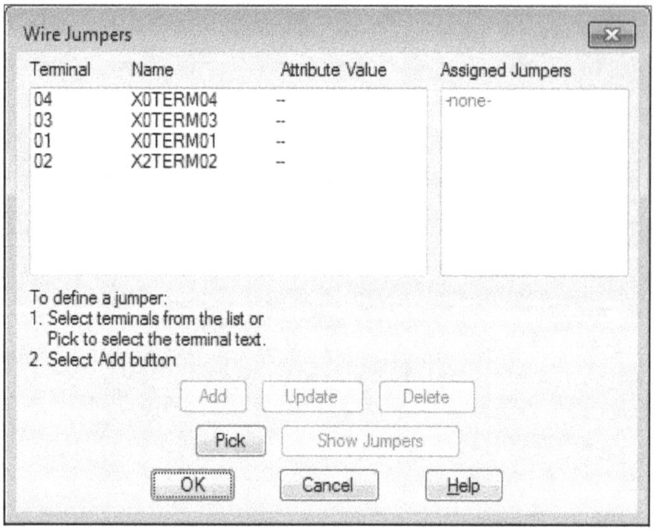

Figure-7. Wire Jumpers dialog box

- Click on the **Pick** button from the dialog box. You are asked to select the jumper points on the component; refer to Figure-8. Note that you can select the terminals directly from the dialog box but in that case you will not be able to identify the wire connections to the jumpers.
- Press **ENTER**. The **Wire Jumpers** dialog box will be displayed with the jumpers selected; refer to Figure-9. Note that the **Add** button is active now.

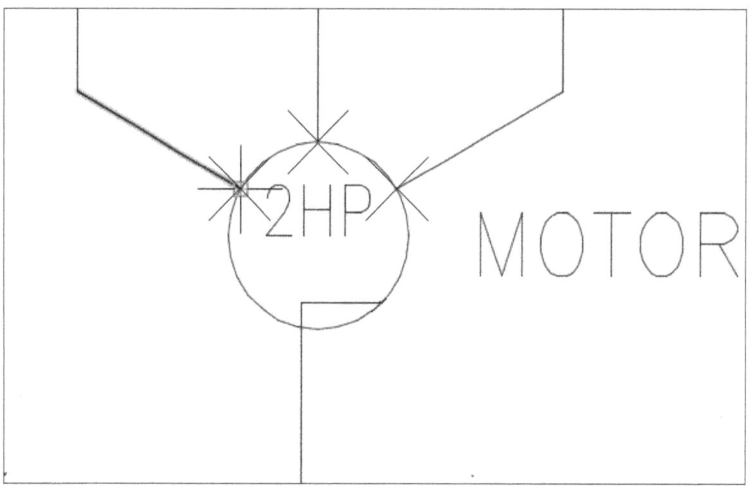

Figure-8. Jumper points selected

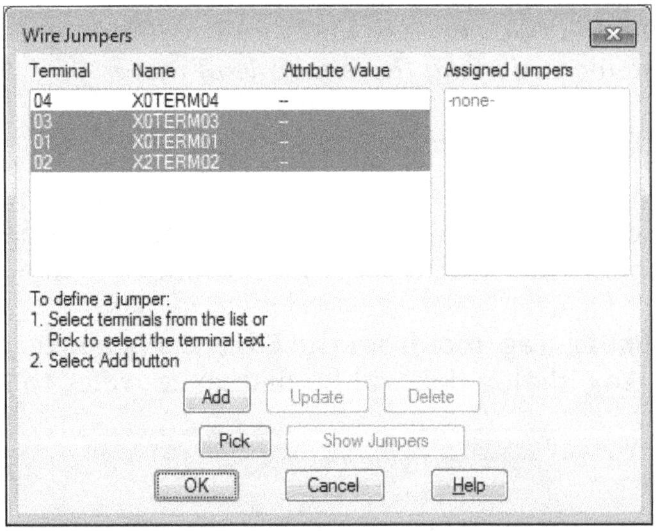

Figure-9. Wire Jumpers dialog box with terminals selected

- Click on the **Add** button from the dialog box. The jumper will be connected to the selected terminals and will be listed in the **Assigned Jumpers** area of the dialog box; refer to Figure-10.

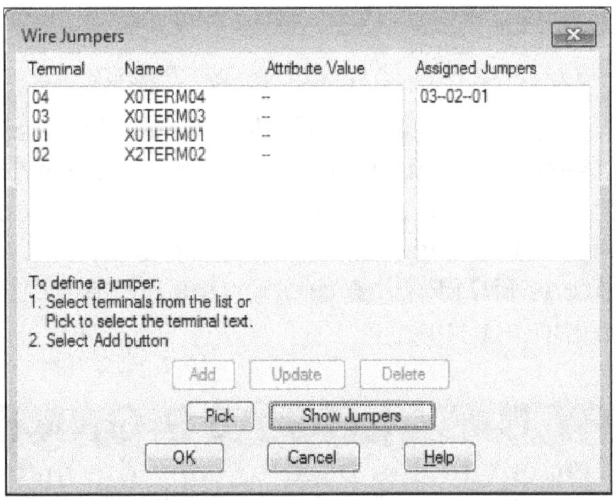

Figure-10. Wire Jumpers dialog box with assigned jumpers

- Click on the **Show Jumpers** button to display the jumper in the drawing area. The jumper will be displayed connecting all the selected terminals; refer to Figure-11.

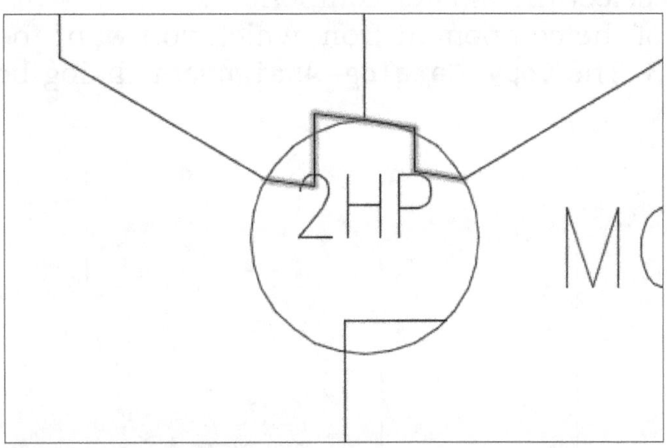

Figure-11. Jumper connecting the terminals

The example used above is just for illustration purpose, in real world you will not be connecting motors like this. We can use PLCs for applying jumpers but they have not been discussed yet so we are using the example of motors.

Fix/UnFix Tag

The **Fix/UnFix Tag** tool is used to fix or unfix the tags assigned to the components. Note that if you fix a component tag then it will not get updated automatically when you re-tag the components in the drawing. The procedure to use this tool is given next.

- Click on the **Fix/UnFix Tag** tool from the **Edit Components** drop-down. The **Fix/UnFix Component Tag** dialog box will be displayed; refer to Figure-12.

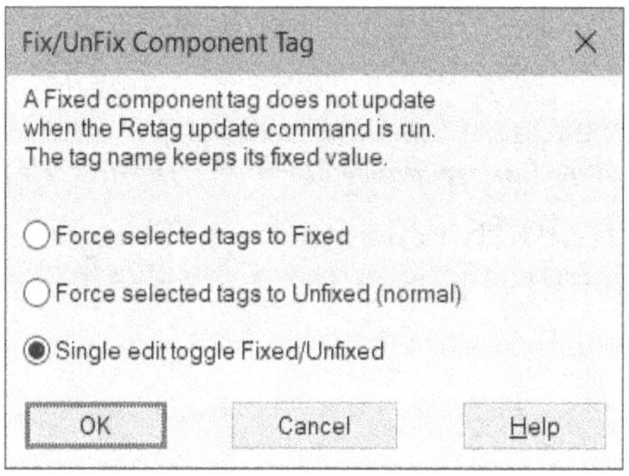

Figure-12. Fix or UnFix Component Tag dialog box

- Select desired radio button from the dialog box. For example, select the **Force selected tags to Fixed** radio button and then all the selected tags will get fixed.
- Click on the **OK** button from the dialog box. You are asked to select the tags that you want to change for their tags fixed/unfixed status.
- Select the tags and press **ENTER**. The properties for selected tags will get modified as per the selected radio button.

COPY CATALOG ASSIGNMENT

The **Copy Catalog Assignment** tool is used to copy the meta data from an existing component and apply it to the other component. The procedure to use this tool is given next.

- Click on the **Copy Catalog Assignment** tool from the **Edit Components** drop-down. You are asked to select the master component.
- Click on the tag of the component from which you want to copy the meta data; refer to Figure-13. The **Copy Catalog Assignment** dialog box will be displayed; refer to Figure-14.

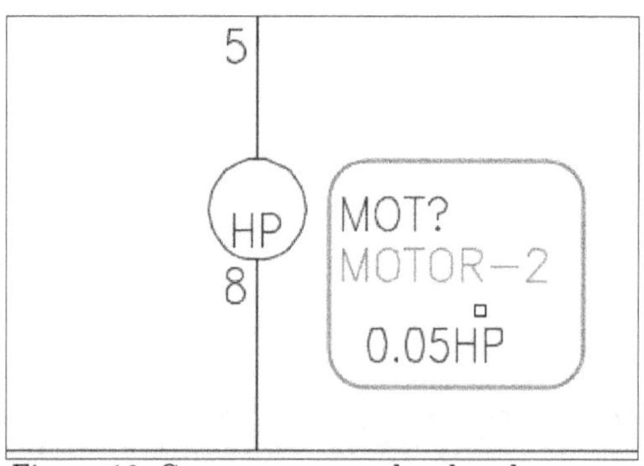

Figure-13. Component tag to be selected

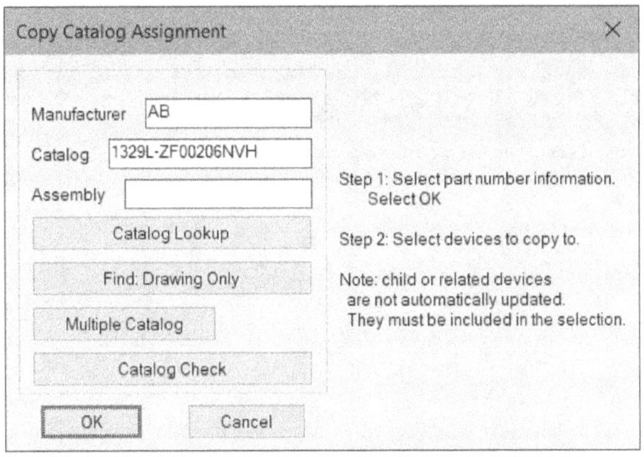

Figure-14. Copy Catalog Assignment dialog box

- Edit the details if you want to and then click on the **OK** button from the dialog box. You are asked to select the target component.
- Select the target component. Some of my friends might get the **Update Related Components** dialog box; refer to Figure-15. Click on the **Yes-Update** button if you want to update all similar components in the drawing or choose the **Skip** button to keep others as they are.
- Few of my friends might also get the **Different symbol block names** dialog box; refer to Figure-16. This dialog box comes if you are assigning properties to an incompatible component. For example, if you are copying properties of a motor to a switch then this dialog box will come. If it is need of your drawing to assign such properties then click on the **OK** button and then **Overwrite** otherwise cancel it and take your decision.

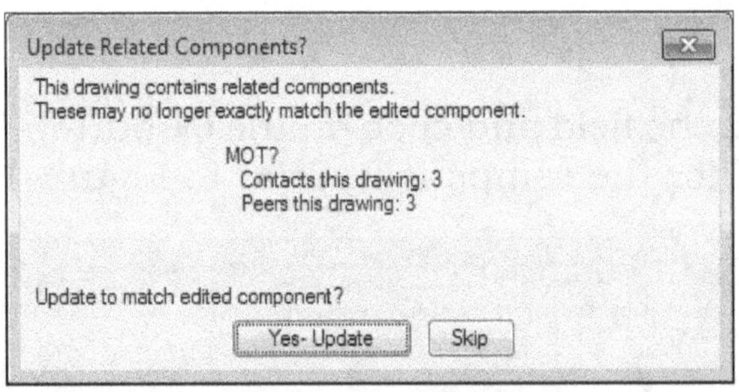

Figure-15. Update Related Components dialog box

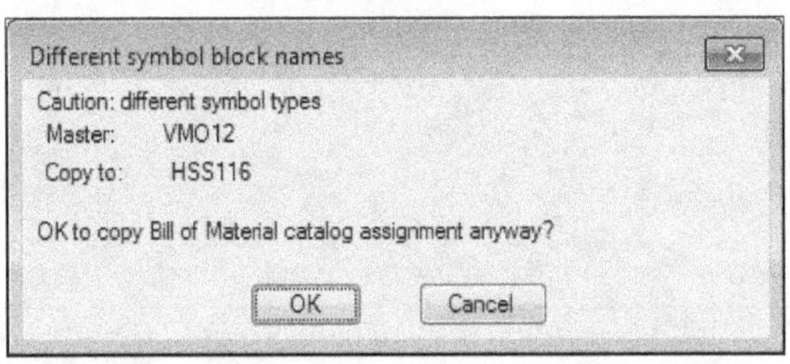

Figure-16. Different symbol block names dialog box

USER TABLE DATA

The **User Table Data** tool allows you to enter any data that you want to include with the components for identifying. The procedure to use this tool is given next.

- Click on the **User Table Data** tool from the **Edit Components** drop-down. The **Edit User Table Data** dialog box will be displayed; refer to Figure-17.

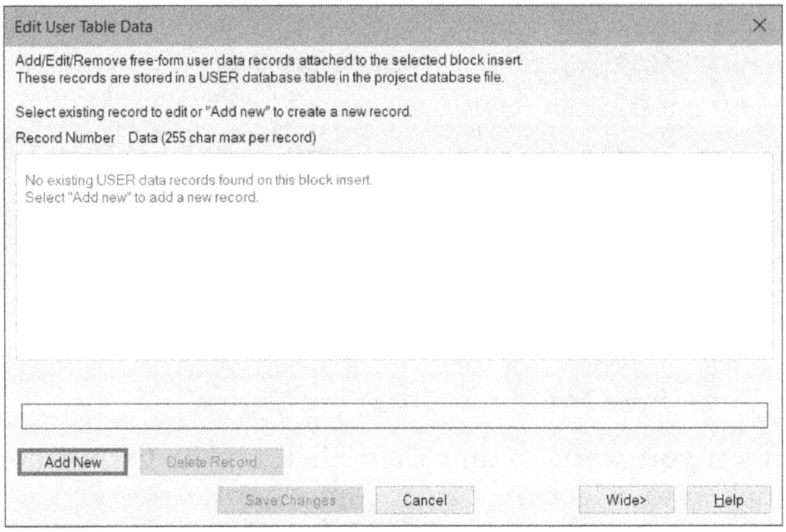

Figure-17. Edit User Table Data dialog box

- Click on the **Add New** button from the dialog box. The **Add New USER data record** dialog box will be displayed; refer to Figure-18.

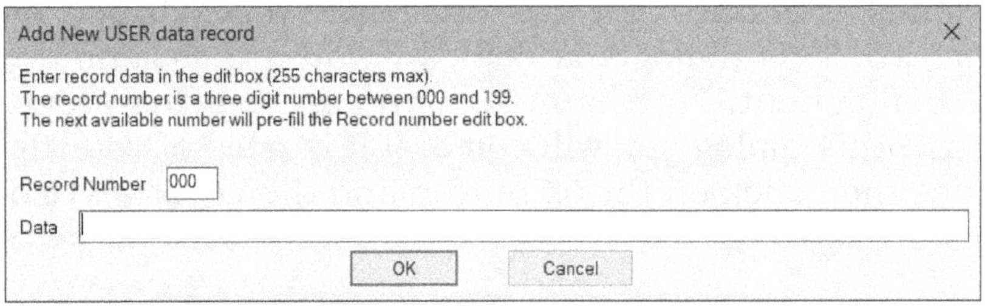

Figure-18. Add New USER data record dialog box

- Enter desired data in the field and click on the **OK** button from the dialog box. The record will be added for the component; refer to Figure-19.

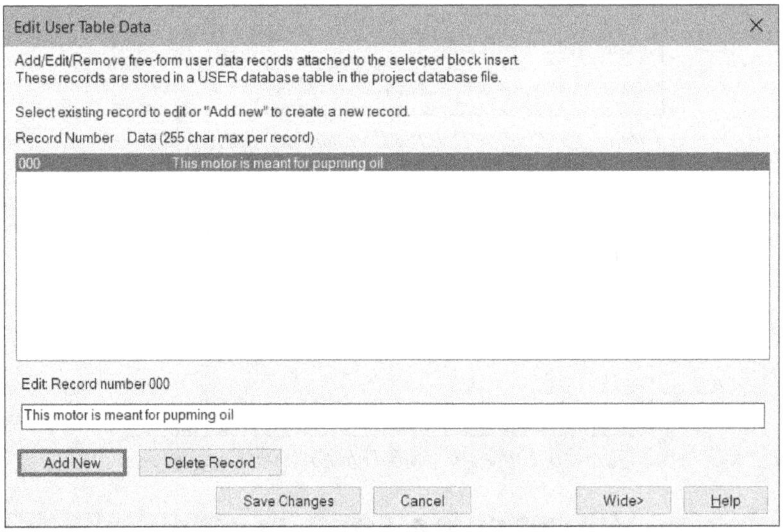

Figure-19. Edit User Table Data dialog box with record

- Click on the **Save Changes** button from the dialog box. The user record will be added to the component.

DELETE COMPONENT

As the name suggests, the **Delete Component** tool is used to delete the components. Note that using this tool to delete components allows the wire to heal. The procedure to do so is given next.

- Click on the **Delete Component** tool from the **Edit Component** panel and click on the components that you want to delete.
- After selecting the components, press **ENTER**. The components will be deleted and the **Search for/ Surf to Children** dialog box will be displayed if there are child components for selected component; refer to Figure-20.

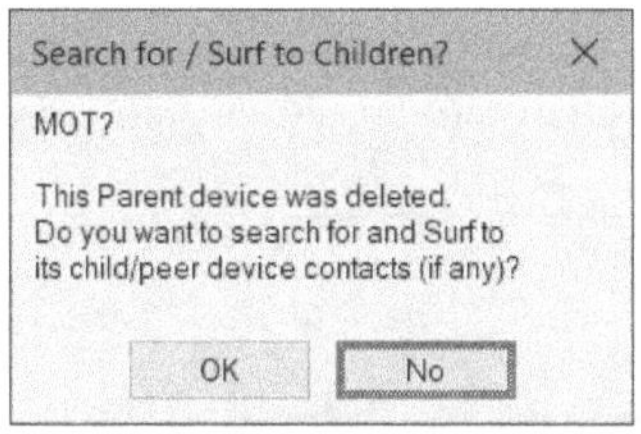

Figure-20. Search for or Surf to Children dialog box

- Click on the **OK** button if you want to search for the child components and want to delete them otherwise select the **NO** button from the dialog box.

COPY COMPONENT

The **Copy Component** tool is used to create copies of an existing component. The procedure to use this tool is given next.

- Click on the **Copy Component** tool from the **Edit Components** panel in the **Ribbon**. You are asked to select the component that you want to copy.
- Select the component. You are asked to specify the insertion points of the copied component.
- Click at desired place in the drawing to create copy and then specify desired parameters in the **Insert/Edit Component** dialog box.
- Click on the **OK** button from the dialog box after specifying the parameters. The copy will be created with the specified parameters.

EDIT CIRCUITS DROP-DOWN

The tools in the **Edit Circuits** drop-down are used to edit the circuits. There are three tools in this drop-down; refer to Figure-21. The procedure to use these tools is discussed next.

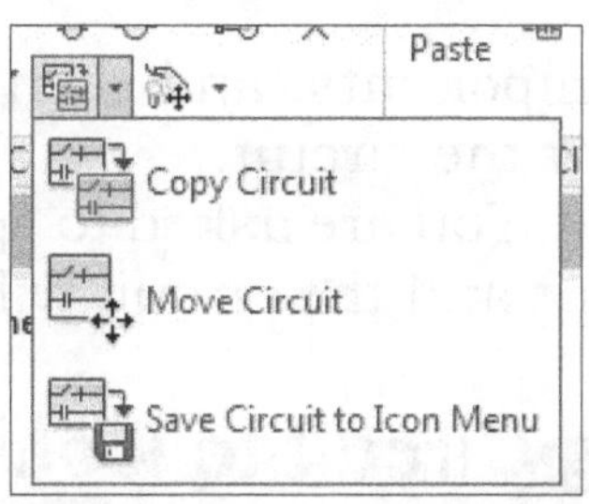

Figure-21. Edit Circuit drop-down

Copying Circuit

- Click on the **Copy Circuit** tool from the **Edit Circuits** drop-down. You are asked to select a circuit.
- Select the circuit and its components and then press **ENTER**. You are asked to specify the insertion point for the copied circuit.
- Click to specify the first point of insertion. You are asked to specify the second point of insertion; refer to Figure-22.

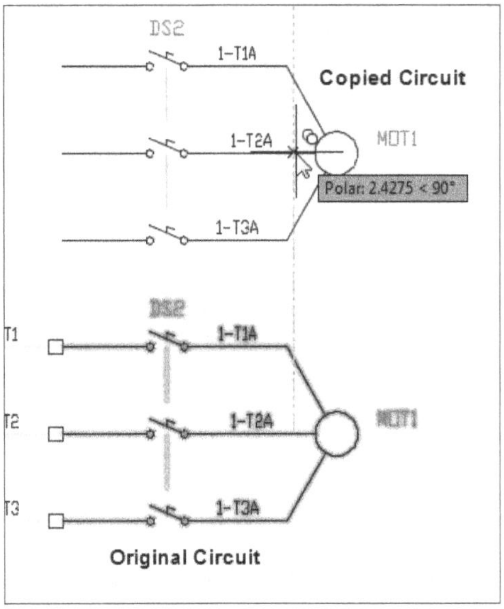

Figure-22. Copying a circuit

- Click to specify the insertion point. The circuit will be placed.
- Note that if you have gaps in the circuit then the **Gapped wire pointer problem** dialog box will be displayed; refer to Figure-23.

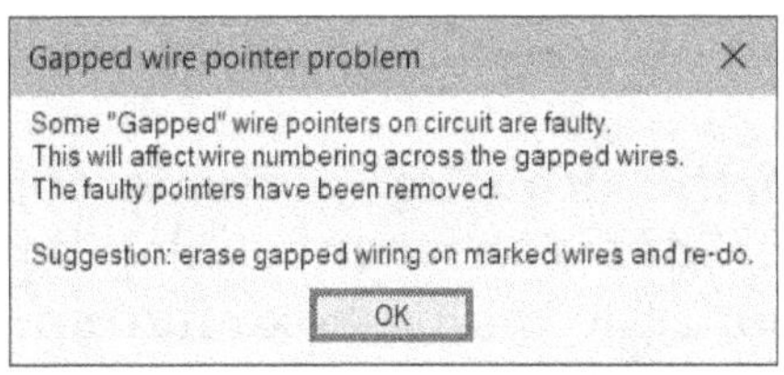

Figure-23. Gapped wire pointer problem dialog box

- Click on the **OK** button from the dialog box and modify the circuit as per your requirement.

Moving Circuit

- Click on the **Move Circuit** tool from the **Edit Circuits** drop-down. You are asked to select a circuit.
- Select the circuit and its components, and then press **ENTER**. You are asked to specify the new base point for the circuit.
- Click to specify the base point. You are asked to specify the insertion point for the circuit. Set the insertion point and the circuit will be placed accordingly.

Saving Circuit to Icon Menu

- Click on the **Save Circuit to Icon Menu** tool from the **Edit Circuits** drop-down. The **Save Circuit to Icon Menu** dialog box will be displayed; refer to Figure-24.

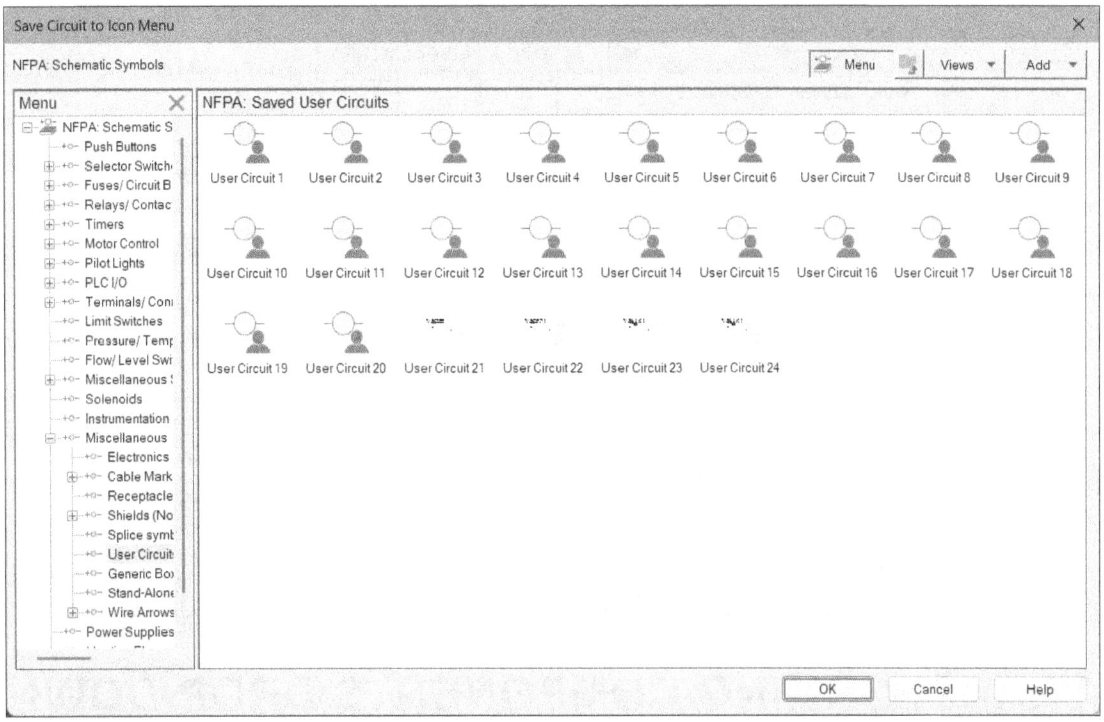

Figure-24. Save Circuit to Icon Menu dialog box

- Right-click on desired slot from the dialog box. A shortcut menu will be displayed.
- Click on the **Properties** button from the shortcut menu. The **Properties-Circuit** dialog box will be displayed; refer to Figure-25.

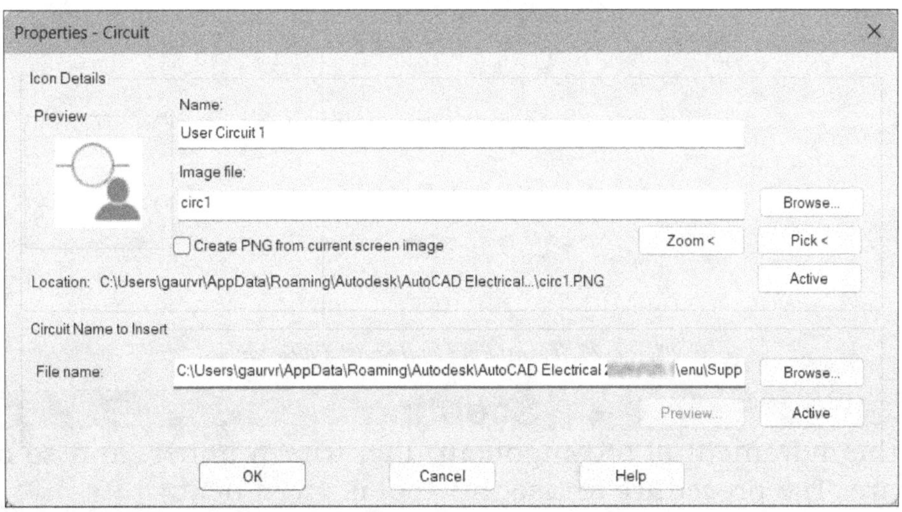

Figure-25. Properties Circuit dialog box

- Click on both the **Active** buttons one by one and click on the **OK** button from the dialog box. The current drawing will be saved as user defined circuit; refer to Figure-26.

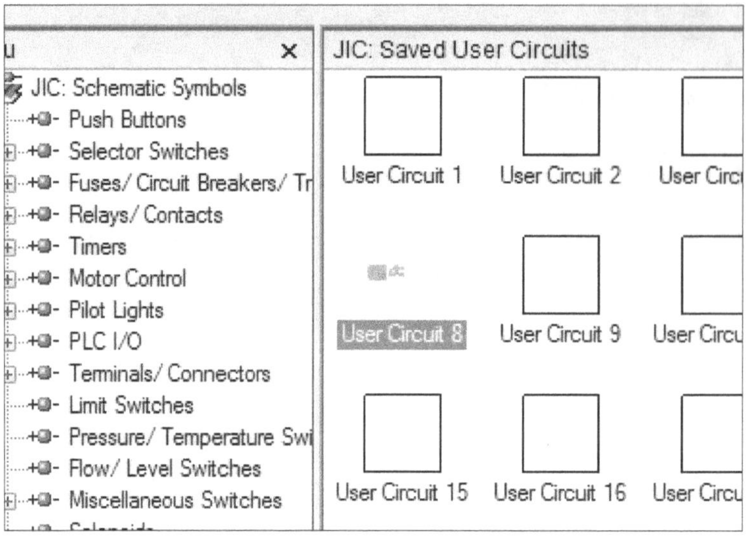

Figure-26. User defined circuit created

- Click on the **OK** button from the **Save Circuit to Icon Menu** dialog box.

TRANSFORMING COMPONENTS DROP-DOWN

The tools in this drop-down are used to transform the components; refer to Figure-27. For example, you can change position of a component on wire, you can align components, and so on. Various tools in this drop-down are discussed next.

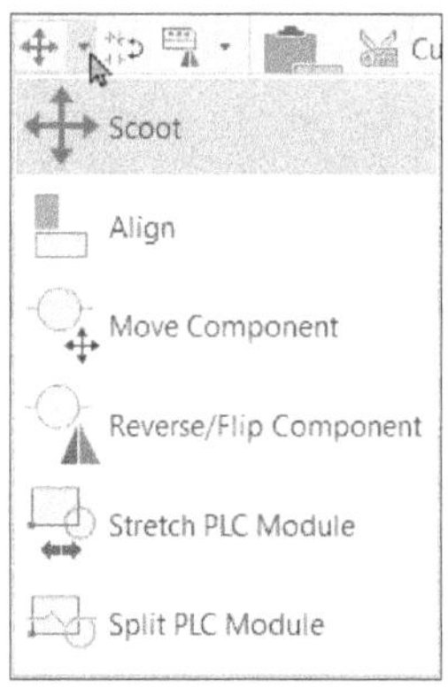

Figure-27. Transform Component dialog box

Scooting

Scooting is the movement of a component, tag, wire number, and so on along the connected wire. The procedure to use this tool is given next.

- Click on the **Scoot** tool from the **Transform Component** drop-down. You are asked to select component, wire, or wire number from the drawing area.
- Click on desired entity. A box will be attached to the cursor representing the component; refer to Figure-28.

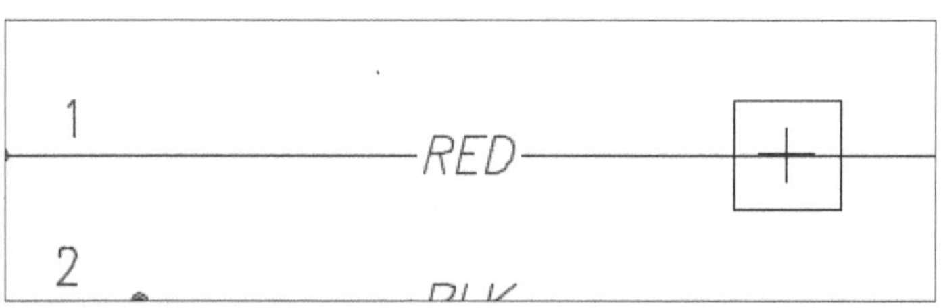

Figure-28. Scooting component

- Click at desired point to specify new location of the entity; refer to Figure-29.

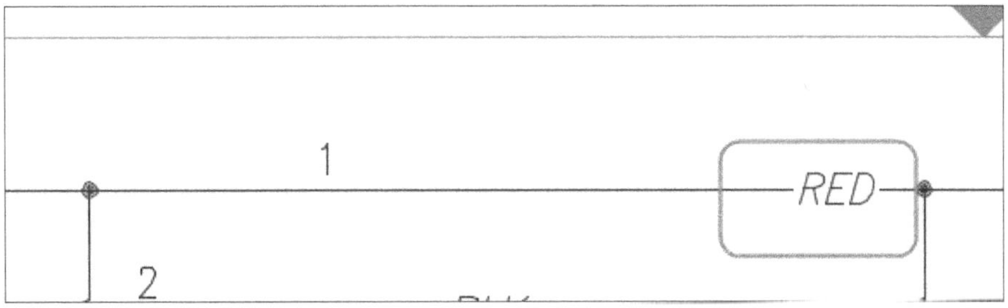

Figure-29. Entity after scooting

Aligning Components

- Click on the **Align** tool from the **Transform Component** drop-down. You are asked to select an entity as the horizontal/vertical reference for the other components.
- Select the component. The reference line will be created from the center of the selected component; refer to Figure-30. Also, you are asked to select the components to be aligned.

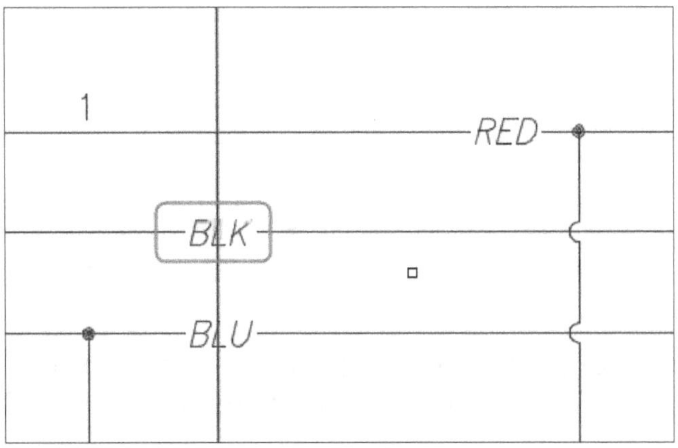

Figure-30. Reference line for selected component

- Click on the entities that you want to align and press **ENTER**. The entities will be aligned; refer to Figure-31.

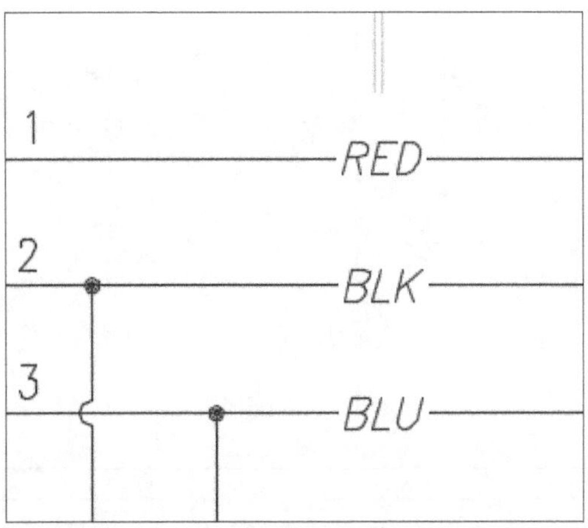

Figure-31. Entities after alignment

Moving Component

You can move components in a circuit without creating gaps in the circuit. The procedure is given next.

- Click on the **Move Component** tool from the **Transform Components** drop-down. You will be asked to select a component.
- Click on the component. The component will get attached to the cursor; refer to Figure-32.

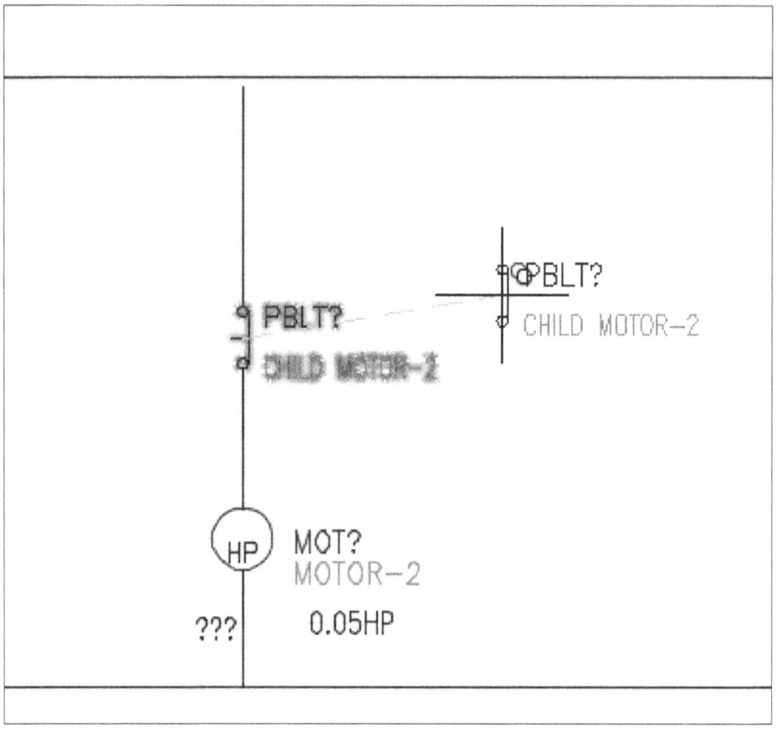

Figure-32. Component attached to the cursor

- Click at new position to place the component. The component will be placed and the circuit will be modified to fill the gap; refer to Figure-33.

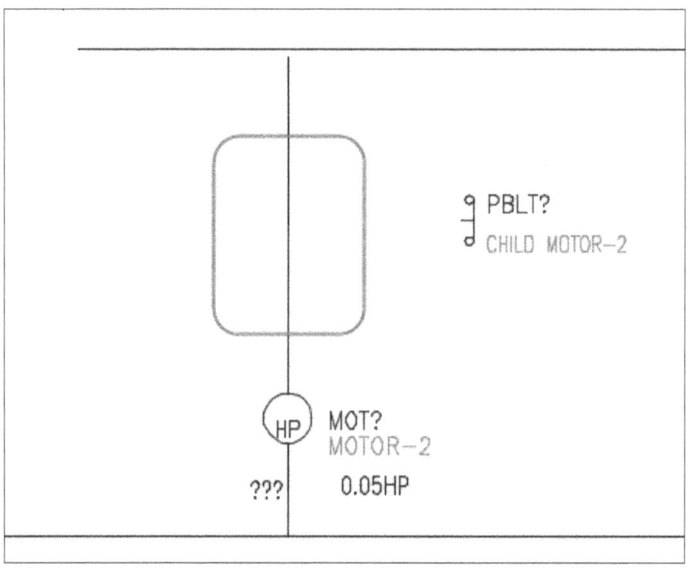

Figure-33. Circuit after moving component

Reversing or Flipping Component

- Click on the **Reverse/Flip Component** tool from the **Transform Components** drop-down. The **Reverse/Flip Component** dialog box will be displayed as shown in Figure-34.

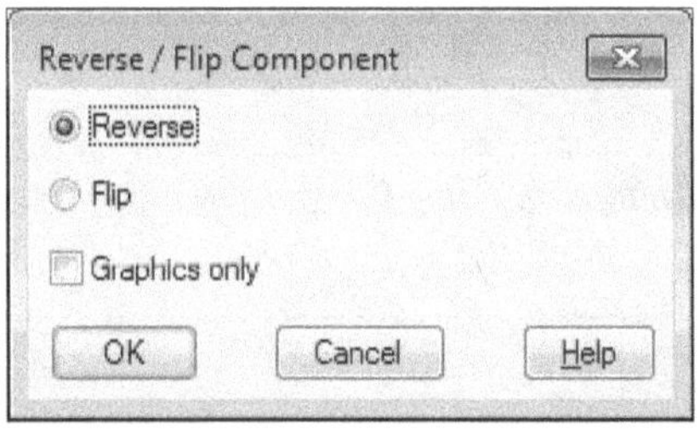

Figure-34. Reverse or Flip Component dialog box

- Select desired radio button for whether you want to reverse the pins of component or you want to flip the component.
- Click on the **OK** button from the dialog box. You are asked to select the component that you want to flip or reverse.
- Select the component, it will get flipped or reversed as per the selected option. Refer to Figure-35.

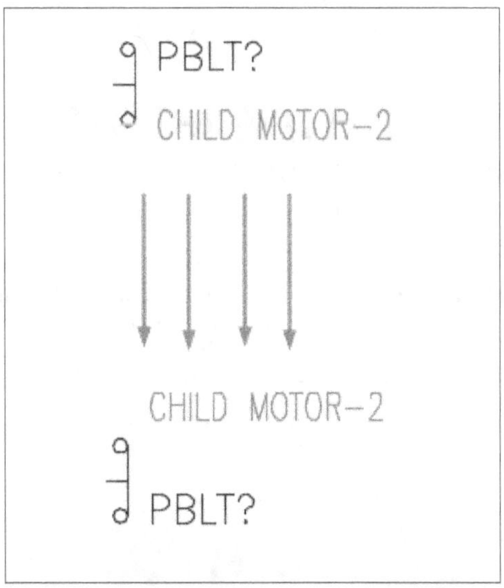

Figure-35. Reverse component

We will discuss about the other two tools in the **Transform Components** drop-down later in the book when we will be discussing PLCs.

RE-TAGGING COMPONENTS

You can assign a new tag to the component by replacing the older one, using the **Retag Components** tool. The procedure to use this tool is given next.

- Click on the **Retag Components** tool from the **Retag Components** drop-down in the **Edit Components** panel in the **Ribbon**; refer to Figure-36. The **Retag Components** dialog box will be displayed; refer to Figure-37.

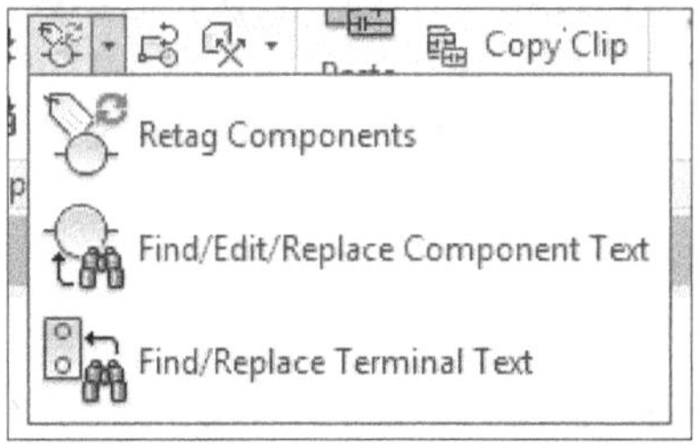

Figure-36. Retag Components drop-down

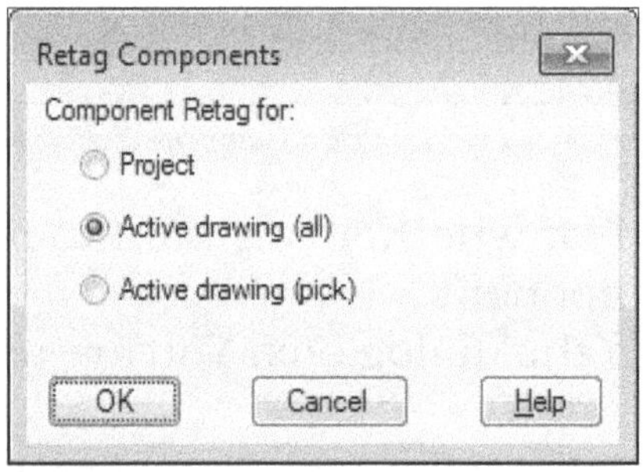

Figure-37. Retag Components dialog box

- Select desired radio button from the dialog box and click on the **OK** button. The components will be re-tagged according to selected radio button.

TOGGLE NO/NC

You can toggle between **Normally Open(NO)** and **Normally Closed(NC)** variants of components by using this tool. The procedure to use this tool is given next.

- Click on the **Toggle NO/NC** tool  from the **Edit Components** panel. You will be asked to select the component for toggling NC/NO.
- Select the component and its status will be changed; refer to Figure-38.

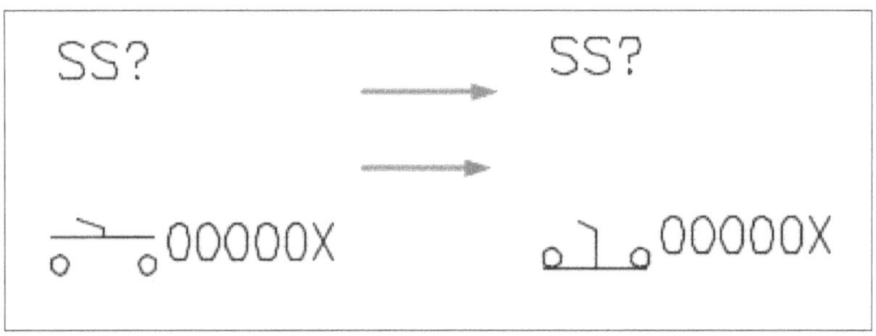

Figure-38. Component after toggling

SWAP/UPDATE BLOCK

The **Swap/Update Block** tool is used to swap or update the block of a component. You need to use this tool when symbol library is updated. The procedure to use this tool is given next.

Swapping

- Click on the **Swap/Update Block** tool from the **Edit Components** panel. The **Swap Block/Update Block/Library Swap** dialog box will be displayed; refer to Figure-39.

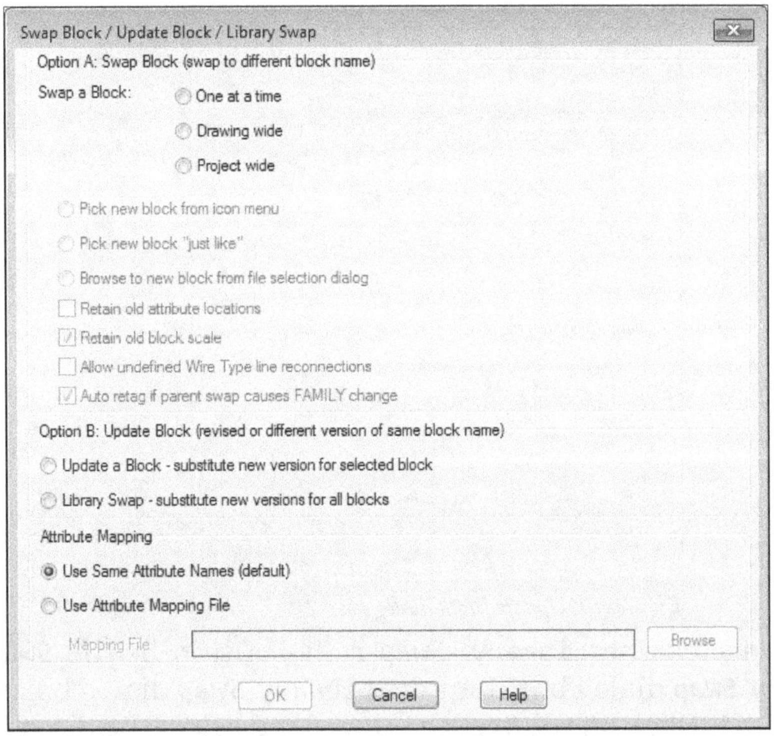

Figure-39. Swap Block or Update Block or Library Swap dialog box

- Click on desired radio button for **Swap a Block** area. The options related to swap will be activated in the dialog box.
- Select desired option for source of swapping by selecting **Pick new block icon menu**, **Pick new block "just like"**, or **Browse to new block from file selection dialog** radio button.
- Click on the **OK** button from the dialog box. If you have selected the **Browse to new block from file selection dialog** radio button, then the dialog box will be displayed prompting you to select the blocks.
- Select the drawing of block. You are asked to select the component.
- Select the component and it will be swapped with the selected block.

Updating

- Click on desired radio button from the **Option B: Update Block** area of the dialog box. If you select the **Update a Block** radio button then new block will be placed for selected component. If you select the **Library Swap** radio button then all the blocks will be updated.
- After selecting desired radio button, click on the **OK** button from the dialog box.
- If you select the **Update a Block** radio button then you will be prompted to select an example of block. Select the block of component (or components) from the drawing area and press **ENTER**. The **Update Block** dialog box will be displayed; refer to Figure-40. Click on the **Browse** button and select desired component block. Click on the **Project** button if you want to update the block for all project drawings or click on the **Active Drawing** button to update the block in current drawing only.

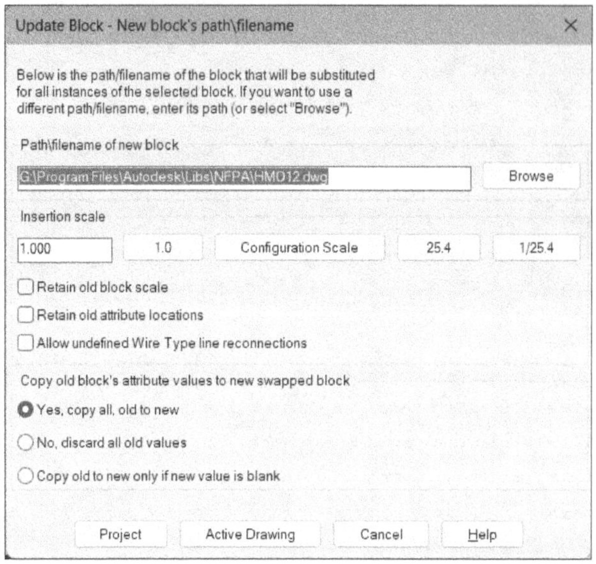

Figure-40. Update Block dialog box

- If you have selected the **Library Swap** radio button in the **Swap Block/Update Block/Library Swap** dialog box then the **Library Swap** dialog box will be displayed; refer to Figure-41. Click on the **Browse** button and select the directory of component blocks. Click on the **Project** button if you want to update the block for all project drawings or click on the **Active Drawing** button to update the block in current drawing only.

Note that this option is very useful if you want to convert all your schematic symbols in drawing to another standard library symbols. Like you want your symbols of JIC to be converted to IEEE standard then you can use this option. Make sure you select the Include subfolder check box in the Path to new block library area and select the Retain old block scale check box so that your symbols do not get bigger or smaller.

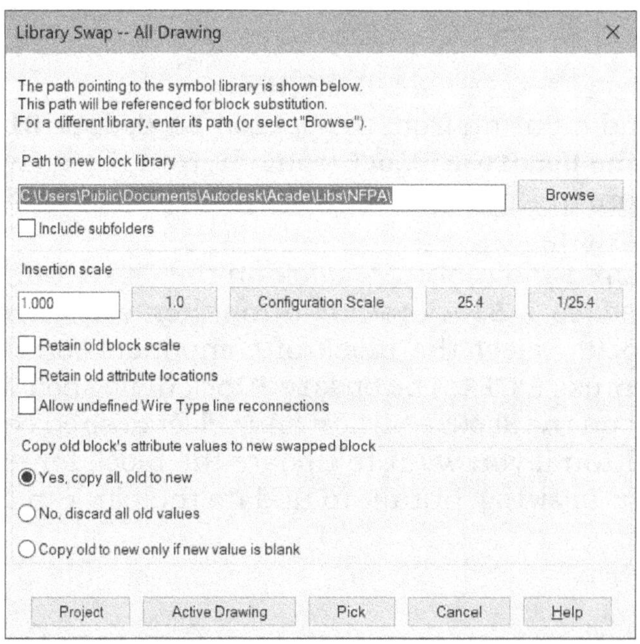

Figure-41. Library Swap dialog box

EDIT ATTRIBUTE DROP-DOWN

The options in the **Edit Attribute** drop-down are used to edit the attributes of the components; refer to Figure-42. These tools are common to all the AutoCAD platform software. An attribute is used to relate the blocks with the real world information. These tools perform the action as their names suggest. You can also perform all these tasks by using the **PROPERTIES Palette**. To use the **PROPERTIES Palette**, click on the component's attribute and enter **PROPERTIES** at the Command Prompt. The **PROPERTIES Palette** will be displayed with the relative options; refer to Figure-43. Click in desired field and change the attributes as required.

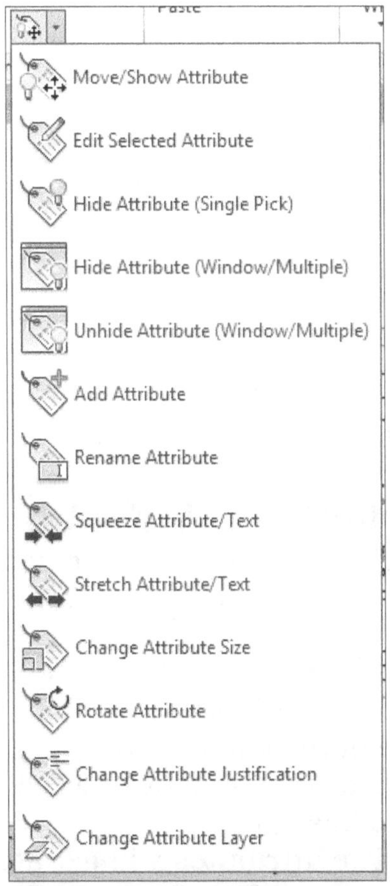

Figure-42. Edit Attribute drop-down

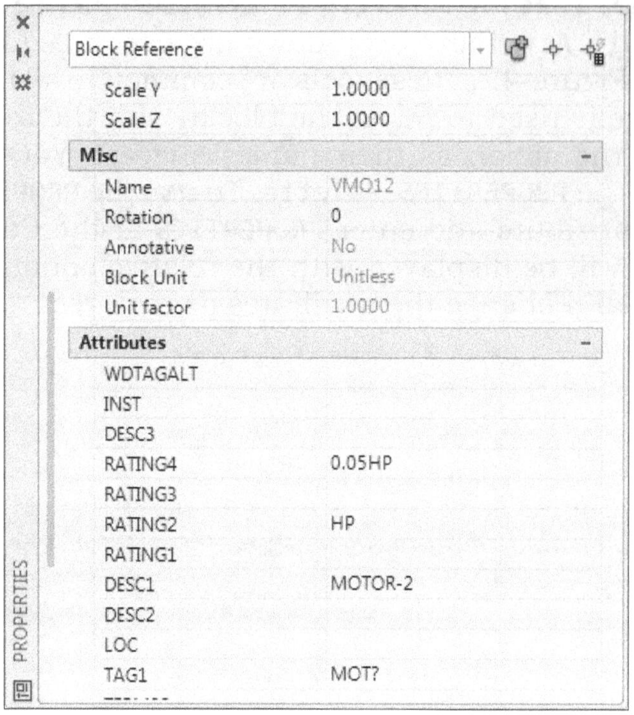

Figure-43. Properties Manager

Various tools in this drop-down are discussed next.

Moving/Showing Attributes

The **Move/Show Attribute** tool is used to move or display/hide attributes of selected component. The procedure to use this tool is given next.

- Click on the **Move/Show Attribute** tool from the **Edit Attribute** drop-down in the **Edit Components** panel of the **Schematic** tab in the **Ribbon**. You will be asked to select block or attribute of the component.
- If you want to move attributes then one by one select them and press **ENTER**. You will be asked to specify the base point.
- Click at desired location. The attributes will get attached to cursor and you will be asked to specify new location of attributes; refer to Figure-44.

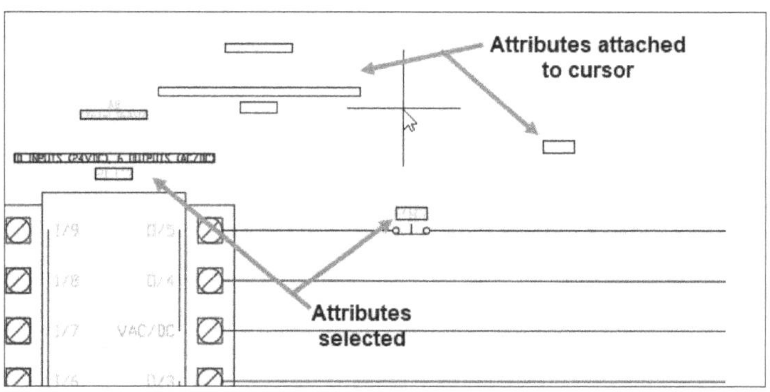

Figure-44. Attributes attached to cursor

- Click at desired location to place the attributes.
- If you want to display or hide attributes of a component then restart the tool and click on the component block instead of attributes. The **SHOW/HIDE Attributes** dialog box will be displayed; refer to Figure-45.

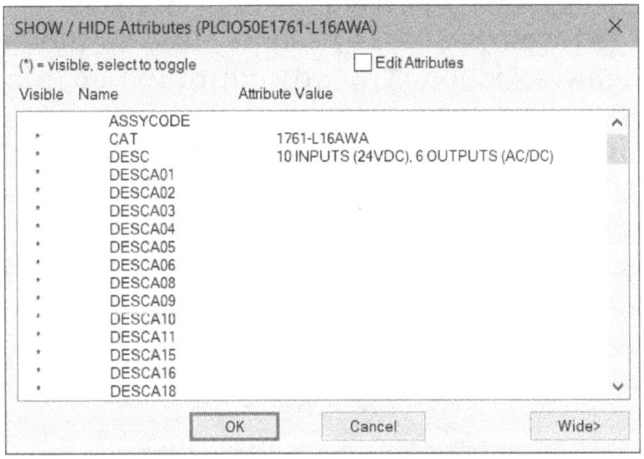

Figure-45. SHOW HIDE Attributes dialog box

- Click on the attribute you want to show/hide from the dialog box. If * is displayed with the attribute in the list then it will be shown in the drawing area.
- After setting desired parameters, click on the **OK** button from the dialog box. Press **ESC** to exit the tool.

Editing Selected Attribute

The **Edit Selected Attribute** tool is used to modify value of selected attribute. The procedure to use this tool is given next.

- Click on the **Edit Selected Attribute** tool from the **Edit Attribute** drop-down in the **Edit Components** panel of **Schematic** tab in the **Ribbon**. You will be asked to select the attribute.
- Select desired attribute from the drawing area. The **Edit Attribute** dialog box will be displayed; refer to Figure-46.

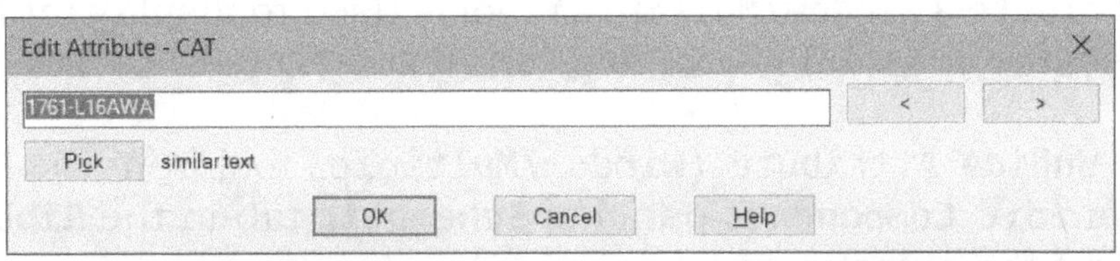

Figure-46. Edit Attribute dialog box

- Specify desired value in the edit box and click on the **OK** button. Press **ESC** to exit the tool.

Hiding Attribute (Single Pick)

The **Hide Attribute (Single Pick)** tool is used to hide selected attribute from drawing area. The procedure to use this tool is given next.

- Click on the **Hide Attribute (Single Pick)** tool from the **Edit Attribute** drop-down in the **Edit Components** panel of the **Schematic** tab in the **Ribbon**. You will be asked to select the attribute or block of the component.
- Click on desired attribute to hide it.
- Click on the block of component to display or hide multiple attributes. The **SHOW/HIDE Attribute** dialog box will be displayed as discussed earlier.
- Press **ESC** to exit the tool.

Hiding Multiple Attributes

The **Hide Attribute (Window/Multiple)** tool is used to hide multiple attributes of component using window selection. The procedure to use this tool is given next.

- Click on the **Hide Attribute (Window/Multiple)** tool from the **Edit Attribute** drop-down of the **Edit Components** panel in the **Schematic** tab of **Ribbon**. You will be asked to select the objects.
- Select desired objects and press **ENTER**. The **Flip Attributes to Invisible** dialog box will be displayed; refer to Figure-47.

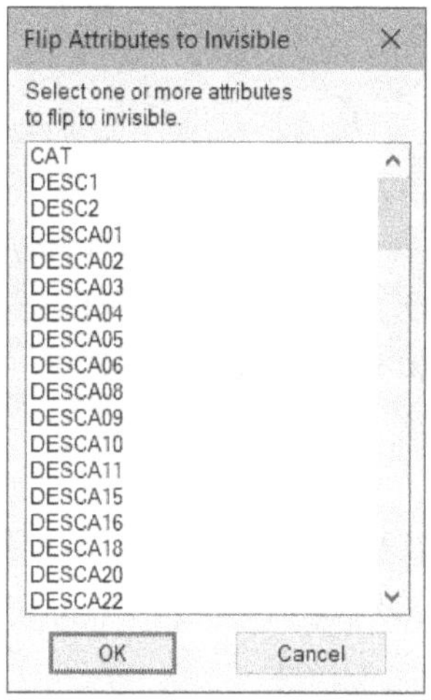

Figure-47. Flip Attributes to Invisible dialog box

- Select the attributes from the list while holding the **CTRL** key to make them invisible and click on the **OK** button.

Unhiding Multiple Attributes

The **Unhide Attribute (Window/Multiple)** tool is used to display multiple attributes again. The procedure to use this tool is given next.

- Click on the **Unhide Attribute (Window/Multiple)** tool from the **Edit Attribute** drop-down of **Edit Components** panel of **Schematic** tab in the **Ribbon**. You will be asked to select the objects.
- Select the components whose attributes are to be displayed again and press **ENTER**. The **Flip Attributes to Visible** dialog box will be displayed; refer to Figure-48.
- Select the attributes you want to display from the list while holding **CTRL** key and click on the **OK** button. The hidden attributes will be visible again.

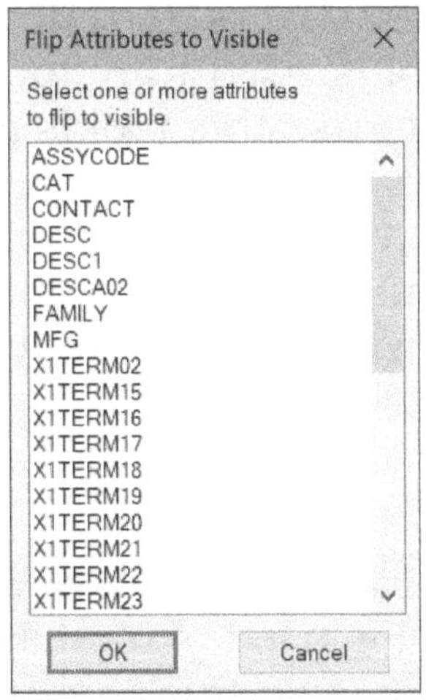

Figure-48. Flip Attributes to Visible dialog box

Adding Attribute to Component

The **Add Attribute** tool is used to add a new attribute to selected component. The procedure to use this tool is given next.

- Click on the **Add Attribute** tool from the **Edit Attribute** drop-down in the **Edit Components** panel of **Schematic** tab in the **Ribbon**. You will be asked to select the component to which you want to add attribute. The **Add Attribute** dialog box will be displayed; refer to Figure-49.

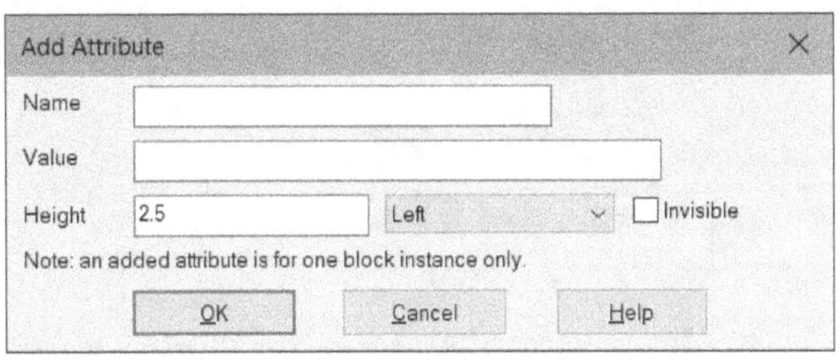

Figure-49. Add Attribute dialog box

- Specify the name, value, and other parameters of the attribute in the dialog box. Click on the **OK** button from the dialog box to create the new attribute. The attribute will get attached to cursor.
- Click at desired location to place the new attribute.

Renaming Attribute

The **Rename Attribute** tool is used to change the name of attribute. Note that this tool does not change the value of attribute. The procedure to use this tool is given next.

- Click on the **Rename Attribute** tool from the **Edit Attribute** drop-down of **Edit Components** panel in the **Schematic** tab of **Ribbon**. You will be asked to select the attribute to be renamed.
- Click on desired attribute. You will be asked to specify new name for the attribute in **Command Bar**. Also, old name of attribute will be displayed in **Command Bar**.
- Click in the **Command Bar** and enter the new name. The name of tab will be changed.

Squeezing Attribute/Text

The **Squeeze Attribute/Text** tool is used to reduce the width of text and attribute in drawing area. The procedure to use this tool is given next.

- Click on the **Squeeze Attribute/Text** tool from the **Edit Attribute** drop-down of **Edit Components** panel in the **Schematic** tab of the **Ribbon**. You will be asked to select the attribute whose width is to be reduced.
- Select desired text. Its width will be reduced by 5% each time you click on it.
- Press **ESC** to exit the tool.

Stretching Attribute/Text

The **Stretch Attribute/Text** tool is used to increase the width of text and attribute in drawing area. The procedure to use this tool is similar to **Squeeze Attribute/ Text** tool.

Changing Attribute Text Size

The **Change Attribute Size** tool is used to change the height and width of attribute text. The procedure to use this tool is given next.

- Click on the **Change Attribute Size** tool from the **Edit Attribute** drop-down in the **Edit Components** panel of the **Schematic** tab in the **Ribbon**. The **Change Attribute Size** dialog box will be displayed; refer to Figure-50.

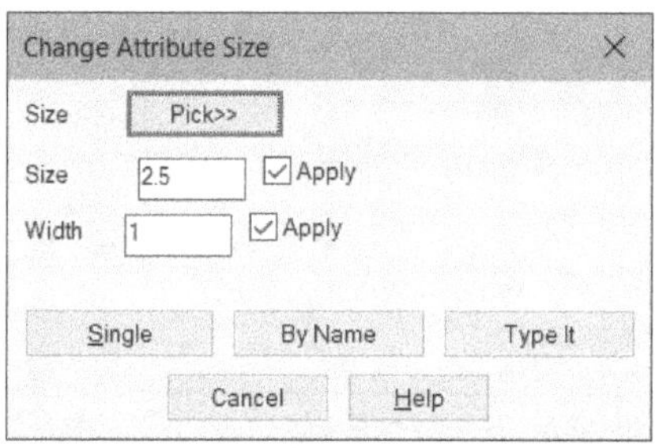

Figure-50. Change Attribute Size dialog box

- Specify desired parameters in the edit boxes of the dialog box and click on the **Single** button to change the size of an attribute. You will be asked to select the attributes.
- One by one click on the attributes whose sizes are to be changed.
- Press **ESC** to exit the tool. You can change the size using **By Name** button if you remember the names of attributes. In that you will be asked to specify name of attribute in **Command Bar**. Similarly, you can use the **Type It** button if you remember the name of attribute.

Rotating Attribute

The **Rotate Attribute** tool is used to rotate selected attribute by 90 degree. The procedure to use this tool is given next.

- Click on the **Rotate Attribute** tool from the **Edit Attribute** drop-down in the **Edit Components** panel of **Schematic** tab in the **Ribbon**. You are asked to select the attribute.
- Click on the attribute you want to rotate. It will be rotated by 90 degree.
- Press **ESC** to exit the tool.

Changing Attribute/Text Justification

The **Change Attribute Justification** tool is used to change the justification of attribute text. The procedure to use this tool is given next.

- Click on the **Change Attribute Justification** tool from the **Edit Attribute** drop-down in the **Edit Components** panel of the **Schematic** tab in the **Ribbon**. The **Change Attribute/Text Justification** dialog box will be displayed; refer to Figure-51.

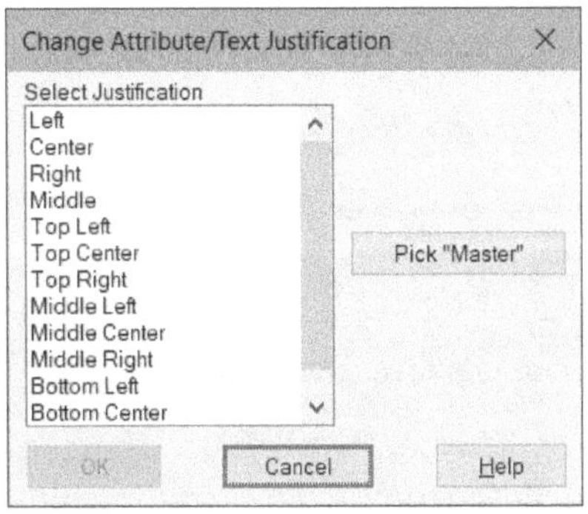

Figure-51. Change Attribute Text Justification dialog box

- Select desired option from the **Select Justifications** list of the dialog box and click on the **OK** button. You will be asked to select the attributes/text whose justification is to be changes.
- You can select the attribute/text one by one or using the window selection. To use window selection, type **W** in the **Command Bar** and press **ENTER**. Now, select the attributes using window selection and then press **ENTER** to apply modification.

Changing Attribute Layer

The **Change Attribute Layer** tool is used to change the layer of attributes. The procedure to use this tool is given next.

- Click on the **Change Attribute Layer** tool from the **Edit Attribute** panel of the **Edit Components** panel of the **Schematic** tab in the **Ribbon**. The **Force Attribute/Text to a Different Layer** dialog box will be displayed; refer to Figure-52.

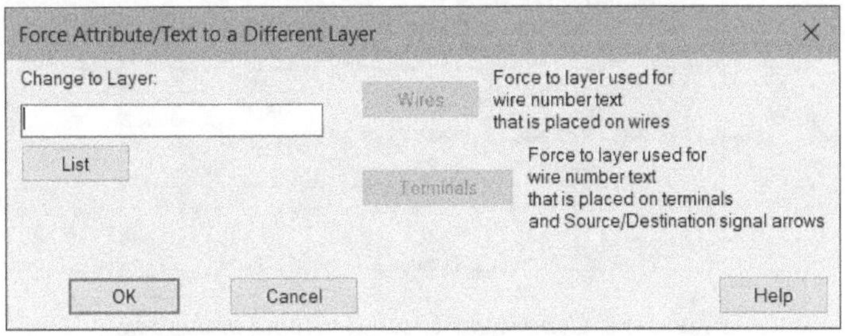

Figure-52. Force Attribute Text to a Different Layer dialog box

- Click on the **List** button from the dialog box and select desired layer on which you want to move the attributes from the **Layers in Drawing** dialog box. After selecting layer, click on the **OK** button from the **Layers in Drawing** dialog box.
- Click on the **OK** button from the dialog box to apply modifications. You will be asked to select the attributes to be moved to selected layer.
- Press **ESC** to exit the tool.

CROSS REFERENCES DROP-DOWN

The tools in this drop-down are used to manage cross-references for the components; refer to Figure-53. The tools in this drop-down are discussed next.

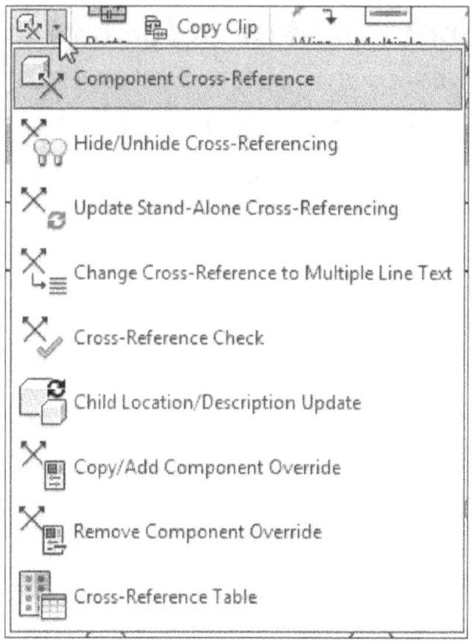

Figure-53. Cross References drop-down

Component Cross-Reference

This tool is used to update cross-reference text of a component. The procedure is given next.

- Click on the **Component Cross-Reference** tool from the **Cross References** drop-down. The **Component Cross-Reference** dialog box will be displayed; refer to Figure-54.

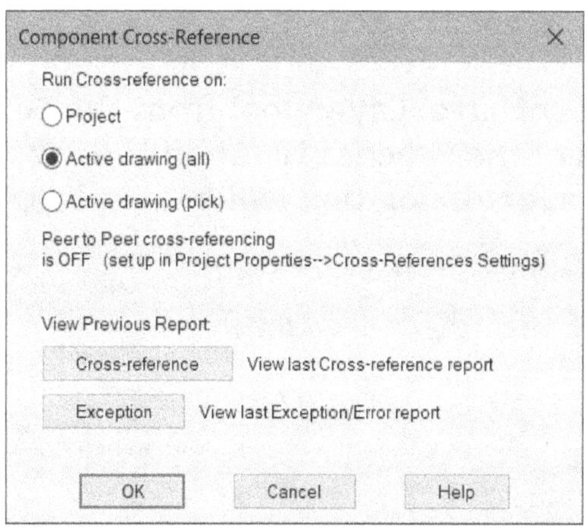

Figure-54. Component Cross-Reference dialog box

- Select desired radio button from the dialog box and click on the **OK** button. The **Error/Exception Report** dialog box will be displayed; refer to Figure-55.

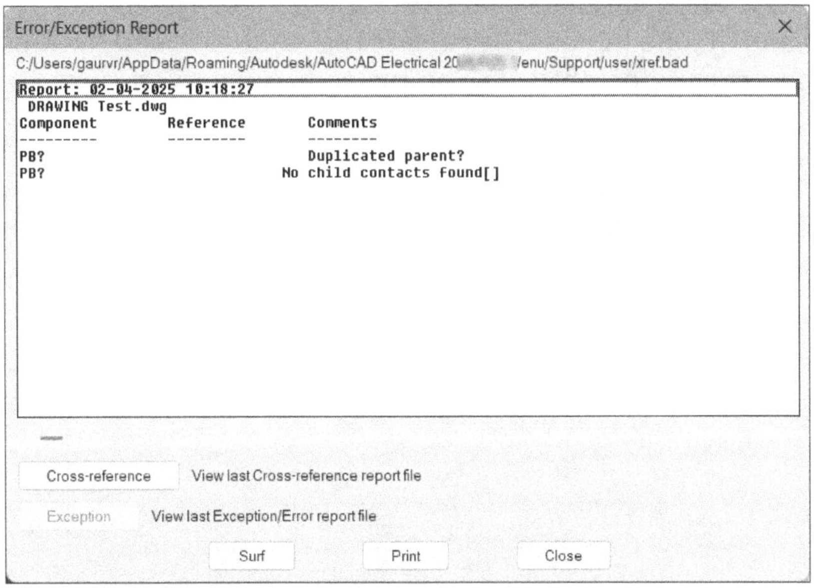

Figure-55. Error or Exception Report dialog box

- Click on the **Cross-reference** button from the dialog box. The **Cross-Reference Report** dialog box will be displayed; refer to Figure-56.

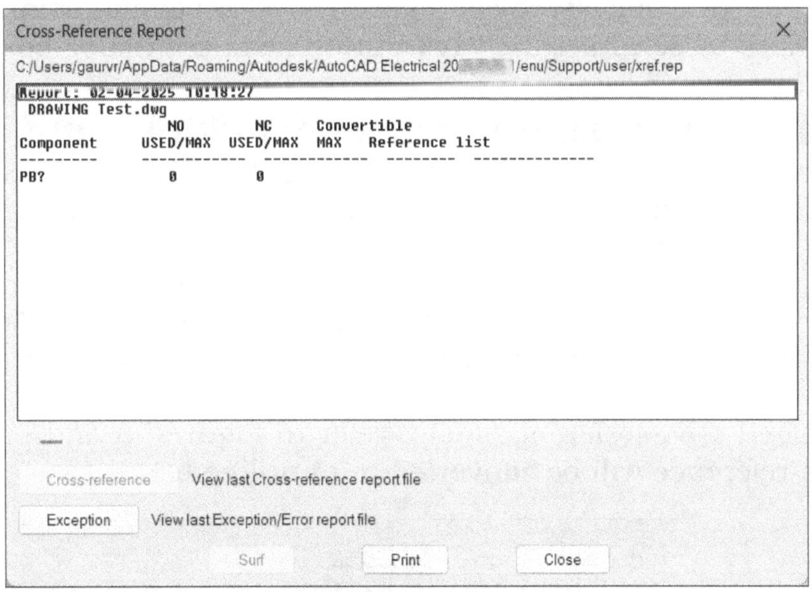

Figure-56. Cross Reference Report dialog box

- From the above result, we can find the components and their respective references in the drawing/project.
- After checking the results, click on the **Close** button to exit.

Hide/Unhide Cross-Referencing

The **Hide/Unhide Cross-Referencing** tool is used to hide or unhide the cross reference of a component in the drawing. Click on this tool from the **Cross References** drop-down and select the component for which you want to change the hide/unhide status.

Update Stand-Alone Cross-Referencing

The **Update Stand-Alone Cross-Referencing** tool is used to update the cross-references in the drawing or project. The procedure is given next.

- Click on **Update Stand-Alone Cross-Referencing** tool from the **Cross References** drop-down. The **Update Wire Signal and Stand-Alone Cross-Reference** dialog box will be displayed; refer to Figure-57.

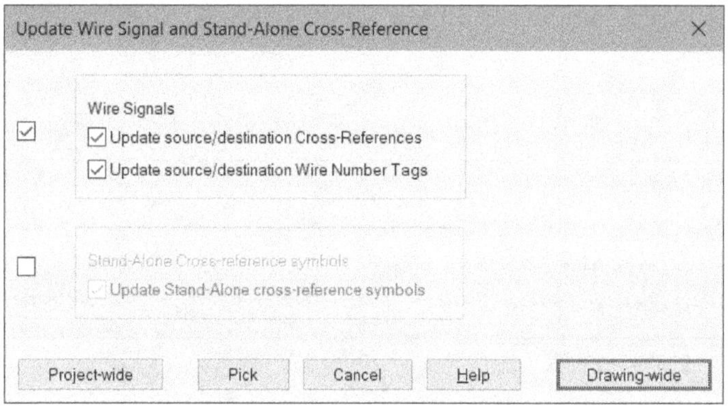

Figure-57. Update Wire Signal and Stand-Alone Cross-Reference dialog box

- Click on the **Drawing-wide** or **Project-wide** button to update all the cross-references in the drawing or project respectively. You can individually update the cross-references by selecting the **Pick** button from the dialog box.

Changing Cross-Reference to Multiple Line Text

The **Change Cross-Reference to Multiple Line Text** tool is used to convert attribute text to simple multiline text. The procedure to use this tool is given next.

- Click on the **Change Cross-Reference to Multiple Line Text** tool from the **Cross-References** drop-down in the **Edit Components** panel of **Schematic** tab in the **Ribbon**. You will be asked to select the cross-reference text to be converted to multiline text.
- Click on desired cross-reference texts; refer to Figure-58 and press **ENTER**. The text of cross-reference will be converted to multiline text.

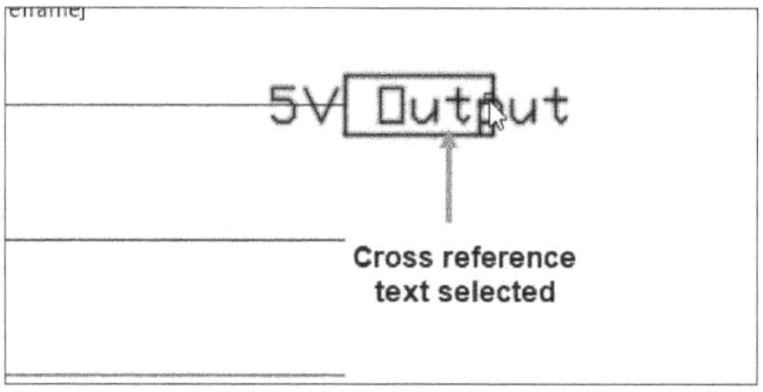

Figure-58. Cross-reference text selected

Cross-Reference Check

The **Cross-Reference Check** tool is used to check cross-references of selected components. The procedure to use this tool is given next.

- Click on the **Cross-Reference Check** tool from the **Cross-References** drop-down in the **Edit Components** panel in the **Ribbon**. You will be asked to select a component whose cross-references are to be checked.

- Select desired component. The **Reference Listing** dialog box will be displayed; refer to Figure-59.

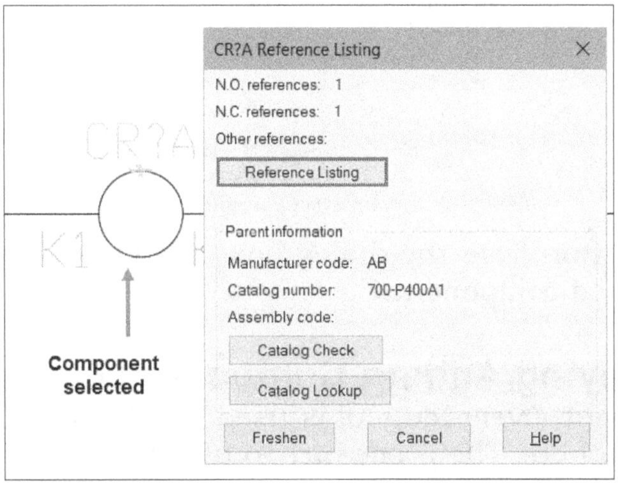

Figure-59. Reference Listing dialog box

- Click on the **Reference Listing** button from the dialog box to check references.
- Click on the **Cancel** button from the dialog box to exit and press **ESC** to exit the tool.

Child Location/Description Update

The **Child Location/Description Update** tool is used to match description and location of child components with schematic symbols. The procedure to use this tool is given next.

- Click on the **Child Location/Description Update** tool from the **Cross-References** drop-down in the **Ribbon**. The **Child Contact and Panel Update from Schematic Parent** dialog box will be displayed.

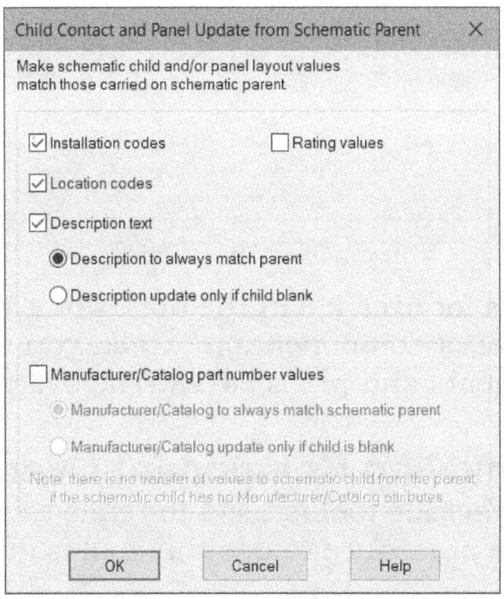

Figure-60. Child Contact and Panel Update from Schematic Parent dialog box

- Set desired parameters in the dialog box and click on the **OK** button. You will be asked to select the components.
- Select desired components to be updated and press **ENTER**. If there are any updates required then **Update other drawings** dialog box will be displayed; refer to Figure-61.

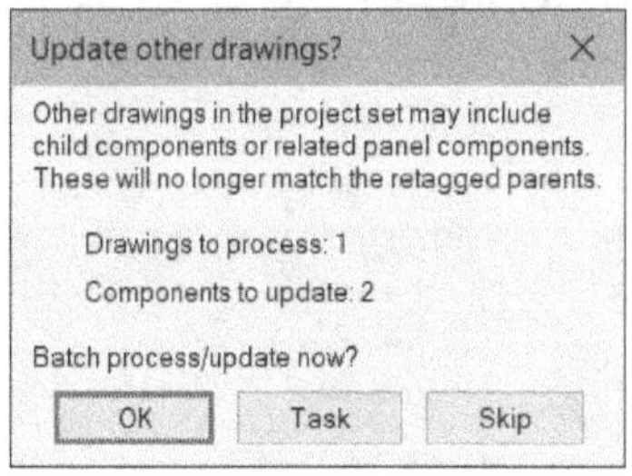

Figure-61. Update other drawings dialog box

- Click on the **OK** button from the dialog box to update cross-references in other drawings for selected components.

Copying/Adding Component Override

The **Copy/Add Component Override** tool is used to add/apply component overrides of one component to others. The procedure to use this tool is given next.

- Click on the **Copy/Add Component Override** tool from the **Cross-References** drop-down in the **Ribbon**. You will be asked to select the component whose overrides are to be copied.
- Select desired component. The **Cross-Reference Component Override** dialog box will be displayed; refer to Figure-62.

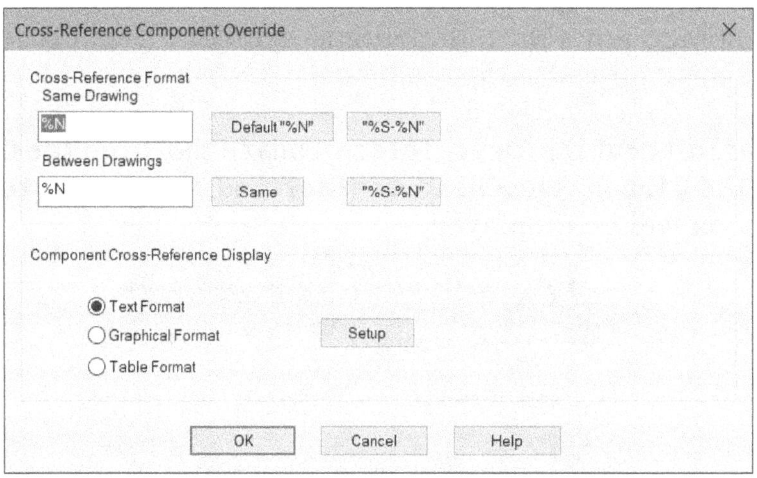

Figure-62. Cross Reference Component Override dialog box

- Set desired parameters for override in the dialog box and click on the **OK** button. You will be asked to select components to which you want to apply override.
- Select desired components and press **ENTER**. The overrides will be applied.

Removing Component Override

The **Remove Component Override** tool is used to remove cross-reference overrides of selected components. The procedure to use this tool is given next.

- Click on the **Remove Component Override** tool from the **Cross-Reference** drop-down in the **Edit Components** panel of **Schematic** tab in the **Ribbon**. The **Remove Component Overrides** dialog box will be displayed; refer to Figure-63.

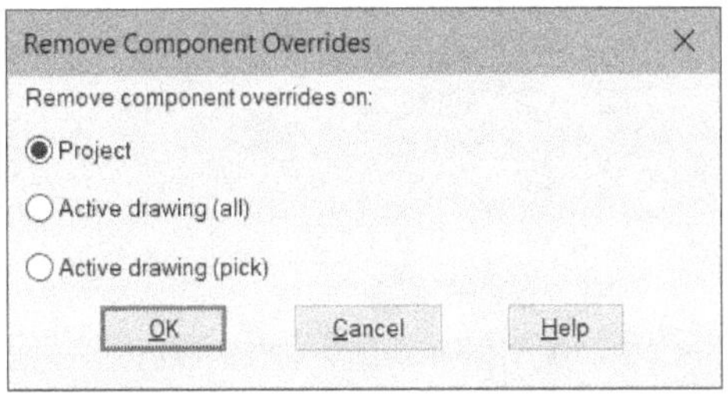

Figure-63. Remove Component Overrides dialog box

- Select desired radio button from the dialog box to define the scope within which component overrides will be removed.
- Click on the **OK** button from the dialog box to remove overrides. If you have selected the **Active drawing (pick)** radio button then you will be asked to select the components whose overrides are to be removed.

Generating Cross-Reference Table

The **Cross-Reference Table** tool is used to generated cross-reference table for selected component. The procedure to use this tool is given next.

- Click on the **Cross-Reference Table** tool from the **Cross-Reference** drop-down in the **Ribbon**. You will be asked to select the component.
- Select desired component. The **Report Generator** dialog box will be displayed; refer to Figure-64.

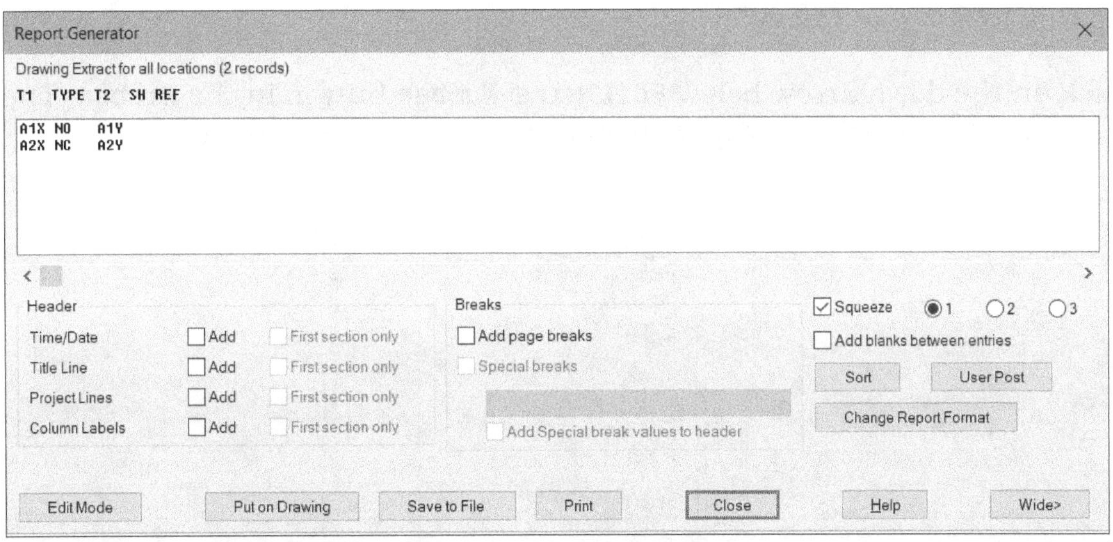

Figure-64. Report Generator dialog box

- Set desired parameters in the dialog box and click on the **Close** button to exit. You will learn more about this dialog box later.

CIRCUIT CLIPBOARD PANEL

The tools in this panel are used to create copy of circuits; refer to Figure-65. Click on the **Cut** or **Copy Clip** tool to copy any circuit in the drawing and using the **Paste** tool, you can paste the copied or cut circuits.

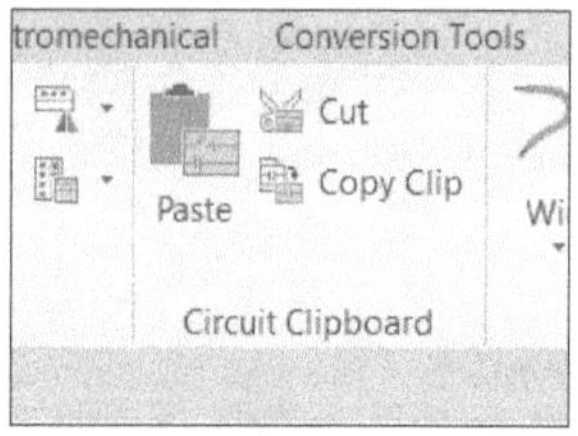

Figure-65. Circuit Clipboard panel

EDITING WIRES OR WIRE NUMBERS

The tools to edit wires/wire numbers are available in the **Edit Wires/Wire Numbers** panel in the **Ribbon**; refer to Figure-66. The tools in this panel are discussed next.

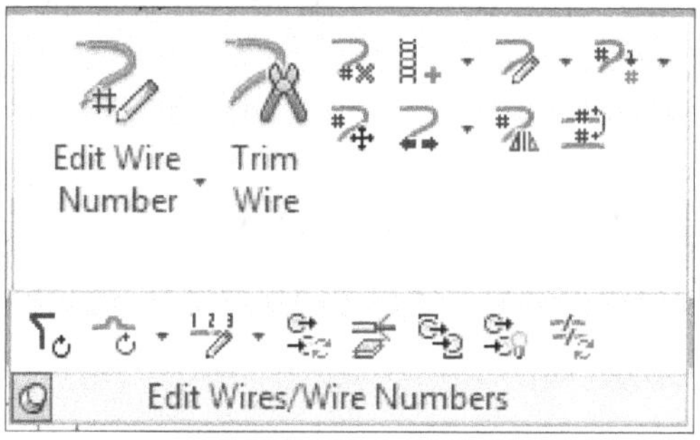

Figure-66. Edit Wires/Wire Numbers panel

Edit Wire Number

The **Edit Wire Number** tool is used to manually change the wire number. The procedure to use this tool is given next.

- Click on the down arrow below **Edit Wire Number** button in the **Ribbon**. The **Edit Wire Number** drop-down will be displayed; refer to Figure-67.

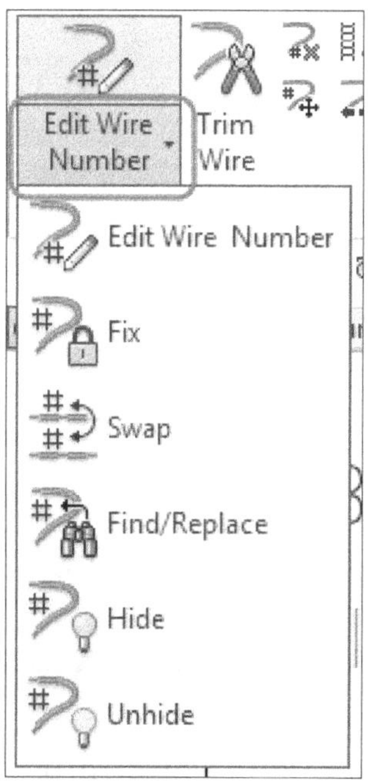

Figure-67. Edit Wire Number drop-down

- Click on the **Edit Wire Number** tool from the drop-down. You are asked to select a wire or wire number.

- Select the wire or wire number that you want to change. The **Edit Wire Numbers/ Attributes** dialog box will be displayed; refer to Figure-68.

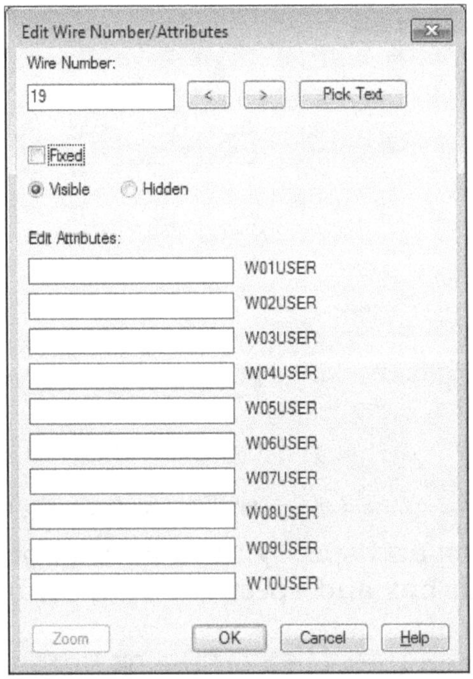

Figure-68. Edit Wire Number or Attributes dialog box

- Click in the **Wire number** edit box and specify desired value. You can use the **<** or **>** buttons to decrease or increase the wire number. Note that if the number is existing then you will be prompted to enter a different wire number.
- Clear the **Fixed** check box, if you want to unfix the wire number.
- Select the **Hidden** radio button if you want to hide the wire number.
- Click on the **OK** button to apply the specified parameters.

Fix

The **Fix** tool is used to make a wire number fixed. To do so, click on the **Fix** tool from the **Edit Wire Number** drop-down. You are asked to select a wire number that you want to be fixed. Select the wire number and it will get fixed. Also, the color of the wire number will change to **Black** color.

Swap

The **Swap** tool is used to swap the wire numbers. Click on this tool from the **Edit Wire Number** drop-down and select two wire numbers one by one. The wire numbers will be swapped.

Find/Replace

The **Find/Replace** tool is available in the **Edit Wire Number** drop-down. This tool is used to find and replace the wire numbers in the drawing. The procedure to use this tool is given next.

- Click on the **Find/Replace** tool from the drop-down. The **Find/Replace Wire Numbers** dialog box will be displayed; refer to Figure-69.

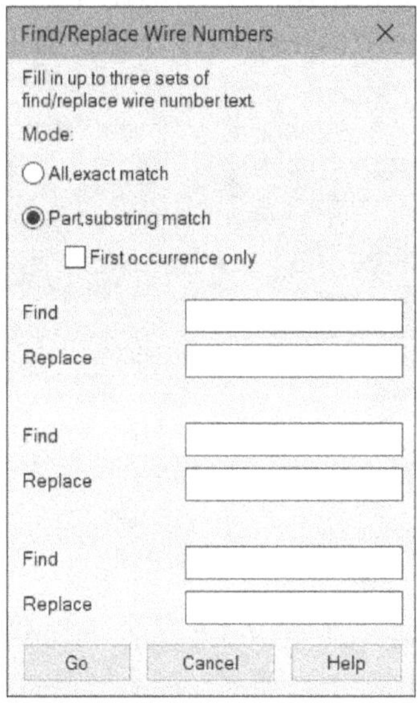

Figure-69. Find or Replace Wire Numbers dialog box

- Click in the **Find** edit box and specify the wire number to be replaced.
- Click in the **Replace** edit box and specify the wire number with which it is to be replaced.
- Click on the **Go** button and then click on the **OK** button, the wire numbers will be replaced.

Hide and Unhide

The **Hide** tool is used to hide the wire numbers and **Unhide** tool is used to unhide the wire numbers. Click on the **Hide** tool and select the wire number that you want to hide. To unhide the wire number, click on the **Unhide** tool from the **Edit Wire Number** drop-down. You are asked to select a wire. Select the wire, the hidden wire number for that wire will be displayed.

Trim Wire

The **Trim Wire** tool is used to trim a wire connected between wire connections. The procedure to use this tool is given next.

- Click on the **Trim Wire** tool from the **Edit Wires/Wire Numbers** panel. You are asked to select the wire that is to be trimmed.
- Select the wire that you want to be trimmed; refer to Figure-70.

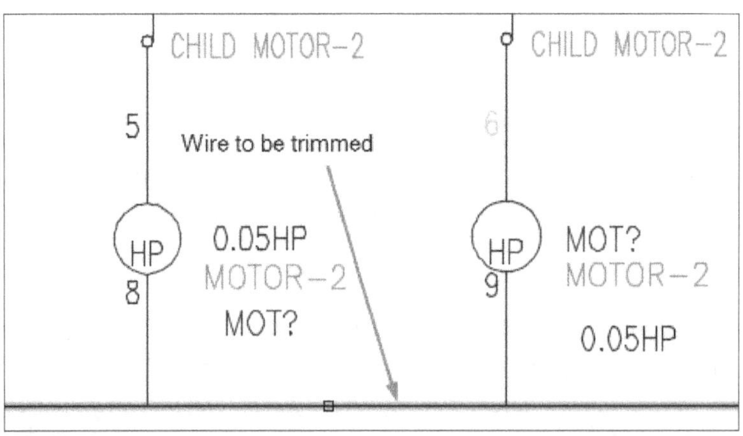

Figure-70. Wire to be trimmed

- On selecting the wire, it will be trimmed between the connecting ends; refer to Figure-71. Also, the wire numbering will be modified accordingly.

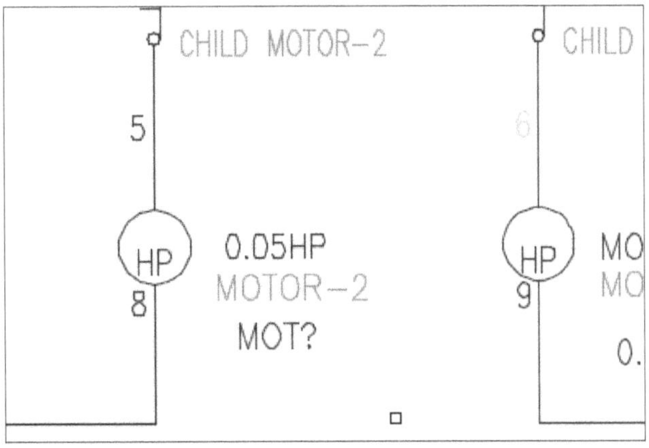

Figure-71. After trimming wire

Delete Wire Numbers

The **Delete Wire Numbers** tool is used to delete the wire numbers. Note that the remaining wire numbers are not modified on using this tool. To delete the wire numbers, click on this tool and then click on the wire numbers that you want to be deleted. The selected wire numbers will be deleted.

Move Wire Number

The **Move Wire Number** tool is used to move the wire numbers. To do so, click on **Move Wire Number** tool from the **Edit Wires/Wire Numbers** panel; you will be prompted to select a new location for the wire number. Click on the wire at desired position. The wire number will move to the location you clicked on the wire.

Add Rung

The **Add Rung** tool is used to add a rung in any circuit or ladder. A rung is a wire connecting two parallel wiring lines. The procedure to use this tool is given next.

- Click on the **Add Rung** tool from the **Ladder Editing** drop-down; refer to Figure-72. You are asked to click at desired location to add rung.
- Click between the two parallel wires. A rung will be created connecting both the wires; refer to Figure-73.

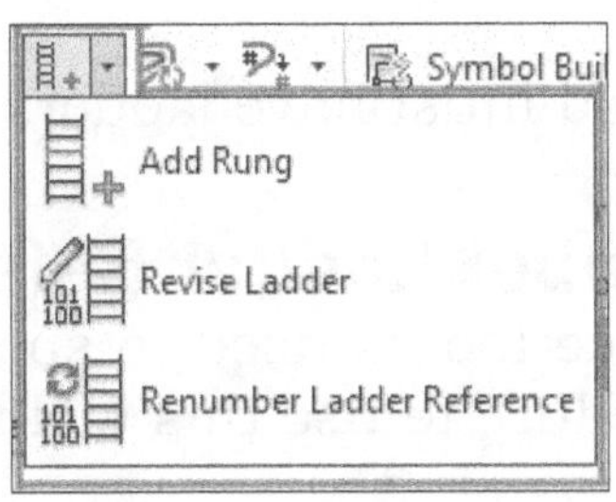

Figure-72. Ladder Editing drop-down

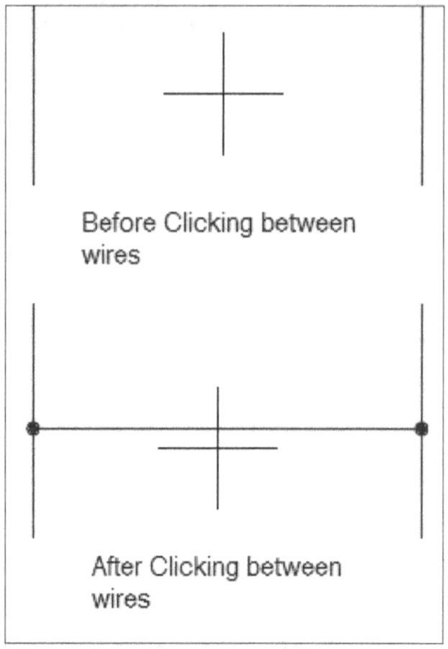

Figure-73. Rung connected

Revise Ladder

The **Revise Ladder** tool is available in the **Ladder Editing** drop-down. This tool is used to modify the ladder numbering. The procedure to use this tool is given next.

- Click on the **Revise Ladder** tool from the **Ladder Editing** drop-down. The **Modify Line Reference Numbers** dialog box will be displayed; refer to Figure-74.

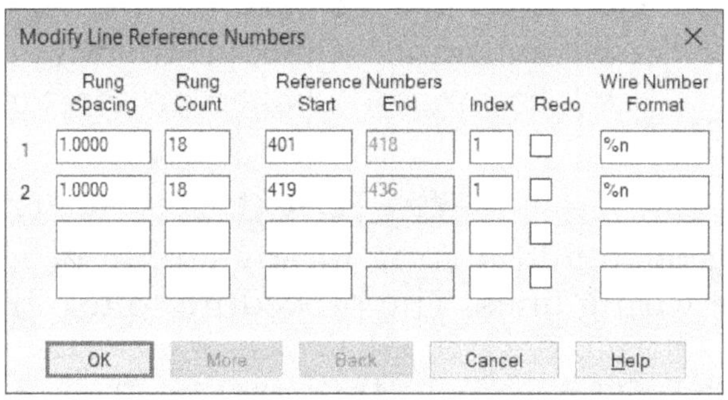

Figure-74. Modify Line Reference Number dialog box

- Specify desired parameters for the ladder numbering in the current drawing and click on the **OK** button from the dialog box. The ladder numbering will be modified accordingly.
- Note that to use this tool, you must have ladders in the drawing area.

Renumber Ladder Reference

The **Renumber Ladder Reference** tool is used to specify the reference point for the ladder renumbering. The procedure to use this tool is given next.

- Click on the **Renumber Ladder Reference** tool from the **Ladder Editing** drop-down. The **Renumber Ladders** dialog box will be displayed; refer to Figure-75.

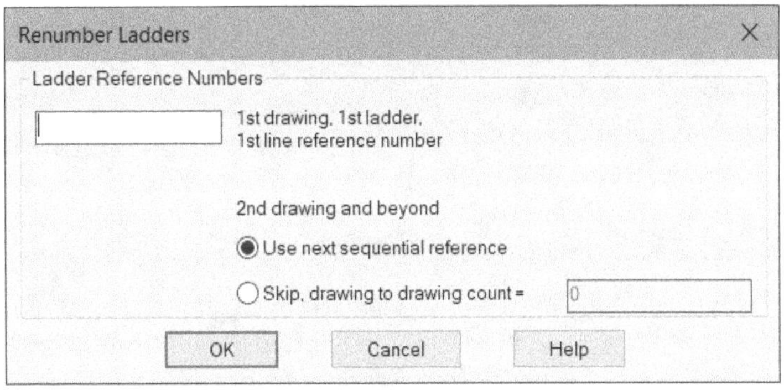

Figure-75. Renumber Ladders dialog box

- Click in the edit box and specify desired reference number. Next, click on the **OK** button from the dialog box. The **Select Drawings to Process** dialog box will be displayed; refer to Figure-76.

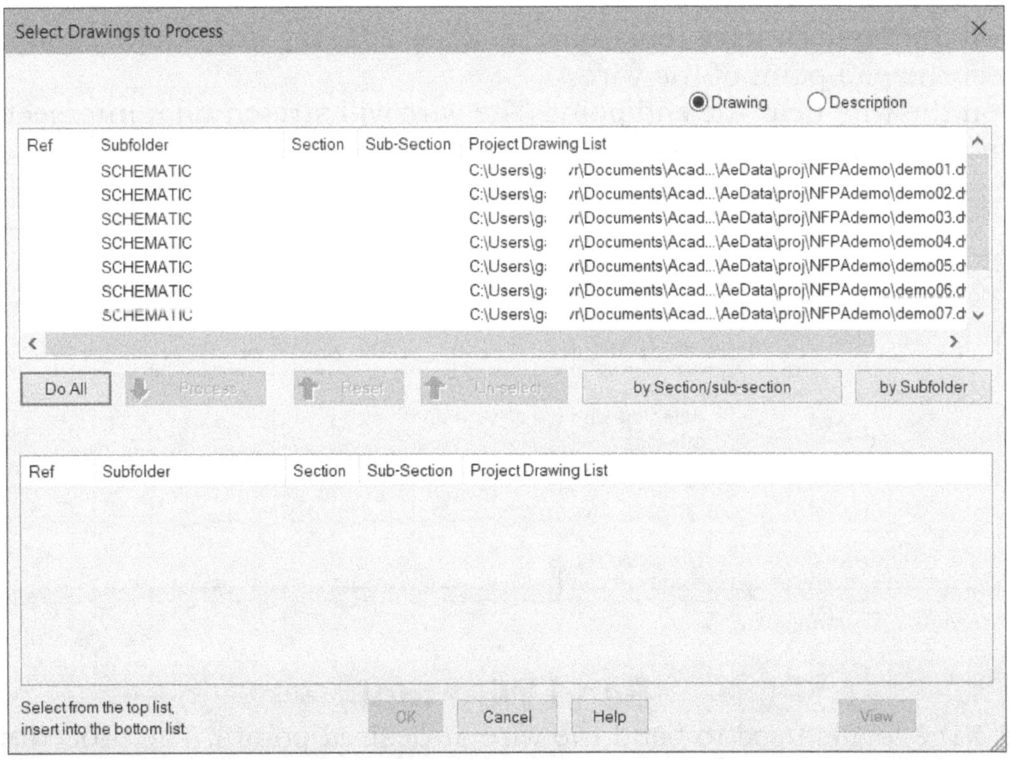

Figure-76. Select Drawings to Process dialog box

- Select the drawing and click on the **Process** button on which you want to apply the numbering. To select all drawings, click on the **Do All** button and then click on the **OK** button from the dialog box. The ladders will be renumbered as per the specified value.

Wire Editing

The tools in the **Wire Editing** drop-down are used to modify wires; refer to Figure-77. The tools in this drop-down are discussed next.

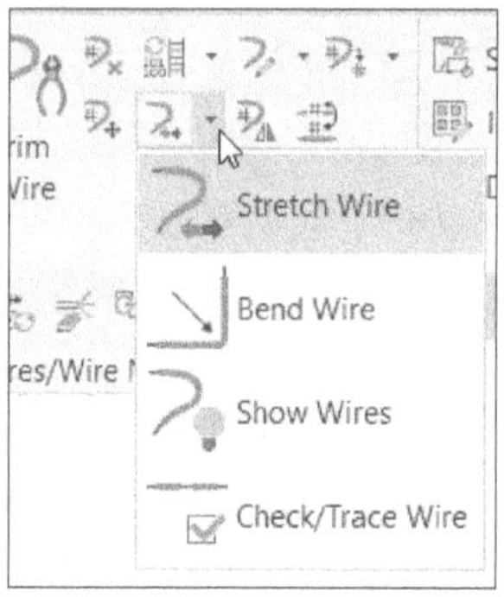

Figure-77. Wire Editing drop-down

Stretch Wire tool

The **Stretch Wire** tool is used to stretch the end point of a wire to connect it to the next intersecting wire. The procedure to use this tool is given next.

- Click on the **Stretch Wire** tool from the **Wire Editing** drop-down. You are asked to select the end point of the wire.
- Click on the wire near the end point. The wire will stretch till it intersect with the next facing wire; refer to Figure-78.

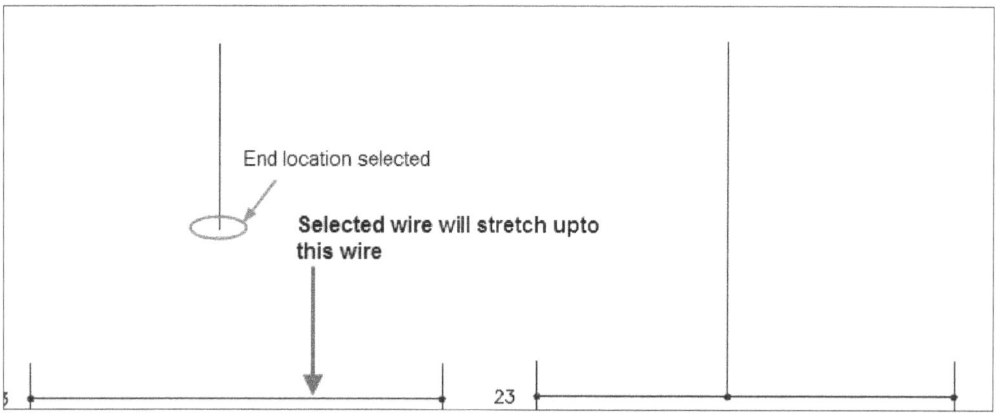

Figure-78. Stretching wire

Bend Wire tool

The **Bend Wire** tool is used to bend the wire at desired points. The procedure to use this tool is given next.

- Click on the **Bend Wire** tool from the **Wire Editing** drop-down. You are asked to select the point on first wire from which the bending will start.
- Click on the wire at desired point; refer to Figure-79.

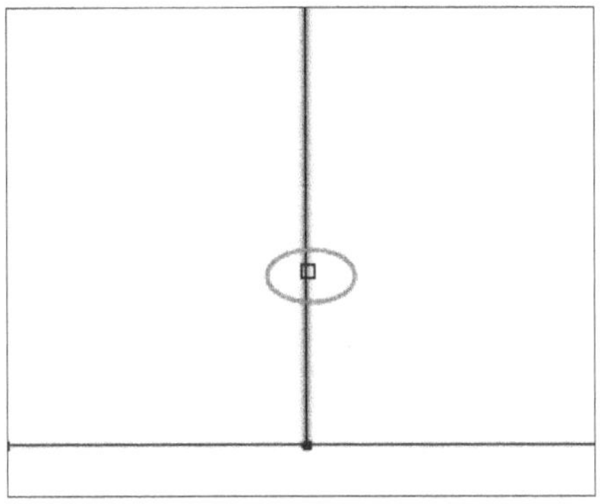

Figure-79. First point of bending

- Click on the second wire to specify the end point of bend. The wire bend will be created refer to Figure-80.

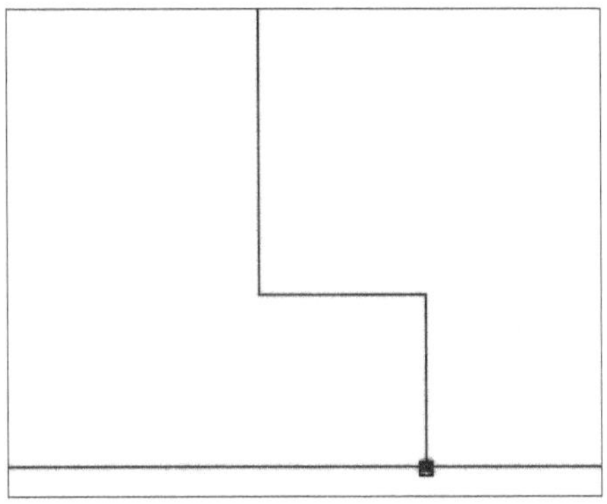

Figure-80. Wire bend after selecting second point

Show Wires

The **Show Wire** tool is used to display all the wires in the drawing. The procedure to use this tool is given next.

- Click on the **Show Wires** tool from the **Wire Editing** drop-down. The **Show Wires and Wire Number Pointers** dialog box will be displayed; refer to Figure-81.

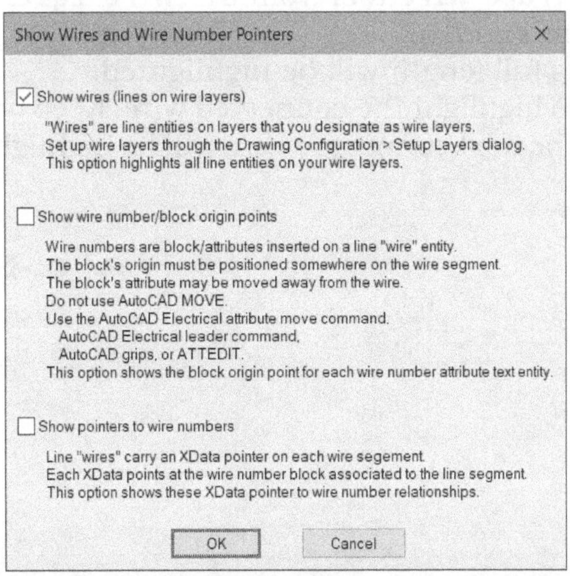

Figure-81. Show Wires and Wire Number Pointers dialog box

- Select desired check boxes to highlight the respective entities in the drawing.

- Click on the **OK** button from the dialog box. The **Drawing Audit** dialog box will be displayed; refer to Figure-82.

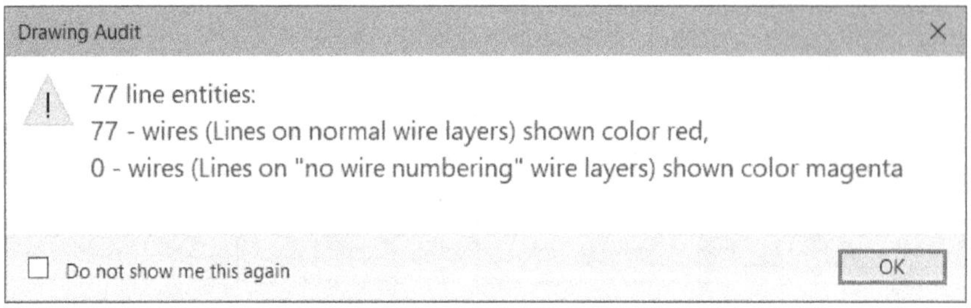

Figure-82. Drawing Audit dialog box

- Click on the **OK** button from the dialog box. The wires will be highlighted; refer to Figure-83.

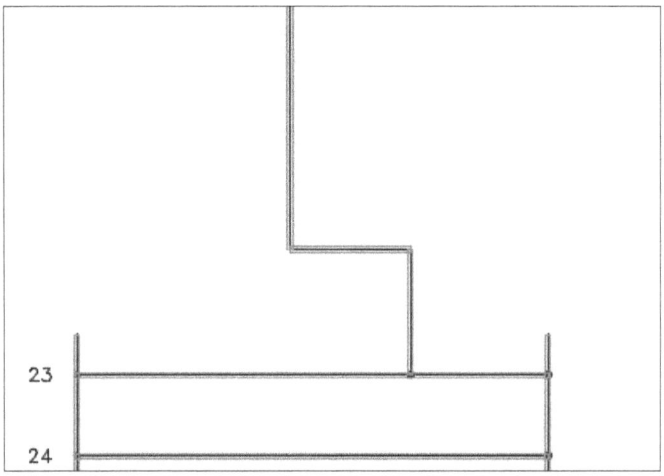

Figure-83. Highlighted wires

Check or Trace Wire tool

The **Check/Trace Wire** tool is used to highlight a wire in the drawing. The procedure to use this tool is given next.

- Click on the **Check/Trace Wire** tool from the **Wire Editing** drop-down. You are asked to select a wire segment.
- Click on the wire, its full length will be highlighted.
- Press the **SPACEBAR** to highlight the connected wire. Keep on pressing the **SPACEBAR** from keyboard and the next connected wire will be highlighted; refer to Figure-84.

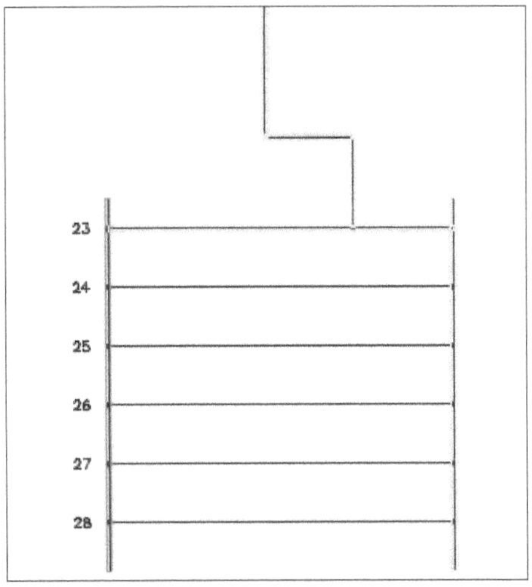

Figure-84. Wire highlighted

Wire Type Editing drop-down

The tools in the **Wire Type Editing** drop-down are used to edit the wire types; refer to Figure-85. These tools and their operations are discussed next.

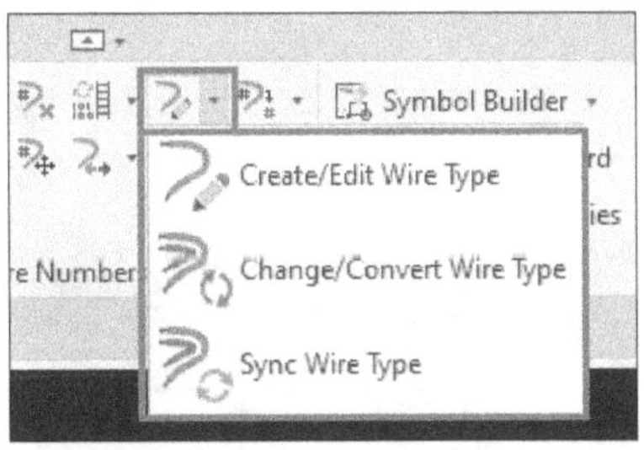

Figure-85. Wire Type Editing drop-down

Create/Edit Wire Type

The **Create/Edit Wire Type** tool is used to create or edit wire types. The procedure to use this tool is given next.

- Click on the **Create/Edit Wire Type** tool from the **Wire Type Editing** drop-down. The **Create/Edit Wire Type** dialog box will be displayed as shown in Figure-86.
- The fields in the table of this dialog box are used to specify parameters of common type of wires. If you want to modify the wire type then click on desired field and specify desired values. Note that **Layer Name** is automatically assigned on the basis of **Wire Color** and **Size** values.
- To add a new wire type, slide down the bottom of table and click in the **Wire Color** field; refer to Figure-87.

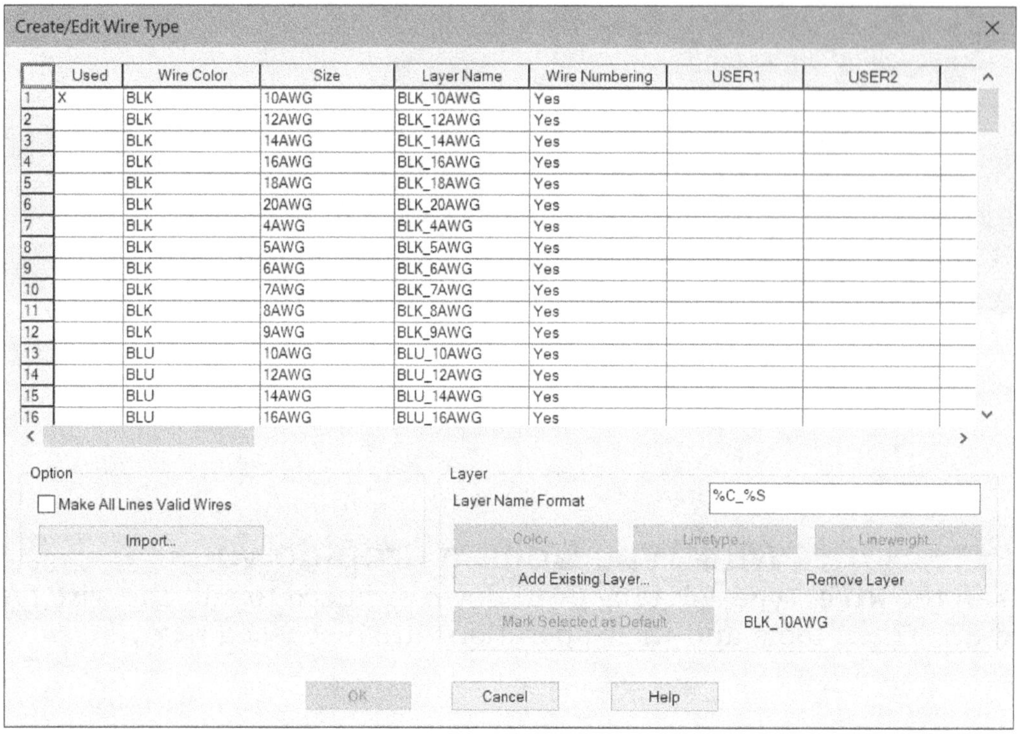

Figure-86. Create or Edit Wire Type dialog box

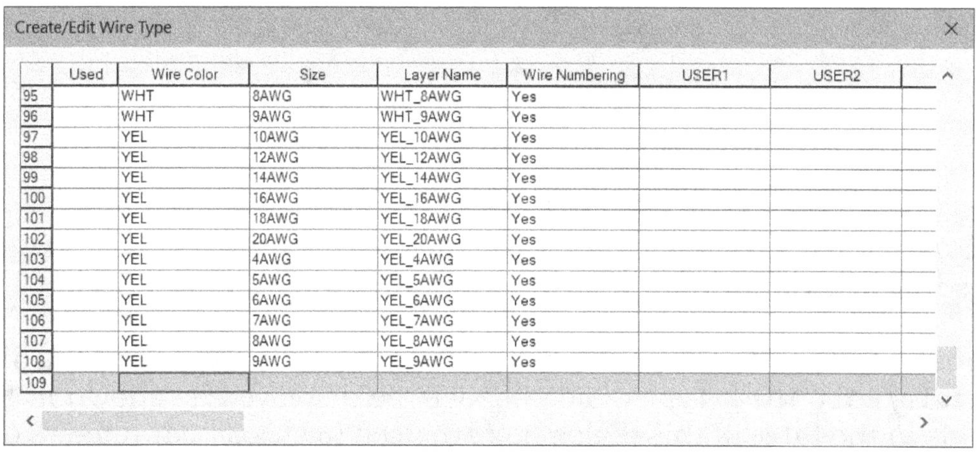

Figure-87. Wire Color field

- Specify desired color. For example, **ORNG**.
- Click in the **Size** field next to it and specify desired size. For example, **4AWG**.
- Click on desired button in the **Layer** area and specify the related properties for the layers.
- After setting desired parameters, click on the **OK** button from the dialog box. The wire type will be created.

Change/Convert Wire Type

The **Change/Convert Wire Type** tool is used to change the wire type of wires in the drawing. The procedure to use this tool is given next.

- Click on the **Change/Convert Wire Type** tool from the **Wire Type Editing** drop-down. The **Change/Convert Wire Type** dialog box will be displayed as shown in Figure-88.

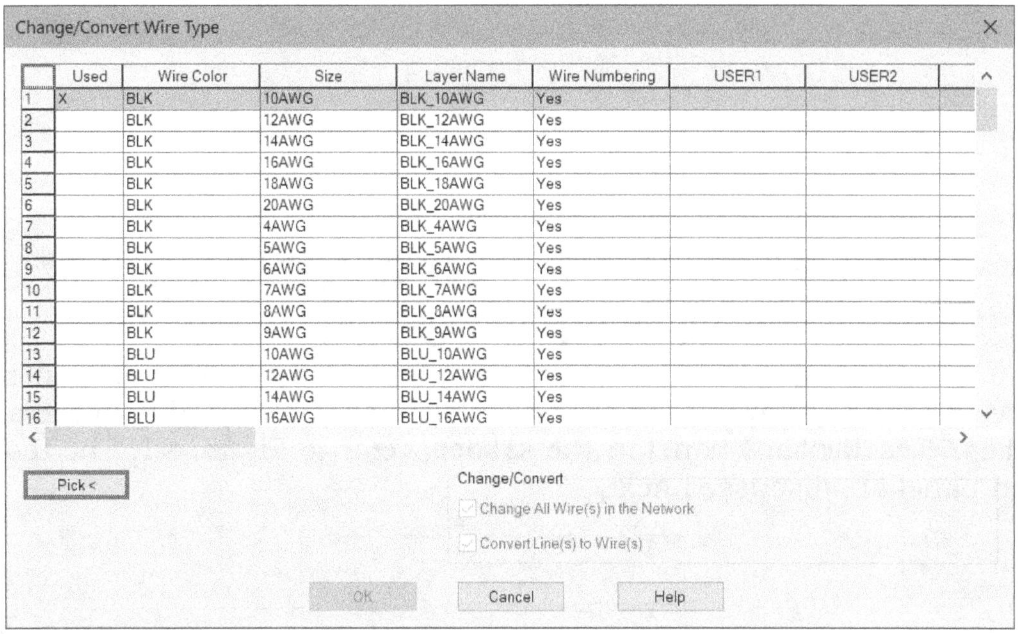

Figure-88. Change/Convert Wire Type dialog box

- Select desired wire type from the list and click on the **OK** button from the dialog box. You are asked to select a wire.
- Select the wire from the drawing area. All the similar wires will be converted to the wire type selected in the dialog box.

Sync Wire Type

The **Sync Wire Type** is used to sync all wires in the network with the same signal code across the drawings.

Flip Wire Number

The **Flip Wire Number** tool is used to flip the wire number on the other side of the wire. The procedure to use this tool is given next.

- Click on the **Flip Wire Number** tool from the **Edit Wires/Wire Numbers** panel in the **Ribbon**. You are asked to select a wire number.
- Select the wire number and it will be flipped to the other side of wire; refer to Figure-89.

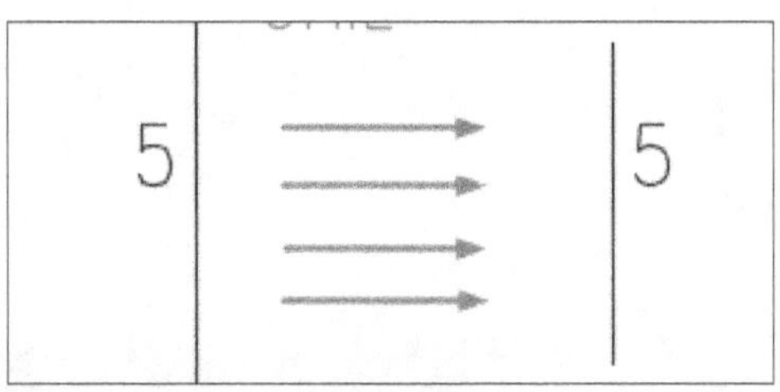

Figure-89. Wire number flipped

Toggle Wire Number In-Line

The **Toggle Wire Number In-Line** tool works in the same way as the **Flip Wire Number** tool works. After using this tool, the wire number gets inserted in the middle of the wire; refer to Figure-90.

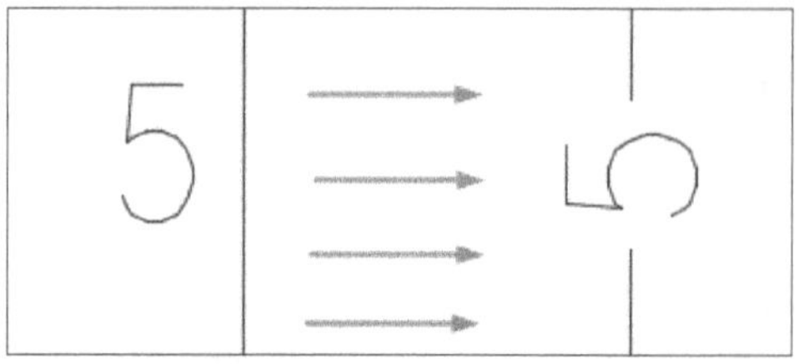

Figure-90. Wire number after toggling

ADVANCED WIRE AND WIRE NUMBER EDITING TOOLS

There are a few advanced wire and wire number editing tools that can make big difference in wiring and wire numbering. These tools are available in the expanded **Edit Wires/Wire Numbers** panel in the **Ribbon**; refer to Figure-91. The tools in the expanded panel are discussed next.

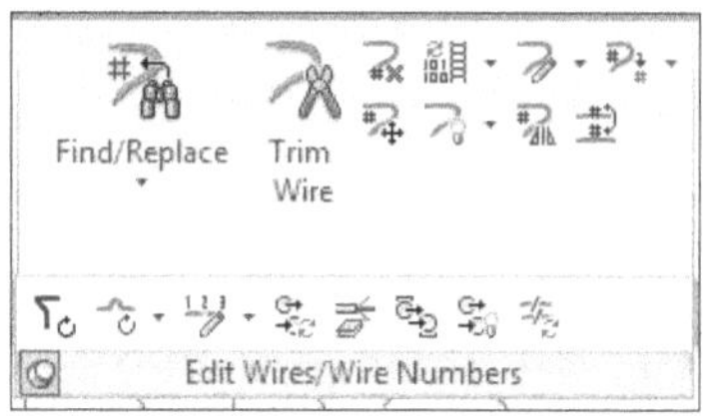

Figure-91. Expanded Edit Wires panel

Toggle Angled Tee Markers

Yes, you can toggle the Tee markers but why?? If we use an angled tee instead of a dot where the wires connect, AutoCAD Electrical can tell exactly how we want it wired. The angle tee orientation indicates the sequence; refer to Figure-92. Note that the default information about Tee markers can be changed from the **Style** tab of **Drawing Properties** dialog box; refer to Figure-93. The procedure to open the **Drawing Properties** dialog box has already been discussed.

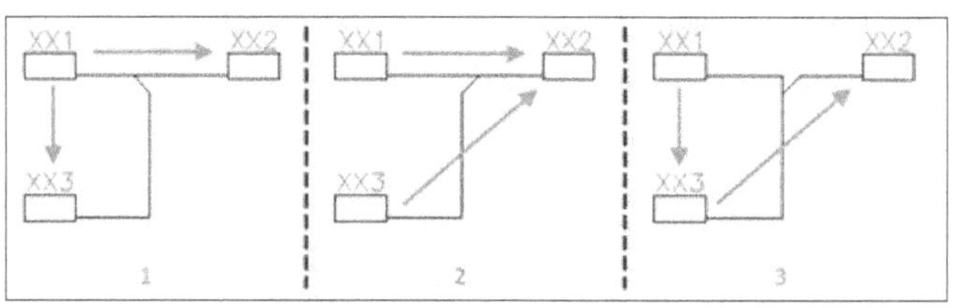

Figure-92. Wiring sequence

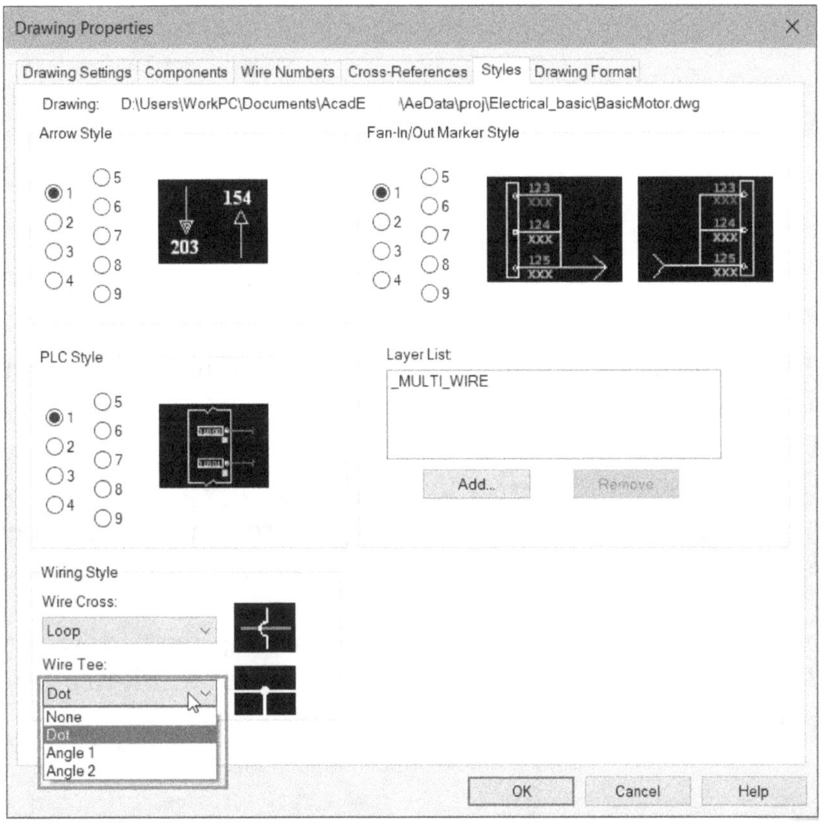

Figure-93. Wire Tee drop-down

The procedure to toggle between dot tee marker and angled tee marker is discussed next.

- Click on the **Toggle Angled Tee Marker** tool from the expanded **Edit Wires/Wire Numbers** panel in the **Schematic** tab of the **Ribbon**. You are asked to select the Tee connections.
- Click on the tee connection that you want to toggle; refer to Figure-94.

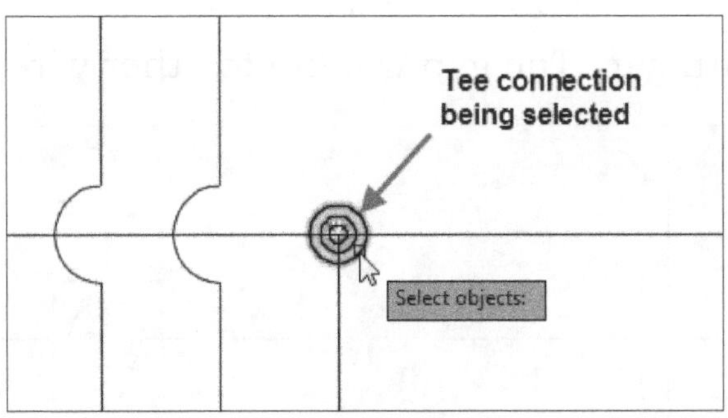

Figure-94. Tee connection being selected

- After selecting the connection(s), press **ENTER** from the keyboard. Preview of angled tee connection will be displayed; refer to Figure-95.

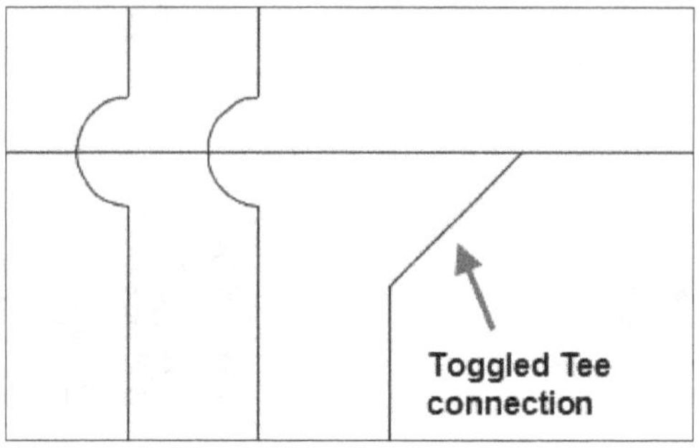

Figure-95. Toggled Tee Connection

- Press **SPACEBAR** from keyboard to toggle position of angled tee connection. Press **ESC** to create the connection.

Flip Wire Gap

The **Flip Wire Gap** tool is used to flip the wire gap of intersecting wires. The procedure to use this tool is given next.

- Click on the **Flip Wire Gap** tool from the **Wire Gap** drop-down in the expanded **Edit Wires/Wire Numbers** panel in the **Ribbon**; refer to Figure-96. You are asked to select the wire with gap.

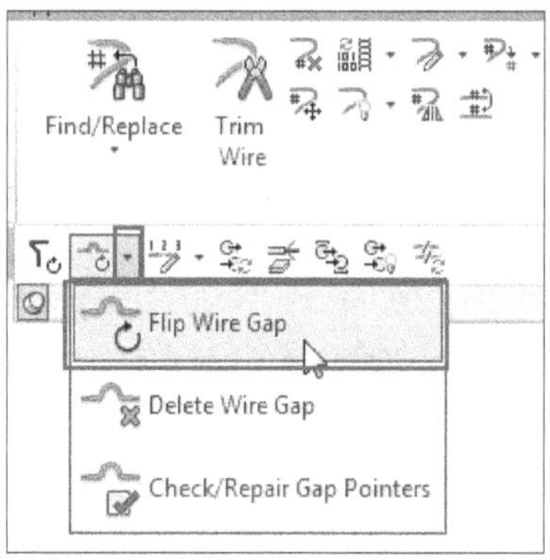

Figure-96. Wire Gap drop-down

- Click on the wire with gap. The gap will flip to other wire; refer to Figure-97.

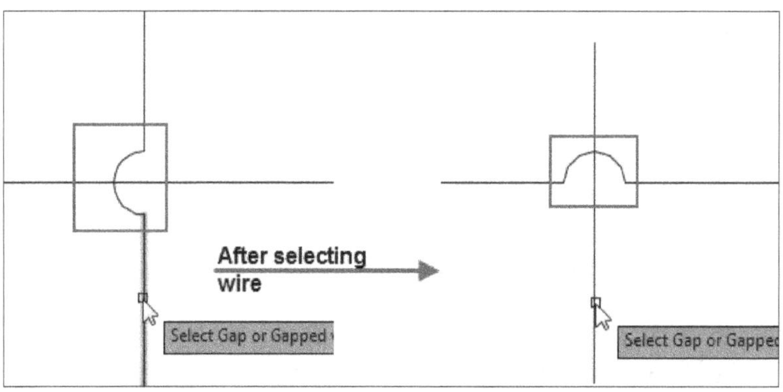

Figure-97. Flipping wire gap

- Press **ESC** to exit the tool.

Delete Wire Gap

The **Delete Wire Gap** tool is used to delete unnecessary gaps in the wires. Some times we delete one of the intersecting wires but wire gap remains there. To delete such wire gaps, we need this tool. The procedure to use this tool is given next.

- Click on the **Delete Wire Gap** tool from the **Wire Gap** drop-down in the expanded **Edit Wires/Wire Numbers** panel of the **Ribbon**. You are asked to select the objects.
- Select the wire with unnecessary gap and press **ENTER**. The extra gaps will be deleted; refer to Figure-98.

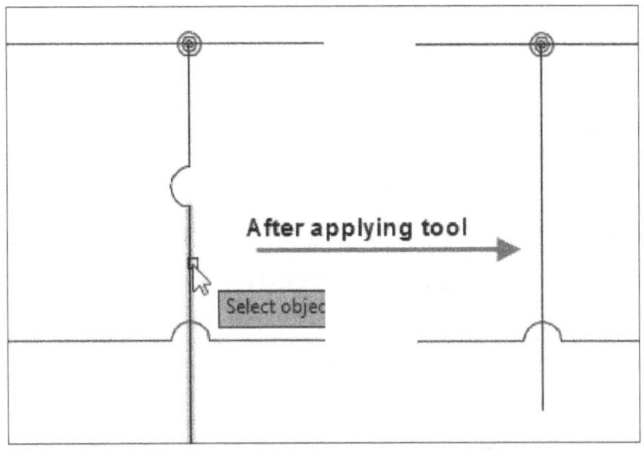

Figure-98. Deleting wire gaps

Check/Repair Gap Pointers

The **Check/Repair Gap Pointers** tool is used to check and repair the error in wiring due to gap. Every wire stores the information about connecting wire in its Xdata. Xdata pointers identify which wire number insert goes with which wire segment. If there is some breakage in the link between wires then you can check it in the command prompt after using this tool. The procedure to use this tool is given next.

- Click on the **Check/Repair Gap Pointers** tool from the **Wire Gap** drop-down in the **Edit Wires/Wire Numbers** panel in the **Ribbon**. You are asked to select one wire segment.
- Select the first segment of wire. You are asked to select the other segment of the wire.
- Select the other segment. Status of the wire gap will be displayed in the command prompt; refer to Figure-99.

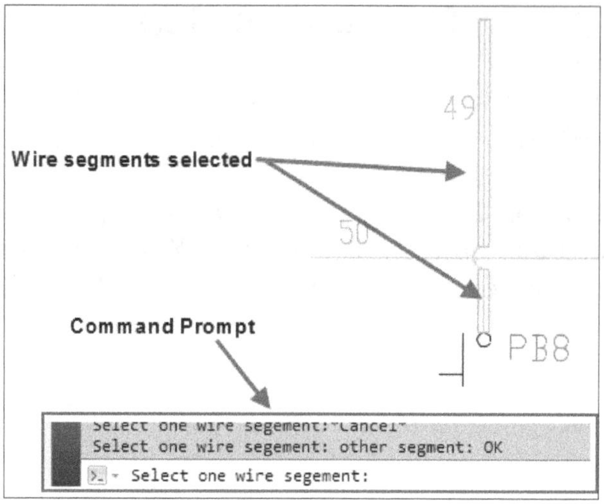

Figure-99. Wire gap checking

You can also connect two wires together by using this tool. Like if you select wire number 50 and wire number 49 after choosing this tool then they will be counted as one wire, and single wire number will be assigned to them. Note that you need to retag all the wire numbers using the **Wire Numbers** tool to check the result.

Editing Wire Sequence

Editing wire sequence directly affects the Wire from/to reports which are generated once the project is completed. This report helps to connect the components in real-world. The **Edit Wire Sequence** tool is used to edit the wire sequence in drawing. The procedure to use this tool is given next.

- Click on the **Edit Wire Sequence** tool from the **Wire Sequence** drop-down in the **Edit Wires/Wire Numbers** panel of the **Ribbon**; refer to Figure-100. You are asked to select a wire network.

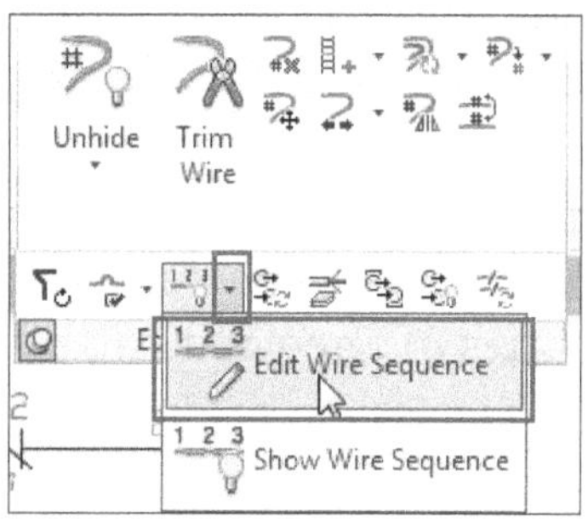

Figure-100. Edit Wire Sequence tool

- Click on the wire in desired wire network. The **Edit Wire Connection Sequence** dialog box will be displayed; refer to Figure-101.

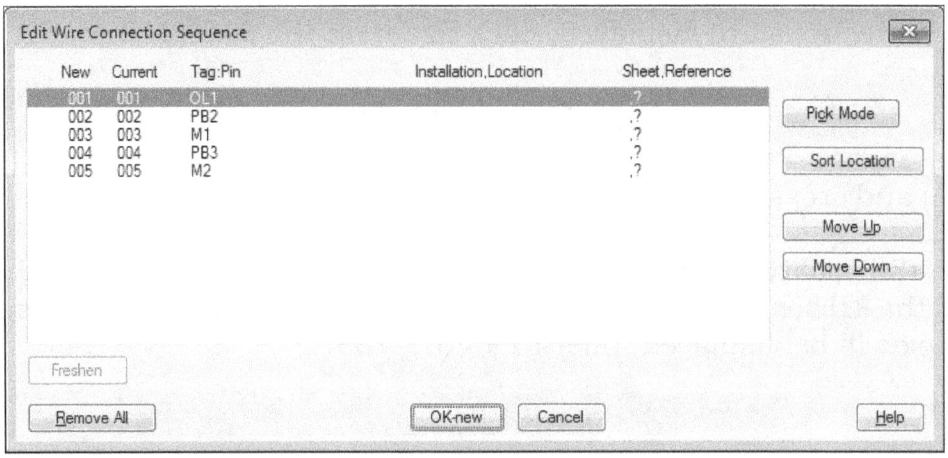

Figure-101. Edit Wire Connection Sequence dialog box

- Select desired component from the list and move it up or down by using the **Move Up** and **Move Down** buttons respectively. The component connections will be as per the sequence in the table of this dialog box.
- After performing the changes, click on the **OK-new** button from the dialog box to apply the changes.

Show Wire Sequence

The **Show Wire Sequence** tool is used to check the sequence of wire connections in the wire network. You can check the results of previous tool by using this tool. The procedure to use this tool is given next.

- Click on the **Show Wire Sequence** tool from the **Wire Sequence** drop-down in the **Edit Wires/Wire Numbers** panel of the **Ribbon**. You are asked to select the wire network.
- Select a wire segment of the network. Preview of connection will be displayed; refer to Figure-102.

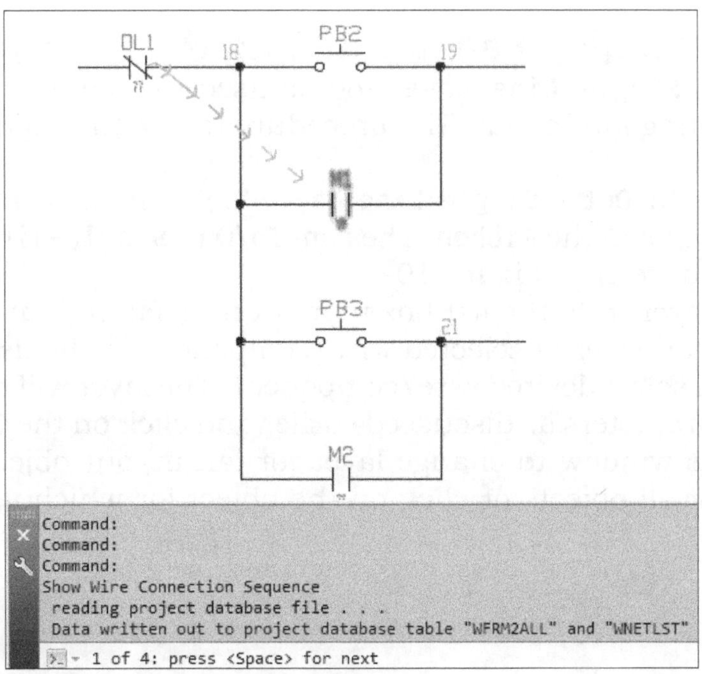

Figure-102. Preview of wire connection

- Press **SPACEBAR** from keyboard to check the next connection in the network. Once all the connections are displayed, click again on the **SPACEBAR** to exit the tool.

Update Signal References

The **Update Signal References** tool is used to update the wire signals (source and destination) and cross-references (tags). The procedure to use this tool is given next.

- Click on the **Update Signal References** tool from the **Edit Wires/Wire Numbers** panel in the **Ribbon**. The **Update Wire Signals and Stand-Alone Cross-Reference** dialog box will be displayed; refer to Figure-103.

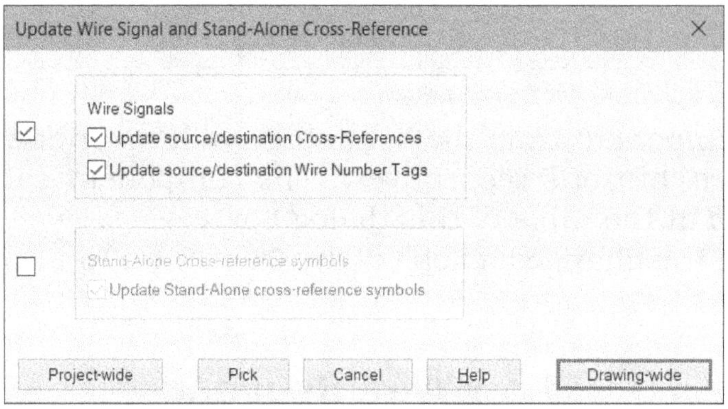

Figure-103. Update Wire Signal and Stand-Alone Cross-Reference dialog box

- If you want to update wire signals then select the **Wire Signals** check box and select the related check boxes underneath.
- To update the cross-reference symbols, select the **Stand-Alone Cross-reference symbols** check box and select the related check box underneath.
- Click on the **Project-wide** button to update the wire signals/cross-references throughout the project.
- Click on the **Drawing-wide** button to update the wire signals/cross-references throughout the drawing.

Changing Fan In/Out Single Line Layer

The **Fan In/Out - Single Line Layer** tool is used to change the layer of selected line/wire representing fan in/out. The procedure to use this tool is given next.

- Click on the **Fan In/Out - Single Line Layer** tool from the expanded **Edit Wires/ Wire Numbers** panel in the **Ribbon**. The **Fan-In/Out Single-line Layer** dialog box will be displayed; refer to Figure-104.
- Select desired layer from the list box to mark it as fan in/out layer.
- If you want to add layer of selected wire/component in the list then click on the Pick button and select desired wire/component. The layer will be added in the list.
- Set the other parameters as discussed earlier and click on the **OK-Change/Convert** button from the window to change layer for fan in/out objects. Press **ENTER** to change layer for all objects or click on the object for which you want to change the layer.

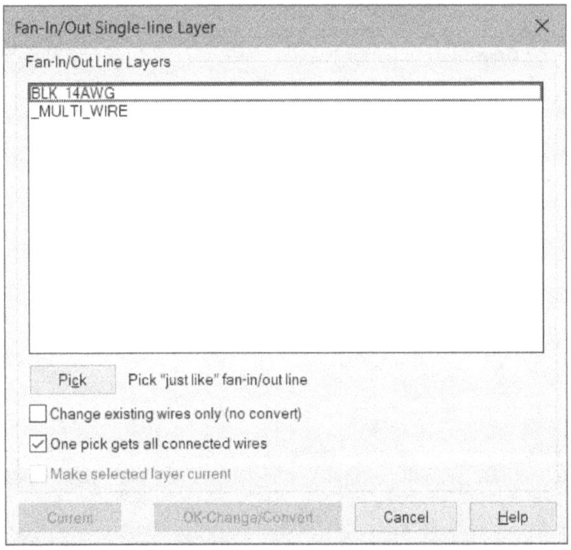

Figure-104. Fan-In/Out Single-line Layer dialog box

Checking Signal Code Flow

The **List Signal Code** tool is used to check the origin and destination of selected signal code. The procedure to use this tool is given next.

- Click on the **List Signal Code** tool from the expanded **Edit Wires/Wire Numbers** panel in the **Ribbon**. You will be asked to select the signal market to checked.
- Select desired signal code from drawing. The **Signal Code** dialog box will be displayed; refer to Figure-105. You can check the source and destination of signal code in this dialog box.
- Select desired option from the list box and click on the **Surf** button to display selected reference in the drawing. Press **ESC** to exit the tool.

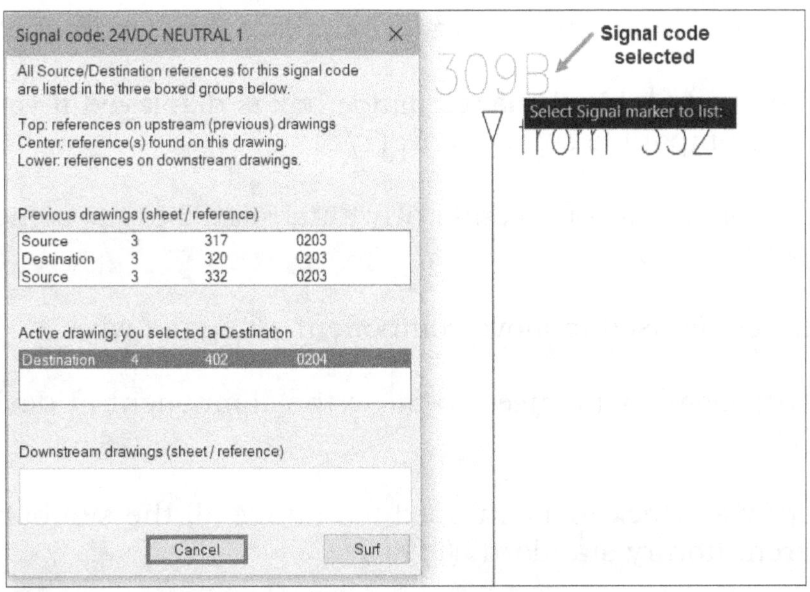

Figure-105. Signal code dialog box

Showing Signal Path in Current Drawing

The **Show Signal Path** tool in expanded **Edit Wires/Wire Numbers** panel of the **Ribbon** is used to highlight signal destination and source in the drawing using yellow and red dashed lines, respectively; refer to Figure-106.

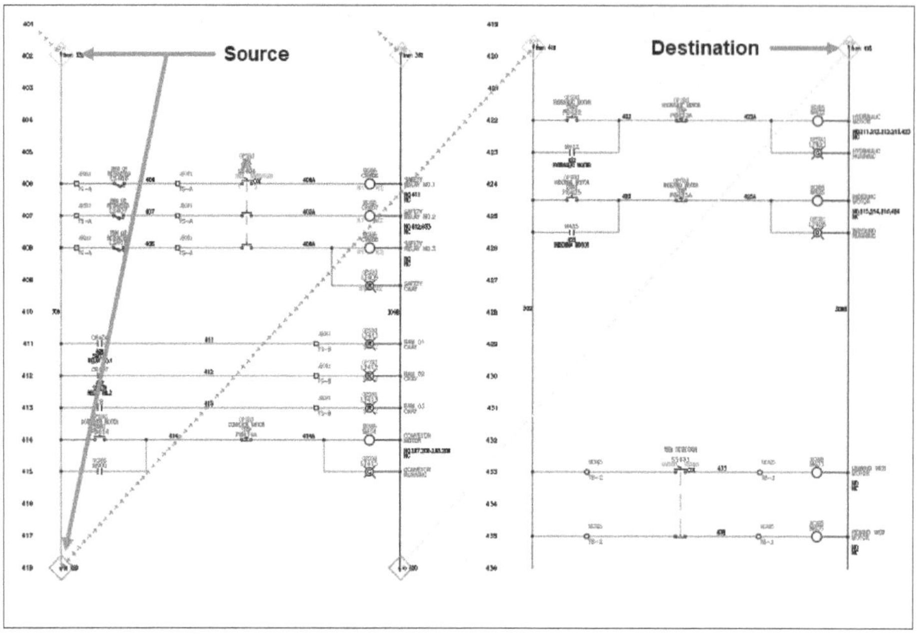

Figure-106. Showing signal path

Similarly, you can use **Multiple Cable Markers Update** tool in the panel to update all the cable markers that have been changed and need updation (drawing-wide or project-wide).

Till this point, we have discussed all the general tools used in editing components, circuits and wires. In the next chapter, we will discuss PLC and its components.

SELF-ASSESSMENT

Q1. .... is used to interconnect terminals of the electrical components.

Q2. The **Different symbol block names** dialog box is displayed if you are assigning properties to an incompatible component. (T/F)

Q3. The .... tool allows you to enter any data related to the component for their identity or purpose.

Q4. The .... tool is used to move component along the wire.

Q5. The **Move Component** tool is used to move the component at desired location in the drawing. (T/F)

Q6. The **Swap/Update Block** tool can used to change all the symbols in drawing to symbols of different library standard. (T/F)

Q7. The **Revise Ladder** tool in the **Ladder Editing** drop-down is used to re-create the ladder previously created. (T/F)

Q8. The **Show Wire** tool is used to display all the wires near the cursor position in the drawing. (T/F)

REVIEW QUESTIONS

Q1. What is the primary function of the Edit tool in the Edit Components drop-down?
A. Insert a new component into the drawing
B. Modify wire numbers
C. Edit the details of an inserted component
D. Delete components from the drawing

Q2. Which button should you click to repeatedly use a component with specified settings in the Insert/Edit Component dialog box?
A. OK
B. Cancel
C. OK-Repeat
D. Apply

Q3. What does the Internal Jumper tool allow you to do?
A. Change the component's tag
B. Interconnect terminals of electrical components
C. Move components in the drawing
D. Delete jumpers

Q4. After selecting jumper points using the Internal Jumper tool, which button must you click to connect the jumper?
A. Apply
B. Add
C. OK
D. Save

Q5. What happens when you click on the Show Jumpers button in the Wire Jumpers dialog box?
A. All jumpers are hidden
B. Jumpers are renamed
C. Jumpers are displayed in the drawing area
D. Jumper data is deleted

Q6. What is the function of the Fix/UnFix Tag tool?
A. To delete component tags
B. To update all tags in the drawing
C. To fix or unfix component tags from being updated automatically
D. To convert tags into labels

Q7. What is the result of choosing "Force selected tags to Fixed" in the Fix/UnFix Component Tag dialog box?
A. All component tags are removed
B. Tags will be editable in bulk
C. The selected tags will not be updated automatically
D. Tags will be hidden in the drawing

Q8. The Copy Catalog Assignment tool allows you to:
A. Copy a component to another layer
B. Apply metadata from one component to another
C. Change component symbols
D. Create new circuits

Q9. What should you do if you get the “Different symbol block names” dialog box while copying a catalog assignment?
A. Cancel the process immediately
B. Change wire numbers
C. Click OK to overwrite if necessary
D. Edit the drawing’s dimensions

Q10. What is the purpose of the User Table Data tool?
A. Delete user-defined data from circuits
B. Change wire label styles
C. Enter custom data for identifying components
D. Edit circuit wire tags

Q11. What happens when you click Save Changes in the Edit User Table Data dialog box?
A. The dialog box closes
B. Data is copied to another component
C. The user record is saved to the component
D. Wire settings are reset

Q12. Which tool is used to remove components from the drawing?
A. Delete Component
B. Copy Component
C. Move Component
D. Reverse Component

Q13. What does the Copy Component tool allow you to do?
A. Change properties of a component
B. Move a component to another sheet
C. Create a duplicate of an existing component
D. Flip a component horizontally

Q14. What happens after selecting a circuit and pressing ENTER using the Copy Circuit tool?
A. The circuit is deleted
B. You specify insertion points for the copied circuit
C. All tags are fixed
D. Circuit is converted into a symbol

Q15. What dialog box appears if there are gaps in the circuit during the copy process?
A. Insert/Edit Component
B. Save Circuit
C. Gapped wire pointer problem
D. Error Message

Q16. What is the function of the Move Circuit tool?
A. Change tag names of components
B. Move a circuit and its components to a new position
C. Add user data to a component
D. Copy wire numbers between circuits

Q17. How do you save a custom circuit to the icon menu?
A. Use the Save Circuit to Icon Menu tool and configure the slot properties
B. Copy the circuit and paste it to a new location
C. Delete the circuit and recreate it
D. Edit user data of all components

Q18. Which tool moves a component along the wire without breaking the connection?
A. Align
B. Scoot
C. Flip
D. Save Circuit

Q19. What does the Align tool require first?
A. The wire to be deleted
B. Component tags
C. A reference entity for alignment
D. Jumper data

Q20. What happens when you use the Move Component tool?
A. The component is flipped
B. The component is deleted
C. The component is moved and the circuit is adjusted
D. The wire number is updated

Q21. What does the Reverse/Flip Component tool allow you to do?
A. Add new circuits
B. Delete component tags
C. Reverse or flip the component's orientation or pins
D. Rename components

Q22. What is the primary purpose of the "Retag Components" tool?
A. To update the block symbol of a component
B. To assign a new tag to a component by replacing the older one
C. To display component attributes
D. To modify component attributes

Q23. Which tool allows switching between Normally Open (NO) and Normally Closed (NC) variants of components?
A. Retag Components
B. Toggle NC/NO
C. Swap/Update Block
D. Edit Selected Attribute

Q24. What happens when the "Library Swap" radio button is selected in the Swap/ Update Block tool?
A. Only a single block is updated
B. All blocks are swapped with identical ones
C. All blocks are updated using a new library
D. Attributes are renamed automatically

Q25. Which tool allows you to hide multiple attributes using window selection?
A. Hide Attribute (Single Pick)
B. Hide Attribute (Window/Multiple)
C. Squeeze Attribute/Text
D. Edit Selected Attribute

Q26. How do you add a new attribute to a component?
A. Use Rename Attribute tool
B. Use Add Attribute tool
C. Use Hide Attribute tool
D. Use Move/Show Attribute tool

Q27. Which tool helps in changing the name of an attribute?
A. Rename Attribute
B. Edit Selected Attribute
C. Retag Components
D. Move/Show Attribute

Q28. What does the Squeeze Attribute/Text tool do?
A. Rotates the attribute by 90 degrees
B. Hides the attribute
C. Reduces the width of the attribute text
D. Adds a new attribute to a block

Q29. To change the height and width of attribute text, which tool should be used?
A. Change Attribute Size
B. Edit Selected Attribute
C. Stretch Attribute/Text
D. Change Attribute Layer

Q30. What is the effect of using the Rotate Attribute tool?
A. Deletes attribute
B. Rotates attribute by 90 degrees
C. Moves attribute to a different component
D. Renames attribute

Q31. Which tool allows modification of text justification of attributes?
A. Change Attribute Layer
B. Change Attribute Justification
C. Move/Show Attribute
D. Edit Selected Attribute

Q32. What is the function of the "Component Cross-Reference" tool?
A. Move attributes across layers
B. Update cross-reference text of a component
C. Rotate attribute text
D. Hide component block

Q33. To update child component locations and descriptions based on schematic symbols, which tool is used?
A. Change Cross-Reference to Multiple Line Text
B. Child Location/Description Update
C. Cross-Reference Check
D. Add Attribute

Q34. What does the Copy/Add Component Override tool do?
A. Copies component blocks
B. Adds new layers
C. Applies component overrides to other components
D. Deletes cross-references

Q35. How does the Remove Component Override tool function?
A. Updates cross-reference reports
B. Removes overrides from selected components
C. Adds new override parameters
D. Changes text justification

Q36. Which tool allows converting cross-reference text into multiline format?
A. Component Cross-Reference
B. Change Cross-Reference to Multiple Line Text
C. Add Attribute
D. Retag Components

Q37. What does the Cross-Reference Check tool help in doing?
A. Updates blocks
B. Matches attributes with schematic
C. Checks references of selected components
D. Moves component across layers

Q38. To update all cross-references in a project, which tool should be used?
A. Hide/Unhide Cross-Referencing
B. Update Stand-Alone Cross-Referencing
C. Swap/Update Block
D. Add Attribute

Q39. Which tool is used to generate a reference table for a selected component?
A. Generate Component Report
B. Cross-Reference Table
C. Cross-Reference Check
D. Rename Attribute

Q40. What is the primary function of the Circuit Clipboard panel?
A. To create new circuits from scratch
B. To edit existing wire types
C. To copy and paste circuit elements
D. To renumber ladders

Q41. What does the Edit Wire Number tool allow you to do?
A. Automatically delete all wire numbers
B. Modify wire numbers manually
C. Fix a wire number permanently
D. Move a wire from one circuit to another

Q42. What does clearing the "Fixed" check box do in the Edit Wire Numbers dialog box?
A. Deletes the wire number
B. Hides the wire number
C. Allows wire number to be changed
D. Shows a preview of the wire number

Q43. What color indicates a fixed wire number after using the Fix tool?
A. Red
B. Blue
C. Green
D. Black

Q44. What does the Swap tool do?
A. Deletes a wire
B. Copies a wire number
C. Swaps two wire numbers
D. Reverses the direction of a wire

Q45. Which dialog box is used in the Find/Replace operation for wire numbers?
A. Edit Wire Number Attributes
B. Find/Replace Wire Numbers
C. Drawing Properties
D. Wire Type Editor

Q46. Which tool is used to trim a wire between connections?
A. Swap
B. Stretch Wire
C. Trim Wire
D. Move Wire Number

Q47. What happens when you use the Hide tool on a wire number?
A. It gets deleted
B. It becomes visible in red
C. It becomes hidden
D. It flips sides

Q48. What is a rung in a circuit or ladder diagram?
A. A fixed wire number
B. A type of component
C. A wire connecting two parallel lines
D. A highlighted wire number

Q49. What must be present in the drawing for the Revise Ladder tool to work?
A. Wire types
B. Hidden wires
C. Ladders
D. Circuit blocks

Q50. What does the Stretch Wire tool do?
A. Removes wire numbering
B. Extends wire to meet another wire
C. Cuts the wire in half
D. Adds a bend to the wire

Q51. What is the purpose of the Bend Wire tool?
A. Rename wires
B. Change wire numbers
C. Connect components
D. Create wire bends at selected points

Q52. Which tool allows you to check the connection of wire segments one by one using the SPACEBAR?
A. Trim Wire
B. Show Wire
C. Check/Trace Wire
D. Move Wire Number

Q53. What dialog box is used for editing or creating new wire types?
A. Ladder Editing
B. Create/Edit Wire Type
C. Edit Wire Numbers
D. Signal Code List

Q54. What does the Flip Wire Number tool do?
A. Reverses the wire type
B. Changes wire size
C. Flips wire number to the other side of wire
D. Moves wire number across drawing

Q55. What visual aid does the Show Signal Path tool use?
A. Blue and green lines
B. Solid black lines
C. Yellow and red dashed lines
D. Flashing circles

Q56. What does the Toggle Angled Tee Markers tool change?
A. Connection color
B. Marker type at wire connections
C. Wire thickness
D. Component type

Q57. Which tool is used to flip the intersection gap between wires?
A. Flip Wire Number
B. Flip Wire Gap
C. Trim Wire
D. Edit Wire Sequence

Q58. What does the Delete Wire Gap tool do?
A. Joins disconnected wires
B. Adds a new gap
C. Removes unwanted gaps
D. Reverses gap direction

Q59. What is the function of the Check/Repair Gap Pointers tool?
A. Move wire number
B. Check and fix errors in wire connections
C. Add tee markers
D. Hide wires

Q60. Which tool is used to modify the sequence of component connections in a wire network?
A. Move Wire Number
B. Check/Trace Wire
C. Edit Wire Sequence
D. Flip Wire Gap

Q61. What does the Update Signal References tool do?
A. Adds a new wire signal
B. Hides all signals
C. Updates wire signals and cross-references
D. Deletes signals across project

Q62. How do you mark a line as Fan In/Out using the Fan In/Out - Single Line Layer tool?
A. Hide the line
B. Flip the layer color
C. Select desired layer from the list
D. Select component type manually

Q63. What is shown in the Signal Code dialog box after using the List Signal Code tool?
A. Only the destination
B. Wire number and size
C. Origin and destination of signal
D. Total voltage

Q64. What happens when you click the Surf button in the Signal Code dialog box?
A. Opens a drawing audit
B. Displays selected reference in the drawing
C. Changes the wire type
D. Flips the signal code

FOR STUDENT NOTES

FOR STUDENT NOTES

Answer for Self-Assessment:

Ans1. Internal Jumper, Ans2. T, Ans3. User Table Data, Ans4. Scoot, Ans5. T, Ans6. T, Ans7. F, Ans8. F

Chapter 7

PLCs and Components

Topics Covered

The major topics covered in this chapter are:

- ***Introduction***
- ***Inserting Parametric PLCs***
- ***Inserting PLCs (Full Unit)***
- ***Inserting Connectors***
- ***Inserting Terminals***

INTRODUCTION

PLC is a solid state/computerized industrial control system that performs discrete or sequential logic in a factory environment. It was originally developed to replace mechanical relays, timers, counters, and so on. PLCs are used successfully to execute complicated control operations in a plant. Its purpose is to monitor crucial process parameters and adjust process operations accordingly. A sequence of instructions is programmed by the user to the PLC memory and when the program is executed, the controller operates based on specified instructions.

PLC consists of three main parts: CPU, memory and I/O units.

CPU is the brain of PLC. It reads the input values from inputs port, runs the program existing in the program memory and writes the output values to the output register. Memory is used to store different types of information in the binary structure form. The I/O units provide communication between PLC control systems.

Application of PLCs in manufacturing process

In an industrial setup PLCs are used to automate manufacturing and assembly processes. By 'Process' we mean a step-by step procedure by which a product is manufactured and assembled. It is the responsibility of the product engineering department to plan for manufacturing of new or modified products. Other processes might involve the filling and capping of bottles, the printing of newspapers, or the assembly of automobiles. In many such manufacturing situations, PLC plays an important role in carrying out the various processes.

Examples of PLC Applications

Internal Combustion Engine Monitoring
A PLC acquires data recorded from sensors located at the internal combustion engine. Measurements taken include water temperature, oil temperature, RPMs, torque, exhaust temperature, oil pressure, manifold pressure, and timing.

Carburetor Production Testing
PLCs provide on-line analysis of automotive carburetors in a production assembly line. The systems significantly reduce the test time, while providing greater yield and better quality carburetors. Pressure, vacuum, and fuel and air flow are some of the variables tested.

Monitoring Automotive Production Machines
The system monitors total parts, rejected parts, produced parts, machine cycle time, and machine efficiency. Statistical data is available to the operator anytime or after each shift.

Power Steering Valve Assembly and Testing
The PLC system controls a machine to ensure proper balance of the valves and to maximize left and right turning ratios.

CHEMICAL AND PETROCHEMICAL

Ammonia and Ethylene Processing
Programmable controllers monitor and control large compressors used during ammonia and ethylene manufacturing. The PLC monitors bearing temperatures, operation of clearance pockets, compressor speed, power consumption, vibration, discharge temperatures, pressure, and suction flow.

Dyes
PLCs monitor and control the dye processing used in the textile industry. They match and blend colors to predetermined values.

Chemical Batching
The PLC controls the batching ratio of two or more materials in a continuous process. The system determines the rate of discharge of each material and keeps inventory records. Several batch recipes can be logged and retrieved automatically or on command from the operator.

Fan Control
PLCs control fans based on levels of toxic gases in a chemical production environment. This system effectively removes gases when a preset level of contamination is reached. The PLC controls the fan start/stop, cycling, and speeds, so that safety levels are maintained while energy consumption is minimized.

Gas Transmission and Distribution
Programmable controllers monitor and regulate pressures and flows of gas transmission and distribution systems. Data is gathered and measured in the field and transmitted to the PLC system.

Pipeline Pump Station Control
PLCs control mainline and booster pumps for crude oil distribution. They measure flow, suction, discharge, and tank low/high limits. Possible communication with SCADA (Supervisory Control and Data Acquisition) systems can provide total supervision of the pipeline.

Oil Fields
PLCs provide on-site gathering and processing of data pertinent to characteristics such as depth and density of drilling rigs. The PLC controls and monitors the total rig operation and alerts the operator of any possible malfunctions.

GLASS PROCESSING

Annealing Lehr Control
PLCs control is used to remove the internal stress from glass products. The system controls the operation by following the annealing temperature curve during the reheating, annealing, straining, and rapid cooling processes through different heating and cooling zones. Improvements are made in the ratio of good glass to scrap, reduction in labor cost, and energy utilization.

Glass Batching

PLCs control the batch weighing system according to stored glass formulas. The system also controls the electromagnetic feeders for in-feed to and out-feed from the weigh hoppers, manual shut-off gates, and other equipment.

Cullet Weighing

PLCs direct the cullet system by controlling the vibratory cullet feeder, weight-belt scale, and shuttle conveyor. All sequences of operation and inventory of quantities weighed are kept by the PLC for future use.

Batch Transport

PLCs control the batch transport system, including reversible belt conveyors, transfer conveyors to the cullet house, holding hoppers, shuttle conveyors, and magnetic separators. The controller takes action after the discharge from the mixer and transfers the mixed batch to the furnace shuttle, where it is discharged to the full length of the furnace feed hopper.

MANUFACTURING/MACHINING

Production Machines

The PLC controls and monitors automatic production machines at high efficiency rates. It also monitors piece-count production and machine status. Corrective action can be taken immediately if the PLC detects a failure.

Transfer Line Machines

PLCs monitor and control all transfer line machining station operations and the interlocking between each station. The system receives inputs from the operator to check the operating conditions on the line-mounted controls and reports any malfunctions. This arrangement provides greater machine efficiency, higher quality products, and lower scrap levels.

Wire Machine

The controller monitors the time and ON/OFF cycles of a wiredrawing machine. The system provides ramping control and synchronization of electric motor drives. All cycles are recorded and reported on demand to obtain the machine's efficiency as calculated by the PLC.

Tool Changing

The PLC controls a synchronous metal cutting machine with several tool groups. The system keeps track of when each tool should be replaced, based on the number of parts it manufactures. It also displays the count and replacements of all the tool groups.

Paint Spraying

PLCs control the painting sequences in auto manufacturing. The operator or a host computer enters style and color information and tracks the part through the conveyor until it reaches the spray booth. The controller decodes the part information and then controls the spray guns to paint the part. The spray gun movement is optimized to conserve paint and increase part throughput.

MATERIALS HANDLING

Automatic Plating Line

The PLC controls a set pattern for the automated hoist, which can traverse left, right, up, and down through the various plating solutions. The system knows where the hoist is at all times.

Storage and Retrieval Systems

A PLC is used to load parts and carry them in totes in the storage and retrieval system. The controller tracks information like a lane numbers, the parts assigned to specific lanes, and the quantity of parts in a particular lane. This PLC arrangement allows rapid changes in the status of parts loaded or unloaded from the system. The controller also provides inventory printouts and informs the operator of any malfunctions.

Conveyor Systems

The system controls all of the sequential operations, alarms, and safety logic necessary to load and circulate parts on a main line conveyor. It also sorts products to their correct lanes and can schedule lane sorting to optimize pelletizer duty. Records detailing the ratio of good parts to rejects can be obtained at the end of each shift.

Automated Warehousing

The PLC controls and optimizes the movement of stacking cranes and provides high turnaround of materials requests in an automated, high-cube, vertical warehouse. The PLC also controls aisle conveyors and case pelletizers to significantly reduce manpower requirements. Inventory control figures are maintained and can be provided on request.

METALS

Steel Making

The PLC controls and operates furnaces to produce metal in accordance with preset specifications. The controller also calculates oxygen requirements, alloy additions, and power requirements.

Loading and Unloading of Alloys

Through accurate weighing and loading sequences, the system controls and monitors the quantity of coal, iron ore, and limestone to be melted. It can also control the unloading sequence of the steel to a torpedo car.

Continuous Casting

PLCs direct the molten steel transport ladle to the continuous- casting machine, where the steel is poured into a water-cooled mold for solidification.

Cold Rolling

PLCs control the conversion of semi finished products into finished goods through cold-rolling mills. The system controls motor speed to obtain correct tension and provide adequate gauging of the rolled material.

Aluminum Making

Controllers monitor the refining process, in which impurities are removed from bauxite by heat and chemicals. The system grinds and mixes the ore with chemicals and then pumps them into pressure containers, where they are heated, filtered, and combined with more chemicals.

POWER

Plant Power System

The programmable controller regulates the proper distribution of available electricity, gas, or steam. In addition, the PLC monitors powerhouse facilities, schedules distribution of energy, and generates distribution reports. The PLC controls the loads during operation of the plant, as well as the automatic load shedding or restoring during power outages.

Energy Management

Through the reading of inside and outside temperatures, the PLC controls heating and cooling units in a manufacturing plant. The PLC system controls the loads, cycling them during predetermined cycles and keeping track of how long each should be on or off during the cycle time. The system provides scheduled reports on the amount of energy used by the heating and cooling units.

Coal Fluidization Processing

The controller monitors how much energy is generated from a given amount of coal and regulates the coal crushing and mixing with crushed limestone. The PLC monitors and controls burning rates, temperatures generated, sequencing of valves, and analog control of jet valves.

Compressor Efficiency Control

PLCs control several compressors at a typical compressor station. The system handles safety interlocks, startup/shutdown sequences, and compressor cycling. The PLCs keep compressors running at maximum efficiency using the nonlinear curves of the compressors.

PULP AND PAPER

Pulp Batch Blending

The PLC controls sequence operation, ingredient measurement, and recipe storage for the blending process. The system allows operators to modify batch entries of each quantity, if necessary, and provides hardcopy printouts for inventory control and for accounting of ingredients used.

Batch Preparation for Paper-Making Processing

Applications include control of the complete stock preparation system for paper manufacturing. Recipes for each batch tank are selected and adjusted via operator entries. PLCs can control feedback logic for chemical addition based on tank level measurement signals. At the completion of each shift, the PLC system provides management reports on materials use.

Paper Mill Digester

PLCs control the process of making paper pulp from wood chips. The system calculates and controls the amount of chips based on density and digester volume. Then, the percent of required cooking liquors is calculated and these amounts are added to the sequence. The PLC ramps and holds the cooking temperature until the cooking is completed.

Paper Mill Production

The controller regulates the average basis weight and moisture variable for paper grade. The system manipulates the steam flow valves, adjusts the stock valves to regulate weight, and monitors and controls total flow.

RUBBER AND PLASTIC

Tire-Curing Press Monitoring

The PLC performs individual press monitoring for time, pressure, and temperature during each press cycle. The system alerts the operator of any press malfunctions. Information concerning machine status is stored in tables for later use. Report generation printouts for each shift include a summary of good cures and press downtime due to malfunctions.

Tire Manufacturing

Programmable controllers are used for tire press/cure systems to control the sequencing of events that transforms a raw tire into a tire fit for the road. This control includes molding the tread pattern and curing the rubber to obtain road-resistant characteristics. This PLC application substantially reduces the space required and increases reliability of the system and the quality of the product.

Rubber Production

PLCs provide accurate scale control, mixer logic functions, and multiple formula operation of carbon black, oil, and pigment used in the production of rubber. The system maximizes utilization of machine tools during production schedules, tracks in-process inventories, and reduces time and personnel required to supervise the production activity and the shift-end reports.

Plastic Injection Molding

A PLC system controls variables, such as temperature and pressure, which are used to optimize the injection molding process. The system provides closed-loop injection, where several velocity levels can be programmed to maintain consistent filling, reduce surface defects, and shorten cycle time.

I know that the above list of applications is quite long but it is important to know the applications of PLC thoroughly before we start working on PLCs in AutoCAD Electrical.

Specifications of PLCs

In real-world, the PLCs are specified by the following parameters.

- Input Voltages
- Output Voltages
- Number of I/O terminals

Now, we are ready to work on the electrical drawings containing PLCs.

INSERTING PLCS (PARAMETRIC)

In market, you can find PLCs in modules (which can be increased/decreased) as well as single built unit. There are two tools in AutoCAD Electrical to insert PLCs; **PLC (Parametric)** and **PLC (Full Units)**. These tools are available in the **Insert PLCs** drop-down in the **Insert Components** panel; refer to Figure-1. The **PLC (Parametric)** tool is used to insert the PLCs based on your inputs whereas the **PLC (Full Units)** is used to insert the ready made PLC modules in the drawing. The procedure to use the **PLC (Parametric)** tool is given next.

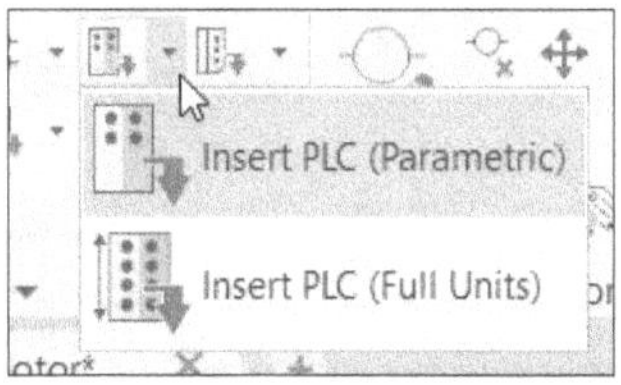

Figure-1. Insert PLCs drop-down

- Click on the **Insert PLC (Parametric)** tool from the **Insert PLCs** drop-down in the **Insert Components** panel. The **PLC Parametric Selection** dialog box will be displayed; refer to Figure-2.
- Click on the **+** signs to expand the categories and select desired type for your PLC.
- Select desired PLC from the table displayed at the bottom in the dialog box.
- Click on desired radio button from the **Graphic Style** area displayed on the left in the dialog box. Preview of the graphic style is displayed in the thumbnail.

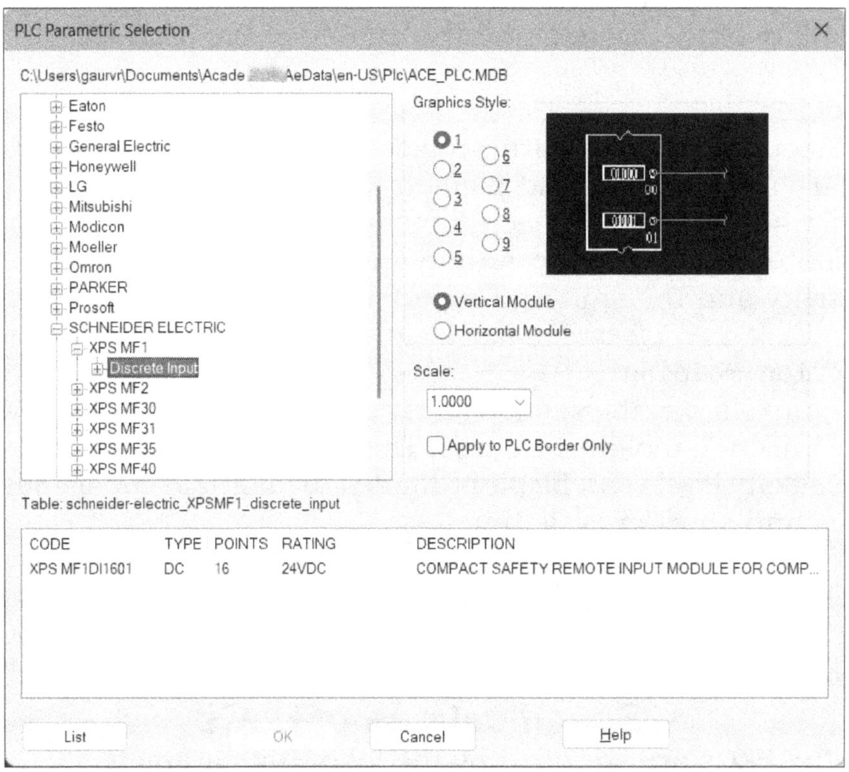

Figure-2. PLC Parametric Selection dialog box

- Click in the **Scale** edit box and specify desired scale factor to make the PLC smaller or larger.

- Click on the **OK** button from the dialog box. The PLC will get attached to the cursor; refer to Figure-3.

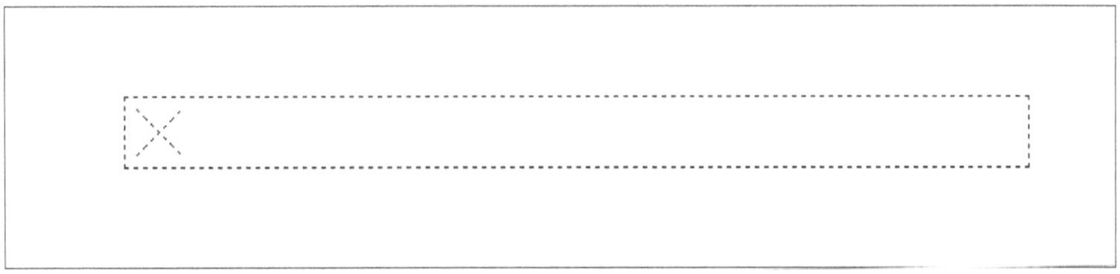

Figure-3. PLC attached to the cursor

- Click in the drawing area to place the PLC. The **Module Layout** dialog box will be displayed; refer to Figure-4.

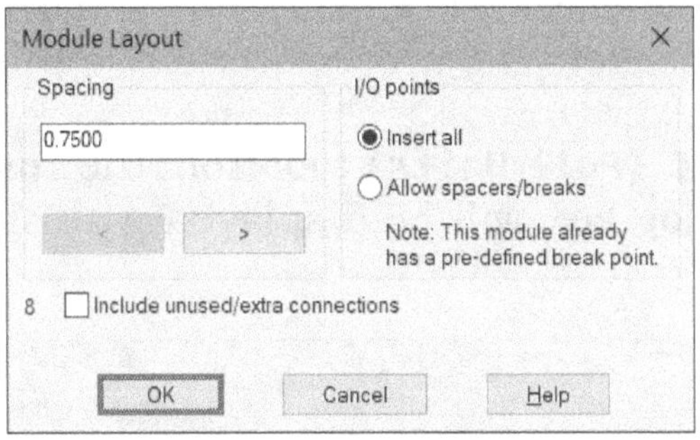

Figure-4. Module Layout dialog box

- Specify desired spacing between the consecutive input and output points in the PLC by using the **Spacing** edit box in this dialog box.
- Select the **Allow spacers/breaks** radio button if you want to add spacers or break lines at the pre-defined break points in the PLC.
- If you want to display all the terminals whether they are used or unused then select the **Include unused/extra connections** check box.
- Click on the **OK** button from the dialog box. The **I/O Point** dialog box will be displayed; refer to Figure-5.

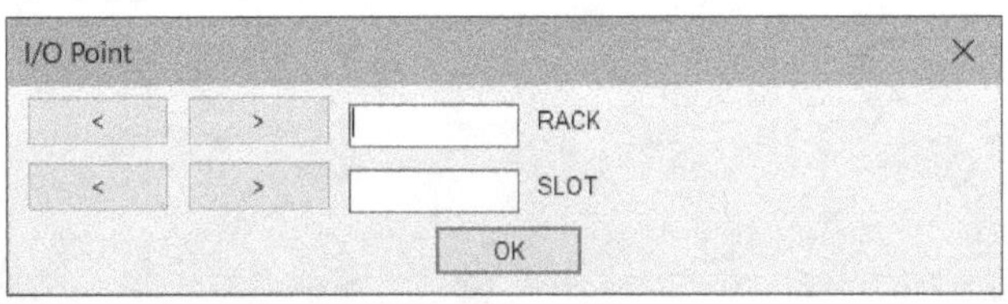

Figure-5. I/O Point dialog box

- Specify desired module identifier value in the **MODULE** edit box of this dialog box and click on the **OK** button from the dialog box. The PLC will be placed. If you have selected the **Allow spacers/breaks** radio button then **Custom Breaks/ Spacing** dialog box will be displayed; refer to Figure-6. Select desired buttons to create customized PLC. Note that AutoCAD Electrical stores preferences of PLC insertion and next time when you will choose this tool, it will prompt you whether to continue previous session or not.

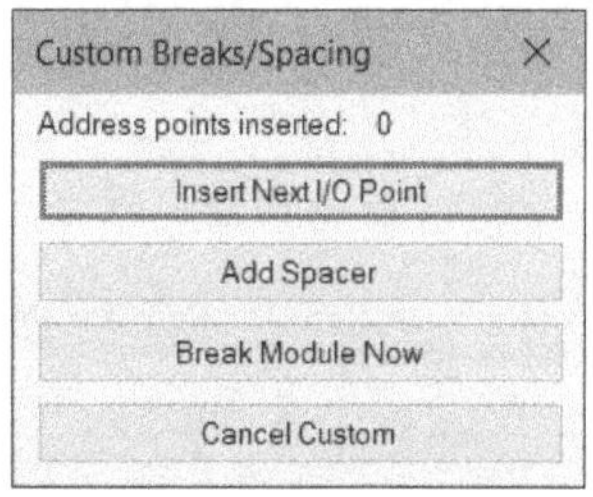

Figure-6. Custom Breaks Spacing dialog box

INSERT PLC (FULL UNITS)

The **Insert PLC (Full Units)** tool is used to insert the full PLC units as per the requirement. Note that in this case, you cannot control the number of connections, display style and so on for the selected PLC. These are some ready to use representations of PLCs available in the market. The procedure to use this tool is given next.

- Click on the **Insert PLC (Full Units)** tool from the **Insert PLCs** drop-down. The **Insert Component** dialog box will be displayed with the PLC components; refer to Figure-7.

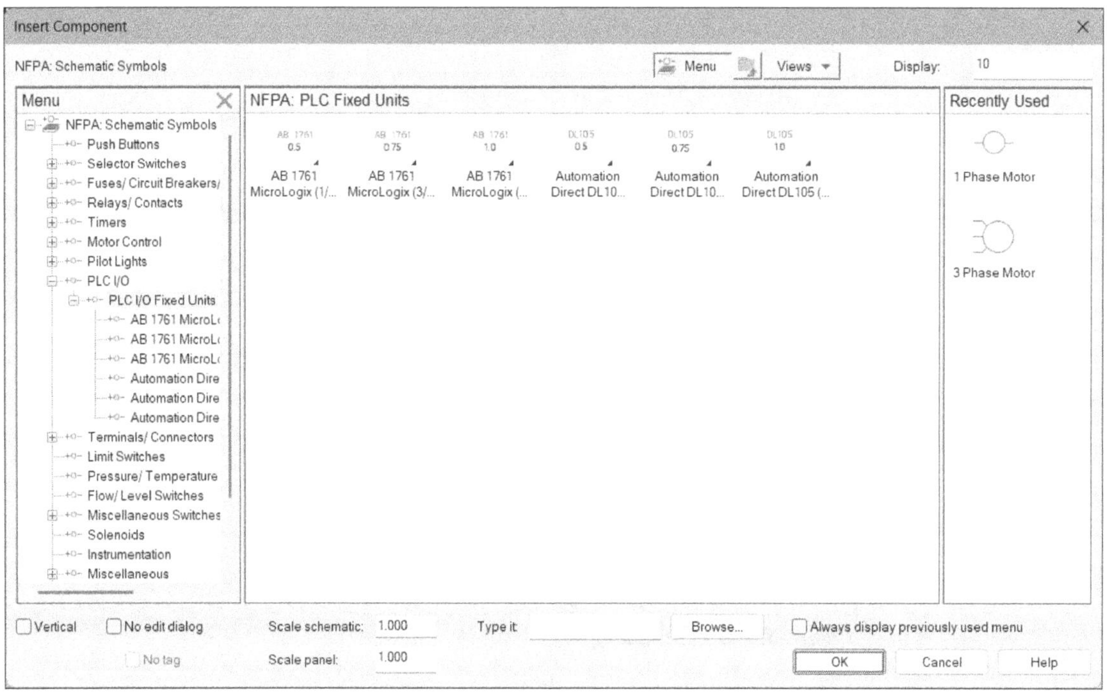

Figure-7. Insert Component dialog box

- Click on desired category and select a plc of your requirement. The PLC will get attached to the cursor; refer to Figure-8.
- Click in the drawing area to place the PLC. The **Edit PLC Module** dialog box will be displayed; refer to Figure-9.

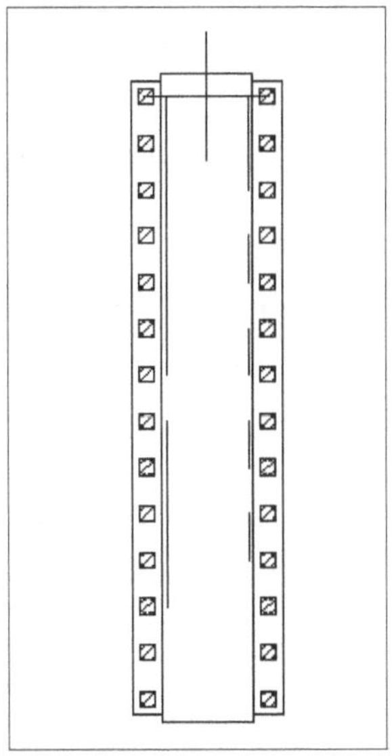

Figure-8. PLC full unit attached to the cursor

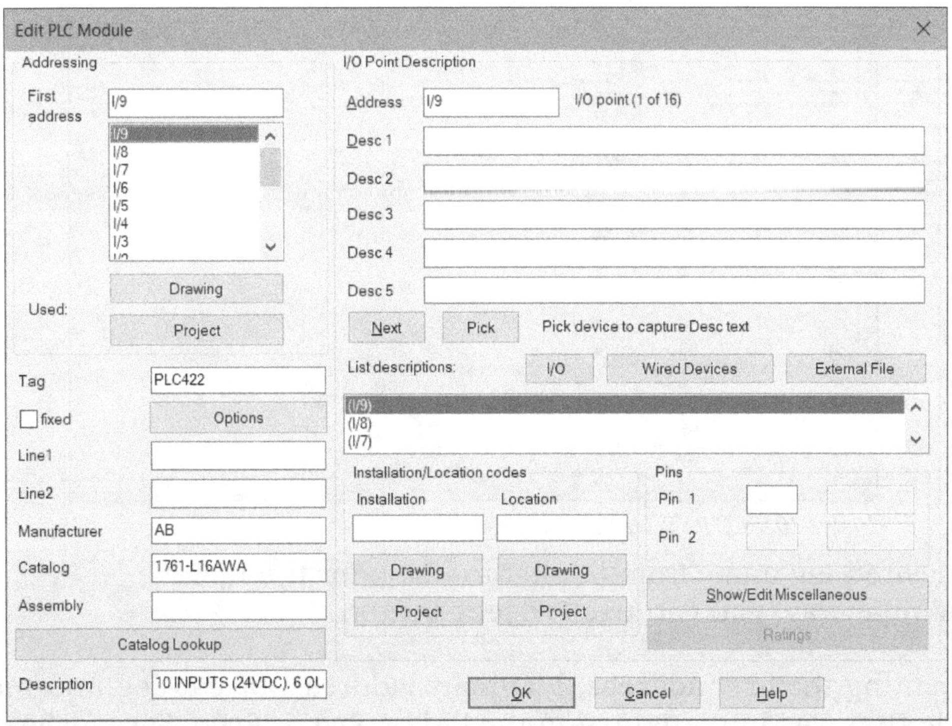

Figure-9. Edit PLC Module dialog box

- Specify the parameters in the dialog box and click on the **OK** button from the dialog box. The options in this dialog box are discussed next (its very important to understand the options in this dialog box).

Addressing Area

The options in this area are used to manage input or output addresses of the PLC.

First Address

Specifies the first I/O address for the PLC module. The addresses that start with I are meant for input like I/9,I/8, and so on. The addresses that start with O are meant for output like O/3, O/4, and so on. In some PLCs, you will get X for input and Y for output. Select desired address from the list to specify it as first I/O address.

Used Area

The buttons in this area are used to select I/O addresses used in earlier drawings or projects. The procedure to use these options is given next.

- Select the **Drawing** button from the area. The **PLC I/O Point List (this drawing)** dialog box will be displayed with the list of I/O points assigned already; refer to Figure-10.

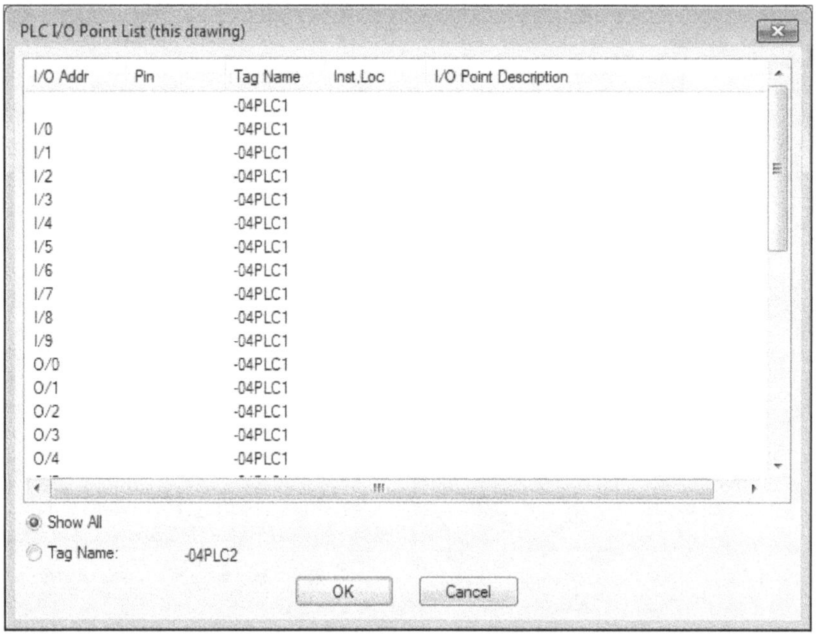

Figure-10. PLC IO Point List dialog box

- Select desired tag name from the list to use it for the PLC.
- In the similar way, you can use **Project** button.

Note that editing the first address, I/O point address, or catalog information for a PLC module that was imported using the **Unity Pro** or **From Spreadsheet** tool may result in problems when you export the data back to Unity Pro. An alert displays to ask whether you want to proceed with the changes.

Tag

The **Tag** edit box is used to specify a unique identifier assigned to the module. The tag value can be manually typed in the edit box.

Options

The **Options** button is used to change the default format of tags for PLCs. On selecting this button, the **Option: Tag Format "Family" Override** dialog box is displayed; refer to Figure-11. Note that you can insert a fixed text string for the %F part of the tag format. Retag Component can then use this override format value to calculate a new tag for the PLC module. For example, a certain PLC module must always have an "IO" family tag value instead of "PLC" so that retag, for example, assigns IO-100 instead of PLC100. To achieve this tag override you would enter "IO-%N" for the tag override format.

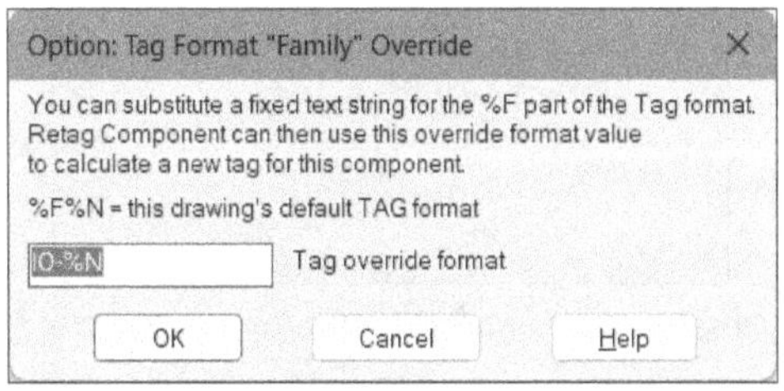

Figure-11. Option dialog box

Line1/Line2

These edit boxes are used to specify optional description text for the module. May be used to identify the relative location of the module in the I/O assembly (example: Rack # and Slot #).

Manufacturer

The **Manufacturer** edit box is used to list the manufacturer number for the module. Enter a value or select one by using the **Catalog Lookup** tool.

Catalog

Lists the catalog number for the module. Enter a value or select one from the **Catalog Lookup**.

Assembly

Lists the assembly code for the module. The Assembly code is used to link multiple part numbers together.

Catalog Lookup

The **Catalog Lookup** tool is used to display the data in catalog related to the selected component. Click on the **Catalog Lookup** tool and the **Catalog Information** dialog box will be displayed; refer to Figure-12. The options of this dialog box have already been discussed.

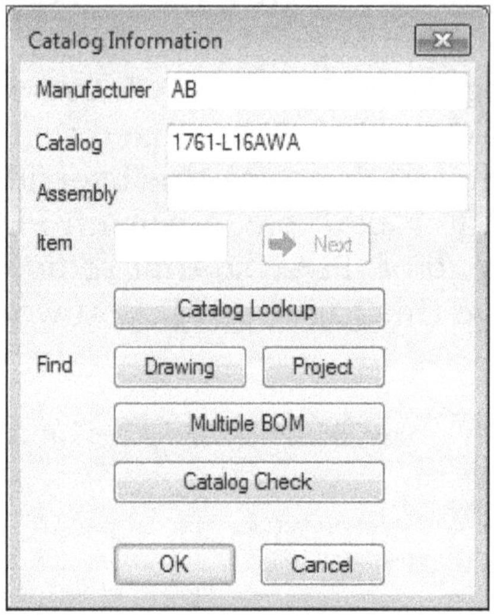

Figure-12. Catalog Information dialog box

Description

Specify the optional line of description text in the **Description** edit box. May be used to identify the module type (for example, "16 Discrete Inputs - 24VDC")

I/O Point Description Area

The options in the I/O Point Description area of the dialog box are used to specify descriptions about various input and output addresses of PLC.

Address

This edit box is used to specify the I/O address assignment.

Description 1-5

These edit boxes are used to specify the optional description text. Enter up to five lines of description attribute text.

Next/Pick

The **Next** button is used to select the next I/O point for specifying the description. **Pick** button is used to copy description of other components in the drawing.

List descriptions

The options in this area are used to list the I/O point descriptions currently assigned to each I/O point on the module or connected wired devices in a pick list.

Pins

The edit boxes in this area are used to assign pin numbers to the pins that are physically located on the module.

Show/Edit Miscellaneous

The **Show/Edit Miscellaneous** button is used to view or edit any attributes that are not predefined by AutoCAD Electrical attributes.

Ratings

This edit box is used to specify values for each ratings attribute. You can enter up to 12 ratings attributes on a module. Select **Defaults** to display a list of default values. Note: If ratings are unavailable, the module you are editing does not carry rating attributes.

MODIFYING PLC MODULE SIZE

There are two tools available in the **Scoot** drop-down of **Edit Components** panel of the **Schematic** tab in the **Ribbon** to modify size of plc modules: **Stretch PLC Module** and **Split PLC Module**; refer to Figure-13. These tools are discussed next.

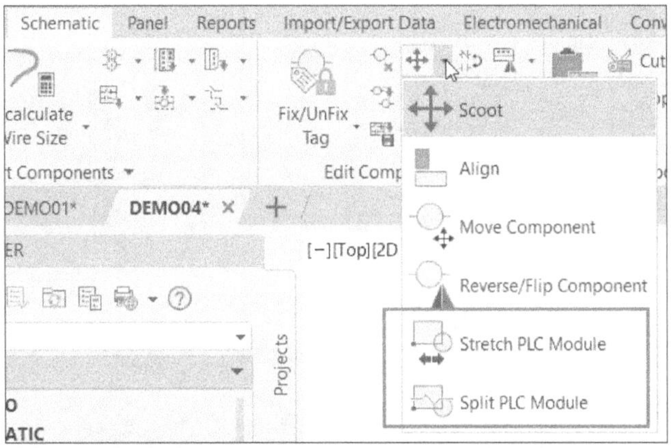

Figure-13. PLC Modification tools

Stretching PLC Module

The **Stretch PLC Module** tool is used to expand selection portion of a plc module. The procedure to use this tool is given next.

- Click on the **Stretch PLC Module** tool from the **Scoot** drop-down of **Edit Components** panel of the **Schematic** tab in the **Ribbon**. You will be asked to select the objects to be stretched.
- Create a rectangular selection box covering the portion to be stretched; refer to Figure-14.

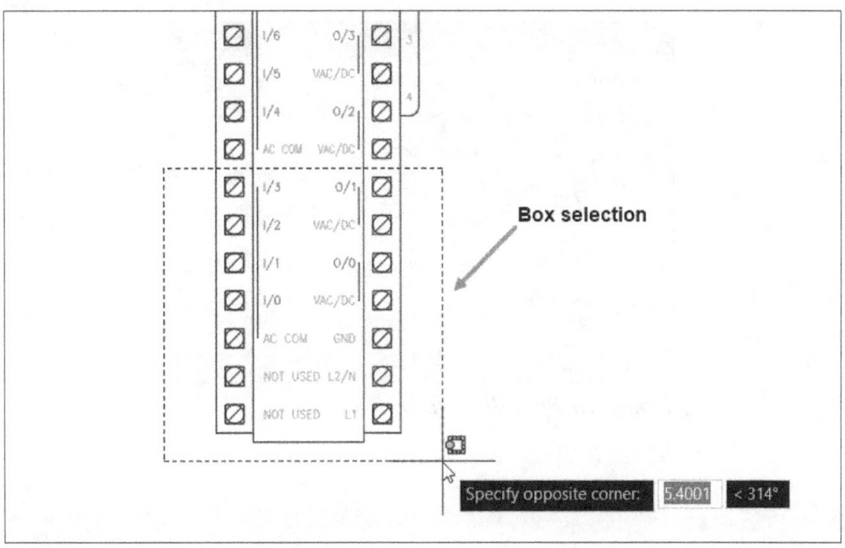

Figure-14. Selecting PLC for stretching

- After performing box selection, press **ENTER**. You will be asked to specify base point to be used for stretching.
- Click at desired location on the PLC to define base point and then click at desired location to define displacement by which PLC will be stretched. The stretch operation will be applied to PLC; refer to Figure-15.

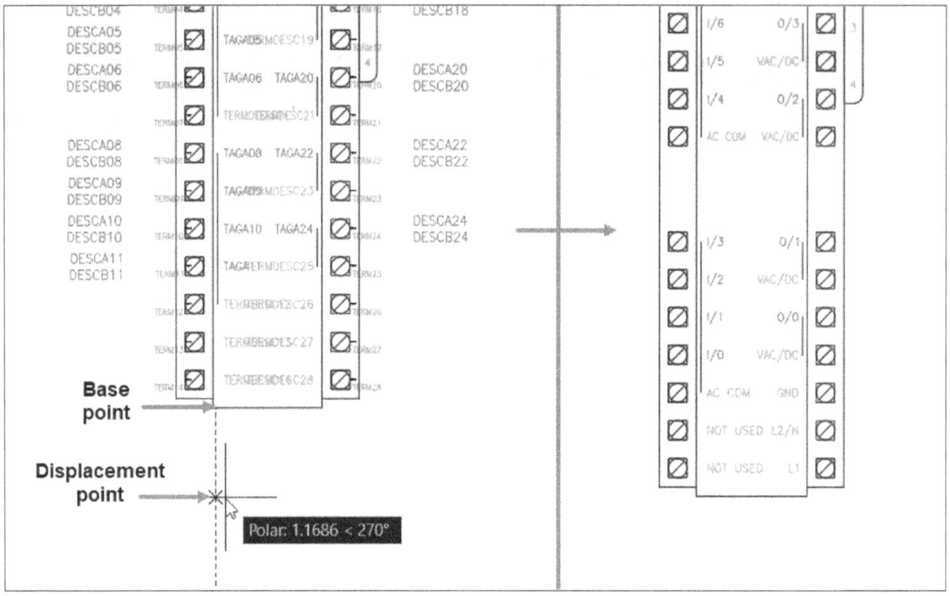

Figure-15. Stretching PLC

Splitting PLC Module

The **Split PLC Module** tool is used to split selected PLC module at specified point to accommodate large PLC in smaller size drawing. The procedure to use this tool is given next.

- Click on the **Split PLC Module** tool from the **Scoot** drop-down of **Edit Components** panel of the **Schematic** tab in the **Ribbon**. You will be asked to select the object to be split.
- Select the PLC to be split. You will be asked to specify split point.
- Click at desired location on PLC where you want to split it. The **Split Block** dialog box will be displayed; refer to Figure-16.

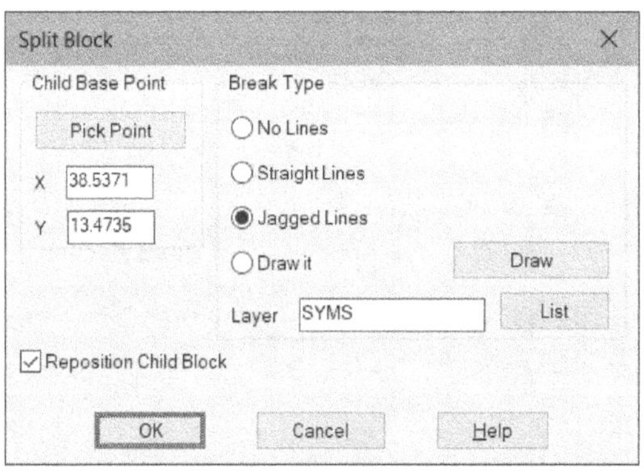

Figure-16. Split Block dialog box

- Select desired radio button from the **Break Type** area of dialog box to define how split will be created. Set the other parameters and click on the **OK** button. The split PLC will be attached to cursor; refer to Figure-17. Click at desired location to place the split PLC.

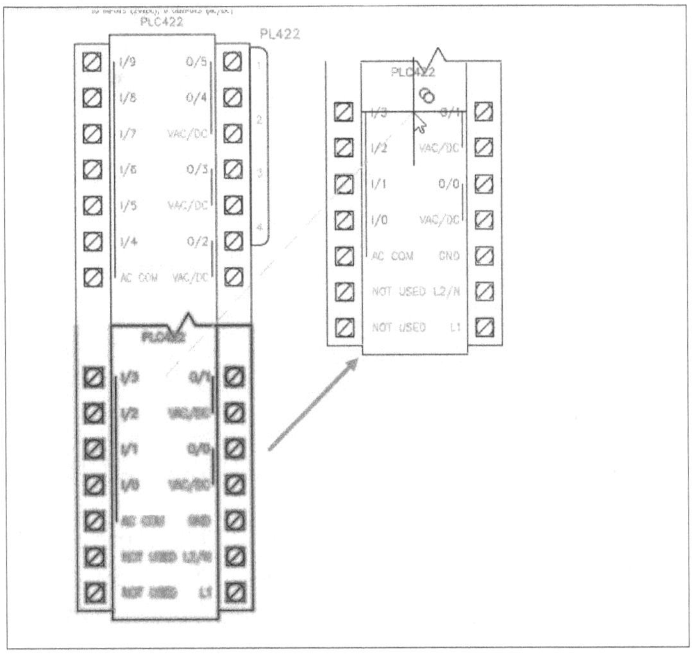

Figure-17. Moving split PLC

CONNECTORS

Connectors are the parts that are used to connect wirings from sources to equipment. The tools to add or modify the connectors are available in the **Insert Connectors** drop-down in the **Insert Components** panel; refer to Figure-18. Various tools in this drop-down are discussed next.

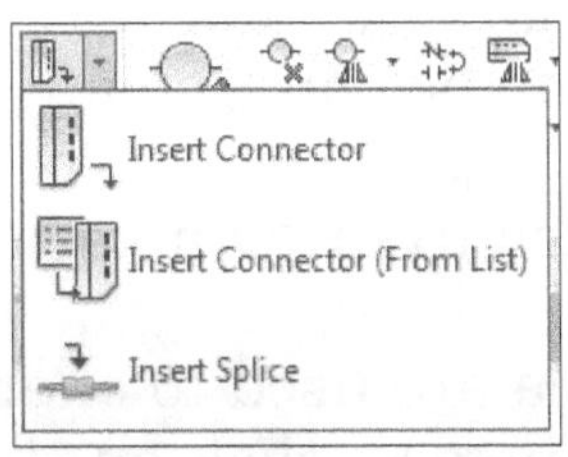

Figure-18. Insert Connector drop-down

Insert Connector

The **Insert Connector** tool is used to insert connectors in the drawing. The procedure to use this tool is given next.

- Click on the **Insert Connector** tool from the **Insert Connectors** drop-down. The **Insert Connector** dialog box will be displayed; refer to Figure-19.

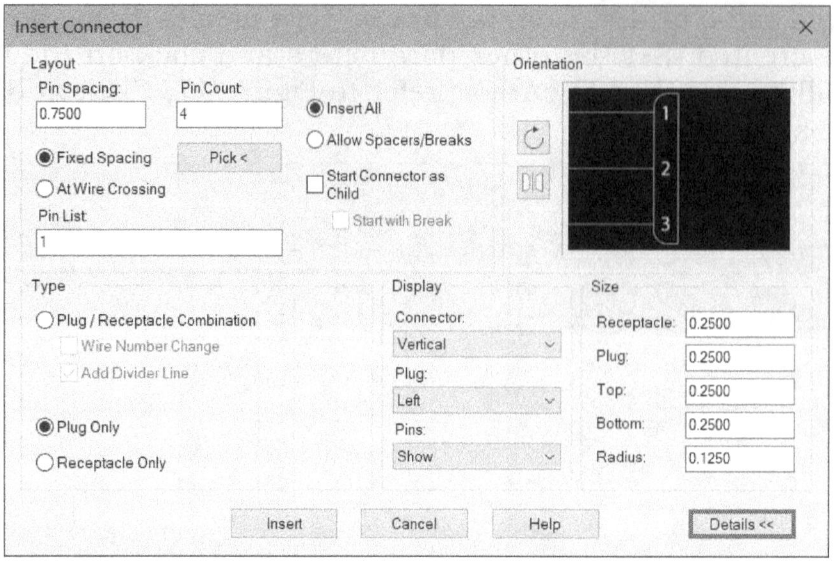

Figure-19. Insert Connector dialog box

- Click in the **Pin Spacing** edit box and specify the space between two consecutive pins of the connector.
- Click in the **Pin Count** edit box and specify the number of pins. Alternatively, click on the **Pick<** button and draw a construction line of desired length. The number of pins will be automatically assigned based on the length of line and space between two pins.
- Select the **At Wire Crossing** radio button if you want to set the pins of connector at the crossing of wires with the connector.
- Click on the **Details>>** button to expand the dialog box for accessing advanced options.
- From the **Type** area, select desired radio button to change the type of connector. On selecting the **Plug/Receptacle Combination** radio button both the parts of connectors will be created. On selecting the **Plug Only** or **Receptacle Only** radio button, the respective part will be created.
- If you want the wire numbers to be changed after the insertion of the connectors, then select the **Wire Number Change** check box for plug and receptacle combination.
- Select desired options from the **Display** area to change the display style of the connector.
- The edit boxes in the **Size** area are used to change the size of connector.
- After specifying desired parameters, click on the **Insert** button from the dialog box. The connector will get attached to the cursor.
- Click in the drawing area to place the connector. If you click on the ladder's wire then the pins of connector will automatically adjust according to the ladder; refer to Figure-20.

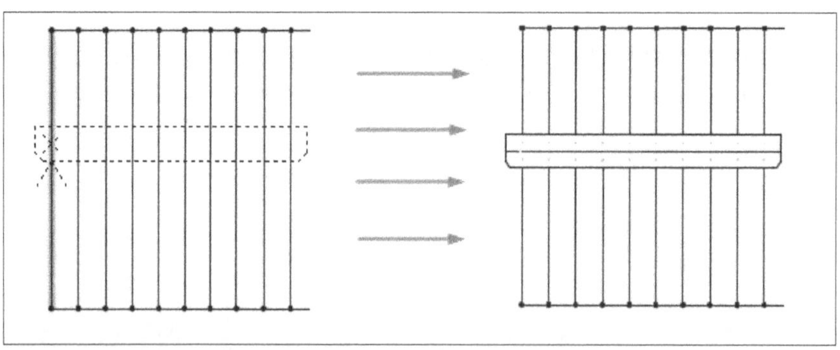

Figure-20. Placing connectors on a ladder

Insert Connector (From List)

The **Insert Connector (From List)** tool is used to insert the connectors from an XML file created by Inventor. In Inventor, you can create model of the connector and then you can export it into XML format. The procedure to use this tool is given next.

- Click on the **Insert Connector (From List)** tool from the **Insert Connectors** drop-down. The **Autodesk Inventor Professional Import File Selection** dialog box will be displayed; refer to Figure-21.

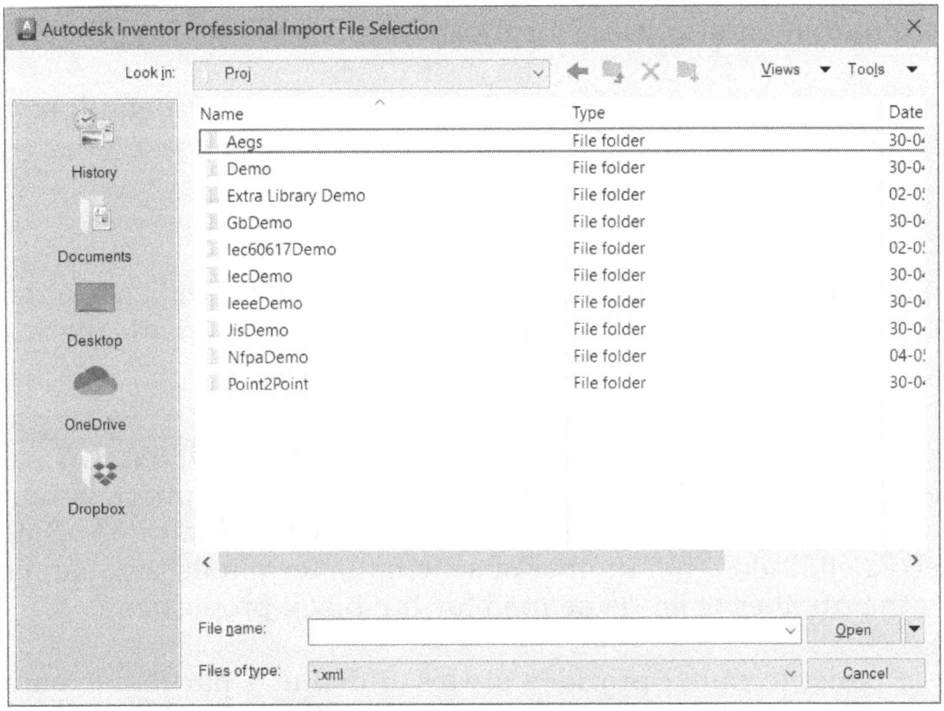

Figure-21. Autodesk Inventor Professional Import File Selection dialog box

- Select desired file and click on the **Open** button. List of connectors will be displayed.
- Select desired connector and place it in the drawing.

Insert Splice

The **Insert Splice** tool is used to insert splice in the wiring of connectors. Click on this tool to insert splice, the **Insert Component** dialog box will be displayed. The procedure to use this dialog box have already been discussed.

TERMINALS

Terminals are used to connect wires to the devices. In real world, terminals are used to connect wires so that they do not cause sparking at high load. You can use the **Insert Component** dialog box to insert terminals in the drawing.

The tools to edit terminals are available in the expanded **Edit Components** panel. Various tools related to terminals are discussed next.

Inserting Terminals from Catalog Browser

- Click on the **Catalog Browser** tool from the **Insert Components** panel in the **Ribbon**. The **Catalog Browser** will be displayed.
- Select the **TRMS** category or any category with terminals.
- Enter a search criteria and click on the **Search** button.
- Click on a field in the results pane.
- Click one of the symbols associated to the catalog value.

OR

- Click **Browse** to locate the symbol on disk. The symbol selected is automatically associated to that catalog value for future insertions.

You can double-click the row in the results pane to insert the default or only symbol associated to the catalog value.

- Specify the insertion point in the drawing.

The symbol orientation matches the underlying wire. If there is no underlying wire, the selected orientation is inserted. The wire breaks automatically if the symbol lands on it.

Associate Terminals on the Same Drawing

You can use the **Associate Terminals** tool to associate two or more terminal symbols together. Associating schematic terminals combine the terminals into a single terminal block property definition. The number of schematic terminals that can be combined is limited to the number of levels defined for the block properties.

Associating a panel terminal provides a way to define a particular panel footprint to represent a schematic block property definition. The procedure to use this tool is given next.

- Click **Schematic** tab > expanded **Edit Components** panel > **Associate Terminals** tool().
- Select a terminal symbol to use as the master. It is used as the basis for any terminal property definition. Make sure the master terminal has more than one layers selected in the **Terminal Block Properties** dialog box; refer to Figure-22.

Note that your terminal symbol must have block properties defined. To define block properties, right-click on the symbol and select **Edit Component** button. In the **Insert/Edit Terminal Symbol** dialog box, click **Block Properties**. The **Terminal Block Properties** dialog box will be displayed.

- On selecting the master terminal, you are asked to select the terminals.
- Select the terminal to be associated. A confirmation message will be displayed in the **Command Prompt** to confirm associativity; refer to Figure-23.

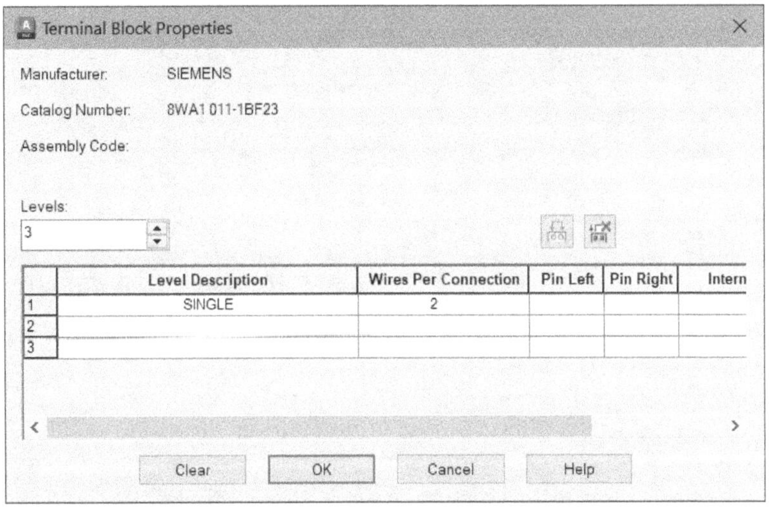

Figure-22. Additional terminal levels

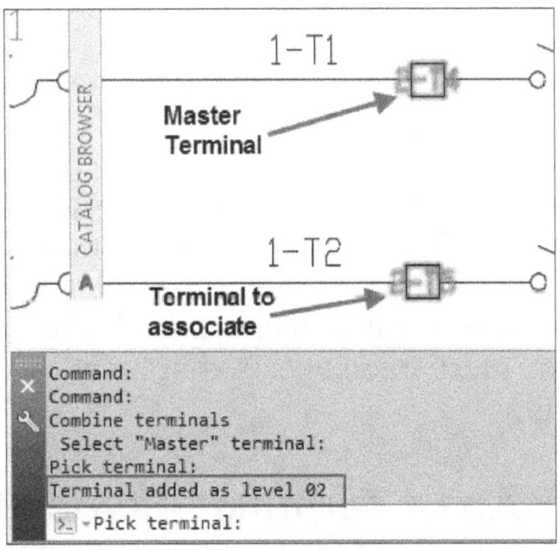

Figure-23. Associating Terminal

- Select the other terminals if required and you have enough levels created in the master terminal.
- Press **ENTER** to exit the tool.
- To check whether the terminals are associated or not, select the master terminal and right-click on it. A contextual menu will be displayed; refer to Figure-24.

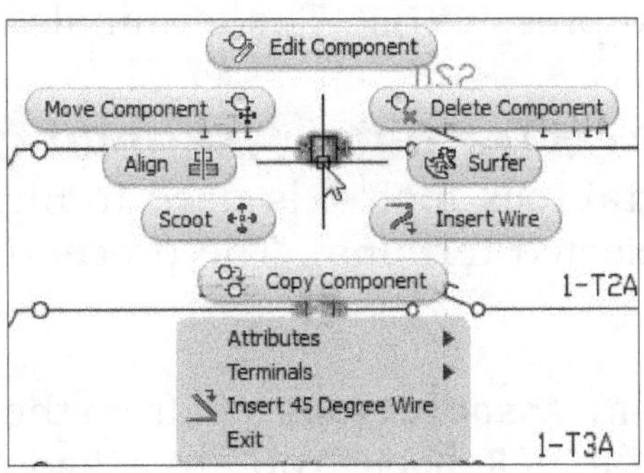

Figure-24. Contextual menu on selecting component

- Click on the **Edit Component** tool. The **Insert/Edit Terminal Symbol** dialog box will be displayed with associated terminals details; refer to Figure-25.

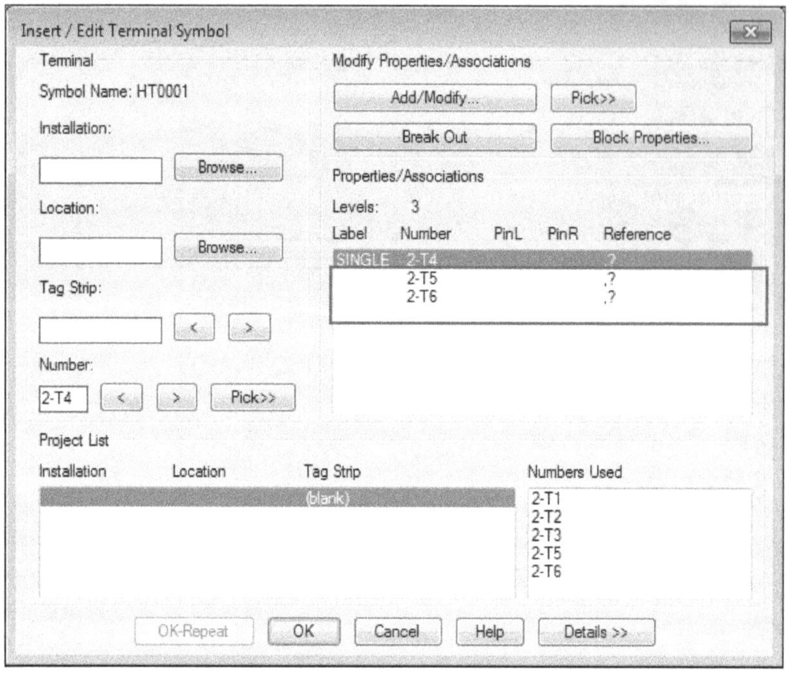

Figure-25. Associations of terminals

- You can check the associated terminals in the **Properties/Associations** area of the dialog box. Click on the **OK** button from the dialog box.

Note: If the number of selected terminals exceeds the total number of levels defined in the block properties, an alert message is displayed and the extra terminals are not added to the association.

Break Apart Terminal Associations

The **Break Apart Terminal Associations** tool does what its name says. Using this tool, you can disassociate the terminals which are associated. The procedure to use this tool is given next.

- Click on the **Break Apart Terminal Associations** tool from the expanded **Edit Components** panel in the **Schematic** tab of the **Ribbon**. You are asked to select the terminals to be disassociated.
- Select the terminals and press **ENTER**. The terminals will be disassociated.

Show Terminal Associations

The **Show Terminal Associations** tool is used to highlight the terminals which are associated with the selected terminal. The procedure to use this tool is given next.

- Click on the **Show Terminal Associations** tool from the expanded **Edit Components** panel in the **Schematic** tab of **Ribbon**. You are asked to pick a terminal.
- Select the terminal whose association is to be checked. The associated terminals will be displayed with a red line connecting them; refer to Figure-26.

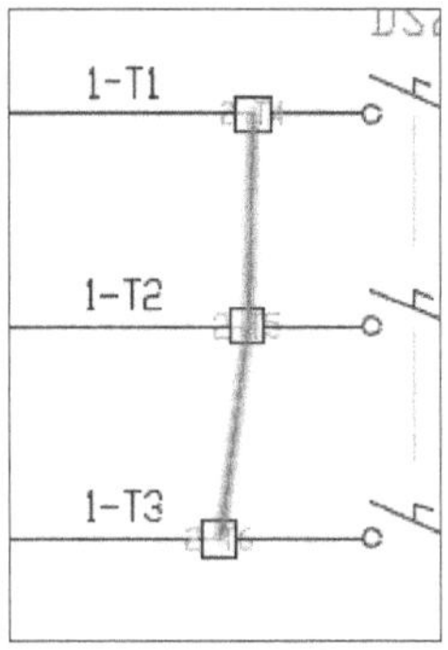

Figure-26. Associated terminals

Edit Jumper

The **Edit Jumper** tool is used to add, edit, and show jumpers connecting the terminals in drawings. The procedure to use this tool is given next.

- Click on the **Edit Jumper** tool from the expanded **Edit Components** panel in the **Ribbon**. You are asked to select a terminal.
- You can either select the terminal or press **ENTER** from keyboard to browse terminals in the drawing. Both the ways are discussed next.

On Pressing ENTER

- On pressing **ENTER**, the **Select Terminals To Jumper** dialog box will be displayed; refer to Figure-27.

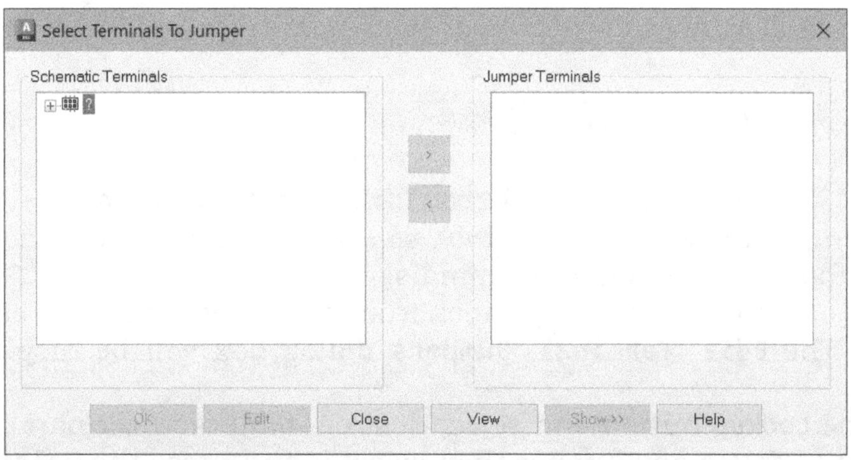

Figure-27. Select Terminals To Jumper

- Expand the node in **Schematic Terminals** area of the dialog box. All the terminals in the drawing will be displayed; refer to Figure-28.

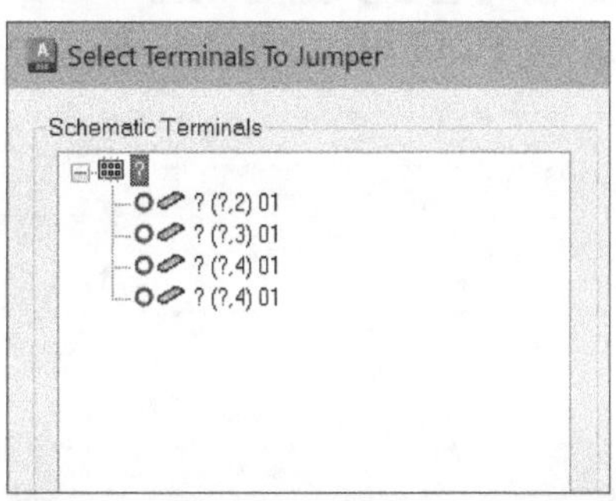

Figure-28. Terminals in schematic drawing

- Select the terminals you want to jumper while holding the **CTRL** key and press the **>** button from the dialog box. The selected terminals will be added in the **Jumper Terminals** area; refer to Figure-29.

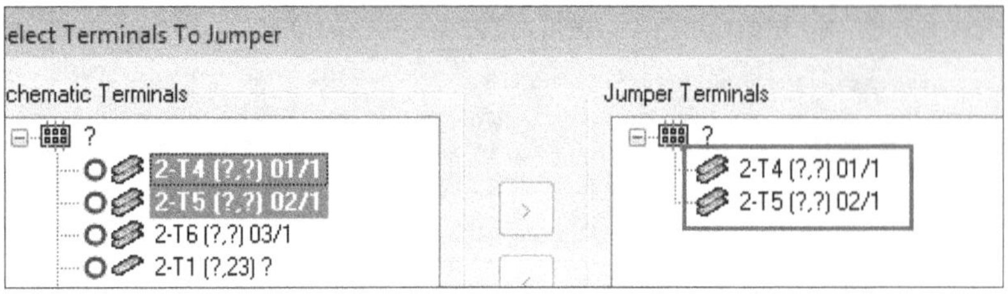

Figure-29. Terminals added in Jumper Terminals area

- Click on the **Edit** button to specify the manufacturer details. The **Edit Terminal Jumpers** dialog box will be displayed; refer to Figure-30.

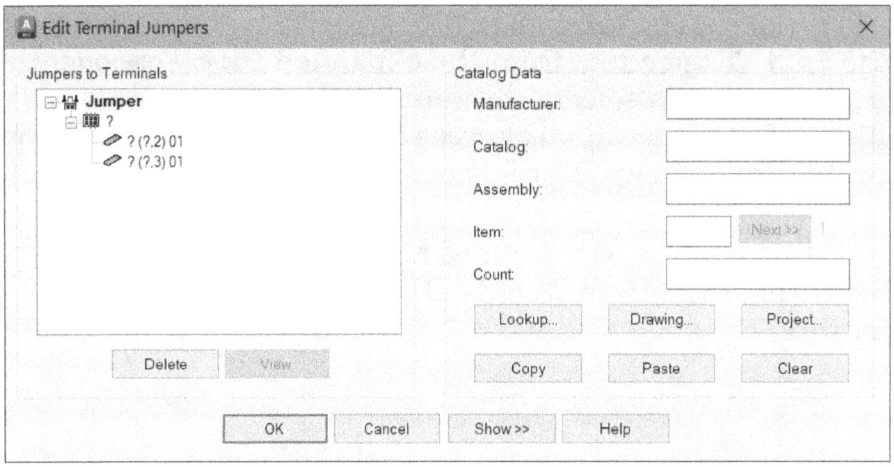

Figure-30. Edit Terminal jumpers dialog box

On Selecting Terminals from Drawing

- After clicking on the **Edit Jumper** tool, select the terminal from the drawing. You are asked to select the jumper terminals.
- Select the other terminals to jumper from the drawing and press **ENTER** from keyboard. The **Edit Terminal Jumpers** dialog box will be displayed; refer to Figure-30.
- Click on the **Lookup** button and select desired manufacturer part.
- If you want to delete any jumper then select it and press the **Delete** button.
- Click on the **OK** button to create the jumper.

PRACTICAL 1

Create the schematic drawing of PLC circuit as shown in Figure-31.

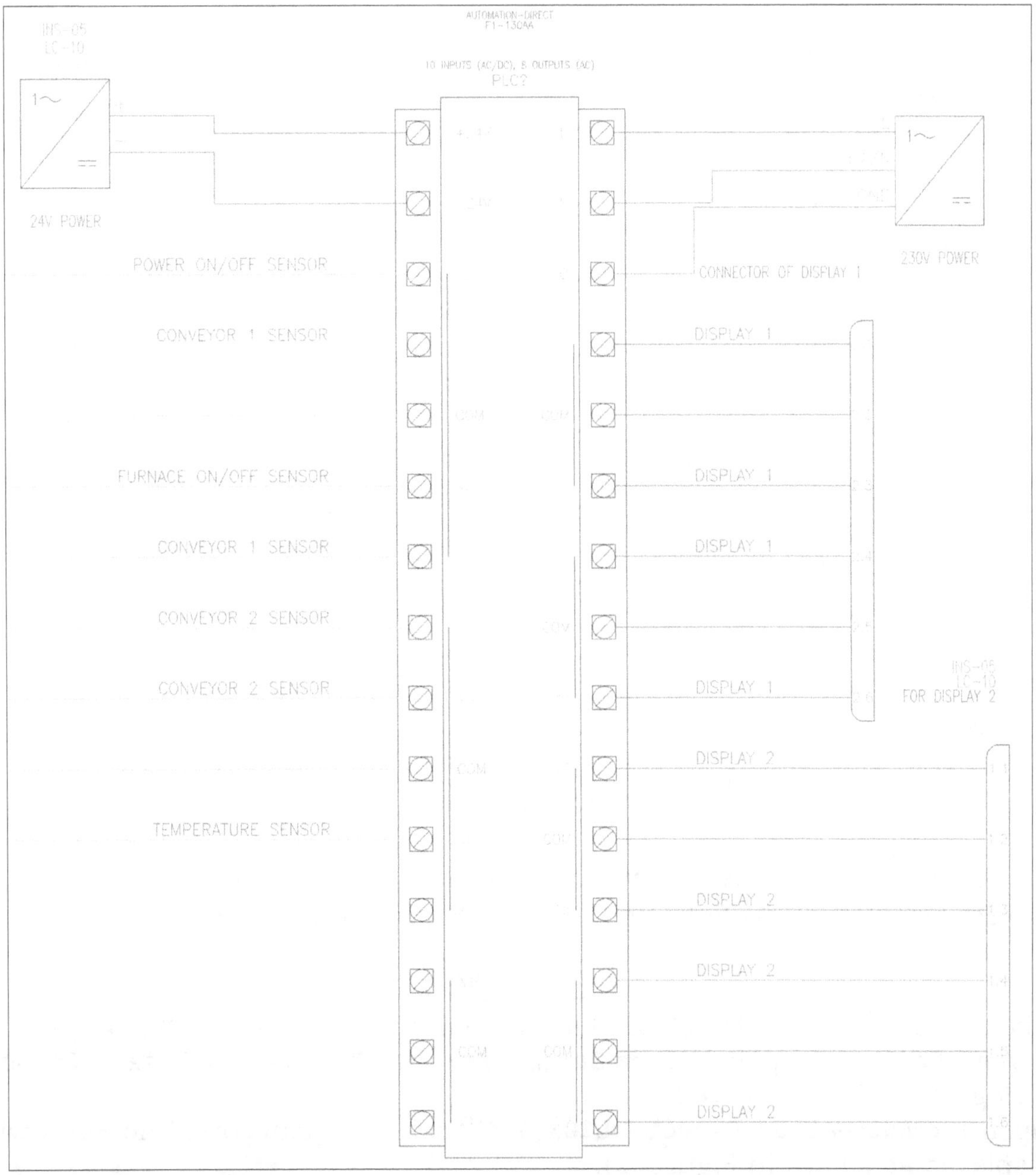

Figure-31. Schematic for PLC

To do List:

- Start a new drawing in current project.
- Place the PLC (Full Unit) of specified description
- Define the parameters of input and output points.
- Create multiple wire buses.
- Place the power supplies and connectors.

The procedure to create this drawing is given next.

Starting a New File

- Click on the **New Drawing** tool from the **Project Browser**. The **Create New Drawing** dialog box will be displayed.
- Specify desired name and other parameters for the file; refer to Figure-32.

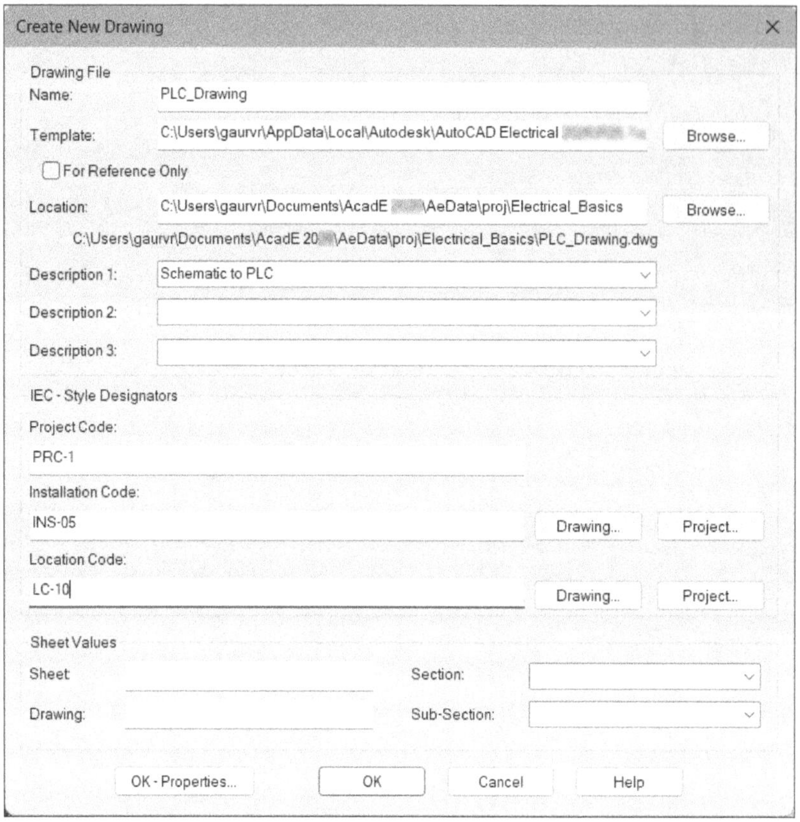

Figure-32. Parameters for new drawing

- Click on the **OK** button from the dialog box. You will be asked whether to specify project defaults. Click on the **Yes** button from the dialog box displayed to apply project defaults to this drawing. The drawing will be created.

Placing PLC

- Click on the **Insert PLC (Full Units)** tool from the **Insert PLC** drop-down of **Insert Components** panel in the **Schematic** tab of **Ribbon**. The **Insert Component** dialog box will be displayed.
- Open the **Automation Direct DL105(3/4" spacing)** category. The PLCs will be displayed as shown in Figure-33.
- Select the F1-130AA PLC from the dialog box. The PLC will get attached to cursor.
- Place PLC at suitable location in drawing where connections can be made to both side. The **Edit PLC Module** dialog box will be displayed; refer to Figure-34.

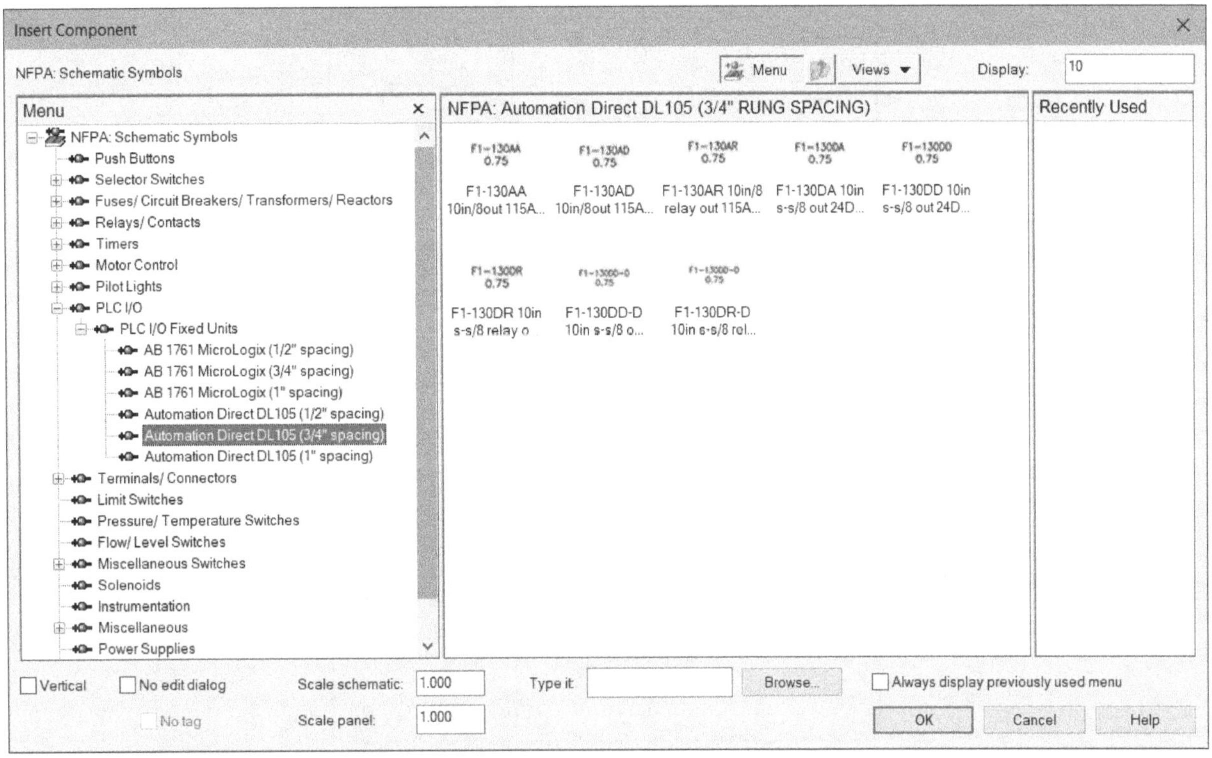

Figure-33. PLCs displayed

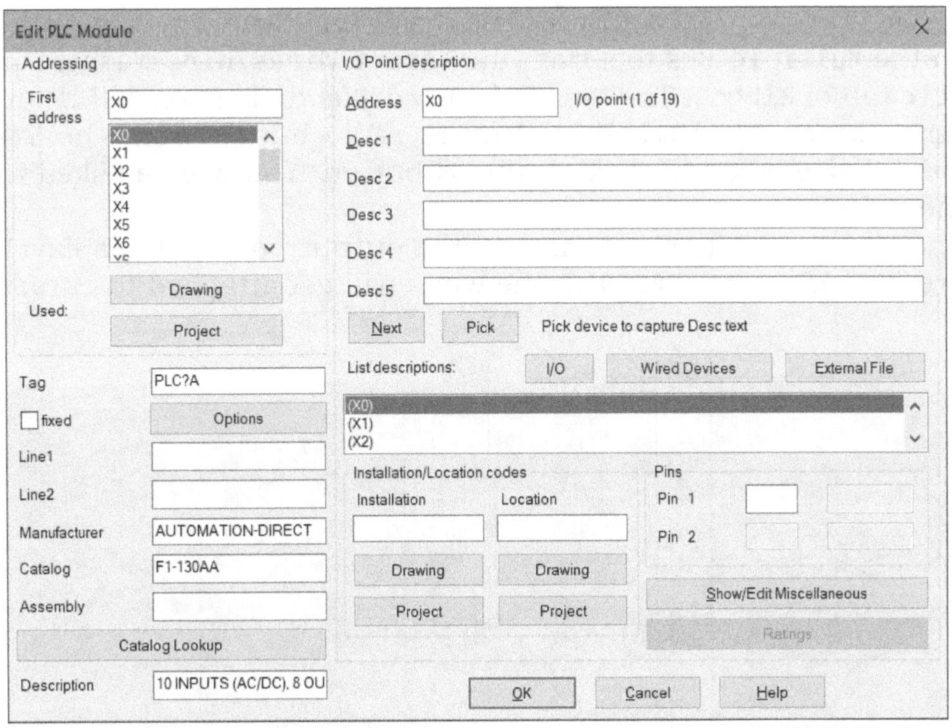

Figure-34. Options to Edit PLC module

- Specify the description of X0 point as **POWER ON/OFF SENSOR**. Click on the **Next** button, you will be asked to specify description for X1 point.
- Similarly, specify descriptions for all the points as per Figure-31. Set the installation and location codes to defaults set for drawing. The drawing should display as shown in Figure-35.

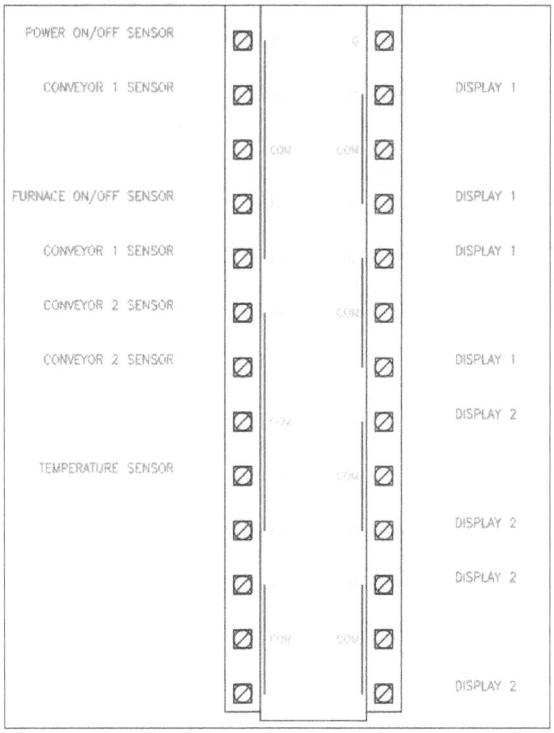

Figure-35. PLC after defining descriptions

Creating Multiple Wire Buses

- Click on the **Multiple Bus** tool from the **Insert Wires/Wire Numbers** panel in the **Schematic** tab of **Ribbon**. The **Multiple Wire Bus** dialog box will be displayed.
- Select the **Component (Multiple Wires)** radio button from the **Starting at** area of the dialog box and click on the **OK** button. You will be asked to select the component for wiring.
- Create a window selection covering all the output points of Display 1 as shown in Figure-36 and press **SPACEBAR**, the wires will get attached to cursor.

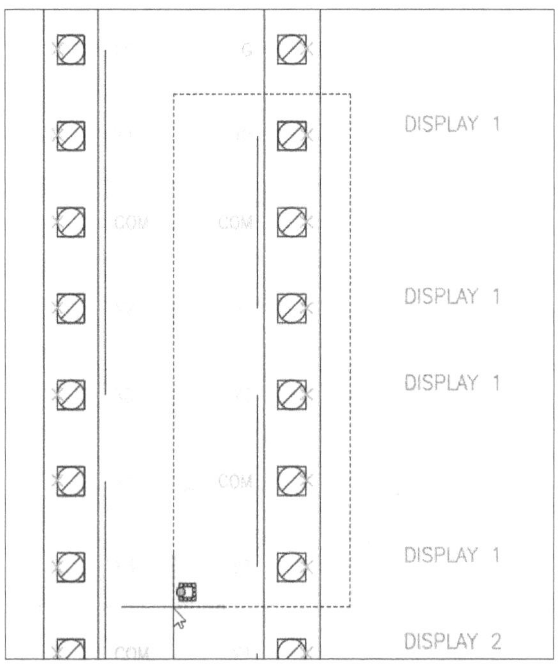

Figure-36. Selection window for Display 1 points

- Click at desired location to specify end point of wires; refer to Figure-37.

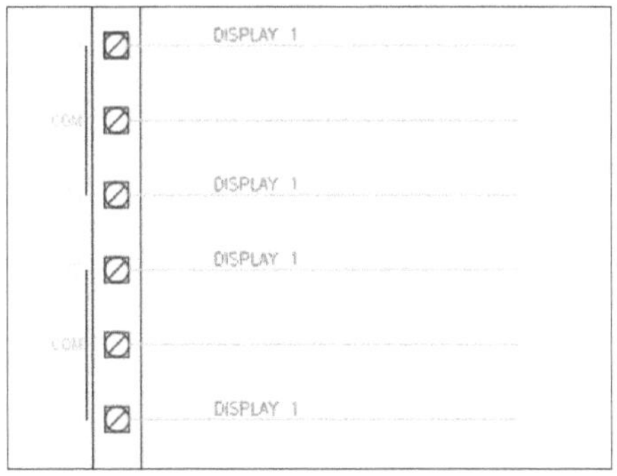

Figure-37. Creating wires for Display 1

- Similarly, create wire bus for Display 2 and sensors on the other side of plc; refer to Figure-38.

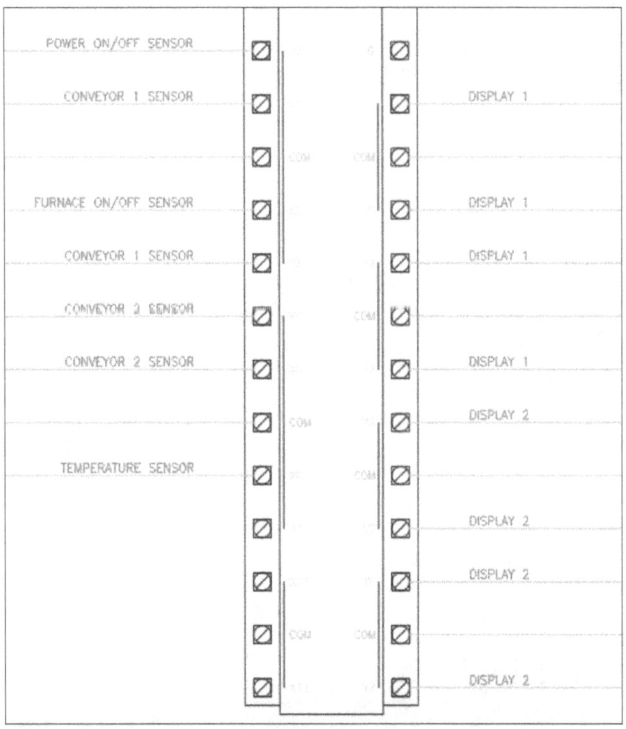

Figure-38. Wires buses created for PLC

Placing Power Supplies

- Click on the **Catalog Browser** tool from the **Icon Menu** drop-down in the **Insert Components** panel in the **Schematic** tab of **Ribbon**. The **CATALOG BROWSER** will be displayed.
- Select the **PW(Power supplies)** option from the drop-down at the top in the **CATALOG BROWSER**. Specify **24VDC** in the **Search** edit box and press **ENTER**. The catalog data for 24 volt DC supplies will be displayed.
- Select the CP-E 24/5 catalog from the catalog and click on the button for inserting symbol; refer to Figure-39. The **Insert Component** dialog box will be displayed.

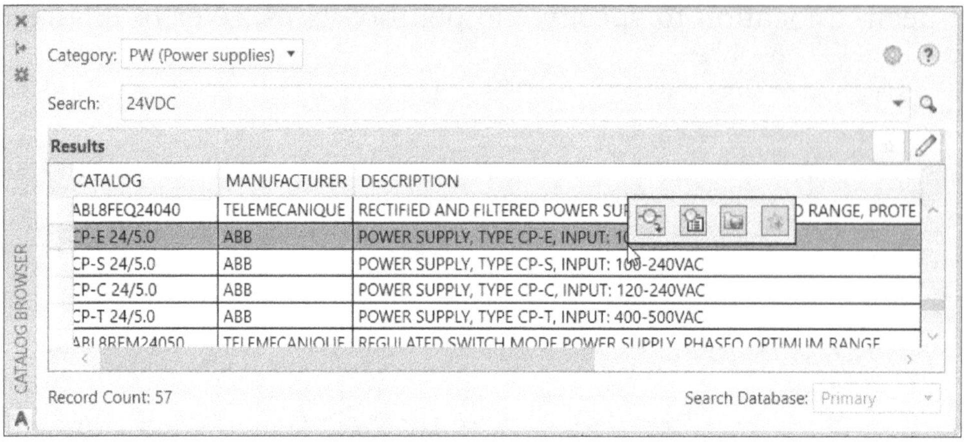

Figure-39. 24 V power supplies in Catalog Browser

- Select the **Power Source 1 Phase** symbol from the **Power Supplies** category in the **Insert Component** dialog box and place the symbol at desired location. The **No Pinlist Defined for Catalog Selection** dialog box will be displayed with **Insert/ Edit Component** dialog box; refer to Figure-40.

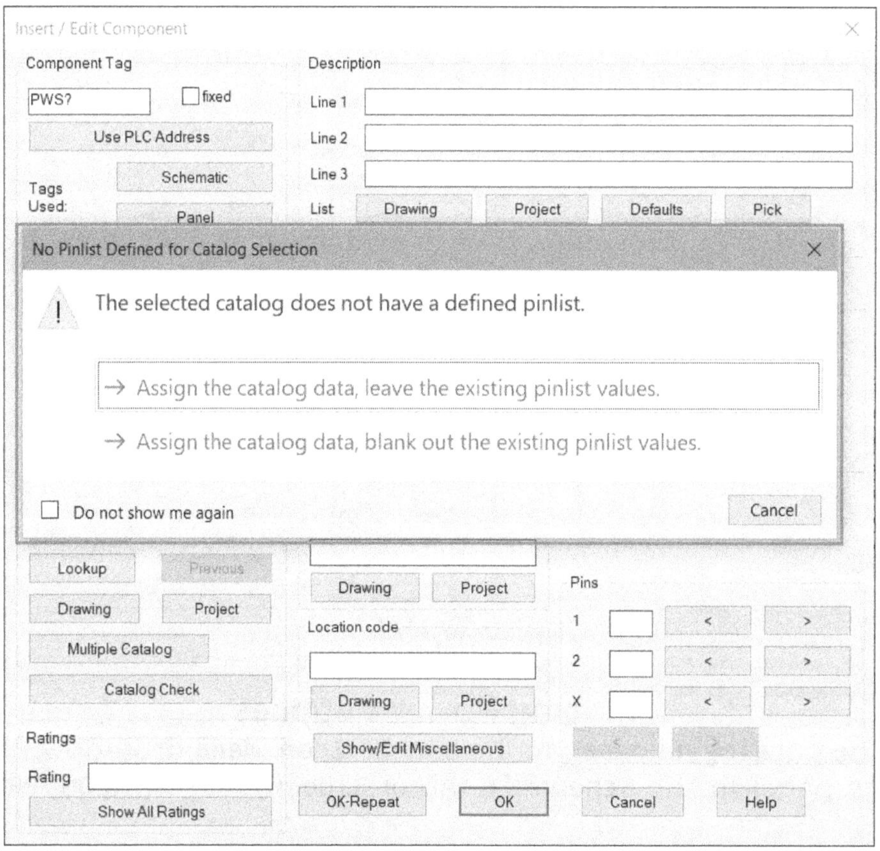

Figure-40. No Pinlist Defined for Catalog Selection dialog box

- Click on the **Assign the catalog data, leave the existing pinlist values** button from the dialog box if you want to leave the pinlist values as they are as per the symbol. Click on the **Assign the catalog data, blank out the existing pinlist values** button from the dialog box if you want to clear the pinlist data and specify the pin data manually. In our case, we have selected the first button. The dialog box will be displayed.
- Clear the pin values for pin 1, 2, and 3; and set the other parameters as shown in Figure-41.

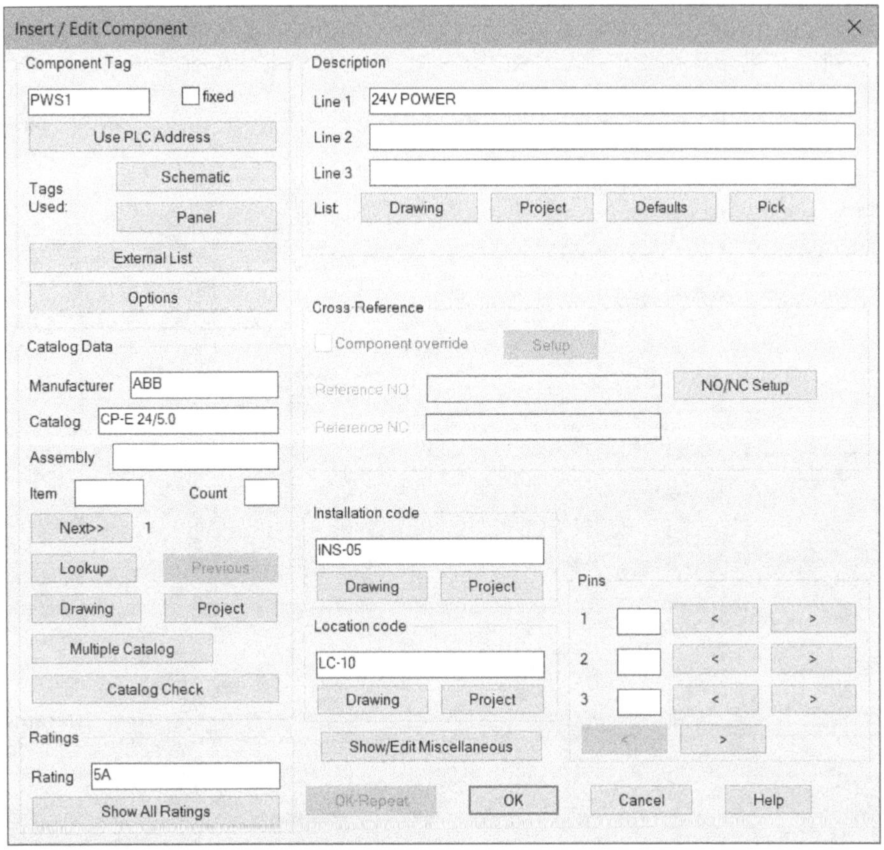

Figure-41. Parameters for 24V power supply

- Click on the **OK** button from the dialog box. The power supply will be created; refer to Figure-42.

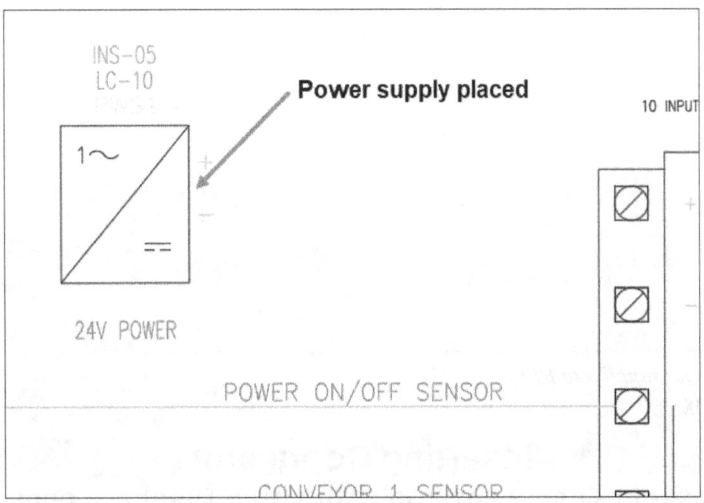

Figure-42. 24VDC power supply placed

- Similarly, place the power supply of 230V with catalog number 1609-P3000A from AB; refer to Figure-43.

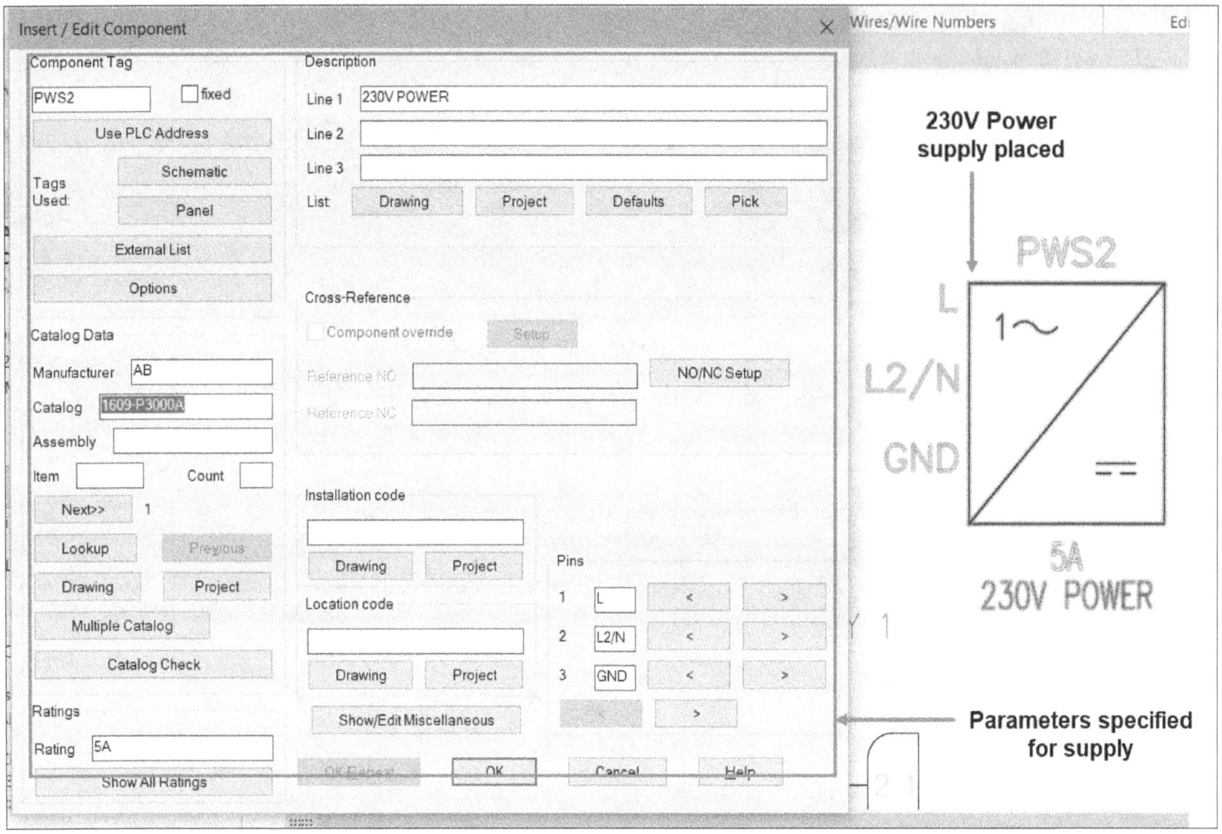

Figure-43. 230V power supply placed

- Connect the power supplies to PLC using **Wire** tool; refer to Figure-44.

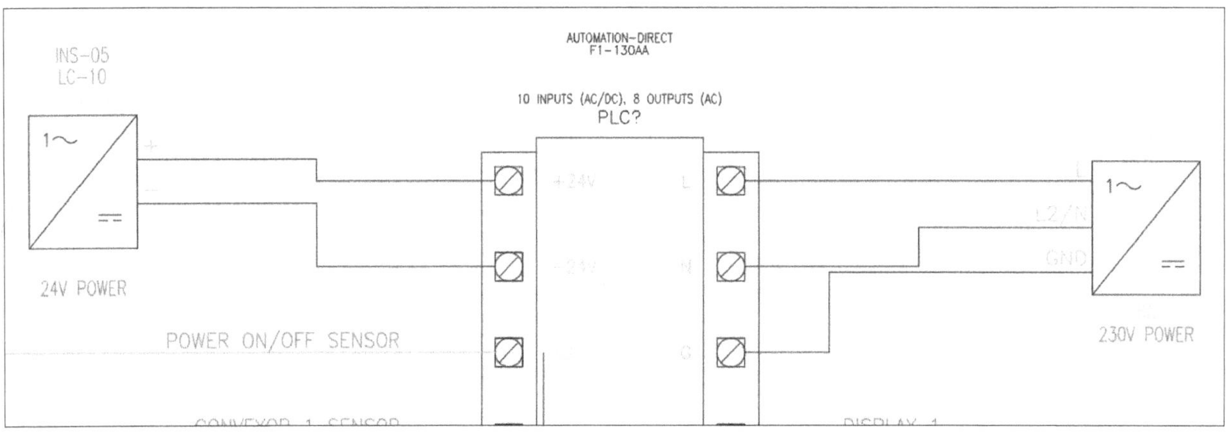

Figure-44. Connecting Power supplies to PLC

Inserting Connectors

- Click on the **Insert Connector** tool from the **Insert Connector** drop-down of **Insert Components** panel of **Schematic** tab in the **Ribbon**. The **Insert Connector** dialog box will be displayed.
- Set the parameters as shown in Figure-45 and place the connector at first wire of **Display 1**; refer to Figure-46. The **Insert/Edit Component** dialog box will be displayed.

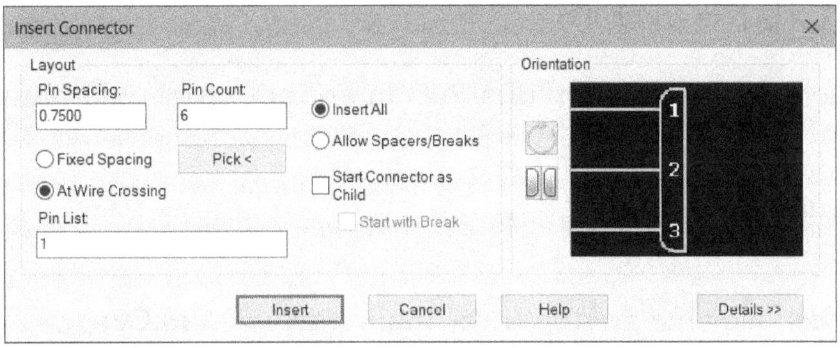

Figure-45. Parameters for Display 1 connector

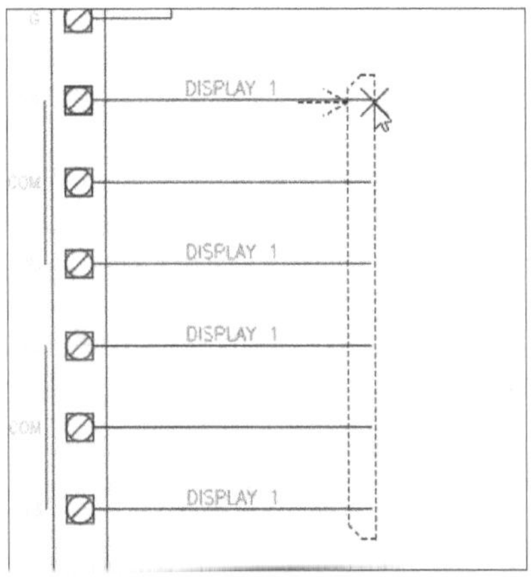

Figure-46. Placing connector for Display 1

- Set desired parameters in the dialog box and click on the **OK** button. The connector will be placed.
- Similarly, you can insert another connector for Display 2.

SELF-ASSESSMENT

Q1. .... is a solid state industrial computer that performs discrete or sequential logic in a factory environment.

Q2. Write down the three main parts of a PLC.

Q3. Discuss various usage of PLCs in different industries.

Q4. The **PLC (Full Units)** tool is used to insert the ready made PLC modules in the drawing. (T/F)

Q5. In some PLCs, the value with X represents .... voltage.

Q6. The **Edit Jumper** tool is used to add, edit, and show jumpers connecting the terminals in drawings. (T/F)

REVIEW QUESTIONS

Q1. What is the primary function of a PLC in an industrial environment?
A. Design circuit boards
B. Perform discrete or sequential logic
C. Replace human workers
D. Conduct mechanical repairs

Q2. Which of the following are the three main parts of a PLC?
A. CPU, RAM, Monitor
B. CPU, Memory, I/O Units
C. RAM, ROM, I/O Boards
D. Sensor, Actuator, Transmitter

Q3. What does the CPU in a PLC do?
A. Only stores data
B. Manages energy usage
C. Reads inputs, executes program, writes outputs
D. Handles internet communication

Q4. In which area is a PLC used to control ammonia and ethylene compressors?
A. Food and Beverage
B. Glass Processing
C. Chemical and Petrochemical
D. Power

Q5. What process does the PLC control in glass batching?
A. Glass cooling
B. Reheating
C. Electromagnetic feeders and weigh hoppers
D. Conveyor belts

Q6. In the context of AutoCAD Electrical, what is the function of the "PLC (Parametric)" tool?
A. Places readymade PLCs without customization
B. Places sensors
C. Inserts PLCs based on user-defined parameters
D. Prints schematics

Q7. What does the "Insert PLC (Full Units)" tool allow you to do?
A. Customize every terminal and connection
B. Insert complete PLC units with fixed configurations
C. Change wiring diagrams
D. Edit program logic

Q8. In PLC addressing, what does a prefix "I/" signify?
A. Instruction
B. Internal address
C. Input
D. Index

Q9. Which PLC application helps in balancing left and right turning ratios in vehicles?
A. Tire Manufacturing
B. Wire Machine Monitoring
C. Power Steering Valve Testing
D. Conveyor System

Q10. In tire curing, what is monitored by the PLC?
A. Road grip
B. Weight
C. Pressure, time, temperature
D. Color

Q11. What is the purpose of the Tag edit box in the PLC module dialog?
A. Assign a color to the module
B. Identify the manufacturer
C. Specify a unique identifier
D. Add a date of installation

Q12. What feature allows the user to maintain previous PLC insertion preferences?
A. Catalog Lookup
B. Tag Format Override
C. AutoCAD Electrical memory
D. Drawing Manager

Q13. Which of the following options in AutoCAD Electrical helps override tag formats?
A. Description box
B. Line1/Line2
C. Options button
D. Next/Pick

Q14. In which application is ramping temperature during the cooking phase crucial?
A. Tire Manufacturing
B. Paper Mill Digester
C. Paint Spraying
D. Rubber Production

Q15. What role do PLCs play in Conveyor Systems?
A. Designing belt sizes
B. Manual control of parts
C. Control sequential operations, alarms, and safety logic
D. Fabricate conveyor rollers

Q16. What is an example of a PLC application in the automotive sector?
A. Glass annealing
B. Chemical batching
C. Carburetor production testing
D. Cold rolling

Q17. What kind of control is used by PLCs in fan operations within chemical production?
A. Manual control only
B. Remote control only
C. Toxic gas level-based control
D. Temperature-based control

Q18. What does the 'Next' button do in the I/O Point Description area in AutoCAD Electrical?
A. Selects the next terminal block
B. Selects the next I/O point for description entry
C. Jumps to the next drawing
D. Saves the current project

Q19. In which application does the PLC assist in controlling molten steel into molds?
A. Annealing Lehr
B. Tire Manufacturing
C. Continuous Casting
D. Plastic Injection Molding

Q20. What PLC-controlled process improves the paint conservation and part throughput?
A. Paint Spraying
B. Tool Changing
C. Wire Machine Control
D. Pulp Batch Blending

Q21. What data is monitored by PLCs in gas transmission and distribution systems?
A. Voltage and current
B. Gas color and density
C. Pressures and flows
D. Air humidity and light

Q22. Which PLC application includes automated handling of hoists in electroplating?
A. Conveyor Systems
B. Storage and Retrieval Systems
C. Automatic Plating Line
D. Tool Changing System

Q23. What component allows linking multiple part numbers in AutoCAD Electrical's PLC dialog?
A. Catalog Lookup
B. Addressing Area
C. Tag
D. Assembly

Q24. Which field in the Insert PLC dialog is used to describe the module type (e.g., "16 Discrete Inputs - 24VDC")?
A. Catalog
B. Description
C. Manufacturer
D. Tag

Q25. What is controlled in a steel plant to calculate oxygen requirements and alloy additions?
A. Annealing Lehr
B. Continuous Casting
C. Steel Making
D. Cold Rolling

Q26. What is the main purpose of the "Allow spacers/breaks" checkbox in PLC (Parametric) insertion?
A. To select spacing for wiring
B. To add manual notes
C. To insert break lines at pre-defined points
D. To highlight errors

Q27. In the context of PLCs, what does the "Used Area" button do?
A. Retrieves part drawings
B. Deletes used addresses
C. Selects I/O addresses used in earlier drawings or projects
D. Displays unused PLC units

Q28. Which of the following is NOT a typical specification parameter for PLCs?
A. Input voltages
B. Output voltages
C. Operating system
D. Number of I/O terminals

Q29. What is the function of the Catalog Lookup tool?
A. Save project file
B. Connect to internet
C. Display catalog data for selected component
D. Print wiring diagrams

Q30. What does a PLC do in a compressor station?
A. Refines oil
B. Optimizes motor load manually
C. Handles safety interlocks and startup/shutdown sequences
D. Filters incoming gas manually

Q31. Which tool is used to expand a portion of a PLC module?
A. Split PLC Module
B. Insert Connector
C. Stretch PLC Module
D. Associate Terminals

Q32. What do you need to define after selecting the objects to be stretched using the Stretch PLC Module tool?
A. Jumper length
B. Terminal symbol
C. Base point and displacement
D. Connector pin spacing

Q33. What is displayed after clicking a point on the PLC when using the Split PLC Module tool?
A. Insert Connector dialog box
B. Catalog Browser
C. Split Block dialog box
D. Edit Component dialog box

Q34. Which radio button should you select in the Insert Connector dialog to insert both parts of the connector?
A. Plug Only
B. Receptacle Only
C. At Wire Crossing
D. Plug/Receptacle Combination

Q35. What happens if you click on a ladder's wire when placing a connector?
A. It adds a jumper
B. It splits the connector
C. The pins automatically adjust
D. The wire is deleted

Q36. What does the Insert Connector (From List) tool use to import connector models?
A. CSV file
B. XML file from Inventor
C. PDF drawing
D. Manual input

Q37. Which dialog box is used when inserting a splice?
A. Split Block
B. Insert Connector
C. Insert Component
D. Edit Jumper

Q38. How do you insert terminals from the Catalog Browser?
A. Select symbol and draw jumper line
B. Right-click and select Insert Terminal
C. Search terminal category and double-click row
D. Use Edit Component tool

Q39. What is the purpose of the Associate Terminals tool?
A. Create wire labels
B. Connect terminal jumpers
C. Combine terminals into a terminal block definition
D. Insert connector from Inventor

Q40. What should you ensure before selecting a terminal as master during association?
A. Wire size matches
B. Manufacturer part number is set
C. Terminal has multiple layers defined
D. Spacing is uniform

Q41. What happens if the number of selected terminals exceeds the defined levels in block properties?
A. Tool crashes
B. Extra terminals are skipped
C. Wire numbers change
D. Pins are misaligned

Q42. What does the Break Apart Terminal Associations tool do?
A. Combines terminals
B. Associates jumpers
C. Splits PLC
D. Disassociates terminals

Q43. What visual indication is given when using Show Terminal Associations?
A. Dotted blue lines
B. Red connecting lines
C. Green dashed lines
D. Yellow box

Q44. What is used to specify jumper terminals if ENTER is pressed during Edit Jumper?
A. Select Connectors Dialog
B. Jumper Settings Dialog
C. Select Terminals To Jumper dialog box
D. Terminal Properties dialog box

Q45. In Edit Jumper, what dialog box allows entering manufacturer details?
A. Edit Component
B. Edit Terminal Jumpers
C. Insert Connector
D. Split Block

For Student Notes

Answer for Self-Assessment:

Ans1. PLC, Ans4. T, Ans5. input, Ans6. T

Chapter 8

Practical and Practice

Topics Covered

The major topics covered in this chapter are:

- ***Introduction***
- ***Practical 1***
- ***Practical 2***
- ***Practical 3***
- ***Practice 1***
- ***Practice 2***
- ***Practice 3***
- ***Practice 4***
- ***Practice 5***

INTRODUCTION

In this chapter, you will create various electrical circuits and at the end of chapter a complete collection of electrical drawings is given for practice.

PRACTICAL 1

Create a schematic for 3 phase motor starter as shown in Figure-1.

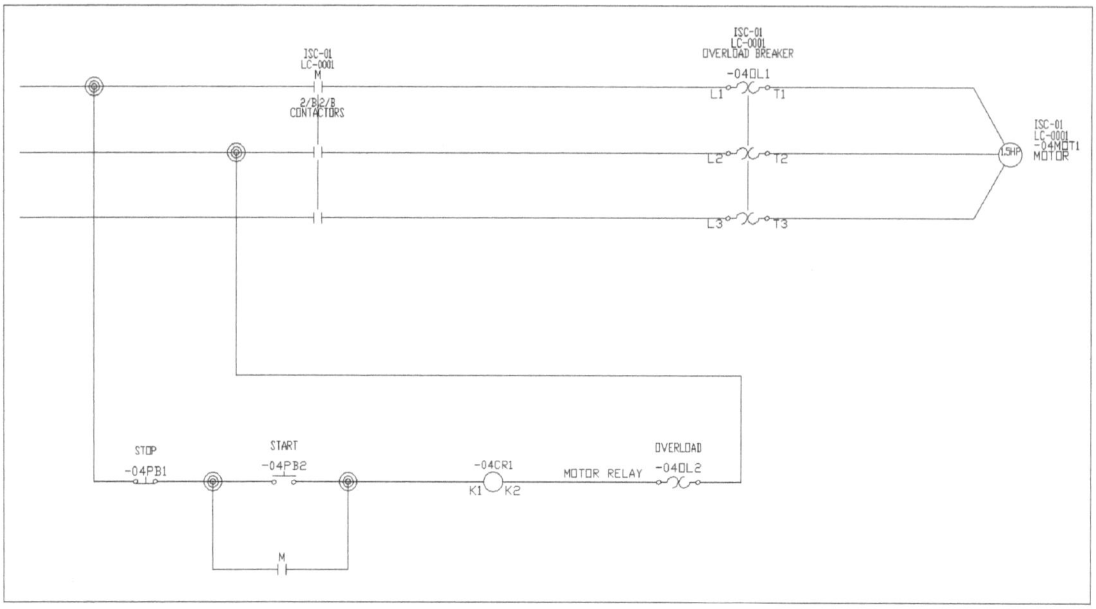

Figure-1. Practical 1

Starting AutoCAD Electrical drawing

- Click on the AutoCAD Electrical icon from the desktop or start AutoCAD Electrical by using the **Start** menu; refer to Figure-2. The interface of AutoCAD Electrical will be displayed.

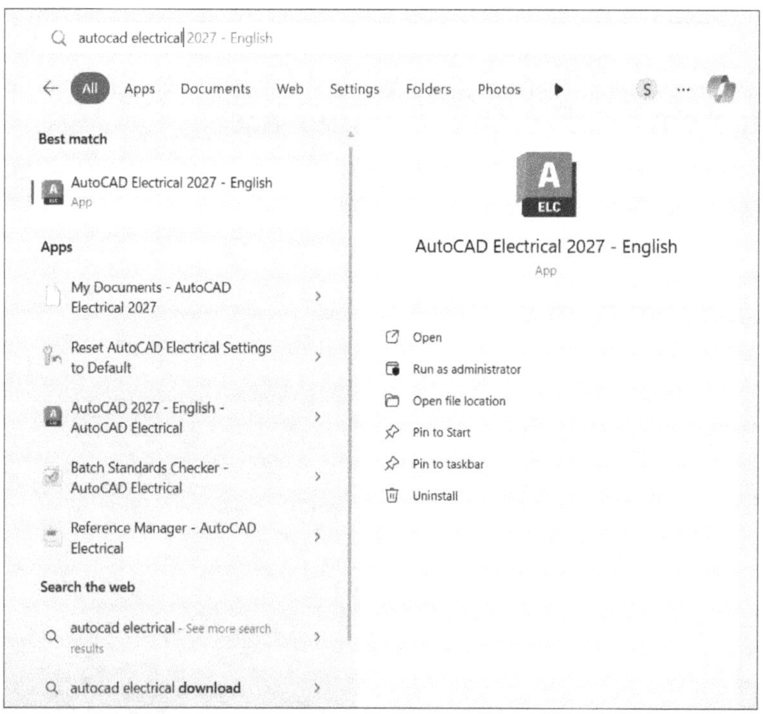

Figure-2. Starting AutoCAD using Start menu

- Click on the down arrow next to **New** button in the first page of interface; refer to Figure-3. Select the **Browse templates** option from the drop-down list. The **Select Template** dialog box will be displayed; refer to Figure-4.

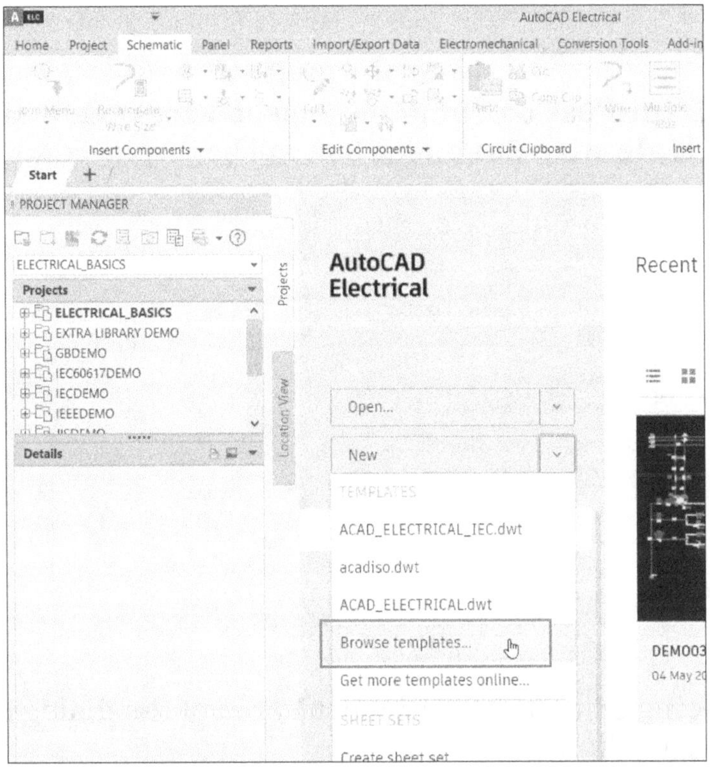

Figure-3. New document options

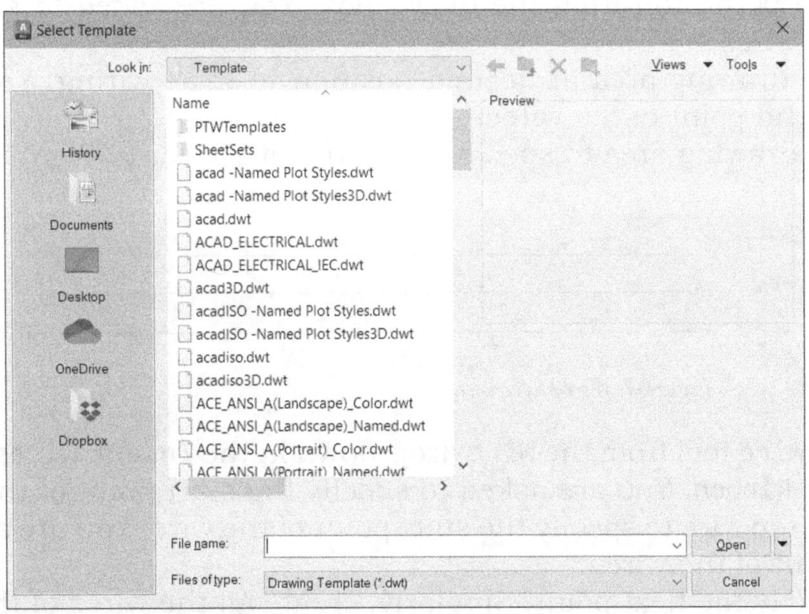

Figure-4. List of templates

- Select the **ACAD ELECTRICAL IEC.dwt** template from the list and click on the **Open** button. A new drawing will be created with the selected template.
- Click on the buttons displayed in Figure-5 to hide the grid and off the grid snap.

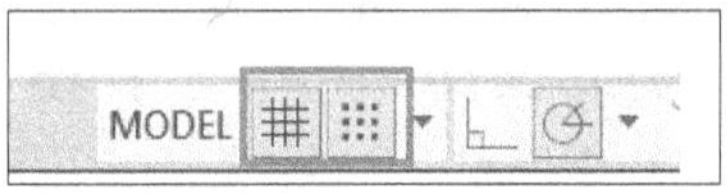

Figure-5. Grid display and snap

Inserting wires

- Click on the **Multiple Bus** tool from the **Insert Wires/Wire Numbers** panel in the **Ribbon**. The **Multiple Wire Bus** dialog box will be displayed as shown in Figure-6.

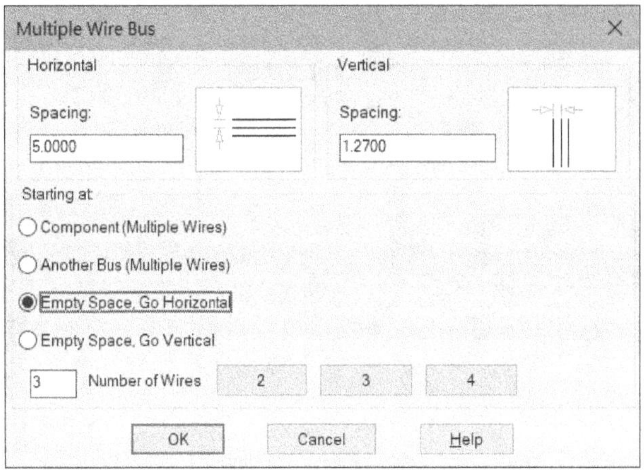

Figure-6. Multiple Wire Bus dialog box

- Click in the **Spacing** edit box in **Horizontal** area of the dialog box and specify the spacing as **5** in the edit box.
- Select the **Empty Space, Go Horizontal** radio button and specify the number of wires as **3** in the **Number of Wires** edit box.
- Click on the **OK** button from the dialog box. You are asked to specify the start point of the wires.
- Click in the drawing area at desired position to start wiring. You are asked to specify the end point of the wires.
- Click in the drawing area to specify the end point; refer to Figure-7.

Figure-7. Wire bus created

- Click in the **Wire** tool from the **Wire** drop-down in the **Insert Wires/Wire Numbers** panel in the **Ribbon**. You are asked to specify the start point of the wire.
- Click on the top wire to specify the start point of the wire. You are asked to specify the next point of the wire.
- Click to specify the next point. Similarly, create all the wires of the circuit; refer to Figure-8.

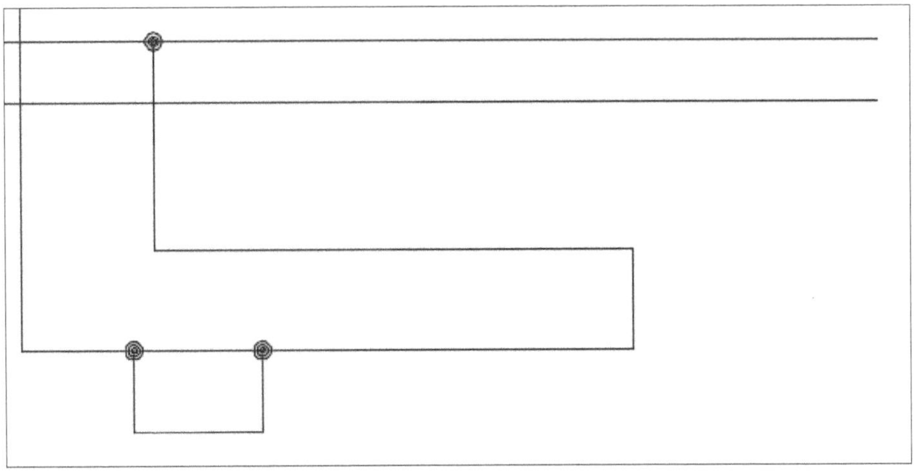

Figure-8. Complete wiring for current circuit

Inserting Components

- Click on the **Icon Menu** button from the **Insert Components** drop-down in the **Insert Components** panel of the **Ribbon**. The **Insert Component** dialog box will be displayed.
- Specify the Scale schematic value as 10 at the bottom in the dialog box and click on the **Motor Control** category from the dialog box. The components related to motor control will be displayed; refer to Figure-9.
- Click on the **3 Phase Motor** component from the dialog box. The component will get attached to the cursor.
- Click at the end point of the center line of the wire bus; refer to Figure-10. The motor will be placed and the **Insert/Edit Component** dialog box will be displayed; refer to Figure-11.

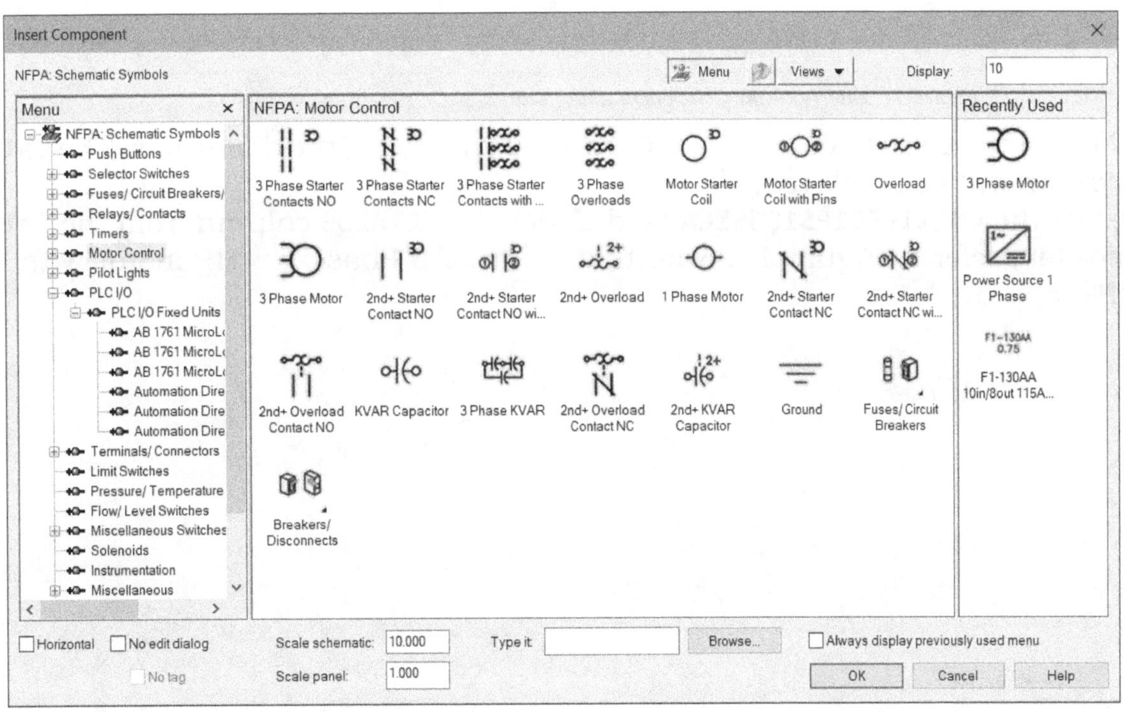

Figure-9. Insert Component dialog box

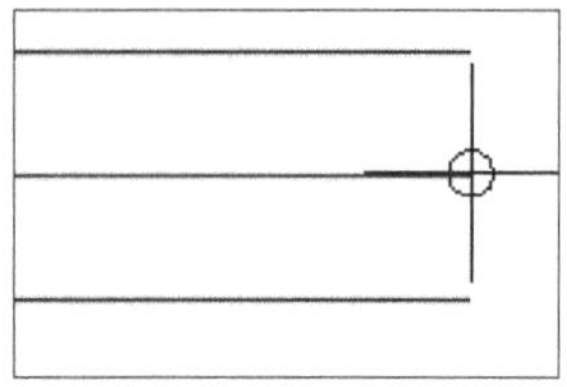

Figure-10. Motor placement

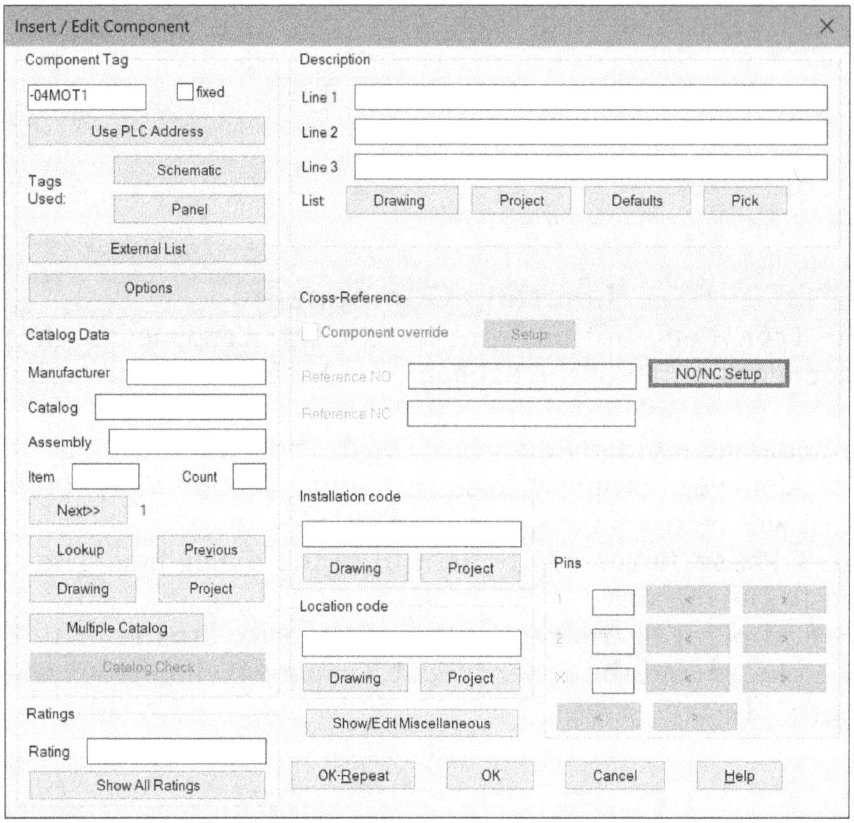

Figure-11. Insert or Edit Component dialog box

- Click on the **Lookup** tool from the **Catalog Data** area of the dialog box. The **Catalog Browser** will be displayed.
- Select the **CM111-FC1F518GSKCA** field under the **CATALOG** column from the **Catalog Browser**; refer to Figure-12. Note that this is a 3 Phase 1.5 HP motor with 1800 RPM speed.

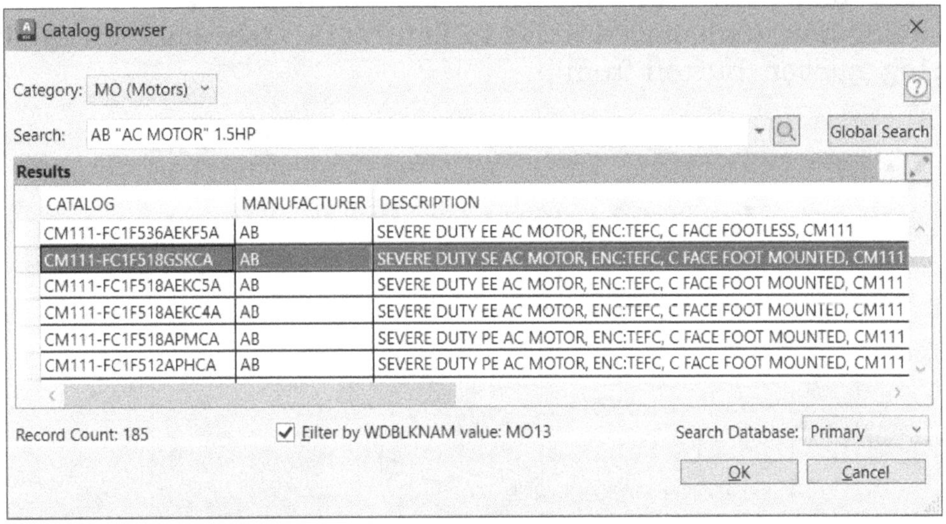

Figure-12. Motor to be selected

- Click on the **OK** button from the **Catalog Browser**.
- The catalog data will be updated automatically in the **Catalog Data** area of the **Insert/Edit Component** dialog box.
- Click on the **Show All Ratings** button from the **Ratings** area of the dialog box. The **View/Edit Rating Values** dialog box will be displayed; refer to Figure-13.
- Click in the **Rating 1** edit box and specify the value as **1.5HP** and remove the **Rating 4** edit box.
- Click on the **OK** button from the dialog box. The value will be displayed in the **Rating** edit box.

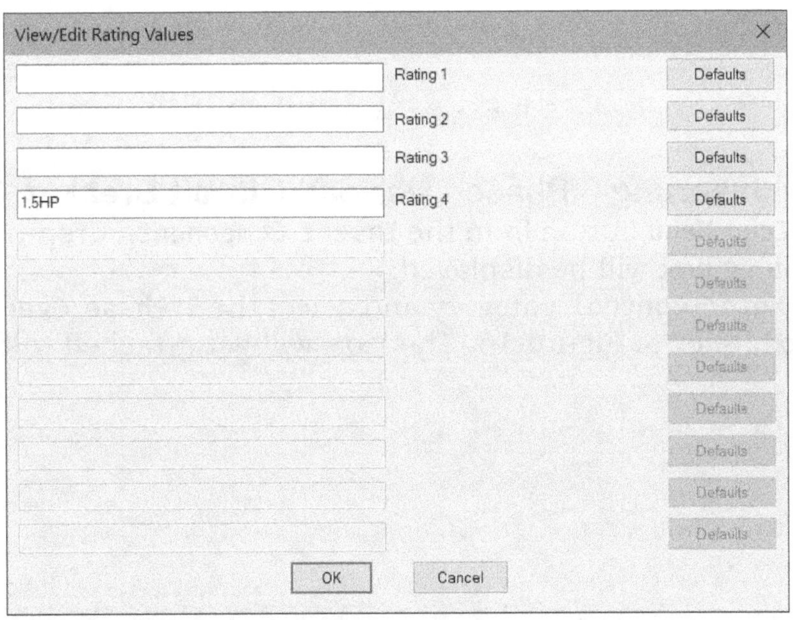

Figure-13. View or Edit Rating Values dialog box

- Click in the **Line 1** edit box in the **Description** area of the dialog box and specify the value as **MOTOR**.
- Click in the **Installation Code** edit box and specify the value as **ISC-01**.
- Click in the **Location Code** edit box and specify the value as **LC-0001**.

- Click on the **OK** button from the dialog box. If the **Assign Symbol To Catalog Number** dialog box is displayed (refer to Figure-14) then click on the **Map symbol to catalog number** button from it.

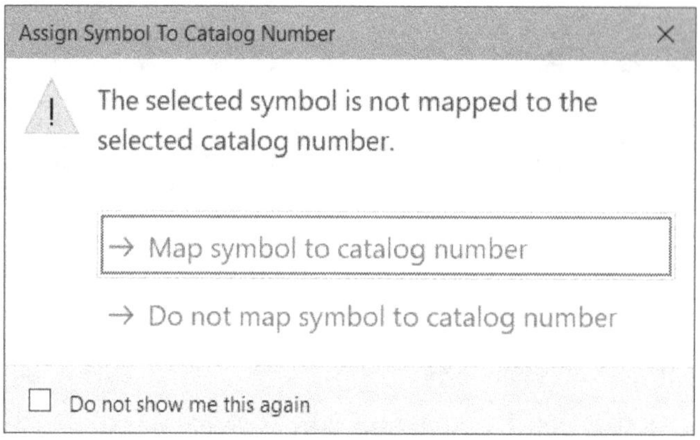

Figure-14. Assign Symbol To Catalog Number dialog box

- The motor after placement will be displayed as shown in Figure-15.

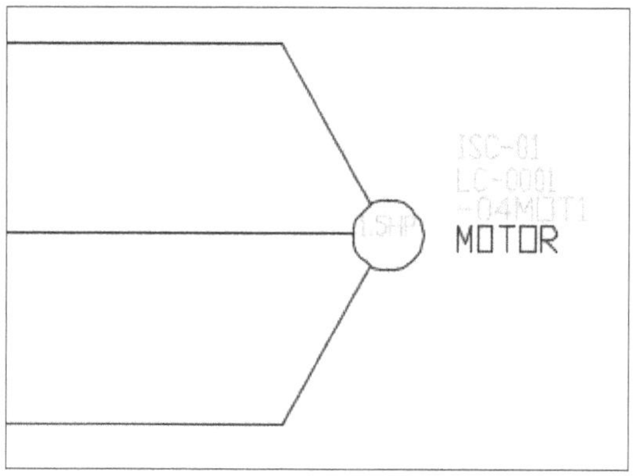

Figure-15. Motor after placement

Inserting 3 Phase Overload Circuit Breaker

- Click on the **Icon Menu** button from the **Insert Components** drop-down. The **Insert Component** dialog box will be displayed.
- Click on the **Motor Control** category and select the **3 Phase Overloads** icon from the dialog box; refer to Figure-16. The icon will get attached to the cursor; refer to Figure-17.

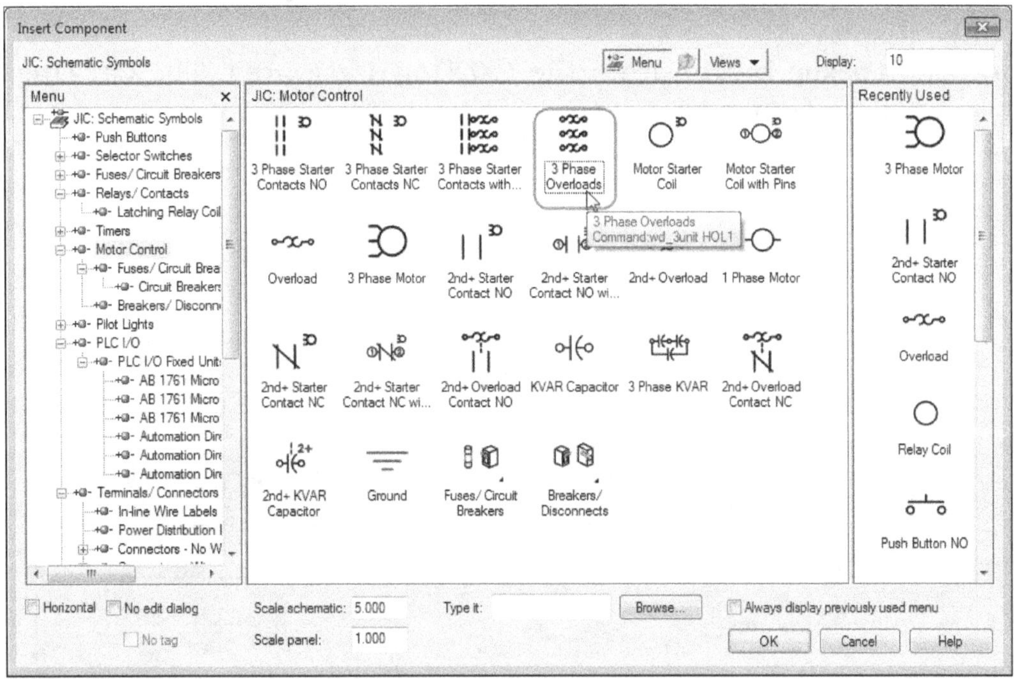

Figure-16. 3 Phase Overloads component

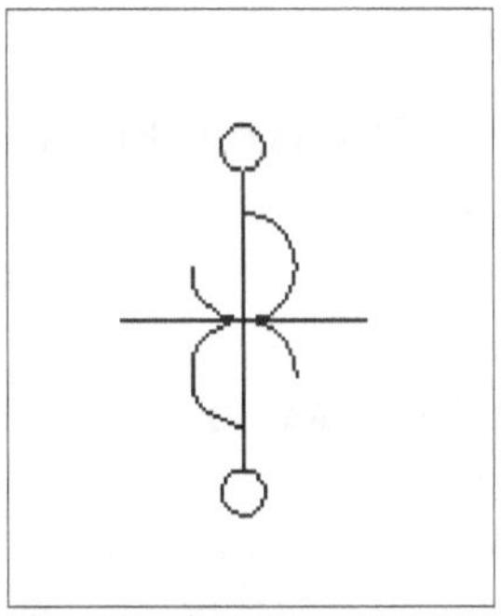

Figure-17. Icon attached to the cursor

- Click on the top wire of the wire bus; refer to Figure-18. The **Build Up or Down?** dialog box will be displayed; refer to Figure-19.

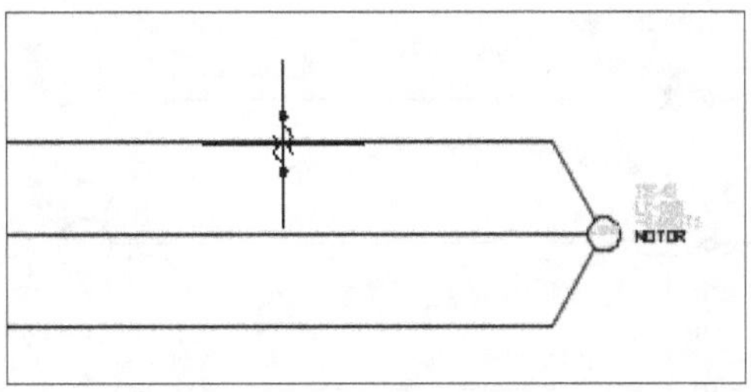

Figure-18. Placement of overload circuit breaker

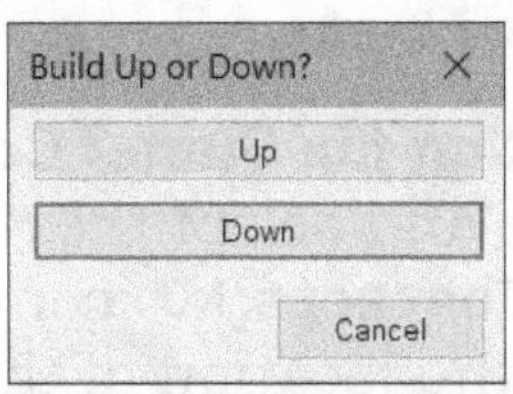

Figure-19. Build Up or Down dialog box

- Click on the **Down** button from the dialog box. The **Insert/Edit Component** dialog box will be displayed.
- Click on the **Lookup** button from the **Catalog Data** area. The **Catalog Browser** dialog box will be displayed.
- Click on the **193-A1A1** field in the **CATALOG** column from the dialog box; refer to Figure-20.

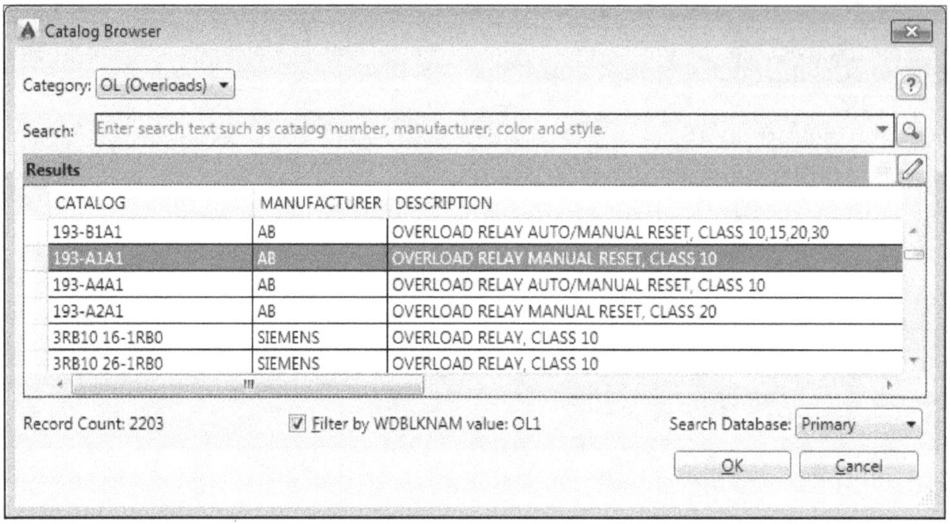

Figure-20. Catalog Browser with Overloads components

- Click on the **OK** button from the **Catalog Browser**. The details will appear in the **Catalog Data** area of the dialog box.
- Click in the **Line 1** edit box in the **Description** area of the dialog box. Specify the value as **OVERLOAD BREAKER**.
- For the **Installation Code** and **Location Code**, click on the **Drawing** button and select the earlier specified code.
- Click on the **OK** button from the dialog box. If the **Assign Symbol To Catalog Number** dialog box is displayed then click on the **Map symbol to catalog number** button. The breaker will be placed in the circuit; refer to Figure-21.

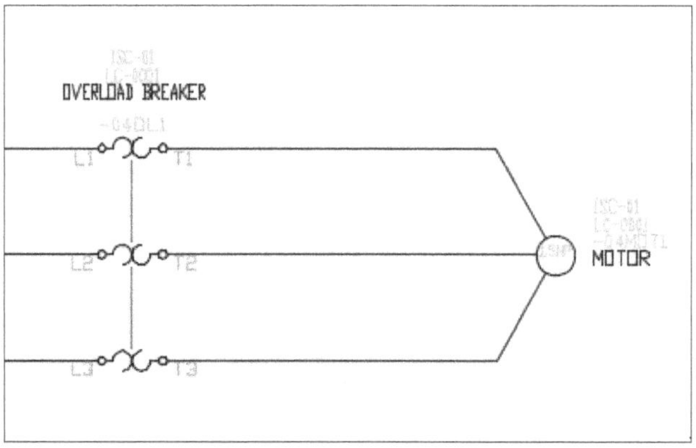

Figure-21. Overload Breaker

Inserting the Starter Contact

- Click on the **Icon Menu** button and select the **Motor Control** category.
- Click on the **3 Phase Starter Contacts NO** icon from the **Insert Component** dialog box. You are asked to specify the location of the contacts.
- Click on the top wire of the bus as did before; refer to Figure-22. The **Build Up or Down?** dialog box will be displayed.

- Click on the **Down** button from the dialog box. The contacts will be placed and the **Insert/Edit Child Component** dialog box will be displayed; refer to Figure-23.
- Click in the **Line 1** edit box in the **Description** area of the dialog box and specify the value as **CONTACTORS**.

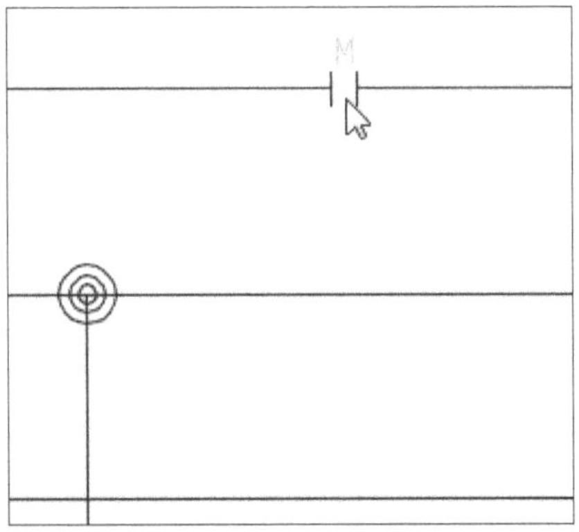

Figure-22. Contact symbol placement

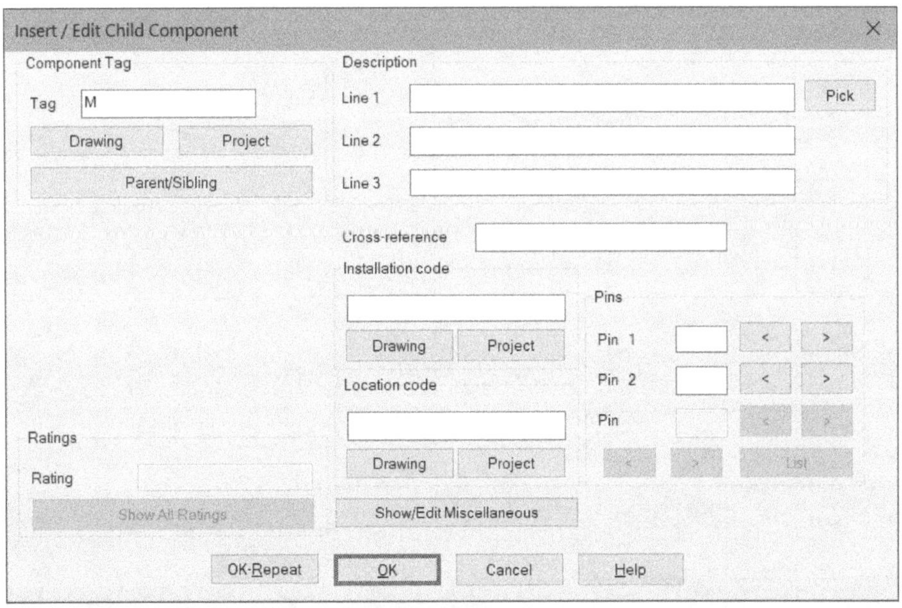

Figure-23. Insert or Edit Child Component dialog box

- For **Installation** and **Location** codes, click on the **Drawing** buttons and select the earlier specified values.
- Click on the **OK** button from the dialog box.

Similarly, insert the other components with the required descriptions. The schematic drawing after inserting all the components will be displayed as shown in Figure-24.

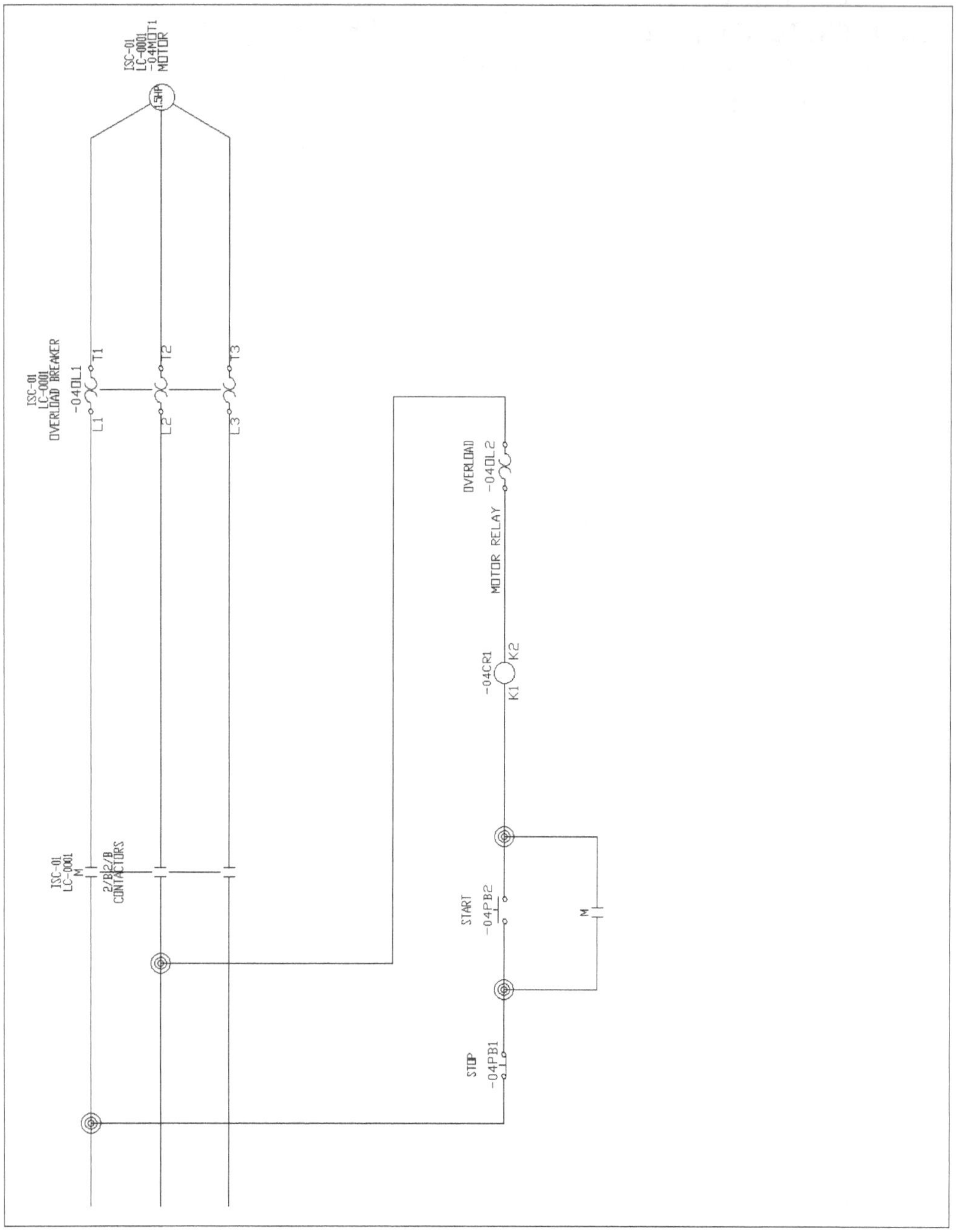

Figure-24. Practical 1

PRACTICAL 2

In this practical, we will create a schematic drawing of circuit as shown in Figure-25. This circuit mainly emphasize on connector.

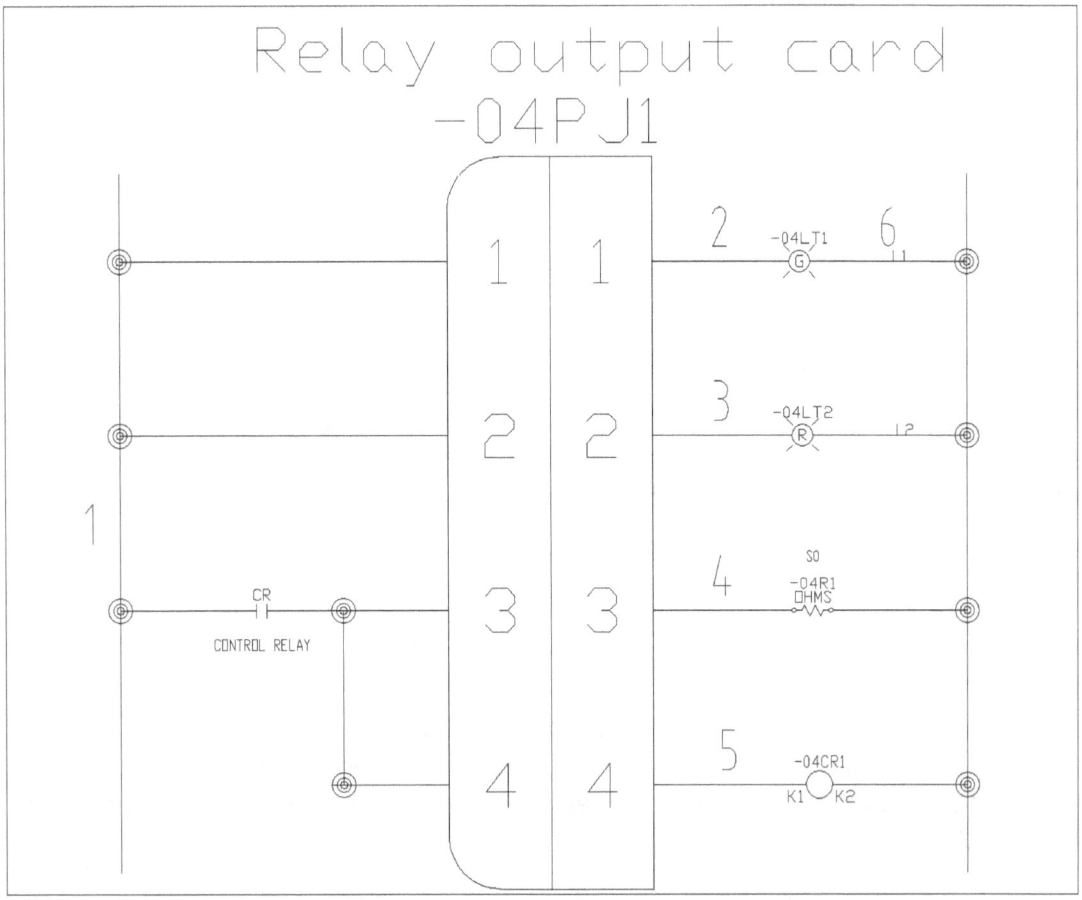

Figure-25. Practica 2

Starting AutoCAD Electrical drawing

- Click on the AutoCAD Electrical icon from the desktop or start AutoCAD Electrical by using the **Start** menu. The interface of AutoCAD Electrical will be displayed.
- Click on the down arrow next to **New** button in the first page of interface; refer to Figure-26. Select the **Browse templates** option from the drop-down list. The **Select Template** dialog box will be displayed.

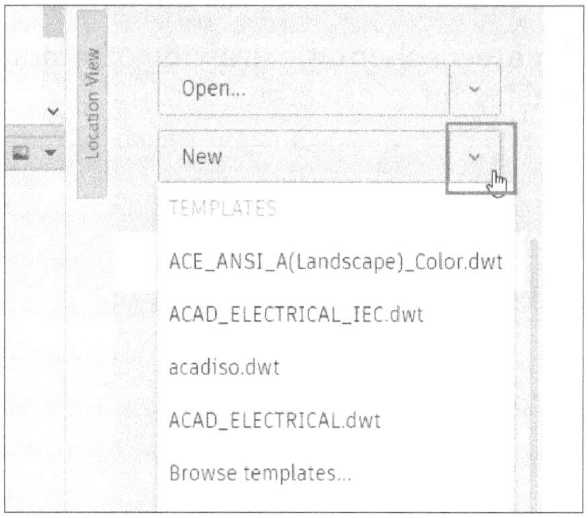

Figure-26. New document option

- Select the **ACAD ELECTRICAL IEC.dwt** template from the list and click on the **Open** button. A new drawing will be created with the selected template.
- Click on the buttons displayed in Figure-27 to hide the grid and off the grid snap.

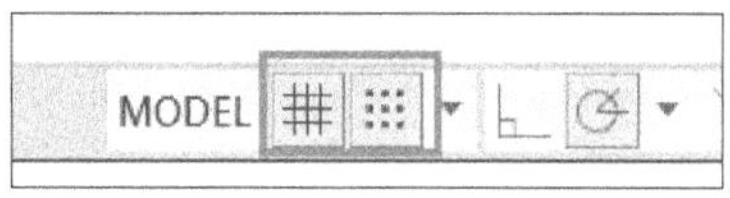

Figure-27. Grid display and snap

Inserting ladder

- Click on the **Insert Ladder** tool from the **Insert Ladders** drop-down in the **Insert Wires/Wire Numbers** panel. The **Insert Ladder** dialog box will be displayed; refer to Figure-28.

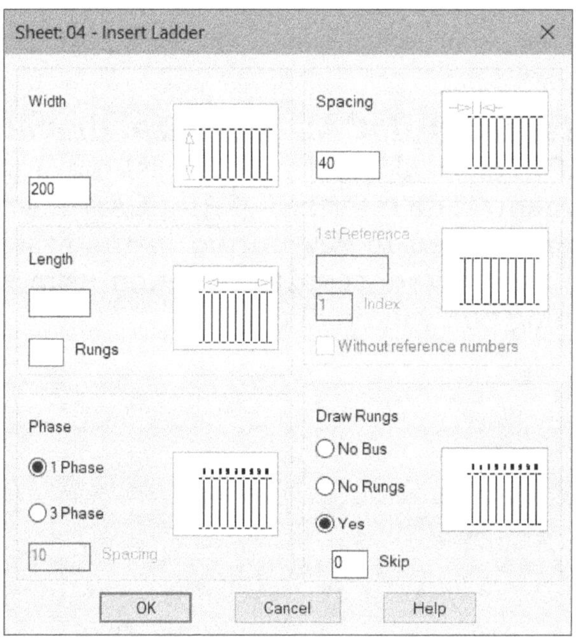

Figure-28. Insert Ladder dialog box

- Specify the **Width** value and **Spacing** value as **50** and **10**, respectively.
- Click in the **Rungs** edit box and specify the value as **4**.
- Click on the **OK** button from the dialog box. You are asked to specify the insertion point for the ladder.

- Click in the drawing area to place the ladder at the desired position. If your ladder is placed with horizontal orientation as shown in Figure-29, then rotate the complete ladder by 90 degree using the **Rotate** tool from the **Modify** panel in the **Home** tab of the **Ribbon**.

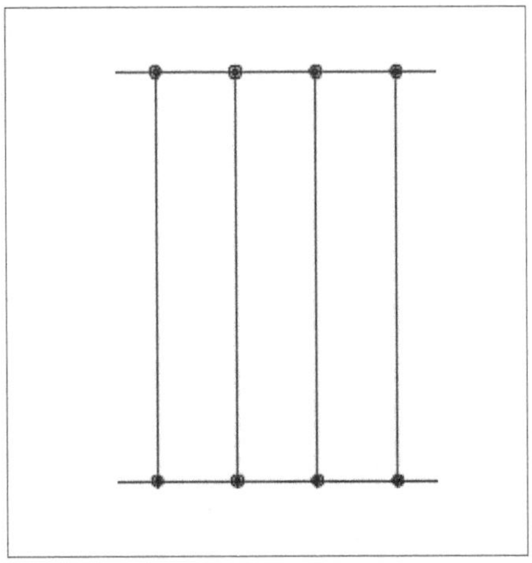

Figure-29. Ladder placed horizontally oriented

- After rotating the ladder, it will be displayed as shown in Figure-30.

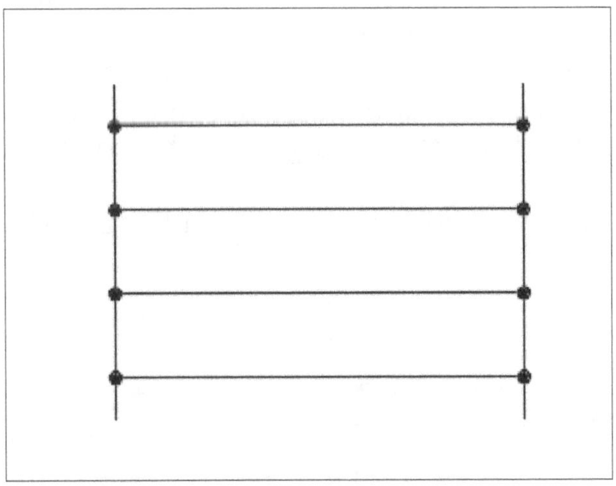

Figure-30. Rotated ladder

Inserting the Connector

- Click on the **Schematic** tab if you are still in the **Home** tab; refer to Figure-31.

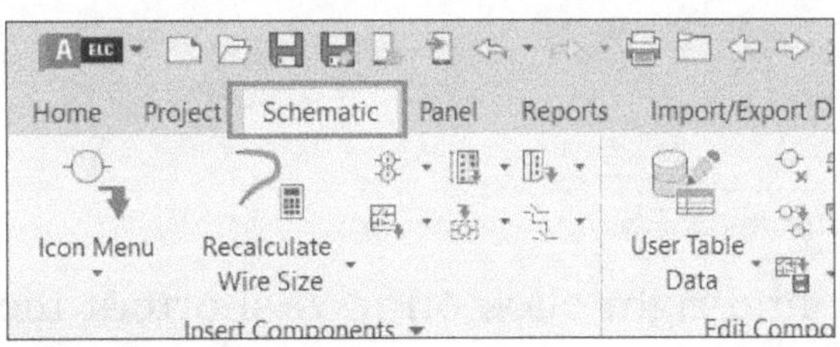

Figure-31. Schematic tab

- Click on the **Insert Connector** tool from the **Insert Connectors** drop-down in the **Insert Components** panel. The **Insert Connector** dialog box will be displayed as shown in Figure-32.

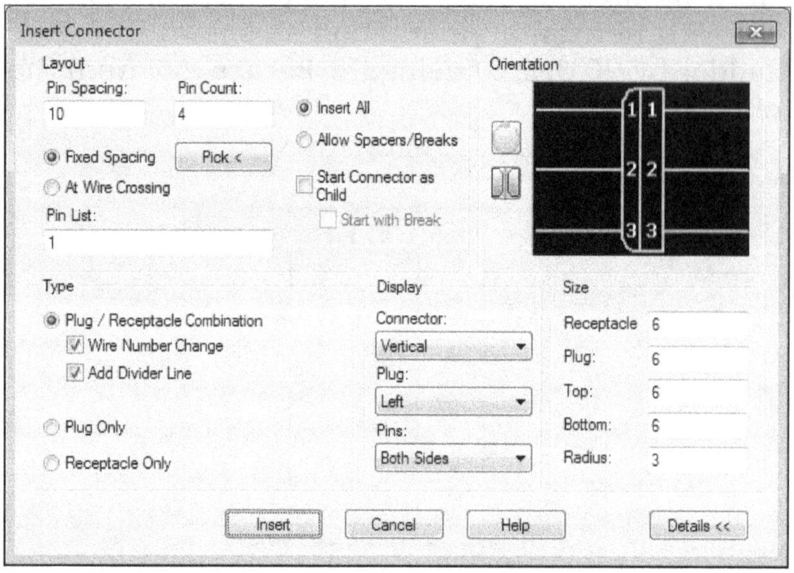

Figure-32. Insert Connector dialog box

- Click in the **Pin Spacing** edit box and specify the value as **10** (this is also the spacing between two consecutive rungs in the ladder earlier created).
- Click in the **Pin Count** edit box and specify the value as **4** (this is also the number of rungs in the ladder earlier created).
- Click in the **Connector** drop-down in the **Display** area of the dialog box and select the **Vertical** option from the list displayed if not selected.
- Click on **Insert** button from the dialog box. The connector will get attached to the cursor.
- Click on the top rung of the ladder as shown in Figure-33 to place the connector.

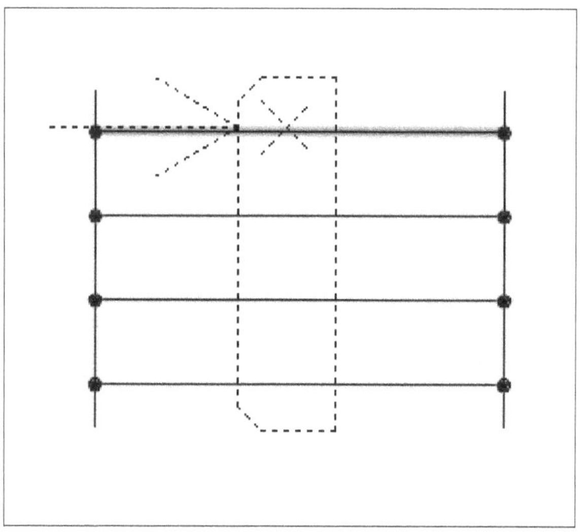

Figure-33. Placing the connector

- Select the connector and right click on it. A shortcut menu will be displayed as shown in Figure-34.

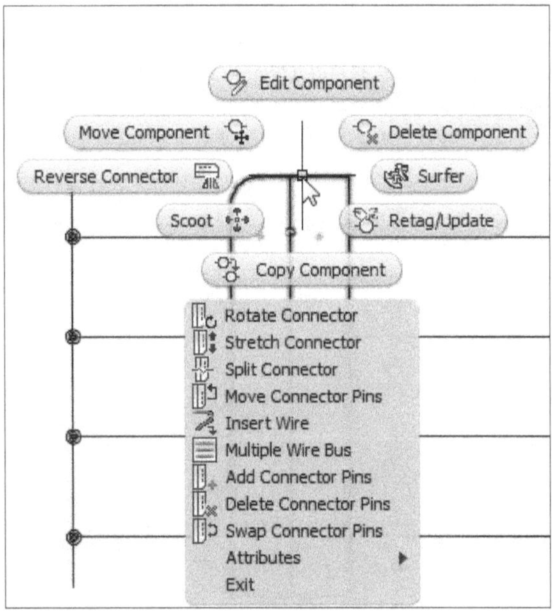

Figure-34. Menu for connector

- Move the cursor on the **Attributes** option in the menu. A cascading menu will be displayed; refer to Figure-35.

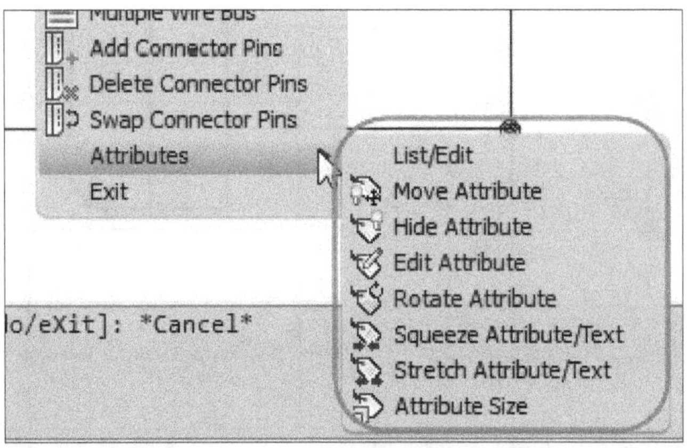

Figure-35. Cascading menu for attributes

- Click on the **Attribute Size** option from the cascading menu. The **Change Attribute Size** dialog box will be displayed; refer to Figure-36.

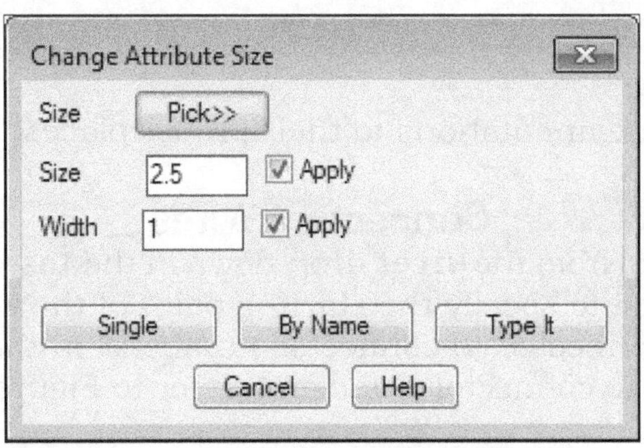

Figure-36. Change Attribute Size dialog box

- Make sure the size is **2.5** and width is **1**. Click on the **Single** button from the dialog box. You are asked to select the annotation objects.
- Click on the connector numbers and tags displayed in very small font to increase their size. Refer to Figure-37.

Figure-37. Numbers and tags in small font

- After increasing the size of annotations, the drawing will be displayed as shown in Figure-38.

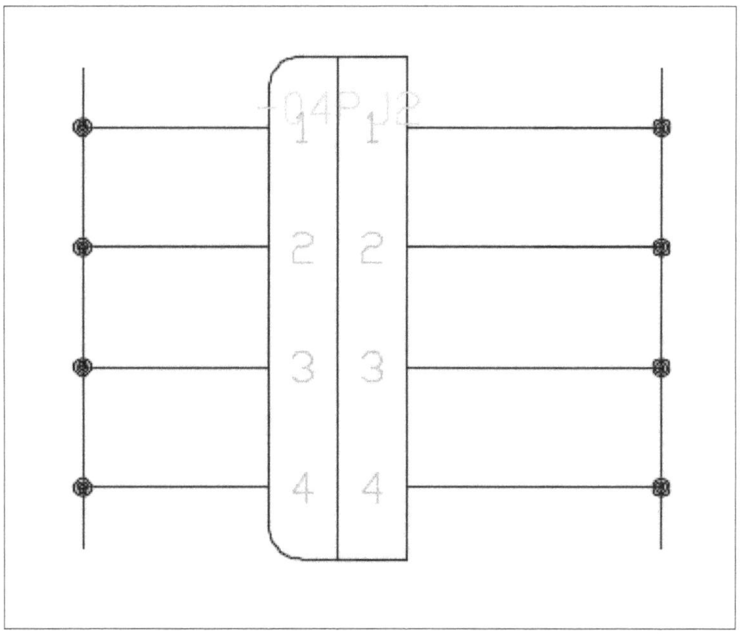

Figure-38. After increasing size of annotations

- Move the overlapping annotations to their proper places.

Connecting Wires

- Click on the **Wire** tool from the **Wires** drop-down in the **Insert Wires/Wire Numbers** panel. You are asked to specify the starting point of the wire.
- Click on the wire connected to connector number **3** in the left and connect it to the wire connected to connector number **4**; refer to Figure-39.

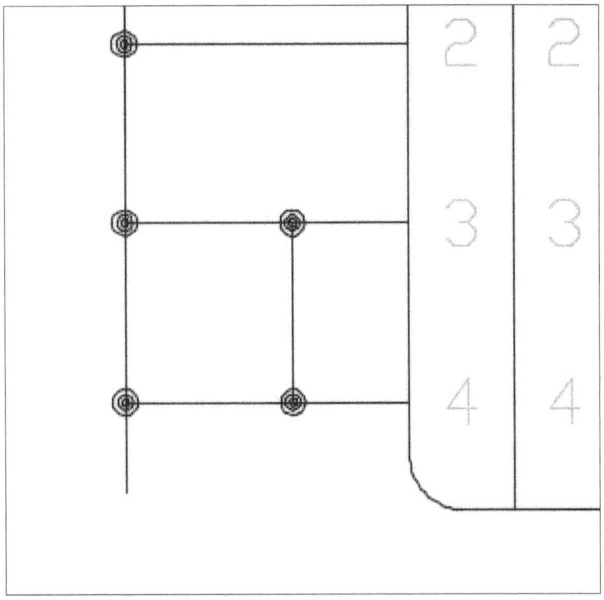

Figure-39. Wire connected

- Click on the **Trim Wire** tool from the **Edit Wires/Wire Numbers** panel in the **Ribbon**. You are asked to select the wire to be trimmed.
- Click on the wire as shown in Figure-40. Press **ENTER** to exit the **Trim Wire** tool.

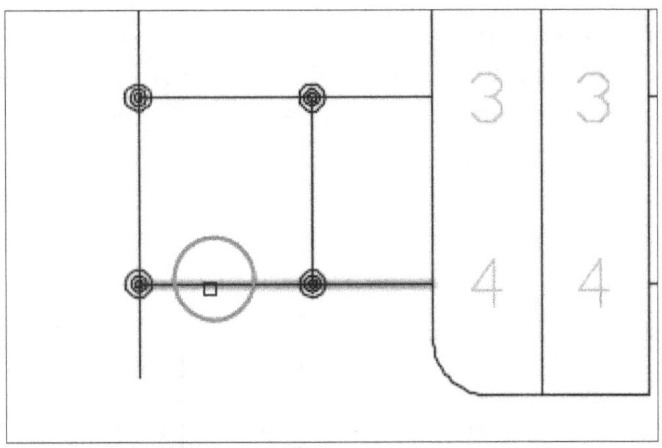

Figure-40. Wire to be trimmed

Inserting Components

- Click on the **Icon Menu** tool from the **Insert Components** drop-down in the **Insert Components** panel. The **Insert Component** dialog box will be displayed.
- Click on the **Relays/Contacts** category from the dialog box. The components related to relays and contacts will be displayed; refer to Figure-41.

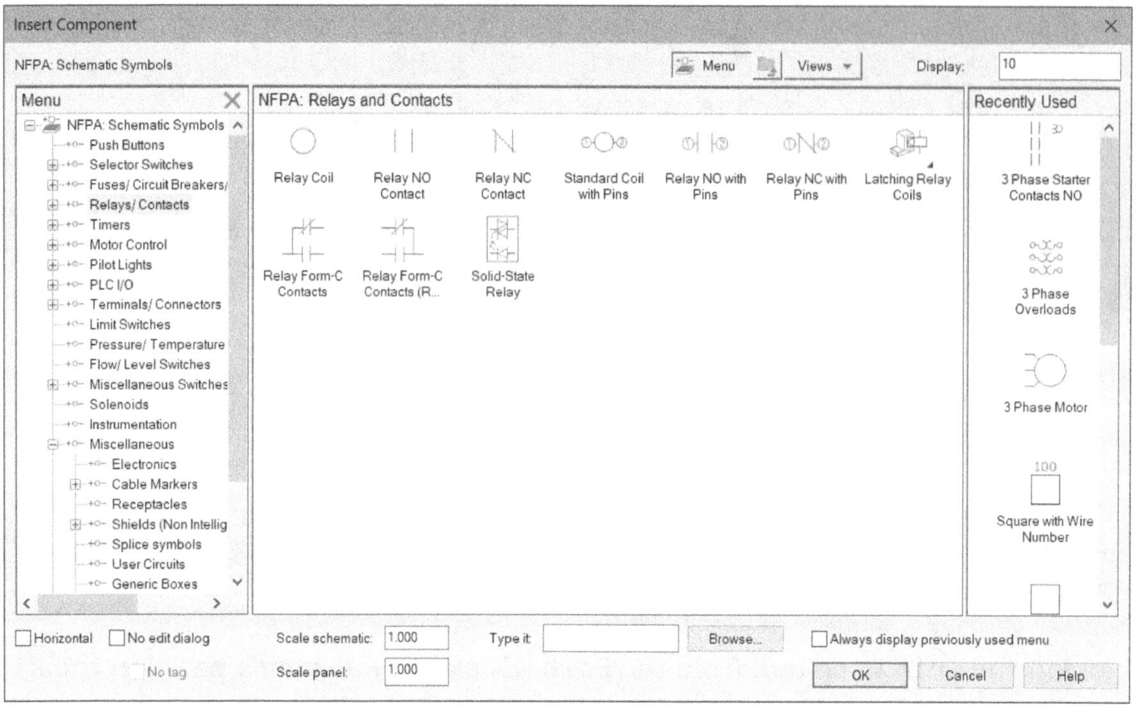

Figure-41. Insert Component dialog box with Relay and Contacts

- Click on the **Relay NO Contact** component from the dialog box. The component will get attached to the cursor.
- Place the component on the wire connected to Connector **3**; refer to Figure-42. The **Insert/Edit Child Component** dialog box will be displayed; refer to Figure-43.

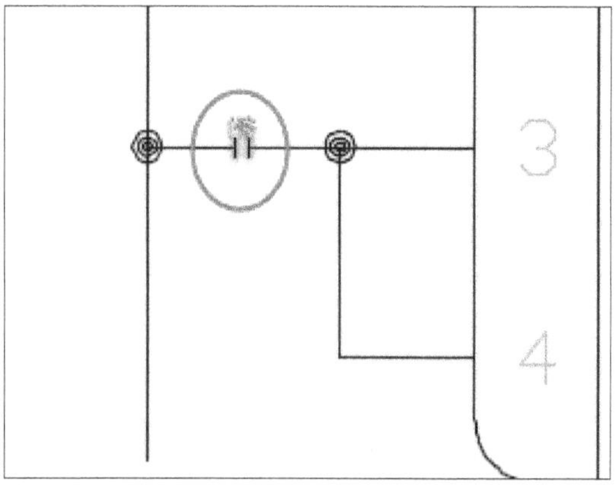

Figure-42. Relay placed

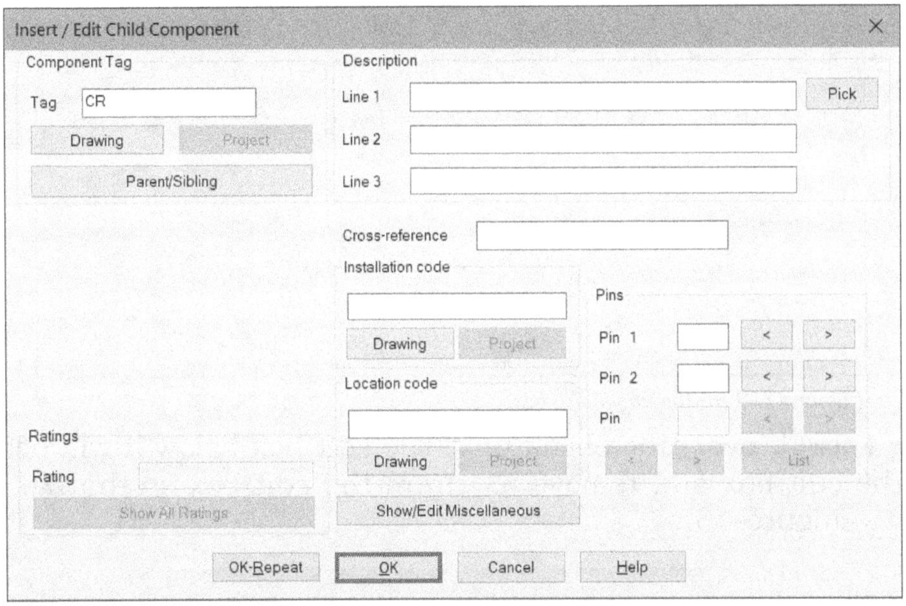

Figure-43. Insert or Edit Child Component dialog box for relay

- Click in the **Line 1** edit box and specify the value as **Control Relay**.
- Click on the **OK** button from the dialog box.
- Similarly, place the other components in the circuit. The schematic after placing all the components will be displayed as shown in Figure-44.

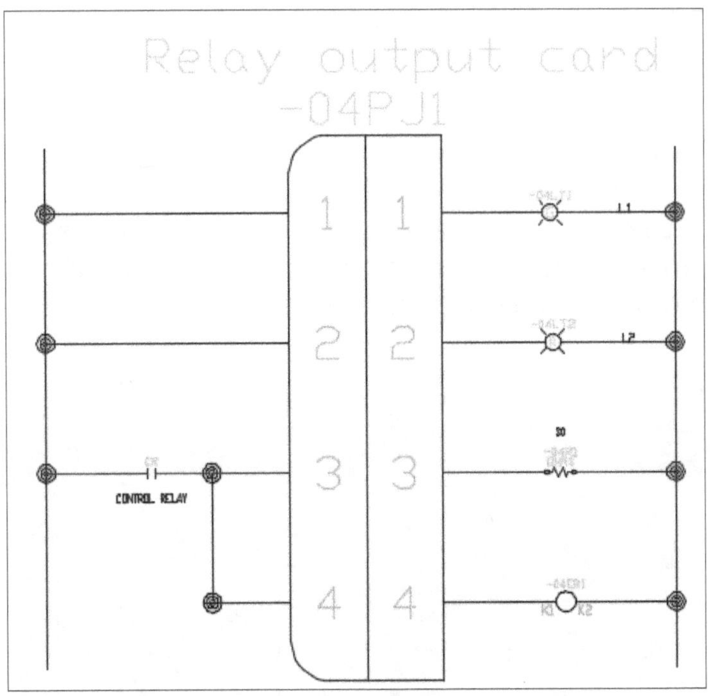

Figure-44. Schematic after placing the components

Assigning Wire Numbers

- Click on the **Wire Numbers** button from the **Wire Numbers** drop-down in the **Insert Wires/Wire Numbers** panel. The **Wire Tagging** dialog box will be displayed; refer to Figure-45.

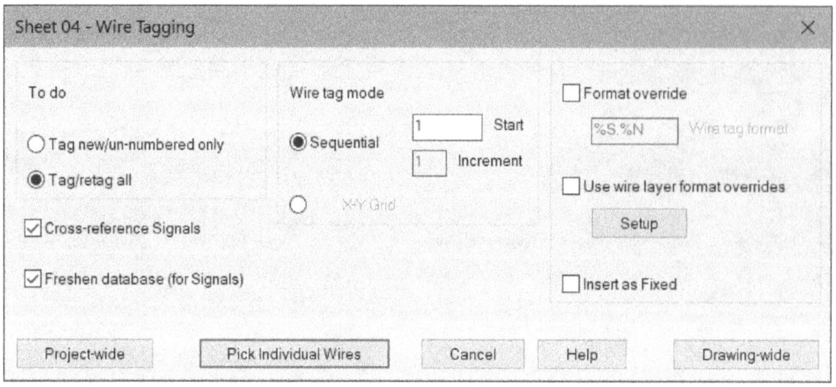

Figure-45. Wire Tagging dialog box

- Select the **Format override** check box. The edit box below it will become active.
- Click in the edit box and remove **%S.** from the edit box so that **%N** is left in the box; refer to Figure-46.

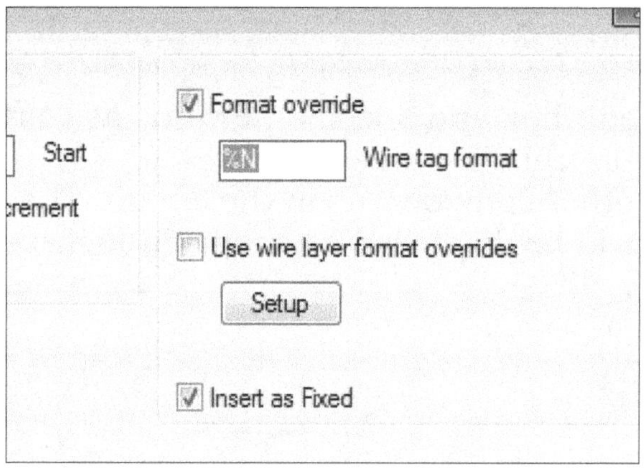

Figure-46. Format override

- Click on the **Drawing-Wide** button from the dialog box to assign the wire numbers to all the wires in the current drawing. The schematic after assigning the wire numbers will be displayed as shown in Figure-47.

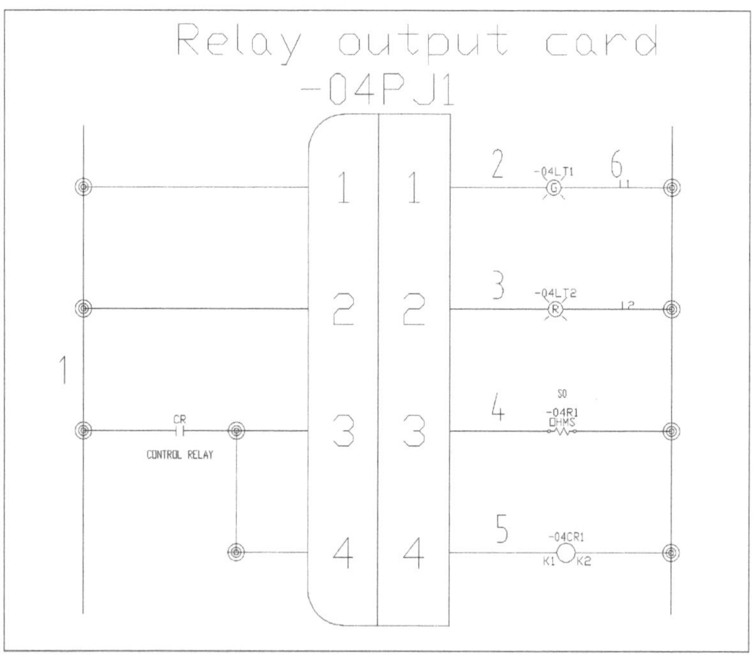

Figure-47. Practical 2

PRACTICAL 3

Create a PLC circuit as shown in Figure-48.

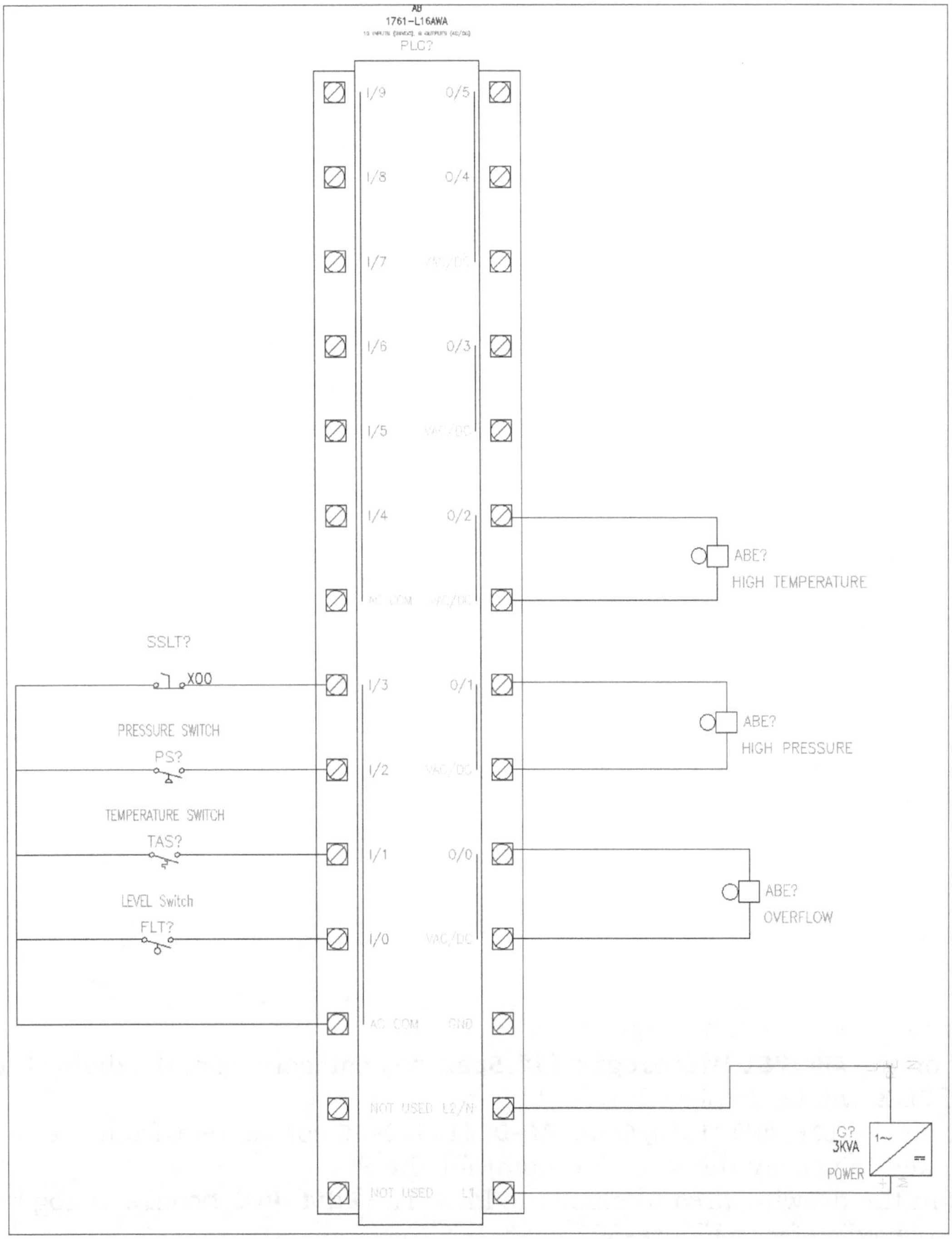

Figure-48. Practical 3

Starting AutoCAD Electrical drawing

- Click on the AutoCAD Electrical icon from the desktop or start AutoCAD Electrical by using the **Start** menu. The interface of AutoCAD Electrical will be displayed.
- Click on the down arrow next to **New** button in the first page of interface. Select the **Browse templates** option from the drop-down list. The **Select Template** dialog box will be displayed.
- Select the **ACAD ELECTRICAL IEC.dwt** template from the list. A new drawing will be created with the selected template.
- Click on the buttons displayed in Figure-49 to hide the grid and off the grid snap.

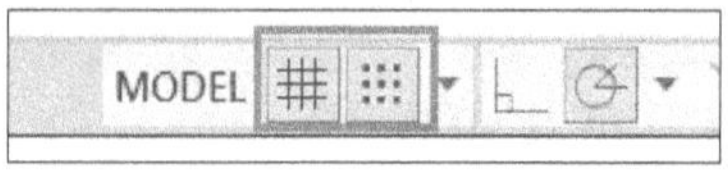

Figure-49. Grid display and snap

Inserting PLC (Full Units)

- In this tutorial, we are going to use Allen Bradley's AB 1761-L16AWA model PLC. We recommend you to know the working of this PLC by browsing through the Allen Bradley's website.
- Click on the **Insert PLC (Full Units)** tool from the **Insert PLCs** drop-down in the **Insert Components** panel. The **Insert Component** dialog box will be displayed; refer to Figure-50.

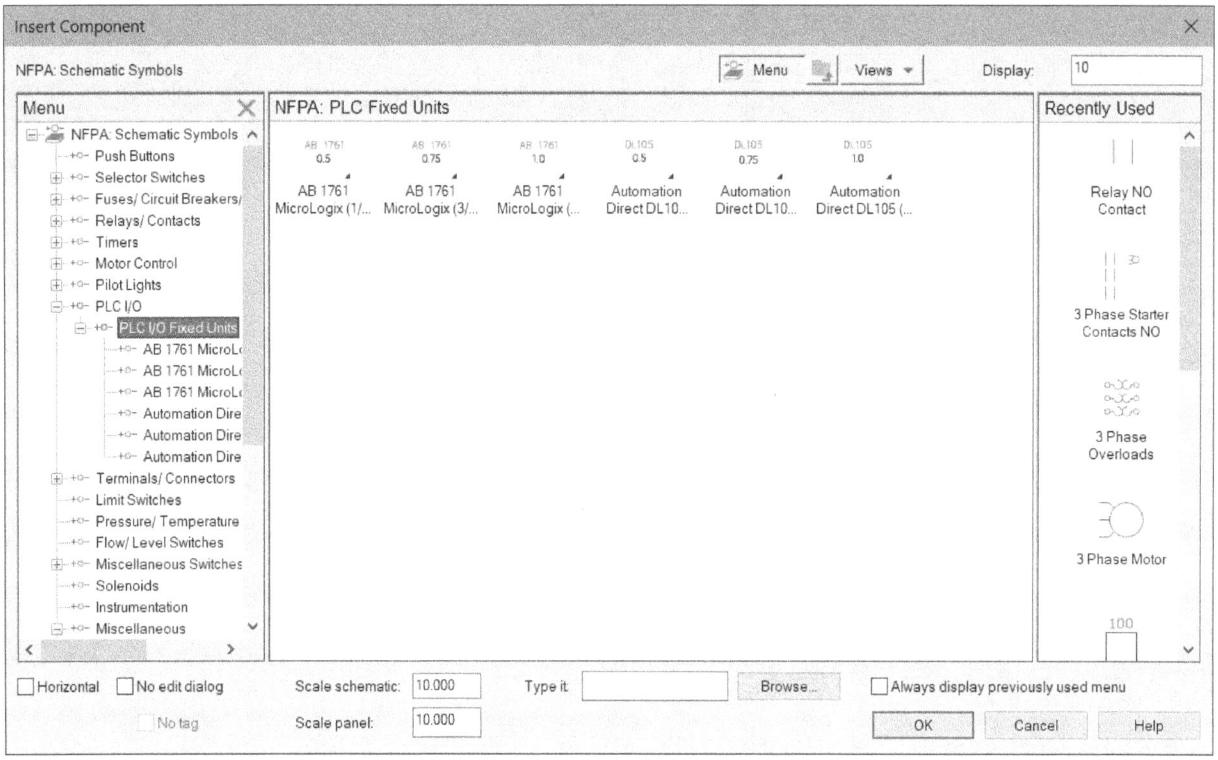

Figure-50. Insert Component with PLC components

- Click on the **AB 1761 Micrologix (1" Spacing)** category from the dialog box. The list of **PLCs** will be displayed; refer to Figure-51.
- Click on the **L16-AWA 10in/6out AC-DC/115AC-DC** component from the list. You are asked to specify the insertion point for the PLC.
- Click in the drawing area to place the PLC. The **Edit PLC Module** dialog box will be displayed; refer to Figure-52.

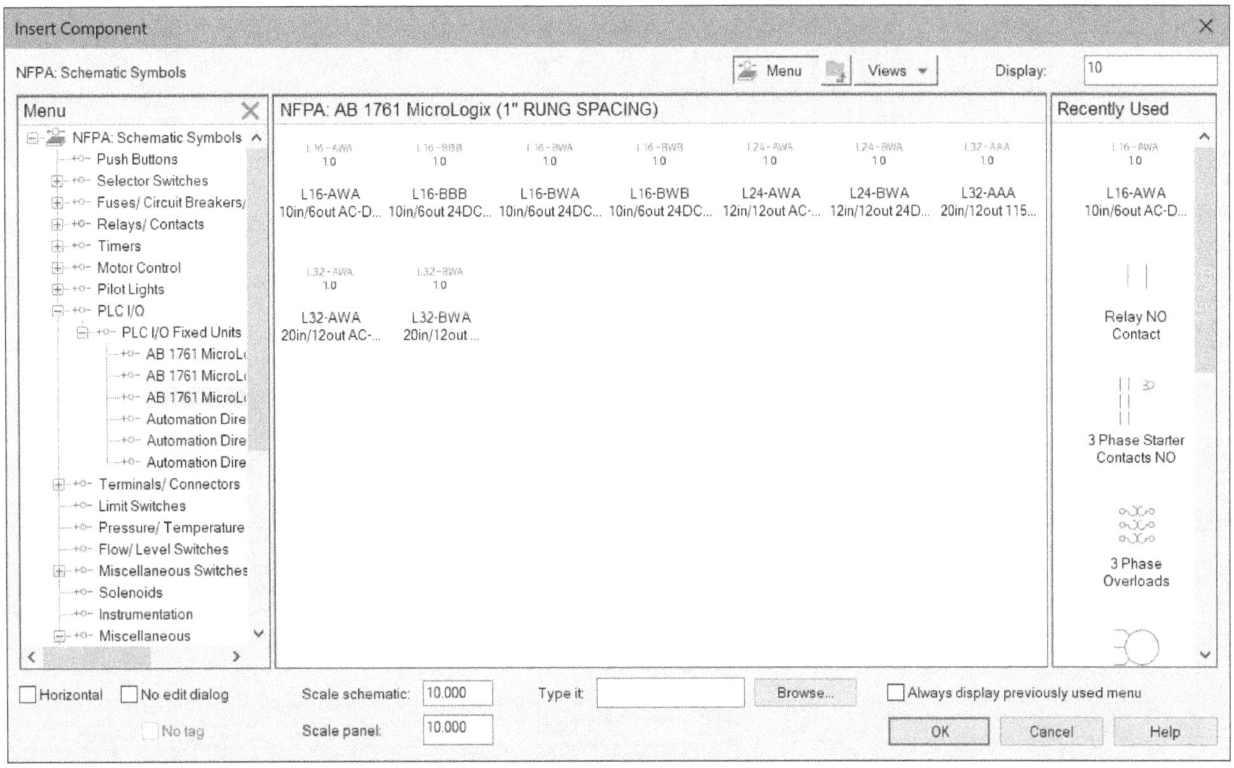

Figure-51. Insert Component with PLCs listed

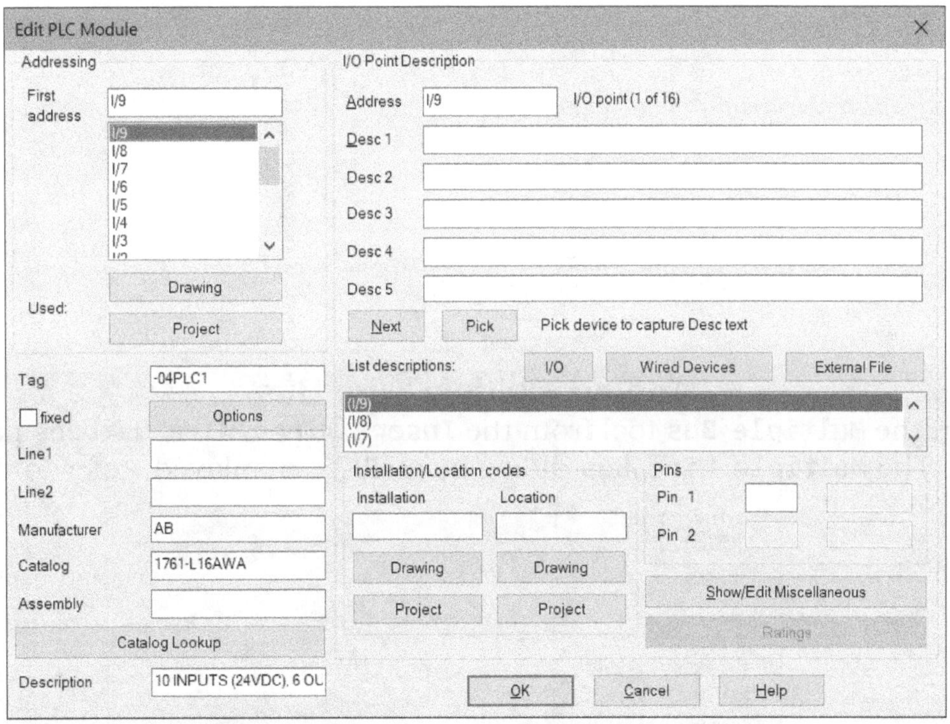

Figure-52. Edit PLC Module dialog box

- You can specify descriptions for each input/output address as desired.
- Click on the **OK** button from the dialog box. The PLC will be placed with specified descriptions; refer to Figure-53.

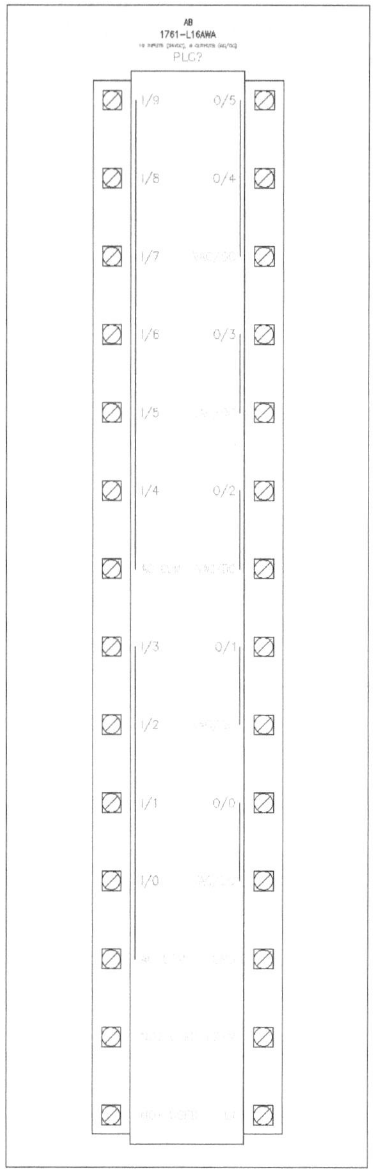

Figure-53. PLC placed in drawing

Adding Multiple Wire Bus

- Click on the **Multiple Bus** tool from the **Insert Wires/Wire Numbers** panel in the **Ribbon**. The **Multiple Wire Bus** dialog box will be displayed; refer to Figure-54.

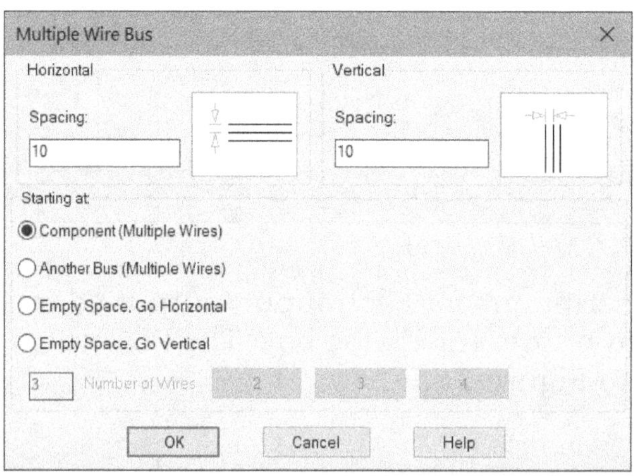

Figure-54. Multiple Wire Bus dialog box with Component radio button selected

- Make sure that the **Component (Multiple Wires)** radio button is selected in the dialog box.
- Click on the **OK** button from the dialog box. You are asked to make a window selection to select the ports of the component for connecting wires.
- Select the Input ports and AC COM as shown in Figure-55.
- Press **ENTER** from the keyboard and move the cursor. End point of the wire bus will get attached to the cursor; refer to Figure-56.
- Click at the desired distance to create the wire bus.

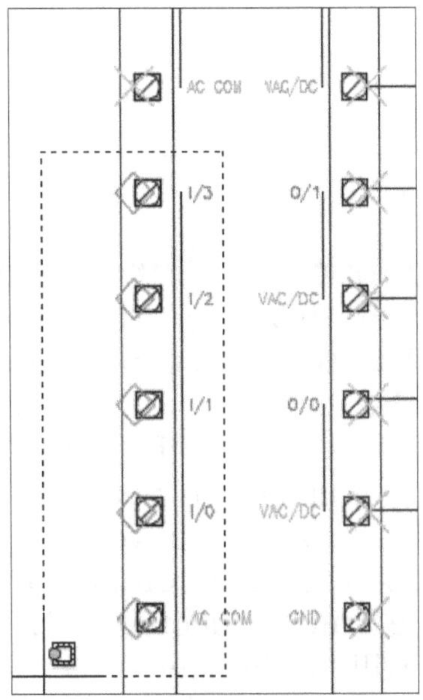

Figure-55. Input ports selected

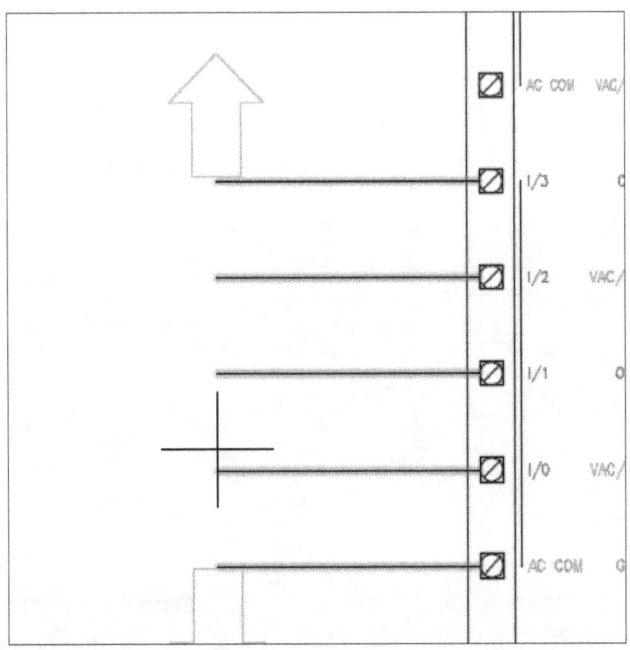

Figure-56. End point of wire bus attached to cursor

- Click on the **Wire** tool from the **Wires** drop-down in the **Insert Wires/Wire Numbers** panel. You are asked to specify the start point of the wire.
- Connect all the end points of the wire bus by creating a wire; refer to Figure-57.

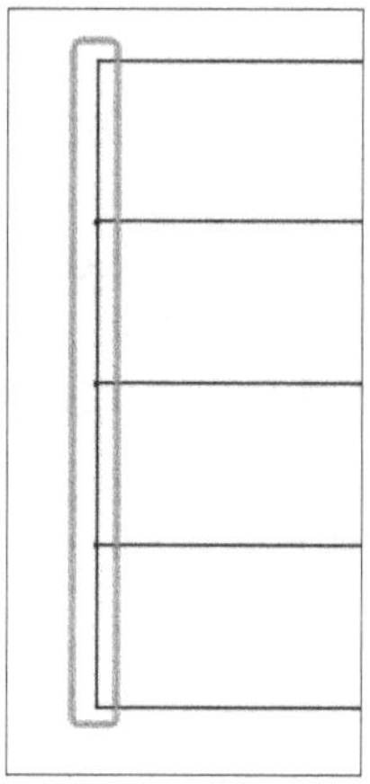

Figure-57. Wire connecting end points

Connecting Components

- Click on the **Icon Menu** button from the **Insert Components** drop-down in the **Insert Components** panel. The **Insert Component** dialog box will be displayed.
- Click on the **Selector Switches** category in the dialog box. The components in the dialog box will be displayed as shown in Figure-58. Note that you might need to change the scale of the component to properly display it in the drawing by using the edit boxes highlighted in a box in Figure-58.
- Select the 3 Position NC component. The component will get attached to the cursor.
- Click on the top wire of wire bus connected to I/3 address of PLC. The **Insert/ Edit Component** dialog box will be displayed.
- Specify the desired parameters and click on the **OK** button from the dialog box.

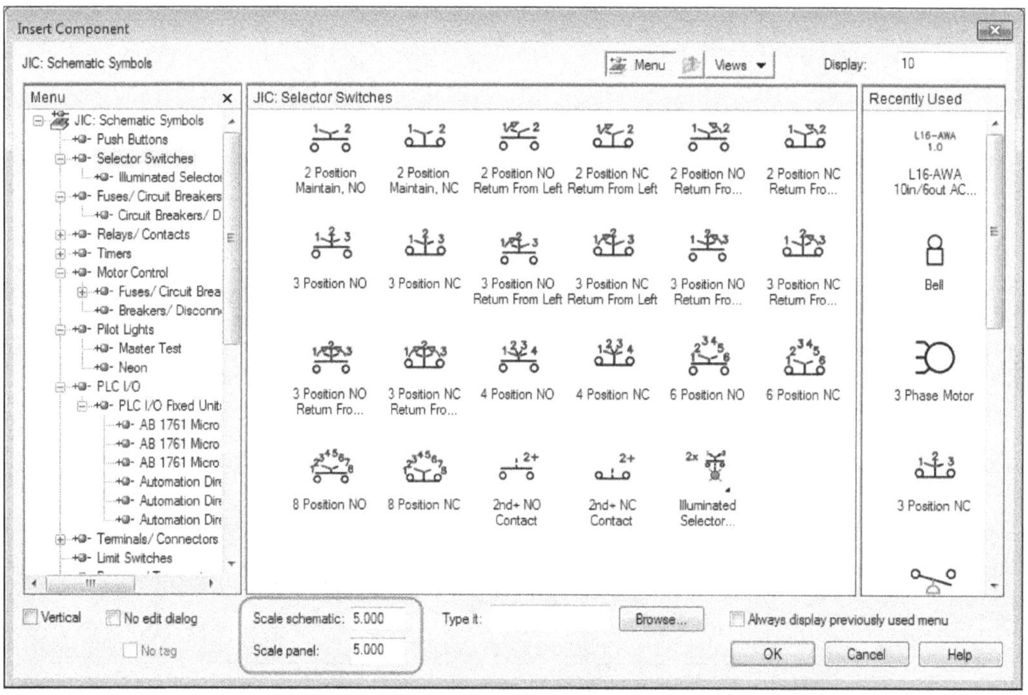

Figure-58. Insert Component dialog box with Selector Switches

- Similarly, you can add the other components in the wire bus. After connecting all the components in the wire bus, the schematic will be displayed as shown in Figure-59.

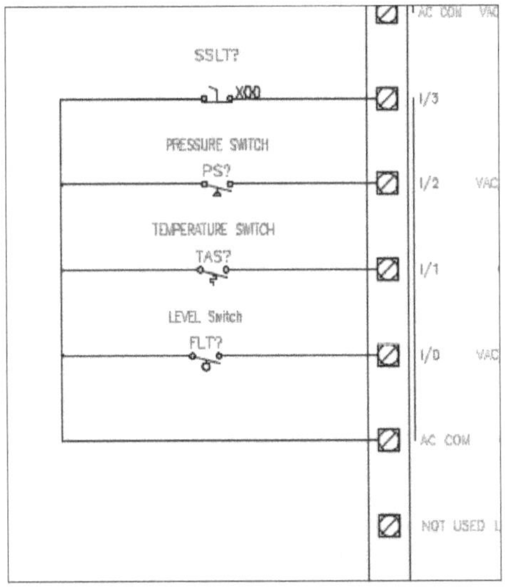

Figure-59. Wirebus after connecting components

Connecting Output Devices

- The right side of PLC is meant for output and power supply. Click on the **Wire** tool and build the circuit as shown in Figure-60.

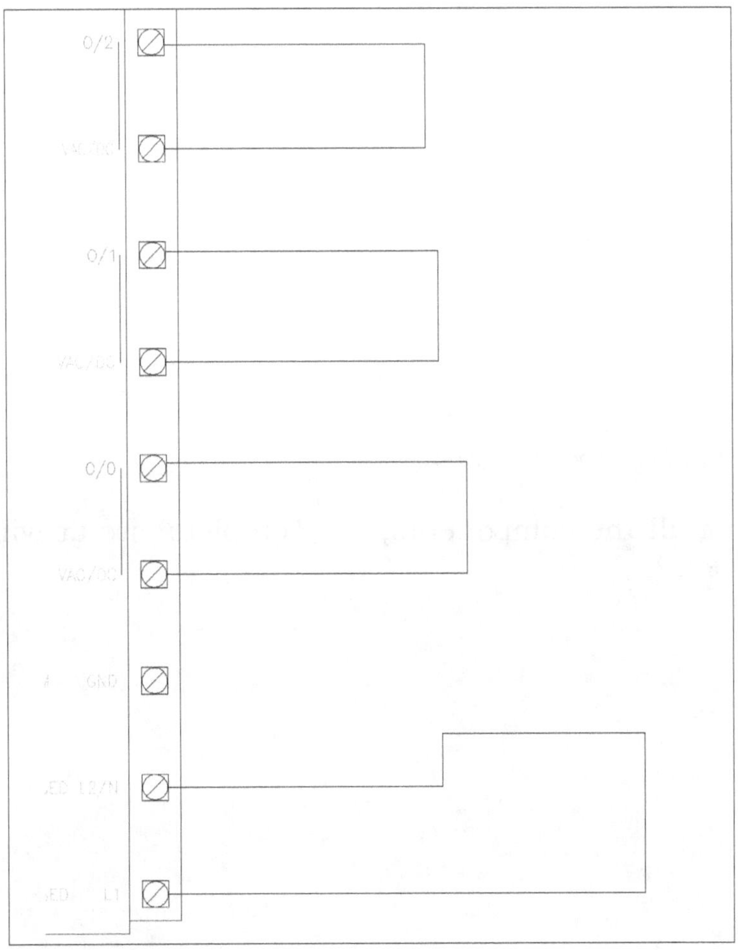

Figure-60. Circuit for output

- Using the **Icon Menu** tool, connect all the components to the circuit; refer to Figure-61.

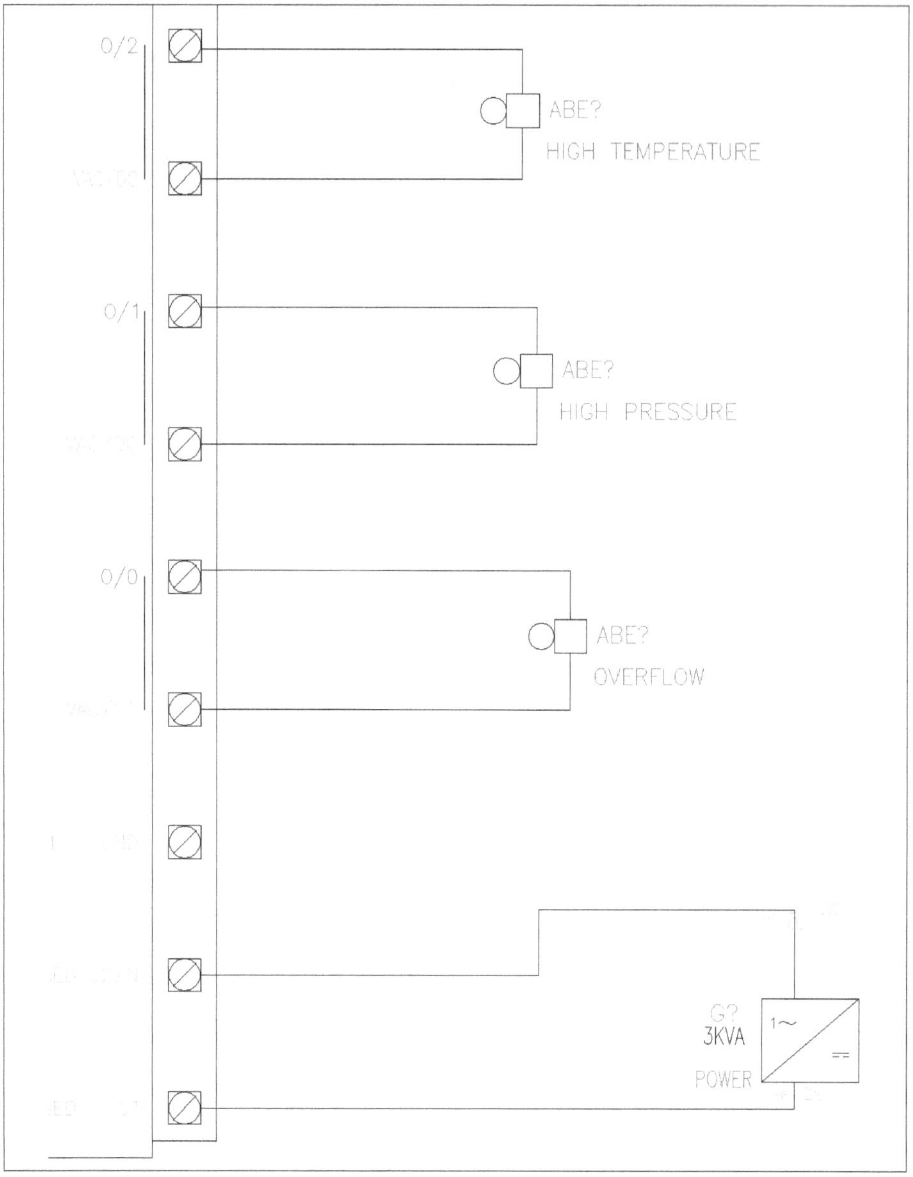

Figure-61. Adding components to the output

- After connecting all the components, the complete circuit will be displayed as shown in Figure-62.

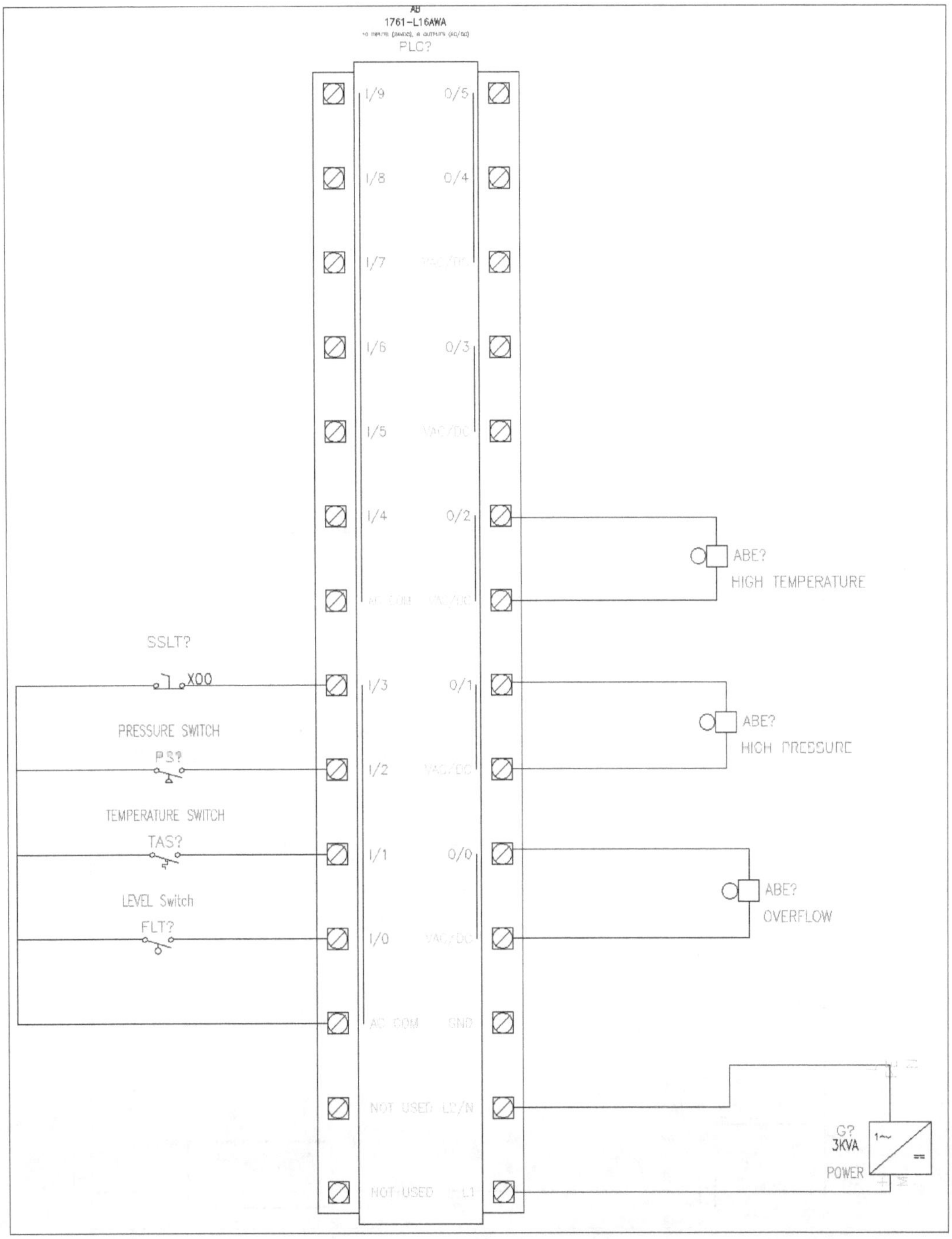

Figure-62. Practical 3

PRACTICE 1

Create the schematic drawing as shown in Figure-63.

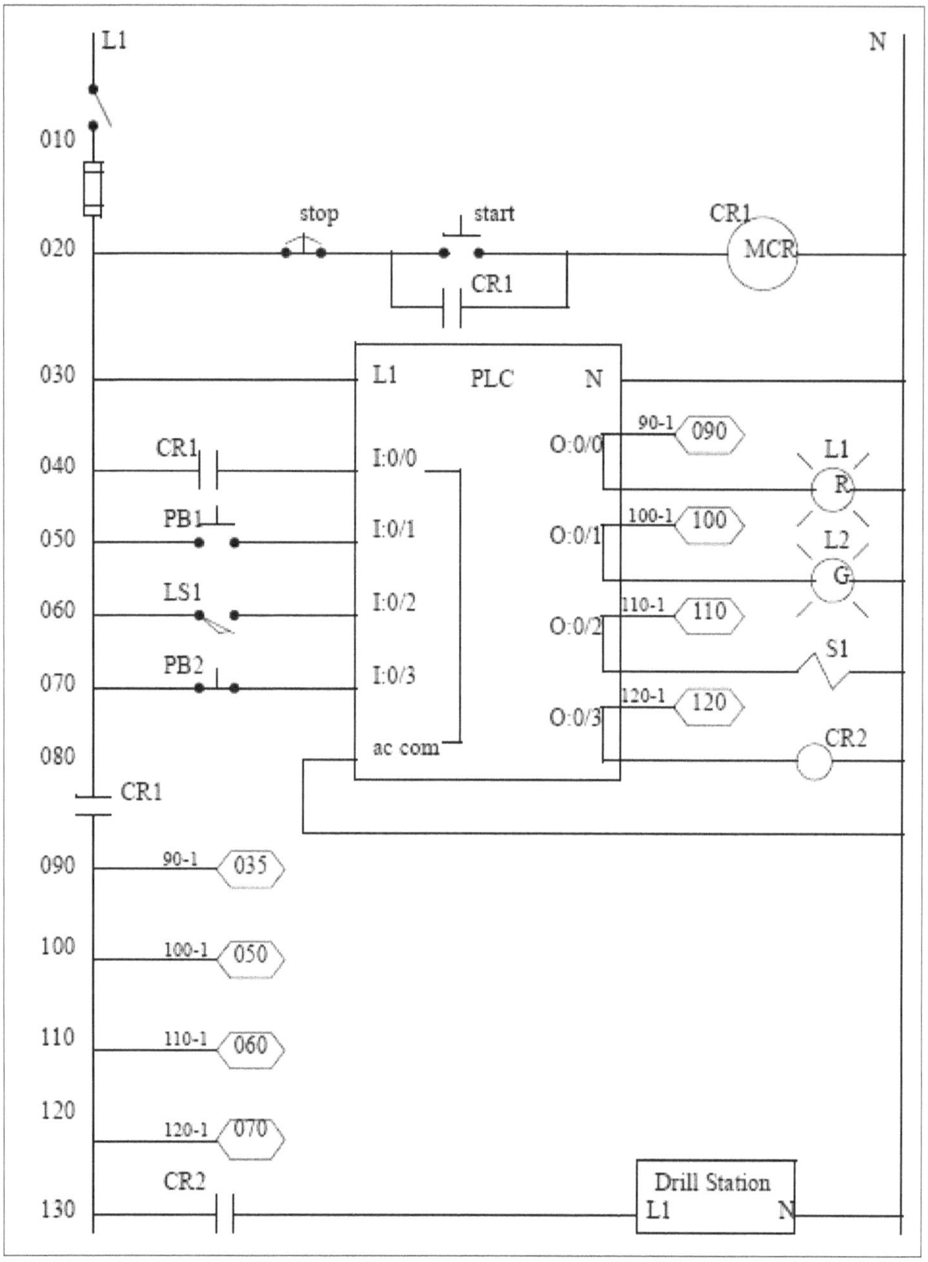

Figure-63. Practice 1

PRACTICE 2

Create the schematic drawing as shown in Figure-64.

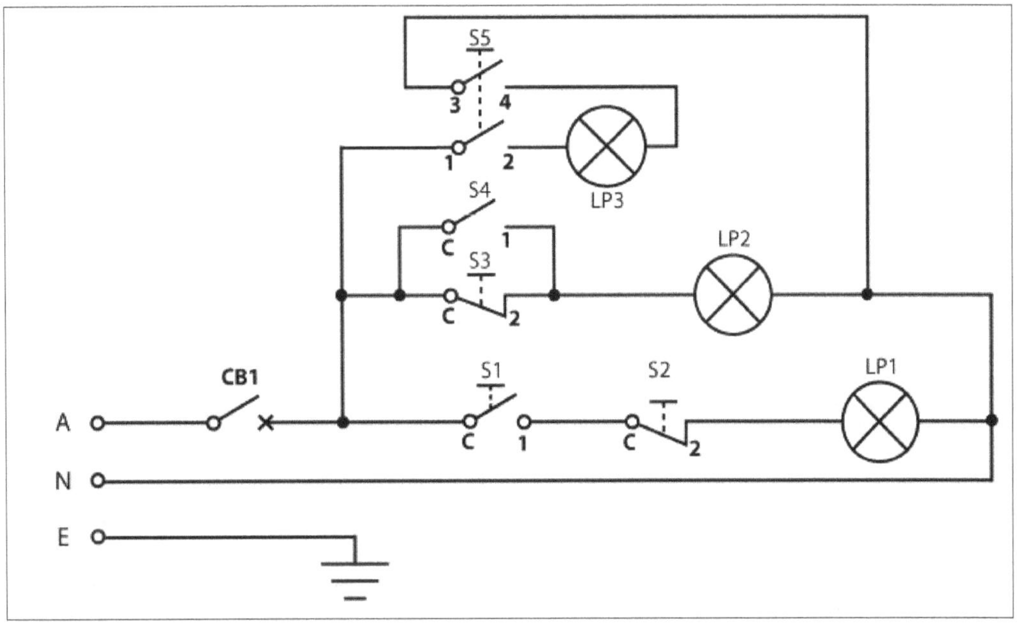

Figure-64. Practice 2

PRACTICE 3

Create the schematic drawing as shown in Figure-65.

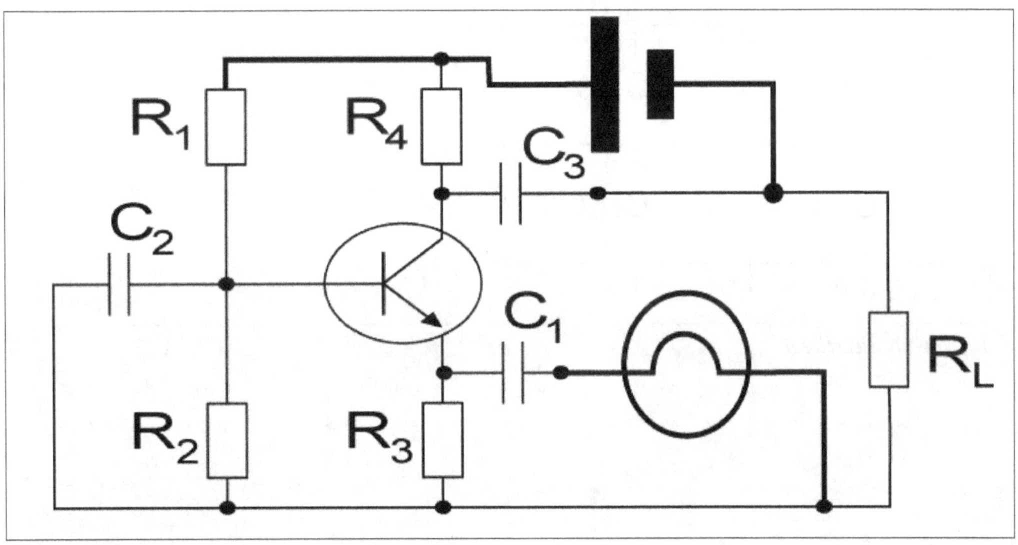

Figure-65. Practice 3

PRACTICE 4

Create the schematic drawing as shown in Figure-66.

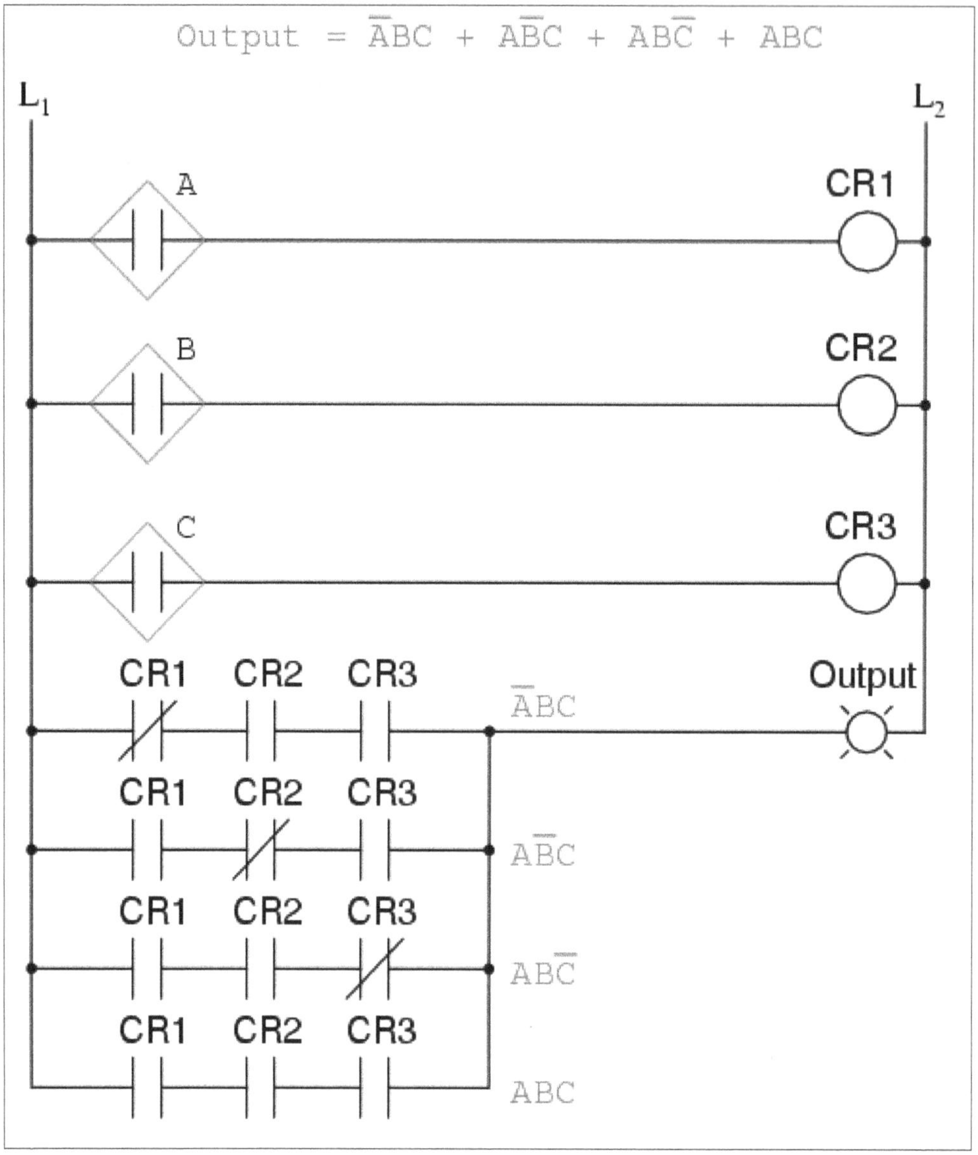

Figure-66. Practice 4

PRACTICE 5

Create the schematic drawing as shown in Figure-67.

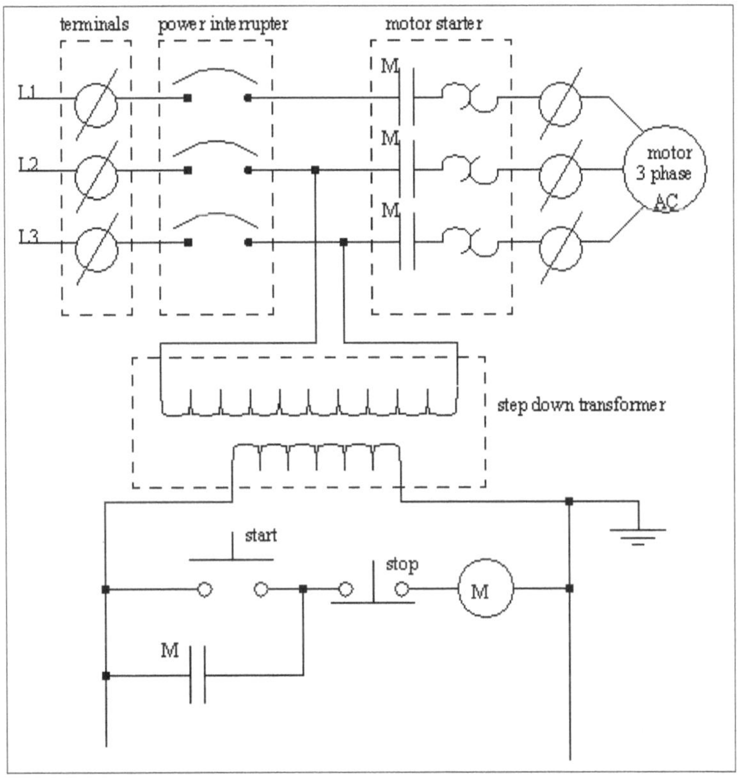

Figure-67. Practice 5

FOR STUDENT NOTES

Chapter 9

Panel Layout

Topics Covered

The major topics covered in this chapter are:

- ***Introduction***
- ***Icon Menu for panels***
- ***Schematic List***
- ***Manual***
- ***Manufacturers Menu***
- ***Balloon***
- ***Wire Annotation***
- ***Panel Assembly***
- ***Editor***
- ***Table Generator***
- ***Terminals***
- ***Editing***

INTRODUCTION

In the previous chapters, you have learnt to create schematic circuit diagrams. After creating those circuit diagrams, the next step is to create panels. A panel is the box consisting of various electrical switches and PLCs to control the working of equipment. Refer to Figure-1. Note that the panel shown in the figure is back side panel of a machine. This panel is generally hidden from the operator. What an operator see is different type of panel; refer to Figure-2. We call this panel as User Control panel. In both the cases, the approach of designing is almost same but the interaction with the user is different. The User Control Panel is meant for Users so it can have push buttons, screen, sensors, key board and so on. On the other hand, the back panel will be having relays, circuit breakers, sensors, connectors, PLCs, switches, and so on for electrician.

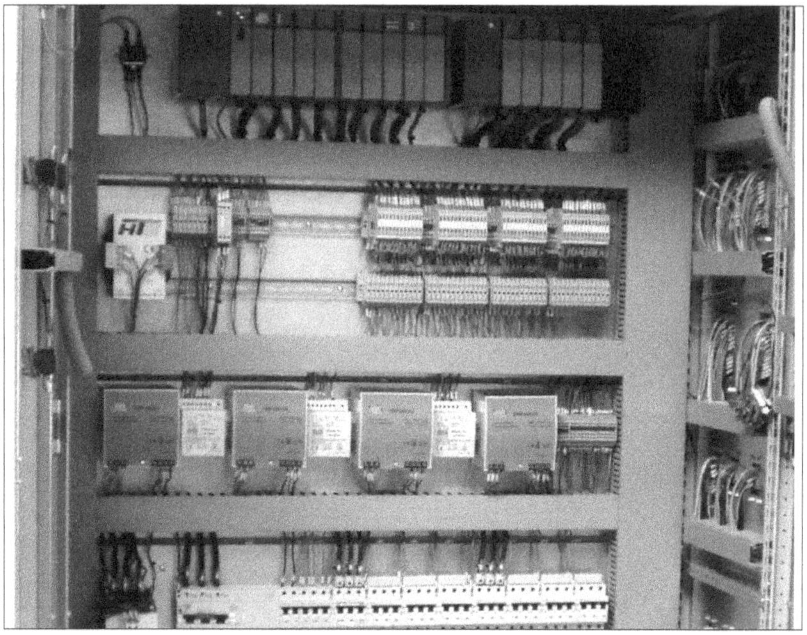

Figure-1. Panel

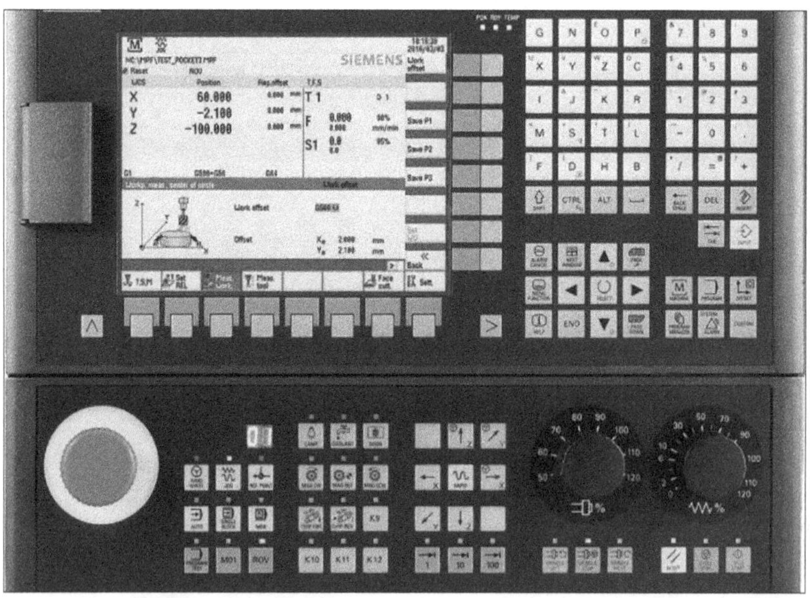

Figure-2. User Control panel

If we start linking the schematic drawings with the panel drawings then the common platform is the component tag and the location code. Suppose, we have created a push button in the schematic with tag -04PB2 then in the panel layout, you should insert the same push button with the same tag. The location of the Push button will be decided on the basis of Location code. The components that have same location code should be placed at the same place in the panel.

In AutoCAD Electrical, there is a separate tab for the tools related to Panel Layout designing with the name **Panel**; refer to Figure-3. The procedures of using various tools of this panel are discussed next.

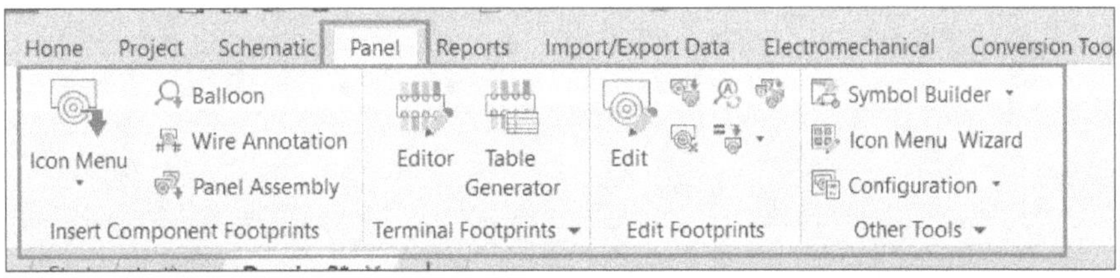

Figure-3. Panel tab

ICON MENU

The **Icon Menu** tool works in a similar way as it does for Schematic drawings. The only difference is representation of components. In panel layouts, footprints are inserted in place of schematic symbols. The procedure to use this tool is given next.

- Click on the **Icon Menu** tool from the **Insert Component Footprints** drop-down in the **Insert Components Footprints** panel in the **Panel** tab of the **Ribbon**; refer to Figure-4. The **Insert Footprint** dialog box will be displayed; refer to Figure-5.

Note that when you start a new panel drawing and activate the **Icon Menu** tool, an Alert message box is displayed asking you to add WD PNLM block in the drawing. Click on the **OK** button to add the block. The WD PNLM block is an invisible block that stores the attributes of components, wires, and other drawing parameters. This block also keeps panel icons and the schematic symbol connected with its panel icon.

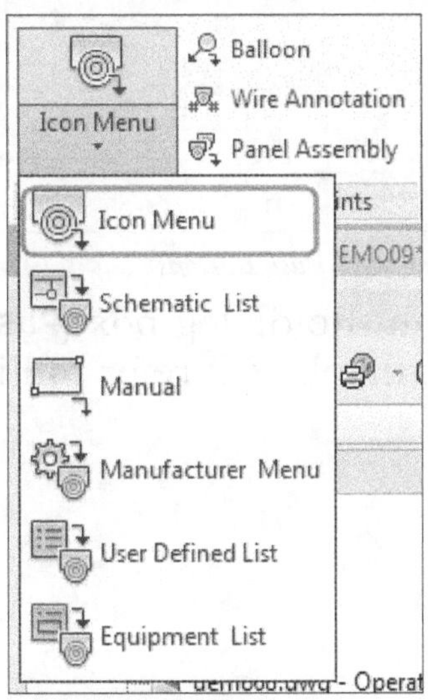

Figure-4. Icon Menu tool

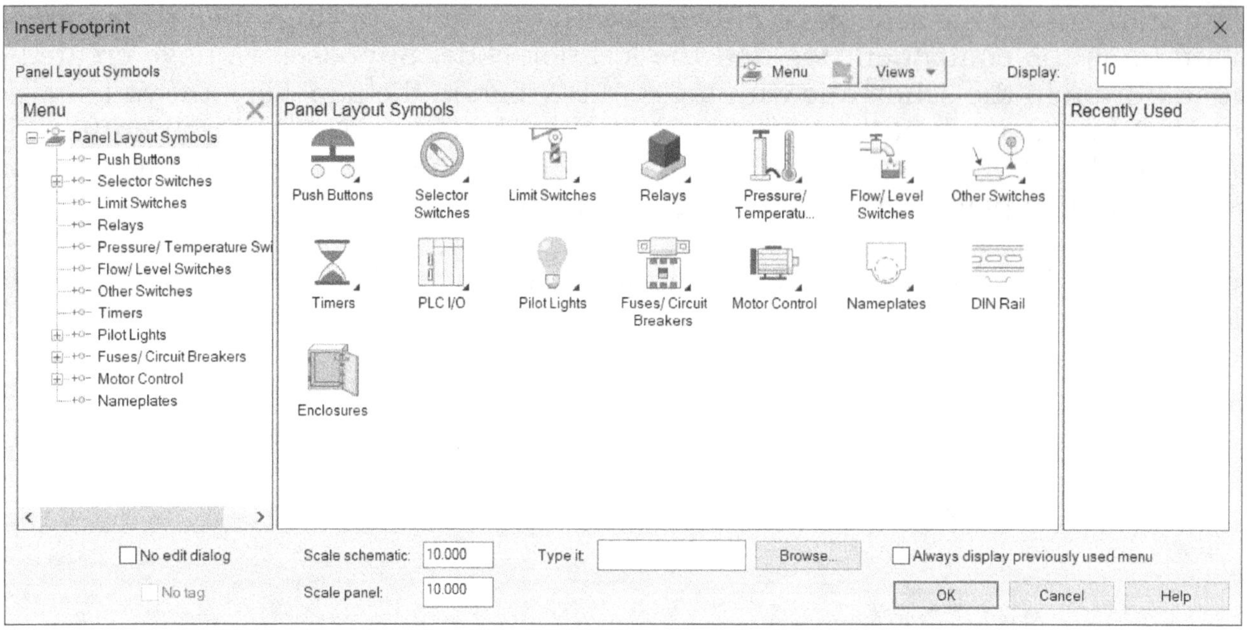

Figure-5. Insert Footprint dialog box

- Click on desired category (Push Buttons in our case). The symbols related to the selected category will be displayed; refer to Figure-6.

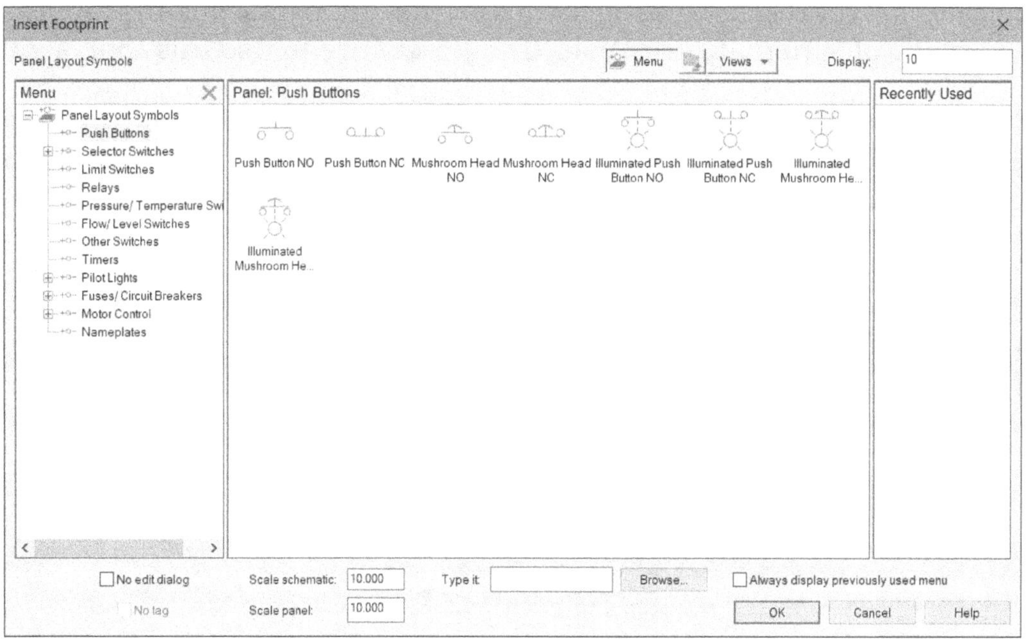

Figure-6. Insert Footprint dialog box with Push Buttons

- Click on desired symbol from the dialog box (**Push Button NC** in our case). The **Footprint** dialog box will be displayed; refer to Figure-7.

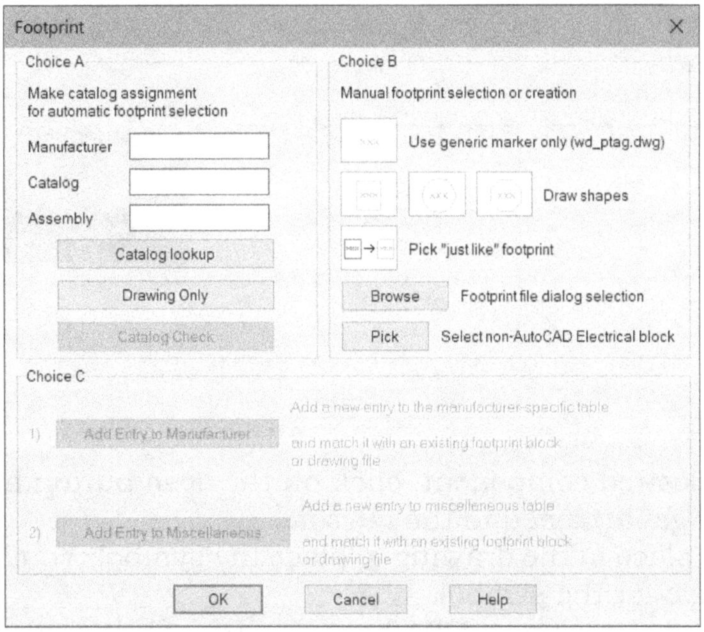

Figure-7. Footprint dialog box

- Click on the **Catalog lookup** button to check the catalog for current symbol. The **Catalog Browser** dialog box will be displayed; refer to Figure-8.

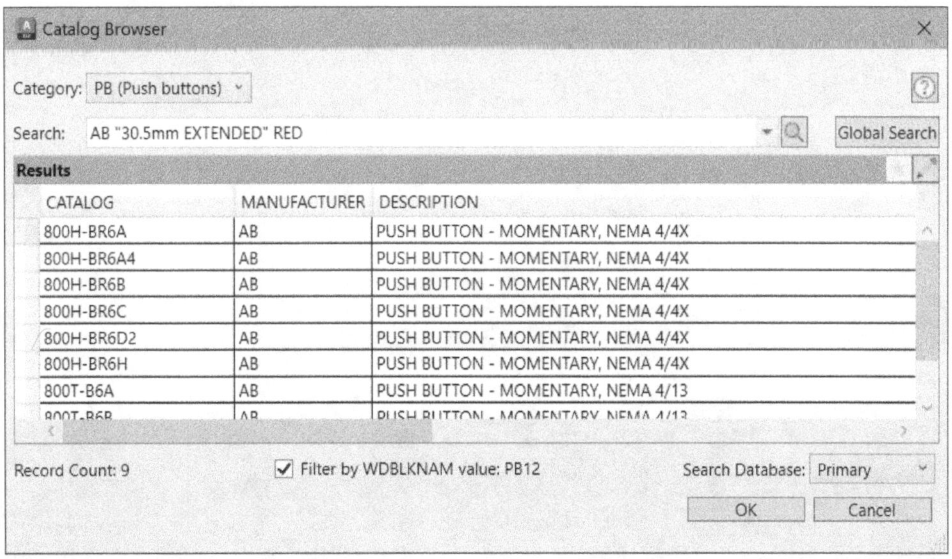

Figure-8. Catalog Browser dialog box

- Select desired component from the catalog and click on the **OK** button from the dialog box.
- If you have created a block for the component in your local drive then click on the **Browse** button and select desired component block from the dialog box; refer to Figure-9.

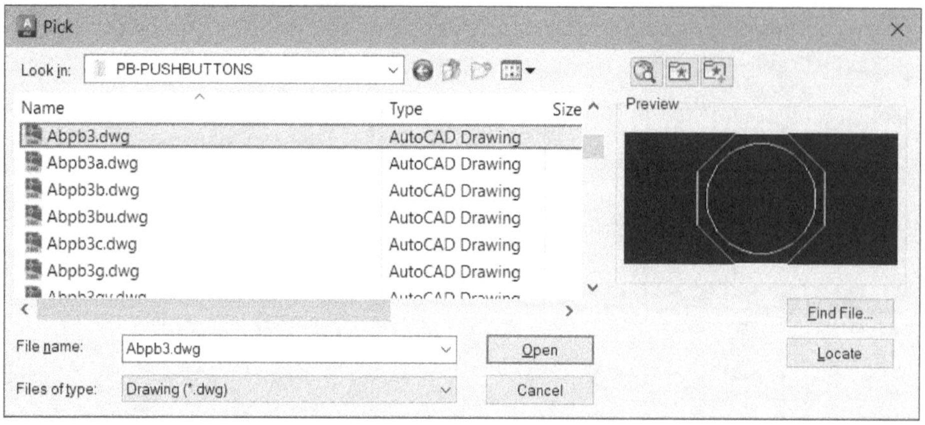

Figure-9. Pick dialog box

- After selecting desired component, click on the **Open** button from the dialog box. The symbol will get attached to the cursor.
- Click at desired place in the drawing to place the symbol. You are asked to specify the rotation angle for the symbol.
- Specify desired angle or press **ENTER** to use the default angle (i.e. 0 degree). The **Panel Layout-Component Insert/Edit** dialog box will be displayed; refer to Figure-10.

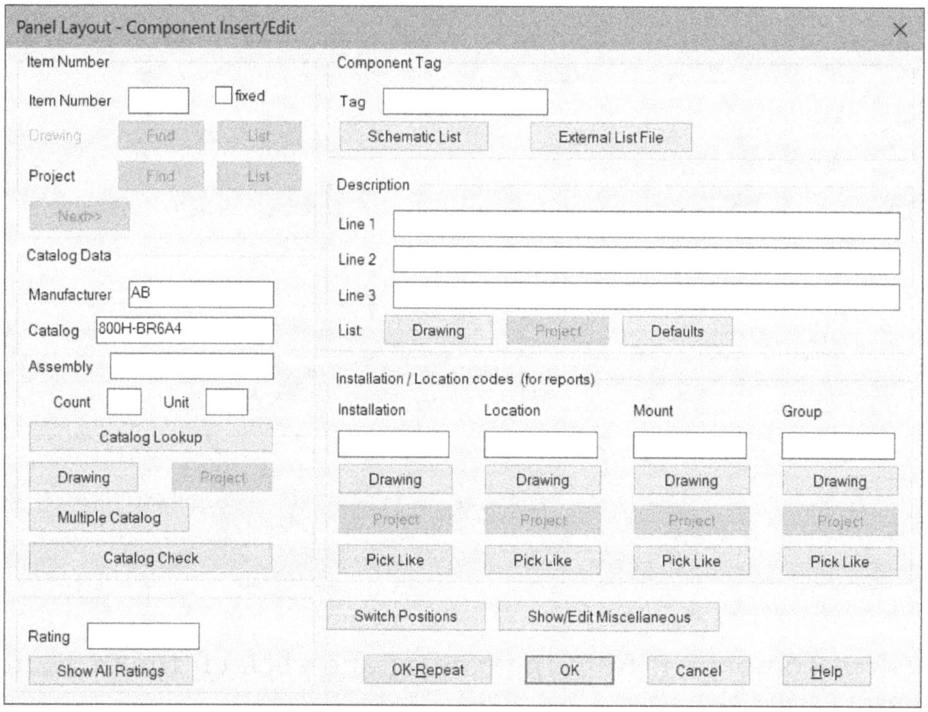

Figure-10. Panel Layout Component Insert or Edit dialog box

- This dialog box is similar to the **Insert/Edit Component** dialog box as discussed earlier. Specify the tags for the component.
- Click on the **Schematic List** button below **Tag** edit box in the **Component Tag** area to display the list of tags used in the drawing/project. The **Schematic Tag List** dialog box will be displayed; refer to Figure-11.
- Select the tag related to the symbol being inserted and click on the **OK** button from the dialog box.

- Click on the **OK** button from the dialog box. The symbol will be placed with the corresponding tag attached. If you want to insert more instances of the component then click on the **OK-Repeat** button from the dialog box.

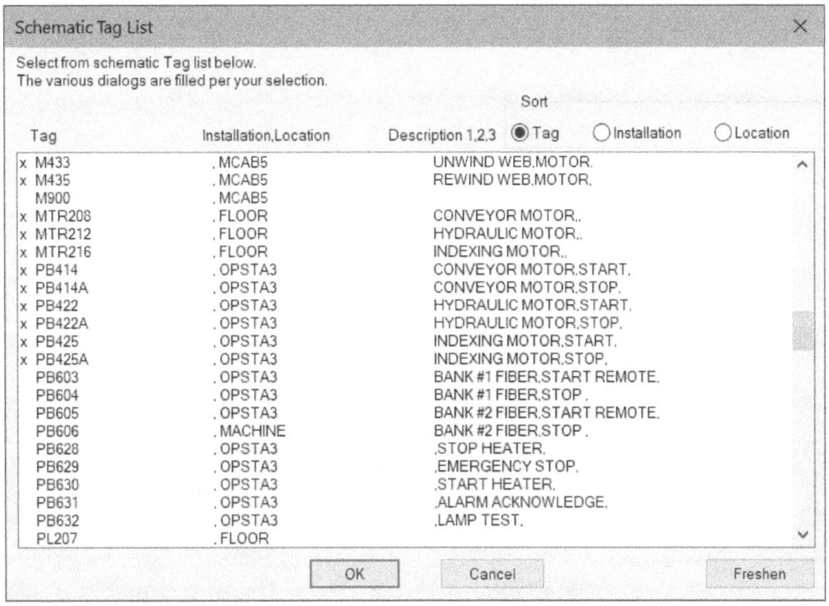

Figure-11. Schematic Tag List dialog box

Before we start inserting components, it is important to understand the basic components of a panel. An electric panel generally has an enclosure to close pack all the components of panel, a DIN rail to support the MCBs (Mounted Circuit Breakers), Nameplates to identify components, and electrical components. We will create a panel combining all of them later in this chapter (under Practical).

SCHEMATIC LIST

The **Schematic List** tool is used to import the list of schematic components from the current project drawings so that the panel representation of those components can be created. The procedure to use this tool is given next.

- Click on the **Schematic List** tool from the **Insert Component Footprints** drop-down. The **Schematic Components List** dialog box will be displayed; refer to Figure-12.

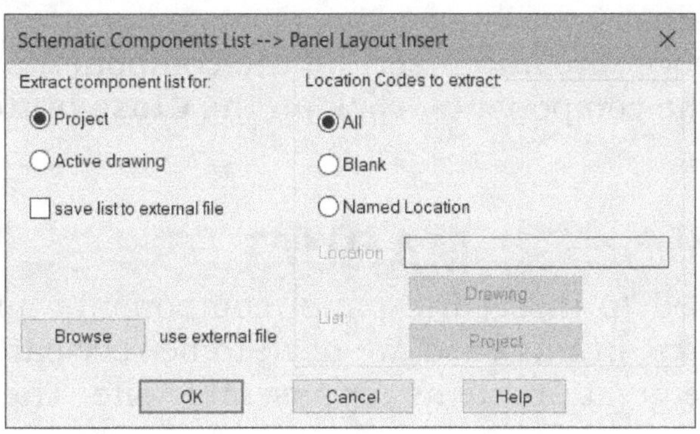

Figure-12. Schematic Components List dialog box

- Select the **Active drawing** if you want to extract the component list of only current drawing (as in our case) or select the **Project** and include the relevant drawings.

- Click on the **OK** button from the dialog box. The **Schematic Components (active drawing)** dialog box will be displayed (in case of selecting **Active drawing** radio button); refer to Figure-13.

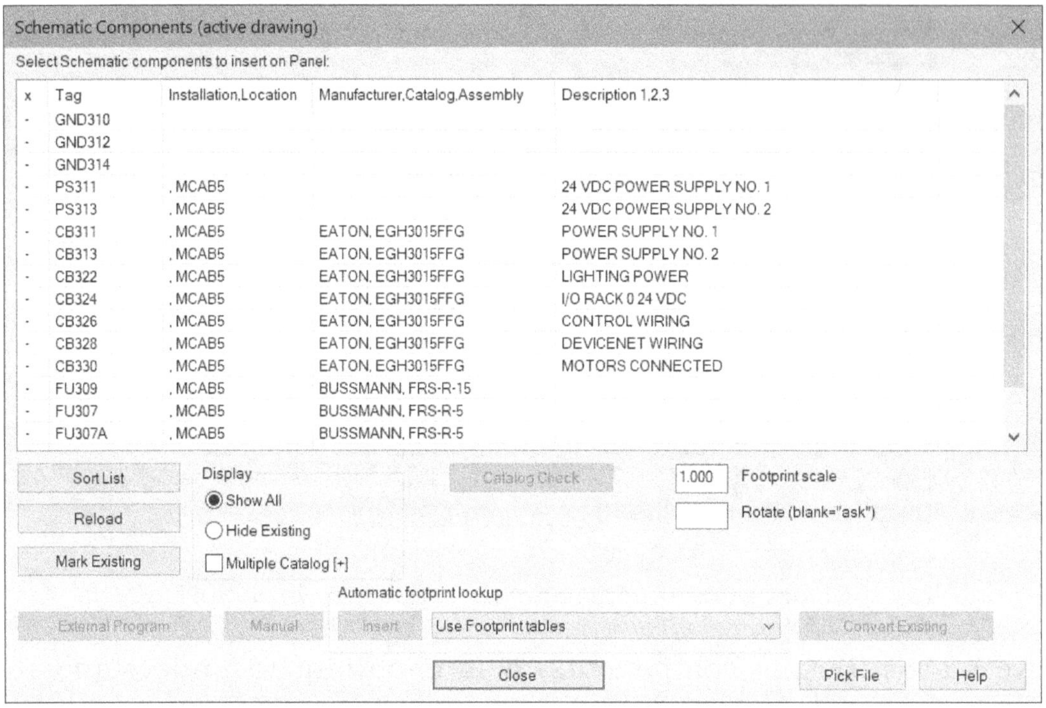

Figure-13. Schematic Components dialog box

- Select any component from the list for which you want to insert symbol in the panel. Click on the **Insert** button at the bottom of the dialog box. If there is symbol for the component then it will get attached to the cursor and you will asked to specify the insertion point for the symbol.
- Click in the drawing area to place the component. You will be asked to specify the rotation angle value for the component.
- Specify desired value in the Command box or press **ENTER** to set the default angle. The **Panel Layout-Component Insert/Edit** dialog box will be displayed as discussed earlier. Note that the tags are defined automatically this time. This is because the tags are automatically extracted as per the schematic list.
- If you want to perform modifications then do that and click on the **OK** button from the dialog box. The symbol will be created and the **Schematic Components** dialog box will appear again.
- Follow the same procedure and insert all the components required in the panel.
- After inserting all the components, click on the **Close** button from the dialog box to exit.

MANUAL

The **Manual** tool is used to insert the panel components as per our requirement directly using the blocks. In this case, we need to be very careful regarding the tags otherwise it can create great problems for assembly site. The procedure to use this tool is given next.

- Click on the **Manual** tool from the **Insert Component Footprints** drop-down. The **Insert Component Footprints -- Manual** dialog box will be displayed; refer to Figure-14.

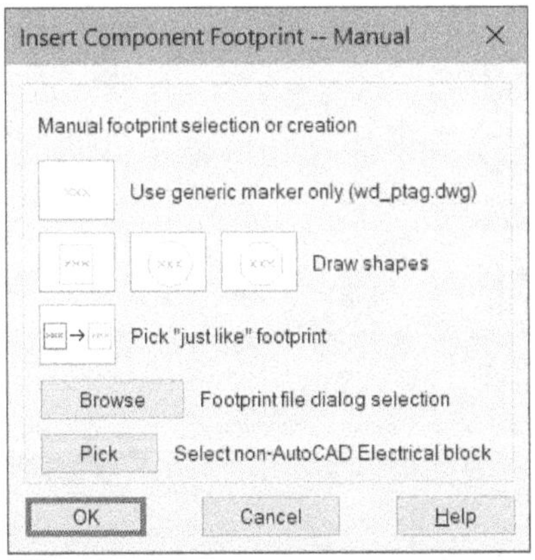

Figure-14. Insert Component Footprint Manual

- This dialog box is similar to the **Choice B** area of the **Footprint** dialog box discussed earlier. Click on the **Browse** button to insert the block of component.
- If you have the component already in the drawing, then click on the **Pick** button and select the component. (Note that you can select only non-AutoCAD Electrical blocks as symbol for using this feature).
- On selecting the block, you are asked to specify the insertion point for the symbol.
- Click in the drawing area and specify the rotation of the symbol. The symbol will be placed and **Panel Layout - Component Insert/Edit** dialog box will be displayed as discussed earlier.
- Specify desired tags and click on the **OK** button from the dialog box.

MANUFACTURER MENU

The **Manufacturer Menu** tool is used to insert the components from the manufacturer's menu. AutoCAD Electrical contains a library of manufacturer's components (like **Allen-Bradley** and **ABECAD**). The procedure to use this tool is given next.

- Click on the **Manufacturer Menu** tool from the **Insert Component Footprints** drop-down. The **Vendor Menu Selection** dialog box will be displayed (for first time); refer to Figure-15.

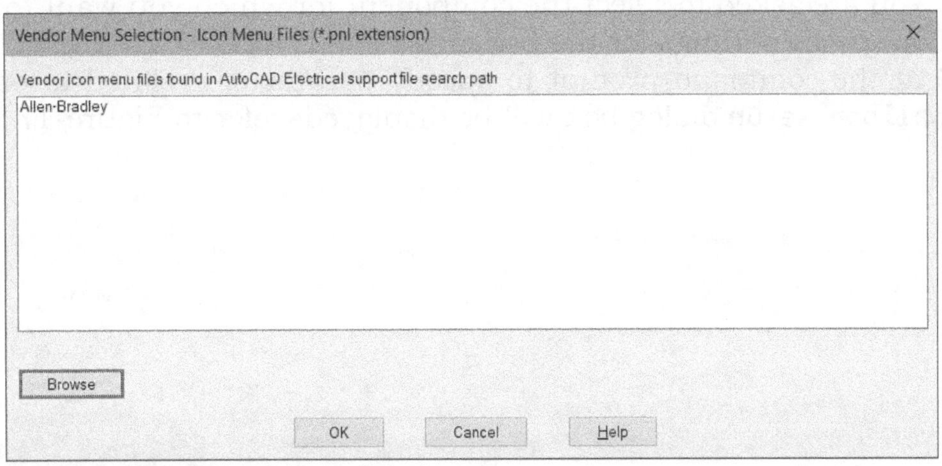

Figure-15. Vendor Menu Selection dialog box

- Select desired vendor from the list and click on the **OK** button from the dialog box. The **Vendor Panel Footprint** dialog box will be displayed; refer to Figure-16.

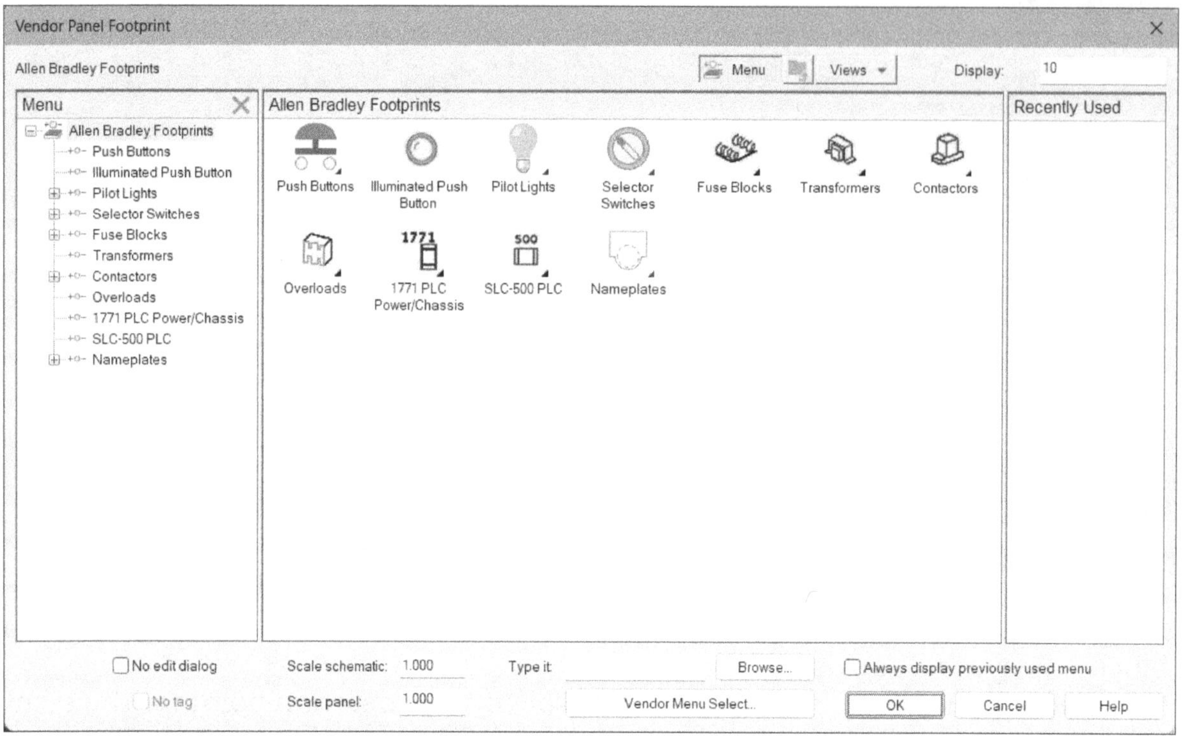

Figure-16. Vendor Panel Footprint dialog box

- The dialog box works in similar way to **Insert Component** dialog box discussed earlier.

Similarly, you can use the **User Defined List** and **Equipment List** tools to insert the **Panel** components.

BALLOON

The **Balloon** tool is used to assign balloons to the components for their identification in the table. The procedure to use this tool is given next.

- Click on the **Balloon** tool from the **Insert Component Footprints** panel in the **Ribbon**. You are asked to select the component for which you want to assign the balloon or specify settings for the balloon.
- Enter **S** at the command prompt to specify parameters related to balloon. The **Panel balloon setup** dialog box will be displayed; refer to Figure-17.

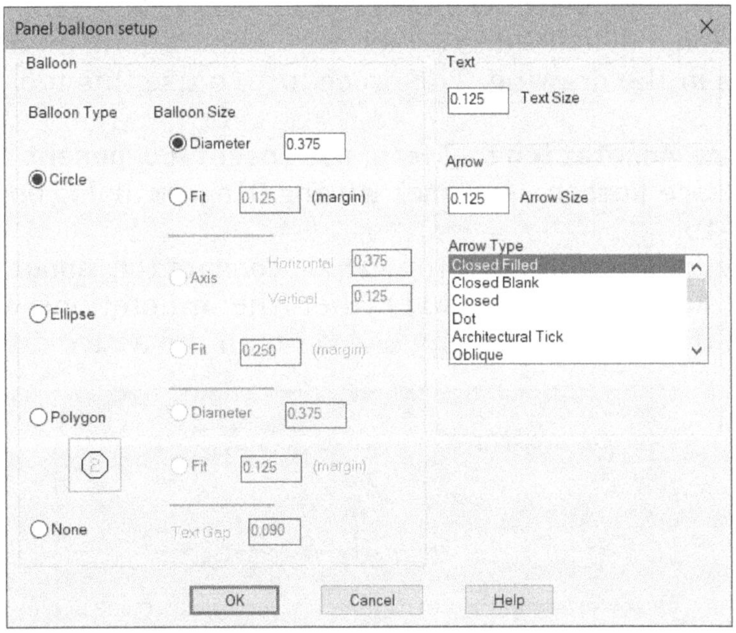

Figure-17. Panel balloon setup dialog box

- Select desired balloon type and specify the parameters in the dialog box.
- Click on the **OK** button from the dialog box to apply the specified parameters.
- Click on the component. You are asked to specify the leader start point or balloon insertion point.
- Click at desired point near the component. A dashed line of leader will be displayed having end point attached to the cursor.
- Click to specify the end point of the leader or press **ENTER** to place the balloon at the earlier selected point; refer to Figure-18.

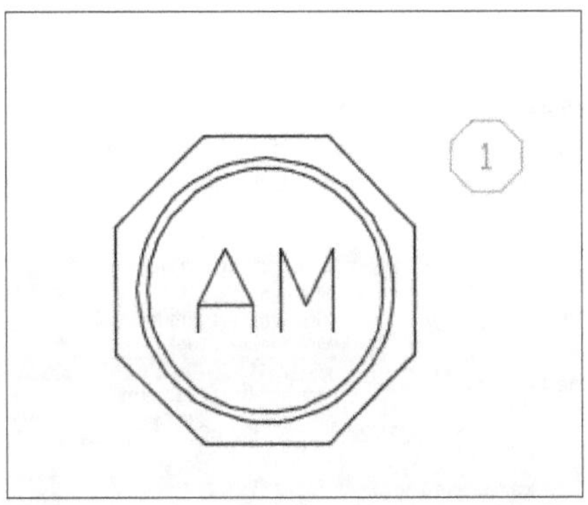

Figure-18. Component with assigned balloon

- If you have specified the end point of the leader then press **ENTER** to place the balloon.

WIRE ANNOTATION

The **Wire Annotation** tool is used to attach wire annotations (marks) to the selected panel components in the drawing. The procedure to use this tool is given next.

- Click on the **Wire Annotation** tool from the **Insert Component Footprints** panel. The **Schematic Wire Number --> Panel Wiring Diagram** dialog box will be displayed; refer to Figure-19.
- Select desired radio button from the **Panel connection annotation for** area. If you select the **Active drawing(all)** then the annotations will be assigned to components in the current drawing based on all the active drawings.

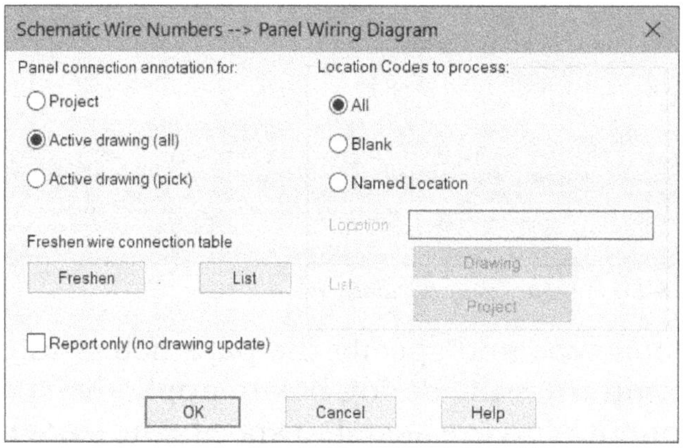

Figure-19. Schematic Wire Number dialog box

- Select desired radio button from the **Location Codes to process** area. The location codes will be assigned to the components as per the selection of the radio button.
- After specifying desired settings, click on the **OK** button from the dialog box. The **Schematic --> Layout Wire Connection Annotation** dialog box will be displayed; refer to Figure-20.

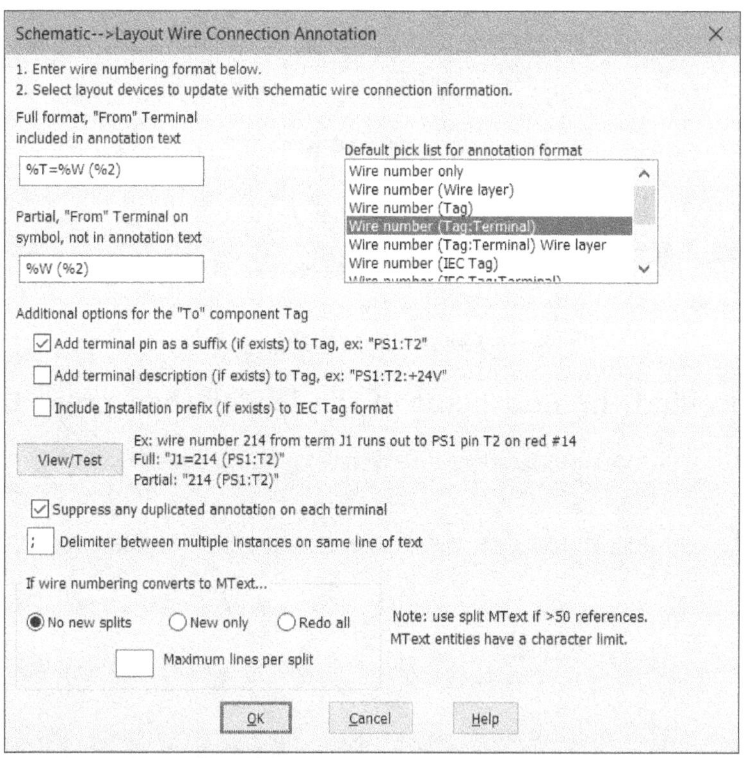

Figure-20. Schematic dialog box

- Set desired annotation format in this dialog box and click on the **OK** button from the dialog box. The annotations will be assigned to the components in the current drawing automatically; refer to Figure-21.

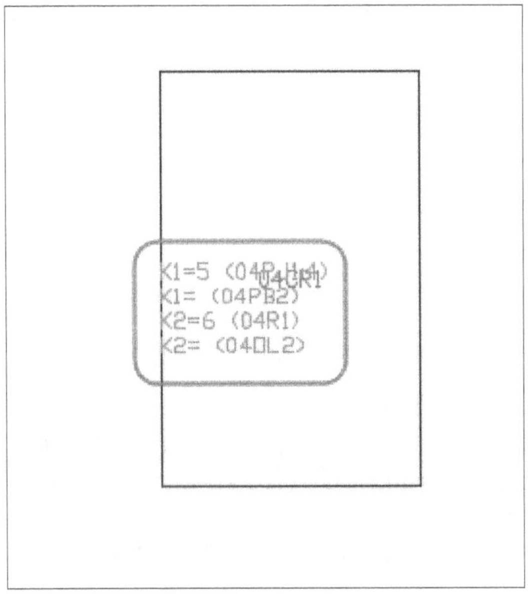

Figure-21. Annotations automatically assigned

PANEL ASSEMBLY

The **Panel Assembly** tool is used to insert an already created assembly of panel in the current drawing. This tool is very useful if you are going to use the same type of panel assemblies in various drawings. The procedure to use this tool is given next.

- Click on the **Panel Assembly** tool from the **Insert Component Footprints** panel in the **Ribbon**. The **Insert Panel Assembly of Blocks** dialog box will be displayed; refer to Figure-22.

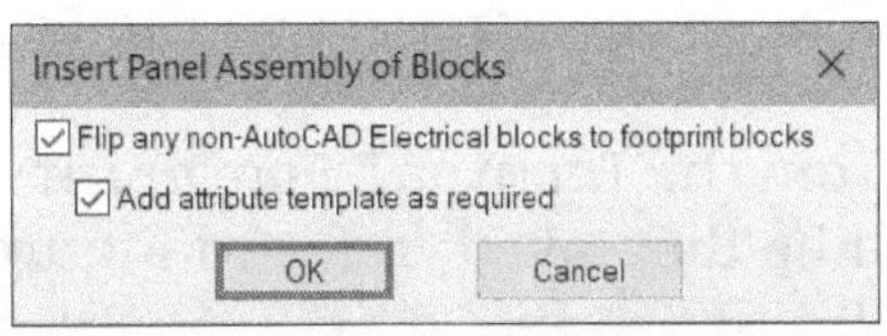

Figure-22. Insert Panel Assembly of Blocks dialog box

- Make sure all the check boxes are selected and then click on the **OK** button from the dialog box. The **WBlocked Assembly to insert** dialog box will be displayed; refer to Figure-23.

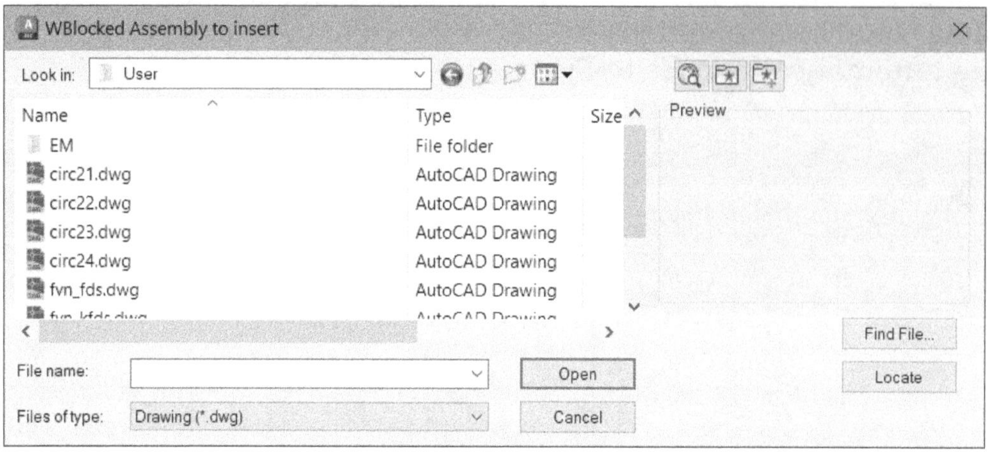

Figure-23. WBlocked Assembly to insert dialog box

- Select the drawing file of earlier created panel block and click on the **Open** button from the dialog box. You are asked to specify the insertion point for the panel block.
- Click in the drawing area to specify the insertion point for the panel block. You are asked to specify the rotation angle value for the block.
- Specify desired angle value and press **ENTER**. The block will be placed.

Till this point, we have discussed about inserting the component footprints in the panel layout. Now, we will learn about terminal footprints. The tools to manage the terminal footprints are available in the **Terminal Footprints** panel in the **Ribbon**. The tools in this panel are discussed next.

EDITOR

The **Editor** tool is used to edit the details of terminals present in the current drawing. Note that the terminal information is collected from all the drawings in the current project. So, it is important to activate the project whose drawing is being used for editing terminal footprints. The procedure to use the **Editor** tool is given next.

- Click on the **Editor** tool from the **Terminal Footprints** panel. The **Terminal Strip Selection** dialog box will be displayed; refer to Figure-24.

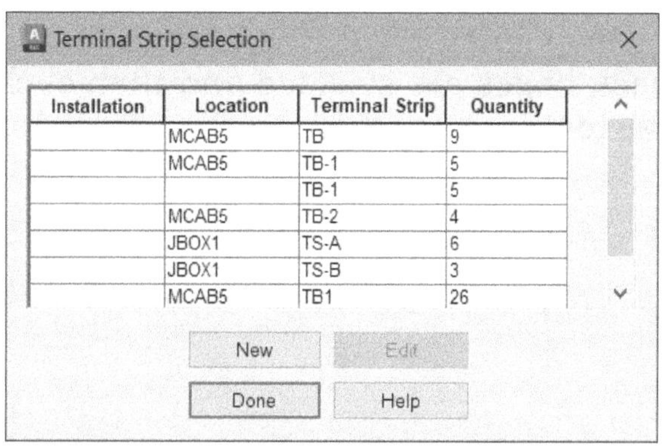

Figure-24. Terminal Strip Selection dialog box

- Select the entry that you want to edit in this table and click on the **Edit** button. If the number of terminals in the current drawing exceed the number of wires then the **Defined Terminal Wiring Constraints Exceeded** dialog box will be displayed; refer to Figure-25.

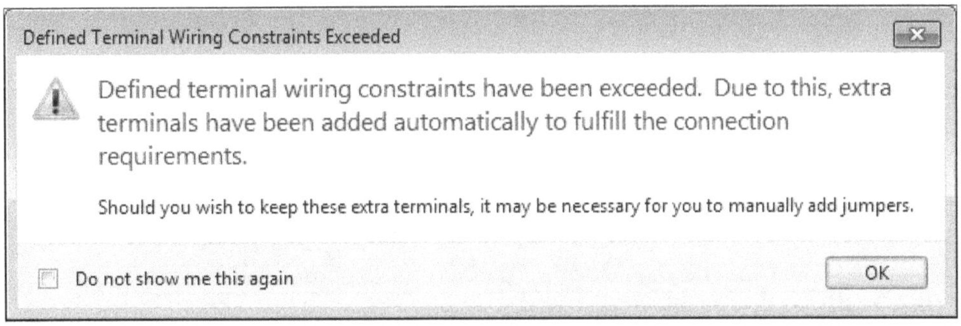

Figure-25. Defined Terminal Wiring Constraints Exceeded dialog box

- Click on the **OK** button from the **Defined Terminal Wiring Constraints Exceeded** dialog box. The **Terminal Strip Editor** dialog box will be displayed; refer to Figure-26.

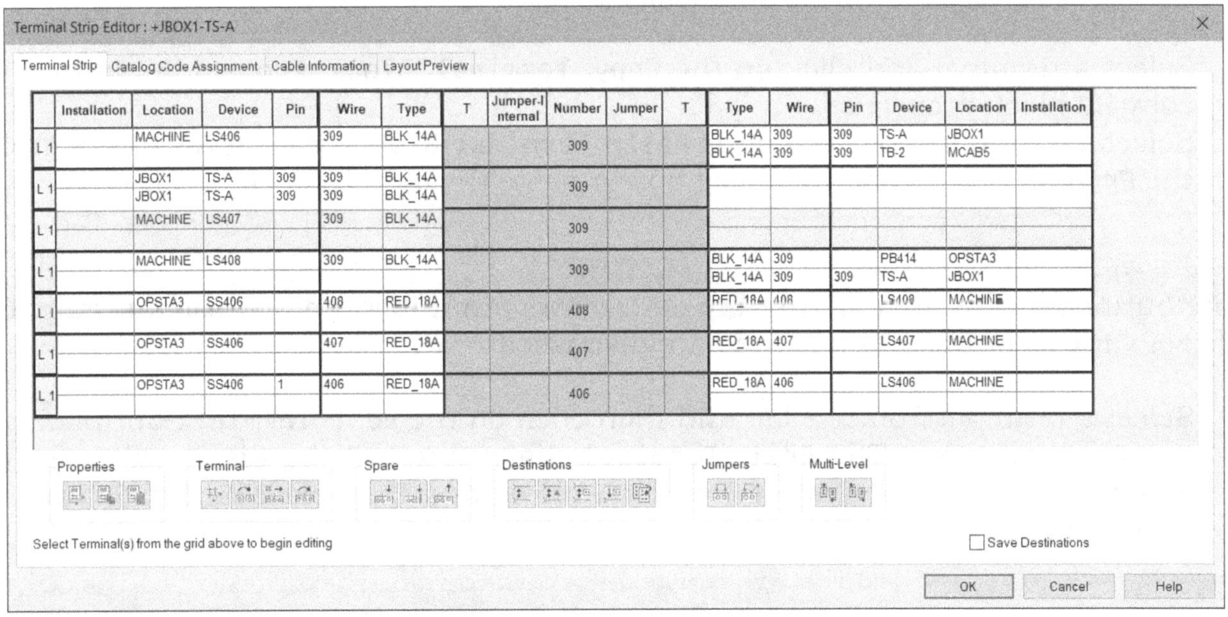

Figure-26. Terminal Strip Editor dialog box

- The selected terminal will be highlighted in yellow color. There are six areas in this dialog box named; **Properties**, **Terminal**, **Spare**, **Destinations**, **Jumpers**, and **Multi-Level**. The tools in these areas are used to edit the respective parameters of the terminals. Hold the cursor on each of the button in these areas to check its function.

Properties

There are three buttons in this area; **Edit Terminal Block Properties**, **Copy Terminal Block Properties**, and **Paste Terminal Block Properties**. The procedures to use these buttons are discussed next.

- Select a terminal from the list and click on the **Edit Terminal Block Properties** button from the **Properties** area. The **Terminal Block Properties** dialog box will be displayed; refer to Figure-27. Options in this dialog box have already been discussed in previous chapters. After performing the modifications, click on the **OK** button from the dialog box.

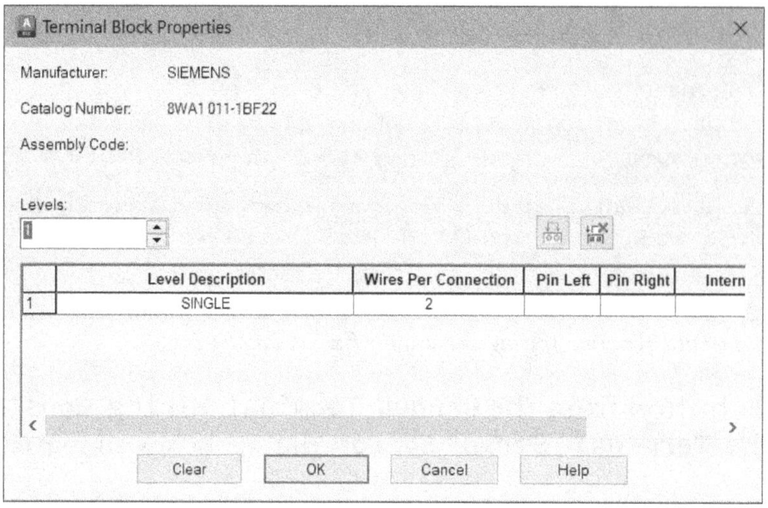

Figure-27. Terminal Block Properties dialog box

- Select a terminal and click on the **Copy Terminal Block Properties** button to copy its block properties.
- Select a terminal and click on the **Paste Terminal Block Properties** button from the **Properties** area of the dialog box to paste the copied block properties.

Terminal

The buttons in this area are used to edit, move, renumber, and reassign terminals. Use of various buttons in this area are discussed next.

- Select a terminal from the list and then click on the **Edit Terminal** button from the **Terminal** area. The **Edit Terminal** dialog box will be displayed; refer to Figure-28. Modify the parameters as required and then click on the **OK** button from the dialog box.

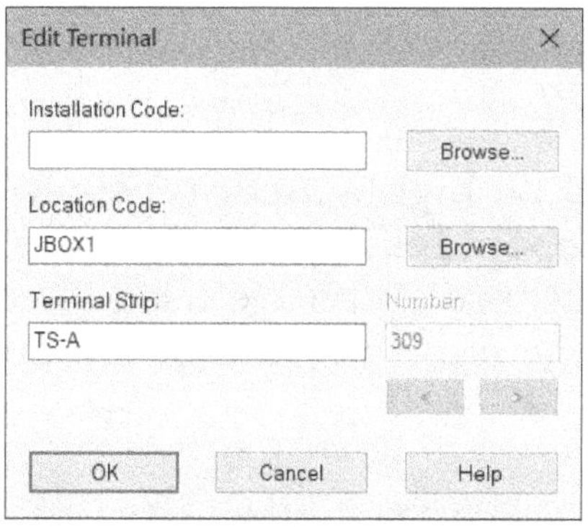

Figure-28. Edit Terminal dialog box

- Select a terminal from the list and click on the **Reassign Terminal** button. The **Reassign Terminal** dialog box will be displayed; refer to Figure-29. Set desired Installation code, Location code, and Terminal strip for the selected terminal and click on the **OK** button from the dialog box.

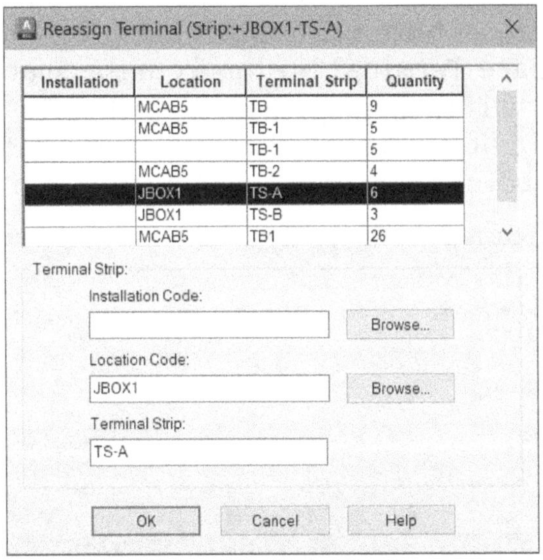

Figure-29. Reassign Terminal dialog box

- Select the terminals and click on the **Renumber Terminal** button from the **Terminal** area of the dialog box. The **Renumber Terminal Strip** dialog box will be displayed; refer to Figure-30. Specify the starting number of terminal, set the other parameters, and click on the **OK** button.

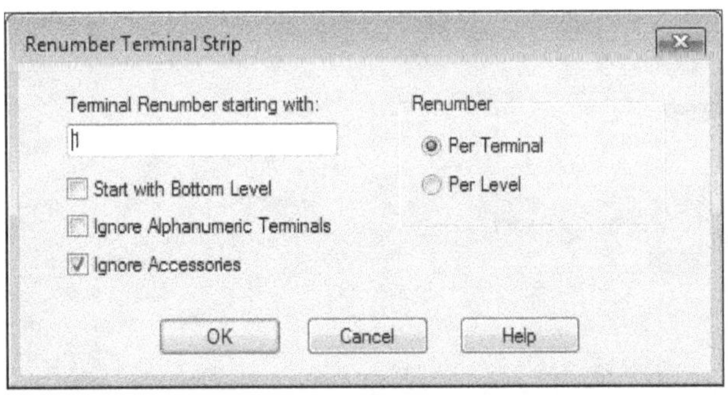

Figure-30. Renumber Terminal Strip dialog box

- Click on the **Move Terminal** button after selecting a terminal if you want to move the terminal in the table. On clicking this button, the **Move Terminal** dialog box will be displayed; refer to Figure-31. Using the buttons in the dialog box, you can move the terminal to desired location. After performing the changes, click on the **Done** button.

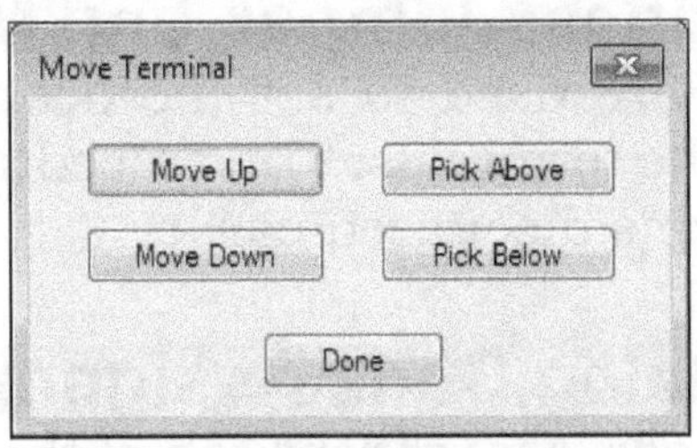

Figure-31. Move Terminal dialog box

Spare

The buttons in this area are used to add or delete spare terminals and accessories. Use of various buttons in this area are discussed next.

- You can insert a spare terminal before or after the selected terminal by using the **Insert Spare Terminal** button from this area. To do so, select a terminal and click on the **Insert Spare Terminal** button from the **Spare** area. The **Insert Spare Terminal** dialog box will be displayed; refer to Figure-32. Specify the terminal number in the **Number** edit box. Set the manufacturer data and quantity, and then click on the **Insert Above** or **Insert Below** button to insert the space terminal.

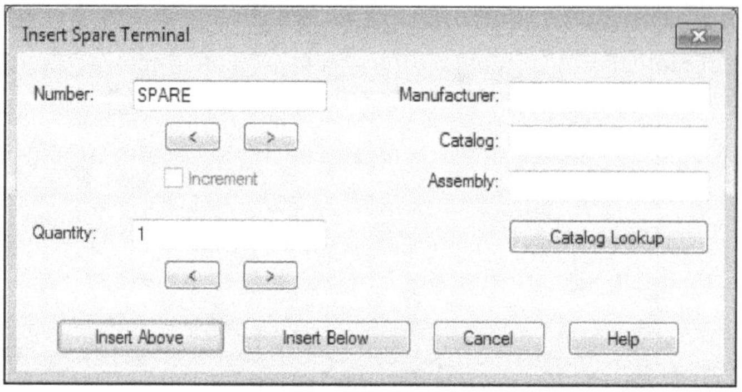

Figure-32. Insert Spare Terminal dialog box

- You can insert accessories in the same way as spare terminals are inserted. To do so, click on the **Insert Accessory** button from the **Spare** area. The **Insert Accessory** dialog box will be displayed; refer to Figure-33. Rest of the procedure is same as discussed for spare terminals.

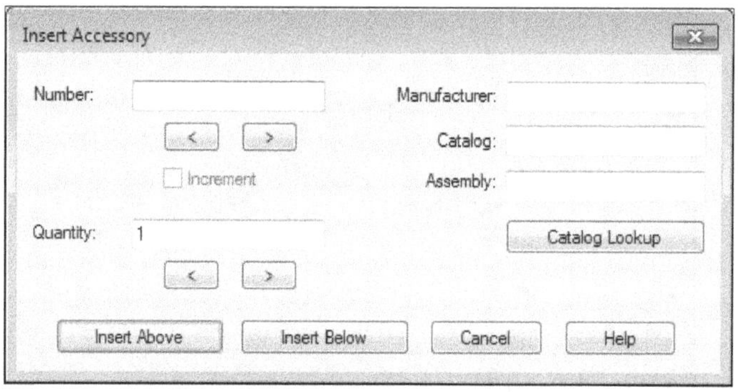

Figure-33. Insert Accessory dialog box

- The **Delete Spare Terminals/Accessories** button is used to delete the spare terminals and accessories. To delete a spare terminal/accessory, select it from the list and then click on the **Delete Spare Terminals/Accessories** button. The selected terminal/accessory will be deleted.

Destinations, Jumper, and Multilevel

The buttons in these areas are used to set the destination values, jumpers and multilevel terminal data for the terminals. The procedure to use the buttons in these areas are similar to the buttons discussed earlier.

Catalog Code Assignment

The options in the **Catalog Code Assignment** tab of the dialog box are used to check and modify catalog data assigned to the terminals; refer to Figure-34.

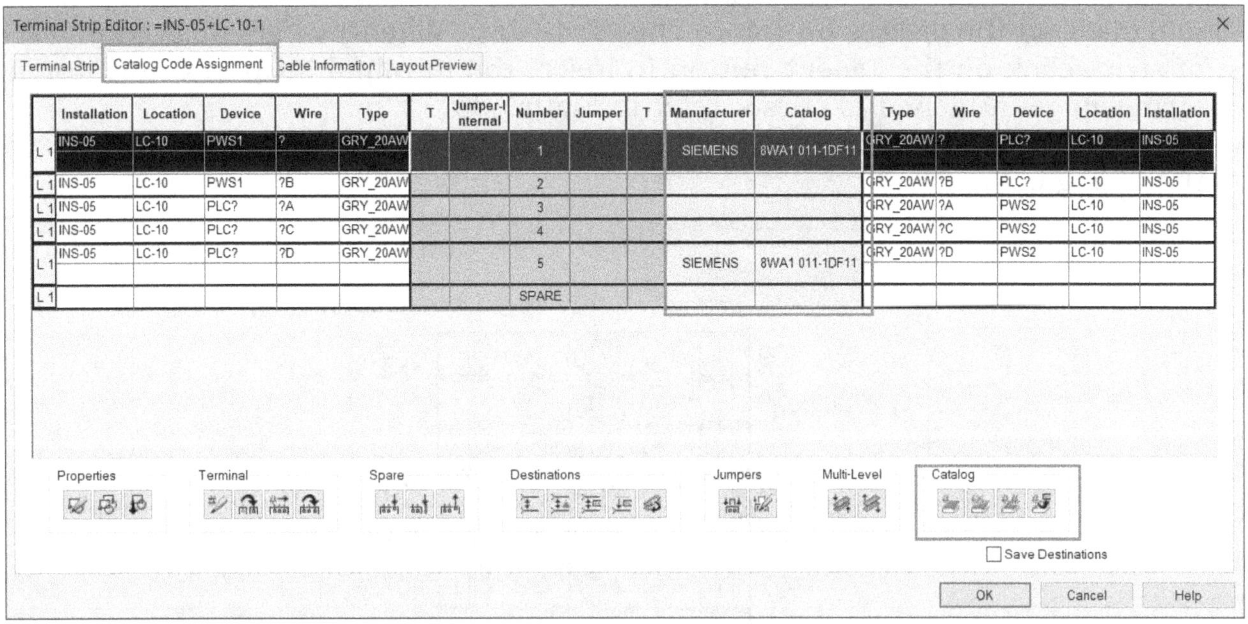

Figure-34. Catalog Code Assignment tab

- Select desired terminal from the table to which you want to assign catalog data and click on the **Assign Catalog Number** button from the **Catalog** area of the dialog box. The **Catalog Browser** will be displayed.
- Select desired catalog number from the browser and click on the **OK** button. Similarly, use the **Delete Catalog Number**, **Copy Catalog Number**, and **Paste Catalog Number** buttons to delete, copy, and paste catalog numbers respectively.
- The options in the **Cable Information** tab are used to check cable and conductor data for terminals.
- Click on the **Layout Preview** tab from the dialog box to check and modify the graphical representation of the terminal strip; refer to Figure-35.

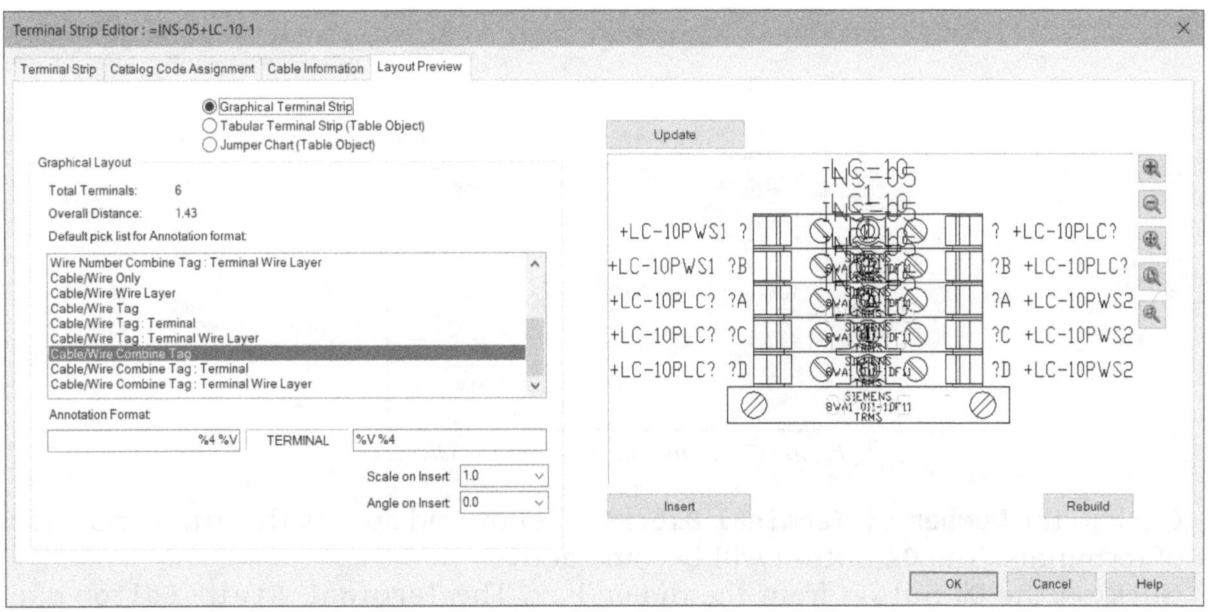

Figure-35. Layout Preview tab

- Select desired options in the dialog box to modify terminal strip representation and click on the **Update** button to check preview. When you get desired preview of strip, click on the **Insert** button to insert the terminal strip in drawing. The strip will get attached to cursor refer to Figure-36.
- Click at desired location to place the strip and then click on the **OK** button from the dialog box.

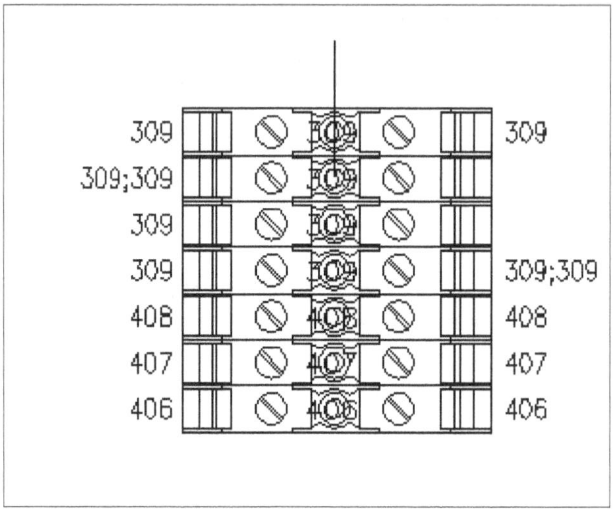

Figure-36. New terminal strip attached to cursor

- In place of selecting the **Insert** button, you can select the **Update** button to update the already existing terminal strip.
- After updating or inserting the strip, the **Terminal Strip Selection** dialog box will be displayed again.
- If you want to add a new terminal strip, then click on the **New** button from the dialog box. The **Terminal Strip Definition** dialog box will be displayed; refer to Figure-37.

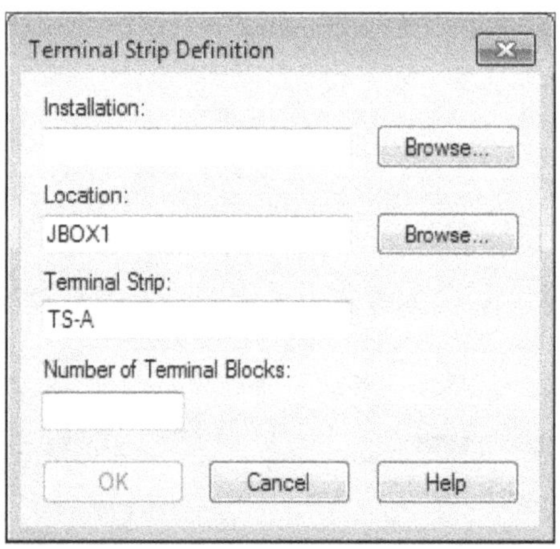

Figure-37. Terminal Strip Definition dialog box

- Click in the **Number of Terminal Blocks** edit box and specify the number of blocks of terminals. The **OK** button will become active.
- Click on the **OK** button from the dialog box. The **Terminal Strip Editor** dialog box will be displayed as discussed earlier. Specify each detail of the terminal as required and then click on the **OK** button from the dialog box. The dialog box will be displayed asking you to insert the graphical representation of the terminal.

- Click on the **Insert** button from the dialog box and place the terminal at desired position in the drawing.
- Click on the **Done** button from the **Terminal Strip Selection** dialog box to exit.

TABLE GENERATOR

The **Table Generator** tool is used to insert a table consisting of details related to the terminals in the drawing. The procedure to use this tool is given next.

- Click on the **Table Generator** tool from the **Terminal Footprints** panel in the **Ribbon**. The **Terminal Strip Table Generator** dialog box will be displayed; refer to Figure-38.

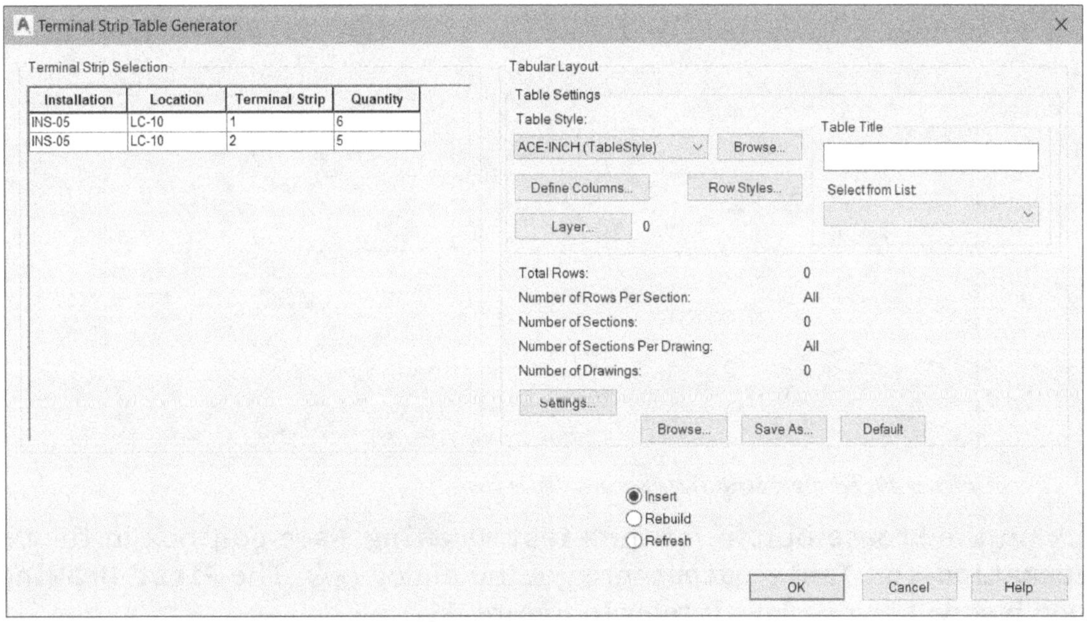

Figure-38. Terminal Strip Table Generator dialog box

- Select desired terminal from the **Terminal Strip Selection** table in the dialog box. For multiple selection, press & hold the **CTRL** key and then select the terminals; refer to Figure-39.

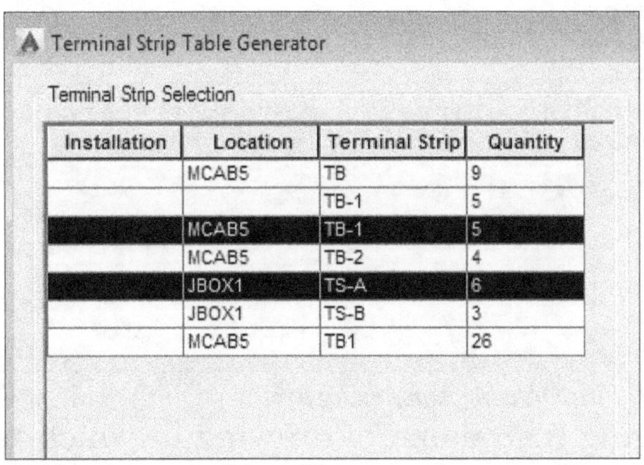

Figure-39. Terminal Selection

- Click on the **Table Style** drop-down in the **Table Settings** area of the dialog box and select desired table style.

- Click in the **Table Title** edit box and specify the suitable title for the terminal table.
- Click on the **Settings** button in the dialog box. The **Terminal Strip Table Settings** dialog box will be displayed; refer to Figure-40.

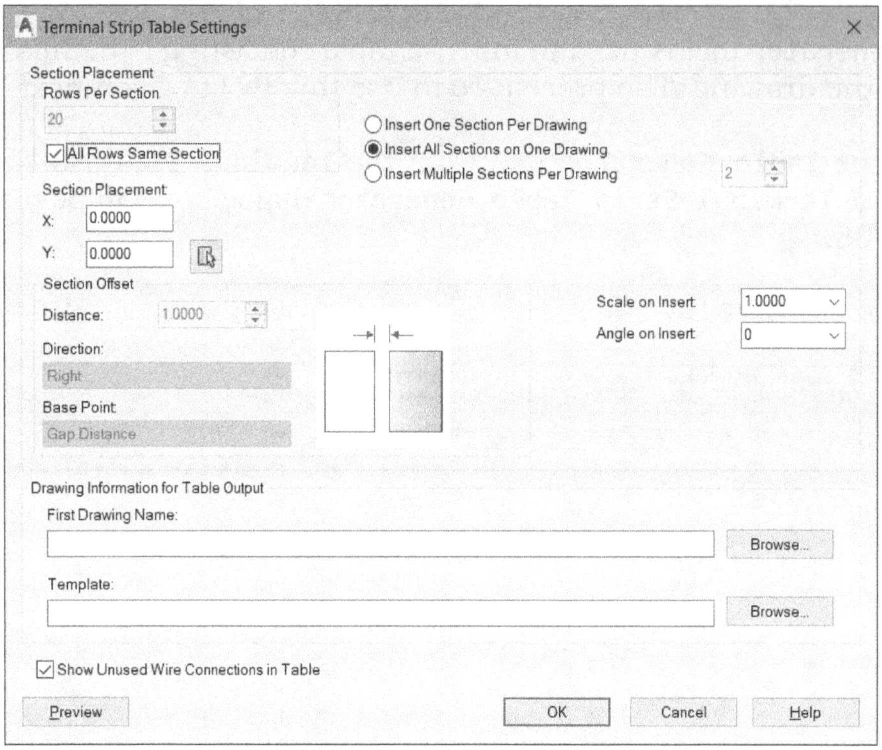

Figure-40. Terminal Strip Table Settings dialog box

- Click on the **Browse** button for the **First Drawing Name** edit box in the **Drawing Information for Table Output** area of the dialog box. The **First Drawing Name** dialog box will be displayed; refer to Figure-41.

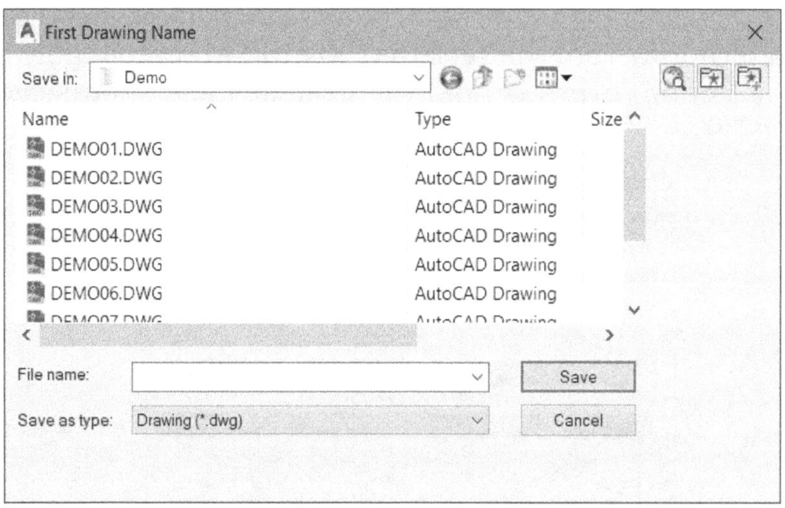

Figure-41. First Drawing Name dialog box

- Select desired drawing (The name of drawing in which table will be inserted is decided on the basis of the selected drawing's name) or you can create a new drawing file by specifying desired name. Click on the **Save** button from the dialog box.
- After specifying the rest of the parameters as required, click on the **OK** button from the dialog box.

- Click on the **OK** button from the **Terminal Strip Table Generator** dialog box. The **Table(s) Inserted** dialog box will be displayed; refer to Figure-42.

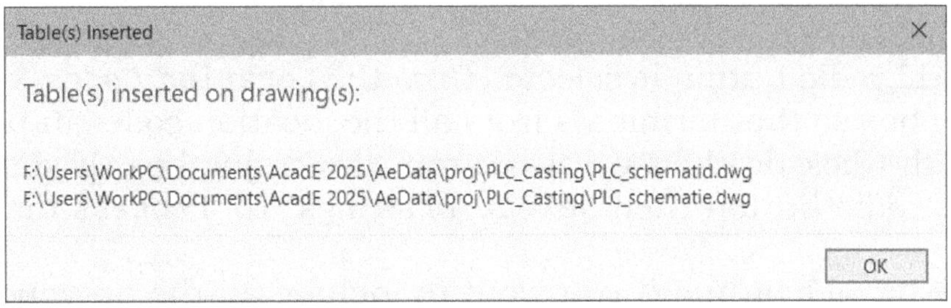

Figure-42. Tables Inserted dialog box

- Click on the **OK** button from the dialog box. Drawing will be added in the current project having tables created recently; refer to Figure-43.

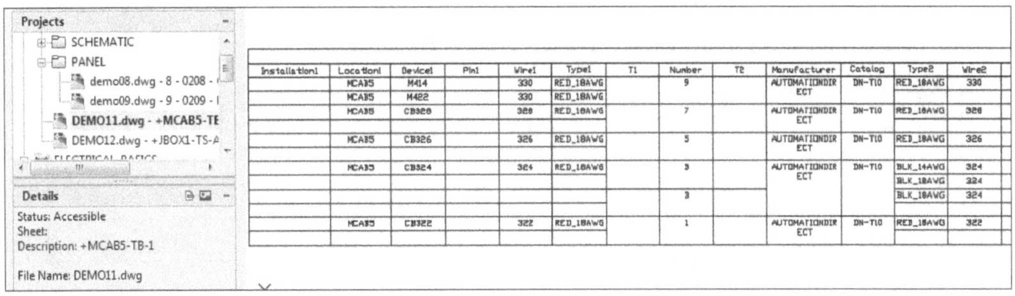

Figure-43. Table created

INSERT TERMINALS

There are two tools to insert terminals in the panel layout; **Insert Terminal (Schematic List)** and **Insert Terminal (Manual)** which work similar to **Schematic List** and **Manual** tool respectively. These tools are available in the **Insert Terminal** drop-down in the expanded **Terminal Footprints** panel of **Panel** tab of the **Ribbon**. The procedure to use these tools are discussed next.

Insert Terminals (Schematic List)

The **Insert Terminals (Schematic List) tool** is used to insert footprints of the terminals which have already been used in a drawing of current project. The procedure to use this tool is discussed next.

- Click on the **Insert Terminal (Schematic List)** tool from the **Insert Terminals** drop-down in the expanded **Terminal Footprints** panel of the **Ribbon**. The **Schematic Terminals List --> Panel Layout Insert** dialog box; refer to Figure-44.

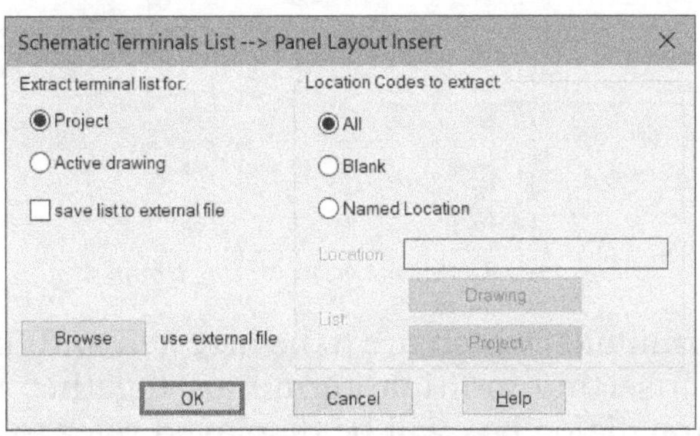

Figure-44. Schematic Terminals List dialog box

- Select the **Project** radio button in **Extract terminal list for** area of the dialog box if you want to extract the complete list of terminals used in current project. If you want the terminals from the current drawing then select the **Active drawing** radio button.
- Make sure **All** radio button is selected from the **Location Codes to extract** area of the dialog box so that terminals from all the location codes in project/drawing are extracted. Click on the **OK** button from the dialog box. If you have selected the **Project** radio button then **Select Drawings to Process** dialog box will be displayed.
- Click on the **Do All** button if you want to include all the drawings or select the drawings while holding the **CTRL** key and click on the **Process** button from the dialog box.
- Click on the **OK** button from the dialog box. The **Schematic Terminals** dialog box will be displayed; refer to Figure-45.

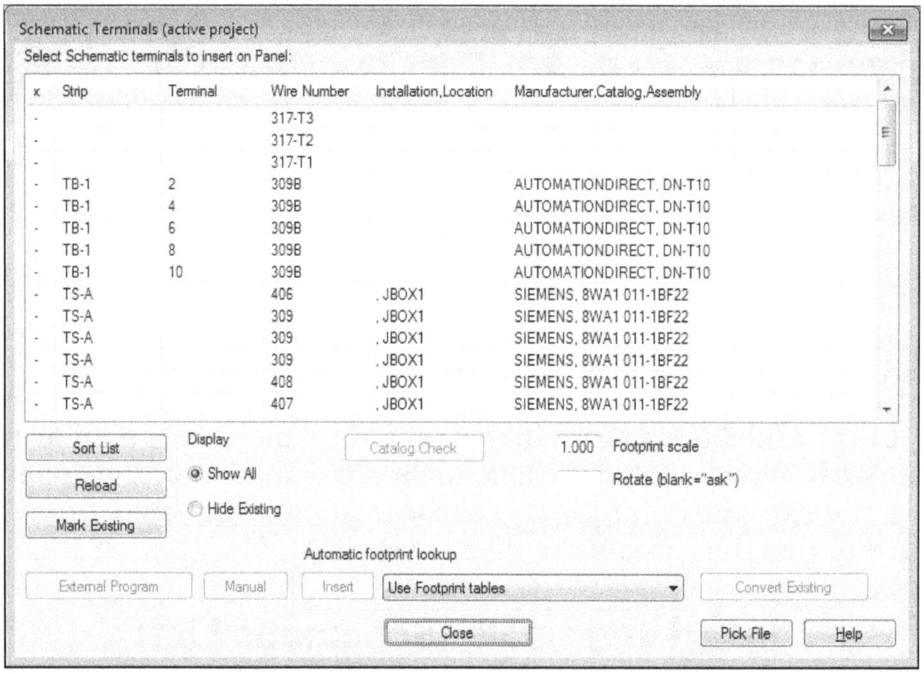

Figure-45. Schematic Terminals dialog box

- Click on the **Mark Existing** button to mark the terminals that have already been inserted in panel drawings of current project. The terminals that are exist in panel drawing are marked with cross (x) before their names; refer to Figure-46.

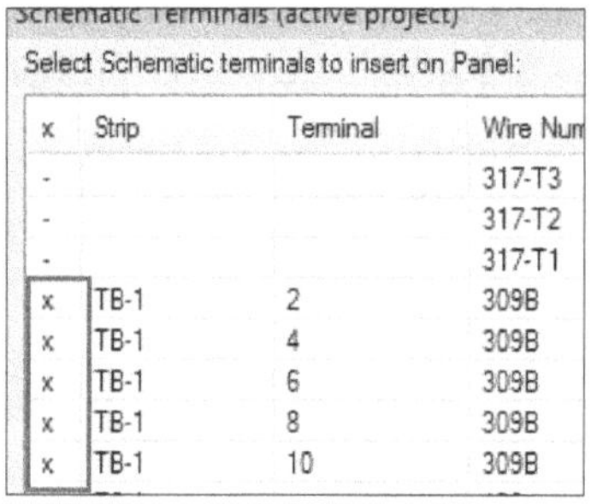

Figure-46. Cross mark before existing terminals

- Select the rest of terminal(s) which are to be inserted and then click on the **Insert** button. If you are inserting multiple terminals together then the **Spacing for Footprint Insertion** dialog box will be displayed refer to Figure-47.

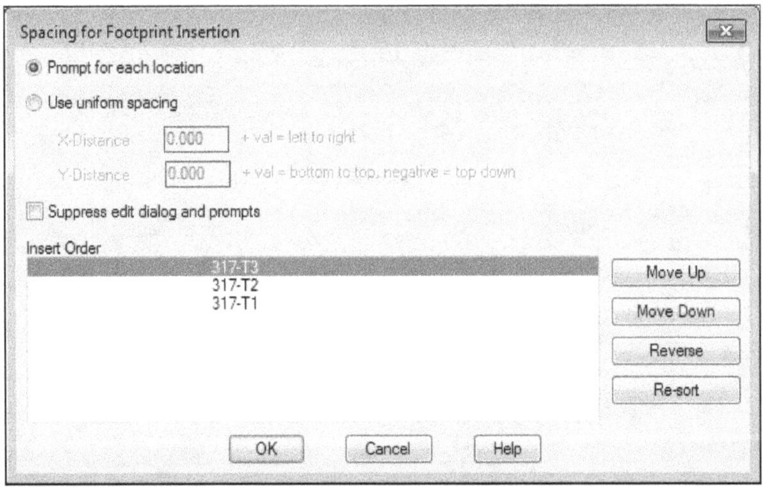

Figure-47. Spacing for Footprint Insertion dialog box

- Arrange the terminals as required by using the buttons along the **Insert Order** area.
- Select the **Use uniform spacing** radio button if you want to specify a uniform space between different terminals or select the **Prompt for each location** radio button so you are asked to specify position of each terminal.
- Click on the **OK** button. The terminals will get attached to cursor and you will be asked to specify insertion point. If the footprints are not assigned to the terminals earlier then the **Terminal** dialog box will be displayed; refer to Figure-48.

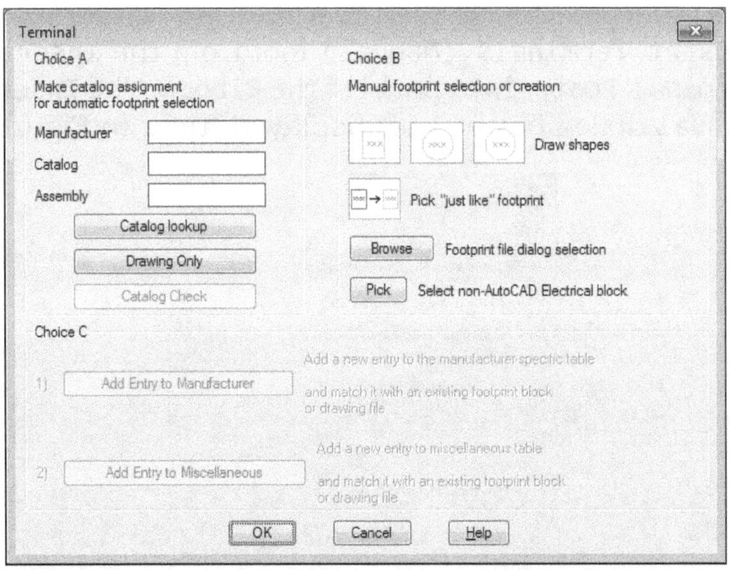

Figure-48. Terminal dialog box

- Set desired manufacturer data and click on the **OK** button from the **Terminal** dialog box.
- Click in the drawing at desired location to place the terminal(s) and specify the orientation. The **Panel Layout- Terminal Insert/Edit** dialog box will be displayed; refer to Figure-49.

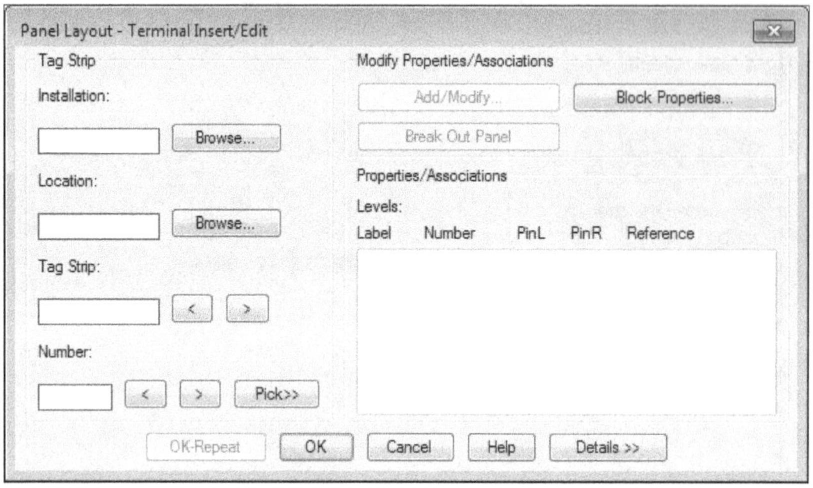

Figure-49. Panel Layout Terminal Insert or Edit dialog box

- Specify the information related to tag strip and block properties and click on the **OK** button. Repeat this step till all the terminals are inserted.
- On doing so, the **Schematic Terminals** dialog box will be displayed again. Insert the other terminals if required and then click on the **Close** button to exit.

Insert Terminals (Manual)

The **Insert Terminals (Manual) tool** is used to insert footprints of the terminals manually. The procedure to use this tool is given next.

- Click on the **Insert Terminals (Manual)** tool from the **Insert Terminals** drop-down in the **Terminal Footprints** panel of the **Ribbon**. The **Insert Panel Terminal Footprint--Manual** dialog box will be displayed; refer to Figure-50.

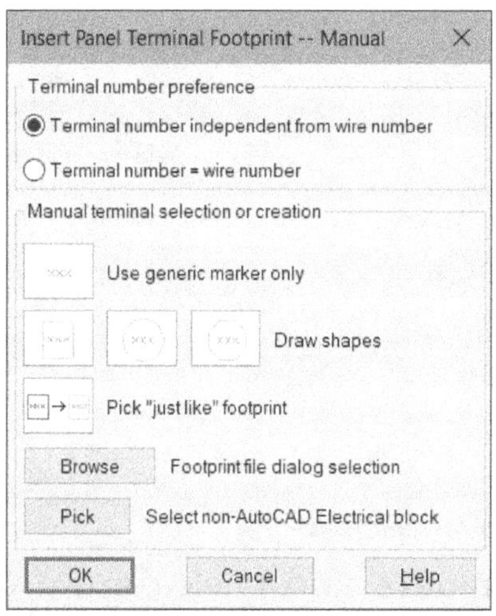

Figure-50. Insert Panel Terminal Footprint-Manual dialog box

- Select the **Terminal number independent from wire number** radio button from the dialog box if you want to create a terminal independent of wire number. If you want the terminal number same as the wire number then select the **Terminal number = wire number** radio button.
- After selecting desired radio button, now we have five options to insert footprints of terminals in the panel layout: by using generic marker, by drawing shapes, by picking just alike, by selecting file from local storage, by picking non-AutoCAD Electrical block. Procedure by using each option is given next.

Use Generic Marker Only

- Click on the white box for the **Use Generic Marker Only** option. You are asked to specify insertion point of the marker.
- Click in the drawing at desired location. You are asked to specify rotation value.
- Press **ENTER** to make it zero. The **Panel Layout-Terminal Insert/Edit** dialog box will be displayed. The options in this dialog box have already been discussed.

Draw Shapes

There are three buttons to draw shapes. You can draw rectangle, circle, or polygon by using the shape button.

- Click on desired button from this area. You are asked to specify the related parameter. If you click on the button with rectangle then you are asked to specify first corner and then opposite corner point of the rectangle to draw it. If you click on the button with circle then you are asked to specify center point and then radius of the circle. If you click on the button with polygon then you are asked to specify number of sides for polygon, then location of center point, then inscribed or circumscribed and then radius of inscribing/circumscribing circle.
- After making desired shape, specify the data in the **Panel Layout-Terminal Insert/Edit** dialog box and click on the **OK** button.

Pick "just like" footprint

The **Pick "just like" footprint** button is used to create a terminal as the selected one. The procedure is given next.

- Click on the white box before **Pick "just like" footprint** option. You are asked to select the existing "just like" footprint.
- Select the footprint whose shape is to be copied. The selected footprint will get attached to the cursor.
- Click at desired location to place the footprint. The **Panel Layout-Terminal Insert/Edit** dialog box will be displayed. Options in this dialog box have already been discussed.

Browse

The **Browse** option is used when you have a footprint for terminal saved in the local storage. The procedure to use this option is given next.

- Click on the **Browse** button from the **Insert Panel Terminal Footprint** dialog box. The **Pick** dialog box will be displayed; refer to Figure-51.

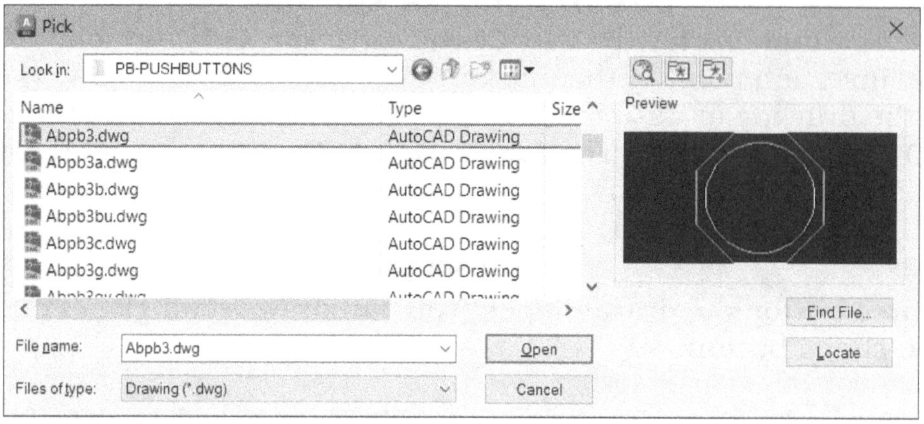

Figure-51. Pick dialog box

- Select the drawing file of terminal block you want to use. Preview will be displayed in the **Preview** area of the dialog box.
- Click on the **Open** button. The selected terminal block will get attached to the cursor.
- Click at desired location to place the terminal. Specify the rotation value and set the properties in the **Panel Layout-Terminal Insert/Edit** dialog box.

Pick

The **Pick** option is used to make any block existing in drawing area as terminal. It can be a non-AutoCAD Electrical object. After selecting this option, select an object from the drawing area. The **Panel Layout-Terminal Insert/Edit** dialog box will be displayed and you will be asked to specify parameters related to terminal.

EDITING FOOTPRINTS

The tools to edit footprints are available in the **Edit Footprints** panel; refer to Figure-52. These tools are discussed next.

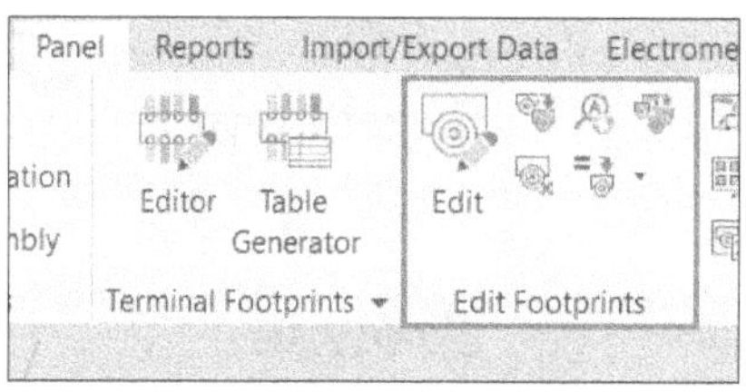

Figure-52. Edit Footprints panel

Edit

The **Edit** tool is used to modify the footprints of components in the panel layout. The procedure to use this tool is given next.

- Click on the **Edit** tool from the **Edit Footprints** panel. You are asked to select the component for editing footprint.
- Select desired component. The **Panel Layout - Component Insert/Edit** dialog box will be displayed; refer to Figure-53.
- Specify desired tags and parameters in the dialog box and click on the **OK** button from the dialog box to apply the specified settings.

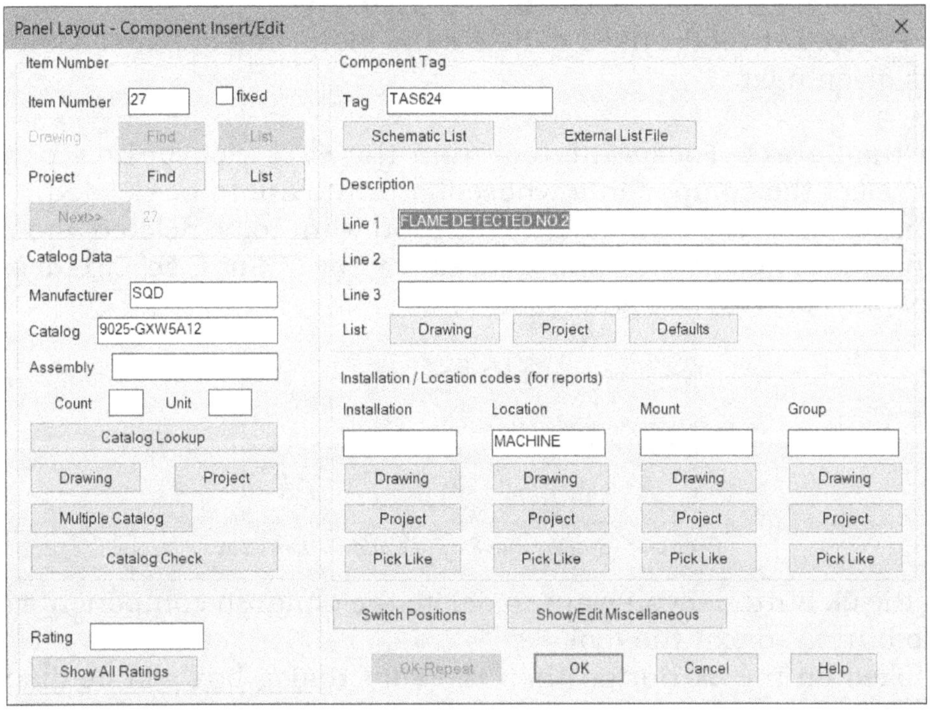

Figure-53. Panel Layout Component dialog box

Copy Footprint

The **Copy Footprint** tool is used to copy the footprint of a component so that you can assign it to other components in the panel layout. The procedure is given next.

- Click on the **Copy Footprint** tool from the **Edit Footprints** panel. You are asked to select the components just alike the component to which you want to assign footprints.
- Select desired component. Its footprint will get attached to the cursor; refer to Figure-54.

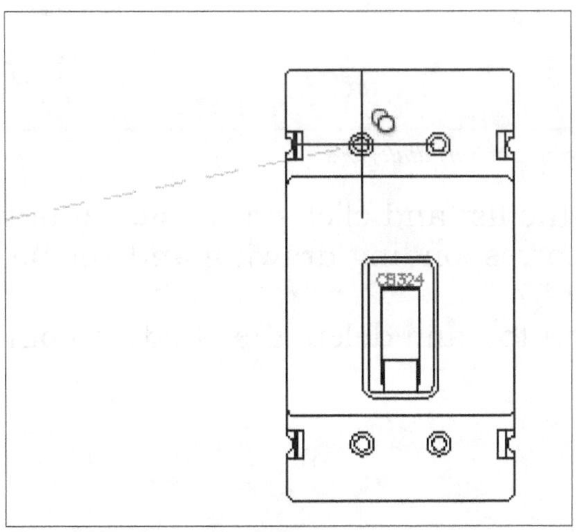

Figure-54. Copied footprint

- Click at desired location to place the footprint. The **Panel Layout - Component Insert/Edit** dialog box will be displayed. The options in this dialog box have already been discussed.
- Follow the procedures as discussed earlier.

Delete Footprint

The **Delete Footprint** tool is used to delete the footprints of panel components. The procedure is given next.

- Click on the **Delete Footprint** tool from the **Edit Footprints** panel. You are asked to select the components whose footprints are to be deleted.
- One by one click on the components that you want to be deleted and press **ENTER**. The footprints will be deleted and the **Search for / Surf to Children** dialog box will be displayed; refer to Figure-55.

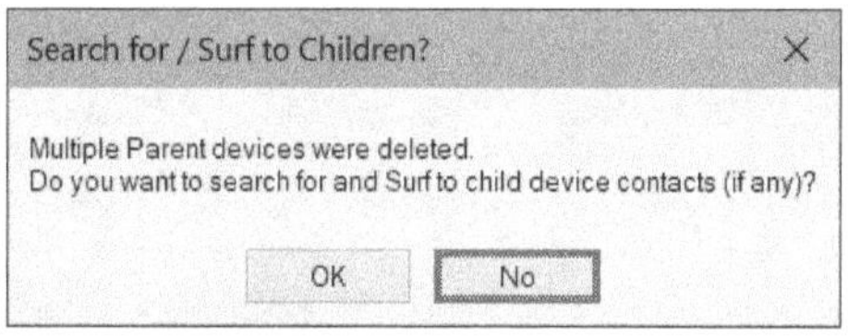

Figure-55. Search for or Surf to Children dialog box

- Click on the **OK** button if you want to delete the children components also or click on the **No** button to exit the tool.
- If you clicked on the **OK** button then the **Surf** dialog box will be displayed; refer to Figure-56.

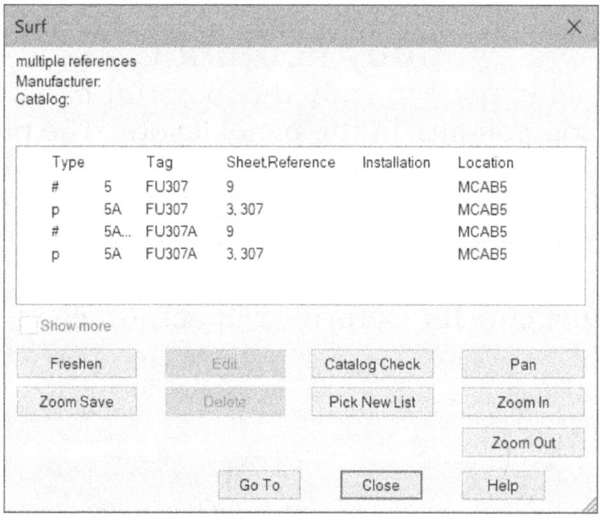

Figure-56. Surf dialog box

- Select the item from the list and click on the **Go To** button. The component will be displayed in the corresponding drawing and the **Delete** button will become active; refer to Figure-57.
- Click on the **Delete** button and delete the children components if you want to.

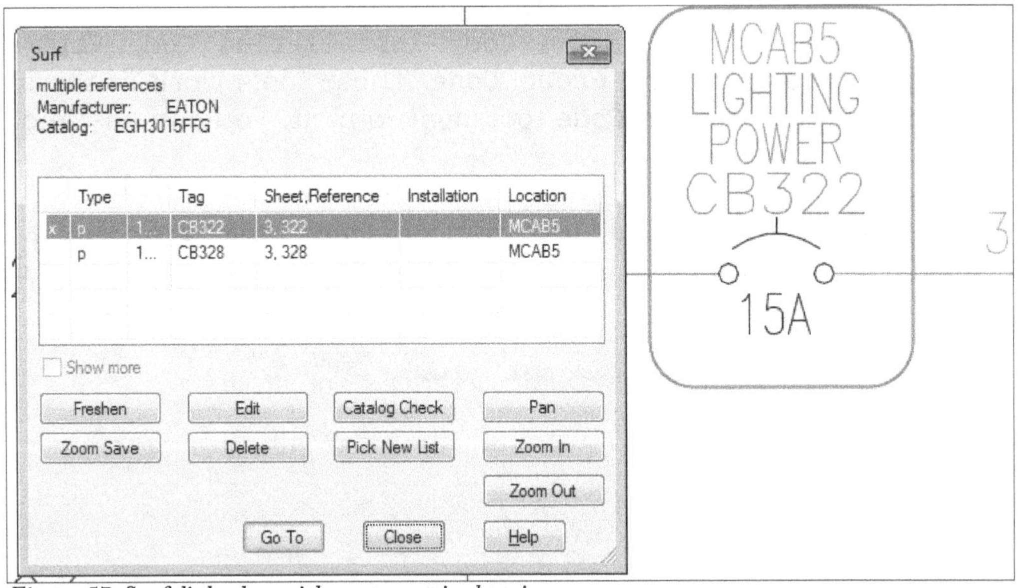

Figure-57. Surf dialog box with component in drawing

Resequence Item Numbers

The **Resequence Item Numbers** tool is used to re-sequence all the items in the current drawing/project. The procedure is given next.

- Click on the **Resequence Item Numbers** tool from the **Edit Footprints** panel. The **Resequence Item Numbers** dialog box will be displayed; refer to Figure-58.

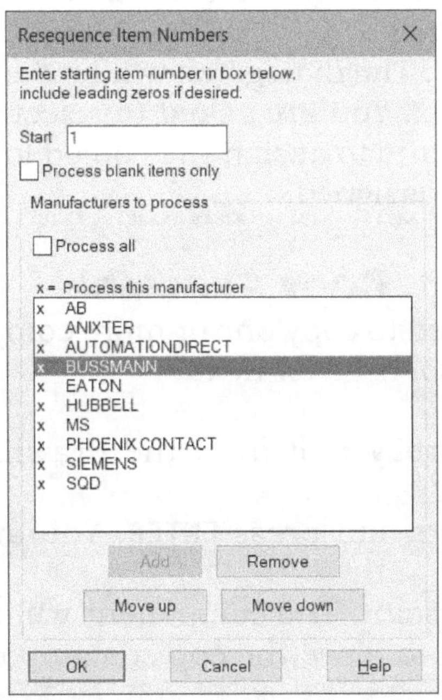

Figure-58. Resequence Item Numbers dialog box

- Specify desired starting number for the items in the **Start** edit box.
- Select the **Process all** check box to apply re-sequencing to all the items in panel drawing. Using the **Move up** and **Move down** buttons after selecting an item from the list, you can apply desired sequence to various items.
- After setting desired parameters, click on the **OK** button from the dialog box. On doing so, the item numbers will be updated automatically.

Copy Codes drop-down

There are four tools in this drop-down; **Copy Installation Code**, **Copy Location Code**, **Copy Mount Code**, and **Copy Group Code**. These tools work in the same way. The procedure for **Copy Location Code** tool is given next. You can assume the same for other tools in the drop-down.

- Click on the **Copy Location Code** tool from the **Copy Codes** drop-down. The **Copy Installation\Location\Mount\Group to Components** dialog box will be displayed; refer to Figure-59.

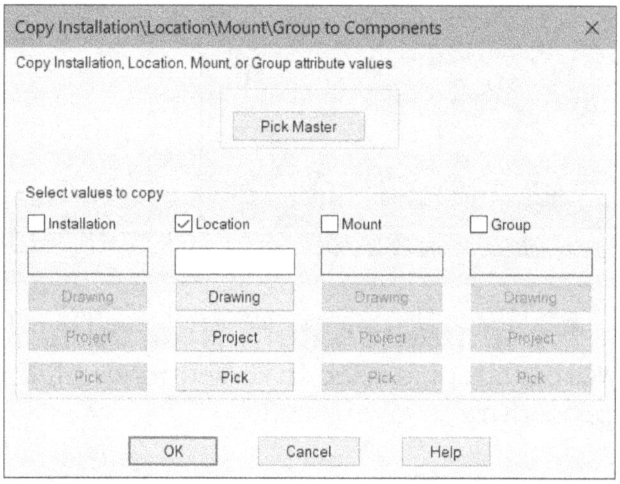

Figure-59. Copy Installation Location Mount Group to Components dialog box

- Click on the **Pick Master** button. You are asked to select the component whose location code is to be copied.
- Select desired component. The dialog box will be displayed again. Click on the **OK** button from the dialog box. You are asked to select the target components.
- Select the target components to assign the copied location code and press **ENTER**. The location code will be assigned.

Copy Assembly

The **Copy Assembly** tool is used to copy one or more components of a panel assembly. The procedure to use this tool is given next.

- Click on the **Copy Assembly** tool from the drawing. You are asked to select components or assembly.
- Select desired components and press **ENTER**. You are asked to specify the base point for copying.
- Click to specify the base point. The component will get attached to the cursor.
- Click in the drawing area to place the copied components/assembly.

MISCELLANEOUS TOOLS

The tools in **Other Tools** panel of **Panel** tab in the **Ribbon** are used to perform various miscellaneous tasks related to panel footprints like creating a new symbol, defining configuration parameters, renaming and updating layers, and so on. Most of the tools in this panel have already been discussed. The rest of the tools of this panel are discussed next.

Setting Panel Footprint Configuration

The **Configuration** tool is used to define default parameters for placing footprints like numbering for balloon, scale for inserting footprint, layer for placing footprint, and so on. The procedure to use this tool is given next.

- Click on the **Configuration** tool from the **Configuration** drop-down in the **Panel** tab of **Ribbon**. The **Panel Drawing Configuration and Defaults** dialog box will be displayed; refer to Figure-60.

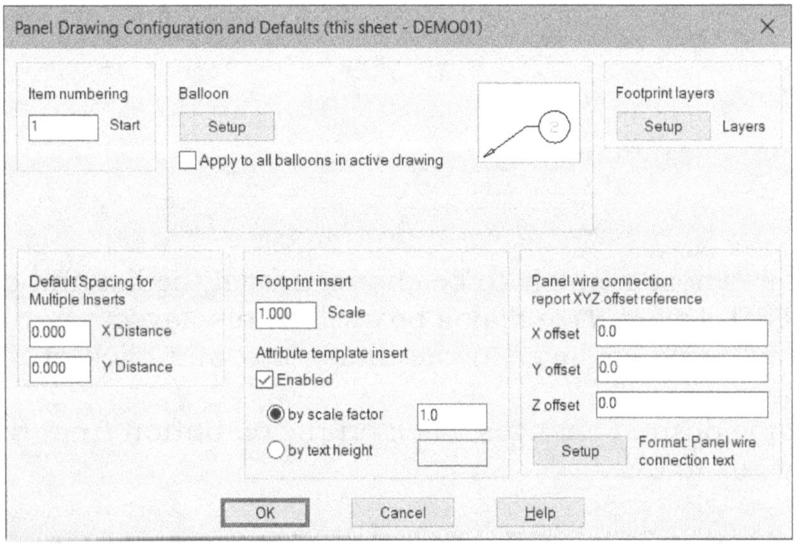

Figure-60. Panel Drawing Configuration and Defaults

- Specify desired value in the **Start** edit box of **Item numbering** area in the dialog box to define the starting number of balloons when marking them in drawing.
- Click on the **Setup** button from the **Balloon** area to define shape, size, and leader style for creating balloons in panel drawing. The **Panel balloon setup** dialog box will be displayed which has been discussed earlier in this chapter.
- If you are inserting multiple balloons then specify desired values in the **X Distance** and **Y Distance** edit boxes of the **Default Spacing for Multiple inserts** area to define distances along X axis and Y axis between consecutive balloons.
- Specify desired value in the **Scale** edit box of the **Footprint insert** area to define default scale at which footprint will be inserted.
- Similarly, you can set the other parameters in dialog box like footprint layers, wire connection offset for panels, and so on in respective fields of the dialog box.
- After setting desired parameters, click on the **OK** button from the dialog box to apply changes.

Renaming Layers

The **Rename Layers** tool is used to change names of layers used for placing various panel drawing components. The procedure to use this tool is given next.

- Click on the **Rename Layers** tool from the **Configuration** drop-down in the **Panel** tab of **Ribbon**. The **Rename Panel Layers** dialog box will be displayed; refer to Figure-61.

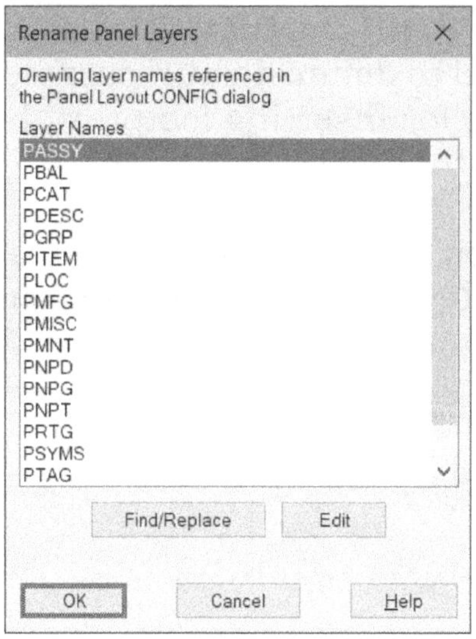

Figure-61. Rename Panel Layers dialog box

- Select the layer whose name is to be changed from the list and click on the **Edit** button. The **Edit Layer Name** dialog box will be displayed.
- Specify desired name for layer in the dialog box and click on the **OK** button to change the name.
- After performing desired changes, click on the **OK** button from the **Rename Panel Layers** dialog box to exit.

Updating Components Based on Layer Configuration

The **Update Layer** tool in **Configurations** drop-down on **Other Tools** panel in the **Panel** tab of **Ribbon** is used to update properties of selected components based on changes applied using Configuration settings.

Making Xdata (External Data) Visible

The **Make Xdata Visible** tool is used to make invisible external data of selected component visible. The procedure to use this tool is given next.

- Click on the **Make Xdata Visible** tool from the **Configuration** drop-down in the **Panel** tab of **Ribbon**. You will be asked to select the footprint for which Xdata is to be made visible.
- Select desired footprint from panel drawing. The **Select XData to Change to a Block Attribute** dialog box will be displayed; refer to Figure-62.
- Select radio button for the attribute which you want to display in drawing with the footprint symbol and click on the **Insert** button. The attribute will be attached to cursor.
- Click at desired location to place the Xdata attribute; refer to Figure-63. After inserting desired data, click on the **Done** button from the dialog box to exit.

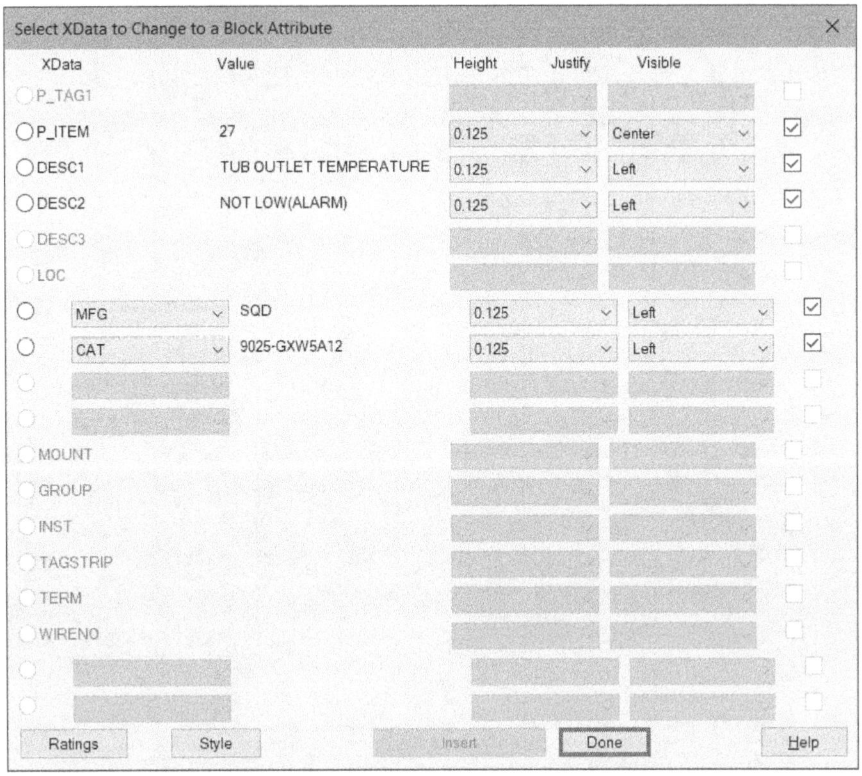

Figure-62. Select XData to Change to a Block Attribute dialog box

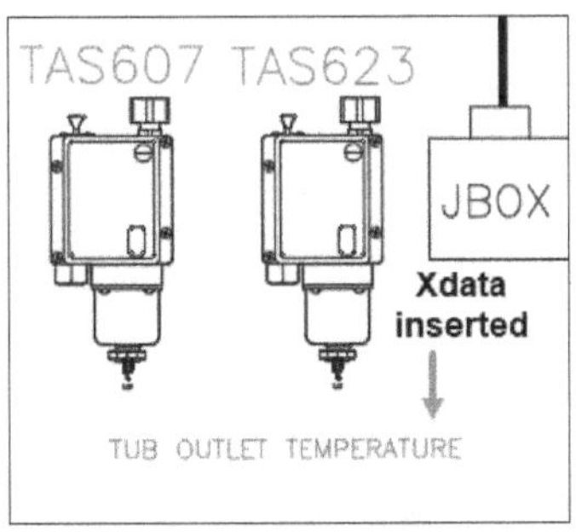

Figure-63. Xdata inserted

PRACTICAL 1

Create the panel layout for schematic diagram created in Practical 1 of Chapter 8; refer to Figure-64.

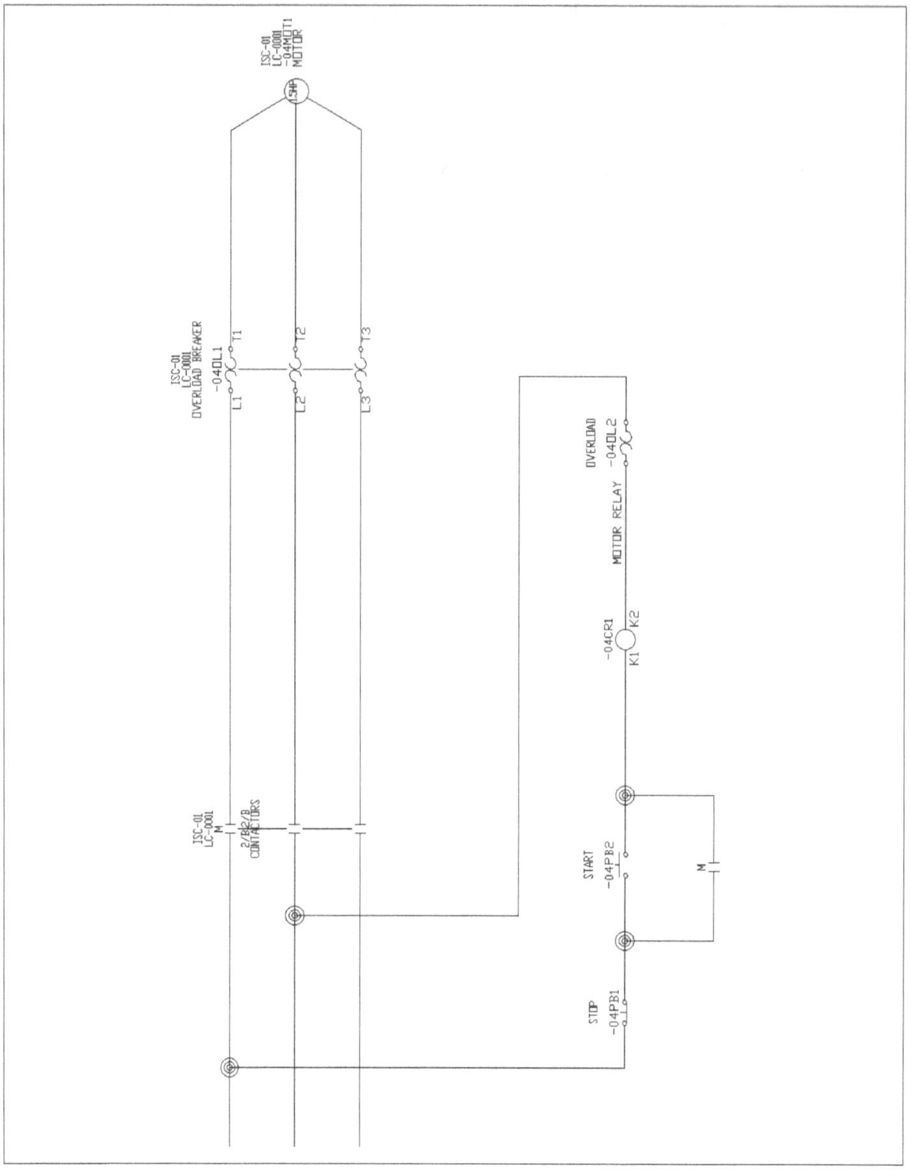

Figure-64. Practical 1 schematic

Starting a drawing

- Select a project having schematic drawing for which you want to design the panel. Note that you need to also add the drawing of Practical 1 of previous chapter in any project.
- Right-click on the project name in the **`Project Manager`**; refer to Figure-65.

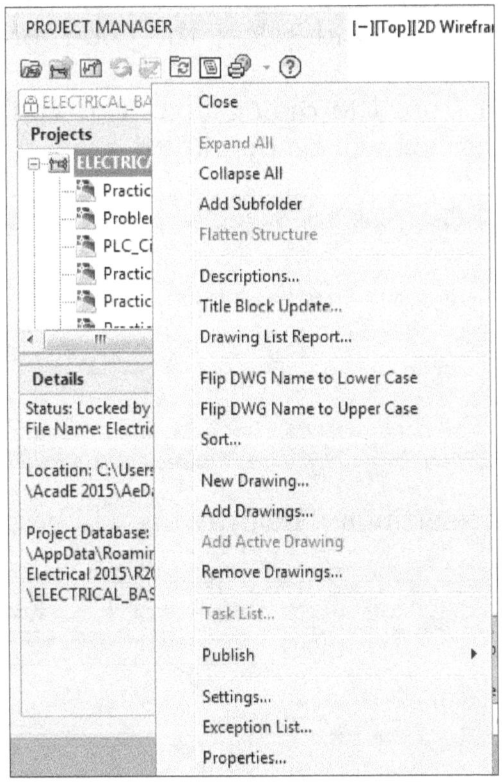

Figure-65. Shortcut menu of Project

- Click on the **New Drawing** button. The **Create New Drawing** dialog box will be displayed; refer to Figure-66.

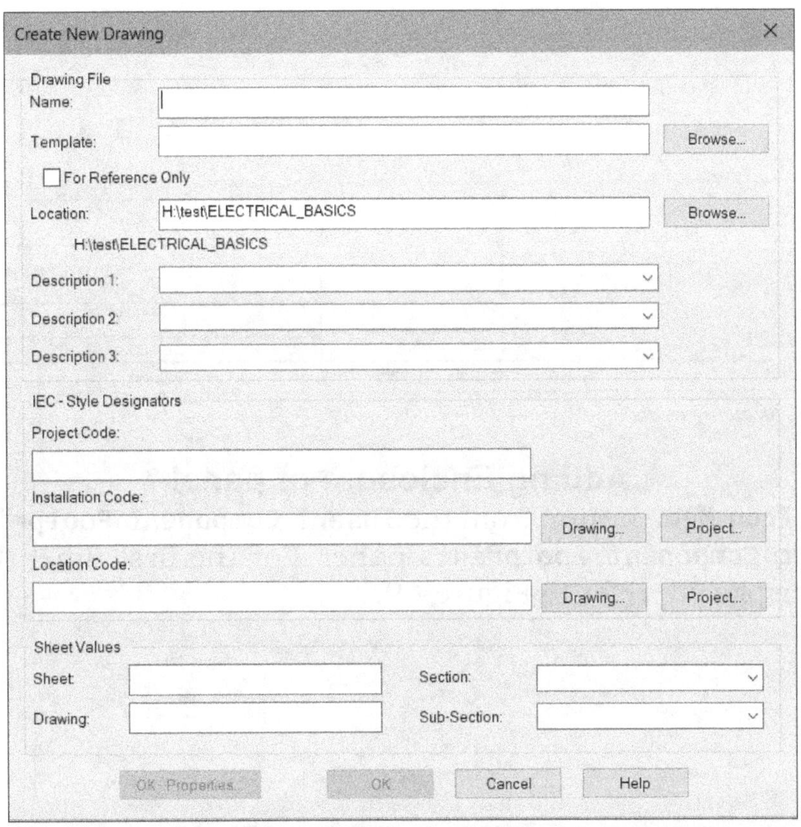

Figure-66. Create New Drawing dialog box

- Click in the **Name** edit box and specify the name as **Panel Layout**.
- Click on the **Drawing** button and select the code for **Installation** as well as **Location** codes.
- Click on the **OK** button from the dialog box. The **Apply Project Defaults to Drawing Settings** dialog box will be displayed; refer to Figure-67.

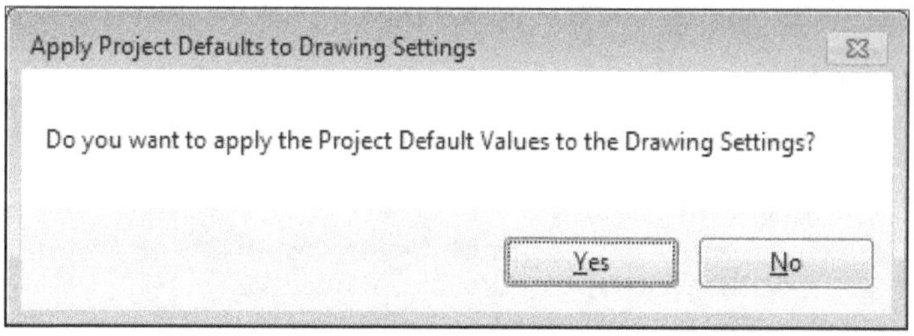

Figure-67. Apply Project Defaults to Drawing Settings dialog box

- Click on **Yes** to apply the settings. The drawing will be created; refer to Figure-68.

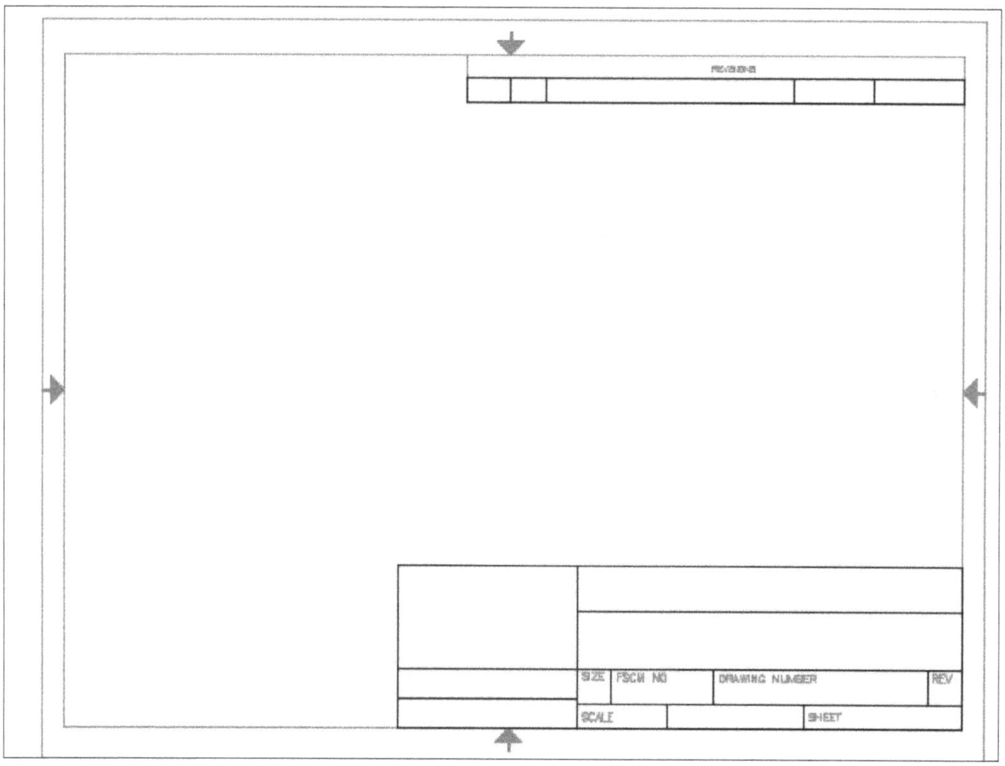

Figure-68. Drawing created

Adding Enclosure of panel

- Click on the **Icon Menu** button from the **Insert Component Footprints** drop-down in the **Insert Component Footprints** panel. For the first time, the **Alert** dialog box will be displayed; refer to Figure-69.

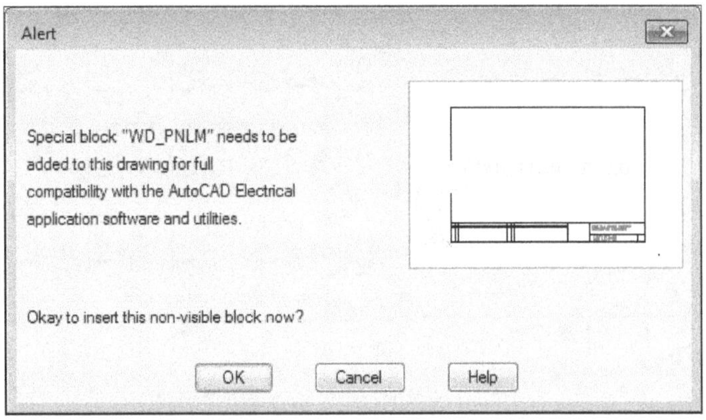

Figure-69. Alert dialog box

- Click on the **OK** button from the dialog box. The **Insert Footprint** dialog box will be displayed; refer to Figure-70.

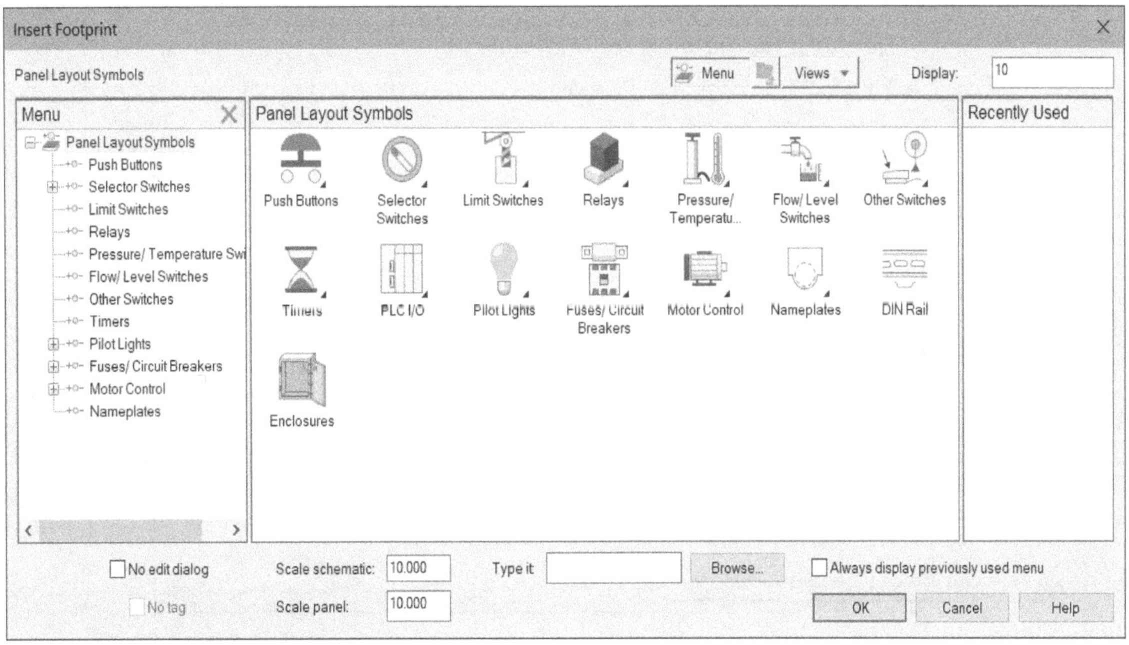

Figure-70. Insert Footprint dialog box

- Click on the **Enclosures** category in the dialog box. The **Footprint** dialog box will be displayed; refer to Figure-71.
- Click on the **Catalog lookup** button to find the manufacturer data or specify desired manufacturer data in edit boxes of the **Choice A** area.
- Click on the **Browse** button from **Choice B** area and select the drawing block for the enclosure.
- Click on the **Open** button from the dialog box, you are asked to specify the insertion point for the footprint.
- Click in the drawing area to specify the insertion point. You are asked to specify the rotation angle for the current footprint. Enter **0** at the command prompt. The enclosure will be placed; refer to Figure-71 and the **Panel Layout** dialog box will be displayed; refer to Figure-73.

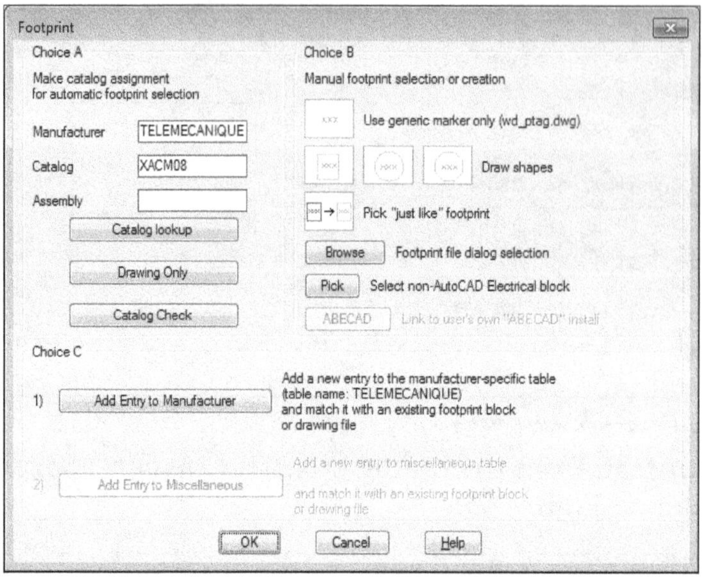

Figure-71. Footprint dialog box for enclosure

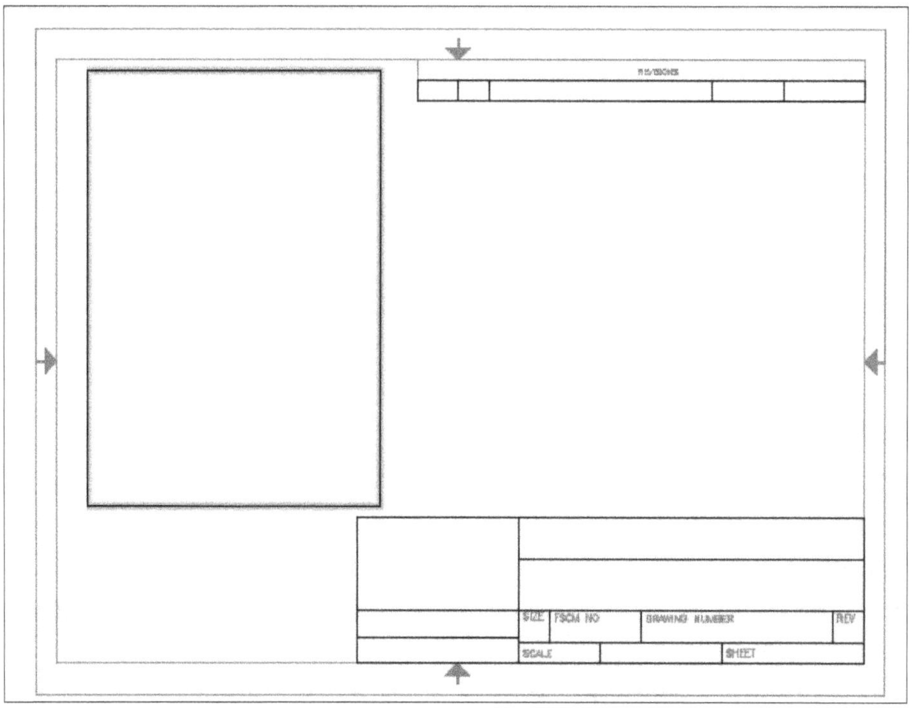

Figure-72. Enclosure placed

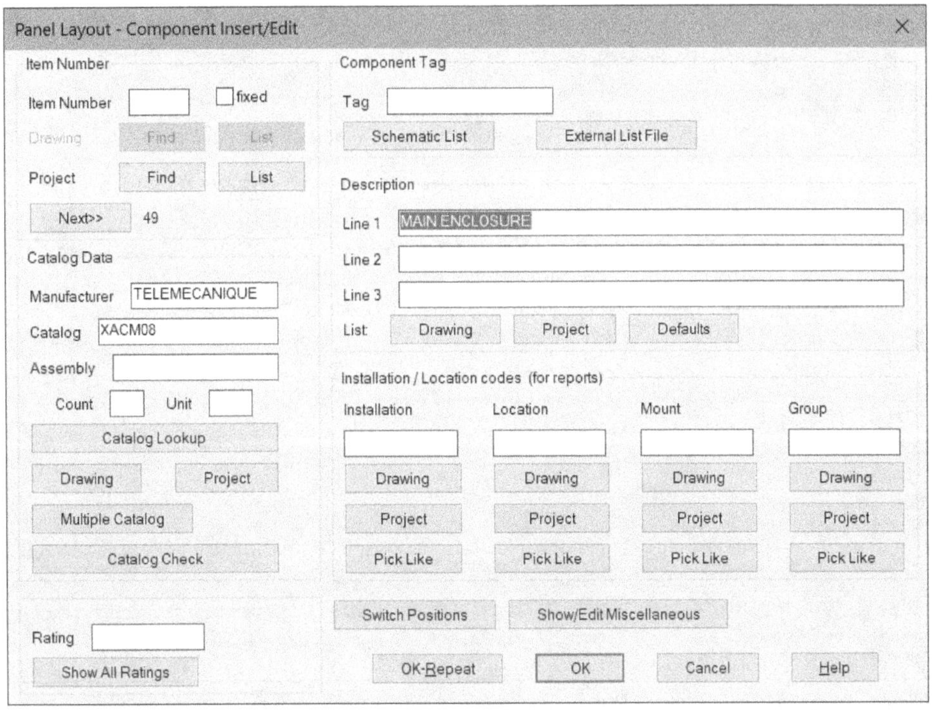

Figure-73. Panel Layout dialog box for enclosure

- Make sure that you have the values specified as in the above figure and then click on the **OK** button from the dialog box.

Inserting other components of the panel

- Click on the **Schematic List** tool from the **Insert Component Footprints** drop-down in the **Insert Component Footprints** panel. The **Schematic Components List** dialog box will be displayed; refer to Figure-74.

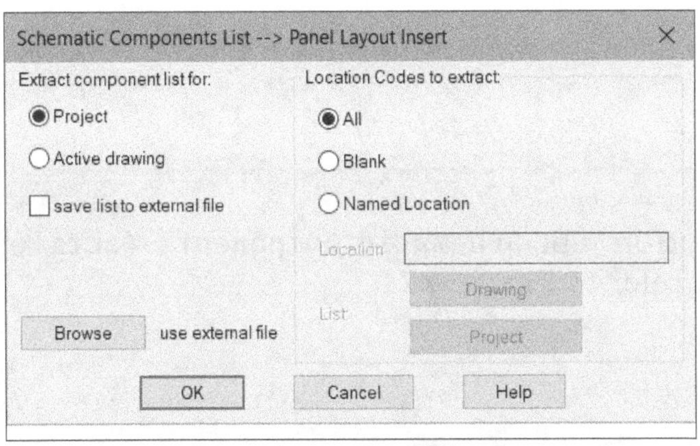

Figure-74. Schematic Components List dialog box

- Make sure that the **Project** radio button is selected and then click on the **OK** button from the dialog box. The **Select Drawings to Process** dialog box will be displayed; refer to Figure-75.

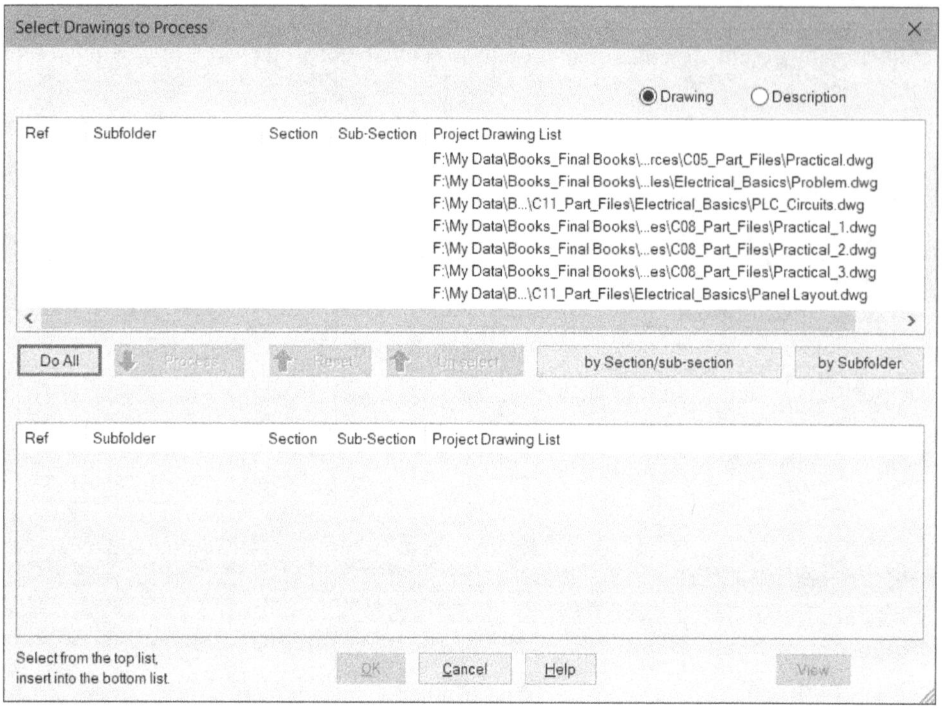

Figure-75. Select Drawings to Process dialog box

- Select the **Practical 1** drawing from the list and click on the **Process** button. The drawing will be included in the schematic list; refer to Figure-76.

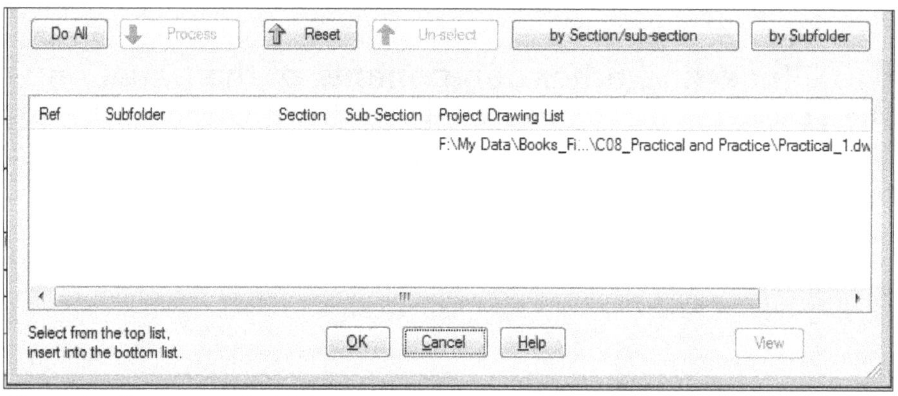

Figure-76. Drawing included for processing

- Click on the **OK** button. The **Schematic Components (active project)** dialog box will be displayed; refer to Figure-77.

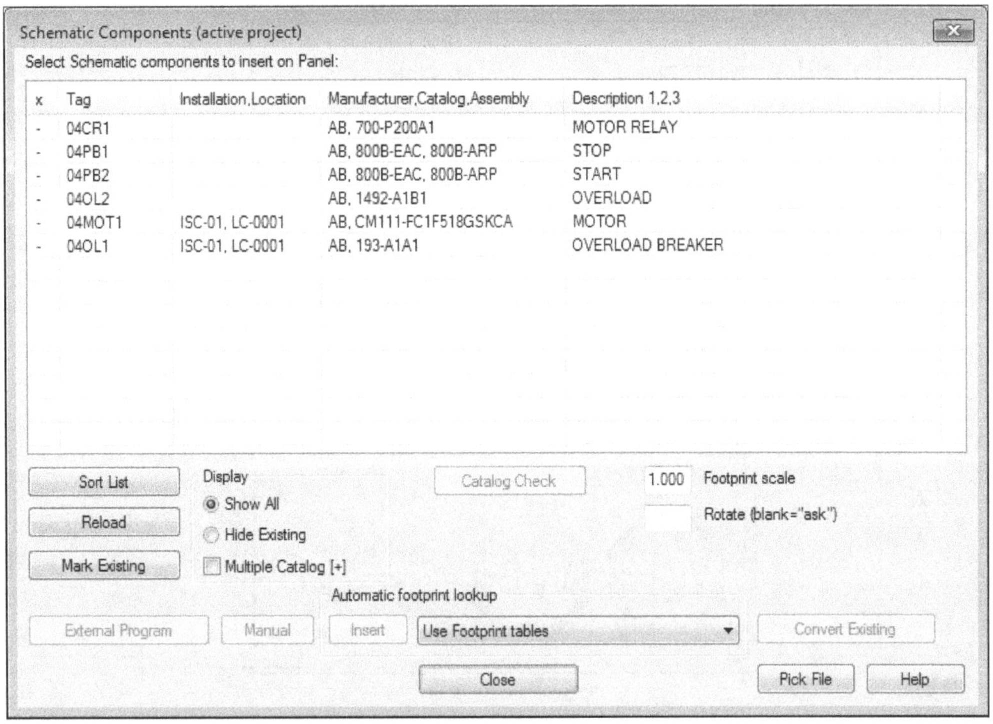

Figure-77. Schematic Components active projects dialog box

- Select the component with description **STOP** and click on the **Insert** button. The icon will get attached to the cursor; refer to Figure-78.

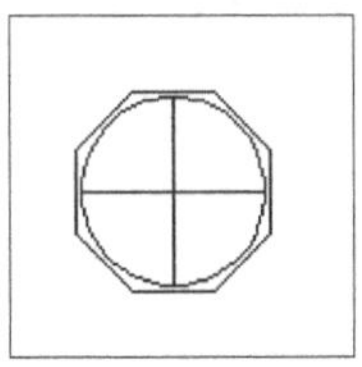

Figure-78. Stop button attached to cursor

- Click in the enclosure to place the button at desired place. You are asked to specify the rotation angle.
- Enter **0** at the command prompt. The component will be placed and the **Panel Layout - Insert/Edit Component** dialog box will be displayed.
- Specify the required parameters and click on the **OK** button. The button will be placed and the **Schematic Components (active project)** dialog box will be displayed again.
- Similarly, insert the other components of the user panel and assign balloons as discussed in the chapter. After adding all the components and balloons, the panel layout will be displayed as shown in Figure-79.

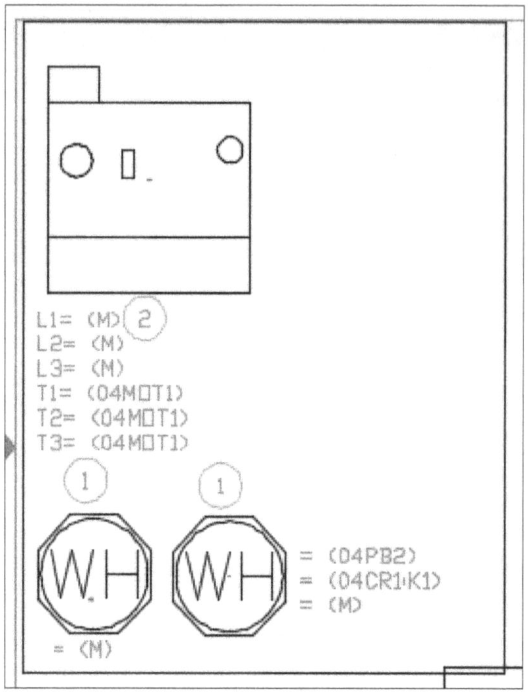

Figure-79. Panel drawing for Practical 1

PRACTICE 1

Create a panel layout for users for the schematic created in Practical 3 of chapter 8; refer to Figure-80. The panel layout is given in Figure-81.

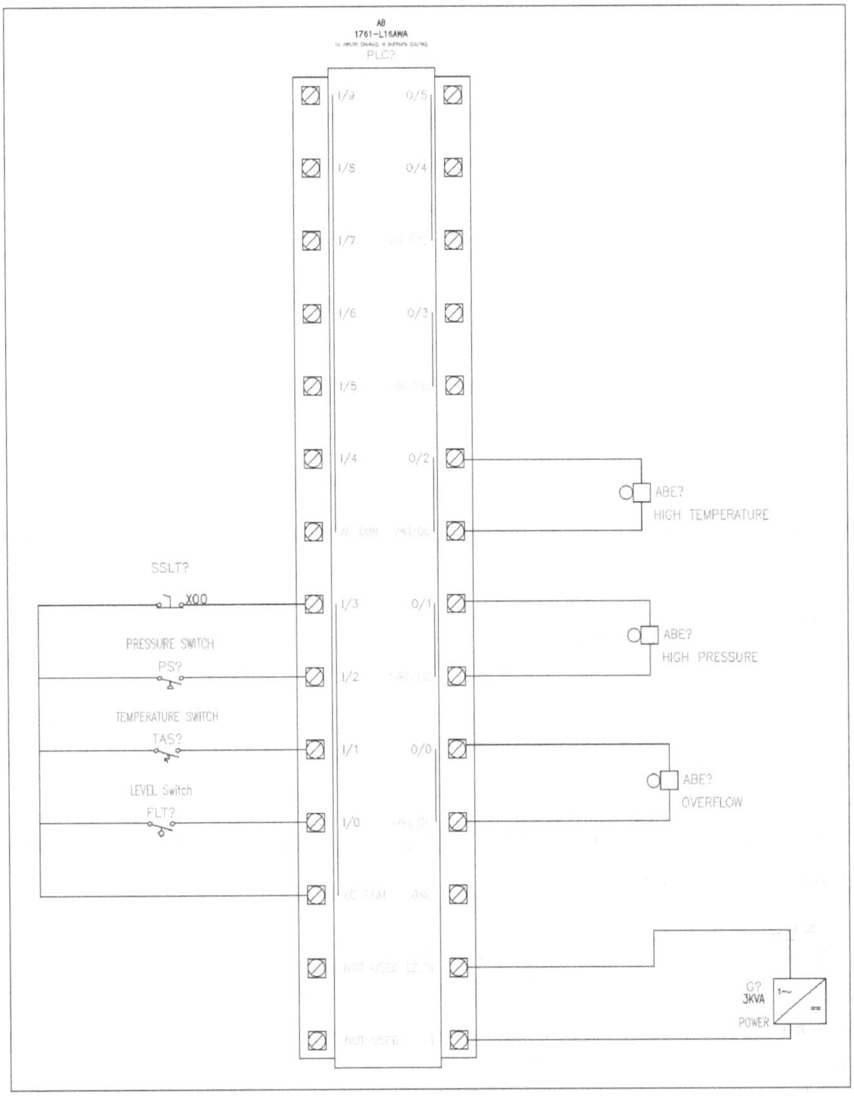

Figure-80. Practical 3

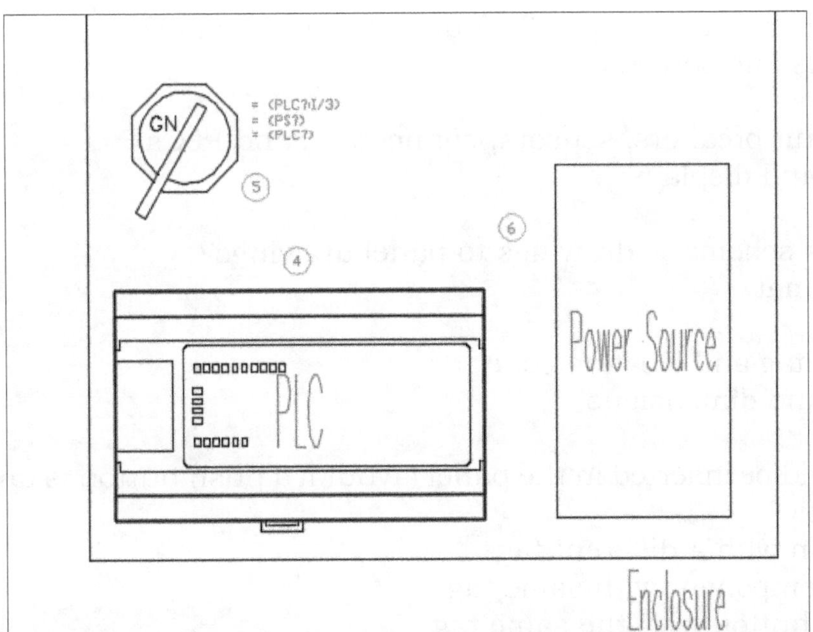

Figure-81. Panel layout for practical 3

SELF-ASSESSMENT

Q1. The **Schematic List** tool is used to insert schematic symbols used in the project, in the form of footprints in panel drawing. (T/F)

Q2. The .... tool is used to insert the components from the manufacturer's menu.

Q3. The ....tool is used to insert an already created assembly of panel in the current drawing.

Q4. The wire piece that internally connects terminals of a component together is called..........

Q5. Write down steps to insert footprint of enclosure in panel drawing.

REVIEW QUESTIONS

Q1. What is the primary purpose of creating a panel after schematic diagrams?
A. To generate reports for engineers
B. To create a 3D model of equipment
C. To control the working of equipment using electrical switches and PLCs
D. To replace old wiring systems

Q2. Which panel is generally visible to the machine operator?
A. Side panel
B. Back panel
C. User Control Panel
D. PLC Cabinet Panel

Q3. What components are typically found in a back panel?
A. Only buttons and screens
B. Only sensors
C. Relays, circuit breakers, sensors, connectors, and PLCs
D. Keyboards and displays

Q4. What links schematic drawings to panel drawings?
A. Drawing format
B. Panel size
C. Component tag and location code
D. Line types and dimensions

Q5. What should be inserted in the panel layout if a push button is tagged as -04PB2 in the schematic?
A. A new button with a different tag
B. Any other component with same tag
C. Same push button with the same tag
D. Only location code

Q6. What does the WD PNLM block do?
A. It prints panel reports
B. It stores attributes and connects panel icons with schematic symbols
C. It creates 3D models
D. It designs cable trays

Q7. In AutoCAD Electrical, which tab contains tools related to panel layout design?
A. Home
B. Insert
C. Panel
D. Annotate

Q8. What does the Icon Menu tool insert in a panel layout?
A. Wiring diagrams
B. Footprints of components
C. Text descriptions
D. Wire marks

Q9. What is the default rotation angle when placing a symbol in the drawing?
A. 90 degrees
B. 180 degrees
C. 45 degrees
D. 0 degrees

Q10. What does the Schematic List tool help with?
A. Creating catalog files
B. Importing schematic components for panel representation
C. Printing layouts
D. Managing layers

Q11. Which tool automatically defines tags while inserting panel components?
A. Manual
B. Icon Menu
C. Schematic List
D. Manufacturer Menu

Q12. What must the user be careful about when using the Manual tool?
A. Symbol color
B. Drawing scale
C. Component tags
D. Wire colors

Q13. What kind of blocks does the Manual tool allow the user to select?
A. AutoCAD Electrical symbols only
B. Only 3D blocks
C. Non-AutoCAD Electrical blocks
D. Wire blocks only

Q14. What does the Manufacturer Menu tool use to insert panel components?
A. Wire tables
B. User drawings
C. Manufacturer's library
D. Panel reports

Q15. What is the function of the Balloon tool?
A. Adjusting balloon inflation
B. Inserting table of contents
C. Assigning balloons to panel components
D. Creating exploded views

Q16. How is a balloon inserted after selecting a component?
A. Automatically appears in the drawing
B. Requires leader point and end point
C. Added from manufacturer's library
D. Generated from tags only

Q17. Which tool attaches wire annotations to selected components?
A. Manual
B. Wire Balloon
C. Wire Annotation
D. Component List

Q18. What does the Wire Annotation tool depend on for annotation format?
A. Vendor type
B. Layout size
C. Settings in Schematic --> Layout Wire Connection Annotation dialog
D. Component color

Q19. What is the purpose of the Panel Assembly tool?
A. Inserting predefined panel assemblies
B. Editing existing wires
C. Assigning tag numbers
D. Checking power ratings

Q20. What file is selected when using the Panel Assembly tool?
A. Vendor PDF
B. Equipment Datasheet
C. WBlocked panel assembly drawing
D. DWG to DXF converter

Q21. What is the primary use of the Editor tool in the Terminal Footprints panel?
A. To assign catalog codes to terminals
B. To insert new terminals in the panel layout
C. To edit terminal details in the current drawing
D. To generate terminal tables

Q22. Which dialog box appears if the number of terminals exceeds the number of wires?
A. Terminal Strip Selection
B. Edit Terminal
C. Defined Terminal Wiring Constraints Exceeded
D. Terminal Block Properties

Q23. What color highlights the selected terminal in the Terminal Strip Editor dialog box?
A. Blue
B. Yellow
C. Green
D. Red

Q24. Which button is used to copy terminal block properties?
A. Edit Terminal Block Properties
B. Copy Terminal Block Properties
C. Paste Terminal Block Properties
D. Duplicate Terminal Block Properties

Q25. What does the Reassign Terminal dialog box allow you to modify?
A. Terminal color and size
B. Manufacturer data
C. Installation code, Location code, Terminal strip
D. Jumper settings

Q26. Which tool allows you to move a terminal within the table?
A. Reassign Terminal
B. Move Terminal
C. Edit Terminal
D. Update Terminal

Q27. Which area of the Terminal Strip Editor dialog box includes the Insert Spare Terminal button?
A. Properties
B. Terminal
C. Spare
D. Destinations

Q28. What does the Insert Accessory button help you do?
A. Add catalog codes
B. Insert new terminals
C. Insert accessories like spare terminals
D. Copy terminal data

Q29. What action does the Delete Spare Terminals/Accessories button perform?
A. Deletes active terminal
B. Deletes associated cable
C. Deletes accessories and spare terminals
D. Deletes terminal block properties

Q30. Which tab is used to assign catalog data to a terminal?
A. Properties
B. Catalog Code Assignment
C. Spare
D. Multilevel

Q31. What does the Layout Preview tab help you do?
A. Assign cable data
B. Generate terminal reports
C. Check and modify graphical representation of terminal strip
D. Access installation settings

Q32. What is the purpose of the Table Generator tool?
A. Insert new terminals
B. Create wiring connections
C. Insert a table with terminal details in the drawing
D. Assign jumper settings

Q33. Which key is used for selecting multiple terminals in the Table Generator dialog box?
A. ALT
B. SHIFT
C. CTRL
D. TAB

Q34. Which button is clicked to create a new drawing file for terminal table output?
A. Insert
B. Save
C. Browse
D. Done

Q35. What tool is used to insert terminals already used in schematic drawings?
A. Insert Terminal (Manual)
B. Insert Terminal (Schematic List)
C. Insert Accessory
D. Insert Spare Terminal

Q36. What is the effect of clicking the Mark Existing button?
A. Removes unused terminals
B. Highlights unmarked terminals
C. Marks terminals already inserted in panel drawing
D. Inserts all available terminals

Q37. What happens when multiple terminals are inserted together?
A. They are grouped as a single block
B. The spacing for insertion dialog box appears
C. Only the first terminal is inserted
D. Terminal names are renamed automatically

Q38. Which radio button allows you to specify a uniform space between terminals during insertion?
A. Use custom spacing
B. Prompt for each location
C. Use uniform spacing
D. Specify manually

Q39. Which method allows you to insert terminal footprints based on an existing one?
A. Browse
B. Draw Shapes
C. Use Generic Marker
D. Pick "just like" footprint

Q40. What is the final step after selecting and placing a terminal using manual method?
A. Exit the tool
B. Save as new drawing
C. Click OK in the Terminal Insert/Edit dialog box
D. Open terminal catalog

Q41. What is the purpose of the "Insert Terminal (Schematic List)" tool?
A. To draw terminal blocks manually
B. To insert terminals from previous projects
C. To insert terminal footprints already used in the current project
D. To delete terminal footprints from the project

Q42. Which dialog box allows arranging terminals and specifying spacing when inserting multiple terminals?
A. Terminal Strip Editor
B. Spacing for Footprint Insertion
C. Terminal Block Properties
D. Edit Terminal

Q43. What does the "Mark Existing" button do in the Schematic Terminals dialog box?
A. Deletes existing terminals
B. Highlights unplaced terminals
C. Marks already inserted terminals with a cross (x)
D. Displays layout view of terminals

Q44. Which radio button should you select to create a terminal with number unrelated to wire number?
A. Terminal number = wire number
B. Wire number = terminal number
C. Terminal number independent from wire number
D. Custom terminal from wire

Q45. What shape can you NOT draw using the Draw Shapes option?
A. Rectangle
B. Circle
C. Ellipse
D. Polygon

Q46. What happens when you select the "Pick 'just like' footprint" option?
A. The terminal is inserted without editing
B. You select a footprint to copy its shape
C. The program auto-generates a footprint
D. You browse the catalog

Q47. Which button is used to select and insert a saved footprint from local storage?
A. Draw
B. Pick
C. Generic Marker
D. Browse

Q48. In the Pick dialog box, what happens after selecting a terminal block drawing file?
A. It gets saved to the catalog
B. It is deleted from the database
C. It gets attached to the cursor for placement
D. It is copied to the clipboard

Q49. What must be specified in the Panel Layout-Terminal Insert/Edit dialog box after placing a shape or footprint?
A. Catalog code only
B. Drawing name
C. Terminal orientation and properties
D. File location

Q50. Which option in Insert Terminals (Manual) allows the use of a symbol that doesn't belong to AutoCAD Electrical?
A. Use Generic Marker Only
B. Browse
C. Pick "just like" footprint
D. Pick non-AutoCAD Electrical block

Q51. What is the primary function of the Resequence Item Numbers tool?
A. To delete items from a drawing
B. To renumber all items in a drawing or project
C. To export item numbers to Excel
D. To lock item numbers for editing

Q52. What must be specified in the Start edit box in the Resequence Item Numbers dialog box?
A. Ending number
B. Total items
C. Desired starting number
D. Layer name

Q53. What does selecting the "Process all" checkbox do?
A. Deletes unused items
B. Updates only visible items
C. Applies re-sequencing to all items in the panel drawing
D. Ignores selected items

Q54. What is the use of Move up and Move down buttons in the Resequence Item Numbers dialog?
A. To zoom in and out of the list
B. To add new items
C. To change sequence of items
D. To lock item numbers

Q55. What does the Copy Location Code tool allow you to do?
A. Move components to a new location
B. Assign a location code from one component to others
C. Rename location folders
D. Export component data

Q56. What is the first step in using the Copy Location Code tool?
A. Select target components
B. Type the location code
C. Click on Pick Master to choose the source component
D. Press ENTER

Q57. What happens after you click OK in the Copy Location Code dialog box?
A. The tool closes without action
B. You are asked to select target components
C. Code is auto-applied to all components
D. The layer is renamed

Q58. What does the Copy Assembly tool allow?
A. Copy the drawing
B. Copy one or more components of a panel assembly
C. Copy an entire panel layout
D. Copy balloon settings

Q59. What is required before placing copied components with Copy Assembly?
A. Export the assembly
B. Assign new item numbers
C. Specify a base point
D. Set panel configuration

Q60. What happens after you specify the base point using Copy Assembly?
A. Component attaches to cursor for placement
B. Component auto-places at origin
C. Panel is reset
D. Dialog box reappears

Q61. What are some uses of tools found in the Other Tools panel?
A. Plotting layouts
B. Managing external files
C. Performing miscellaneous panel footprint tasks
D. Uploading projects

Q62. What can be defined using the Configuration tool?
A. Color settings for AutoCAD
B. Default parameters for placing footprints
C. Auto-save frequency
D. External data link settings

Q63. Which edit box sets the balloon numbering start value in Panel Drawing Configuration?
A. X Distance
B. Start
C. Scale
D. Mount Code

Q64. What is set using the Setup button in the Balloon area?
A. Location code
B. Shape, size, and leader style for balloons
C. Background color
D. Project name

Q65. What are X Distance and Y Distance used for in Configuration settings?
A. Setting up 3D view
B. Panel layout margins
C. Spacing for multiple balloon inserts
D. Exporting balloon positions

Q66. What does the Scale edit box define in the Configuration dialog box?
A. Print scale
B. Zoom level
C. Scale at which footprint will be inserted
D. Balloon spacing

Q67. What is the function of the Rename Layers tool?
A. Rename files in drawing
B. Rename item numbers
C. Change names of layers used in panel drawings
D. Rename components

Q68. What is the function of the Edit button in Rename Layers dialog box?
A. Change panel layout
B. Reset all settings
C. Edit the selected layer name
D. Open the configuration tool

Q69. What does the Update Layer tool do?
A. Prints the current layer structure
B. Updates component properties based on configuration changes
C. Deletes unused layers
D. Saves drawing to a new location

Q70. What is the purpose of the Make Xdata Visible tool?
A. Convert block to Xref
B. Hide attributes from drawing
C. Make invisible external data visible for selected components
D. Replace attribute values

Q71. What is the first step when using Make Xdata Visible tool?
A. Open the Balloon Setup
B. Select the footprint to modify
C. Export the layer configuration
D. Lock the footprint

Q72. After selecting an attribute with Make Xdata Visible, what is the next step?
A. Save drawing
B. Click on Setup
C. Click Insert to attach it to the cursor
D. Type new layer name

Q73. What is the final action in the Make Xdata Visible workflow?
A. Click Done to exit
B. Reopen the dialog box
C. Save file as a block
D. Assign new component number

FOR STUDENT NOTES

Answer to Self-Assessment:
Ans1. T, Ans2. Manufacturer Menu, Ans3. Panel Assembly, Ans4. Jumper

Chapter 10

Reports

Topics Covered

The major topics covered in this chapter are:

- ***Introduction***
- ***Reports***
- ***Missing Data Catalog***
- ***Electrical Audit***
- ***Drawing Audit***
- ***Dynamic Table editing***

INTRODUCTION

After schematics and panel drawing, the next important step is to generate different reports. The reports are the tabulated representation of components being used in the schematic diagrams and panel layouts. The tools to create reports are available in the **Reports** tab of the **Ribbon**; refer to Figure-1. Before we start using the tools to create reports, it is important to note that the reports created here are totally dependent on the schematic diagrams and panel layouts created earlier. The tables generally have Names of components, Manufacturer details of components, quantity of components, and so on. We can include other details also like location code, installation code, and so on in the table. The tools to create tables are discussed next.

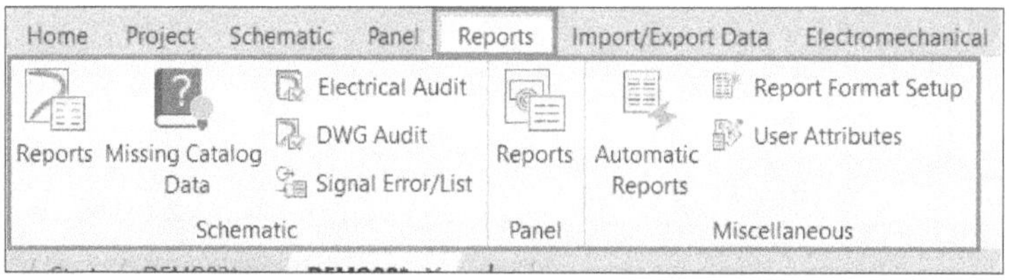

Figure-1. Reports panel

REPORTS (SCHEMATIC)

The **Reports** tool in the **Schematic** panel is used to create reports of schematic components in the project. The procedure to use this tool is given next.

- Click on the **Reports** tool from the **Schematic** panel in the **Reports** tab of **Ribbon**. The **Schematic Reports** dialog box will be displayed; refer to Figure-2.

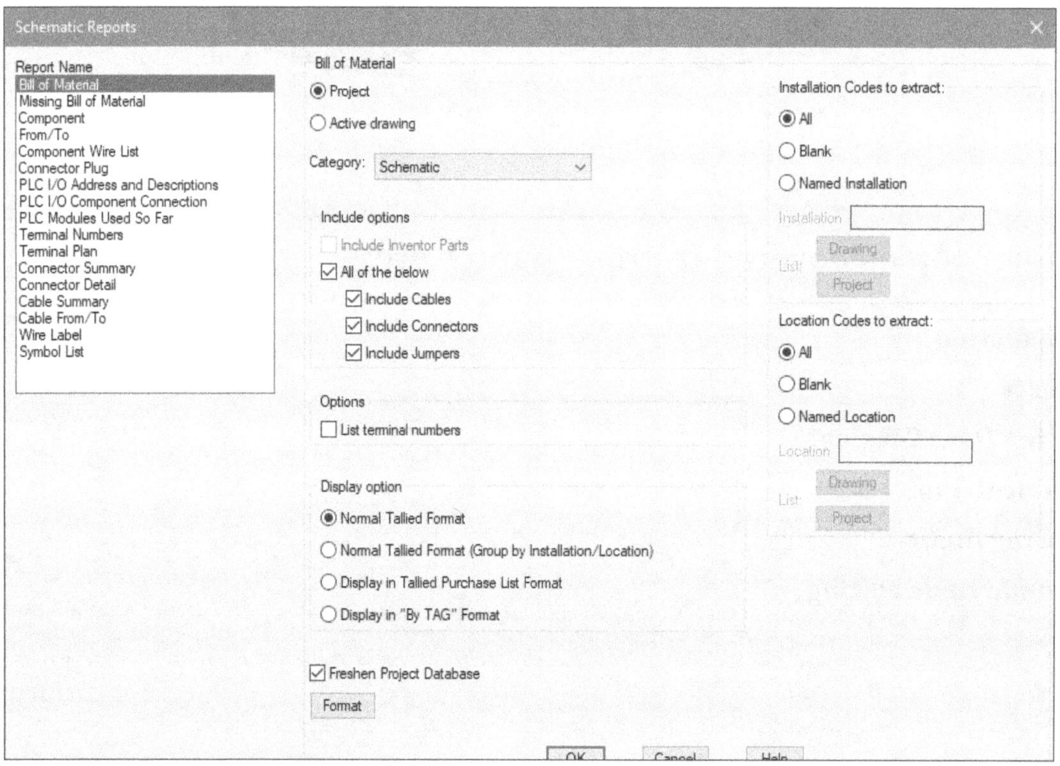

Figure-2. Schematic Reports dialog box

- We have a big list of report types in the **Report Name** selection box at the left; refer to Figure-3. Select desired report type from the selection box, the options in the dialog box will be modified accordingly.

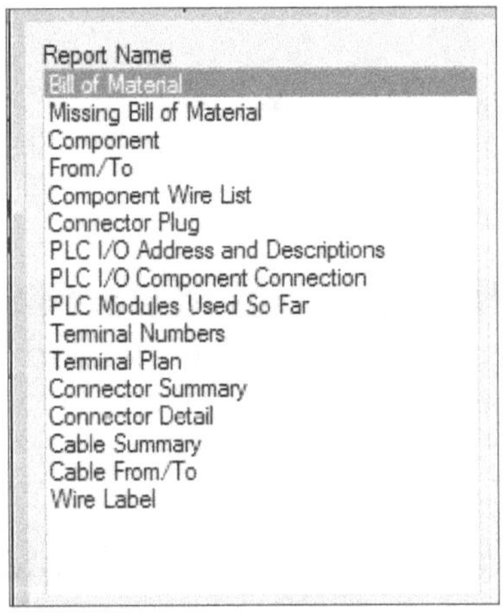

Figure-3. Report Name box

The list of report types available in the dialog box is given next:

Bill of Materials reports

The Bill of Material reports only components with assigned BOM information. These reports provide the following BOM-related features:

- Extract BOM reports on demand, active drawing, or project-wide
- Extract BOM reports on a per-location basis
- Change BOM report format
- Output BOM reports to ASCII report file
- Export BOM data to a spreadsheet or database program
- Insert BOM as a table on an AutoCAD Electrical drawing
- List parent or stand-alone components without catalog information

Procedure to create:

- Select the **Bill of Material** option from the **Report Name** selection box. The options related to Bill of Materials will be displayed in the dialog box.
- Select the **Project** radio button if you want to include the drawings from the current project or select the **Active Drawing** radio button to create Bill of Materials based on the current active drawing.
- Click in the **Category** drop-down. The list of various options will be displayed; refer to Figure-4.

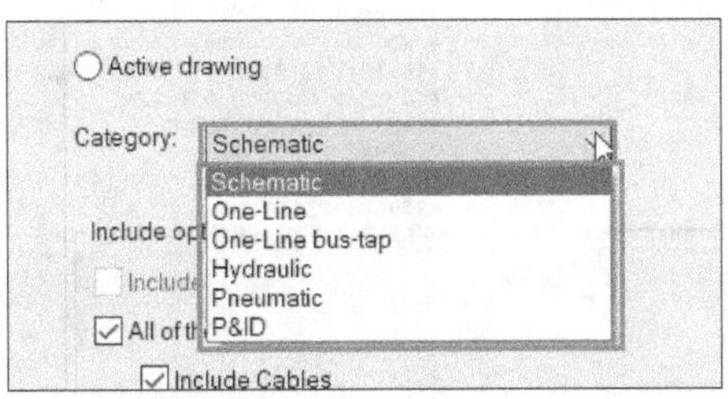

Figure-4. Category drop-down

- Select desired option, the components related to the selected option will be included in the Bill of Materials report.
- Select the check boxes in the **Include options** area of the dialog box to include the corresponding components of the circuit in the report. You can select **All the below** check box to include all the items displayed below this check box.

- Select the **List terminal numbers** check box to include the terminal numbers in the report.
- The four radio buttons in the **Display option** area of the dialog box are used to specify sorting of the components in the table. The effect of these radio buttons is discussed next.

Normal Tallied Format

Identical components or assemblies are tallied and reported as single line items (example: Red push-button operator 800EP-F4 with 800E-A3L latch and two 800E-3X10 N.O. contact blocks).

Normal Tallied Format (Group by Installation/Location)

Identical components or assemblies with the same installation/location codes are tallied and reported as single line items.

Display in Tallied Purchase List Format

Each part becomes its own line item (example: no longer any sub-assembly items) and each is tallied across all component types. For example, all 800E-3X10 N.O. contact blocks for all components are reported as a single line item.

Display in "By TAG" Format

All instances of a given component-ID or terminal tag are processed together and then reported as a single entry.

- Select desired radio button.
- Click on the **Format** button and select desired format for the report.
- Select **All** radio buttons from **Installation Codes to extract** and **Location Codes to extract** areas to include installation codes and location codes in the report.
- Click on the **OK** button from the dialog box. The **Report Generator** dialog box will be displayed; refer to Figure-5.

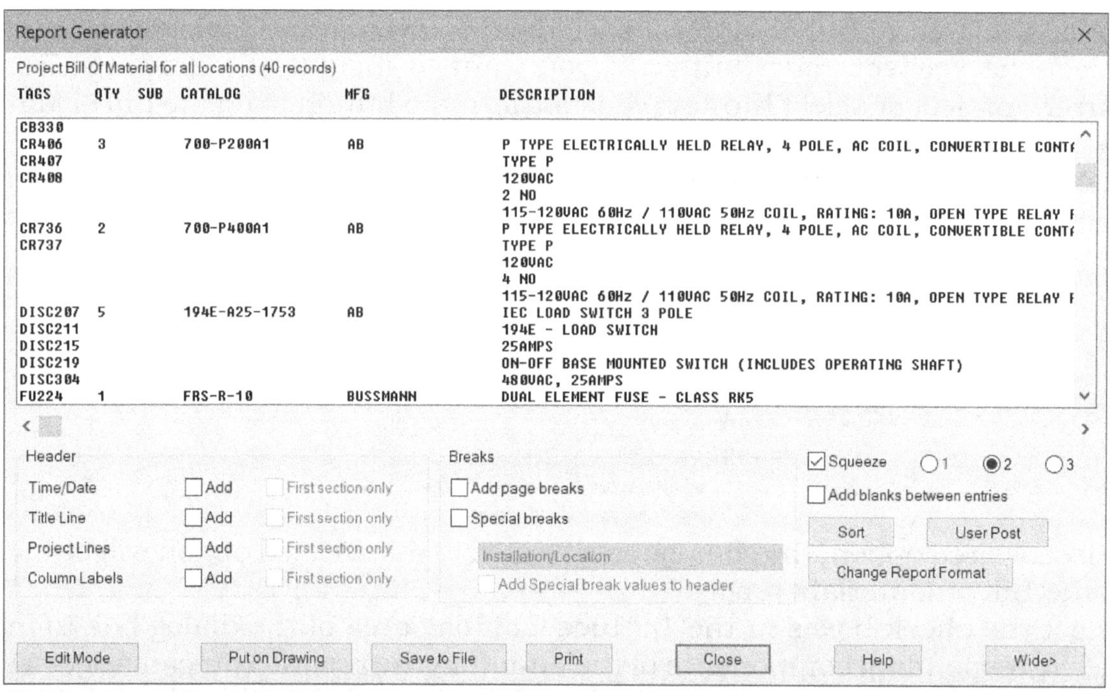

Figure-5. Report Generator dialog box

- Select the **Add** check box for desired data header from the **Header** area of the dialog box.
- Click on the **Edit Mode** button to edit the table data. The **Edit Report** dialog box will be displayed; refer to Figure-6.

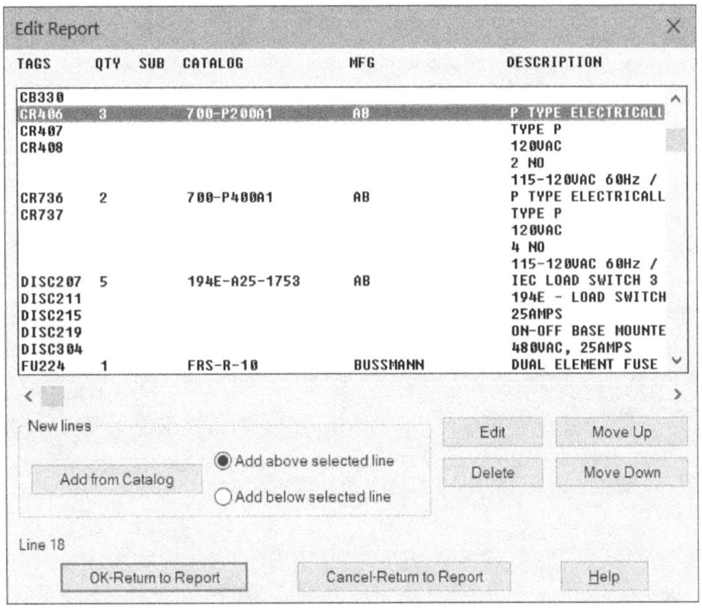

Figure-6. Edit Report dialog box

- Double-click on desired row, the **Edit Line** dialog box will be displayed; refer to Figure-7.

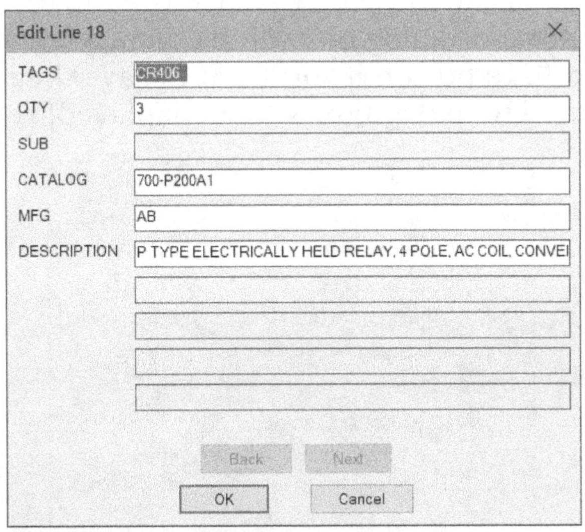

Figure-7. Edit Line dialog box

- Specify desired values in the edit boxes of this dialog box and click on the **OK** button to exit the dialog box.
- You can add a new line by using the **Add from Catalog** button and use the catalog data.
- Click on the **OK-Return to Report** button to return the report creation.
- Now, we are ready to insert the report in the drawing. Click on the **Put on Drawing** button from the **Report Generator** dialog box. The **Table Generation Setup** dialog box will be displayed; refer to Figure-8.

- The options in this dialog box are used to modify the details of the table. Specify desired parameters and click on the **OK** button from the dialog box. A box of table boundary will get attached to the cursor.

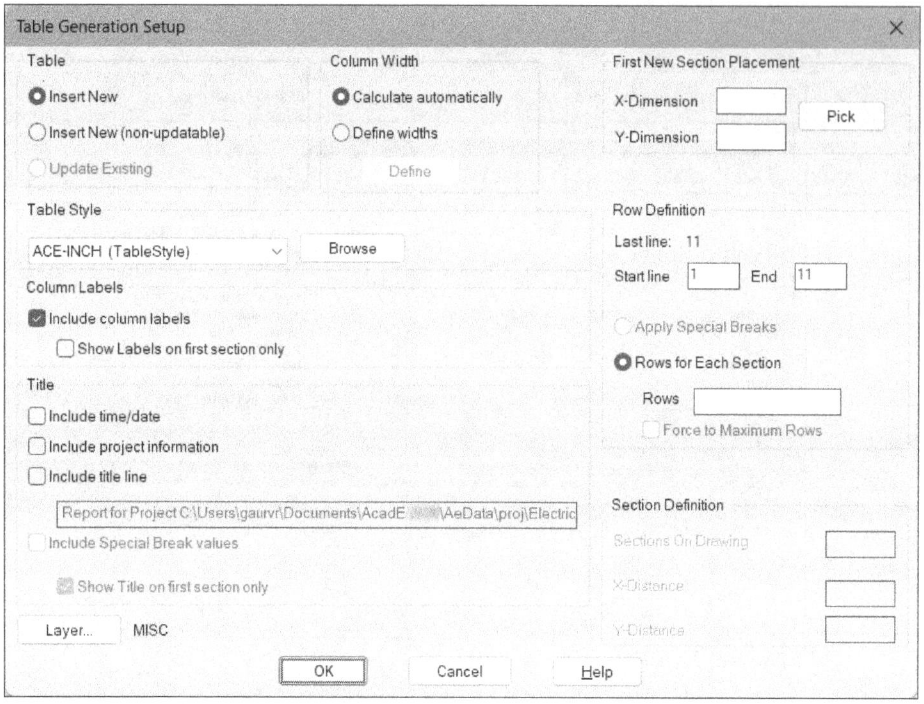

Figure-8. Table Generation Setup dialog box

- Click at desired place in the drawing to place the table. The table will be placed and the **Report Generator** dialog box will be displayed again.
- Click on the **Save to File** button if you want to save the report in a separate file. The **Save Report to File** dialog box will be displayed; refer to Figure-9.

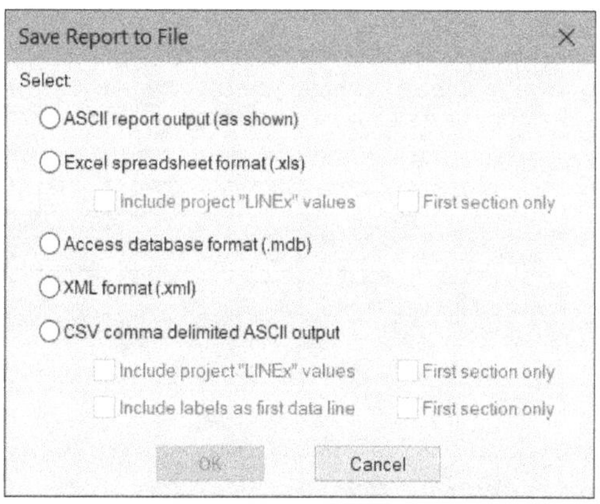

Figure-9. Save Report to File dialog box

- Select desired radio button to specify format for the file. (**Excel spreadsheet format(.xls)** in our case)
- Click on the **OK** button from the dialog box. The **Select file for report** dialog box will be displayed; refer to Figure-10.

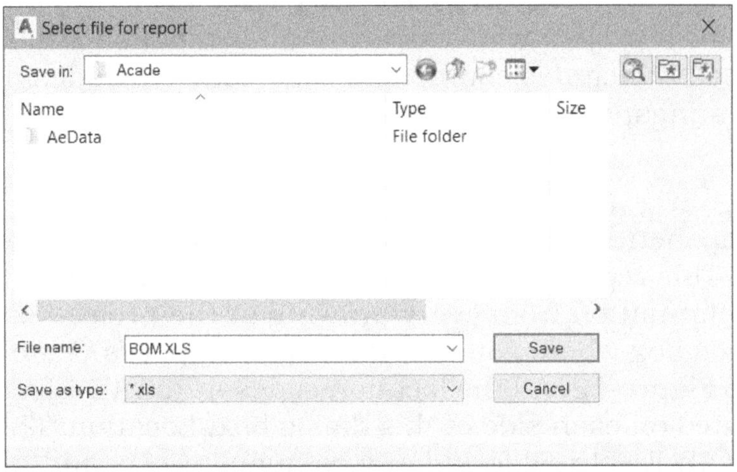

Figure-10. Select file for report dialog box

- Browse to desired location in your local storage drive and save the file. On selecting the **Save** button, the **Optional Script File** dialog box will be displayed; refer to Figure-11.

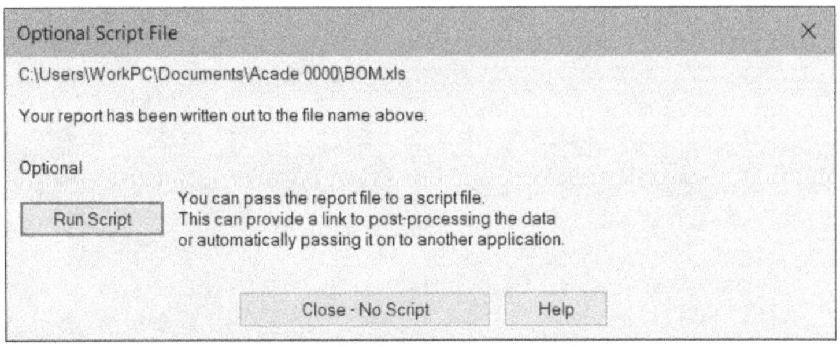

Figure-11. Optional Script File dialog box

- Click on the **Run Script** button to generate the script file containing report or click on the **Close-No Script** button to exit. If you generate the script file then it help the other programs to link the table data to its generation process.
- Click on the **Close** button from the **Report Generator** dialog box to exit. Figure-12 shows a table inserted in the drawing.

TAGS	QTY	SUB	CATALOG	MFG	DESCRIPTION
04CR1	1		700-P200A1	AB	P TYPE ELECTRICALLY HELD RELAY, 4 POLE, AC COIL, CONVERTIBLE CONTACTS TYPE P 120VAC 2 NO 115-120VAC 60Hz / 110VAC 50Hz COIL, RATING: 10A, OPEN TYPE RELAY RAIL MOUNT
04MOT1	1		CM111-FC1F518GSKCA	AB	SEVERE DUTY SE AC MOTOR, ENC:TEFC, C FACE FOOT MOUNTED, CM111 AC MOTOR 1.5HP 208-230/460VAC 3 PHASE, 60Hz, SYNCH. RPM: 1800, FRAME: 56C, INVERTER: CT 4:1
04OL1	1		193-A1A1	AB	OVERLOAD RELAY MANUAL RESET, CLASS 10 MANUAL RESET 0.1-0.32AMPS MOUNTS DIRECTLY TO CONTACTORS 100 AND 104 IEC SOLID STATE RELAY SMP-1
04OL2	1		1492-A1B1	AB	SINGLE PHASE, 80 A, 57 DEVICES PER METER ACCESSORIES 57 DEVICES PER METER 100 A)
04PB1 04PB2	2	*1	800B-EAC	AB	800B MAINTAINED PUSH BUTTON - RECTANGULAR BEZEL 16mm w/800B-ARP
		*1	800B-ARP	AB	REPLACEMENT MOUNTING RING ACCESSORY

Figure-12. Report of BOM

Above procedure is similarly applicable to other type reports discussed next.

Component report

This report performs a project-wide extract of all components found on your wiring diagram set. This data includes component tags, location codes, location reference, description text, ratings, catalog information, and block names.

Wire From/To report

If you marked components and/or terminals with location codes, you can make good use of this report. This report first extracts component, terminal, location code, and wire connection information from every drawing in the project set. Then it displays a location list dialog box where you can make your report's "from" and "to" location selections; refer to Figure-13. All the location codes AutoCAD Electrical found on the drawing set are listed on each side of this dialog box. Location "(??)" is also included in the list if AutoCAD Electrical found any component or stand-alone terminal that did not have an assigned location code. Add desired location codes and click on the **OK** button. The **Report Generator** dialog box will be displayed which has been discussed earlier.

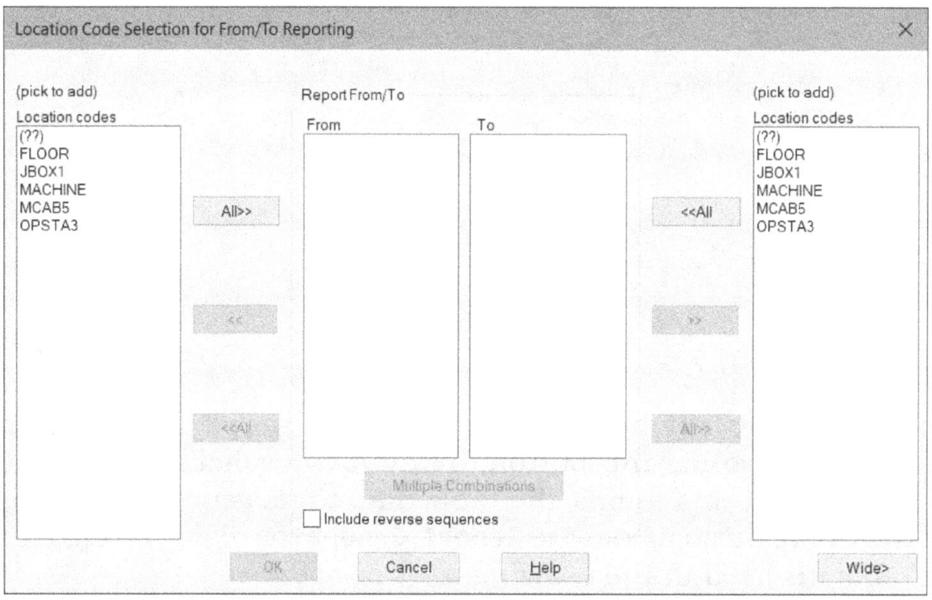

Figure-13. Location Code Selection for From To Reporting

Component Wire List report

This report extracts the component wire connection data and displays it in a dialog box. Each entry shows a connection to a component, the wire number, component tag name, terminal pin number, component location code (if present), and the layer that the connected wire is on.

Connector Plug report

This report extracts plug/jack connection reports and optionally generates pin charts. Since each wire tied to each connector pin displays in the report, each pin has two entries. One entry is for the 'in' wire and the other is for the 'out' wire. To create a more useful report, select the PIN chart 'on' radio button at the bottom of the **Report Generator** dialog box. Make sure the Remove duplicated pin numbers toggle is checked and click **OK**. It reformats the report so each pin is listed only once.

PLC I/O Address and Description report

This report lists each PLC module and its beginning and ending I/O address numbers. It scans your drawing set and returns all individual I/O connection points it finds. It includes up to five lines of description text and the connected wire number for each I/O point.

PLC I/O Component Connection report

This report scans the selected drawings and returns information about any components connected to PLC I/O points. Data for the report starts at each wire connection point and follows the connected wire. The first terminal symbol, fuse symbol, or connector symbol that is hit reports in the column marked with default "TERMTAG" label. The first schematic component reports in the column with the "CMPTAG" label. If there are multiple instances, the one closest to the PLC I/O point is the one that shows up in the report column.

PLC Modules Used So Far report

For this report, AutoCAD Electrical quickly scans the wiring diagram set. It returns in a few moments and displays the I/O modules it finds. Each entry shows the beginning and ending address of the module.

Terminal Numbers report

This project-wide, stand-alone report lists all instances of terminals. Each entry includes information tied directly to the terminal such as terminal number, terminal strip number, and location code.

Terminal Plan report

This project-wide, stand-alone report does a wire network extraction. It takes longer to generate, but the report includes wire number and wire layer name information.

Connector Summary report

This report lists a single line for each connector tag along with pins used, maximum pins allowed (if the parent carries the PINLIST information), and a list of any repeated pin numbers used. You can run this report across the project or for a single component.

Connector Details report

This report extracts the component wire connection data and displays it in a dialog box. Each entry shows a connection to a component, the wire number, component tag name, terminal pin number, component location code (if present), and the layer that the connected wire is on.

Cable Summary report

This project-wide cable conductor report gives a report listing all the cable marker tags (parent tags) found.

Cable From/To report

This project-wide cable conductor report lists the "from / to" for each cable conductor. It also lists the parent cable number of the conductor, conductor color code, and wire number (if present).

Wire Label report

This report lists wire markers/labels and can be used to create physical wire or cable labels.

MISSING CATALOG DATA

The **Missing Catalog Data** tool is used to display a diamond mark on the components in the schematic diagram and panel layout that do not have catalog data attached. Catalog data defines the actual part number and manufacturer details using which part can be ordered physically. The procedure to use this tool is given next.

- Click on the **Missing Catalog Data** tool from the **Schematic** panel. The **Show Missing Catalog Assignments** dialog box will be displayed; refer to Figure-14.

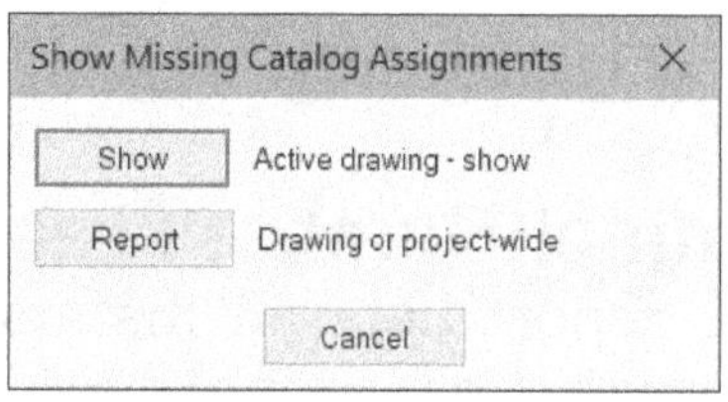

Figure-14. Show Missing Catalog Assignments dialog box

- Click on the **Show** button to highlighted the components not having catalog data attached; refer to Figure-15.

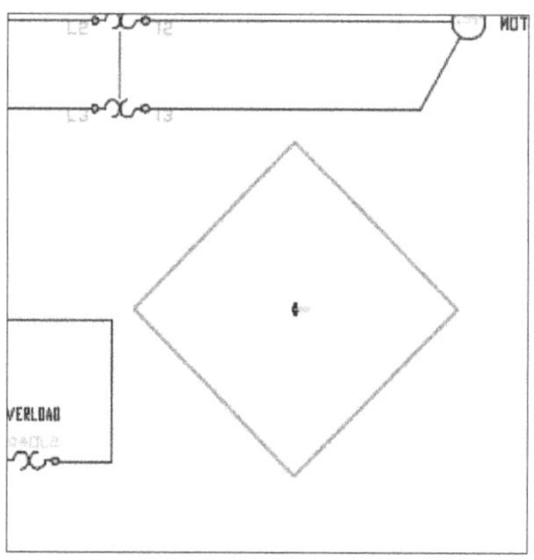

Figure-15. Component not having catalog data attached

- Click on the **Report** button from the **Show Missing Catalog Assignments** dialog box to generate the report. The **Schematic Reports** dialog box will be displayed.
- Click on the **OK** button from the dialog box. The **Select Drawings to Process** dialog box will be displayed.
- Select the **Do All** button to include all the drawings of the project and click on the **OK** button from the dialog box. The **QSAVE** dialog box will be displayed.
- Click on the **OK** button from the dialog box. The **Report Generator** dialog box will be displayed with the missing entries; refer to Figure-16.
- Click on the **Put on Drawing** button or **Save to File** button as discussed earlier to extract the report. You can also print the table by using the **Print** button from the dialog box.

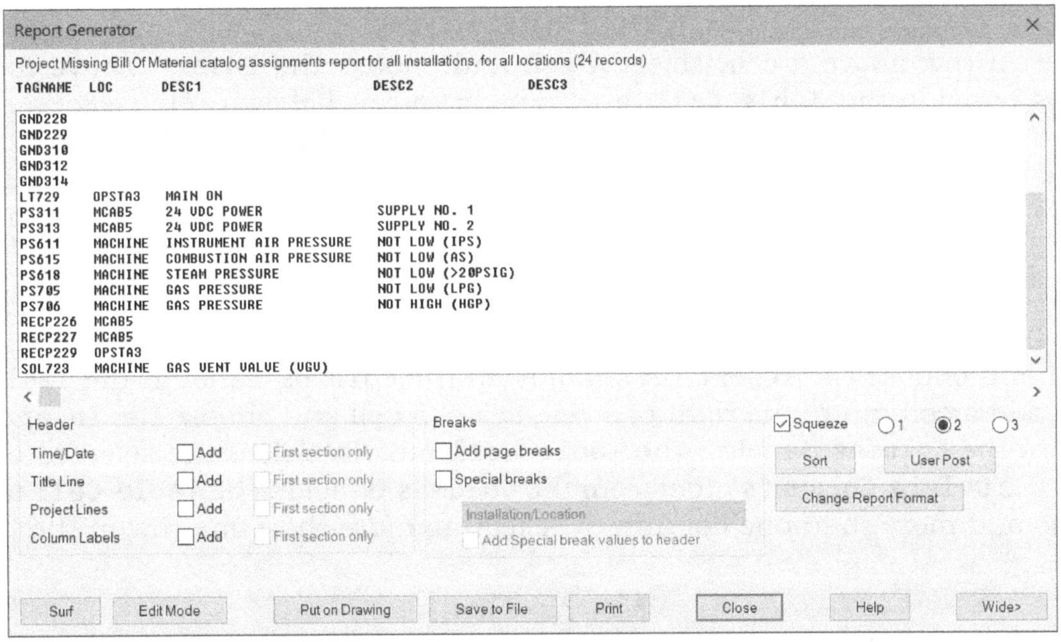

Figure-16. Report Generator dialog box with missing catalog data components

DYNAMIC EDITING OF REPORTS IN DRAWING

You can edit any entry of the reports inserted in the drawing by using the dynamic editing facility of the AutoCAD Electrical. The procedure is given next.

- Click in any cell of the table. The **Table Cell** tab will be displayed; refer to Figure-17.

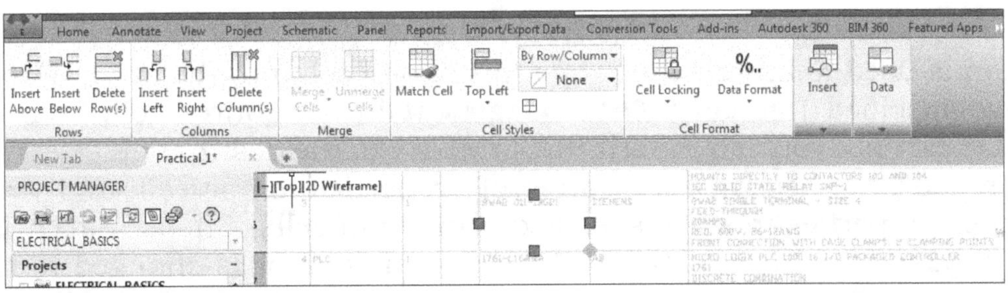

Figure-17. Table Cell tab in Ribbon

- Type the new value for the selected cell if you want to change the value.
- Using the tools in the **Table Cell** tab you can change the format of the table.

The tools in the **Table Cell** tab are discussed next.

MODIFYING TABLES

Select any cell in the table; the **Table Cell** tab will be added to the **Ribbon**, as shown in Figure-18. The options in the **Table Cell** tab are used to modify the table, insert block, add formulas, and perform other operations.

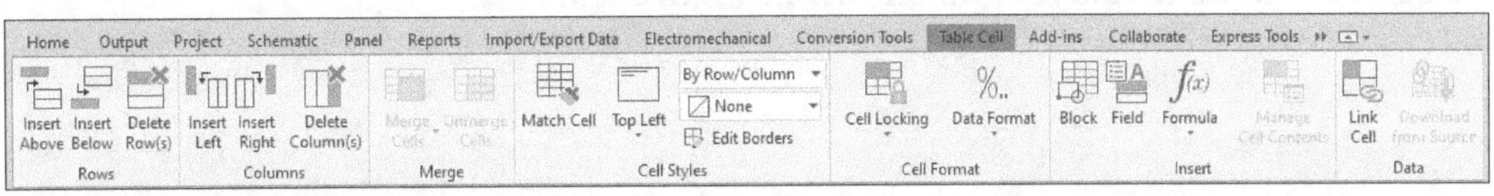

Figure-18. The Table Cell tab added to the Ribbon

Modifying Rows

To insert a row above a cell, select a cell and choose the **Insert Above** tool from the **Rows** panel in the **Table Cell** tab. To insert a row below a cell, select a cell and choose the Insert Below tool from the Rows panel in the **Table Cell** tab. To delete the selected row, choose the **Delete Row(s)** tool from the **Rows** panel in the **Table Cell** tab. You can also add more than one row by selecting more than one row in the table.

Modifying Columns

This option is used to modify columns. To add a column to the left of a cell, select the cell and choose the **Insert Left** tool from the **Columns** panel in the **Table Cell** tab. To add a column to the right of a cell, select a cell and choose the **Insert Right** tool from the **Columns** panel in the **Table Cell** tab. To delete the selected column, choose the **Delete Column(s)** tool from the **Columns** panel in the **Table Cell** tab. You can also add more than one column by selecting more than one row in the table.

Merge Cells

This button is used to merge cells. Choose this button; the **Merge Cells** drop-down is displayed. There are three options available in the drop-down. Select multiple cells using the **SHIFT** key and then choose **Merge All** from the drop-down to merge all the selected cells. To merge all the cells in the row of the selected cell, choose the **Merge By Row** button from the drop-down. Similarly, to merge all cells in the column of the selected cell, select the **Merge By Column** tool from the drop-down. You can also divide the merged cells by choosing the Unmerge Cells button from the Merge panel.

Match Cells

This button is used to inherit the properties of one cell into the other. For example, if you have specified **Top Left Cell Alignment** in the source cell, then using the **Match Cells** button, you can inherit this property to the destination cell. This option is useful if you have assigned a number of properties to one cell, and you want to inherit these properties in some specified number of cells. Choose **Match Cell** from the **Cell Styles** panel in the **Table Cell** tab; the cursor is changed to the match properties cursor and you are prompted to choose the destination cell. Choose the cells to which you want the properties to be inherited and then press **ENTER**.

Table Cell Styles

This drop-down list displays the preexisting cell styles or options to modify the existing ones. Select desired cell style to be assigned to the selected cell. The **Cell Styles** drop-down also has the options to create a new cell style or manage the existing ones.

Edit Borders

Choose the **Edit Borders** button from the **Table Cell** tab; the **Cell Border Properties** dialog box will be displayed, as shown in Figure-19.

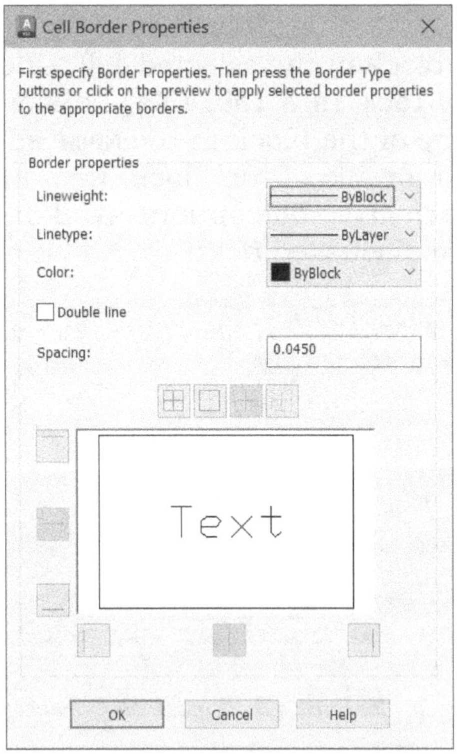

Figure-19. The Cell Border Properties dialog box

Text Alignment

The down arrow with the **Middle-Center** button of the **Cell Styles** panel in the **Ribbon** is used to align the text written in cells with respect to the cell boundary. Choose this button; the **Text Alignment** drop-down will be displayed. Select desired text alignment from this drop-down; the text of the selected cell will get aligned accordingly.

Locking

This button is used to lock the cells so that they cannot be edited by accident. Select a cell and choose **Cell Locking** from the **Cell Format** panel in the **Table Cell** tab; the **Cell Locking** drop-down list will be displayed. Four options are available in the drop-down list. The **Unlocked** option is chosen by default. Choose the **Content Locked** option to prevent the modification in the content of the text, but you can modify the formatting of the text. Choose the **Format Locked** option to prevent the modification in the formatting of the text, but in this case, you can modify the content of the text. Select the **Content and Format Locked** option to prevent the modification of both formatting and content of the text.

Data Format

The display of the text in the cell depends on the format type selected. Choose **Data Format** to change the format of the text in the cell. On doing so, the **Data Format** drop-down list is displayed. Choose the required format from it. You can also select the **Custom Table Cell Format** option and choose the required format from the **Data type** list box in the **Table Cell Format** dialog box. Click on the **OK** button to apply changes.

Block

This tool is used to insert a block in the selected cell. Choose the **Block** tool from the **Insert** panel; the **Insert a Block in a Table Cell** dialog box will be displayed, refer to Figure-20. Enter the name of the block in the **Name** edit box or choose the **Browse** button to locate the destination file of the block. If you have browsed the file path, it will be displayed in the **Path** area. The options available in the **Insert a Block in a Table Cell** dialog box are discussed next.

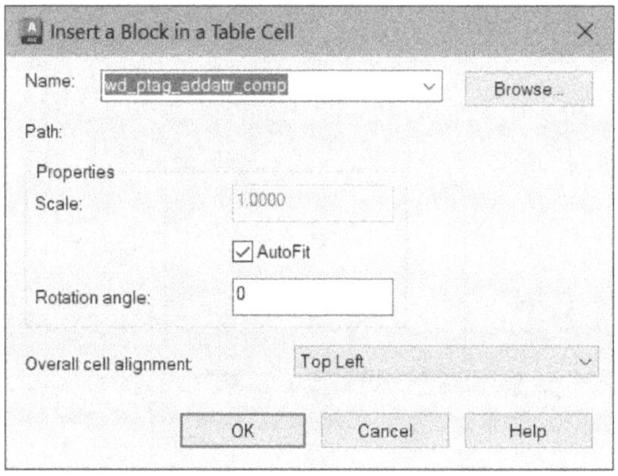

*Figure-20. The **Insert a Block in a Table Cell** dialog box*

Properties Area

The options in this area are discussed next.

Scale. This edit box is used to specify the scale of the block. By default, this edit box is not available because the **AutoFit** check box is selected below this drop-down list. Selecting the **AutoFit** check box ensures that the block is scaled such that it fits in the selected cell.

Rotation angle. The **Rotation angle** edit box is used to specify the angle by which the block will be rotated before being placed in the cell.

Overall cell alignment. This drop-down list is used to define the block alignment in the selected cell.

Field

You can also insert a field in the cell. The field contains the data that is associative to the property that defines the field. For example, you can insert a field that has the name of the author of the current drawing. If you have already defined the author of the current drawing in the **Drawing Properties** dialog box, it will automatically be displayed in the field. If you modify the author name and update the field, the changes will automatically be made in the text. When you choose the **Field** button from the **Insert** panel, the **Field** dialog box will be displayed. You can select the field to be added from the **Field names list** box and select the format of the field from the **Format list box**. Choose **OK** after selecting the field and format. If the data in the selected field is already defined, it will be displayed in the Text window. If not, the field will display dashes (----).

Formula

Choose **Formula** from the **Insert** panel; a drop-down list is displayed. This drop-down list contains the formulas that can be applied to a given cell. The formula calculates the values for that cell using the values of other cell. In a table, the columns are named with letters (like A, B, C, ...) and rows are named with numbers (like 1, 2, 3, ...). The **TABLEINDICATOR** system variable controls the display of column letters and row numbers. By default, the **TABLEINDICATOR** system variable is set to 1, which means the row numbers and column letters will be displayed when Text Editor is invoked. Set the system variable to zero to turn off the visibility of row numbers and column letters. The nomenclature of cells is done using the column letters and row numbers. For example, the cell corresponding to column A and row 2 is A2. For a better understanding, some of the cells have been labeled accordingly to Figure-21.

Formulas are defined by the range of cells. The range of cells is specified by specifying the name of first and the last cell of the range, separated by a colon (:). The range takes all the cells falling between specified cells. For example, if you write A2 : C3, this means all the cells falling in 2nd and 3rd rows, Column A and B will be taken into account. To insert a formula, double click on the cell; Text Editor is invoked. You can now write the syntax of the formula in the cell. The syntax for different formulas are discussed later while explaining different formulas. Formulas can also be inserted by using the Formula drop-down list. Different formulas available in the Formula drop-down list are discussed next.

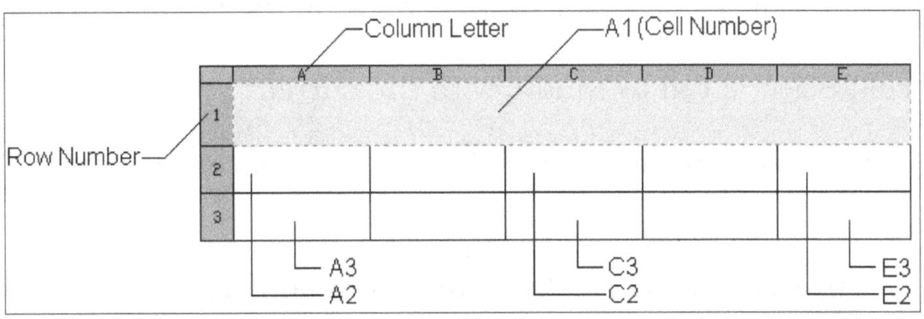

*Figure-21. The **Table** showing nomenclature for **Columns**, **Rows**, and **Cells***

Sum

The **Sum** option gives output for a given cell as the sum of the numerical values entered in a specified range of cells. Choose the **Sum** option from the **Table Cell > Insert > Formula** drop-down; you will be prompted to select the first corner of the table cell range and then the second corner. The sum of values of all the cells that fall between the selected range will be displayed as the output. As soon as you specify the second corner, the Text Editor is displayed and also the formula is displayed in the cell. In addition to the formula, you can also write multiline text in the cell. Choose **Close Text Editor** from the **Close** panel to exit the editor. When you exit the text editor, the formula is replaced by a hash (#). Now, if you enter numerical values in the cells included in the range, the hash (#) is replaced according to the addition of those numerical values. The prompt sequence, when you select the Sum option, is given next.

Select first corner of table cell range: Specify a point in the first cell of the cell range.

Select second corner of table cell range: Specify a point in the last cell of the cell range.

Note that the syntax for the Sum option is: =Sum{Number of the first cell of cell range (for example: A2): Number of the last cell of the cell range (for example: C5)}

Average

This option is used to insert a formula that calculates the average of values of the cells falling in the cell range. Prompt sequence is the same as for the Sum option.

Note that the syntax for the **Average** option is: =Average{Number of the first cell of the cell range (for example: A2): Number of the last cell of the cell range (for example: C5)}

Count

This option is used to insert a formula that calculates the number of cells falling under the cell range. The prompt sequence is the same as for the Sum option.

Note that the syntax for the Count option is: =Count{Number of the first cell of cell range (for example: A2): Number of the last cell of cell range (for example: C5)}

Cell

This option equates the current cell with a selected cell. Whenever there is a change in the value of the selected cell, the change is automatically updated in the other cell. To do so, choose the **Cell** option from **Table Cell > Insert > Formula** drop-down; you will be prompted to select a table cell. Select the cell with which you want to equate the current cell. The prompt sequence for the Cell option is given next.

Select table cell: Select a cell to equate with the current cell.

Note that the syntax for the **Cell** option is: =Number of the cell.

Equation

Using this option, you can manually write equations. The syntax for writing the equations should be the same as explained earlier.

Manage Cell Content

The **Manage Cell Content** button is used to control the sequence of the display of blocks in the cell, if there are more than one block in a cell. Choose **Manage Cell Contents** from the **Insert** panel; the **Manage Cell Content** dialog box will be displayed, see Figure-22. The options in the **Manage Cell Content** dialog box are discussed next.

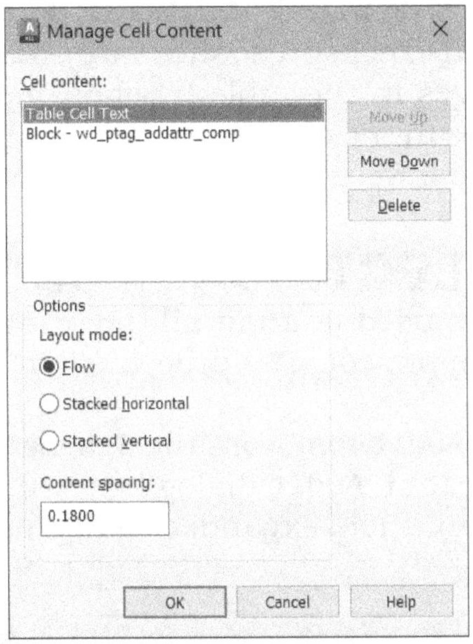

Figure-22. The Manage Cell Content dialog box

Cell content Area

This area lists all blocks entered in the cell according to the order of their insertion sequence.

Move Up

This button is used to move the selected block to one level up in the display order.

Move Down

This button is used to move the selected block to one level down in the display order.

Delete

This button is used to delete the selected block from the table cell.

Options Area

The options in this area are used to control the direction in which the inserted block is placed in the cell. If you select the **Flow** radio button, the direction of placement of the blocks in the cell will depend on the width of the cell. Select the **Stacked horizontal** radio button to place the inserted blocks horizontally. Similarly, select the **Stacked vertical** radio button to place the inserted blocks vertically. You can also specify the gap to be maintained between the two consecutive blocks by entering desired gap value in the **Content** spacing edit box.

Link Cell

To insert data from an excel into the selected table, choose **Data > Link Cell** from the **Table Cell** tab; the **Select a Data Link** dialog box will be displayed. The options in this dialog box are used to link the data in a cell to the external sources.

Download from source

If the contents of the attached excel spreadsheet are changed after linking it to a cell, you can update these changes in the table by choosing this button. AutoCAD will inform you about the changes in the content of the attached excel sheet by displaying an information bubble at the lower-right corner of the screen.

ELECTRICAL AUDIT

The **Electrical Audit** tool is used to audit all the drawings in project and find out the faulty components in the project. The procedure to use this tool is given next.

- Click on the **Electrical Audit** tool from the **Schematic** panel in the **Reports** tab of the **Ribbon**. The **Electrical Audit** dialog box will be displayed.
- Click on the **Details** button. The expanded dialog box will be displayed; refer to Figure-23.

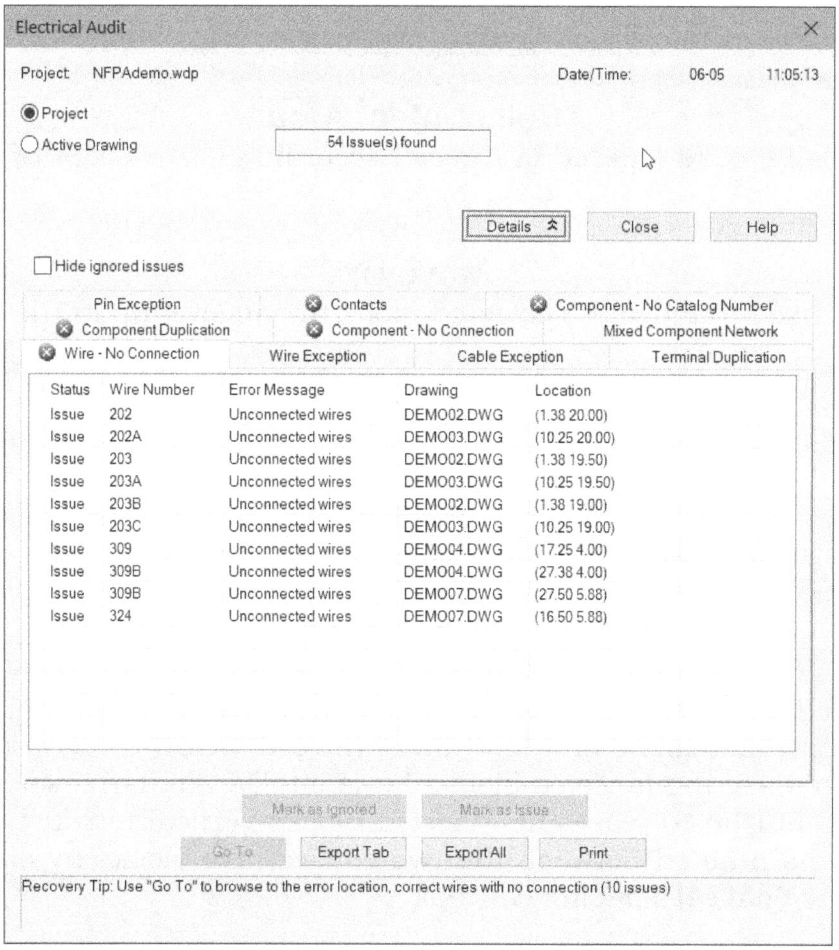

Figure-23. Electrical Audit dialog box

- Click on the **Active Drawing** radio button to display the problems in the current drawing. Click on the tabs in the expanded dialog box which are marked with a cross, the errors will be displayed; refer to Figure-24.

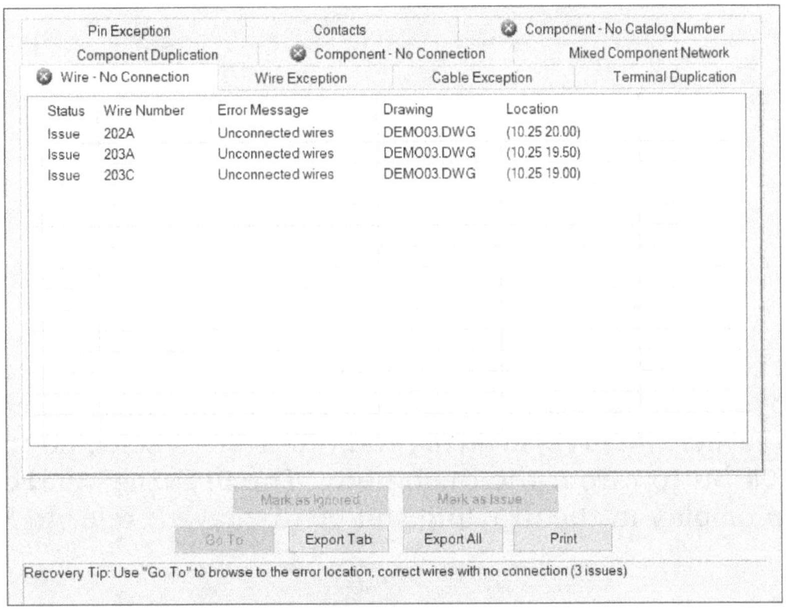

Figure-24. Errors in the drawing

- Select the error from the table and click on the **Go To** button from the dialog box. The component having error will be highlighted in the drawing; refer to Figure-25.

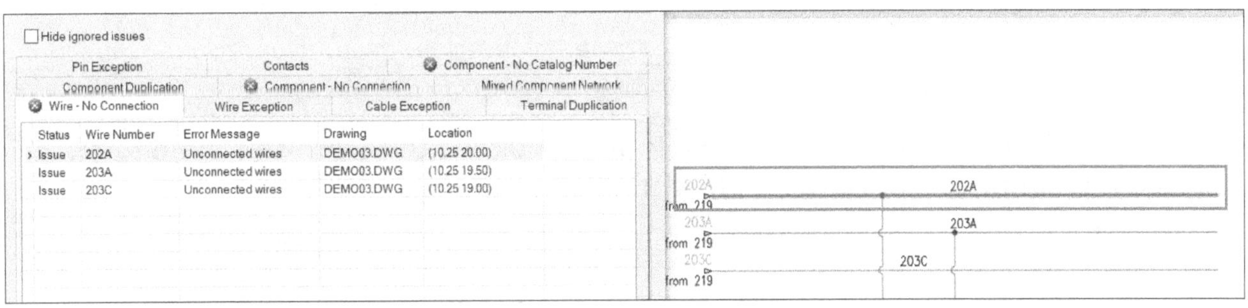

Figure-25. Component of error highlighted

- Modify the component as per the requirement.
- If you have intentionally left the wire unconnected or child component without parent component or similar issues then you right-click on the error message in the table and select **Mark as Ignored** option; refer to Figure-26.

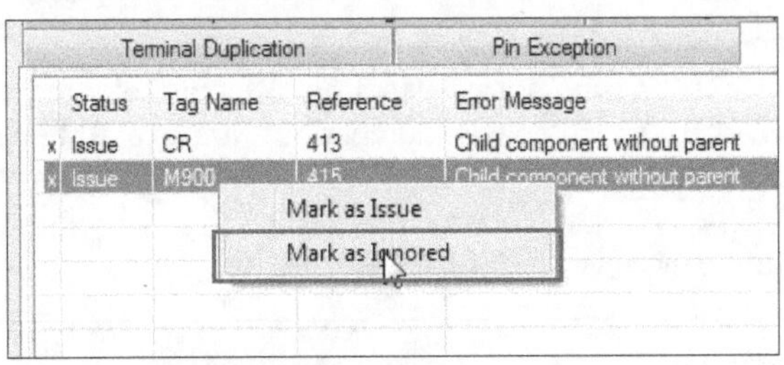

Figure-26. Mark as Ignored option

DRAWING AUDIT

The **DWG Audit** tool is used to audit the drawing for errors of wiring or wire numbers. The procedure to use this tool is given next.

- Click on the **DWG Audit** tool from the **Schematic** panel of the **Ribbon**. The **Drawing Audit** dialog box will be displayed; refer to Figure-27.

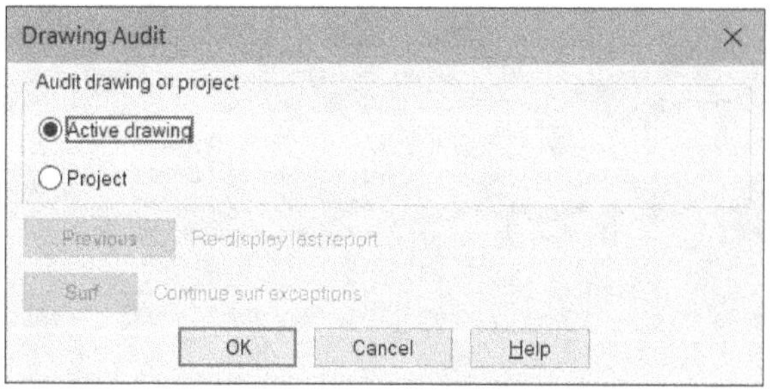

Figure-27. Drawing Audit dialog box

- Select the **Active drawing** or **Project** radio button as per your requirement. If you select the **Project** radio button then you need to include the drawings of project as discussed earlier. (**Active drawing** radio button is selected in our case)
- Click on the **OK** button from the dialog box. The **Drawing Audit** dialog box with the options to display in the drawing will be displayed; refer to Figure-28.

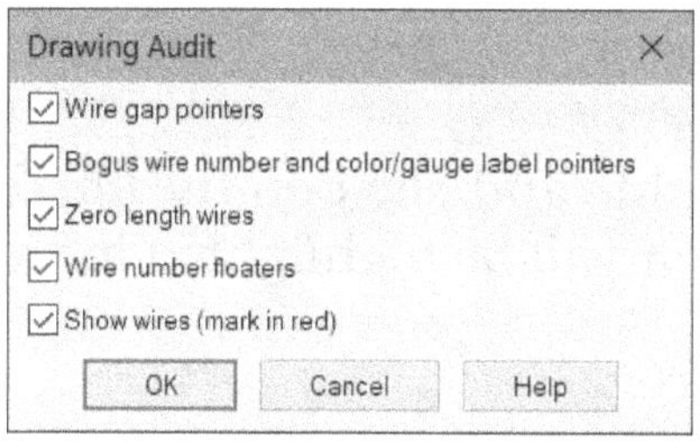

Figure-28. Drawing Audit dialog box with entities to display

- Select desired check boxes and click on the **OK** button from the dialog box. The selected components will be highlighted; refer to Figure-29.

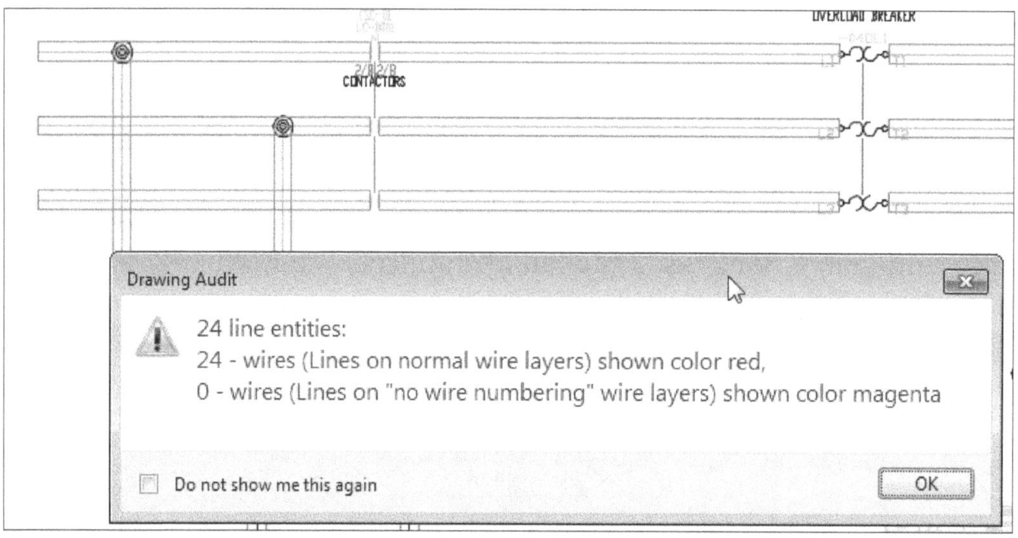

Figure-29. Wires highlighted

- Click on the **OK** button from the dialog box to exit.

SIGNAL ERROR/LIST

The **Signal Error/List** tool is used to display signal reports and find out errors in the destinations of wire signals. The procedure to use this tool is given next.

- Click on the **Signal Error/List** tool from the **Schematic** panel in the **Reports** tab of the **Ribbon**. The **Wire Signal or Stand-Alone References Report** dialog box will be displayed; refer to Figure-30.

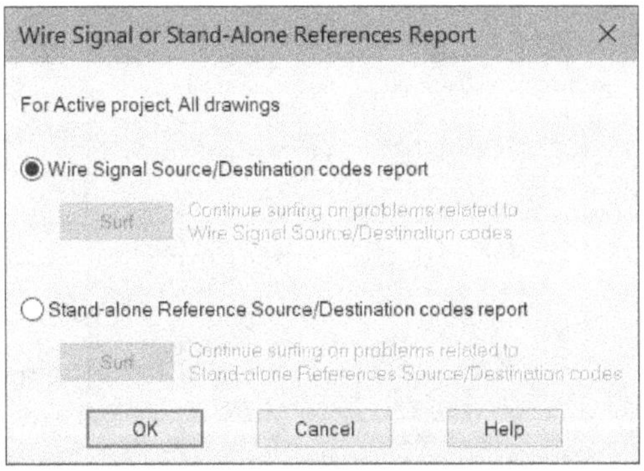

Figure-30. Wire Signal or Stand-Alone References Report dialog box

- If you want to surf wire signal source and destination then select the **Wire Signal Source/Destination codes report** radio button. If you want to surf stand-alone source/destinations in the project then select the **Stand-alone Reference Source/ Destination codes report** radio button.
- After selecting desired radio button, click on the **Surf** button below it. The **Surf** dialog box will be displayed; refer to Figure-31.

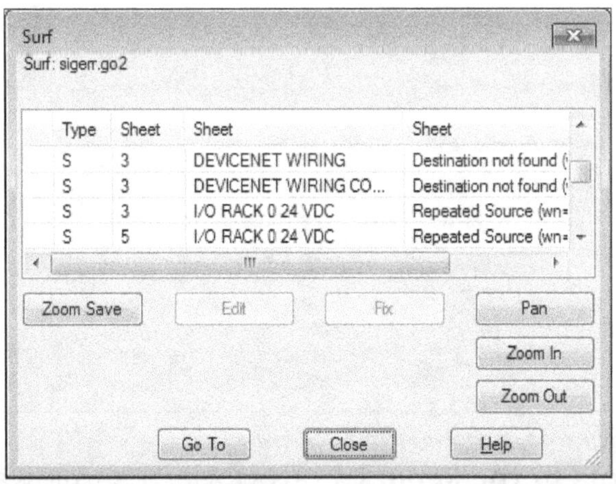

Figure-31. Surf dialog box

- Select the object with error from the list and click on the **Go To** button to check it. Location of error will be displayed; refer to Figure-32.

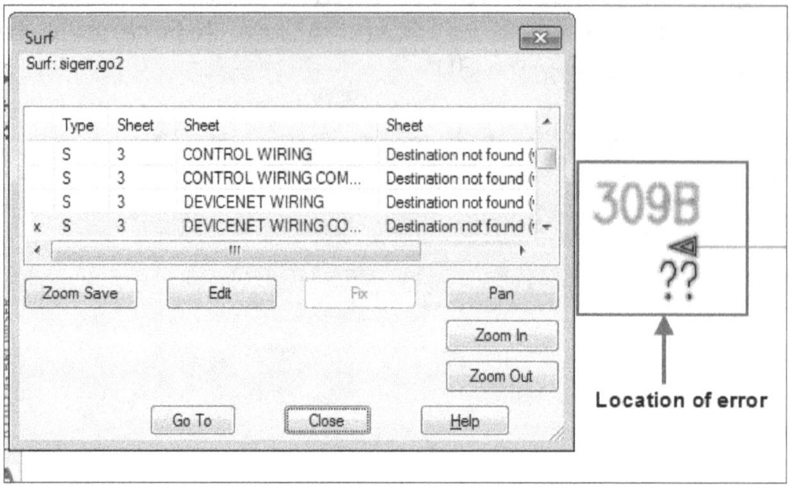

Figure-32. Location of error in destination

- Click on the **Edit** button from the dialog box and specify the missing parameter in the dialog box displayed; refer to Figure-33.

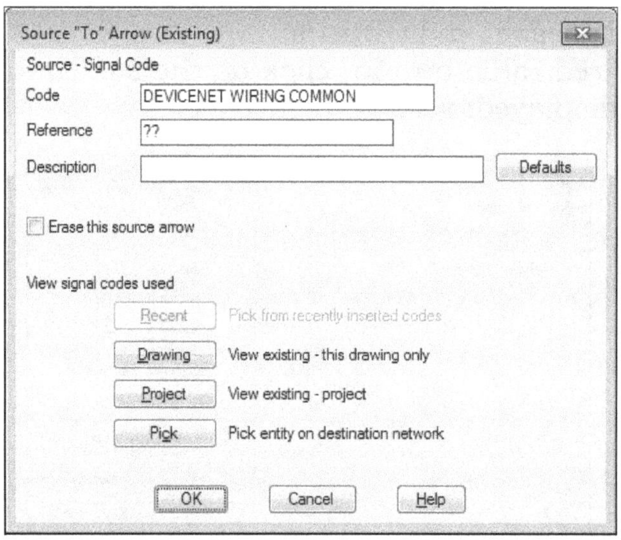

Figure-33. Source to Arrow dialog box

Note that rest of the tools in the **Reports** tab work in same way as discussed earlier.

IMPORTING AND EXPORTING DATA

Earlier in this book, you have learned about importing and exporting files using tools available in the **File** menu. In AutoCAD Electrical, you can also import and export data related to electrical drawings. The tools to do so are available in the **Import/Export Data** tab of the **Ribbon**; refer to Figure-34. Various tools of this tab are discussed next.

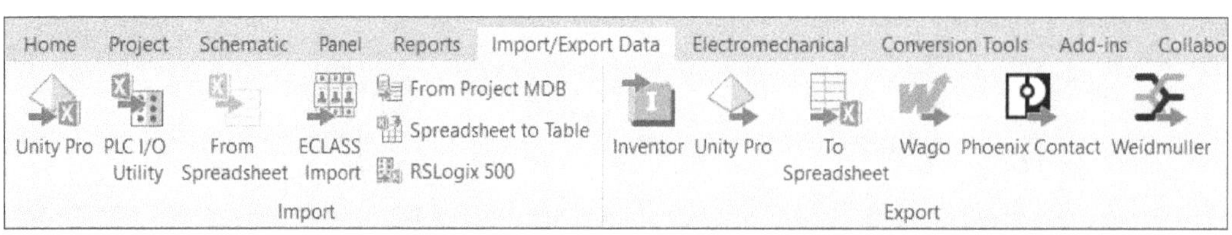

Figure-34. Import Export Data tab

Exporting Data

The tools in **Export** panel are used to export various electrical components related data like terminals, components, wire numbers, etc in different formats compatible to respective software. For example, you can use **Unity Pro** tool in **Export** panel of **Import/Export Data** tab of the **Ribbon** to export electrical drawings of current project in XSY format which can be imported in Unity Pro software (now known as EcoStruxure Control Expert) for further operations. The procedure to use all the tools in **Export** panel are similar. One of the most used format for exporting data is Spreadsheet which allows faster modification of multiple properties of components before importing them back.

Exporting to Inventor

The **Inventor** tool in **Export** panel of **Ribbon** is used to export electrical drawings and terminal data in xml format which can be imported in Autodesk Inventor. The procedure to use this tool is given next.

- Click on the **Inventor** tool from the **Export** panel in the **Import/Export Data** tab of the **Ribbon**. The **Autodesk Inventor Professional Export** dialog box will be displayed; refer to Figure-35.

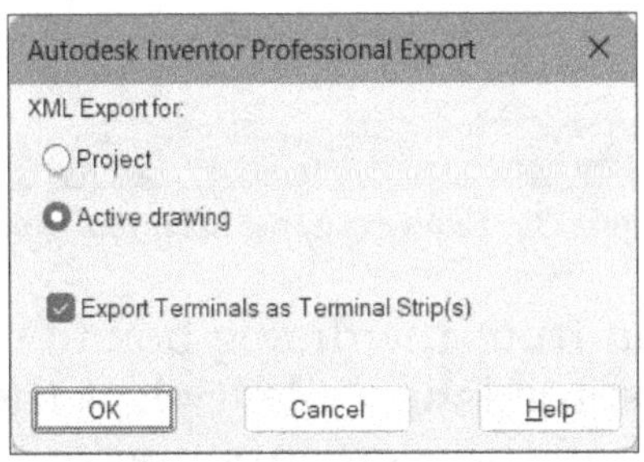

Figure-35. Autodesk Inventor Professional Export dialog box

- Select the **Active drawing** radio button if you want to export the contents of current drawing only. Select the **Project** radio button if you want to select drawing files of current project to be exported.
- Select the **Export Terminals as Terminal Strip(s)** check box to export terminals of current project/drawing as a terminal strip rather then individual terminals.
- If you have selected the **Project** radio button then **Select Drawings to Process** dialog box will be displayed which has been discussed earlier. After setting desired parameters, click on the **OK** button from the **Autodesk Inventor Professional Export** dialog box. The **Autodesk Inventor Professional XML File Export** dialog box will be displayed.
- Specify desired name of the file and click on the **Save** button to save the file.

Similarly, you can use the **Unity Pro** tool from the **Export** panel in the **Import/Export Data** tab of **Ribbon** to export drawing data to unity pro format (now known as EcoStruxure Control Expert).

Exporting to Spreadsheet

The **To Spreadsheet** tool is used to export various elements like components, terminals, wire numbers, PLC I/O connections, and so on to spreadsheet format (like *.xls,*.mdb, *.csv). The procedure to use this tool is given next.

- Click on the **To Spreadsheet** tool from the **Export** panel in the **Import/Export Data** tab of the **Ribbon**. The **Export to Spreadsheet** dialog box will be displayed; refer to Figure-36.

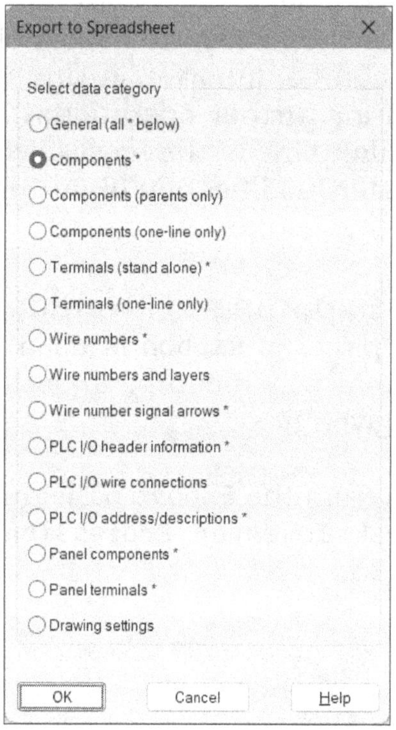

Figure-36. Export to Spreadsheet dialog box

- Select desired radio button from the dialog box to define which type of elements are to be exported in the spreadsheet file. Select the **General (all*below)** radio button to export details of all the elements marked by * in the dialog box.
- Click on the **OK** button from the dialog box. The **General Data Export** dialog box will be displayed; refer to Figure-37.

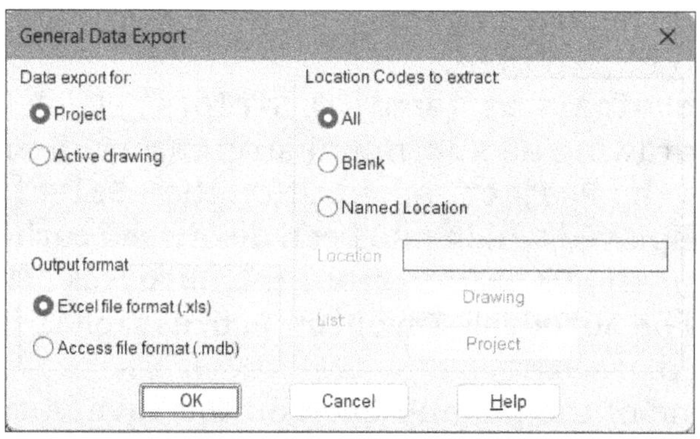

Figure-37. General Data Export dialog box

- Set desired parameters in the dialog box as discussed earlier (Excel file format in our case) and click on the **OK** button from the dialog box. The **Select file name for Project-wide XLS output** dialog box will be displayed; refer to Figure-38.

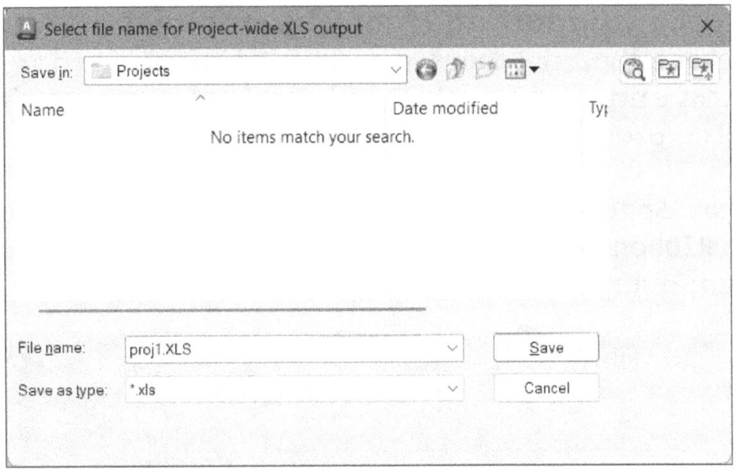

Figure-38. Select file name for Project-wide XLS output dialog box

- Specify desired name of file in the dialog box and click on the **Save** button. The spreadsheet file will be created.

Similarly, you can use other tools of the **Export** panel.

Checking Exported Spreadsheet

Once you have exported the spreadsheet file, open it in Microsoft Excel or any other database management software. Contents of such an exported spreadsheet file are shown in Figure-39. Note that you can modify the contents of this file if needed and then import it back to make changes in drawing. Assume that you want to change manufacturer of these terminals to ABB then you can change SIEMENS to ABB in all fields by simple text editing operations and copy-paste.

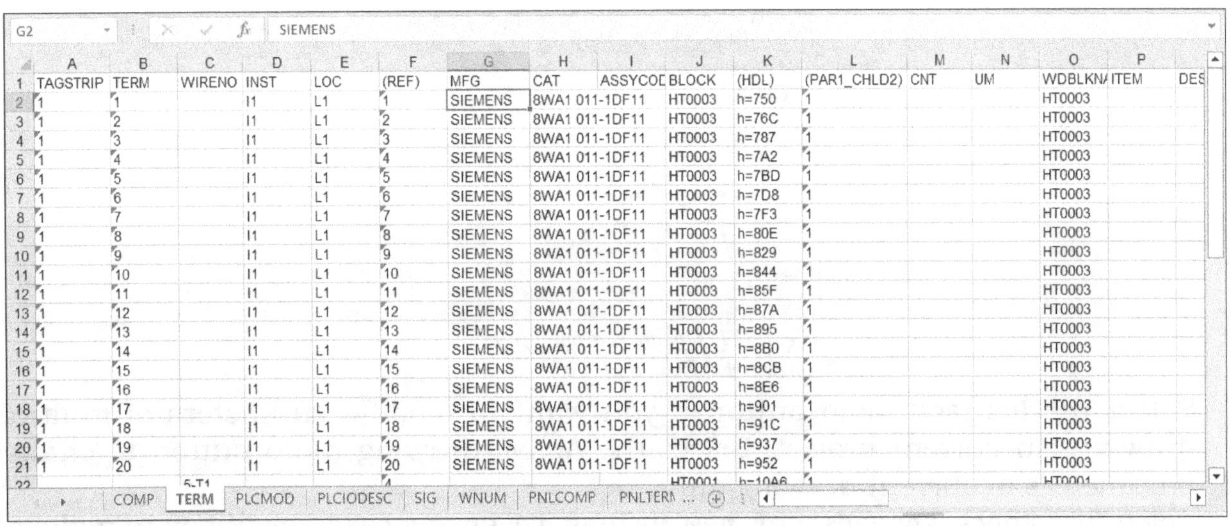

Figure-39. Contents of sample file

Importing Data

The tools in the **Import** panel of **Import/Export Data** tab of the **Ribbon** are used to import electrical data generated by other software in current drawing/project. You can also use the tools in this panel to import manufacturing data and model of electrical components using tools like ECLASS Import, RSLogix 500, and so on. The procedures to use all the tools in **Import** panel are similar.

Importing from Spreadsheet

The **From Spreadsheet** tool in **Import** panel of **Ribbon** is used to import data from selected spreadsheet and apply them to related components/wires present in the current drawing. The procedure to use this tool is given next.

- Click on the **From Spreadsheet** tool from the **Import** panel in the **Import/Export Data** tab of the **Ribbon**. The **Update Drawing from Spreadsheet File** dialog box will be displayed; refer to Figure-40.

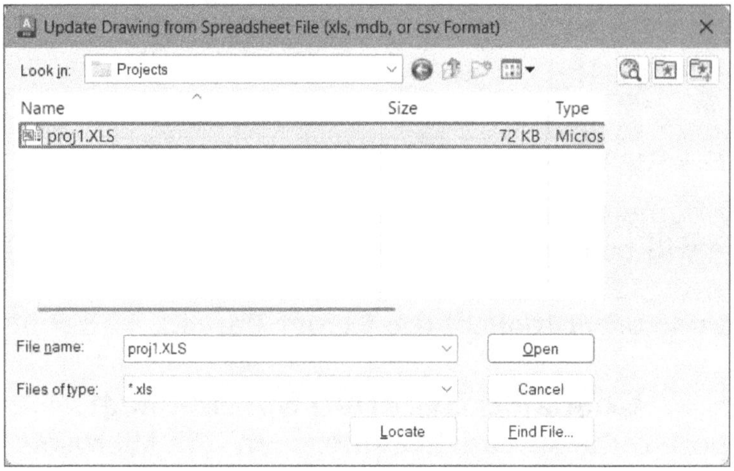

Figure-40. Update Drawing from Spreadsheet File dialog box

- Select desired file from which data will be imported and click on the **Open** button. The **Update Drawings per Spreadsheet Data** dialog box will be displayed; refer to Figure-41.

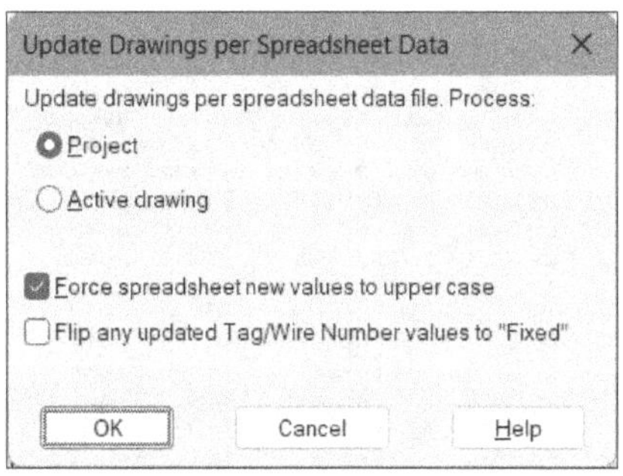

Figure-41. Update Drawing from Spreadsheet Data dialog box

- Select the **Project** radio button if you want to modify parameters of multiple drawings in current project. Select the Active drawing radio button to change parameters of current drawing only.
- Select the **Force spreadsheet new values to upper case** check box to make all the new values uppercase which are imported from spreadsheet.
- Select the **Flip any updated Tag/Wire Number values to "Fixed"** check box to make sure updated values do not change unintentionally.
- After setting desired parameters, click on the **OK** button from the dialog box. The drawing(s) will be updated.

Now, you have learned about the reports which is last piece of information need in the electrical engineering drawing so you are ready for a project given in the next chapter.

SELF-ASSESSMENT

Q1. The .... tool in the .... panel is used to create reports of schematic components in the project.

Q2. The .... report extracts plug/jack connection reports and optionally generates pin charts.

Q3. The .... tool is used to display a diamond mark on the components in the schematic diagram and panel layout that do not have catalog data attached.

Q4. The **Electrical Audit** tool is used to audit all the drawings in project and find out the faulty components in the project. (T/F)

Q5. The **DWG Audit** tool is used to audit the drawing for errors of wiring or wire numbers. (T/F)

REVIEW QUESTIONS

Q1. What is the primary function of the Reports tool in the Schematic panel?
A. To create wiring diagrams
B. To generate schematic reports of components in the project
C. To edit AutoCAD drawings
D. To simulate circuits

Q2. Which report type only includes components with assigned BOM information?
A. Wire Label report
B. Connector Plug report
C. Bill of Materials report
D. Terminal Numbers report

Q3. What option allows you to generate a Bill of Materials for the entire project?
A. Active Drawing radio button
B. Project radio button
C. Category drop-down
D. Installation Codes section

Q4. In the Display options of BOM, which format tallies identical components with the same location/installation code?
A. Normal Tallied Format
B. Display in "By TAG" Format
C. Normal Tallied Format (Group by Installation/Location)
D. Display in Tallied Purchase List Format

Q5. What does the Format button in the BOM report setup allow you to do?
A. Select components
B. Edit wire numbers
C. Choose layout format for the report
D. Import catalog data

Q6. What is the purpose of the 'Put on Drawing' button in the Report Generator dialog box?
A. To open schematic panel
B. To place the report as a table on the drawing
C. To print the report
D. To save the report in .xls format

Q7. Which report includes component tags, location codes, and catalog information across the project?
A. Component report
B. Connector Details report
C. Terminal Plan report
D. Cable Summary report

Q8. What unique function does the Wire From/To report offer?
A. Edits BOM data
B. Displays terminal ratings
C. Allows selection of "from" and "to" locations
D. Generates missing catalog entries

Q9. What detail is NOT included in the Component Wire List report?
A. Wire number
B. Component tag
C. Cable marker tag
D. Terminal pin number

Q10. Which report type provides information for each connector pin showing 'in' and 'out' wires?
A. Cable From/To report
B. Connector Plug report
C. Terminal Plan report
D. PLC Modules Used So Far report

Q11. What does the PLC I/O Address and Description report list?
A. All terminal blocks used
B. Beginning and ending I/O addresses of PLC modules
C. Component location tags
D. Missing wires

Q12. Which report identifies the first terminal or component connected to a PLC I/O point?
A. PLC I/O Address and Description report
B. Component Wire List report
C. PLC I/O Component Connection report
D. Terminal Numbers report

Q13. What information does the Terminal Numbers report provide?
A. Catalog data
B. Terminal strip number and location code
C. Cable summary
D. PLC tags

Q14. Which report includes wire number and layer name and takes longer to generate?
A. Wire Label report
B. Terminal Plan report
C. Cable Summary report
D. Component report

Q15. What does the Connector Summary report list for each connector tag?
A. Complete wire paths
B. Wire marker tags
C. Pins used and repeated pin numbers
D. Terminal numbers

Q16. What is a key feature of the Cable From/To report?
A. Extracts fuse info
B. Lists cable conductor's parent number and wire number
C. Only shows cable colors
D. Only available for active drawing

Q17. What is the purpose of the Missing Catalog Data tool?
A. Delete unused components
B. Assign wire colors
C. Highlight components missing catalog data
D. Replace schematic diagrams

Q18. Which dialog box is displayed when you click 'Put on Drawing' during a missing catalog data report?
A. Format Selection
B. Report Generator
C. Table Generation Setup
D. Catalog Editor

Q19. What is shown in the location list dialog box of the Wire From/To report if a component has no assigned location?
A. "None"
B. "Not Set"
C. "(??)"
D. "Default"

Q20. Which action follows clicking the 'Save to File' button during report generation?
A. Print setup opens
B. Edit Line dialog appears
C. Optional Script File dialog box appears
D. Missing Components report is generated

Q21. What is the primary use of the Table Cell tab in AutoCAD Electrical?
A. To create new tables
B. To modify tables, insert blocks, and add formulas
C. To manage drawing layers
D. To insert symbols in schematic drawings

Q22. How can you insert a row below a selected cell in a table?
A. Select the cell and click Insert Above
B. Click on Merge Cells and select Insert Row
C. Select the cell and choose Insert Below from the Rows panel
D. Right-click and select Add Row

Q23. Which tool would you use to delete a column from a table?
A. Delete Row(s)
B. Insert Right
C. Delete Column(s)
D. Remove Column

Q24. What is the function of the Merge By Column option in the Merge Cells drop-down?
A. Merges all rows of selected cells
B. Merges all cells in the column of the selected cell
C. Merges selected columns into a block
D. Deletes columns of selected cells

Q25. Which button is used to inherit properties from one cell to others?
A. Merge Cells
B. Edit Borders
C. Match Cells
D. Format Painter

Q26. How can you lock both content and formatting of a table cell?
A. Choose Format Locked from Cell Format
B. Choose Content Locked from Cell Format
C. Choose Content and Format Locked from Cell Format
D. Choose Cell Locking > All Locked

Q27. What does the Data Format option control?
A. The position of text within the cell
B. The border style of the table
C. The display format of text in the cell
D. The locking settings of the table cell

Q28. Which dialog box is displayed when using the Block tool to insert a block into a cell?
A. Insert Block
B. Block Editor
C. Insert a Block in a Table Cell
D. Table Block Insert

Q29. What does the AutoFit checkbox in the Insert a Block dialog box do?
A. Automatically rotates the block
B. Automatically fits the block in the table row
C. Scales the block to fit within the cell
D. Aligns block text to the center

Q30. What happens when you modify a field's source property like author name in Drawing Properties?
A. It only updates in the title block
B. Field remains unchanged
C. Field data updates automatically if updated
D. You have to reinsert the field

Q31. What is the correct syntax for inserting a SUM formula?
A. =SUM(A2, C5)
B. SUM(A2:C5)
C. =Sum{A2:C5}
D. =Add(A2 to C5)

Q32. What is the purpose of the Count function in a formula?
A. To display the total of numeric values
B. To display average of values
C. To count number of cells in range
D. To count number of columns

Q33. What will the Cell formula do in a table?
A. Add a new formula
B. Copy data of a selected cell to another
C. Merge selected cells
D. Count the selected range

Q34. What tool is used to reorder blocks in a cell?
A. Insert
B. Manage Cell Content
C. Block Editor
D. Format Cells

Q35. Which layout option in Manage Cell Content places blocks vertically?
A. Flow
B. Stacked horizontal
C. Stacked vertical
D. Inline vertical

Q36. What is the use of the Link Cell feature?
A. To insert a hyperlink
B. To add formula from Excel
C. To link table cell with Excel data
D. To insert an image

Q37. What tool should be used to identify errors in the entire electrical project?
A. Drawing Audit
B. Electrical Audit
C. Signal Error/List
D. Formula Checker

Q38. If a wire is intentionally left unconnected, what can you do in Electrical Audit?
A. Delete the error
B. Ignore the warning
C. Mark as Ignored
D. Save the error log

Q39. What does the Drawing Audit tool check?
A. Component placement
B. Signal flow
C. Wiring or wire numbers
D. Block insertion errors

Q40. What is the first step when using the Signal Error/List tool?
A. Click on Surf
B. Click on Edit
C. Click on Signal Error/List tool
D. Select object from the list

Q41. What happens when you select Surf in Signal Error/List tool?
A. Opens a new drawing
B. Exports the table
C. Displays a dialog box with errors
D. Clears all errors

Q42. Which report displays stand-alone source/destination codes?
A. Wire Signal Source/Destination
B. Block Reference Report
C. Stand-alone Reference Source/Destination
D. Formula Audit

Answer for Self-Assessment:
Ans1. Reports, Schematic, Ans2. Connector Plug report, Ans3. Missing Catalog Data, Ans4. T, Ans5. T

Chapter 11

Project

Topics Covered

The major topics covered in this chapter are:

- ***Schematic Drawing***
- ***Panel Layout***
- ***Report Generation***

PROJECT

Create schematic of the circuit as shown in Figure-1. Design the panel for the user and then generate report for the components.

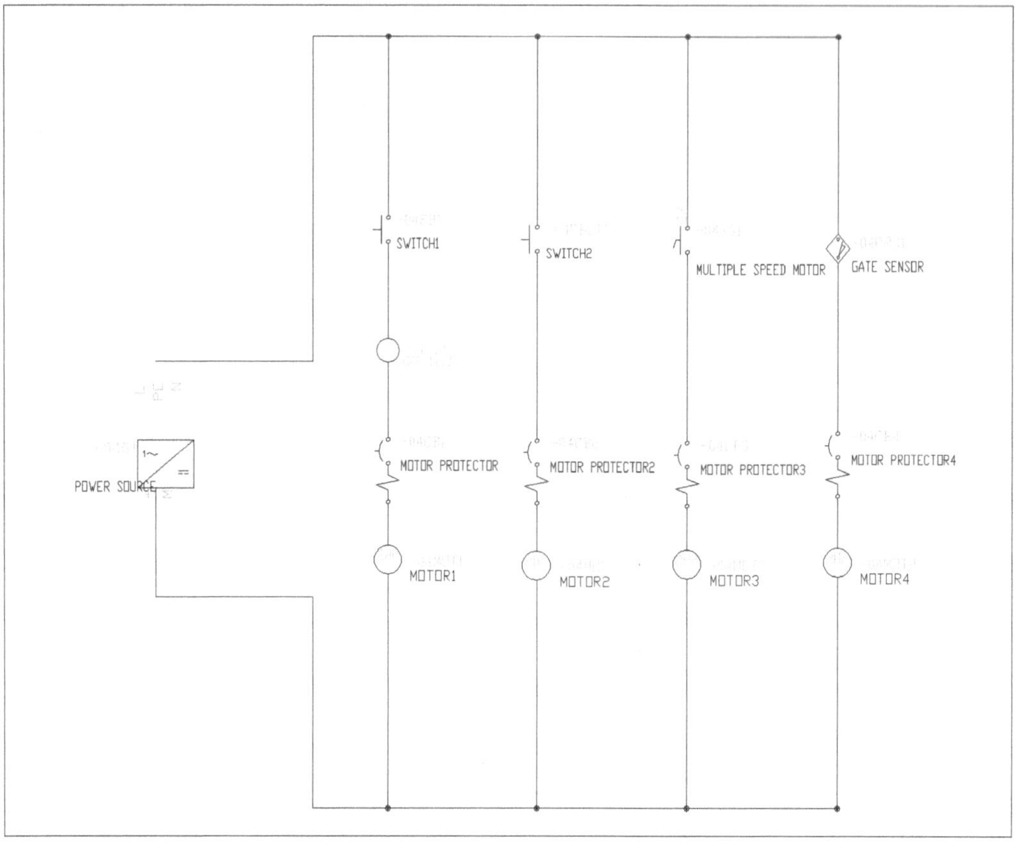

Figure-1. Project-Model

Starting a New Project File

- Click on the AutoCAD Electrical icon from the desktop or Start AutoCAD Electrical by using the Start menu.
- Click on the **New Project** button from the **Project Manager**; refer to Figure-2.

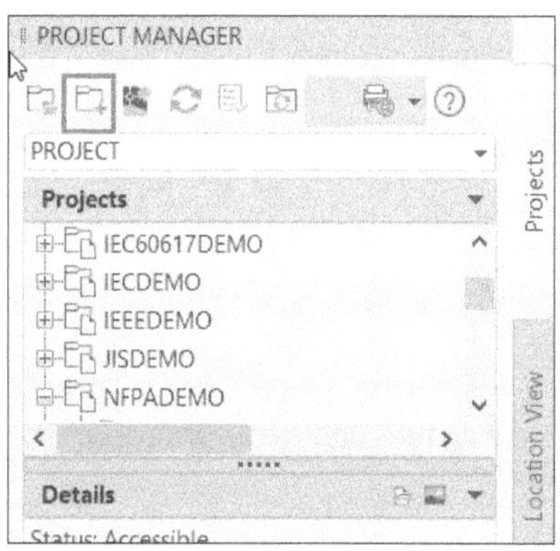

Figure-2. New Project button

- On doing so, the **Create New Project** dialog box will be displayed; refer to Figure-3.

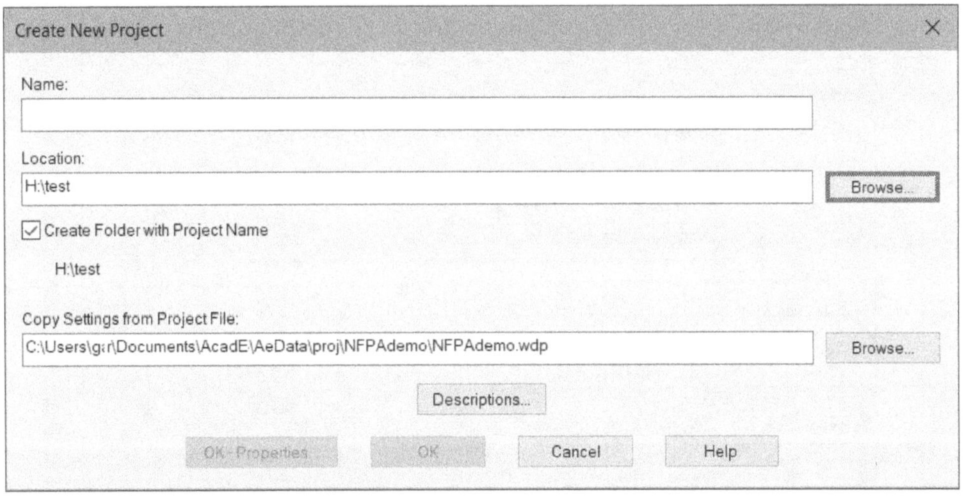

Figure-3. Create New Project dialog box

- Specify the name of the file as **Project** in the **Name** field of the dialog box.
- Click on the **OK-Properties** button from the dialog box. The **Project Properties** dialog box will be displayed.
- Click on the **Drawing Format** tab from the dialog box. The options in the dialog box will be displayed as shown in Figure-4.

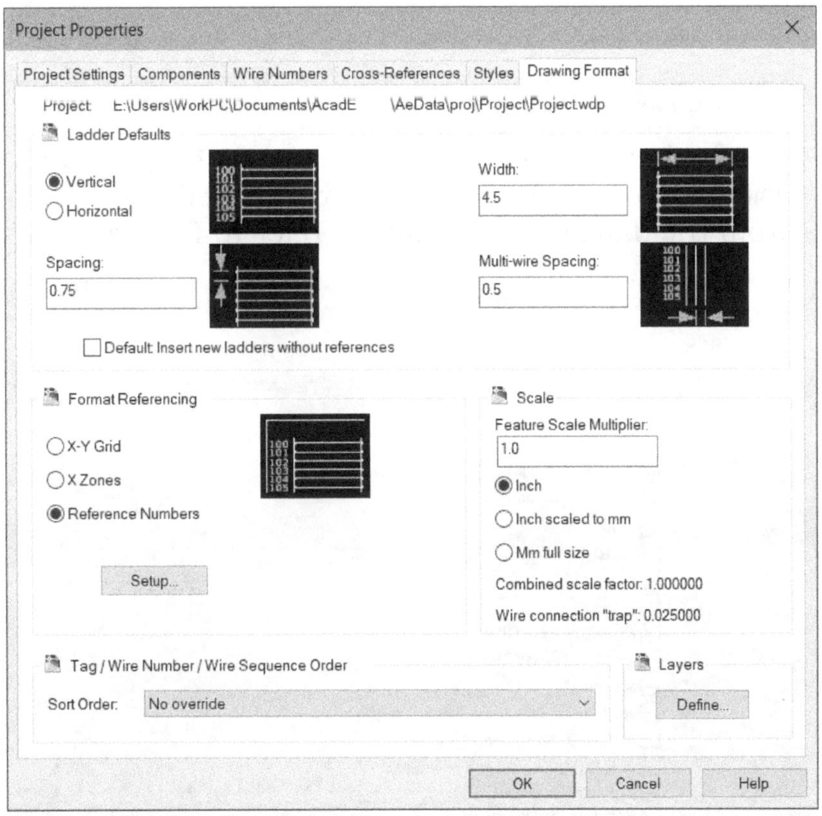

Figure-4. Project Properties dialog box

- Select the **Horizontal** radio button from the **Ladder Defaults** area of the dialog box and click on the **OK** button. A new project file will be added in the **Project Manager** with the name **Project**.

Adding Drawing in the Project

- Click on the **New Drawing** button from the **Project Manager**. The **Create New Drawing** dialog box will be displayed; refer to Figure-5.

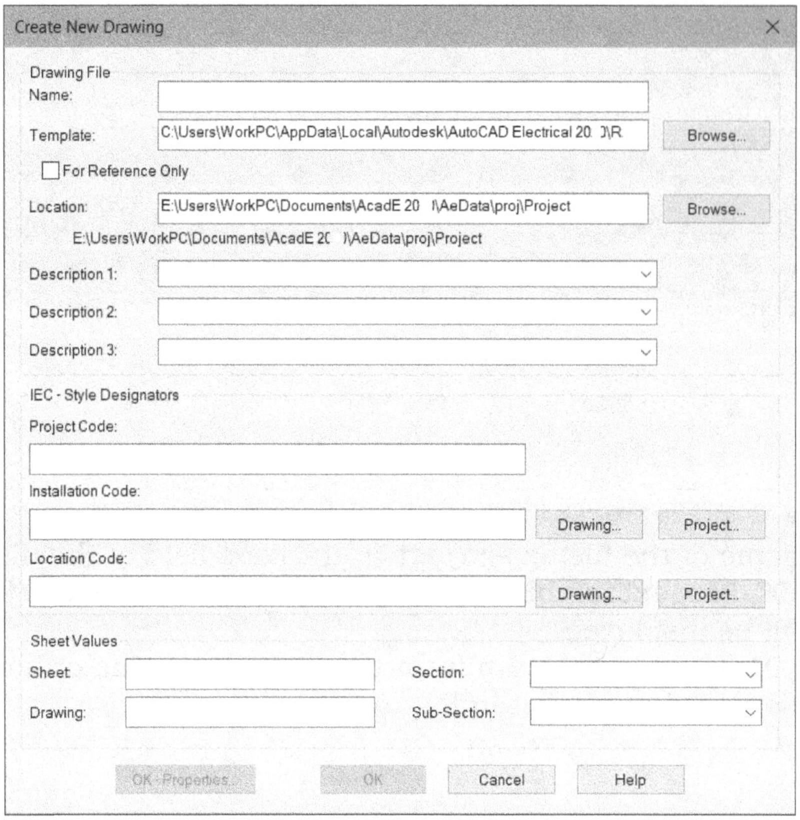
Figure-5. Create New Drawing dialog box

- Click in the **Name** field of the dialog box and specify **Project 1** as the name.
- Click on the **Browse** button for the **Template** edit box and select the **ACAD ELECTRICAL IEC** template from the dialog box displayed; refer to Figure-6.

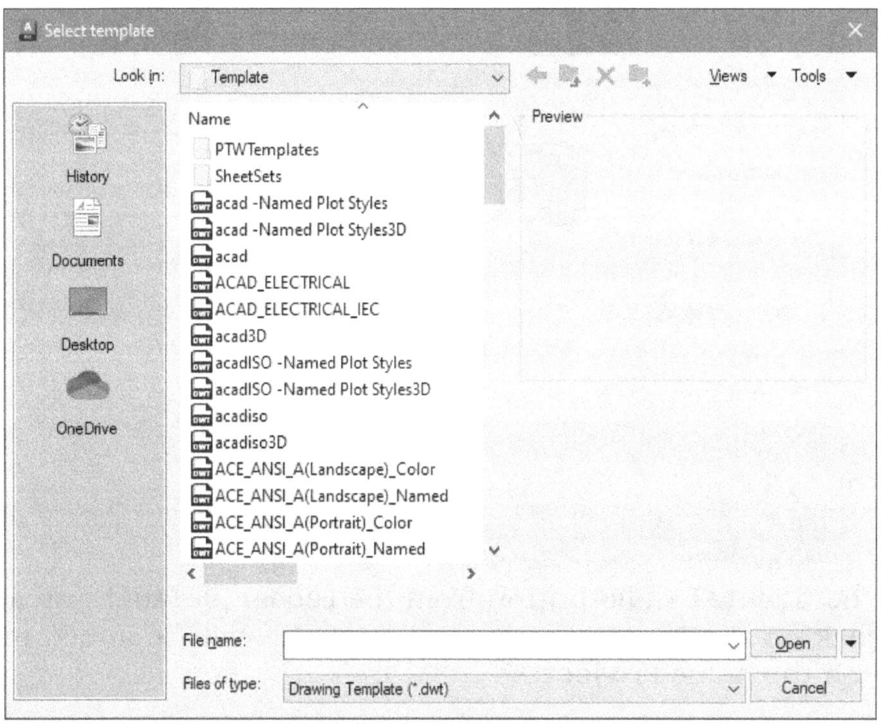
Figure-6. Select Template dialog box

- Click on the **Open** button from the dialog box.

- Click on the **OK** button from the **Create New Drawing** dialog box. The **Apply Project Defaults to Drawing Settings** dialog box will be displayed; refer to Figure-7.

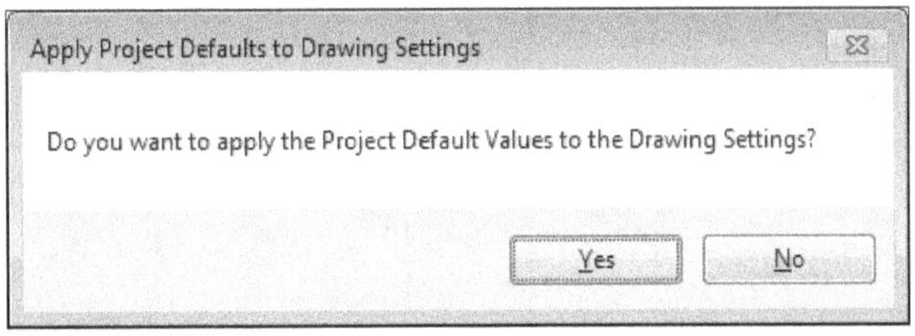

Figure-7. Apply Project Defaults to Drawing Settings dialog box

- Click on the **No** button from the dialog box. New drawing will be added in the Project with the name Project 1.
- Click on the **Grid Snap** and **Grid Display** button to turn off the grid snap and hide the grid from drawing.

Creating Ladder

- Click on the **Insert Ladder** button from the **Ladders** drop-down in the **Insert Wires/Wire Numbers** panel in the **Schematic** tab of the Ribbon. The **Insert Ladder** dialog box will be displayed; refer to Figure-8.

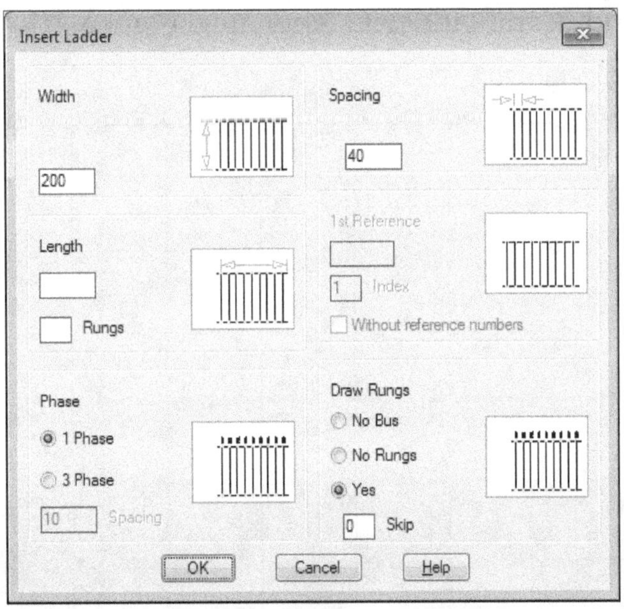

Figure-8. Insert Ladder dialog box

- Click on the **OK** button from the dialog box. You are asked to specify the start position of the first rung.
- Click in the drawing area to specify the starting position. The end point of the ladder gets attached to the cursor and you are prompted to specify the last rung.
- Click in the drawing area to specify the last rung so that the ladders are displayed as shown in Figure-9.

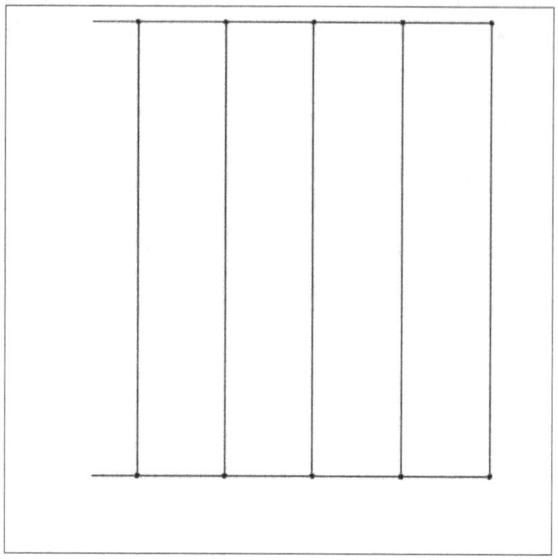

Figure-9. Ladders created

Adding wire

- Click on the **Wire** tool from the **Insert Wires/Wire Numbers** panel in the **Ribbon**. You are asked to specify the start point of the wire.
- Click on the top end of the ladder as shown in Figure-10.

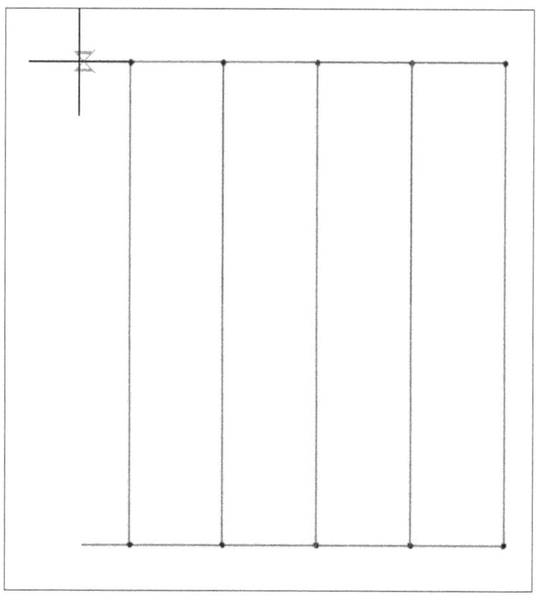

Figure-10. End to be selected

- Drawing the wire as shown in Figure-11.

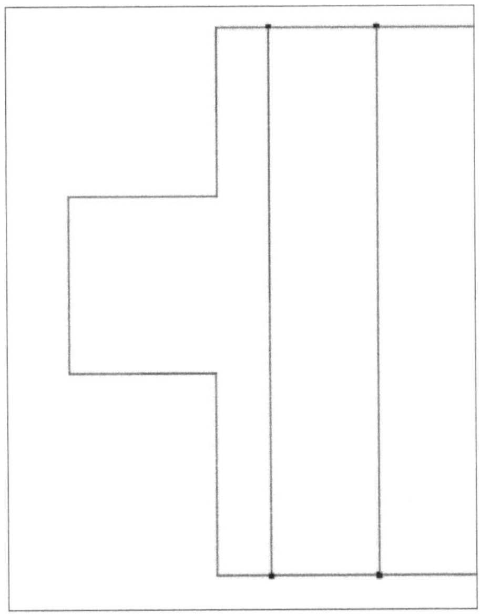

Figure-11. Wire drawn

Adding Components in the circuit

- Click on the **Icon Menu** button from the **Insert Components** panel in the **Ribbon**. The **Insert Component** dialog box will be displayed; refer to Figure-12.

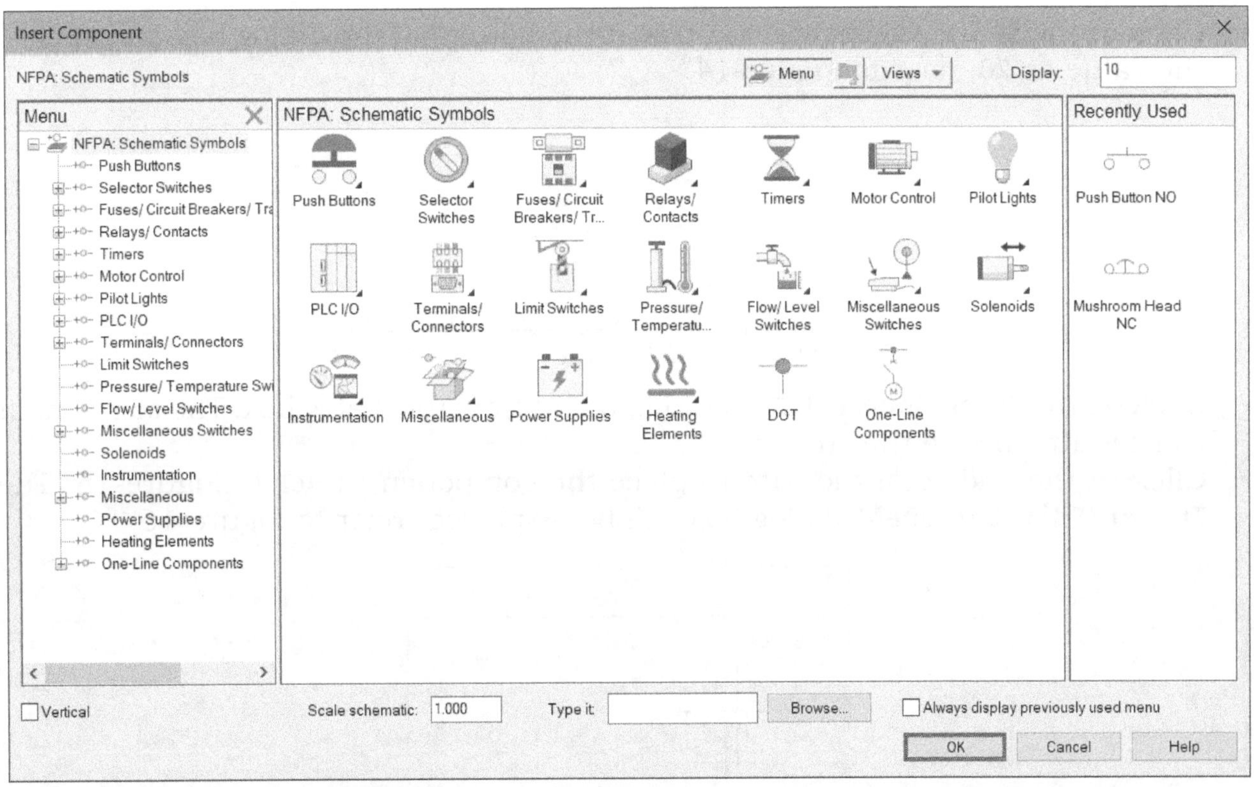

Figure-12. Insert Component dialog box

- Click on the **Power Supplies** category from the dialog box. The components will be displayed as shown in Figure-13.

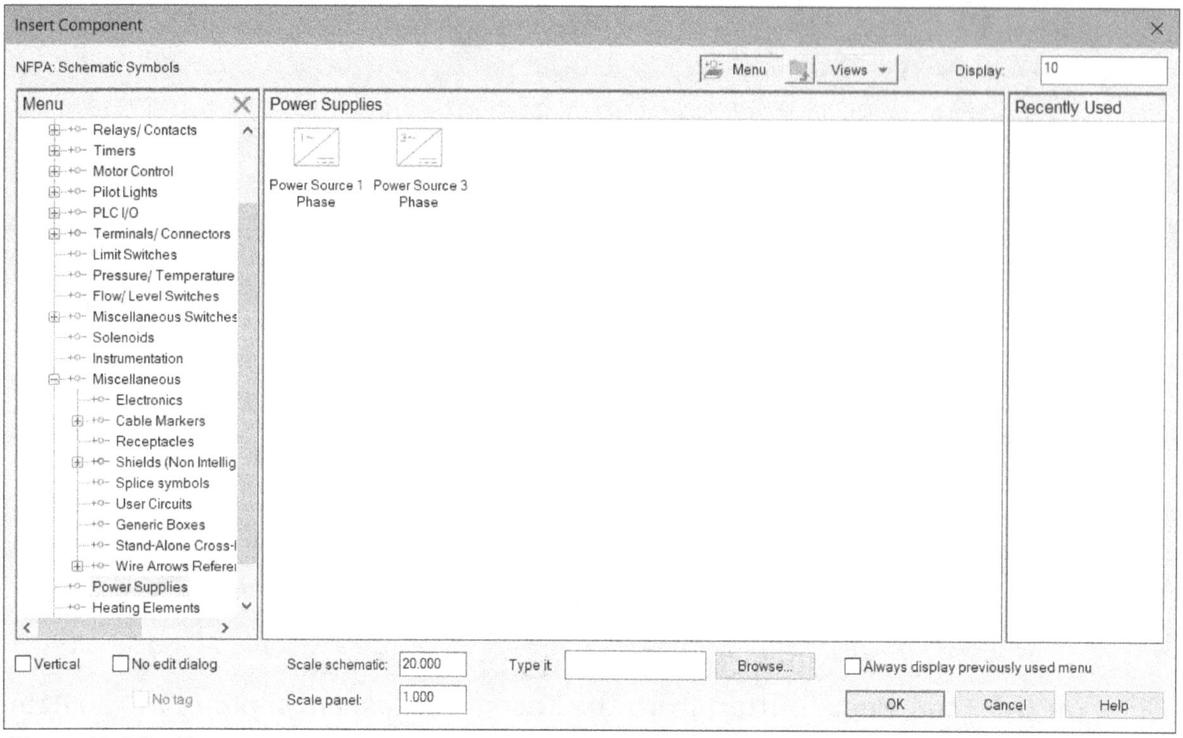

Figure-13. Miscellaneous components

- Click in the **Scale schematic** edit box at the bottom of the dialog box and specify the value as **20**; refer to Figure-14.

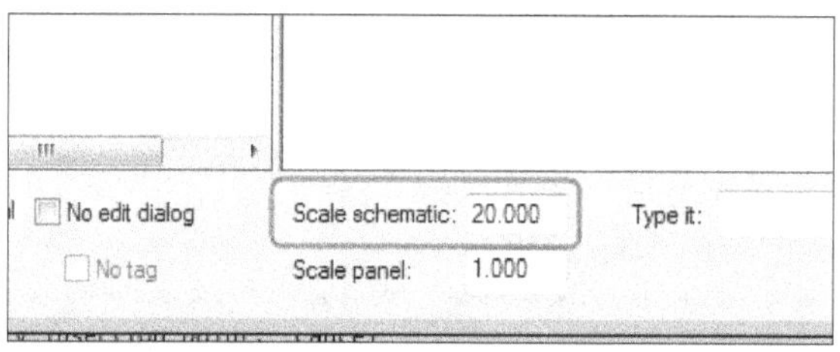

Figure-14. Scale schematic edit box

- Click on the **Power Source 1 Phase** component from the dialog box. The component will get attached to the cursor.
- Click in the middle of the wire to place the component; refer to Figure-15. The **Insert/Edit Component** dialog box will be displayed; refer to Figure-16.

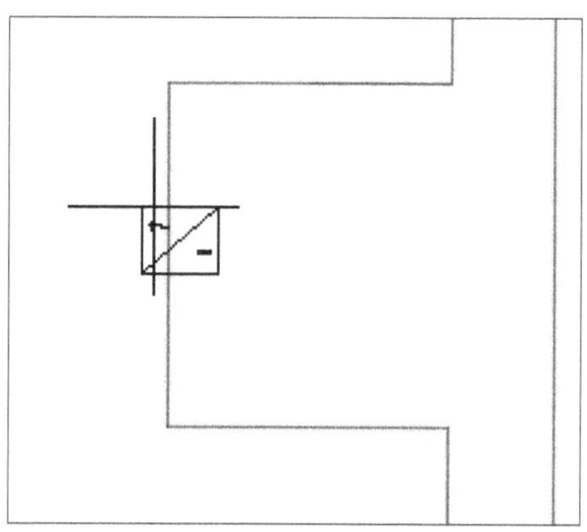

Figure-15. Placing power source

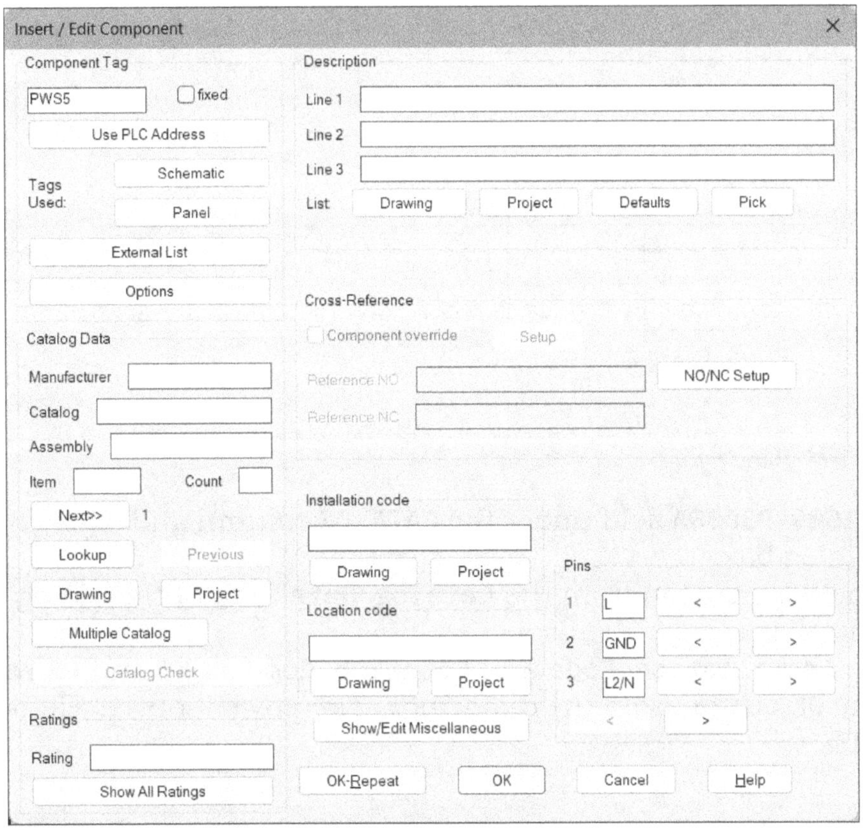

Figure-16. Insert or Edit Component dialog box

- Click in the **Line 1** edit box of the **Description** area of the dialog box and specify **Power Source** in it.
- Click on the **Lookup** button from the **Catalog Data** area of the dialog box. The **Catalog Browser** will be displayed; refer to Figure-17.

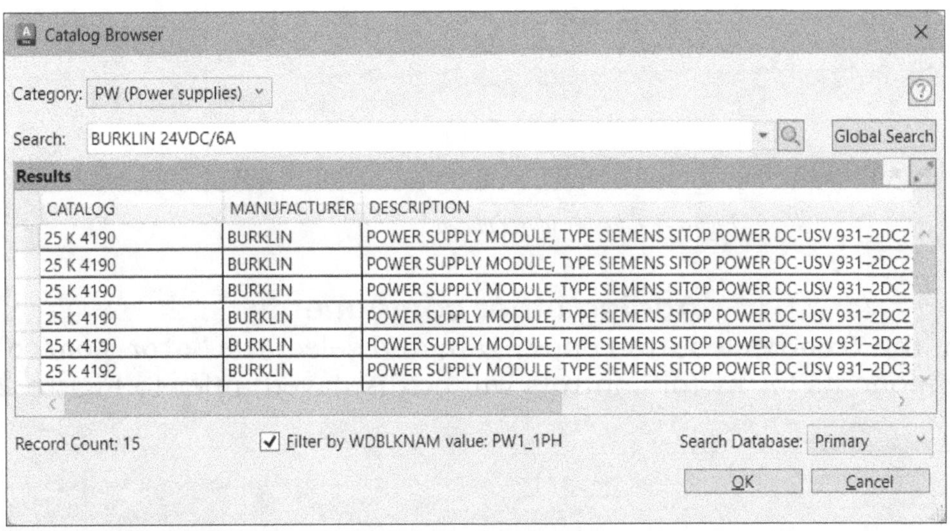

Figure-17. Catalog Browser

- Click in the **Search** field and specify **230VAC**.
- Click on the **Search** button next to **Search** field. The list of power sources will be displayed; refer to Figure-18.

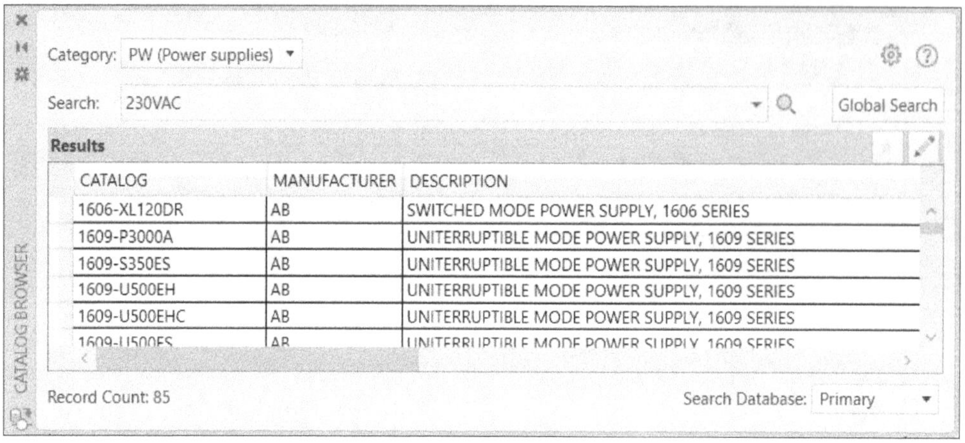

Figure-18. Power supplies

- Select the **1609-P3000A** field under the **CATALOG** column and click on the **OK** button from the dialog box.
- Click on the **Next** button from the **Catalog Data** area of the dialog box to specify the item number.
- Specify the **Installation Code** and **Location Code** as **X1** and **L3** respectively.
- Click on the **OK** button from the dialog box. The power source will be placed; refer to Figure-19.

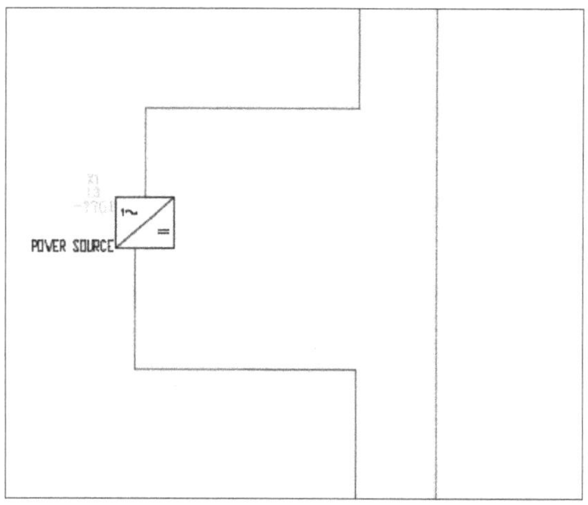

Figure-19. Power source placed

Adding other components

- Click on the **Icon Menu** button once again and select the **Motor Control** category. The components for motor controls will be displayed; refer to Figure-20.

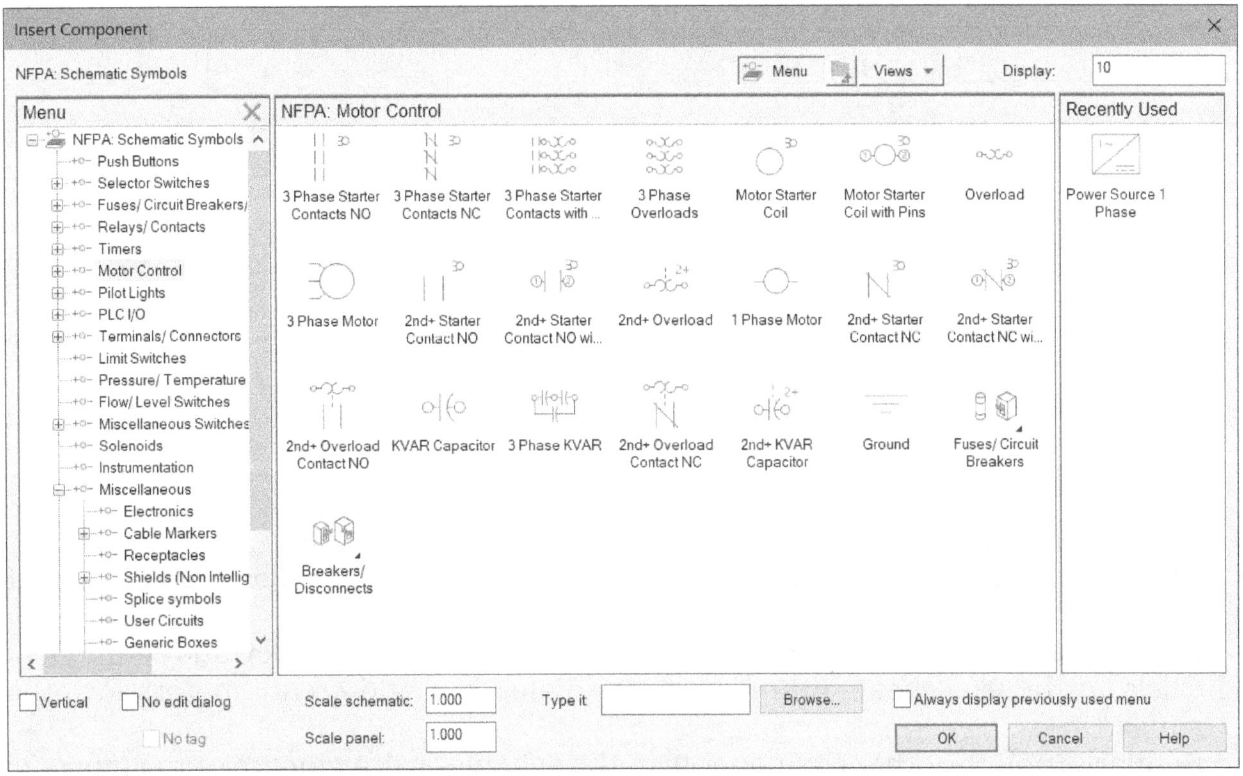

Figure-20. Components of Motor Control

- Click on the **1 Phase Motor** component. It will get attached to the cursor.
- Click on the first rung of the ladder and place the component as shown in Figure-21.

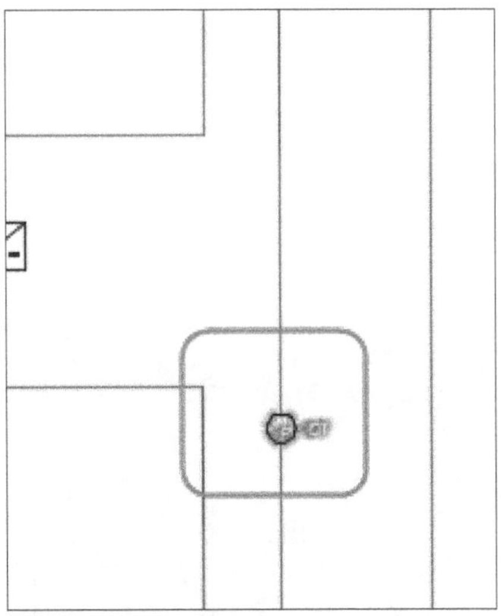

Figure-21. Motor placed

- Click in the **Line 1** edit box in the **Description** area of the dialog box displayed and specify the name as **Motor1**.
- Click on the **Lookup** button from the **Catalog Data** area of the dialog box and select the **1329-ZF00206NVH** field in the **CATALOG** column; refer to Figure-22.

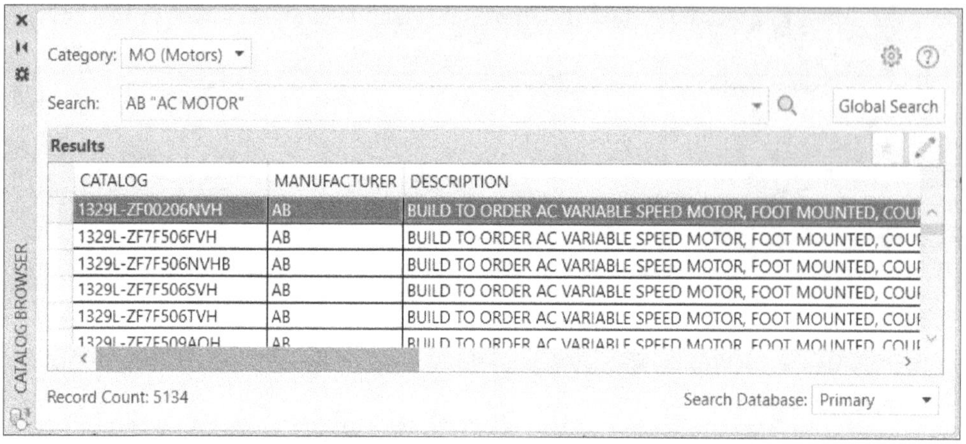

Figure-22. Motor in catalog browser

- Click on the **OK** button from the **Catalog Browser** dialog box.
- Click on the **Next** button from the **Catalog Data** area of the dialog box to specify the item number.
- Specify the same Installation and Location codes as done for previous component.
- Click on the **OK-Repeat** button from the dialog box and create motors for other rungs.
- Similarly, add the other components in the schematic drawing; refer to Figure-23.

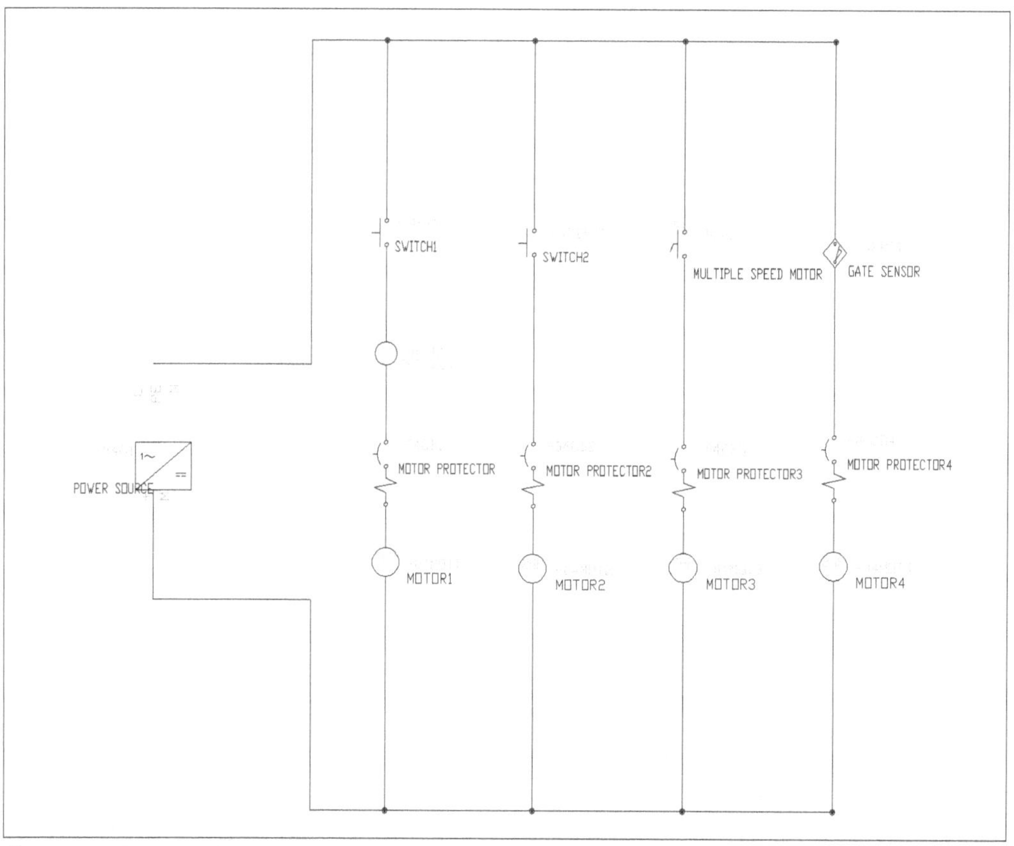

Figure-23. Project-Model

Adding Panel drawing in the Project

- Click on the **New Drawing** button from the **Project Manager**. The **Create New Drawing** dialog box will be displayed as discussed earlier.
- Specify the name as **Panel Drawing** in the **Name** field of the dialog box.

- Specify the same Installation and Location codes, and click on the **OK** button from the dialog box.
- Click on **No** button from the dialog box displayed for applying defaults of project.

Adding panel components

- Click on the **Panel** tab from the **Ribbon**. The tools related to panel will be displayed.
- Click on the **Icon Menu** button from the **Insert Component Footprints** panel. The **Alert** dialog box will be displayed.
- Click on the **OK** button from the dialog box. The **Insert Footprint** dialog box will be displayed; refer to Figure-24.

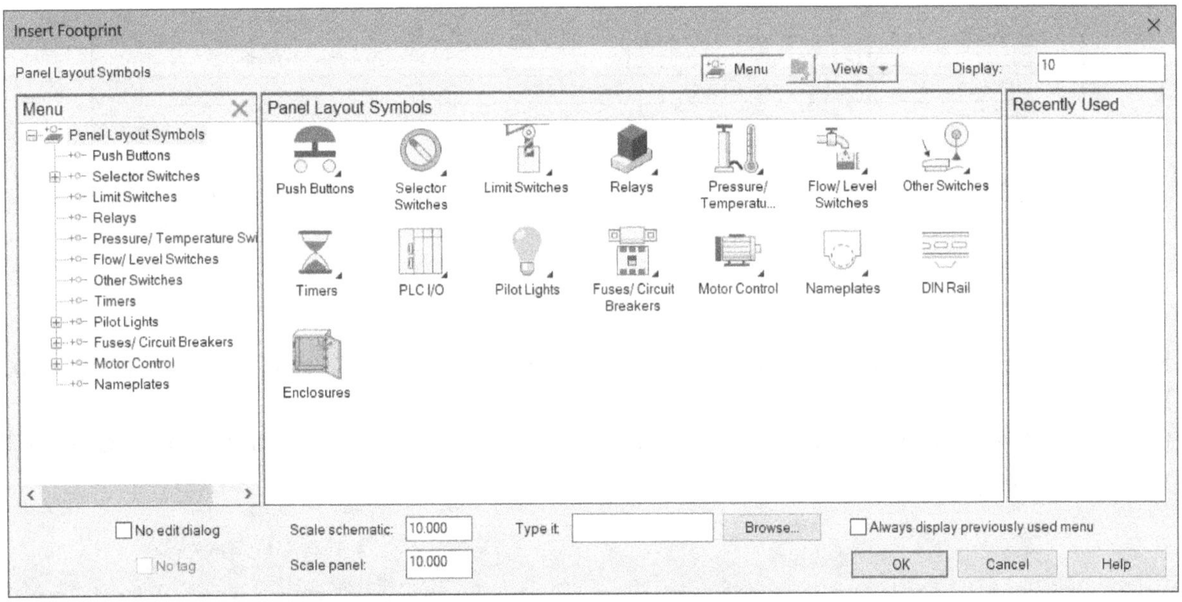

Figure-24. Insert Footprint dialog box

- Click on the **Enclosures** category from the dialog box. The **Footprint** dialog box will be displayed as shown in Figure-25.

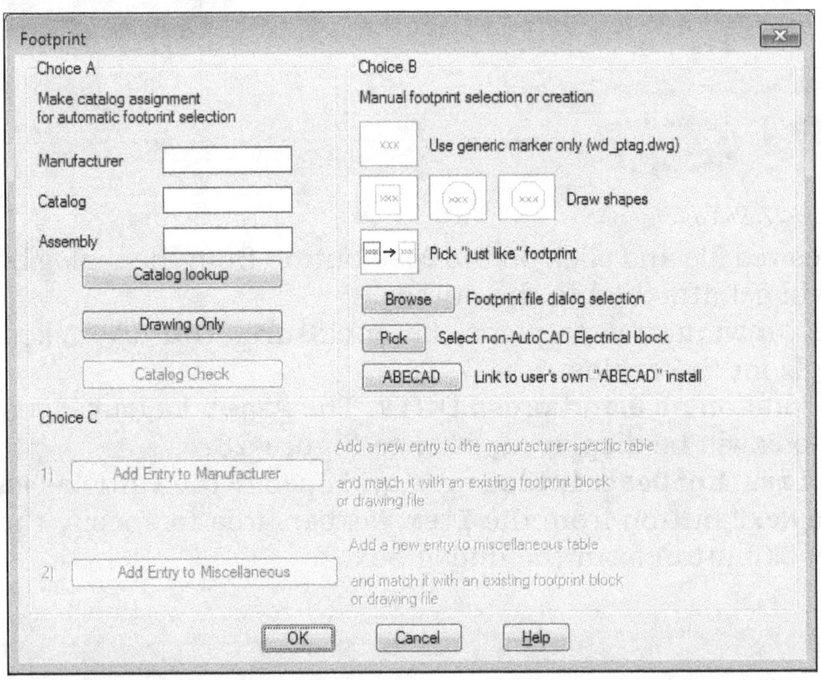

Figure-25. Footprint dialog box

- Click on the **Catalog lookup** button from the dialog box.
- Make the **Search** field empty and click on the **Search** button from the **Catalog Browser**. The list of enclosures will be displayed; refer to Figure-26.

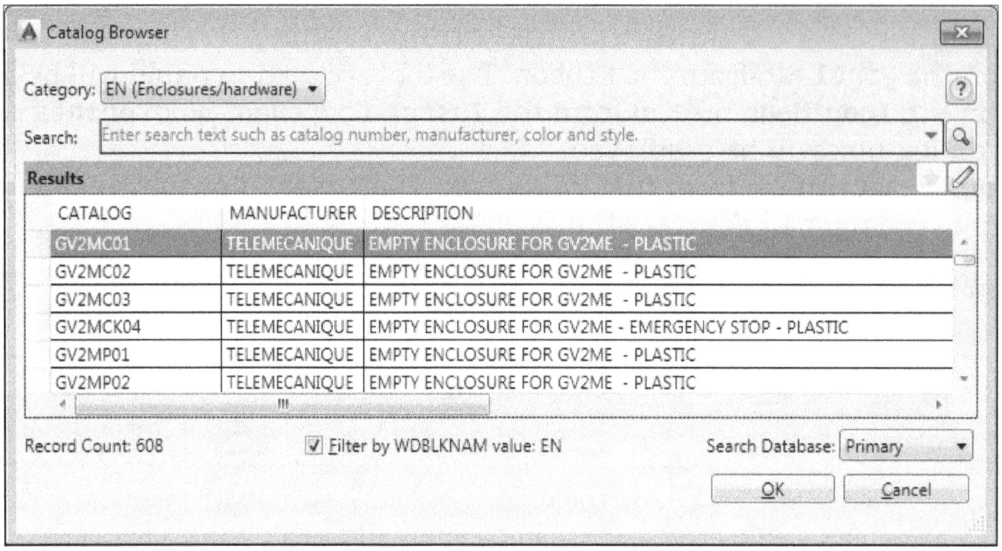

Figure-26. List of enclosures

- Select the first enclosure from the list and click on the **OK** button from the **Catalog Browser**.
- Click on the **Browse** button from the **Footprint** dialog box. The **Pick** dialog box will be displayed; refer to Figure-27.

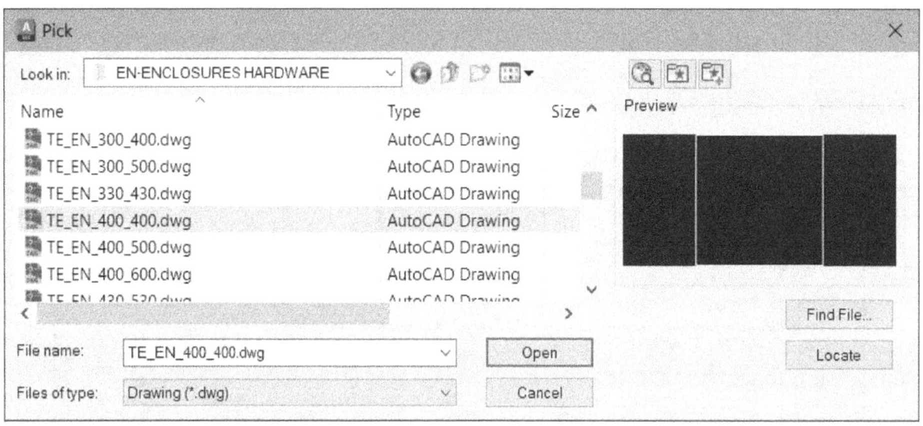

Figure-27. Pick dialog box

- Select the desired file and click on the **Open** button from the dialog box. The selected enclosure will get attached to the cursor.
- Click in the drawing area to place the enclosure. You are asked to specify the rotation angle for the enclosure.
- Specify the rotation as **0** and press **ENTER**. The **Panel Layout-Component Insert/ Edit** dialog box will be displayed; refer to Figure-28.
- Click in the **Line 1** of **Description** area and specify the value as **Panel** enclosure.
- Click on the **Next** button from the **Item Number** area to specify the item number.
- Click on the **OK** button from the dialog box.

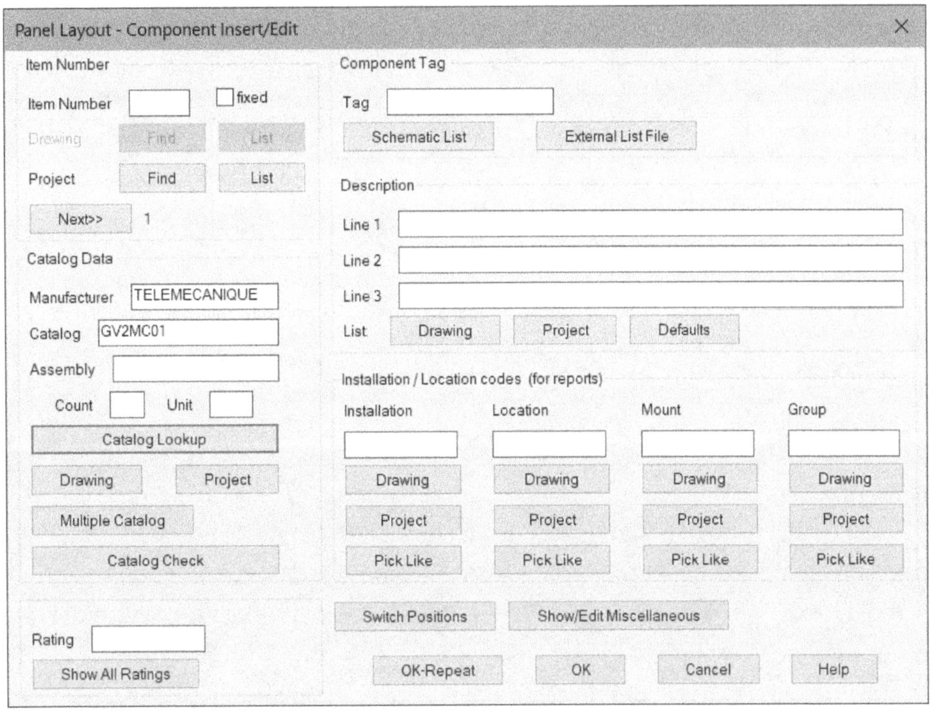

Figure-28. Panel Layout dialog box

Inserting buttons in the Panel layout

- Click on the **Schematic List** button from the **Insert Component Footprints** drop-down in the **Insert Component Footprints** panel in the **Ribbon**. The **Schematic Components List** dialog box will be displayed as shown in Figure-29.

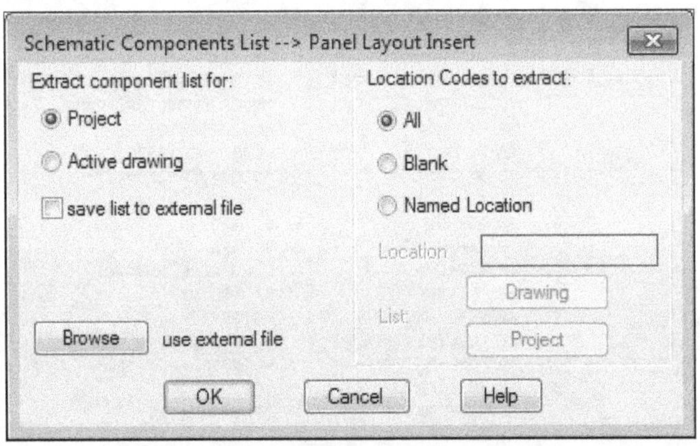

Figure-29. Schematic Components List dialog box

- Click on the **OK** button from the dialog box. The **Select Drawings to Process** dialog box will be displayed; refer to Figure-30.

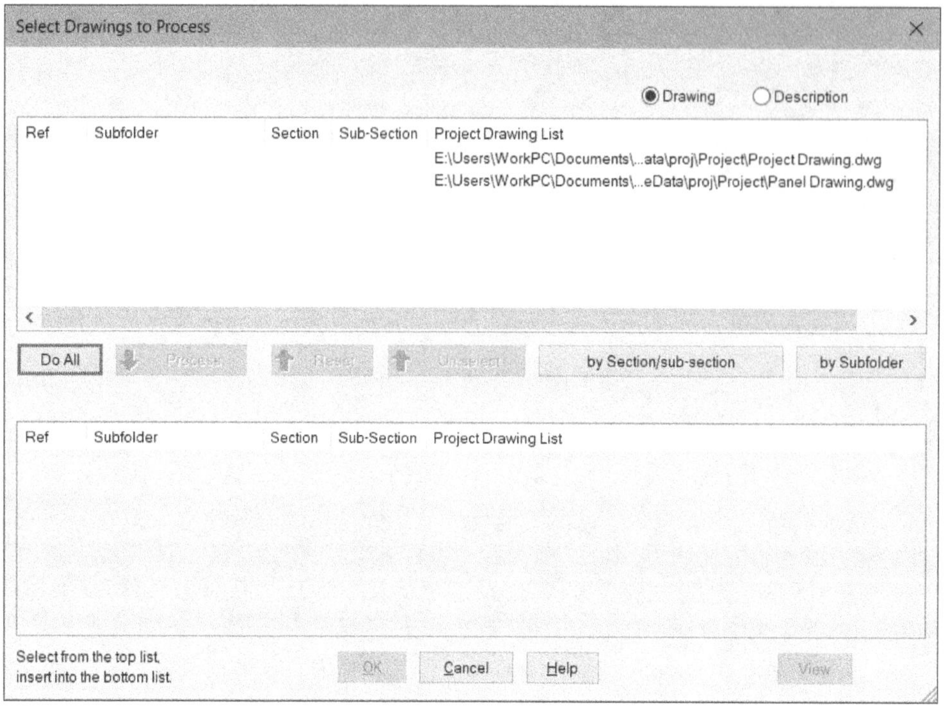

Figure-30. Select Drawings to Process dialog box

- Select the **Project 1** drawing from the list and click on the **Process** button from the dialog box.
- Click on the **OK** button from the dialog box. The **Schematic Components (active project)** dialog box will be displayed; refer to Figure-31.

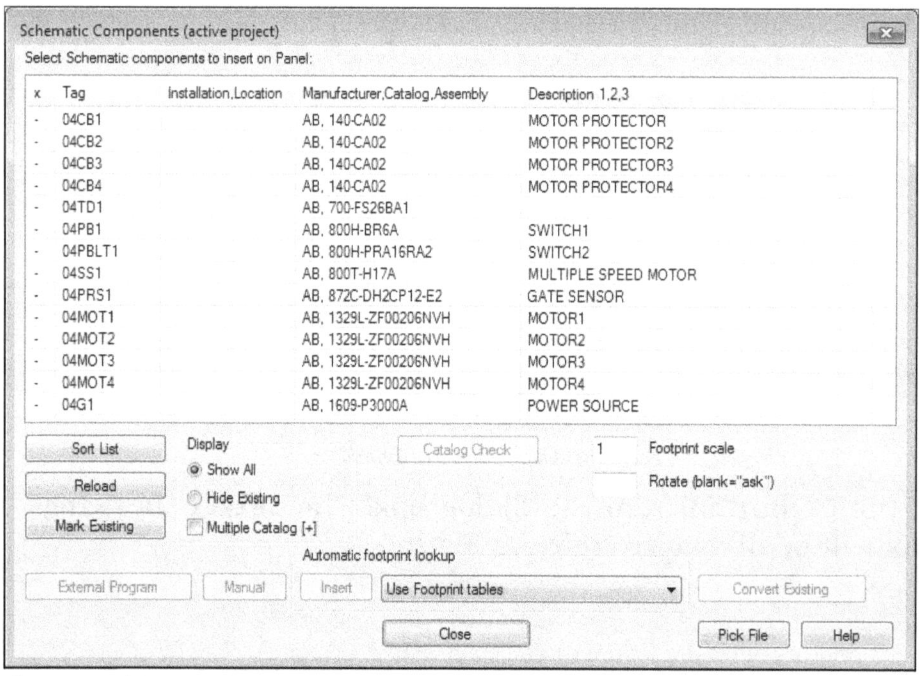

Figure-31. Schematic Components dialog box

- Select the component from the list for which you want to insert the footprint and click on the **Insert** button from the dialog box.
- If the footprint is mapped to component then the footprint will get attached to the cursor and you are asked to place the footprint.

- Place the footprint in the panel enclosure. The **Panel Layout** dialog box will be displayed. Specify the desired parameters and click on the **OK** button from the dialog box.
- If the footprint for the component is not mapped then the **Footprint** dialog box will be displayed and you need to manually select the block for the component by using the **Browse** button as done earlier.
- Similarly, place all the desired components in the panel enclosure. After adding all the components, the panel is displayed as shown in Figure-32.

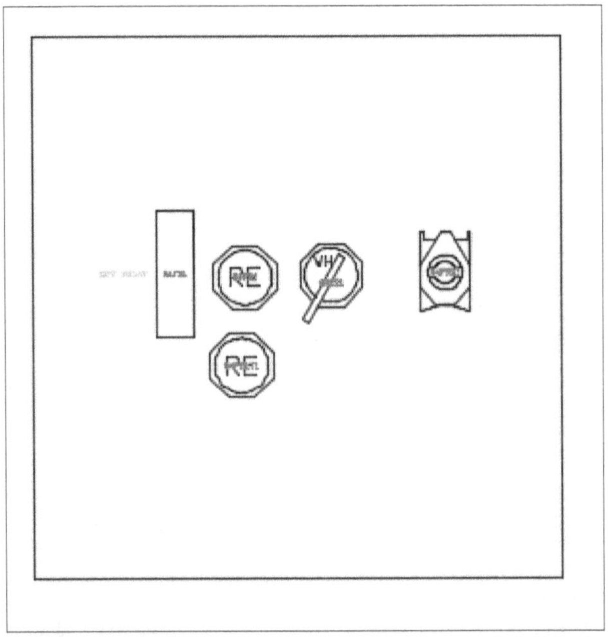

Figure-32. Panel after adding components

- Specify the desired annotation to the buttons to make it easy for interpretation.

Generating Reports

- Click on the **Reports** tab in the **Ribbon**. The tools related to reports will be displayed; refer to Figure-33.

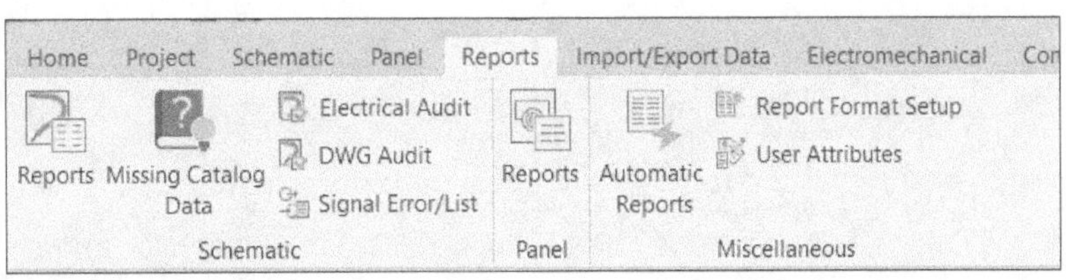

Figure-33. Reports tab

- Click on the **Reports** tool from the **Schematic** panel in the tab. The **Schematic Reports** dialog box will be displayed; refer to Figure-34.

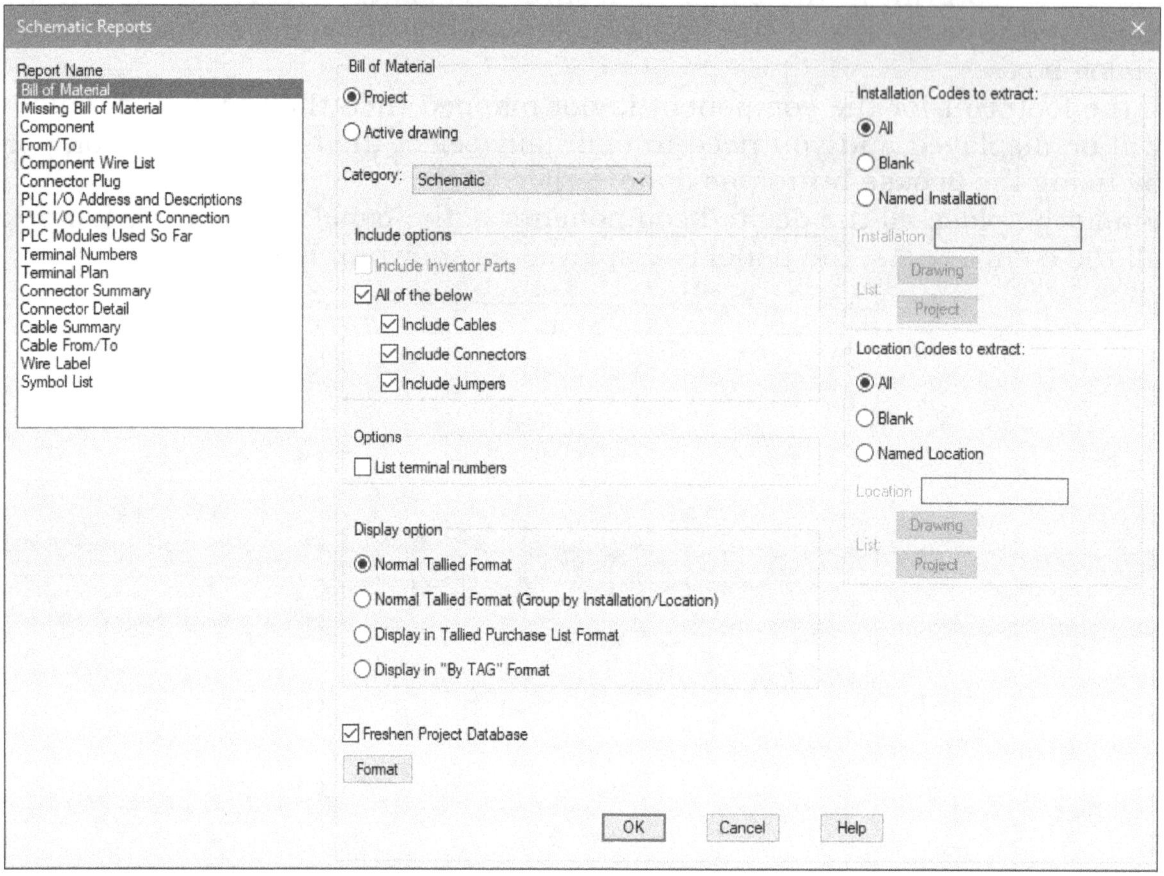

Figure-34. Schematic Reports dialog box

- Select the **Bill of Material** option from the **Report Name** selection box.
- Click on the **OK** button from the dialog box. The **Select Drawings to Process** dialog box will be displayed.
- Click on the **Do All** button and click on the **OK** button from the dialog box. The **QSAVE** dialog box will be displayed if the drawing is not saved.
- Click on the **OK** button from the dialog box. The **Report Generator** dialog box will be displayed; refer to Figure-35.

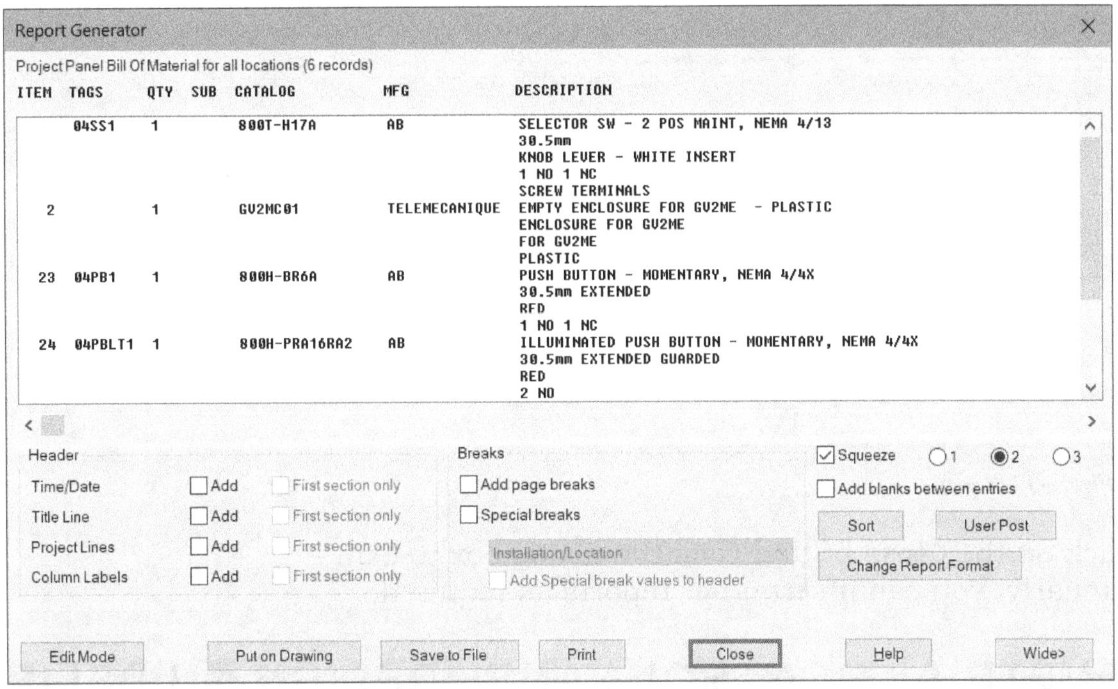

Figure-35. Report Generator dialog box

- Click on the **Put on Drawing** button. The **Table Generation Setup** dialog box will be displayed; refer to Figure-36.

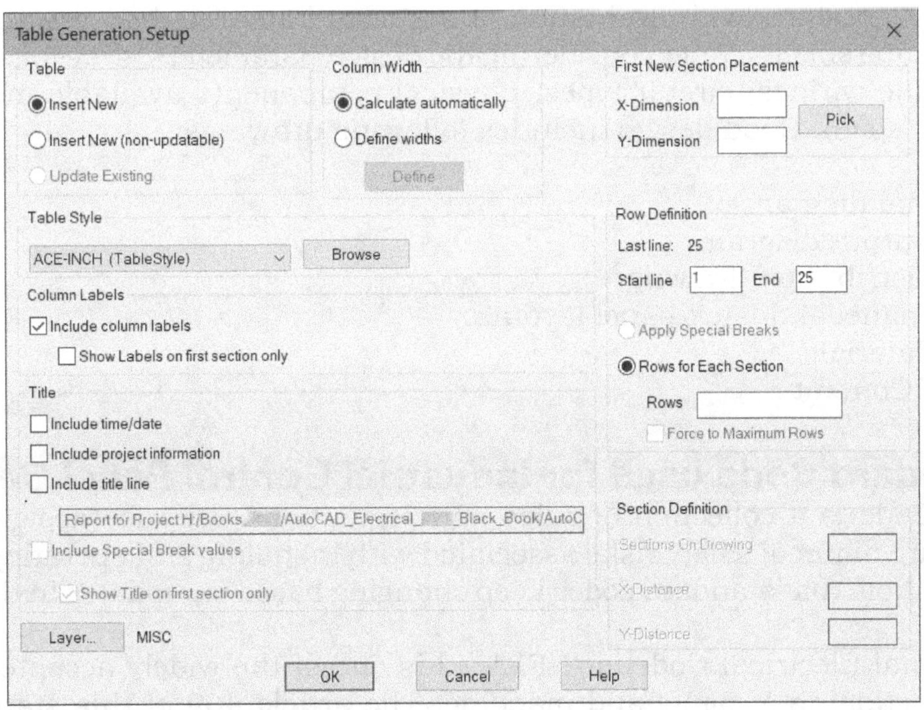

Figure-36. Table Generation Setup dialog box

- Click on the **OK** button from the dialog box. The table will get attached to the cursor.
- Click at the desired location in the drawing area to place the table. Refer to Figure-37.

TAGS	QTY	SUB	CATALOG	MFG	DESCRIPTION
04CB1 04CB2 04CB3 04CB4	4		140-CA02	AB	AUXILIARY CONTACT BLOCK, INTERNAL FRONT MOUNTED, 2 NC AUXILIARY INTERNAL FRONT MOUNTED 2 N.C
04MOT1 04MOT2 04MOT3 04MOT4	4		1329L-ZF00206NVH	AB	BUILD TO ORDER AC VARIABLE SPEED MOTOR, FOOT MOUNTED, COUPLED OR BELTED DUTY PER STANDARD NEMA LIMITS, SERIES A, 1329L AC MOTOR 2HP 230V 650RPM, CONSTANT TORQUE, 1000:1 SPEED RANGE, MOTOR ENCLOSURE: TENV
04PB1	1		800H-BR6A	AB	PUSH BUTTON - MOMENTARY, NEMA 4/4X 30.5mm EXTENDED RED 1 NO 1 NC
04PBLT1	1		800H-PRA16RA2	AB	ILLUMINATED PUSH BUTTON - MOMENTARY, NEMA 4/4X 30.5mm EXTENDED GUARDED RED 2 NO 120 AC XFMR
04G1	1		1609-P3000A	AB	UNITERRUPTIBLE MODE POWER SUPPLY, 1609 SERIES 230VAC INDUSTRIAL SERIES 3 KVA, 230VAC INDUSTRIAL DOUBLE CONVERSION ON-LINE UPS
04PRS1	1		872C-DH2CP12-E2	AB	PROXIMITY SWITCH 2.0MM SENSING RANGE, N.C., PNP
04SS1	1		800T-H17A	AB	SELECTOR SW - 2 POS MAINT, NEMA 4/13 30.5mm KNOB LEVER - WHITE INSERT 1 NO 1 NC SCREW TERMINALS
04TD1	1		700-FS26BA1	AB	TIME DELAY RELAY - TYPE FS OFF DELAY 120VAC 1 NO (CONVERTIBLE) DELAY - 0.15 SEC. TO 3 MIN.

Figure-37. Report

- Click on the **Close** button from the dialog box to exit.
- Similarly, you can insert other reports as required.

INDUSTRIAL CONTROL PANEL DESIGN GUIDELINES

Control panel designing for industrial use addresses various issues like interface design for user panel, load requirements for control panel, wire size calculations, enclosure selection based on environmental standards, and so on. The procedure of panel design starts with evaluation of specifications, regulatory standards, and customer requirements. After considering the design parameters, drawings are created which outline wirings, circuit types, physical components available in market for design, and so on. Good design includes following drawings:

1. Functional diagram
2. Input/Output diagram
3. Power Distribution drawing
4. Control panel and back panel layouts
5. Bill of Materials
6. Table of Content

Standard Code used for Industrial Control Panel Design

Standard Code is a collection of rules and regulations used for designing panels which ensure minimal safety risks associated with installing and operating industrial machinery. Note that standard codes keep changing based on updates in the industry.

NEC (National Electrical Code) or NFPA 70 is one of the widely accepted standard for safe functioning of panel and machine. The article 409 of this standard deals with panels intended for voltage less than or equal to 600V. According to this article, industrial control panels are evaluated and marked for their Short Circuit Current Rating (SCCR). SCCR is calculated by evaluating each feeder individually as well as all branch circuits. The smallest kA value found in calculation is used as the kA value for the panel as a whole.

NFPA (National Fire Protection Association) 79 is section of NEC which deals with size of wires used in panel for safety. The scope of this standard encompasses electrical and electronic elements of all machinery that operates at or below 600V, including injection molding machines, assembly machinery, machine tools, and material handling machinery, among others, as well as inspection and testing machinery.

Design Consideration for Industrial Control Panel

There are various criteria that need to be checked for designing control panel like enclosure and space requirement, wire sizing, control components, circuits, and so on for meeting functional requirements, application specifications, and regulatory standards. These criteria are discussed next.

Enclosures and Space Requirements

Generally the environment dictates the type of enclosure. First thing to consider is space in enclosure for components and wires. There should be enough space in enclosure for placing all the components and wires. Second point to consider is placement location for control panel enclosure. If you are using cabinet enclosure then make sure there is enough space for swing of cabinet door. It is important to know temperature of surroundings where enclosure will be placed. If there is high temperature near control panel then a ventilation fan or air conditioner may be required to keep control panel operating within ideal operating temperature.

Wire size and Component Type

Wire size should be enough to supported required load current. Note that you need to also select circuit protection based on wire size. Selecting the appropriate wire size is essential for ensuring that the circuit is capable of delivering the required load current, while selecting the best circuit protection reduces the risk of fire by preventing wires from overheating. Main supply wires/conductors should be sized in such a way that it can bear sum of all the loads in circuit plus 125% of full load current of highest rated motor.

Components should be selected based on their load capacity and required function. Selecting components of right size is critical as voltage and load currents should match with the carrying capacity of wires. There are two categories of components, power circuit components and control circuit components. Power circuit components are the ones which perform action in the machinery and control circuit components are those which control how and when the actions will be performed. For example, a motor is a power circuit component and a proximity sensor is a control circuit component. Make sure to also add overcurrent protection component in your panel.

FOR STUDENT NOTES

Chapter 12

AutoCAD Electrical with Autodesk Inventor

Topics Covered

The major topics covered in this chapter are:

- ***Electromechanical Link***
- ***Electrical Part creation in Inventor***
- ***Wire Harness in Autodesk Inventor***

INTRODUCTION TO AUTODESK INVENTOR

Autodesk Inventor is a CAD software which provides tools to create 3D models. Once you have created electrical drawings for the circuit, the next step is to create panel drawings. You have learnt these things in earlier chapters. Note that the panel drawing is enough to manufacture the panels in real-world but for representation purpose and for making the things easy to visualize, we use 3D models. We need the 3D model of the panel with all the components placed at their proper position to get complete view of how our panel will look once manufactured. For creating 3D models of panel with electrical properties saved in AutoCAD Electrical, you need Autodesk Inventor. After creating the model, the next step is to connect model with the electrical drawing for easy synchronization. AutoCAD Electrical gives us capability to make connections between Electrical drawing and Inventor models by Electromechanical linking. Creating models in Autodesk Inventor is not in the scope of this book. For that you can use our another book Autodesk Inventor 2027 Black Book. Here, we will start with creating an Electromechanical link and then we will discuss other options related to linking.

CREATING ELECTROMECHANICAL LINK

Electromechanical link is an option to connect the electrical drawing with its 3D model. The 3D model generally includes the wire harness same as wiring in the electrical drawing. The procedure to create an electromechanical link file in AutoCAD Electrical is given next.

- Open any project that you have created earlier. Now, open the panel drawing for which the 3D model has been created in AutoCAD Electrical; refer to Figure-1.

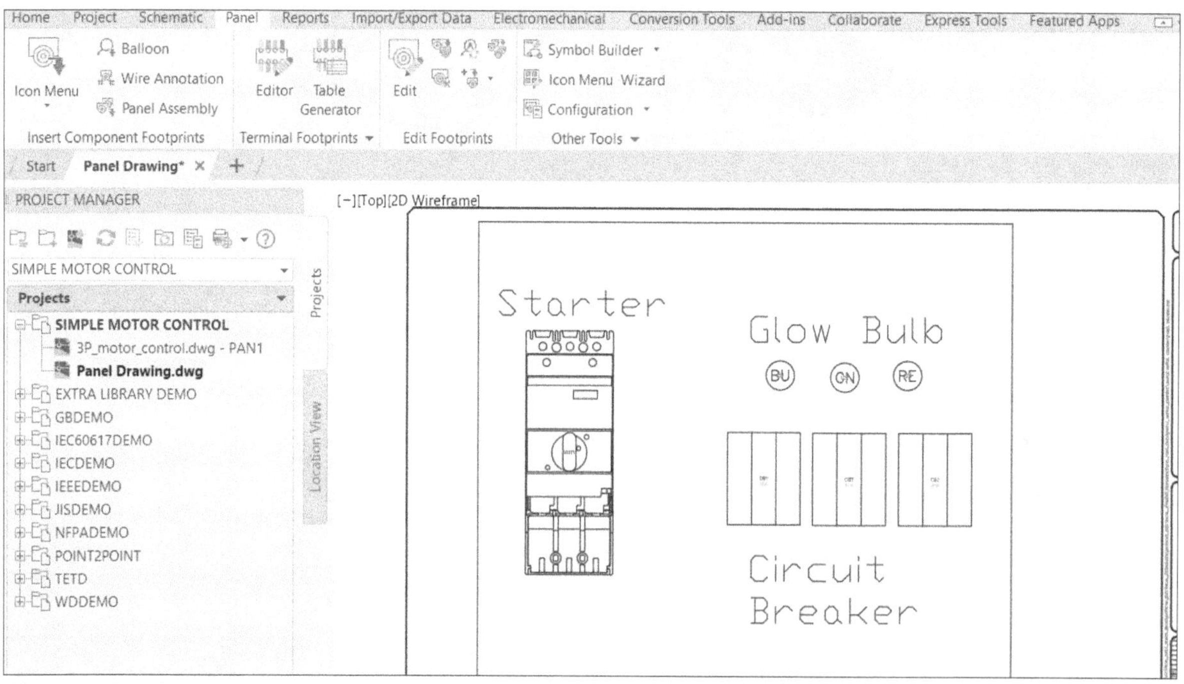

Figure-1. Panel drawing in AutoCAD Electrical

- Click on the **Electromechanical** tab from the **Ribbon**. The tool(s) related to electromechanical linking will be displayed; refer to Figure-2.

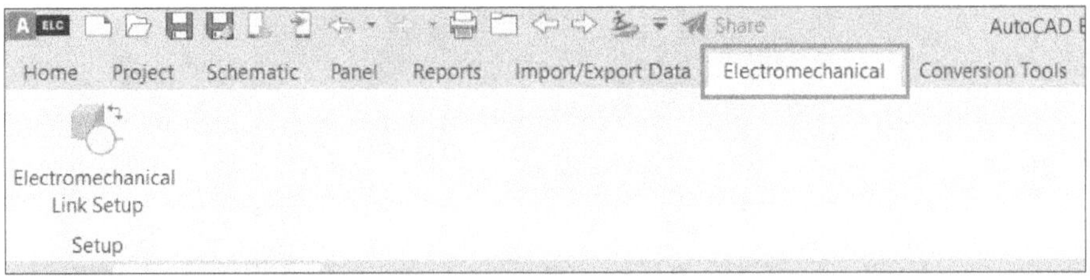

Figure-2. Electromechanical tab in Ribbon

- Click on the **Electromechanical Link Setup** tool from the **Ribbon**. The **Electromechanical Link Setup** dialog box will be displayed; refer to Figure-3.

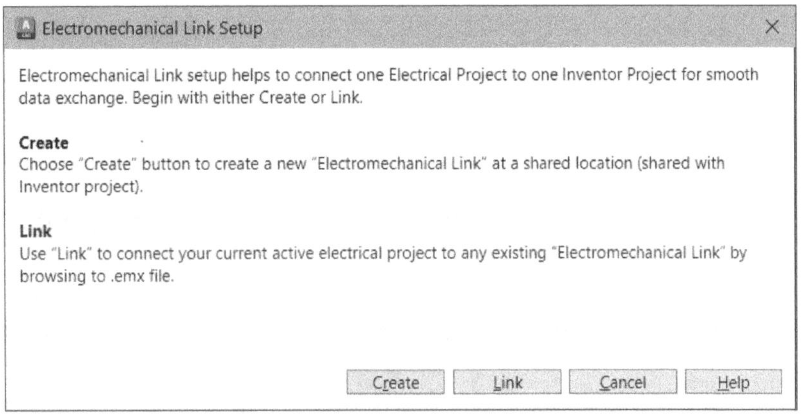

Figure-3. Electromechanical Link Setup dialog box

- There are two ways for linking electrical drawing and inventor model. Either create a new link or use the link earlier created with Autodesk Inventor. Both the ways are discussed next.

Creating New Electromechanical link

- Click on the **Create** button from the **Electromechanical Link Setup** dialog box. The **Create Electromechanical Link** dialog box will be displayed; refer to Figure-4.

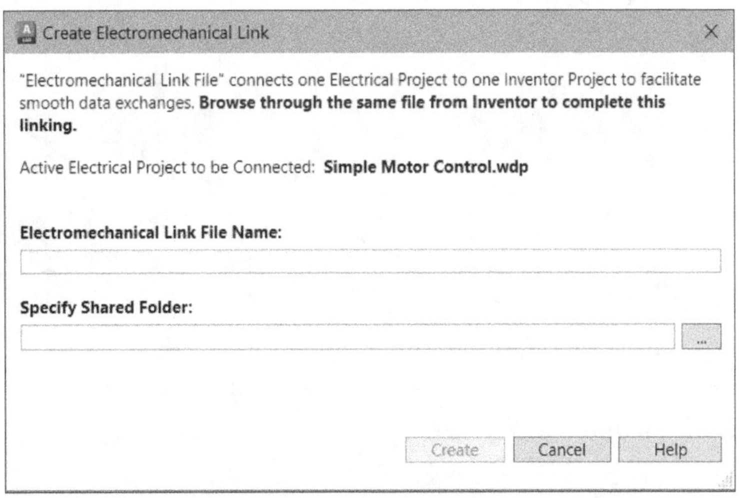

Figure-4. Create Electromechanical Link dialog box

- Specify desired name for the link file in the **Electromechanical Link File Name** edit box.
- Click on the ellipse button (...) to browse for shared folder. The **Browse For Folder** dialog box will be displayed; refer to Figure-5.

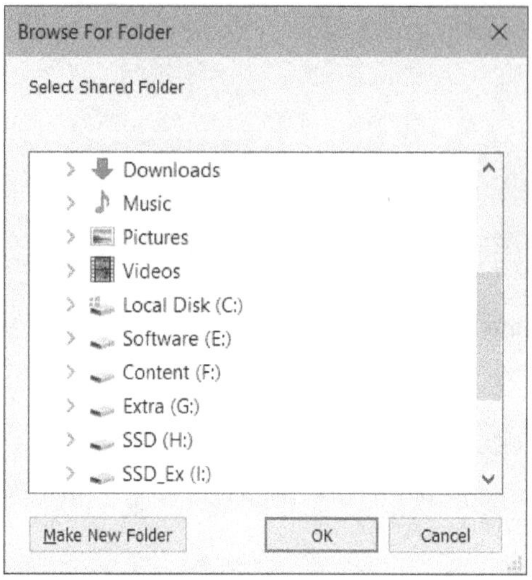

Figure-5. Browse For Folder dialog box

- Browse to desired folder from the dialog box and click on the **OK** button from the dialog box to make it shared folder.
- Click on the **Create** button from the dialog box. The link file will be created and the **Electromechanical Link Creation** dialog box will be displayed; refer to Figure-6.

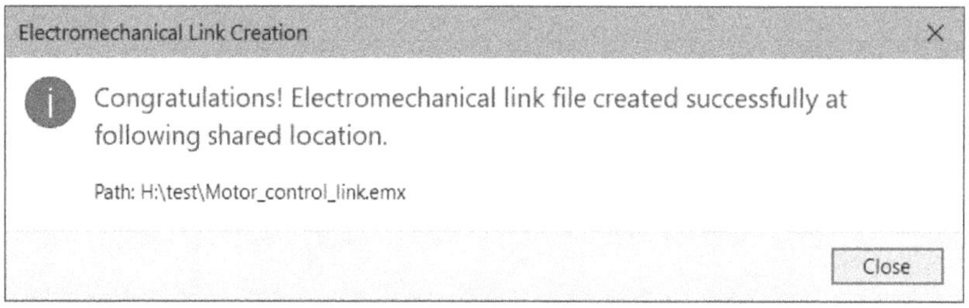

Figure-6. Electromechanical Link Creation dialog box

- Click on the **Close** button from the dialog box. The **Electromechanical Link Setup** dialog box will be displayed; refer to Figure-7.

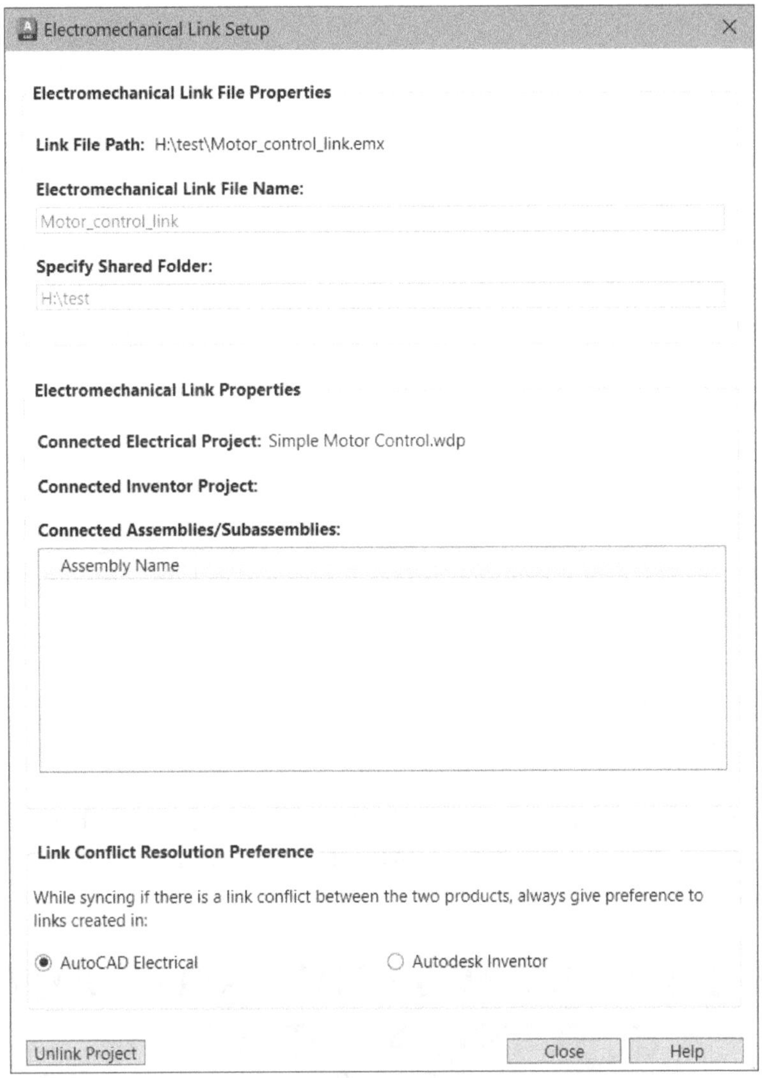

Figure-7. Electromechanical Link Setup dialog box with properties

- Select desired option from the **Link Conflict Resolution Preference** area of the dialog box to set the priority of AutoCAD Electrical or Autodesk Inventor in case of conflict during synchronization. In simple word: If there is any conflict between files of AutoCAD Electrical and Autodesk Inventor then priority will be given to the software for which the radio button is selected in the **Link Conflict Resolution Preference** area.
- Click on the **Close** button from the dialog box to exit the dialog box.

Note that you can revise the properties of the link anytime by using the **Electromechanical Link Setup** button from the **Electromechanical** tab in the **Ribbon**.

Using Existing Electromechanical Link in AutoCAD Electrical

- You can use any existing electromechanical link file instead of creating new file. After selecting the **Electromechanical Link Setup** button from the **Ribbon**, click on the **Link** button from the **Electromechanical Link Setup** dialog box. The **Select Electromechanical Link File** dialog box will be displayed; refer to Figure-8.

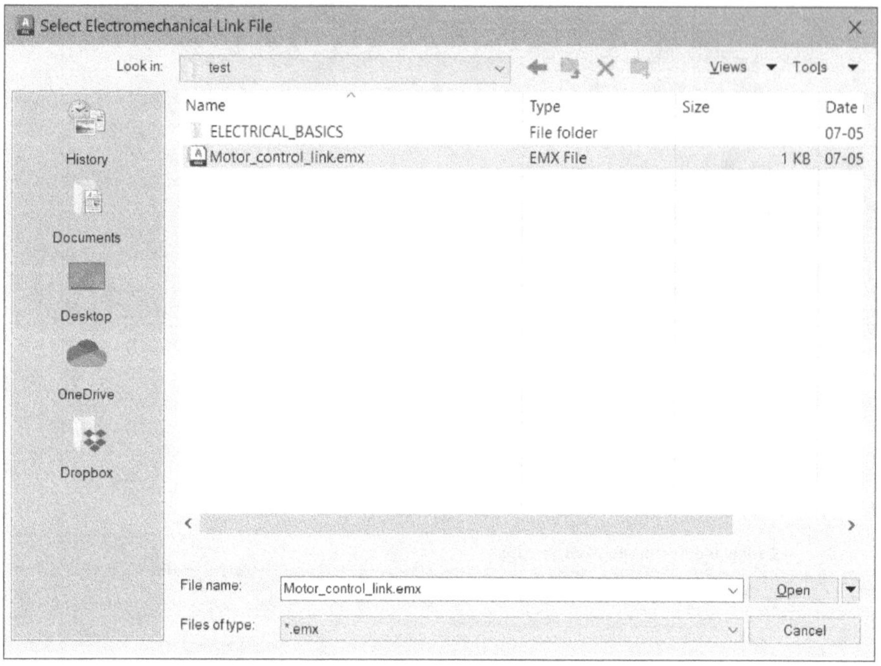

Figure-8. Select Electromechanical Link File dialog box

- Browse to desired folder and select the already existing file for linking the electrical and inventor project.

Note that the Electromechanical link file acts as a bridge between inventor and AutoCAD Electrical so the same file should be linked to projects in both AutoCAD Electrical as well as Autodesk Inventor.

LINKING INVENTOR MODEL WITH AUTOCAD ELECTRICAL DRAWING

I hope that you have already created the Wire harness model of the drawing in Autodesk Inventor. The steps after creating the model are given next.

- Open the assembly file of Electrical model in Autodesk Inventor; refer to Figure-9.

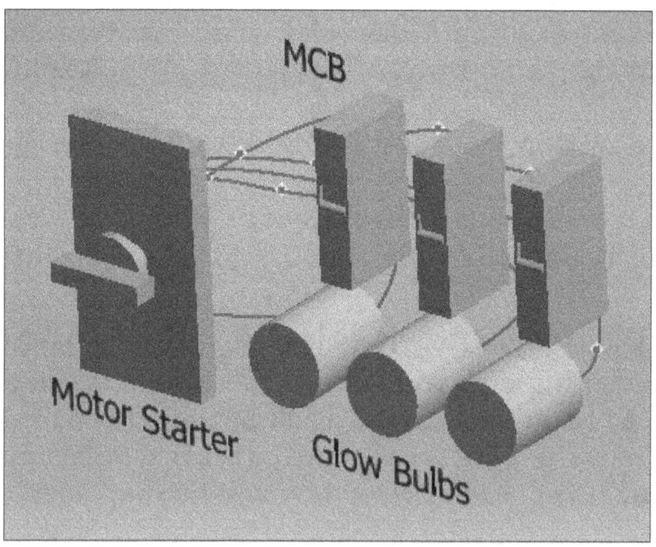

Figure-9. 3D Model of Electrical assembly

- Click on the **Electromechanical** tab from the **Ribbon** in Inventor. The tools related to Electromechanical linking will be displayed; refer to Figure-10.

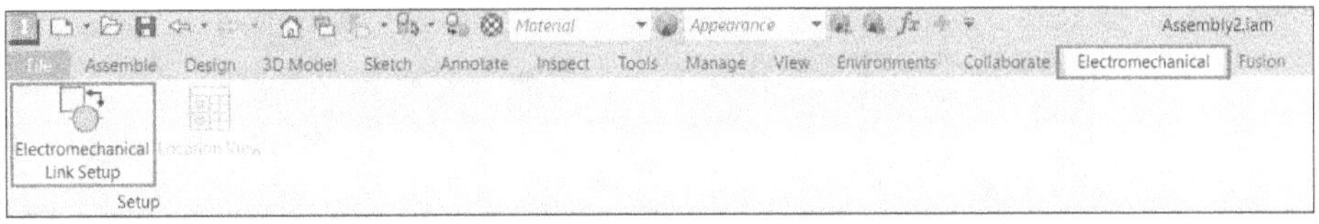

Figure-10. Electromechanical tab in Inventor

- Click on the **Electromechanical Link Setup** button from the **Setup** panel in the **Electromechanical** tab. The **Electromechanical Link Setup** dialog box will be displayed as discussed earlier.
- Click on the **Link to an electromechanical file** button from the dialog box. The **Select Electromechanical File** dialog box will be displayed as shown in Figure-11.

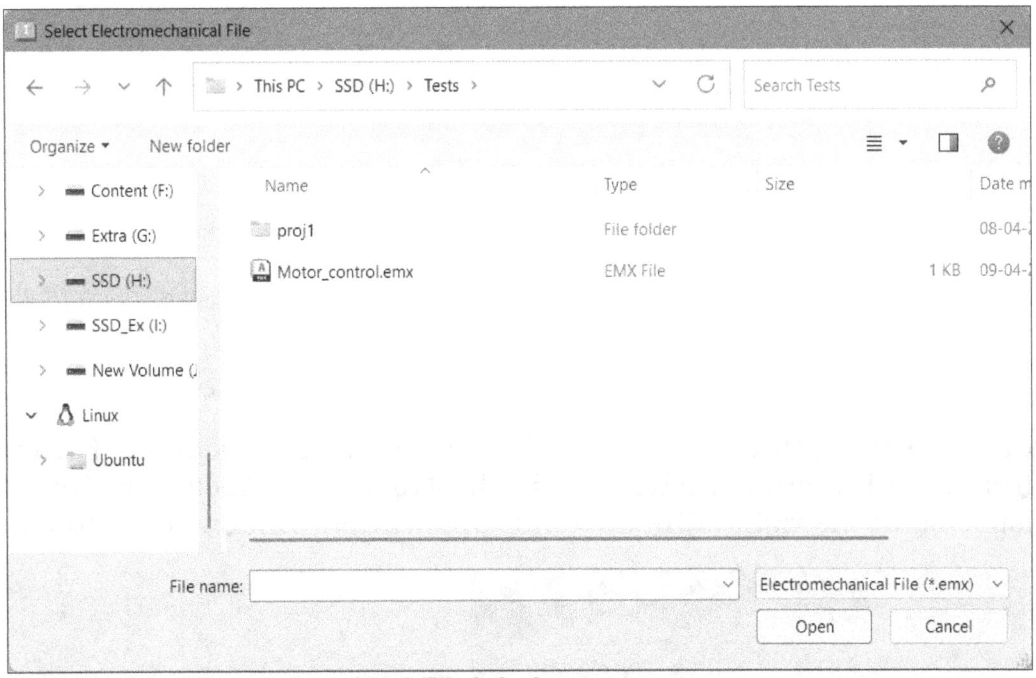

Figure-11. Select Electromechanical Link File dialog box

- Select the link file earlier created by AutoCAD Electrical and click on the **Open** button from the dialog box. If there are unsaved changes in the assembly file then you will be asked to save the unsaved changes; refer to Figure-12.

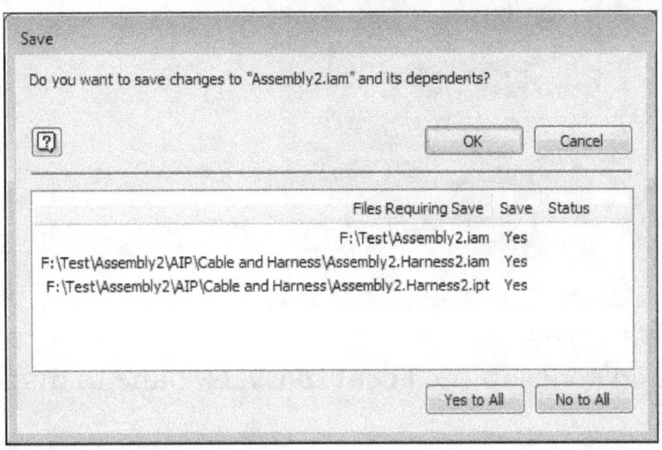

Figure-12. Save dialog box

- Click on the **Yes to All** button and then on the **OK** button from the dialog box to save the changes. The **Electromechanical Link Setup** dialog box will be displayed; refer to Figure-13.

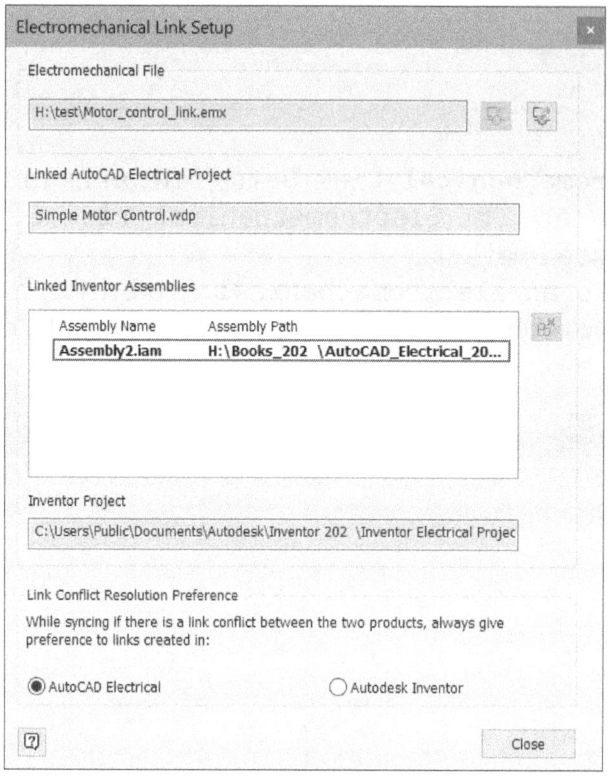

Figure-13. Electromechanical Link Setup dialog box with Inventor project selected

- Click on the **Close** button from the dialog box and then click on the **Location View** tool from the **Electromechanical** tab in the **Ribbon**. The **Location View** pane will be displayed at the right of the Inventor application window; refer to Figure-14.

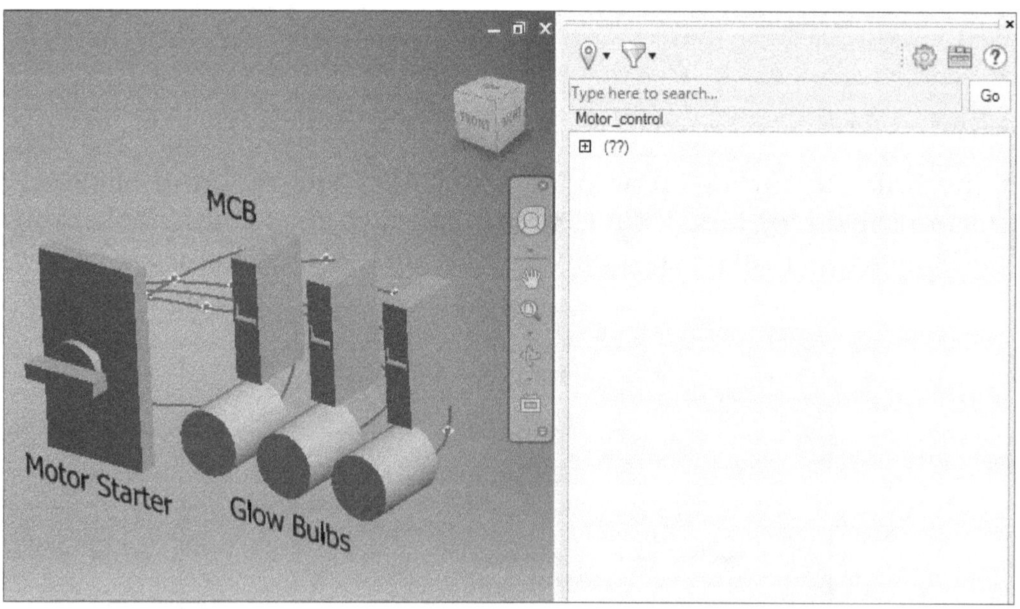

Figure-14. Location View pane

- Expand the location codes in the **Location View** pane to display the components; refer to Figure-15.

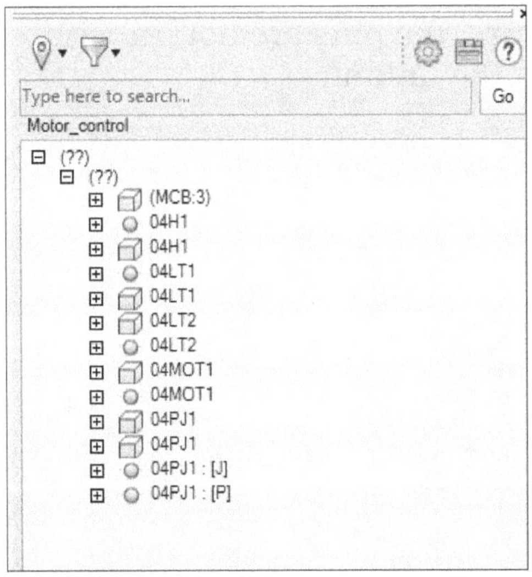

Figure-15. Expanded locations in Location View pane

Note that in Figure-15, the components are displayed with different icons. These icons show the nature and origin of component. There are four type of icons that are displayed in **Location View** pane before the component name, which are:

Inventor Only 

Electrical Only 

Linked 

Linked with Differences 

- To link the electrical component to the Autodesk Inventor component, right-click on desired electrical only component. A shortcut menu will be displayed as shown in Figure-16.

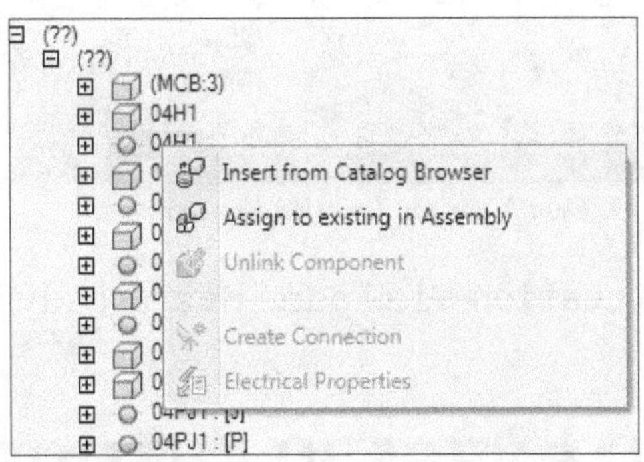

Figure-16. Shortcut menu for linking file

- Click on the **Assign to existing in Assembly** option if you have the component already created in the model space. You will be asked to select the component from the model space.
- Click on the component that you want to link with the selected electrical component. On selecting the Inventor component, both the components will be replaced with a common component with symbol of linking; refer to Figure-17.

- If we expand the link, we can find out the differences between the two objects like in Figure-18. In this figure, the pins are not properly connected and hence cause the differences between the linking.

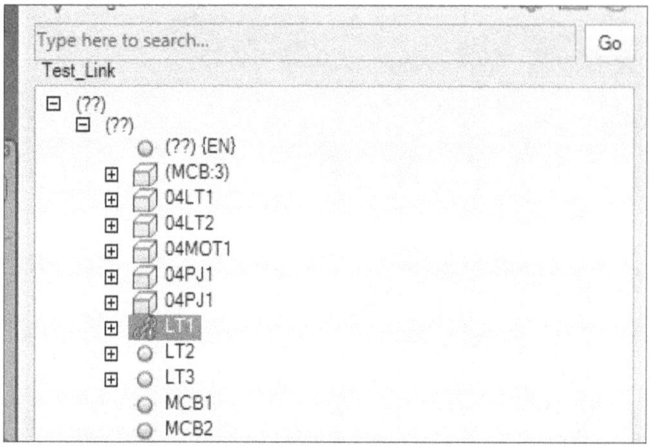

Figure-17. Linked light with differences

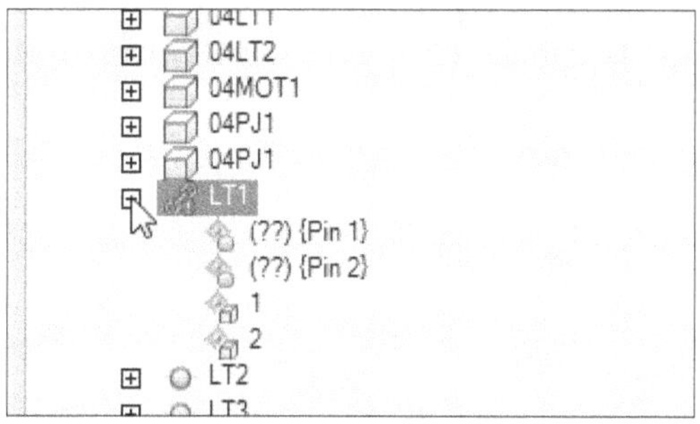

Figure-18. Differences in the link

- We can resolve such problems by using the **Details** pane of **Location View** tab; refer to Figure-19. (Click on **Display Details and Connections** button in the **Location View** pane to show these options).

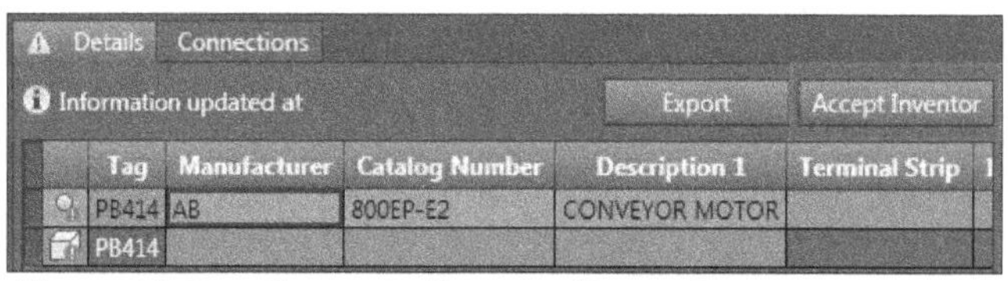

Figure-19. Details pane in Location View tab

Note that the functions of **Location View** pane has already been discussed in Chapter 3 of this book.

ELECTRICAL HARNESS IN AUTODESK INVENTOR FOR AUTOCAD ELECTRICAL

This topic is not related to AutoCAD Electrical directly. In this topic, we will be discussing the functions of Autodesk Inventor for Electrical harnessing. So, those who are not willing to work on Autodesk Inventor for 3D electrical model of panel can skip this topic.

Creating Electrical components in Autodesk Inventor

After creating the Electrical drawing some of the client may demand the 3D model of panel from you. For creating 3D model of the panel, you will need the components of the panel. Most of the time, you will not be able to get all the components in the standard library. In those cases, you will need a procedure to create custom components which have electrical properties. Note that you can use grabcad.com traceparts.com, and component manufacturer's website to download 3D models of standard components. The procedure to create such components in Autodesk Inventor is given next.

- After starting Autodesk Inventor, click on the **Part** option from the **New** cascading menu in the **Application** menu; refer to Figure-20. The modeling environment will be displayed.

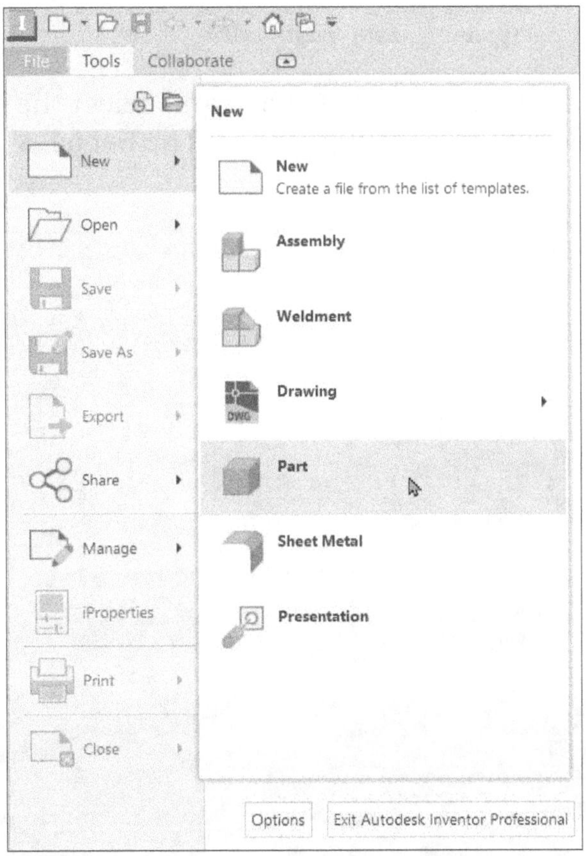

Figure-20. Part option from Application menu

- Create a solid model by using the 3D modeling tools of Autodesk Inventor; refer to Figure-21.

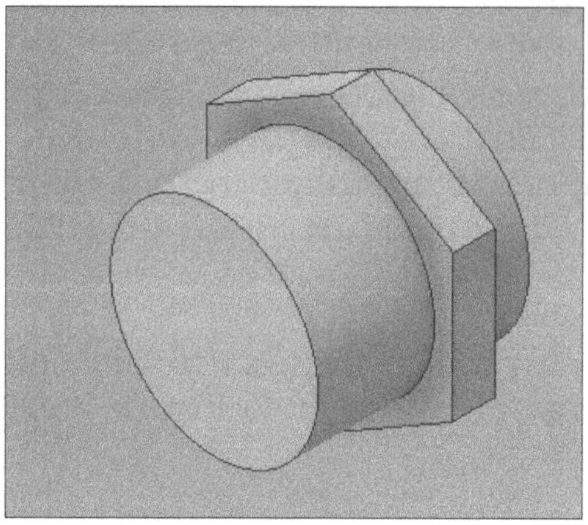

Figure-21. Model of a glow bulb

- Right-click on any panel name in the **Ribbon** and select the **Harness** option from the shortcut menu displayed; refer to Figure-22. The **Harness** panel will be displayed, see Figure-23.

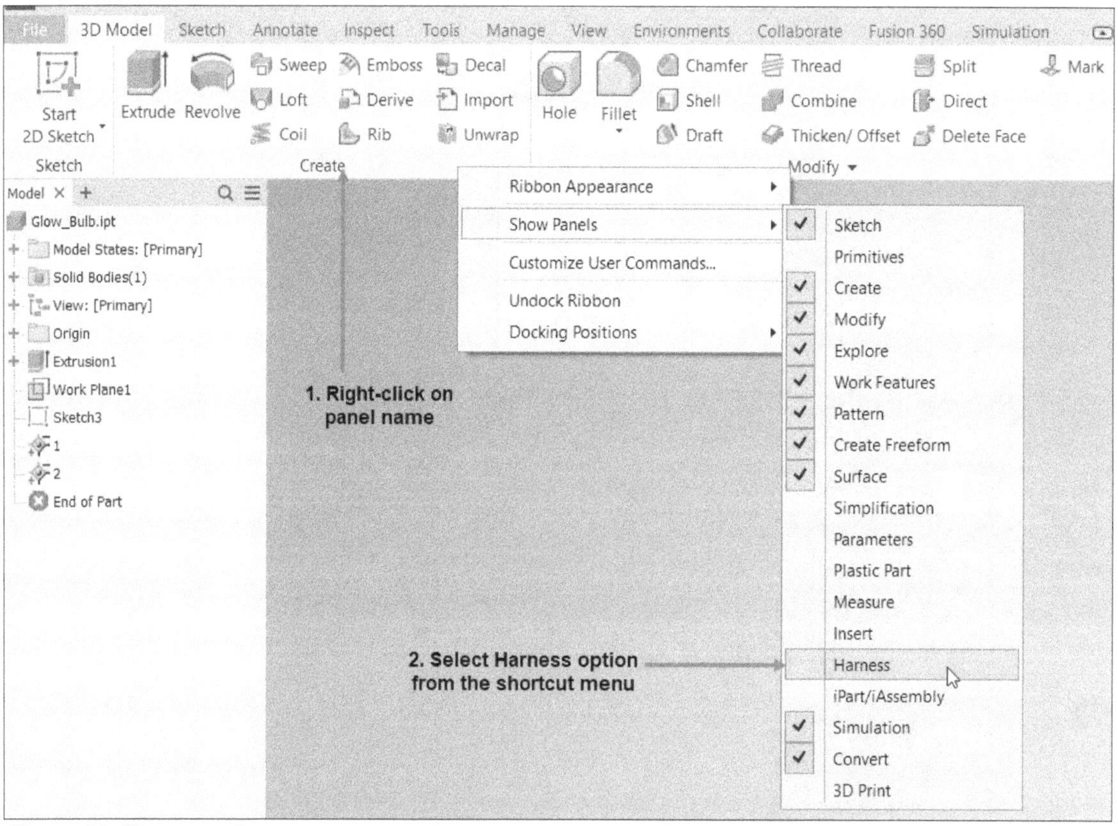

Figure-22. Harness option

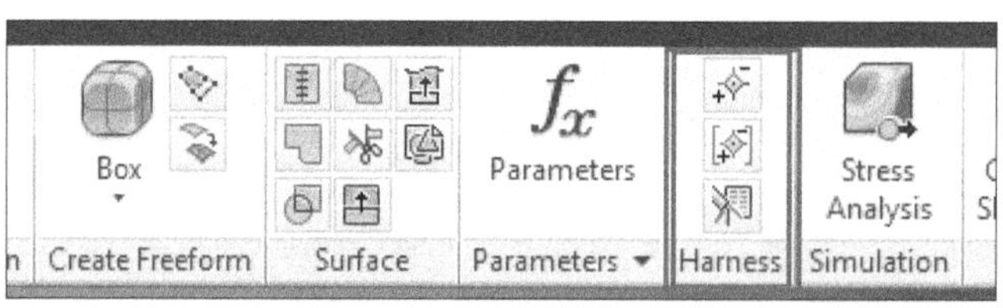

Figure-23. Harness panel

- Click on the **Place Pin** tool from the **Harness** panel in the **Ribbon**. You are asked to specify position of the pins.
- Click at desired location on the face of model. The **Place Pin** dialog box will be displayed; refer to Figure-24.

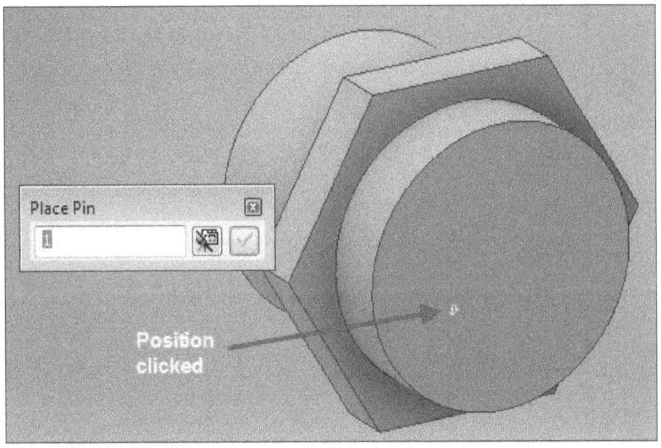

Figure-24. Placing pin

- Specify the pin number in the dialog box and click on the **OK** button. You will be asked to specify position of the next pin. Continue the same procedure till you get desired number of pins.
- After creating desired number of pins, right-click in the empty area of the modeling space and select the **Done** option from the shortcut menu; refer to Figure-25. You can also press the **ESC** button from keyboard in place of selecting **Done** option from the shortcut menu.

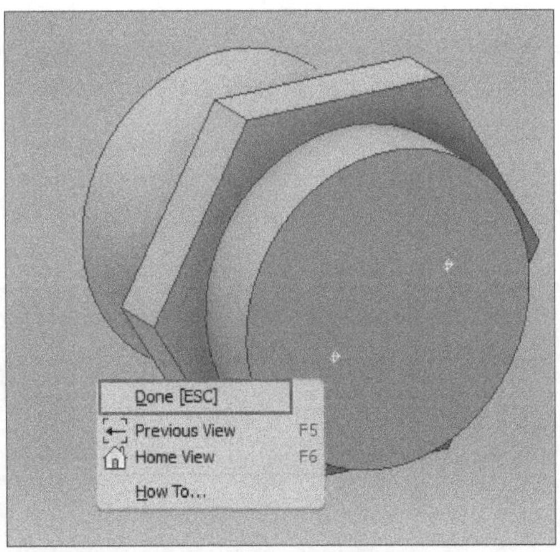

Figure-25. Completing pin creation

- Save the part and use it in assembly for harness.

Placing Multiple pins on part

Instead of placing pins individually, you can place a group of pins on the component to make it useful for connectors. The procedure of placing multiple pins is given next.

- Click on the **Place Pin Group** tool from the **Harness** panel in the **Ribbon**. The **Place Pin Group** dialog box will be displayed; refer to Figure-26.
- Specify desired number of pins in the **Pins Per Row** edit box in the dialog box.

Similarly, specify the number of rows, distance between two pins (pitch), starting number of pins and prefix.

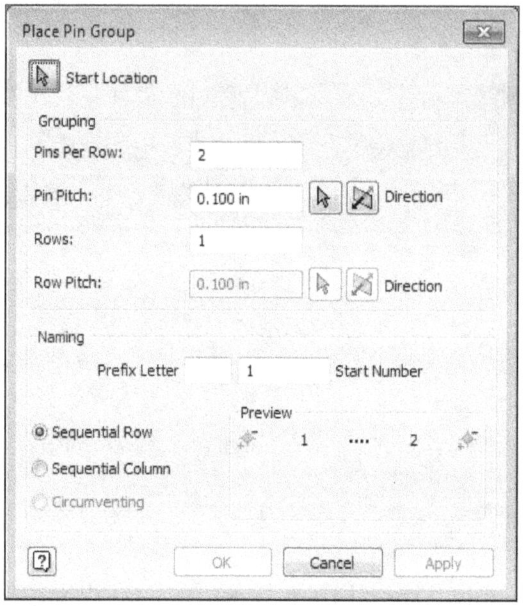

Figure-26. Place Pin Group dialog box

- Select desired option from the **Sequential Row**, **Sequential Column**, and **Circumventing** radio buttons.
- Click at desired location on the model to create the pin group. Note that you need to have a reference point at the location where you want to place pin group. In Figure-27, we have already created a sketch point on the surface of model, which we are going to select.

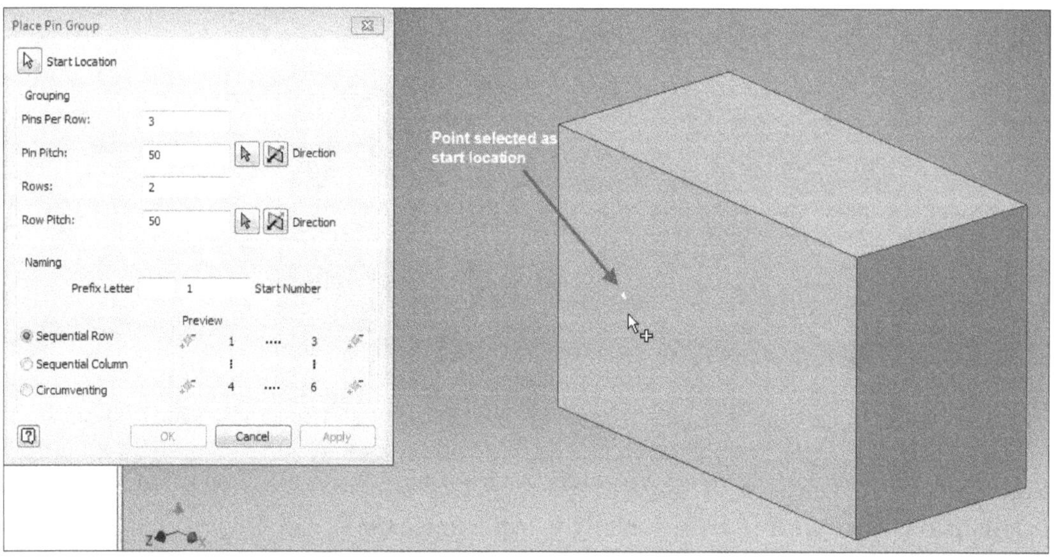

Figure-27. Specifying start location for pin group

- Click on the **Select reference** button next to **Pin Pitch** edit box in the dialog box to select reference for column direction; refer to Figure-28. On doing so, you will be asked to select a reference entity.

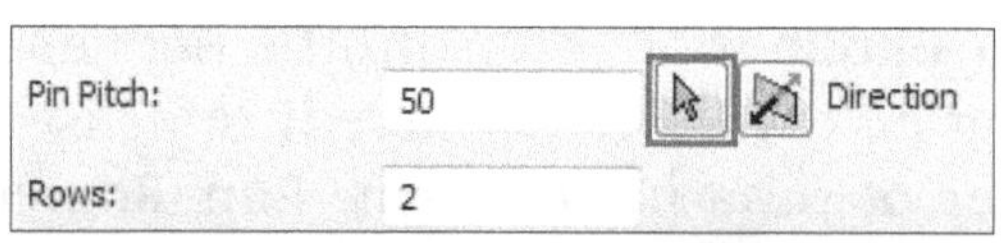

Figure-28. Button selecting Pin Pitch reference

- Select a reference like in Figure-29. You can flip direction of pin pitch by using the **Flip** button next to **Select reference** button.

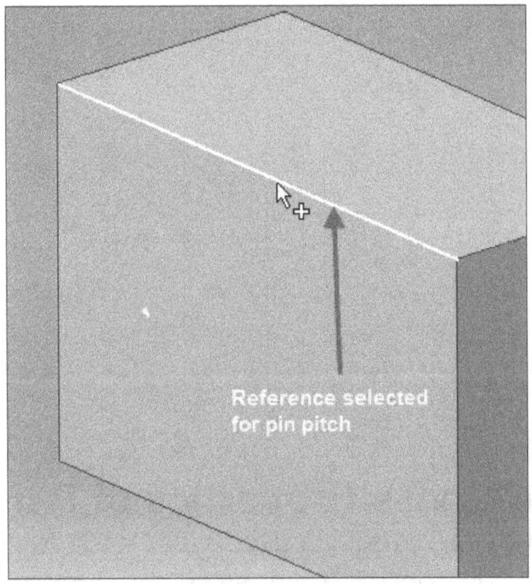

Figure-29. Reference selected for pin pitch

- Similarly, select the row pitch; refer to Figure-30.

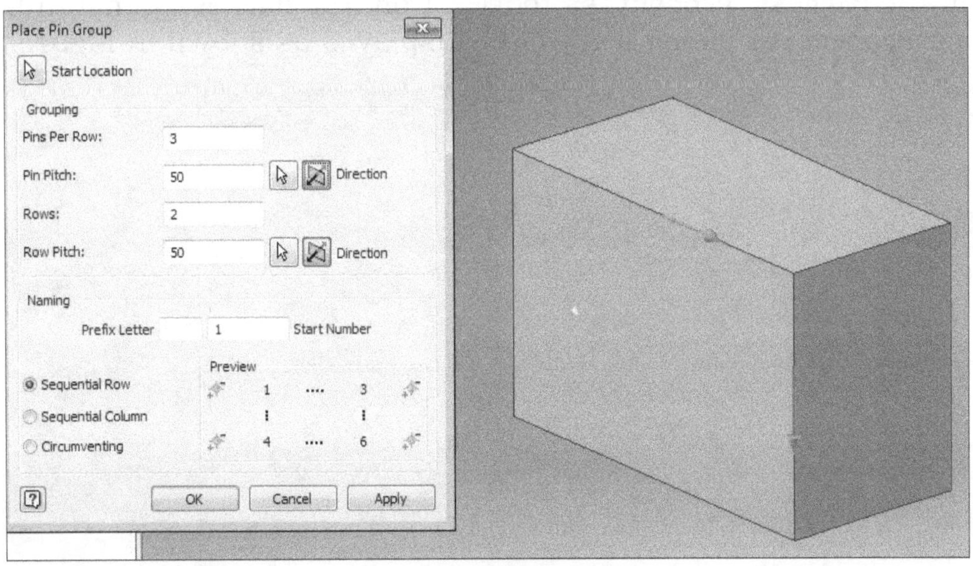

Figure-30. Directions of row pitch and pin pitch references

- Click on the **OK** button to create the pin group; refer to Figure-31.

Figure-31. Pin group created

Harness Properties of Part

To match schematic symbol of AutoCAD Electrical with Autodesk Inventor model, we need to specify some properties of the harness part. The procedure to apply properties to harness part are given next.

- Click on the **Harness Properties** tool from the **Harness** panel in the **Ribbon**. The **Part Properties** dialog box will be displayed as shown in Figure-32.

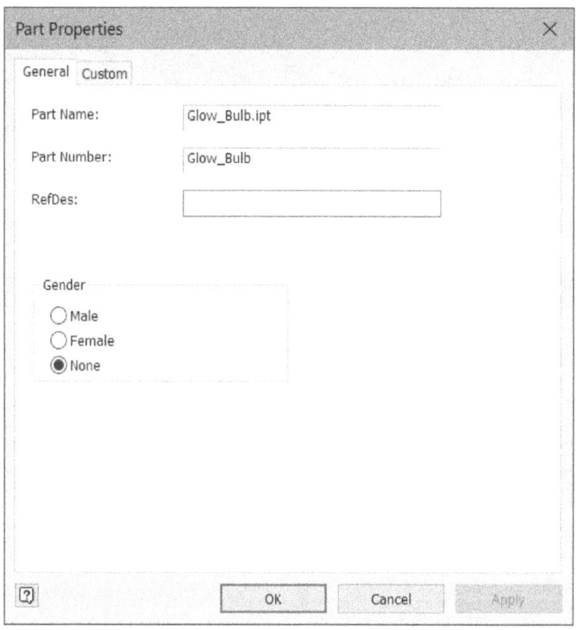

Figure-32. Part Properties dialog box

- Specify the reference description in the **RefDes** edit box in the dialog box. Note that this reference data is matched with the description in AutoCAD Electrical catalog for cross-referencing.
- You can also specify whether the part is Male or Female. If the part is not to act as connector then select the **None** radio button from the **Gender** area of the dialog box.
- Click on the **Custom** tab in the dialog box to define custom properties. Using the options in the **Custom** tab, you can create various properties of part and set their values; refer to Figure-33.

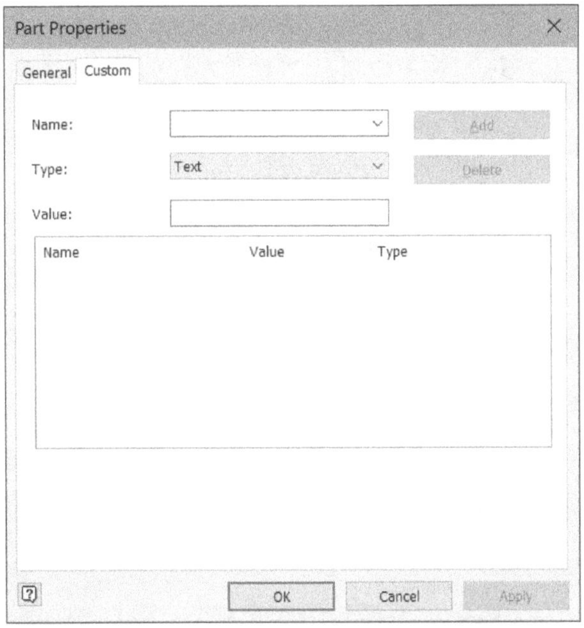

Figure-33. Custom tab of Part Properties dialog box

- Specify desired parameters; refer to Figure-34 and click on the **Add** button to add it in the table; refer to Figure-35.

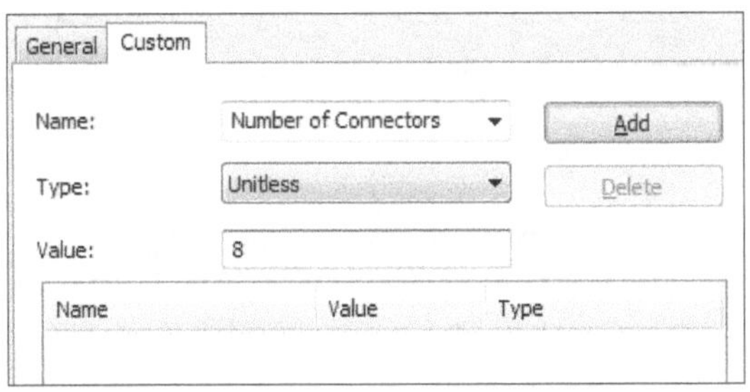

Figure-34. Specifying parameters for part

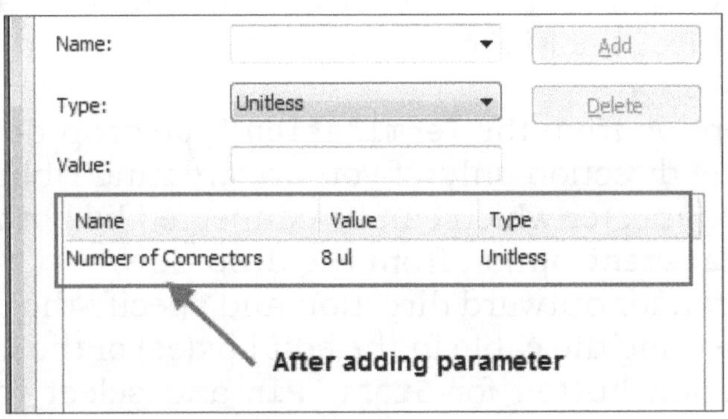

Figure-35. After adding parameter in table

- Similarly, add desired parameters and click on the **OK** button from the dialog box.

Converting the part to Connector

- After creating the electrical part, click on the **Manage** tab in the **Ribbon** and select the **Connector** tool from the **Connections** drop-down in **Author** panel as shown in Figure-36. The **Connector Authoring** dialog box will be displayed; refer to Figure-37.

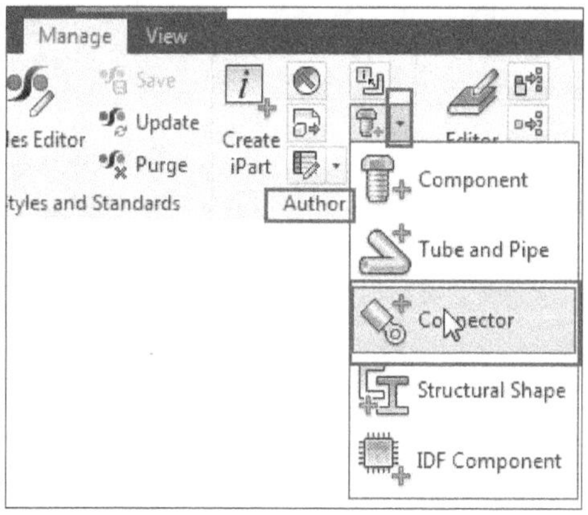

Figure-36. Connector tool

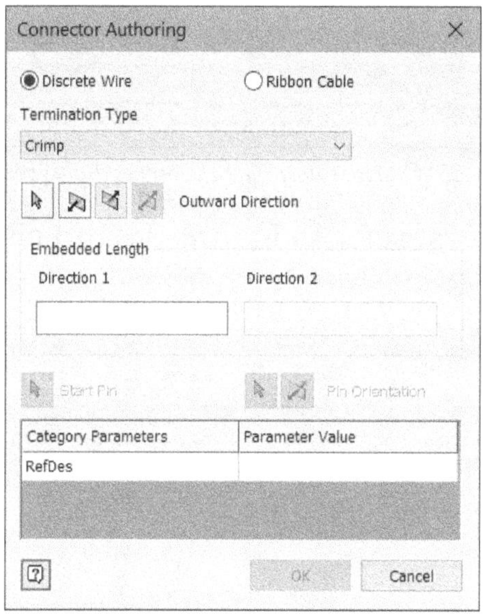

Figure-37. Connector Authoring dialog box

- Select the **Discrete Wire** radio button if you want to make the connector for discrete wires or select the **Ribbon Cable** radio button to make the connector for ribbon cable connection.
- Select the **Crimp** option from the **Termination Type** drop-down if the wire is to be connected from one direction only. If you are creating ribbon cable connector or any other type of connecter which can be connected by both sides then select the **Insulation Displacement** option from the drop-down.
- Select desired button for outward direction and specify the distance which will be spared while connecting the cable in the edit box(es) of the **Embedded Length** area.
- Click on the selection button for **Start Pin** and select the start pin from the component.
- Similarly, click on the selection button for **Pin Orientation** and select a face to specify the orientation of pins; refer to Figure-38.

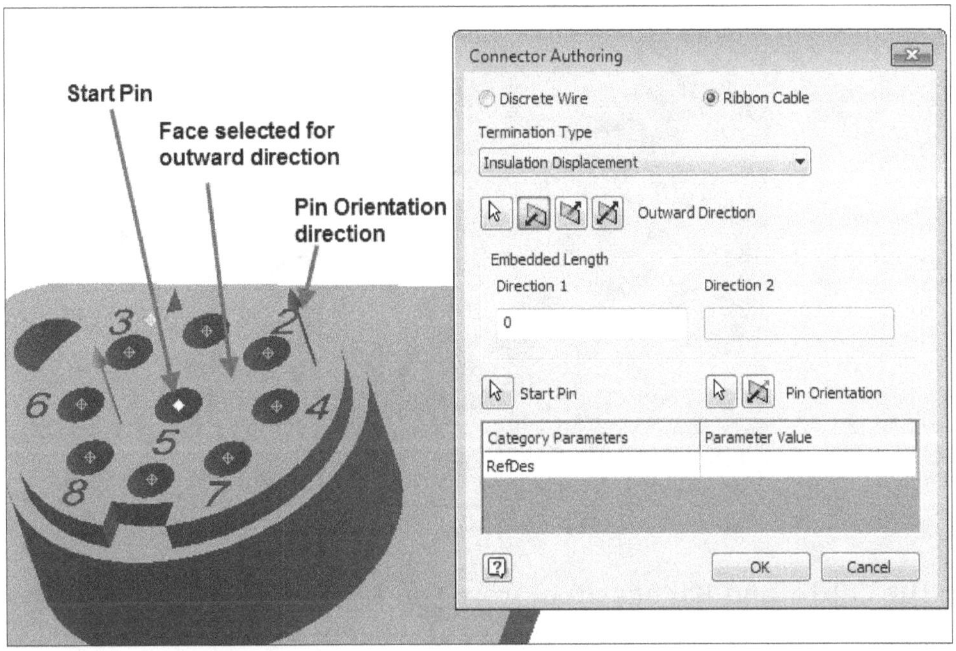

Figure-38. Creating connector

- Specify desired reference value in the **RefDes** edit box and click on the **OK** button. The **Authoring Result** message box will be displayed; refer to Figure-39.

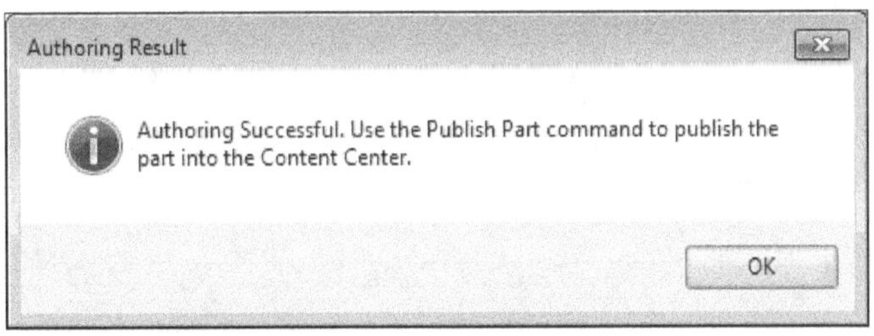

Figure-39. Authoring Result message box

- Click on the **OK** button and Save the part.

Now, we have the components in the form of 3D models. The next step is to create harness assembly using these components. Since this book is dedicated to AutoCAD Electrical only, so we will discuss small part of the harness assembly.

Creating Wiring in Harness Assembly

The benefit of wiring in harness assembly is that you can create the bill of material for wiring. In this way, you can get the idea of total length of various wires required in your project. The procedure to create wiring is given next.

- Open the assembly file of the model that you want to use for harness; refer to Figure-40. Note that pins are must on components for creating wiring. Without pins, you cannot create wiring.

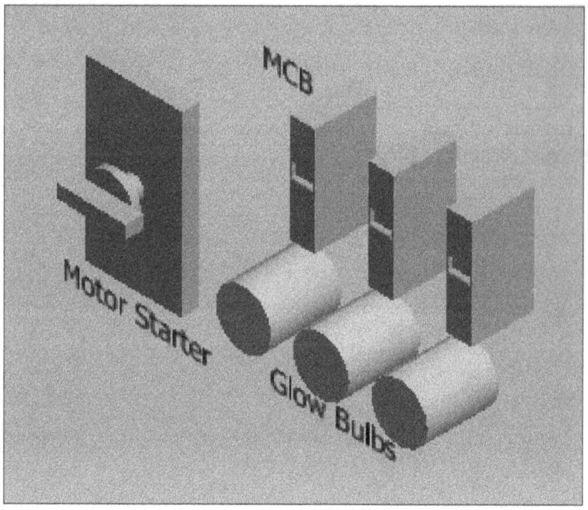

Figure-40. Model used for harness

- Click on the **Cable and Harness** tool from the **Begin** panel in the **Environments** tab of the **Ribbon**; refer to Figure-41. The **Create Harness** dialog box will be displayed; refer to Figure-42.

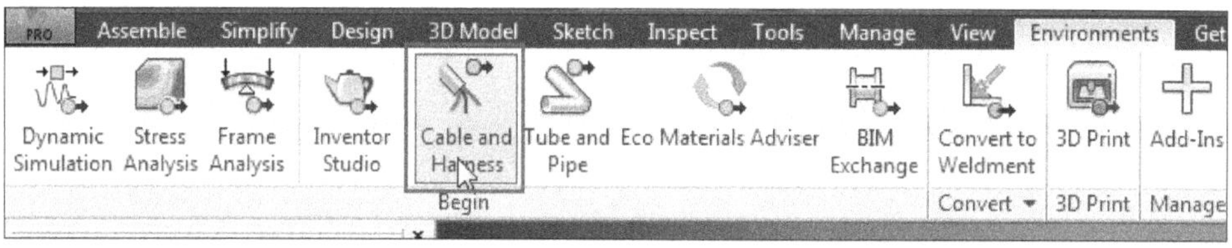

Figure-41. Cable and Harness button

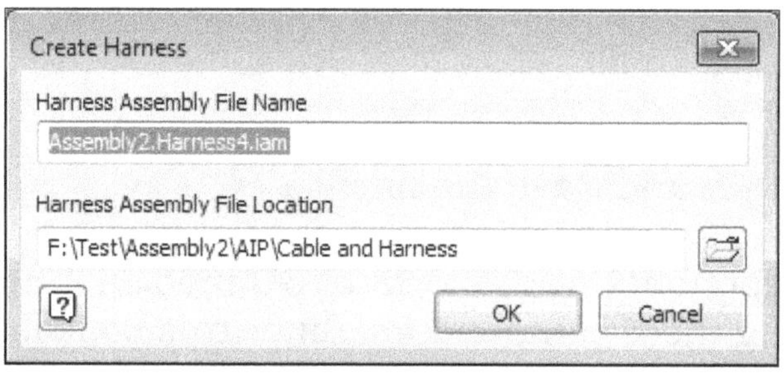

Figure-42. Create Harness dialog box

- Specify desired name and location for harness model. Click on the **OK** button from the dialog box. The **Cable and Harness** contextual tab will be added in the **Ribbon** with options related to cable and harness; refer to Figure-43.

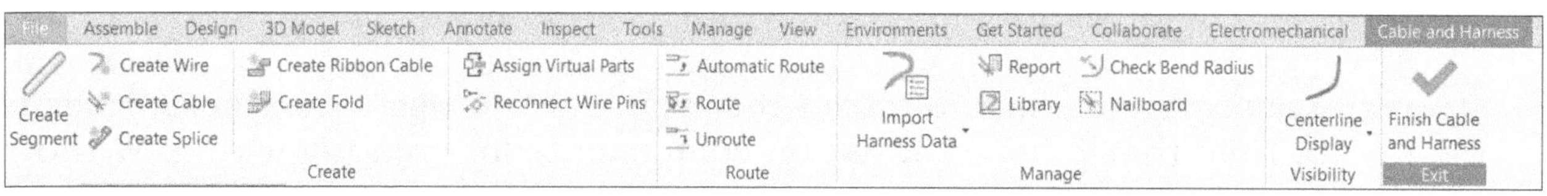

Figure-43. Cable and Harness contextual tab

- Click on the **Create Wire** tool from the **Create** panel in the **Cable and Harness** contextual tab of the **Ribbon**. The **Create Wire** dialog box will be displayed; refer to Figure-44. Also, you are asked to select a pin from the component.

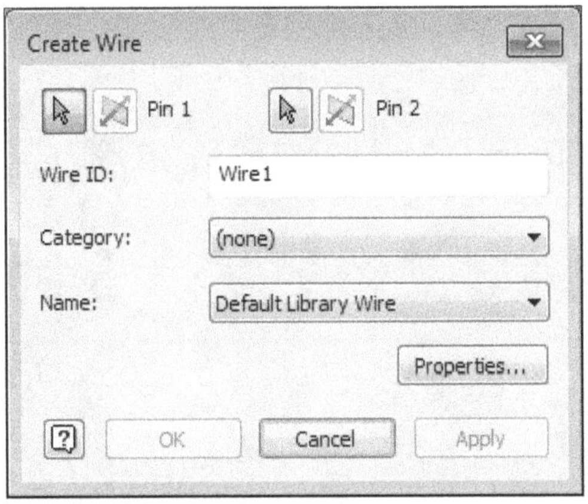

Figure-44. Create Wire dialog box

- Select the first pin to be connected by wire; refer to Figure-45.

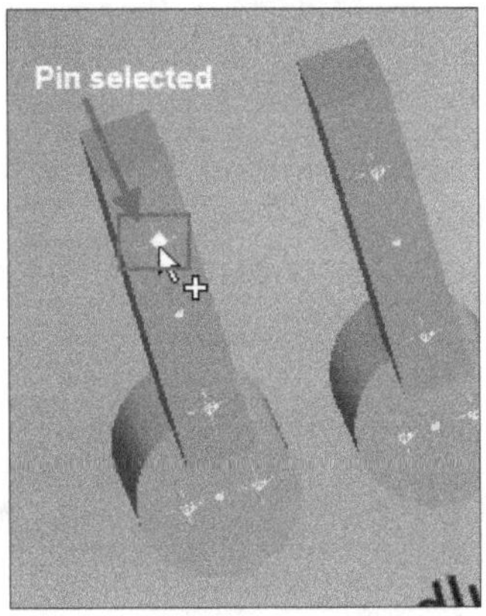

Figure-45. First pin selected

- Specify the wire id in the **Wire ID** edit box of the dialog box.
- Select desired category from the **Category** drop-down and then select the wire type from the **Name** drop-down.
- Now, select the other pin to specify end point of the wire.
- Click on the **Apply** button from the dialog box and keep on creating other wires.
- After creating all desired wires, click on the **Cancel** button from the dialog box to exit.

Modifying Shape of wires

You will get the wires created as straight line but straight wires are not desired at every place; refer to Figure-46. To change the shape of wires, follow the steps given next.

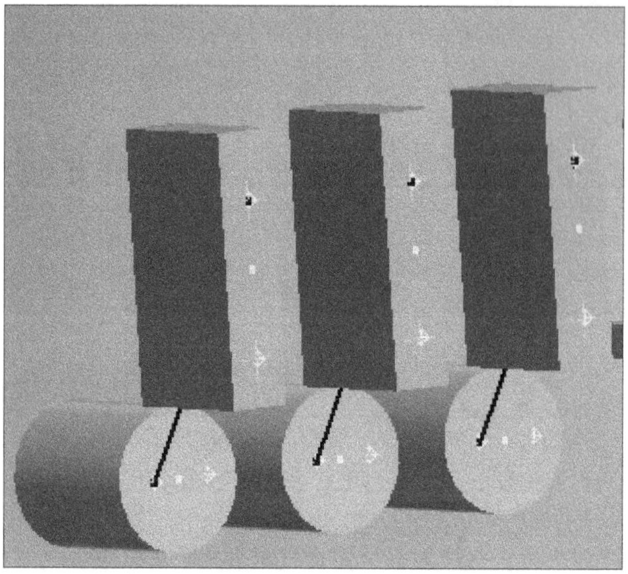

Figure-46. Straight wires created

- Select the wire to be modified. (Note that we will create points on the wire by which we can change shape of wire.)
- Right-click on the wire. A shortcut menu will be displayed; refer to Figure-47.

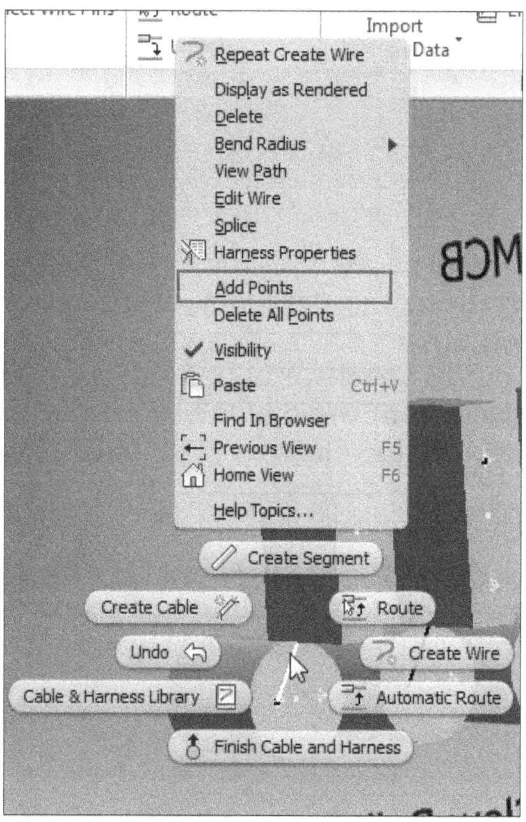

Figure-47. Shortcut menu for wires

- Click on the **Add Points** option from the shortcut menu. You are asked to specify position of the point on wire.
- Click at desired location and press **ESC**; refer to Figure-48.
- Right-click at the new point created and select the **3D Move/Rotate** option from the shortcut menu displayed. A 3D dragger will be displayed on the point; refer to Figure-49.

Figure-48. Specifying position of point

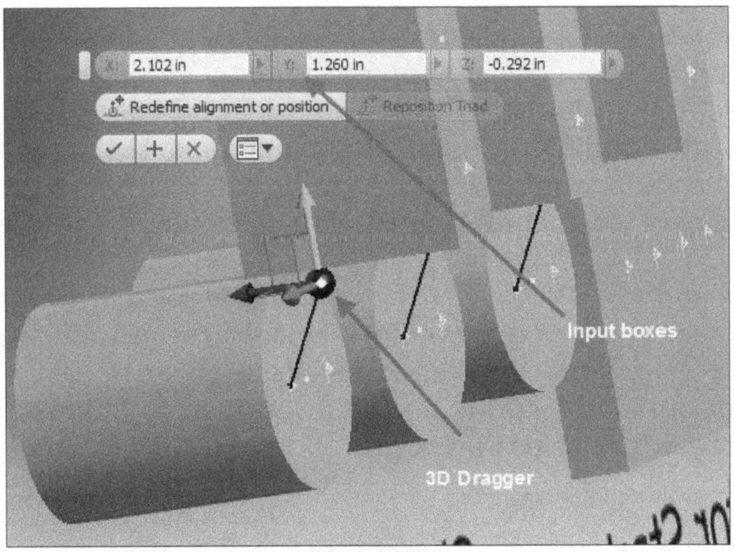

Figure-49. 3D Dragger and Input boxes

- Move the dragger in desired direction and click on the **OK** button from the Input boxes to apply the shape modification.

Creating Cable

In place of wires, you can create cables to connect various components. Like, it would be better to use cable for connecting a system to the power supply which allows sophistication of connectors. The procedure to add cable is given next.

- Click on the **Create Cable** tool from the **Create** panel in the **Cable and Harness** tab of the **Ribbon**. The **Create Cable** dialog box will be displayed as shown in Figure-50.

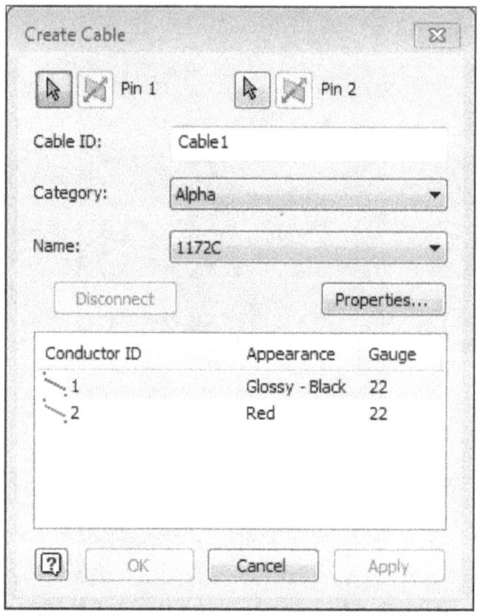

Figure-50. Create Cable dialog box

- Click in the drop-down for **Category** and select desired category. Note that select a category changes the gauge of wire and type.
- Click in the drop-down for **Name** and select desired cable. The related wires will be displayed in the conductors list in the dialog box. Also, you are asked to select first pin for conductor number 1 in the cable.
- Click on the pin of first component. You are asked to select pin of second component.
- Click on the second pin. Preview of the conductor in cable will be displayed; refer to Figure-51. Also, you are prompted to select first pin for second conductor of the cable.
- Similarly, select pins for second conductor and then click on the **OK** button. The cable will be created; refer to Figure-52.

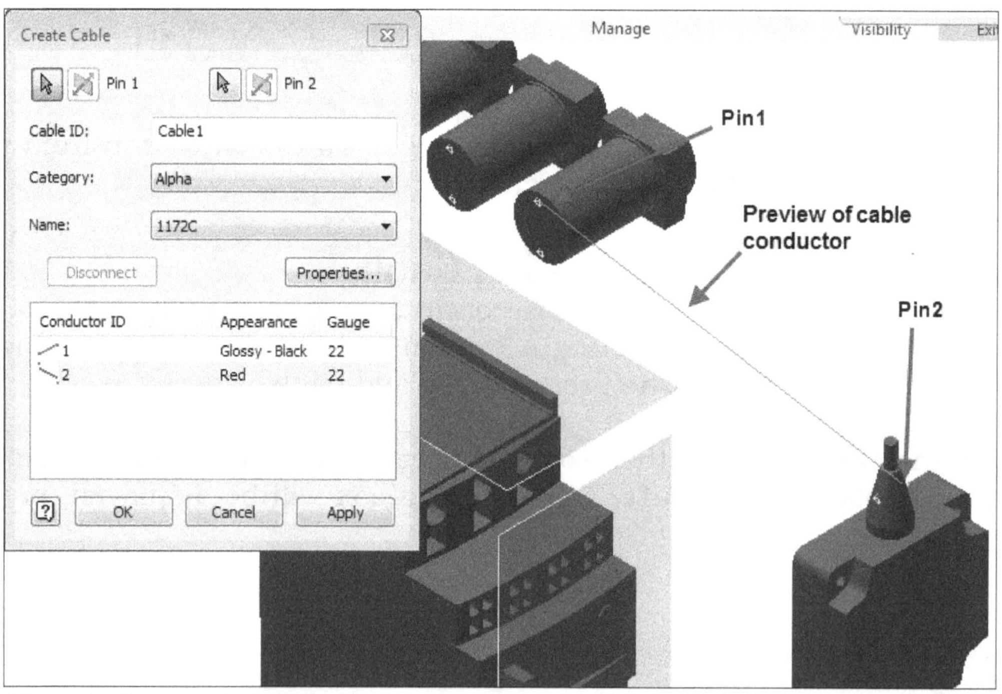

Figure-51. Preview of cable

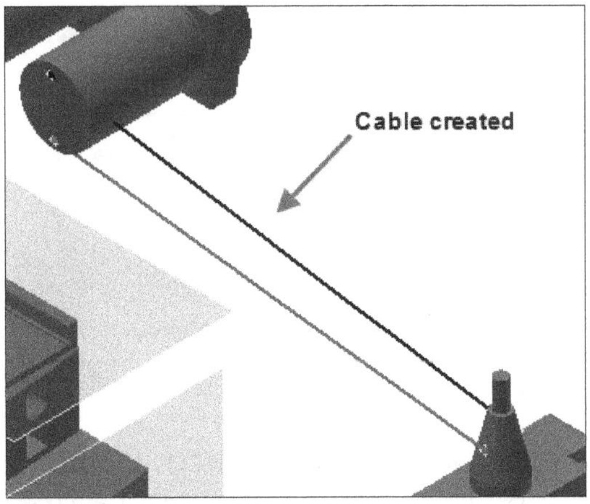

Figure-52. Cable created

Creating Ribbon Cable

Most of the time in DC circuits, you will find connectors for power supply through SMPS. These connectors are connected by ribbon cables. In Autodesk Inventor also, we can create ribbon cables to connect these connectors. The procedure to create ribbon cable is discussed next.

- Click on the **Create Ribbon Cable** tool from the **Create** panel in the **Cable and Harness** tab of the **Ribbon**. The **Create Ribbon Cable** dialog box will be displayed; refer to Figure-53. Also, you are asked to select the starting connector.

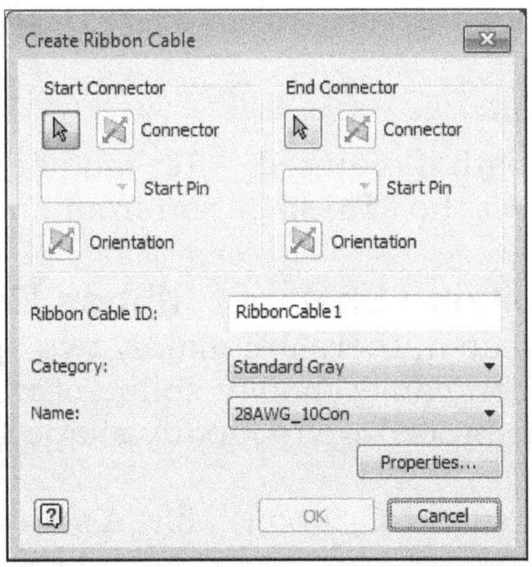

Figure-53. Create Ribbon Cable dialog box

- Click on the first connector. The **Start Pin** drop-down will get activated and preview of start pin on selected connector will be displayed; refer to Figure-54. Also, you are asked to select the end connector.

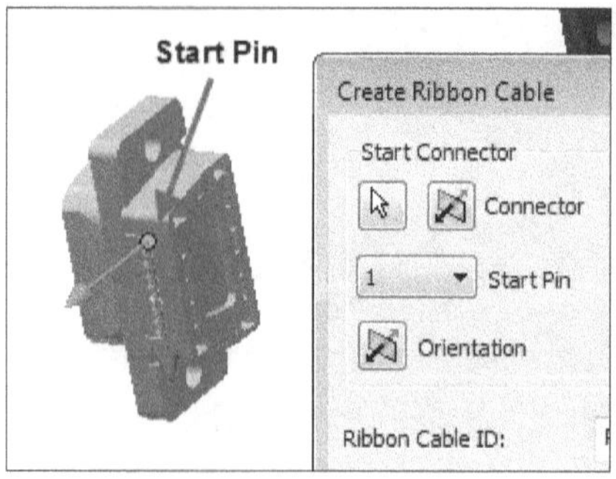

Figure-54. First connector on selection

- Select the second connector and set the starting pins for both the connectors.
- Click on the **OK** button. Preview of the cable will be displayed; refer to Figure-55. Also, you are asked to select vertex, point, edge, or face to modify path of cable.

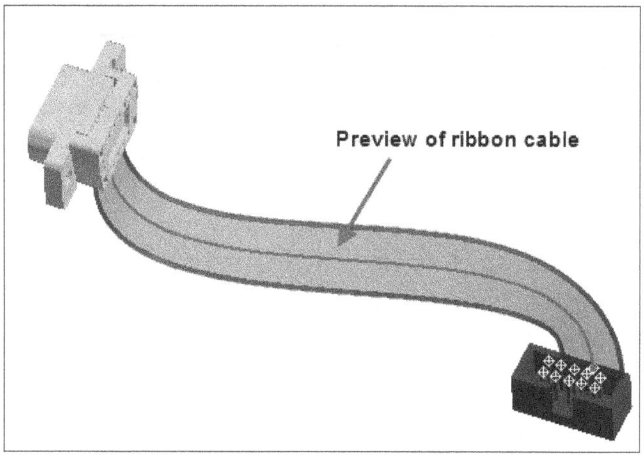

Figure-55. Preview of ribbon cable

- Select points to modify path if required. After modifications, right-click in empty area of drawing and select the **Finish** option from shortcut menu displayed.

Modifying Shape of Ribbon Cable

Once the ribbon cable is created, there are mainly two operations that can be done to modify the shape of ribbon cable; modify spline of ribbon cable to shape it and create fold in the cable. The procedure to modify shape is given next.

- Right-click on the center spline of ribbon cable. A shortcut menu will be displayed; refer to Figure-56.

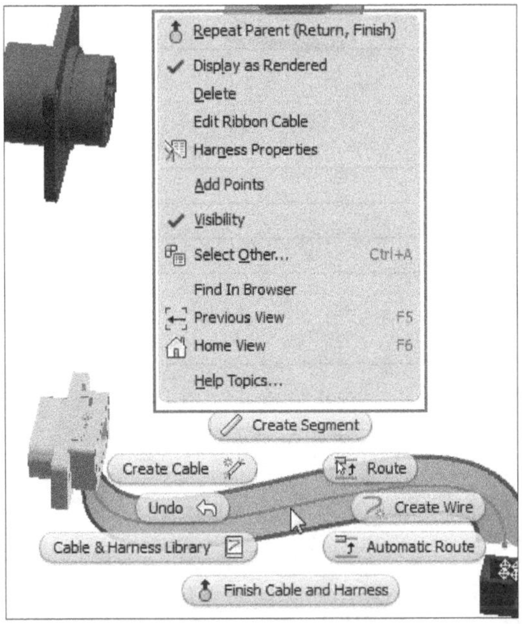

Figure-56. Shortcut menu for ribbon cable

- Click on the **Add Points** option from the shortcut menu. Now, if you hover the cursor over the ribbon line, preview of the point will be displayed.
- Click at desired location on the cable. A point will be created; refer to Figure-57. You can create as many points as you want.

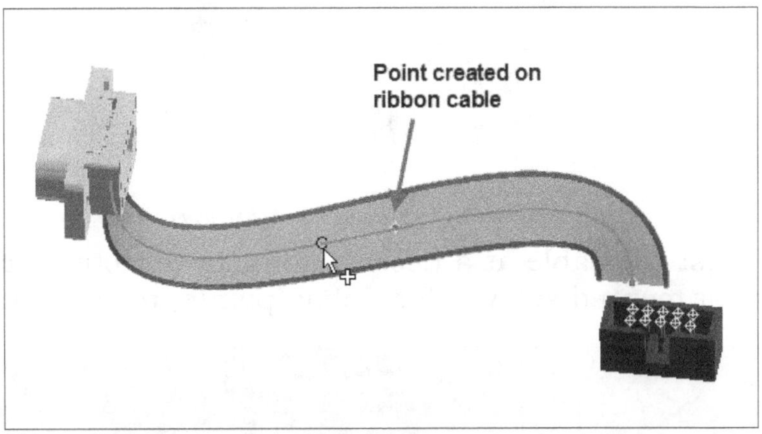

Figure-57. Point created on ribbon cable

- Once you have created the required number of points, press **ESC** from the keyboard to exit the tool.
- Now, select a point by which you want to change the shape of ribbon cable and right-click on it. A shortcut menu will be displayed as shown in Figure-58.

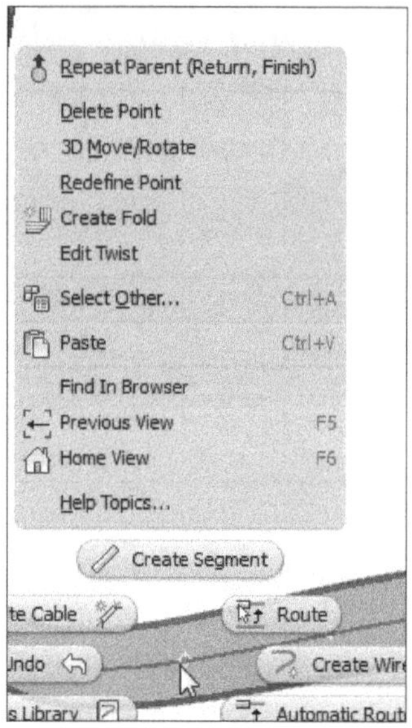

Figure-58. Shortcut menu for point on ribbon cable

- Click on the **3D Move/Rotate** option from the menu and using the handles move the point to desired location; refer to Figure-59.

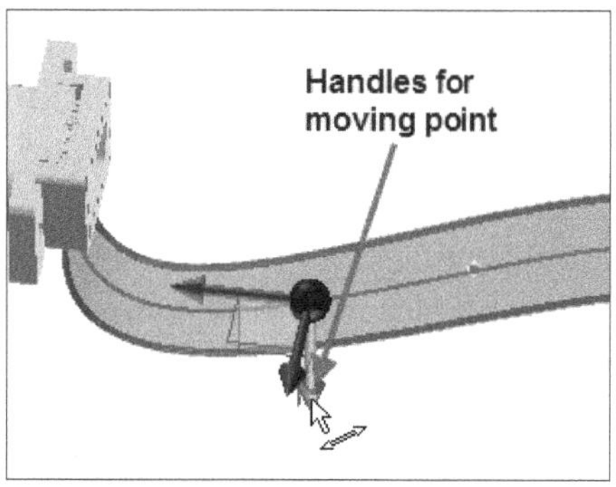

Figure-59. Handles for moving point of ribbon cable

- You can also twist the cable at a desired point by selecting **Edit Twist** option from the shortcut menu displayed for ribbon points; refer to Figure-60.

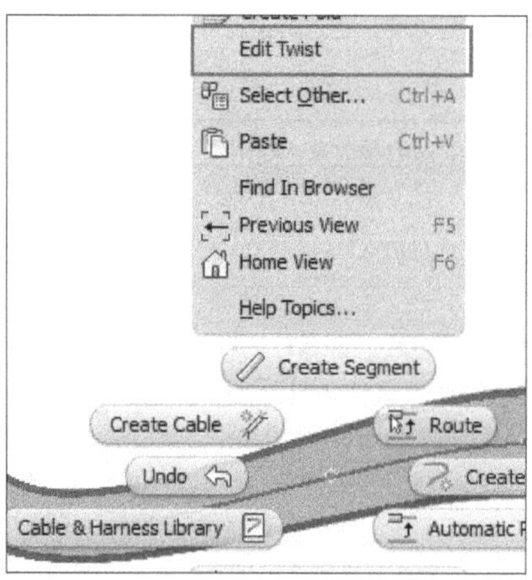

Figure-60. Edit Twist option

- On selecting the **Edit Twist** option, arrows for twisting are displayed; refer to Figure-61.

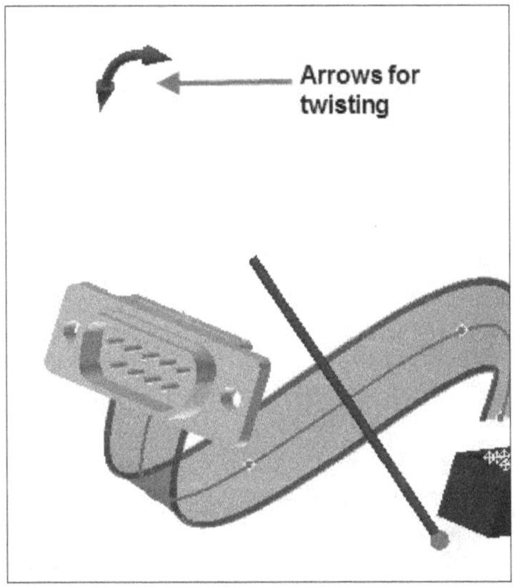

Figure-61. Arrows for twisting

- Rotate the arrows to desired direction, the ribbon will be twisted accordingly.
- After specifying desired twist, right-click and select the **Apply** option.

Creating Fold in Ribbon Cable

The **Create Fold** tool is used to create single or double fold in the ribbon cable. The procedure to use this tool is given next.

- Click on the **Create Fold** tool from the **Create** panel in the **Cable and Harness** tab of the **Ribbon**. The **Create Fold** dialog box will be displayed; refer to Figure-62. Also, you are asked to select a point.

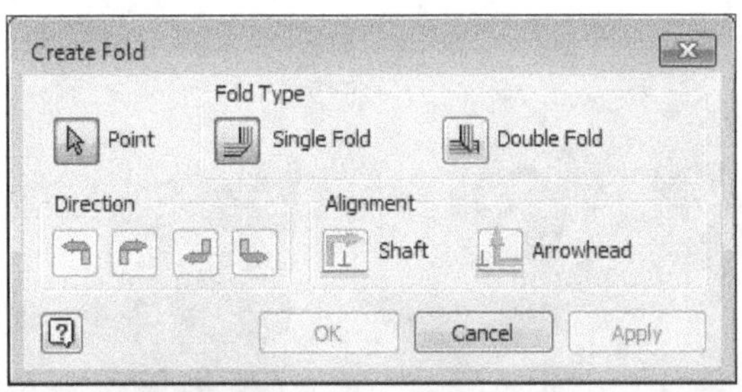

Figure-62. Create Fold dialog box

- Select a point on the ribbon cable (if you don't have the point then create it using the procedure discussed in previous topic). An arrow mark will be displayed on the point and the tools in **Create Fold** dialog box will become active; refer to Figure-63.

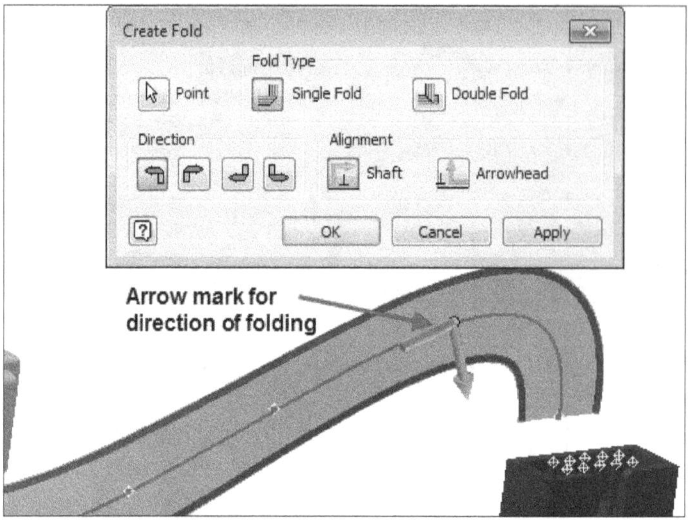

Figure-63. Arrow mark for folding direction

- Select desired button for direction of folding.
- Click on desired button for alignment and then click on the **OK** button from the dialog box. The folded ribbon cable will be displayed.

Automatic Route

The **Automatic Route** tool is used to automatically route all the unrouted wires through continuous segments. Make sure you have created a segment and wires before using this tool; refer to Figure-64. The procedure to use this tool is given next.

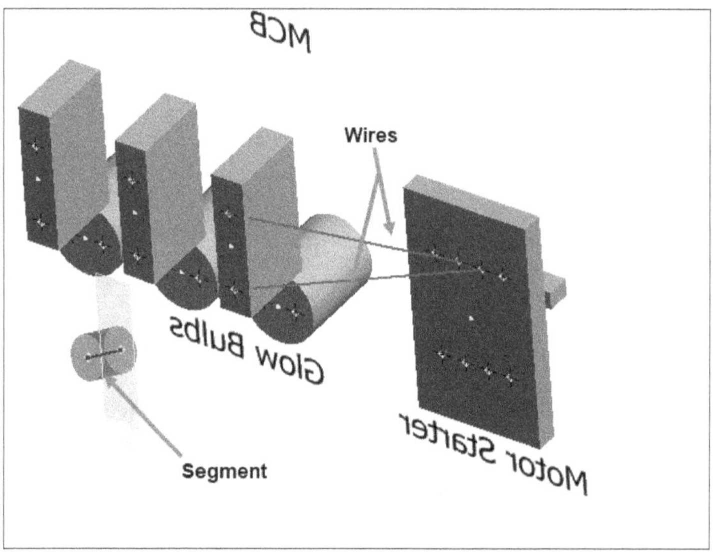

Figure-64. Wires and segment for auto route

- Click on the **Automatic Route** tool from the **Route** panel in the **Cable and Harness** contextual tab of **Ribbon**. The **Auto Route** dialog box will be displayed; refer to Figure-65.

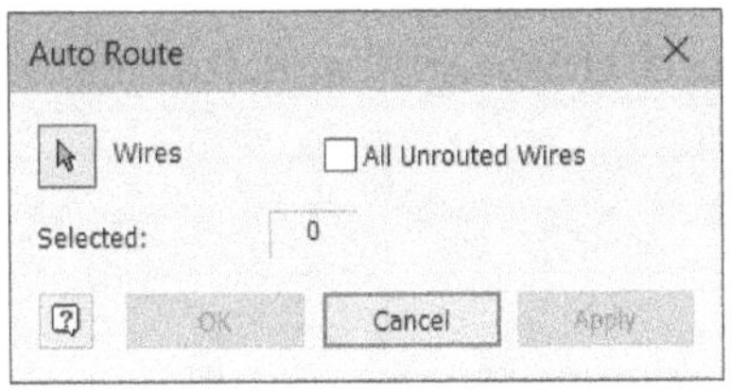

Figure-65. Auto Route dialog box

- If you want to select all the unrouted wires in assembly for auto routing then select the **All Unrouted Wires** check box from the dialog box otherwise, select the wires one by one to be routed.
- After selecting wires, click on the **OK** button from the dialog box. The wires will get routed through segment; refer to Figure-66.

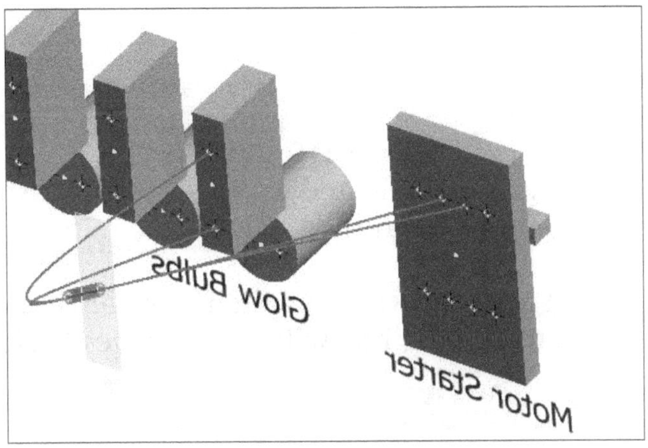

Figure-66. Wires routed through segment

Manual Routing

The **Route** tool is used to manually route wires through selected segments. The procedure to use this tool is given next.

- Click on the **Route** tool from the **Route** panel in the **Cable and Harness** contextual tab of **Ribbon**. The **Route** dialog box will be displayed; refer to Figure-67.

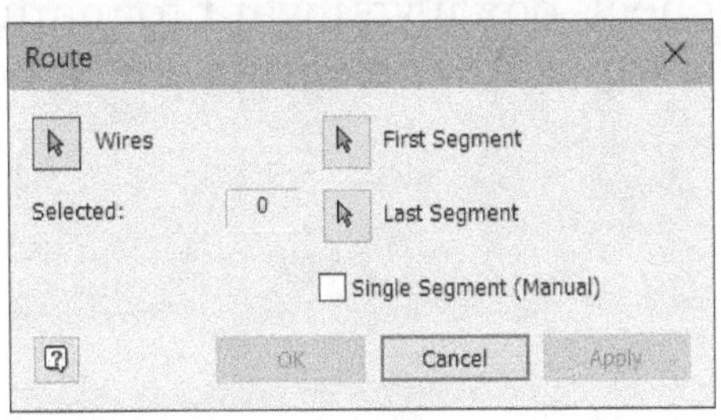

Figure-67. Route dialog box

- One by one select all the wires to be routed manually.
- Click on the **First Segment selection** button and select first segment through which the wires will enter. You will be asked to select the last segment.
- Select the segment through which wires will exit.
- If you are using single segment then select the **Single Segment (Manual)** check box.

- After selecting entities and parameters, click on the **OK** button from the dialog box. The wires will be routed through segments; refer to Figure-68.

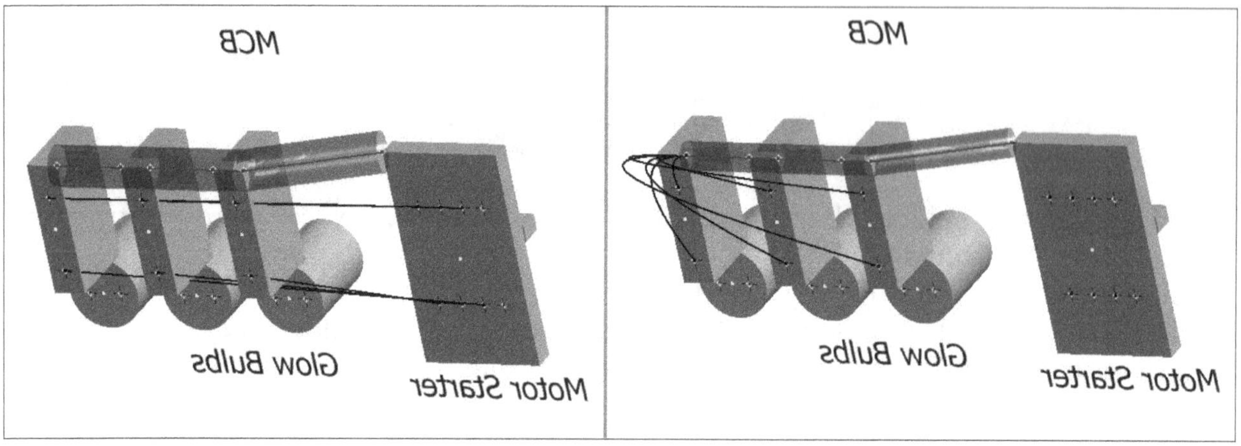

Figure-68. Wires after routing

Unrouting wires

The **Unroute** tool is used to unroute selected wires for route path of selected segment. The procedure to use this tool is given next.

- Click on the **Unroute** tool from the **Route** panel in the **Cable and Harness** contextual tab of **Ribbon**. The **Unroute** dialog box will be displayed; refer to Figure-69.

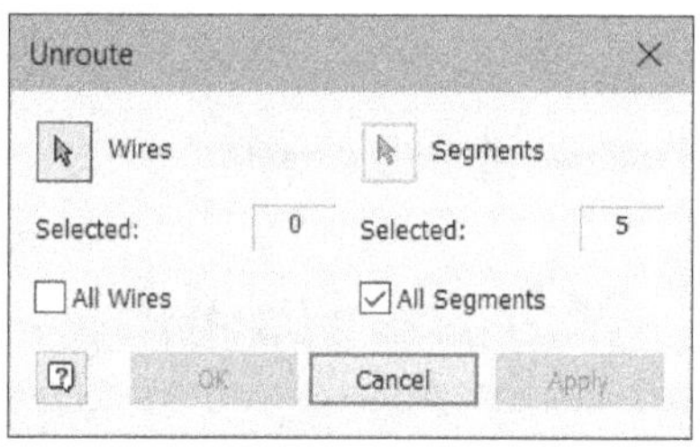

Figure-69. Unroute dialog box

- Select all the wires that you want to unroute. If you want to select all the wires then select the **All Wires** check box.
- Clear the **All Segments** check box if you want to manually select the segments to be skipped for routing the wires. After clearing check box, select desired segments; refer to Figure-70.

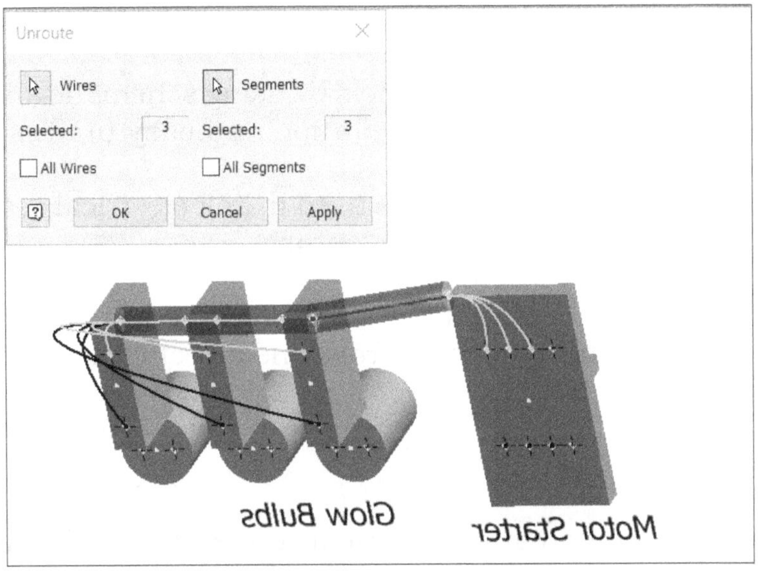

Figure-70. Wires and segments selected for unroute

- Click on the **OK** button from the dialog box to unroute the wires; refer to Figure-71.

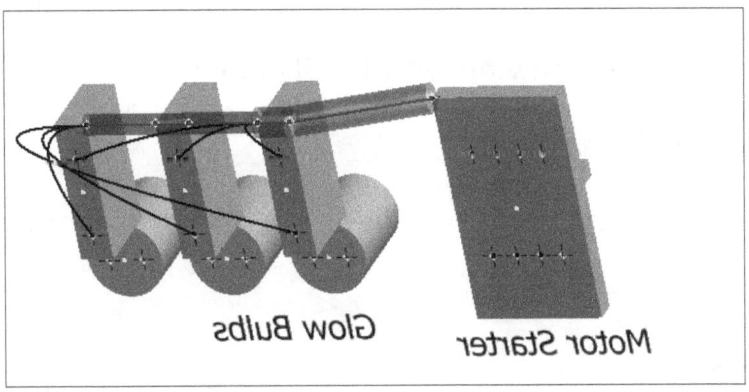

Figure-71. Unrouted wires

Self-Assessment

Q1. You can manufacture the electrical panels by using schematic and panel drawings. The 3D model of panel is only for representation and easy understanding. (T/F)

Q2. The .... button in the **Ribbon** is used to link electrical drawing in AutoCAD Electrical to the 3D model in Autodesk Inventor.

Q3. While authoring connector, select the .... option from the **Termination Type** drop-down if the wire is to be connected from one direction only.

Q4. The **Create Fold** tool can be used to create only single fold in the Ribbon cable. (T/F)

Q5. Write down the steps for creating electrical connector part in Autodesk Inventor.

Review Questions

Q1. What is the purpose of an electromechanical link in AutoCAD Electrical?
A. To connect multiple projects together
B. To connect the electrical drawing with its 3D model
C. To create schematic diagrams
D. To simulate electrical circuits

Q2. Which tab must be accessed to initiate the Electromechanical Link Setup in AutoCAD Electrical?
A. Insert
B. Home
C. Electromechanical
D. Annotate

Q3. What is displayed after clicking the Electromechanical Link Setup button in the Ribbon?
A. Wire Harness Manager
B. Electromechanical Link Setup dialog box
C. Project Explorer
D. Electrical Panel Editor

Q4. What are the two ways to create an electromechanical link?
A. Create a new schematic or import a file
B. Use cloud sync or local backup
C. Create a new link or use an existing one
D. Use the Drawing Setup or Link Manager

Q5. What does the link file act as between AutoCAD Electrical and Autodesk Inventor?
A. A configuration file
B. A 3D render file
C. A bridge
D. A CAD extension

Q6. What should you do before using an existing electromechanical link in AutoCAD Electrical?
A. Rename the link file
B. Open the schematic view
C. Browse and select the existing link file
D. Reinstall Inventor

Q7. What does the Link Conflict Resolution Preference option control?
A. Which software starts first
B. Which model gets updated first in case of conflicts
C. Network synchronization speed
D. File size reduction

Q8. In Autodesk Inventor, where is the Electromechanical Link Setup button located?
A. Project tab
B. Harness tab
C. Setup panel in the Electromechanical tab
D. Assembly panel

Q9. What is the final step after selecting the electromechanical link file in Inventor?
A. Save the project
B. Click the Cancel button
C. Click on the Location View tool
D. Restart the software

Q10. What is displayed in the Location View pane?
A. Only electrical components
B. Only Inventor components
C. All linked files
D. Components with their icons indicating type and origin

Q11. What does the "Linked with Differences" icon indicate?
A. That the link is broken
B. That the components are identical
C. That there are unresolved mismatches
D. That the model is updated

Q12. How can differences between components be resolved?
A. By restarting both programs
B. By using the Details pane of Location View
C. By relinking all components
D. By recreating the harness

Q13. What is the first step to create an electrical component in Autodesk Inventor?
A. Create a wiring diagram
B. Click on the New Part option
C. Link it with AutoCAD
D. Start the Harness Environment

Q14. Which tool is used to add individual pins on a component in Inventor?
A. Place Wire
B. Add Terminal
C. Place Pin
D. Insert Symbol

Q15. What should you do after placing all pins individually?
A. Click Save All
B. Select the Done option or press ESC
C. Restart Inventor
D. Create a new file

Q16. What tool is used to add a group of pins for connectors?
A. Wire Harness
B. Group Pin Creator
C. Pin Matrix Tool
D. Place Pin Group

Q17. Which of the following is NOT a configuration option for placing pin groups?
A. Sequential Row
B. Circumventing
C. Grid Layout
D. Sequential Column

Q18. What must be selected to define the column direction while placing a pin group?
A. Base point
B. Reference entity
C. Harness tab
D. Project file

Q19. What is the purpose of the Flip button next to the reference selection in pin placement?
A. To zoom in on the reference
B. To rotate the model
C. To change the direction of the pin pitch
D. To duplicate the pins

Q20. What is necessary before placing the pin group on the model surface?
A. Selecting an icon
B. Creating a sketch point
C. Rebooting the system
D. Installing additional plugin

Q21. What is the primary purpose of the Harness Properties tool in Autodesk Inventor?

A. To create a wire from a component
B. To convert the harness to a ribbon cable
C. To match schematic symbol with 3D model
D. To define cable path

Q22. In the Part Properties dialog box, what does the RefDes field do?

A. Specifies the number of pins
B. Matches reference with AutoCAD Electrical catalog
C. Defines wire length
D. Sets harness color

Q23. What happens when you select the None radio button in the Gender area of the Part Properties dialog box?

A. The part acts as both male and female
B. The part is excluded from wiring
C. The part does not act as a connector
D. The part becomes a power source

Q24. Which tab in the Part Properties dialog box allows defining custom parameters?

A. General
B. Options
C. Custom
D. Settings

Q25. What tool is used to convert a part into a connector?

A. Harness Converter
B. Cable Builder
C. Connector Authoring
D. Pin Tool

Q26. What option should be selected if a connector connects discrete wires?

A. Ribbon Cable
B. Insulation Displacement
C. Crimp
D. Discrete Wire

Q27. What tool is used to initiate harness creation in the assembly?

A. Place Pin
B. Create Cable
C. Create Harness
D. Add Component

Q28. Which field must be filled when using the Create Wire tool?

A. Segment Name
B. Category
C. Wire ID
D. Orientation

Q29. What is the purpose of the Add Points option in wire editing?

A. Change the wire color
B. Create routing paths for wire shaping
C. Add pins to wire
D. Add insulation

Q30. Which option is used to manually move wire points in 3D?

A. 3D Drag Tool
B. 3D Sketch
C. 3D Move/Rotate
D. Wire Bend

Q31. Why would you use a cable instead of individual wires?

A. To reduce resistance
B. To simplify power supply connections
C. To avoid pin usage
D. For temporary connections

Q32. What must be done first when placing a cable?

A. Set wire length
B. Choose pin orientation
C. Select wire category and type
D. Create a spline

Q33. What is the benefit of using ribbon cable in DC circuits?

A. Easier to twist
B. Suitable for SMPS connections
C. Lower cost
D. Requires fewer pins

Q34. What is the function of the Edit Twist tool?

A. Change cable color
B. Flip wire direction
C. Apply twisting at a point on ribbon
D. Add connector

Q35. What must be selected before applying a fold in ribbon cable?

A. Wire ID
B. Harness tab
C. A point on the ribbon
D. Cable route

Q36. What does the Automatic Route tool do?

A. Routes one wire through a single pin
B. Routes all unrouted wires automatically
C. Unroutes all wires
D. Connects ribbon cables only

Q37. What should be created before using Automatic Route?

A. Harness tab
B. Ribbon cable
C. Wire categories
D. A segment and wires

Q38. What is the function of the Route tool?

A. Create pin connection
B. Edit wire shape
C. Manually route wires
D. View wire details

Q39. Which checkbox should be selected if routing through only one segment?

A. Single Segment (Manual)
B. Auto Select
C. Segment Lock
D. Segment Filter

Q40. What does the Unroute tool allow you to do?

A. Delete the wire
B. Unroute selected wires from segments
C. Rename all segments
D. Group multiple wires together

For Student Notes

Answer for Self-Assessment:

Ans1. T, Ans2. Electromechanical Link Setup, Ans3. Crimp Ans4. F,

Annexure I

Basics of Electrical System

This chapter is an introduction to electrical system so we have given brief introduction here. To learn more about electrical systems, you can check YouTube channel of **Mr. Jim Pytel** | Electro-Mechanical Technology Instructor | Columbia Gorge Community College at **https://www.youtube.com/user/bigbadtech** link.

Topics Covered

The major topics covered in this chapter are:

- ***Basics of Electrical Circuits***
- ***Types of Electricity***
- ***Resistance Coding and Calculation***
- ***Star & Delta Connections***
- ***Voltage and Current Source***
- ***Kirchhoff's Laws***
- ***DC Circuits***
- ***Electromagnetism***
- ***Alternating Current***
- ***Single AC Circuits***
- ***Three Phase AC circuits***
- ***Balanced and Unbalanced Circuits***

INTRODUCTION

In the previous part of this book, you have learned about AutoCAD Electrical commands and tools to create various schematics. This part of the book is about electrical engineering concepts that are important to understand before you start designing electrical panels. Depending on your need, you can start with this Part-II first and then you can work with Part-I of the book. We will start with basic terminology of electrical engineering.

WHAT IS ELECTRICITY?

Electrical engineering revolves around the fundamentals of electricity. It is concerned about how electricity is produced, transmitted, and manipulated to run various machines. Here, the basic question is what is electricity? Electricity is a form of energy caused by the presence or flow of electrons. If free electrons are present in a metal in charges state then they produce static current. If the electrons are flowing from one point to another point in a metal conductor then it is called live current. So, current is one of the parameter used to measure electricity. There are few more parameters used to measure electricity which are discussed next.

How Electricity is Measured?

Every material around us is made up of elements. Every element is made up of atoms and every atom is composed of electrons, protons and neutrons (At least till writing this book!!). The electrons have negative charge, the protons have equal positive charge, and neutrons as name suggests are neutral. Due to external influence, an electron can be freed from its atom and forced to move in desired direction. The charge induced by one such electron is measures as 1.602×10^{-19} C (coulomb). So, we can say that the quantity by which we can measure electricity is **coulomb**.

Current

The coulomb value of a metal is not feasible to measure as there can be millions of electrons in small section of metal and we might need supercomputers to measure the charge in this small section. So, we derived a smart parameter to measure charge called **current**. Current can be defined as rate of flow of charge particle through a point on the conductor (A conductor is a material which allow the current to pass through due to availability of free electrons). In mathematical terms, it can be defined as :

$$I = \frac{dq}{dt} = -nev_d$$

Here, I is current, dq/dt is the rate of charge per second
n is the number of electrons, -e is the charge on electrons, and vd is drift velocity of electron.

The unit of current is ampere denoted by A. One ampere current is the flow of 1 C (coulomb) charge from a conductor at a point per second.

There are two types of current; direct current (DC) and alternating current (AC). Graphically, we can define these currents as shown in Figure-1.

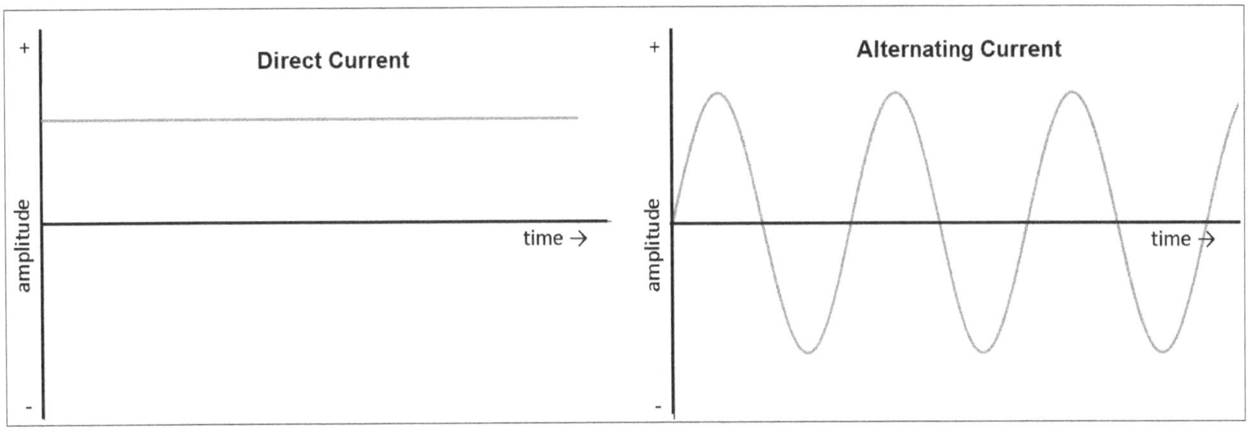

Figure-1. Types of currents

As you can see from the figure, in direct current, the amplitude of current remains constant with respect to time and flow is unidirectional. In case of alternating current, the amplitude of current changes both value as well as direction. Generally, AC current follows two wave forms, sinusoidal wave and square wave. There can be other forms of AC currents but we will not delve deeper here.

Current Density

The current density if the amount of current carried by conductor per unit area. The area should be perpendicular to the direction of current flow. Mathematically, it can be given by formula:

$$J = \frac{I}{A}$$

Since, current density is dependent of current flow direction so it is a vector quantity and its unit is A/m^2.

Voltage or Potential Difference

The **Voltage** or potential difference is the amount of work required to move per coulomb charge from one point to another point in conductor. Unit of voltage is Volt or simply V. Mathematically, it can be given as:

$$V = \frac{W}{Q}$$

So, 1 volt = 1 Joule/1 Coulomb.

When representing voltage in electrical drawing, it should always be defined with + and - signs as shown Figure-2.

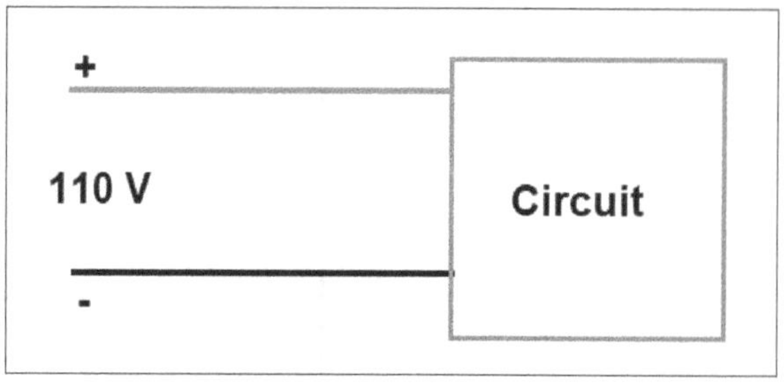

Figure-2. Voltage representation

Electrical Field

Electrical field is the pseudo force field generated in a conductor when voltage source is connected at its two ends. Electrical field is measures by volt per meter and mathematically it can be defined as:

$$E = \frac{V}{L}$$

Electrical Power

The electrical power is defined as product of Voltage and current, P = V x I. The unit of power is watt (W). You can also define power in horsepower (hp). One hp is equal to 746 W.

Efficiency

The electrical efficiency of a component, circuit, or system can be defined as useful power output divided by total power consumed. The efficiency is generally denoted in percentage (%). Mathematically, it can be given as:

$$Efficiency\ (\%) = \frac{Useful\ Power\ Output}{Total\ Power\ Consumed}$$

OHM'S LAW

Ohm's law is the most basic rule in electrical engineering for circuits. As per the Ohm's law, the Voltage across two points of a conductor is directly proportional to the current flowing through it.

$$V \propto I$$

Or, say V = R x I where R is constant of proportionality (Also called Resistance)

Resistance

The resistance to flow of current in a material is called Resistance. The unit of resistance is ohm (Ω). Mathematically, R is calculated by:

$$R = \rho \frac{L}{A}$$

Here, ρ is resistivity defined by unit ohm meter.
L is the length of conductor
A is the cross-section area of conductor.

There are mainly two types of resistors you will come across during practical use; wire wound resistors and carbon molded resistors. There is also another category of resistors called metal film resistors. The metal film resistors have very low power rating and can be easily damaged due to power surge.

Wire Wound Resistors

The wire wound resistors are those in which insulated resistance wire is wound around heat-resistant ceramic tubes; refer to Figure-3. These resistors can be of fixed resistance or variable resistance. The resistance value is based on the length of wire and its resistivity. The value of resistance goes from fraction of ohm to kilo ohms. You can find the resistance value and rated power value printed on the resistors. The power rating of these resistors can be from few watts to 1000 watts.

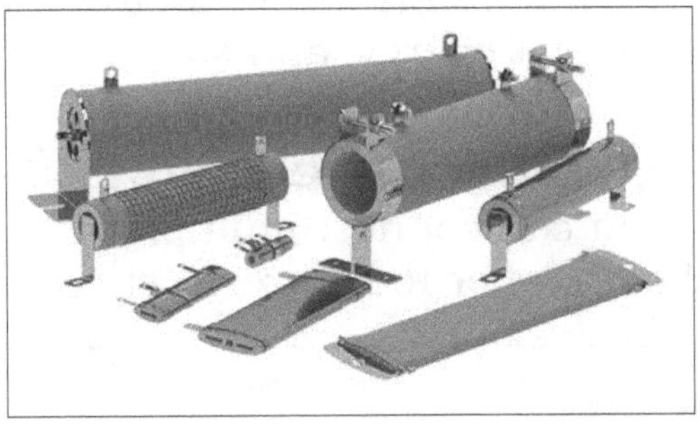

Figure-3. Wire wound resistors

Carbon Molded Resistors

In electronic circuits and other precise electrical circuits where high resistance is required, the carbon molded resistors are the best solution. The cost of creating wire wound resistor for mega ohm resistance will be very high whereas comparatively, you can get a carbon molded resistor of same rating at very low cost. Note that although resistance value of carbon molded resistors is high but their power rating is generally low (in the range of 1/4 W to 5 W). The carbon molded resistors follow a color coding to denote their resistance value; refer to Figure-4. (Want to learn more follow the link https://www.wikihow.com/Identify-Resistors). Note that metal film resistor might look similar to carbon molded resistors but you can identify the two by their background color. The metal film resistors generally have blue background color whereas carbon molded resistors have light brown color background.

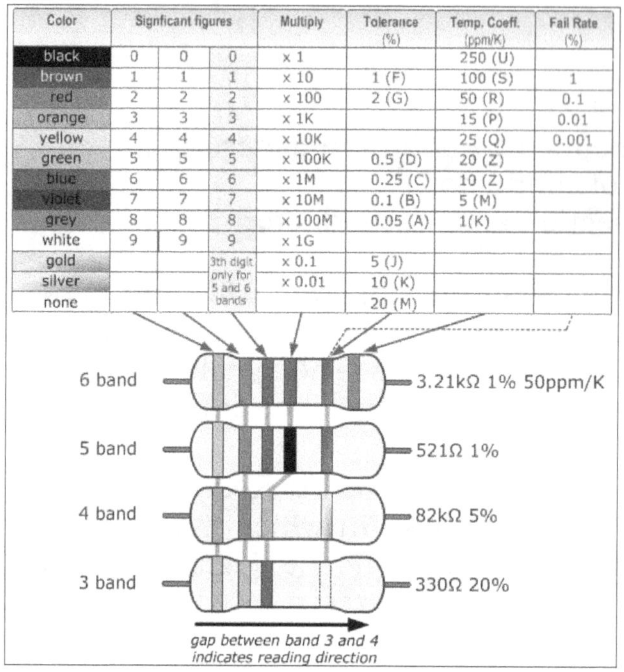

Color	Signficant figures			Multiply	Tolerance (%)	Temp. Coeff. (ppm/K)	Fail Rate (%)
black	0	0	0	x 1		250 (U)	
brown	1	1	1	x 10	1 (F)	100 (S)	1
red	2	2	2	x 100	2 (G)	50 (R)	0.1
orange	3	3	3	x 1K		15 (P)	0.01
yellow	4	4	4	x 10K		25 (Q)	0.001
green	5	5	5	x 100K	0.5 (D)	20 (Z)	
blue	6	6	6	x 1M	0.25 (C)	10 (Z)	
violet	7	7	7	x 10M	0.1 (B)	5 (M)	
grey	8	8	8	x 100M	0.05 (A)	1(K)	
white	9	9	9	x 1G			
gold			3th digit only for 5 and 6 bands	x 0.1	5 (J)		
silver				x 0.01	10 (K)		
none					20 (M)		

Figure-4. Resistor color codes chart

Most of the time, resistance is intentional in circuit to achieve desired objectives like reducing current, generating heat energy, and so on.

Series and Parallel Connections

Based on the application, you can connect multiple resistors in a circuit in series, parallel or combination of both; refer to Figure-5.

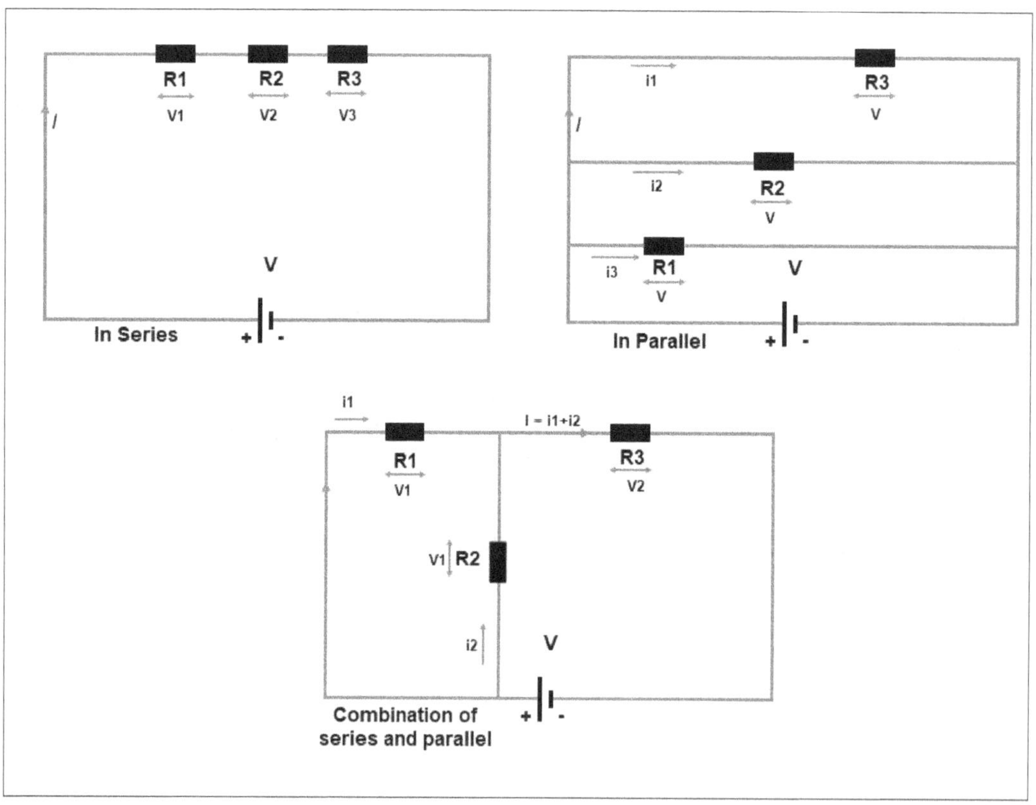

Figure-5. Resistor connections

Note that when we connect resistors in series in a circuit then the total voltage is divided among the components but current flowing through them is same.

Using this information in equation, we get

$$V = V1+V2+V3$$

$$IR = IR1+IR2+IR3 = I(R1+R2+R3)$$

Total Resistance $R = (R1+R2+R3);$

So, we can say that when resistors are connected in series then we can add all the values of resistors to get equivalent resistance value.

When we connect resistors in parallel in a circuit then voltage across each resistor is same but current gets divided among the components so we can say that,

Total current $I = I1+I2+I3$

$$V/R = V/R1+V/R2+V/R3 = V(1/R1+1/R2+1/R3)$$

$$1/R = 1/R1+1/R2+1/R3$$

Star and Delta Connections

While working with circuits, you will not find all the resistors in series or parallel, there can be some devil configurations as shown in Figure-6.

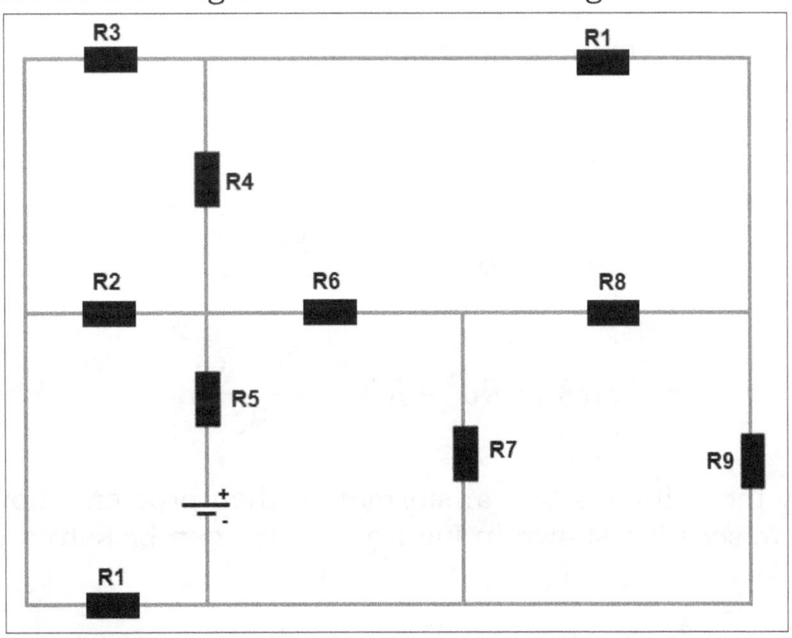

Figure-6. Complex connections of resistors

In such cases, we are looking at delta connections like R7, R8, and R9 form one delta connection and R2, R3, and R4 form another delta connection. There resistors are neither in parallel and not in series. The solutions to such problems are star-delta and delta-star transformation equations; refer to Figure-7.

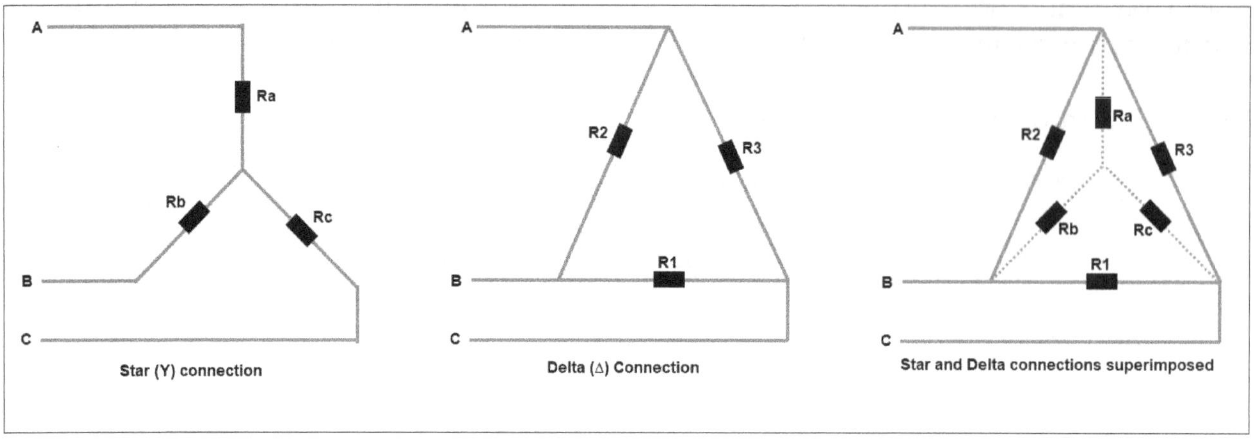

Figure-7. Star and Delta connections

For Delta to Star Transformation, the values are :

$$Ra = \frac{R2\ .\ R3}{R1 + R2 + R3}$$

$$Rb = \frac{R3\ .\ R1}{R1 + R2 + R3}$$

$$Rc = \frac{R1\ .\ R2}{R1 + R2 + R3}$$

For Star to Delta transformation, the values are:

$$R1 = Rb + Rc + \frac{Rb\,.Rc}{Ra}$$

$$R2 = Rc + Ra + \frac{Rc\,.Ra}{Rb}$$

$$R3 = Ra + Rb + \frac{Ra\,.Rb}{Rc}$$

Now, if we apply the delta to star transformation then problem shown in Figure-6 can be reduced to solution shown in Figure-8 which can be solved easily.

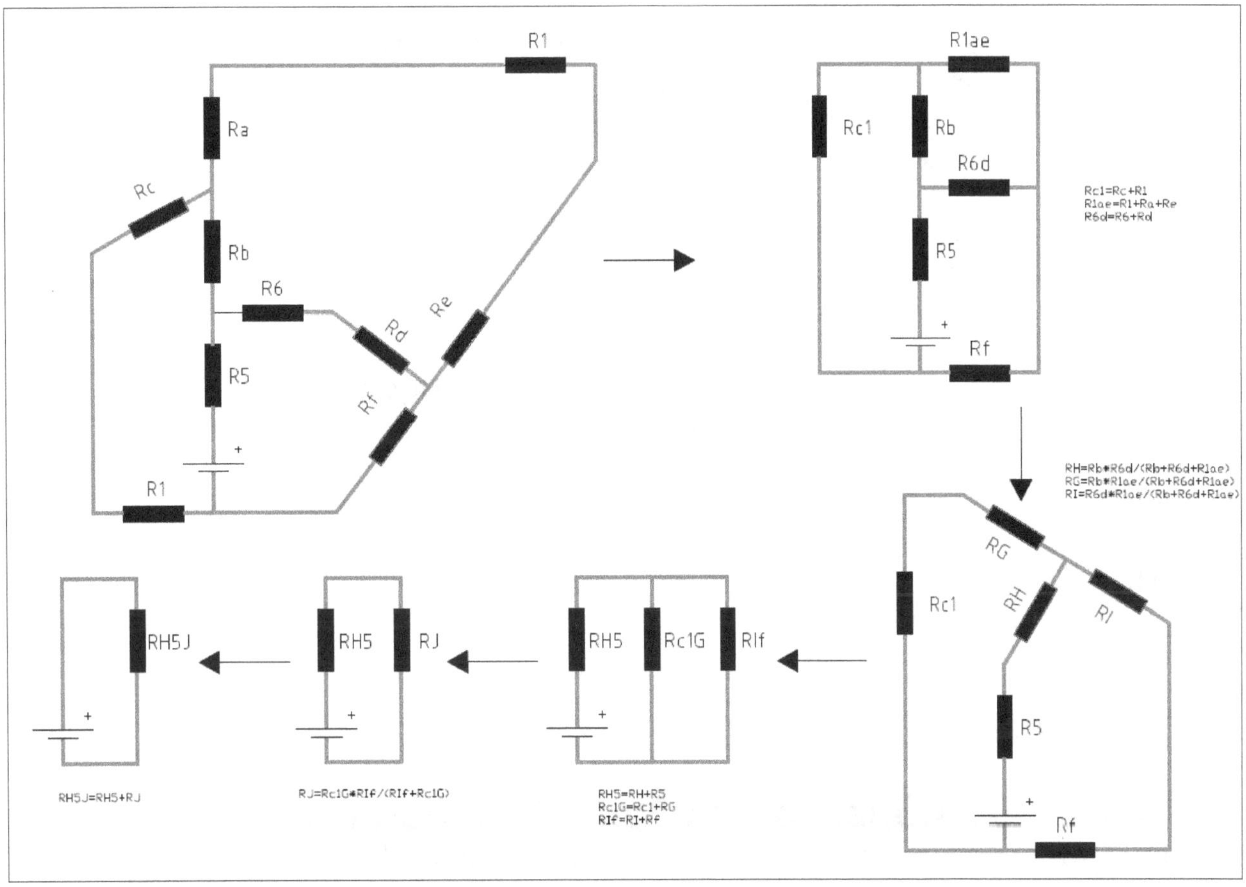

Figure-8. Solution approach for complex connection

CAPACITORS AND INDUCTORS

Capacitors and Inductors, both draw power from the source, store it and then discharge it when there is a sudden change in current/voltage in the circuit. Both the components are widely used in filtering signals and AC circuits. The main difference between the two is how they store the energy. A capacitor stores the energy in the form of electrical field in a dielectric material trapped between two conductor plates. However, an inductor stores the energy in the form of magnetic field in coils and loops. Also, a capacitor activates on change in potential difference whereas inductor acts when there is a change in current. The unit of capacitance is farad (F) which describes the amount of current (in ampere seconds) a capacitor can store/discharge per volt change. The symbol for capacitor is C in a schematic. The unit of inductance is henry (H) which describes the amount of voltage (in volt seconds) it can provide per ampere in case of change in current of circuit. The symbol for inductor is L in a schematic. You can easily identify capacitors and inductors by looking at them; refer to Figure-9.

Figure-9. Capacitors and Inductors

Inductors in Series and Parallel

The inductors combine in the same way as resistors do in series and parallel.

In series:

$L = L_1+L_2+L_3+$...............

In parallel:

$1/L = 1/L_1+1/L_2+1/L_3+$................

Capacitors in Series and Parallel

The capacitors combine in the way opposite to resistors do.

In series:

$1/C = 1/C_1+1/C_2+1/C_3+$................

In Parallel:

$L = C_1+C_2+C_3+$...............

COMBINATION OF ENERGY SOURCES

The are mainly two energy sources in electrical circuits; voltage source and current source. For simplification of circuits when there are two voltage sources in series then we can sum their values to express total voltage; refer to Figure-10. Note that you can combine two voltage sources (mainly batteries) in parallel if both the sources have same voltage values to increase total current storage capacity but you cannot combine sources with different voltage values in parallel as they will cause short circuit in battery of lower voltage. Similarly, when there are two current sources in parallel then we can sum their values to express total current; refer to Figure-11. Note that current sources should not be combined in series because no one can predict what will be the output current, it can be combination of both current value, it can be value of one current source. Keep in mind that practically batteries are voltage sources not the current sources.

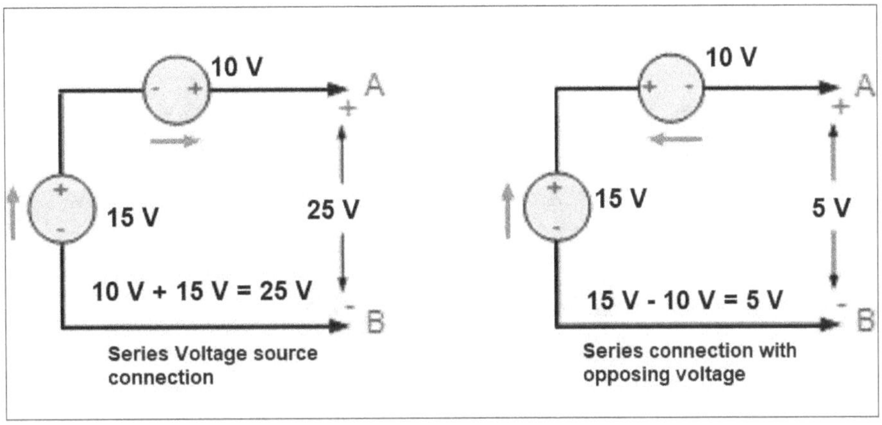

Figure-10. Voltage source combinations

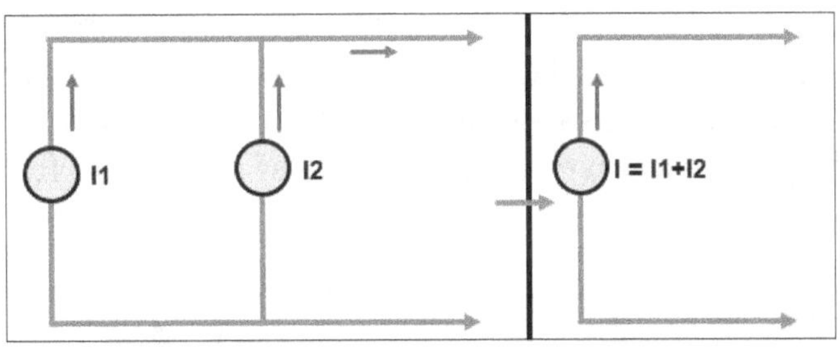

Figure-11. Current source combination

KIRCHHOFF'S LAWS

Gustav Robert Kirchhoff stated two laws for circuit equations which can be used to determine current and voltage values at any section of circuit. These two laws are called Kirchhoff's Current Law (KCL) and Kirchhoff's Voltage Law (KVL).

The Kirchhoff's Current Law (KCL) states that the algebraic sum of current incoming to a point in circuit and the current outgoing from that point in circuit is zero; refer to Figure-12.

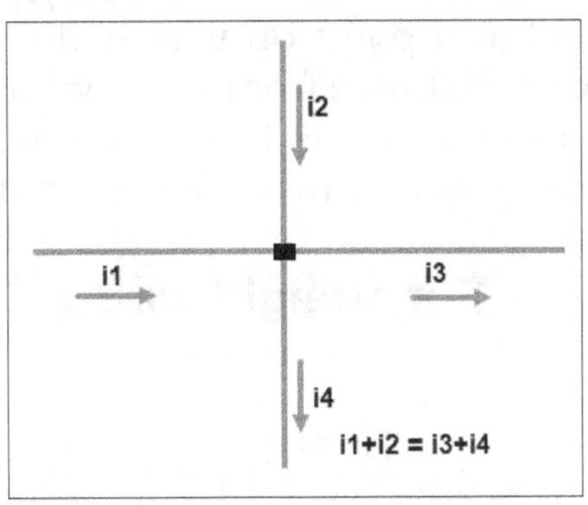

Figure-12. Kirchhoff's Current Law

The Kirchhoff's Voltage Law (KVL) states that the algebraic sum of voltages around a closed loop in circuit is zero; refer to . Note that when you encounter + sign before a voltage source then you need to add it in total voltage and when you encounter - sign before a voltage source then you need to subtract it from total voltage in the loop.

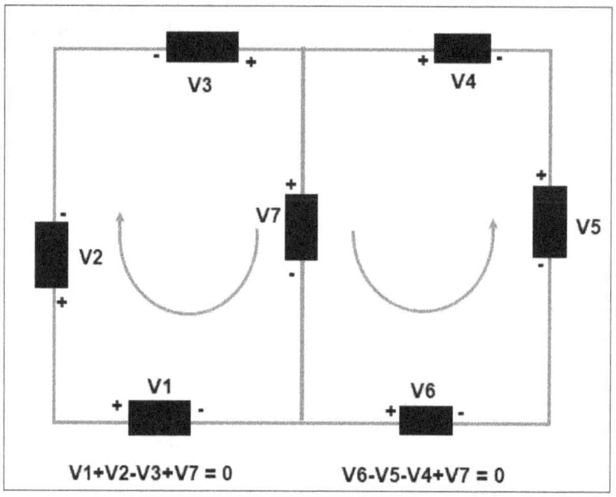

Figure-13. Kirchhoff's Voltage Law

ELECTROMAGNETISM

When direct current passes through a wire then a magnetic field is setup in its vicinity in the form of concentric circles. The direction of magnetic field can expressed by right-hand thumb rule which goes like stretch the thumb of your right hand along the current. The curl of fingers will give the direction of magnetic field; refer to Figure-14.

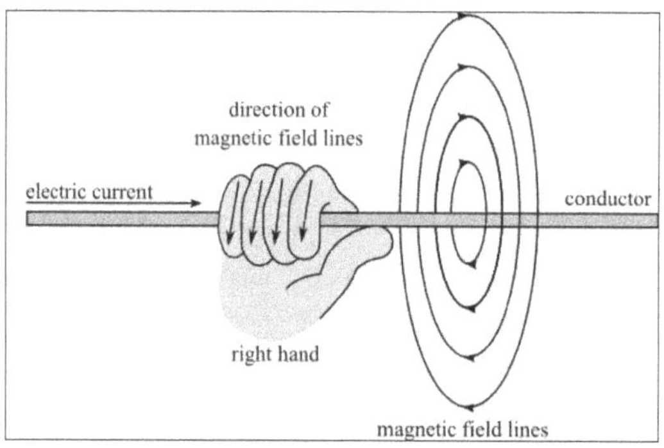

Figure-14. Thumb rule for electromagnetism

Magnetic Field Strength

The strength of magnetic field at a point on wire is directly proportional to current and inversely proportional to its distance from start point. The magnetic field strength parameter is denoted by B and its unit is tesla (T). The formulas for magnetic field strength of different wiring arrangements are given next.

For straight wire

$$B = \frac{\mu_0 I}{2\pi x}$$

Here, I is current, x is distance from start point and μ_0 is permeability of air which has a value of $4\pi \times 10^{-7}$ Tm/A.

For Circular Loop

$$B = \frac{\mu_0 I}{2r}$$

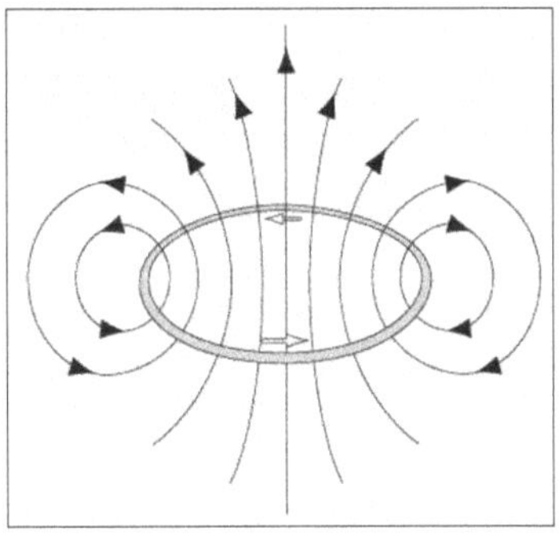

Here, r is the radius of circular loop.

For Solenoid

At center of solenoid, the magnetic field strength is given by,

$$B = \mu_0 n I$$

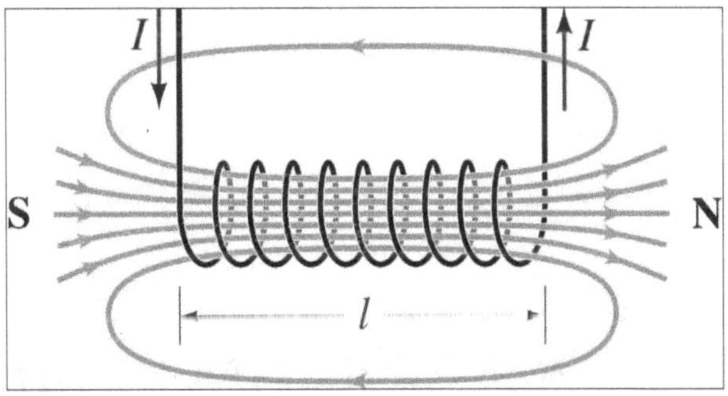

Here, n is the number of turns per unit length

At the either end of solenoid, the magnetic field strength gets halved.

For Toroid

$$B = \frac{\mu_0 n I}{2\pi r}$$

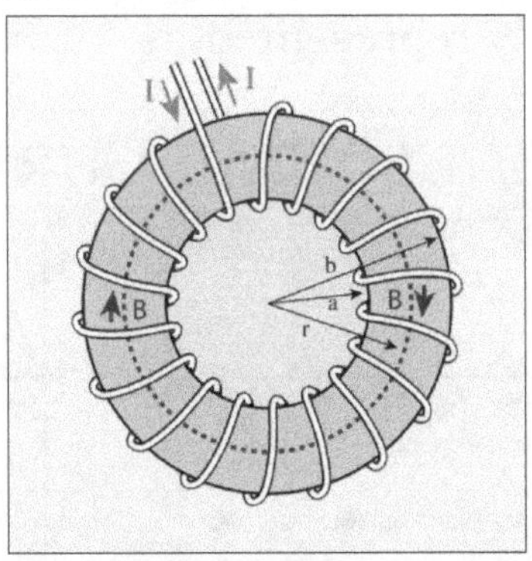

DC CIRCUITS

DC circuits are mostly used to electronic systems and logic based systems like computers, calculators, smart phones, and so on. In these circuits, current flows in one direction and mostly low voltage in the range of 0 to 24 V is used. In DC circuits, loads can be connected in series or in parallel as discussed earlier in this chapter.

AC CIRCUITS

AC circuits are used to run heavy electrical motors and appliances. In these circuits, magnitude and direction of voltage & current in circuit are not fixed. A sinusoidal waveform is used in transmission of current. For half time of waveform, the voltage is positive and rest half time of waveform, the voltage is negative; refer to Figure-15.

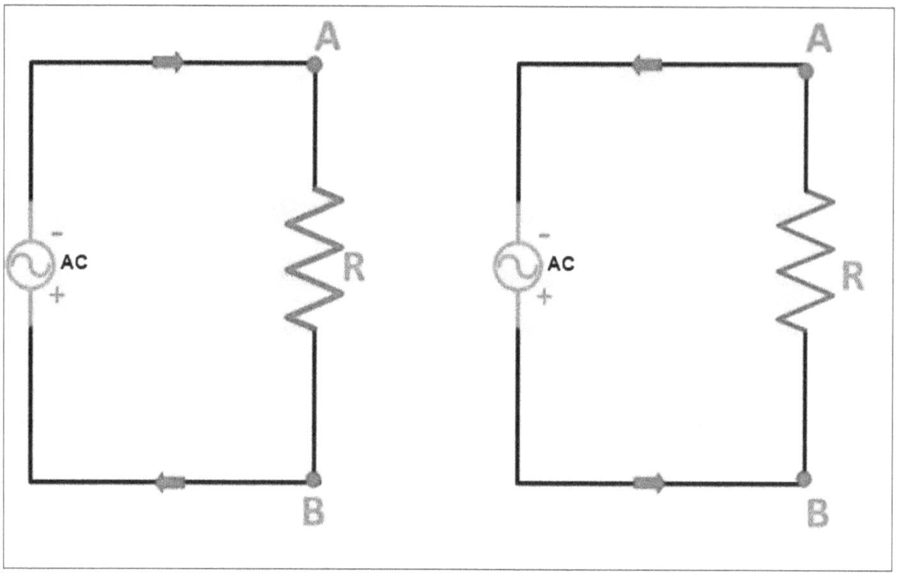

Figure-15. AC circuit

ALTERNATING CURRENT

Alternating means changing direction so the current and voltage that changes direction is called alternating current and alternating voltage, respectively. In alternating current supply, the current varies sinusoidally and can be mathematically represented by;

$$i = I_m \sin\omega t$$

which can be graphically represented as:

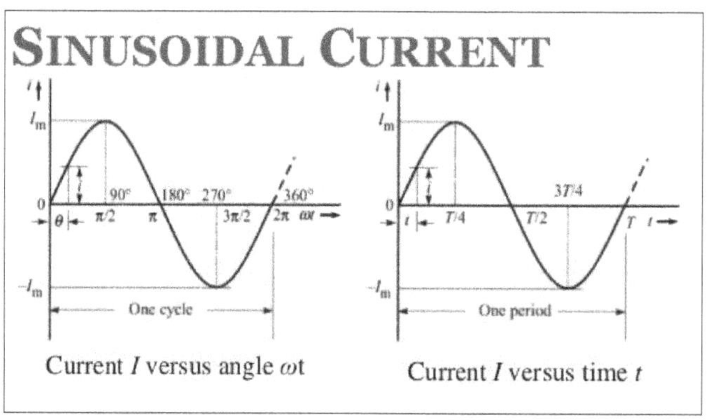

Common Terms for ac supply:

Frequency : Frequency of alternating current is the number of cycles that occur per second. In simple words, it is the number of times current value becomes positive in a second. The unit for frequency is Hz (hertz) which is cycles per second. Frequency is denoted by *f*. In USA, the frequency of alternating current supply is 60 Hz.

Cycle : One complete set of positive and negative values of waveform is called cycle.

Peak or Maximum Value : It is the maximum value of current generated by waveform. It is also called **amplitude** and denoted by I_m.

Time Period : It is the time in second required to complete one cycle. It is denoted by T and is reciprocal of Frequency (F).

Angular Frequency : It is the number of radians covered per second. Angular frequency is denoted by ω which equal to 2π*f*.

Phase : Phase is used to identify if there a mismatch in two waveforms of alternating current and if there is a mismatch then what is its value. Figure-16 shows two alternating current that are in same phase as the two currents attain their maximum, minimum, and zero values at the same time. If there is a difference between two waveforms then the difference is called phase difference or phase shift.

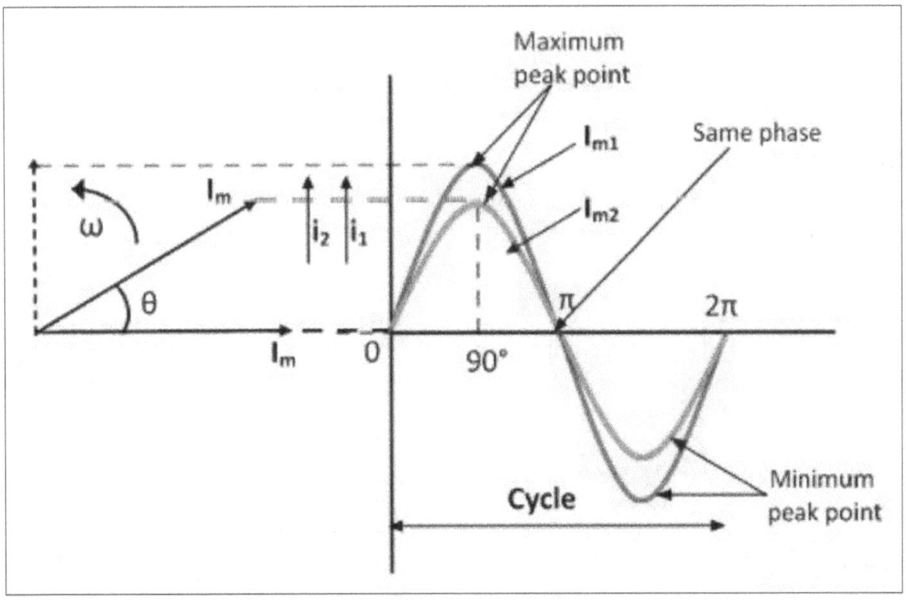

Figure-16. Phase of alternating current

Three Phase Voltage

In a three phase voltage system, three emf of same magnitude and frequency are applied which are 120 degree phase displaced to each other; refer to Figure-17. The line voltage for three phase supply is 440 V (i.e. when you use two live lines for connection) and phase voltage for three phase supply is 220 V (i.e. voltage between neutral and any one live line).

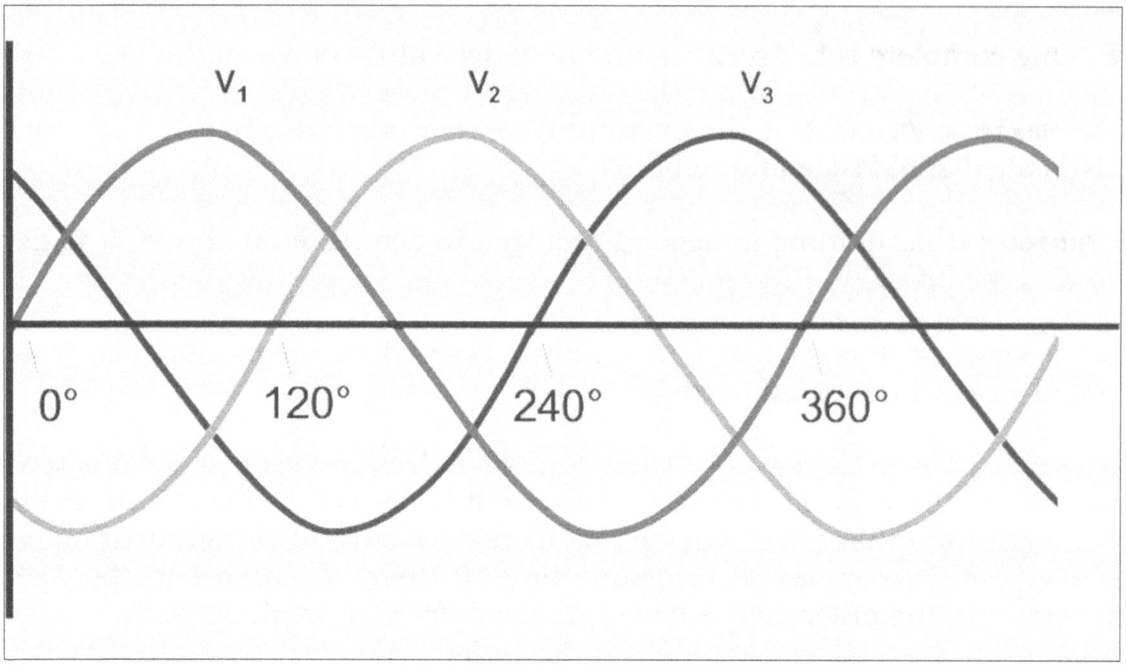

Figure-17. Three phase voltage waveform

Three Phase Loads

The three phase loads are connected to the three phase supply in either Star form or Delta form; refer to Figure-18.

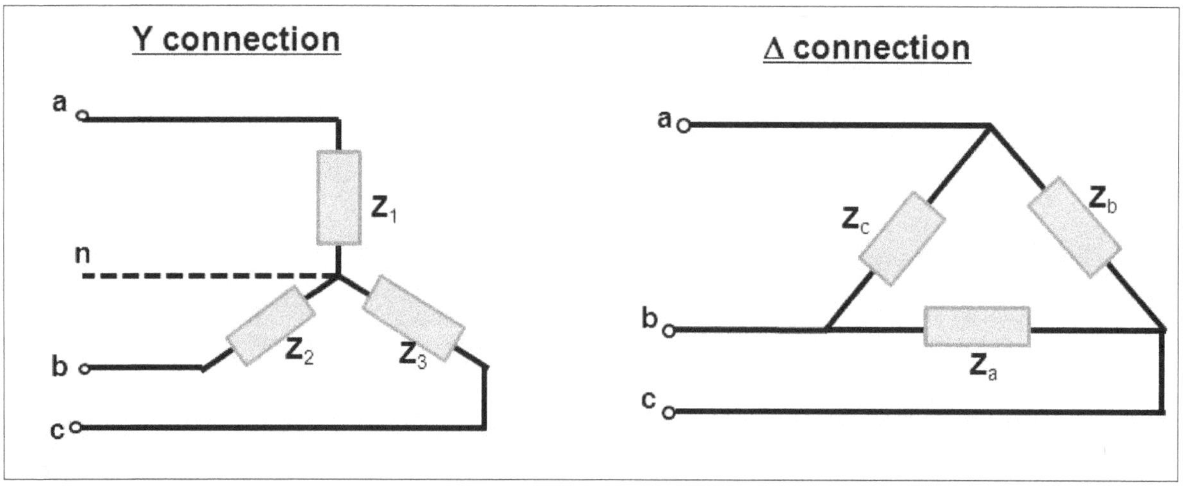

Figure-18. Three phase load system

In a three phase load system if all the three impedances (loads) equal then it is called balanced load system. If the three loads are of different values then it is called unbalanced load system. Generally, three phase supply is provided for both industrial and domestic use. In industries, most of the machines run on balanced load but in domestic application, the loads are not always balanced on supply lines. If an unbalanced load is connected in star configuration to the supply then voltage drops for different branches will be different causing situation called neutral shift. In such cases, voltage in one line can be significantly high whereas in another line it might be very low. So, it is suggested to avoid unbalanced load and if it is not possible then add pseudo loads so that circuit gets balanced. If possible use 3 line + 1 neutral configuration for three phase system as extra current will pass through neutral line and will keep the system balanced at cost of some extra energy.

Review Questions

Q1. What is electricity?
A. A magnetic field created by protons
B. A form of energy caused by electrons' motion or presence
C. The movement of neutrons in a conductor
D. A static field produced by a battery

Q2. Which unit is used to measure electric charge?
A. Ampere
B. Volt
C. Coulomb
D. Watt

Q3. What does one ampere of current represent?
A. 1 V flowing per second
B. 1 C of charge passing a point per second
C. 1 W of power consumed
D. 1 Ω resistance per meter

Q4. What kind of current flows in one direction and has constant amplitude?
A. Alternating Current
B. Static Current
C. Direct Current
D. Live Current

Q5. What is the unit of current density?
A. A
B. V/m
C. A/m^2
D. C/s

Q6. Voltage is defined as:
A. Rate of change of current
B. Work done per unit charge
C. Resistance across two points
D. Charge per second

Q7. What is the formula for electrical power?
A. V × R
B. I × R
C. V × I
D. V/I

Q8. Which resistor type is most cost-effective for high resistance values?
A. Wire wound
B. Metal film
C. Carbon molded
D. Ceramic

Q9. In a parallel resistor connection, what remains the same across all resistors?
A. Current
B. Resistance
C. Voltage
D. Power

Q10. What transformation is used to simplify complex resistor arrangements?
A. AC to DC
B. Star-Delta and Delta-Star
C. Parallel to Series
D. Power to Efficiency

Q11. What do capacitors store energy in the form of?
A. Electric Field
B. Magnetic Field
C. Heat
D. Sound

Q12. What is the unit of inductance?
A. Farad
B. Ampere
C. Tesla
D. Henry

Q13. How are inductors combined in series?
A. L = L1 + L2 + L3
B. 1/L = 1/L1 + 1/L2
C. L = L1 × L2
D. L = L1 - L2

Q14. How are capacitors combined in parallel?
A. 1/C = 1/C1 + 1/C2
B. C = C1 + C2
C. C = C1 - C2
D. 1/C = C1 + C2

Q15. Why can't voltage sources of different values be connected in parallel?
A. Voltage will drop
B. They cancel each other
C. Short circuit may occur
D. Current will stop flowing

Q16. Kirchhoff's Current Law (KCL) states:
A. Total voltage in loop is zero
B. Current at node is equal to resistance
C. Sum of incoming and outgoing current is zero
D. Voltage equals current times resistance

Q17. What is the unit of magnetic field strength?
A. Tesla
B. Henry
C. Farad
D. Ampere

Q18. What determines the direction of magnetic field around a current-carrying wire?
A. Fleming's left-hand rule
B. Ohm's law
C. Right-hand thumb rule
D. Lenz's law

Q19. In AC circuits, what is one complete set of positive and negative values called?
A. Frequency
B. Amplitude
C. Phase
D. Cycle

Q20. What is the frequency of AC supply in the USA?
A. 50 Hz
B. 60 Hz
C. 100 Hz
D. 120 Hz

Q21. What is angular frequency represented by?
A. α
B. θ
C. ω
D. f

Q22. What are the line and phase voltages in a 3-phase supply system?
A. 110 V and 220 V
B. 220 V and 440 V
C. 440 V and 220 V
D. 330 V and 110 V

Q23. What type of three-phase load system has equal impedances?
A. Unstable
B. Linear
C. Unbalanced
D. Balanced

Q24. What can happen if an unbalanced load is connected in star configuration?
A. Current will become zero
B. Voltage will be equally distributed
C. Neutral shift will occur
D. Current will double

Q25. What is generally suggested for unbalanced 3-phase loads?
A. Use less voltage
B. Avoid AC power
C. Use 3-line + 1 neutral
D. Increase resistance

Q26. What does the symbol 'I' represent in electrical equations?
A. Resistance
B. Voltage
C. Current
D. Power

Q27. What does the unit "volt" represent?
A. One coulomb per ampere
B. One joule per coulomb
C. One watt per ampere
D. One ampere per second

Q28. What is the resistivity unit?
A. Ohm
B. Ohm/meter
C. Ohm-meter
D. Volt-meter

Q29. Which resistor is made by winding resistance wire on a ceramic tube?
A. Carbon molded
B. Wire wound
C. Metal film
D. Variable resistor

Q30. Which resistor typically has a blue background color?
A. Carbon molded resistor
B. Wire wound resistor
C. Metal film resistor
D. Ceramic resistor

Q31. What happens when resistors are connected in series?
A. Voltage remains same across each
B. Current gets divided
C. Total resistance is sum of individual resistances
D. Total resistance is lower than smallest resistor

Q32. In a parallel connection of resistors, what can be said about total resistance?
A. It equals the largest resistor
B. It's more than the largest resistor
C. It's less than the smallest resistor
D. It remains constant

Q33. What does the unit "farad" measure?
A. Voltage
B. Capacitance
C. Resistance
D. Inductance

Q34. In a circuit, what happens when a capacitor faces a sudden change in voltage?
A. It ignores the change
B. It stores magnetic energy
C. It reacts and stores electrical energy
D. It overheats

Q35. What is the formula to combine capacitors in series?
A. C = C1 + C2 + C3
B. 1/C = 1/C1 + 1/C2 + 1/C3
C. C = C1 × C2
D. C = C1 - C2 + C3

Q36. What is not true about combining current sources?
A. Can be combined in parallel
B. In series, result is unpredictable
C. Series combination is recommended
D. Batteries are voltage sources, not current sources

Q37. Which law states the total voltage around a closed loop is zero?
A. KCL
B. Ohm's Law
C. Kirchhoff's Voltage Law (KVL)
D. Maxwell's Law

Q38. Which rule helps determine magnetic field direction around a wire?
A. Left-hand rule
B. Kirchhoff's Rule
C. Right-hand thumb rule
D. Ampere's Rule

Q39. What is the formula for magnetic field at center of a circular loop?
A. µ□I/2πx
B. µ□nI
C. µ□I/2r
D. µ□nI/2

Q40. What happens to magnetic field at ends of a solenoid?
A. It doubles
B. It becomes zero
C. It gets halved
D. It stays constant

Q41. What is the usual voltage range in DC circuits?
A. 0 to 12 V
B. 0 to 24 V
C. 0 to 50 V
D. 0 to 100 V

Q42. What waveform is commonly used in AC power transmission?
A. Square wave
B. Triangle wave
C. Sine wave
D. Pulse wave

Q43. What does "Im" represent in an AC waveform?
A. Current over time
B. Frequency
C. Peak or maximum value of current
D. Power

Q44. What is the time period of an AC waveform?
A. $T = f$
B. $T = 1/f$
C. $T = 2\pi f$
D. $T = f^2$

Q45. Which of the following describes phase shift?
A. Difference in amplitude
B. Difference in frequency
C. Difference in zero crossing points
D. Difference in voltage

Q46. What is the typical phase difference between lines in 3-phase system?
A. 90°
B. 120°
C. 180°
D. 240°

Q47. What happens in an unbalanced three-phase star load?
A. Power is lost completely
B. Voltage remains the same
C. Neutral shift occurs
D. Inductance increases

Q48. What's a method to stabilize unbalanced 3-phase systems?
A. Increase frequency
B. Use 2-phase supply
C. Add pseudo loads
D. Remove all loads

Annexure II

Basics Components of Electrical Control System

Topics Covered

The major topics covered in this chapter are:

- ***Switches***
- ***Relays and Timer Relays***
- ***Contactors***
- ***Fuses, Circuit Breakers and Disconnectors***
- ***Terminals and Connectors***
- ***Solenoids***
- ***Motors and Transformers***
- ***Sensors and Actuators***

INTRODUCTION

There are various components involved in designing an electrical control circuit like push buttons, relays, contactors, disconnectors and so on. When designing an electrical control system, it is important to understand the parameters of these components and how they fit in your circuit. In this chapter, we will learn about some common electrical components used in these circuits.

PUSH BUTTONS

The push buttons are used to start and stop motors and other electrical machines. There are mainly two type of push buttons based on operation; Normally Open (NO) and Normally Closed (NO); refer to Figure-1. These push buttons are generally coupled with contacts to run motor or other electrical machines. So, once you have pressed NO push button to start a circuit, the contact will keep on supplying current until NC push button is pressed; refer to Figure-2. The contact points in good quality push buttons are made of silver to provide long life and better conductivity. Gold contacts are typically necessary when switching at logic level, generally defined as covering 1 to 100 mA. The momentary voltage on push buttons is very high so you should make sure that proper insulation is provided in the push button depending on your application. The factors involved in selection of push button are load, voltage, contact materials, circuit type, terminal type, and mounting.

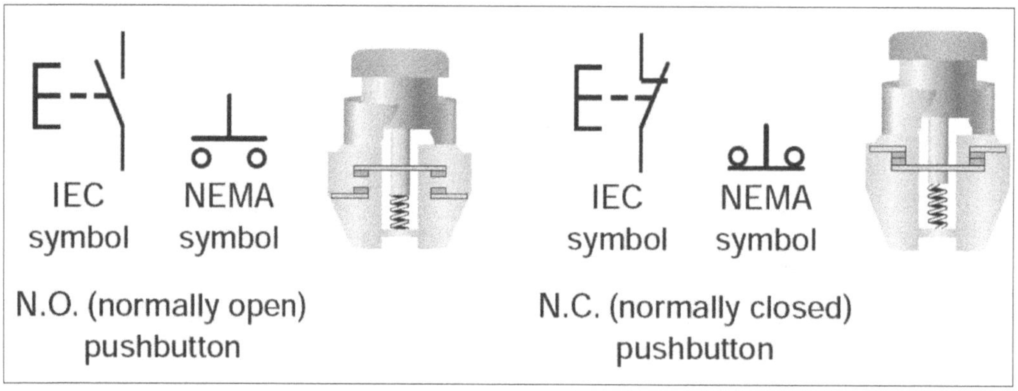

Figure-1. Push buttons

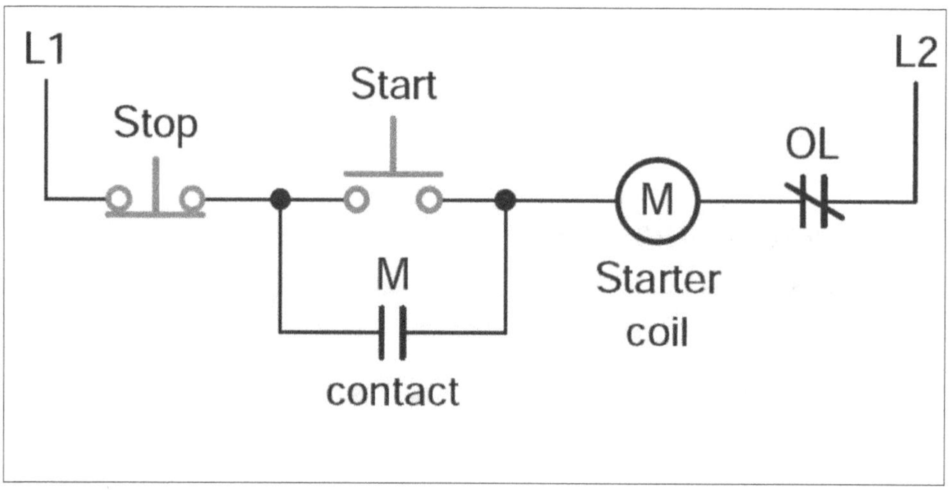

Figure-2. Circuit showing use of push button

PILOT LIGHTS

The pilot lights provide visual indication of circuit status. If a pilot light is glowing then it means the circuit is connected and power is running through it. If you have controls install at one place and connected motors are half a mile away then there is no way to instantaneously confirm whether motor is running or not after you have pressed the push button for it. In such case, pilot lights are connected in motor circuit. The pilot lights are available in different colors which are mainly Red, Yellow, Green, Blue, White, Grey, and Black. As per IEC standards, the use of these lights is described in Figure-3.

Pilot Light Color and its meaning

Color	Meaning	Use
RED	Emergency	It is used to show emergency stop of a machine
YELLOW	Abnormal condition	It is used to show there is some problem in circuit
GREEN	Normal	It is used to show the circuit is working properly
BLUE	Action required	It is used to show that manual interventaion is required for machine
WHITE, GREY, BLACK, CLEAR	Neutral	These lights are used for custom indications for example if a specific part of machine is to be monitored then assign a pilot light to it.

Figure-3. Pilot color coding

Legend Plate

Legend plate is installed around push buttons and pilot lights to identify their purpose. The legend plate text includes text like START, STOP, RESET, RUN, UP, DOWN, and so on.

SWITCHES

Various types of switches are used in electrical systems like selector switch, toggle switch, drum switch, limit switch, temperature switch, pressure switch, float & flow switches, and so on. These switches are discussed next.

Selector Switch

The selector switches are used where you need to select from a broad range of contacts; refer to Figure-4. The selector switch use rotating cam for their operation. You can easily find the use of selector switch in washing machines, dishwashers and other home appliances where you need to select a mode.

Figure-4. Selector switch

The selector switches can be illuminated, non-illuminated, and non-illuminated with key. Selector switch with key are used when machine being operated by that switch can dangerous for people around it if not handled properly. So, the person authorized with key can only operate the machine. The specifications for selection of selector switch are lighted/non-lighted/key-type, physical dimensions, color coding, and current rating.

Toggle Switch

The toggle switch is used to manually toggle between On and Off position of the circuit. The switches on switchboards found in our home to on/off fans, lights, and other appliances are a type of toggle switches. There are mainly four types of toggle switches; SPST (Single Pole Single Throw), SPDT (Single Pole Double Throw), DPST (Double Pole Single Throw), and DPDT (Double Pole Double Throw); refer to Figure-5.

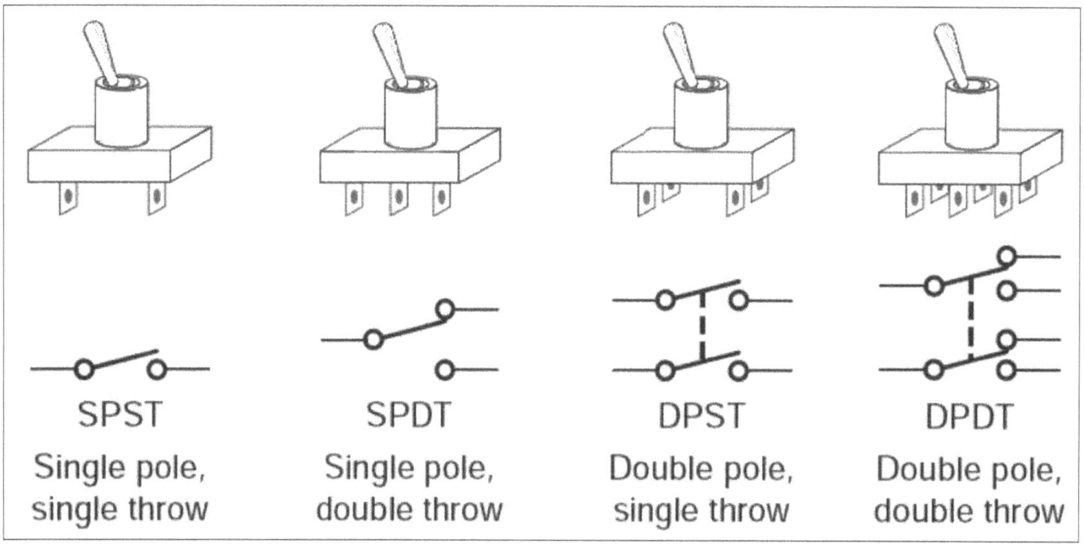

Figure-5. Toggle switches

Single pole single throw (SPST) toggle switches make or break the connection of a single conductor in a single branch circuit. This switch type typically has two terminals and is referred to as a single-pole switch.

Single pole double throw (SPDT) toggle switches make or break the connection of a single conductor with either of two other single conductors. These switches usually have three terminals and are commonly used in pairs. SPDT switches are sometimes called three-way switches.

Double pole single throw (DPST) toggle switches make or break the connection of two circuit conductors in a single branch circuit. They usually have four terminals.

Double pole double throw (DPDT) toggle switches make or break the connection of two conductors to two separate circuits. They usually have six terminals are available in both momentary and maintained contact versions.

Specifications for selection of toggle switch are dimensions, electrical ratings, terminal types, materials, and features.

Drum Switch

In drum switches, you can open or close contacts by moving center shaft; refer to Figure-6. The drum switches can be used to forward, reverse and stop the motor. The parameters for selection drum switch are number of poles, current rating, Switch type (reversing/non-reversing), voltage rating, AC/DC, and so on. You will find the connection settings for reversing motor in the manual supplied with the drum switch. The general schematic for forward and reverse operation by drum switch can be given by Figure-7.

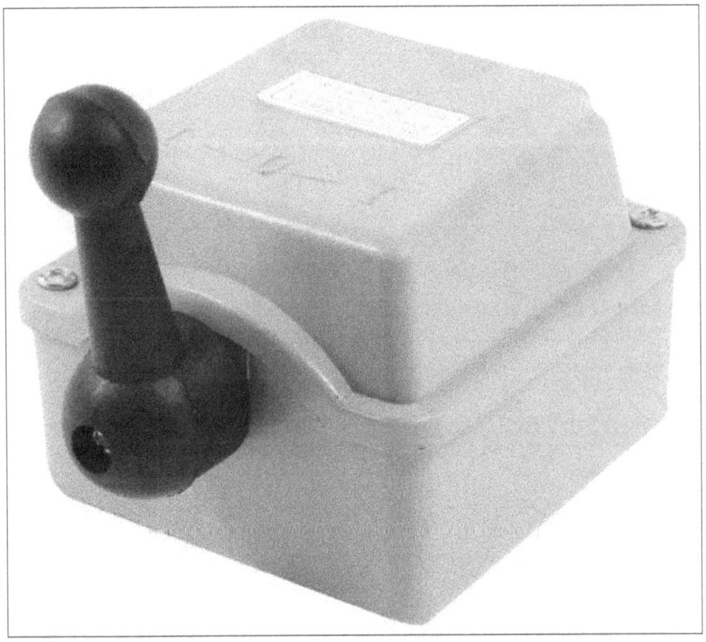

Figure-6. Drum switch

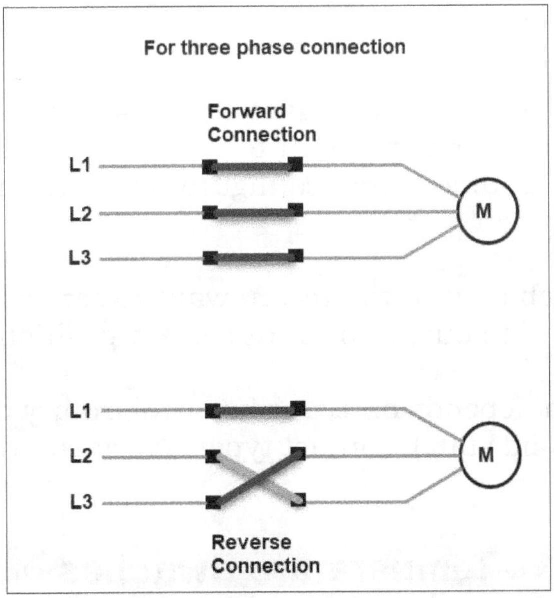

Figure-7. Drum switch connections

Limit Switch

The limit switches work automatically based on various factors like pressure, temperature, position and so on; refer to Figure-8. When the governing factor for limit switch reaches a certain threshold then the switch starts functioning. For example, there is a normally closed limit switch which will keep closed until the pressure reaches 30 bars inside a container. When pressure inside the container reaches 30 bar then the switch will automatically open. There are mainly two parts of a limit switch; body and actuator. The body part contains contacts that open or close based on movement of actuator. The symbols for limit switches are shown in Figure-9.

Figure-8. Limit switch

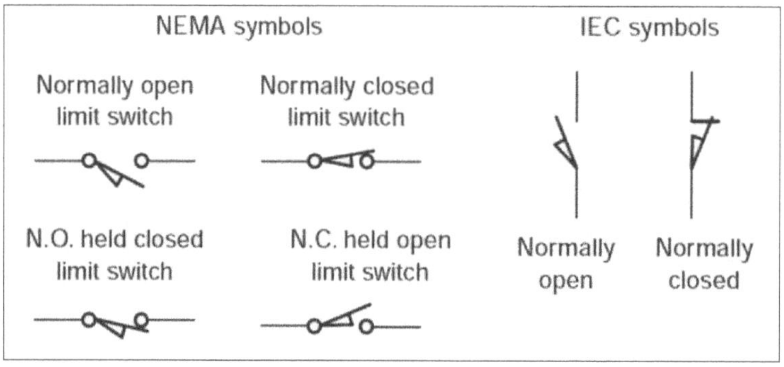

Figure-9. Limit switch symbols

Micro-limit switch is a variation of limit switches in which snap action switch is used to quickly transfer contact from one point to another point. Micro limit switches are available in single pole double throw configuration and one terminal for both NC and NO contacts is common.

Rotating cam limit switch is used where you want to control to rotation of a shaft in machine. They are useful to cutoff power at certain position or rpm threshold.

Selection of limit switch depends on type of control (rotary or linear), type of switch (electromechanical or solid state), contact types, poles and throw configuration, and so on.

Temperature Switches

Temperature switches are actuated by temperature or change in temperature. The symbol for temperature switch is given in Figure-10. The most common mechanism used in temperature switch is bimetallic strip which has two different metals having different thermal expansion properties joined together. Heating or cooling this bimetallic strip causes it to bend causing switch to close or open.

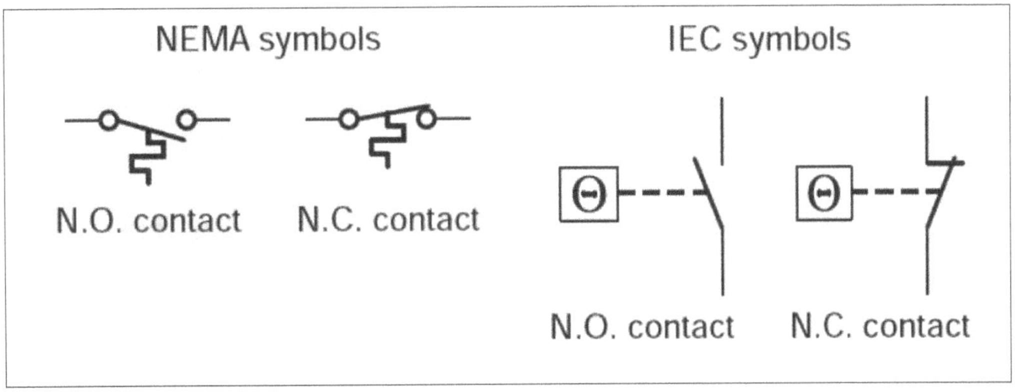

Figure-10. Temperature switch symbols

There is another type of thermal switch which works on expansion of fluid due to temperature called liquid filled temperature switch. It comprises a brass bulb filled with a chemical fluid (sometimes gas). It includes a small tube which hooks up the bulb to a pressure sensing mechanism consisting of bellows, bourdon tube or diaphragm. The functioning of both switches is shown in Figure-11.

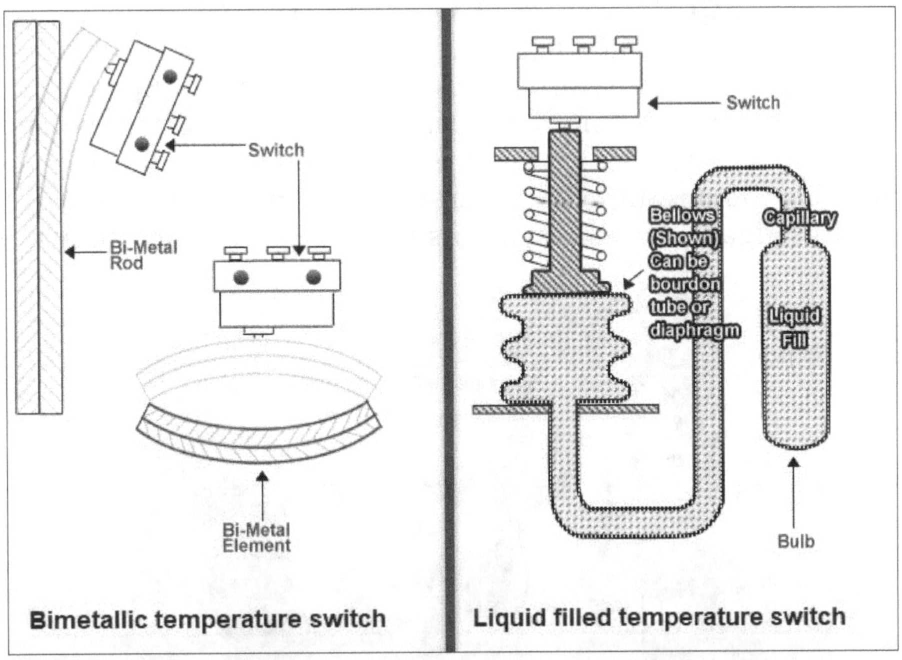

Figure-11. Temperature switches

The selection criteria for temperature switch includes temperature range, current and voltage rating, type of temperature switch required, and so on.

Pressure Switches

The pressure switch is used to control and monitor pressure. The most common application of pressure switch is found in air compressor system where the pressure switch causes to stop motor after pressure in the tank has reached a certain threshold. The symbols for pressure switches are given in Figure-12. The stiffness of spring defines the limit at which pressure switch will be activated.

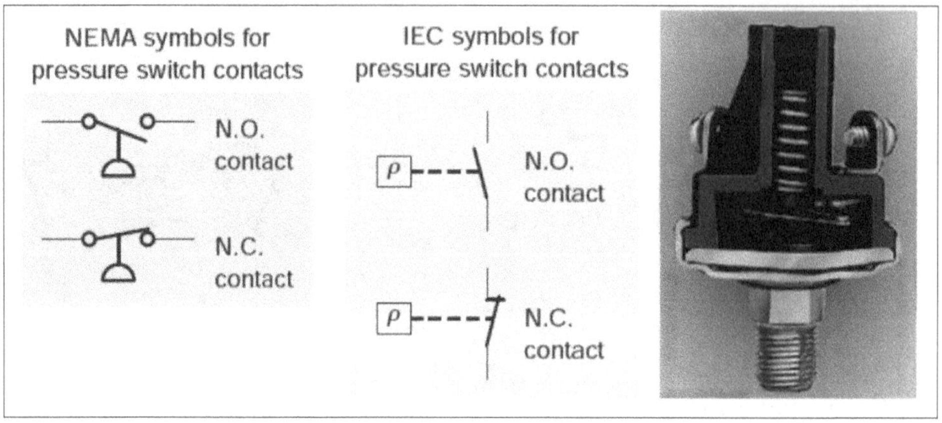

Figure-12. Pressure switch symbols

The selection criteria of pressure switch includes fixed or adjustable type pressure limit, type of fluid, fatigue, electrical output, pressure and electrical connections, response time, and ratings.

Liquid Level and Flow Switches

A liquid level switch is also called float switch as the bulb of switch floats on liquid. This switch is used when you want to cut off power once liquid in container has reached certain level. There is another variation of this switch in which two wires of low current are suspended in the liquid tank and when a conductive liquid submerges both the wire ends, the circuit gets closed causing the switch to activate. Both the types are shown in Figure-13. The symbols for float switch are shown in Figure-14.

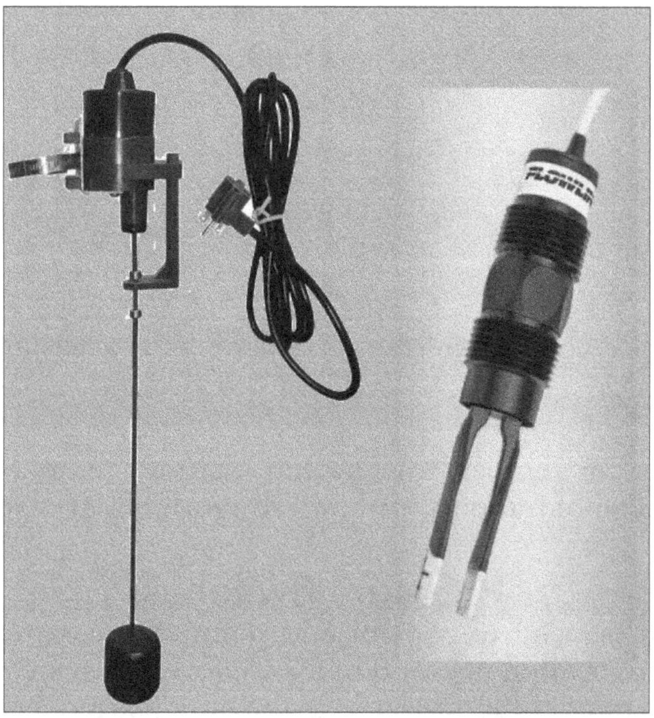

Figure-13. Float or liquid level switches

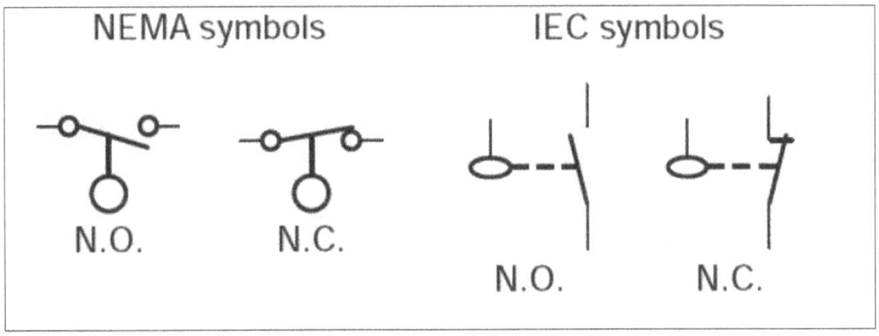

Figure-14. Symbols for float switch

The flow switches are used to monitor flow of fluid and control opening/closing of contacts accordingly. The symbols for flow switches are shown in Figure-15. Figure-16 shows different types of flow switches.

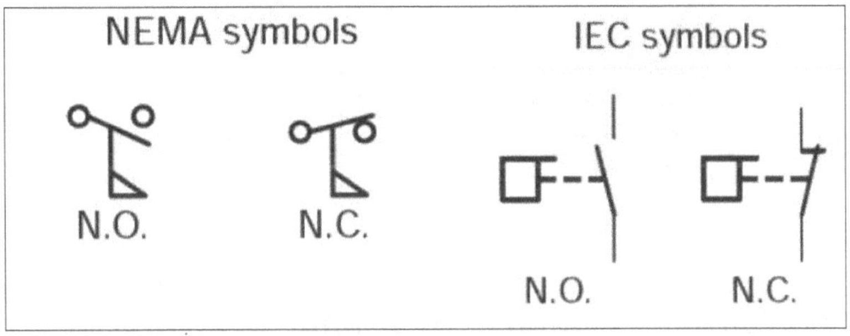

Figure-15. Symbols for flow switches

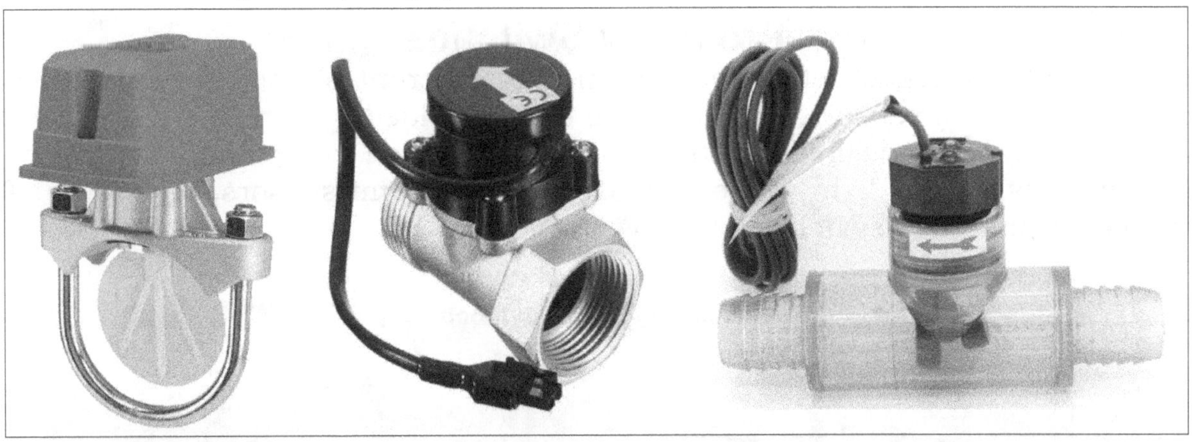

Figure-16. Flow switches

Joystick Switch

A joystick switch is used to control movement of a machine part in more than one direction. They are manually operated control switches and you can easily find their applications in backhoe loaders, cranes, remote control devices and so on. The symbol of joystick can be given as :

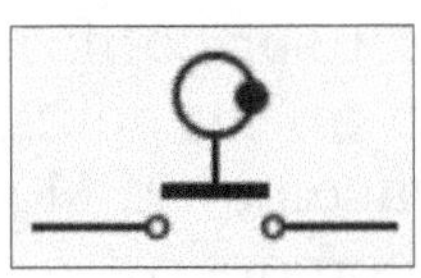

There can be upto 8 actuation directions with different switch combinations on a joystick; refer to Figure-17.

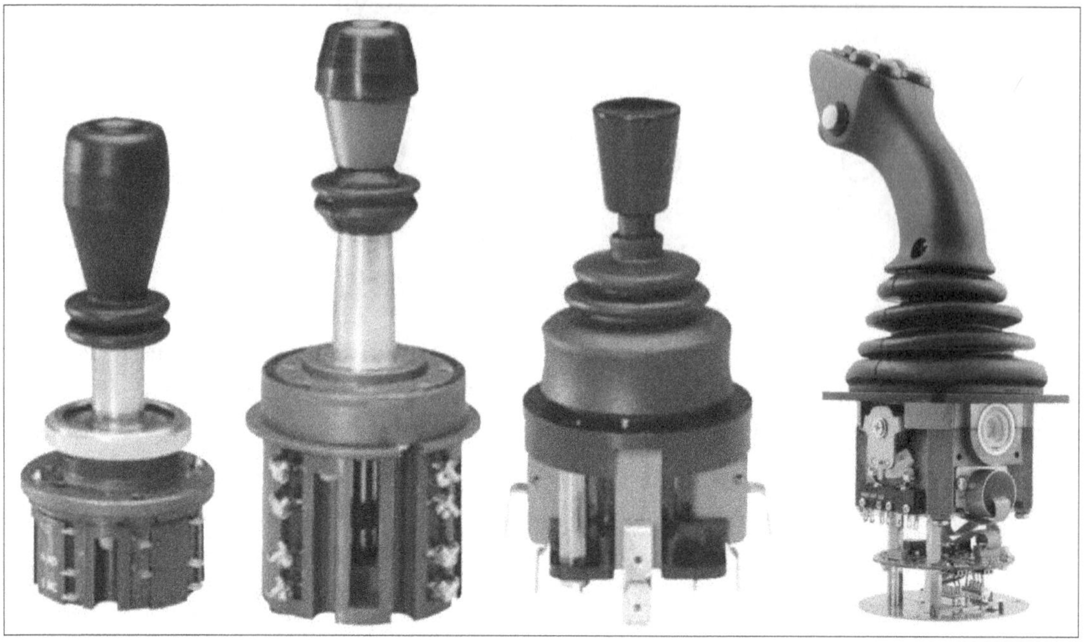

Figure-17. Joystick switches

The selection criteria for joystick includes number of contacts, current & voltage ratings, material of controller, contact types, and so on.

Proximity Switches

The proximity switches detect the presence of different shape, size, and type of materials based on sensors. Based on the data provided by sensors, the proximity switches activate the desired function. The sensors used in proximity switches can be different types like light sensor, pressure sensor, emf sensors, and so on. The symbols for proximity switches are shown in Figure-18.

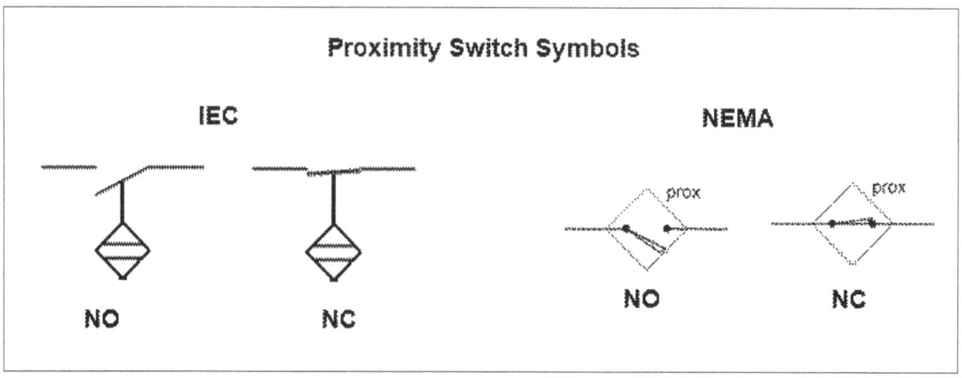

Figure-18. Proximity switch symbols

The point at which the proximity sensor confirms an incoming target is called the operating point. The point at which an outgoing target makes the switch to return to normal position is called the release point. Most proximity sensors also have LED indicator the status. The area between operating and release points is known as the hysteresis zone. The sensors in proximity switches generally work on either 24 V DC or 120 V AC power supply. Selection criteria for proximity switches include range of sensor, type of material to be sensed, space for sensor, environment, output type, contact type, and so on.

Sensors

Sensors are use to transmit digit or analog signals based on detected parameter. There are various types of sensors available for control designers like proximity sensor, temperature sensor, pressure sensor, humidity sensor, radiation sensor, and so on; refer to Figure-19. Since there is a long list of sensors now available in market, it is not feasible to discussed every sensor here. As an electrical control designer, your task will be to select appropriate sensor for your job. For example, you need to run a boom barrier motor at toll plaza based on information received through an id scanner. Now, you need to check the list of sensors and find out which one suits you the best. Like in this case, an RFID sensor will be best choice when each car has an RFID installed in it. The next parameters for electrical engineer will be voltage and current ratings of sensor, range of sensor, environmental conditions for sensor, contact types for sensor, and so on.

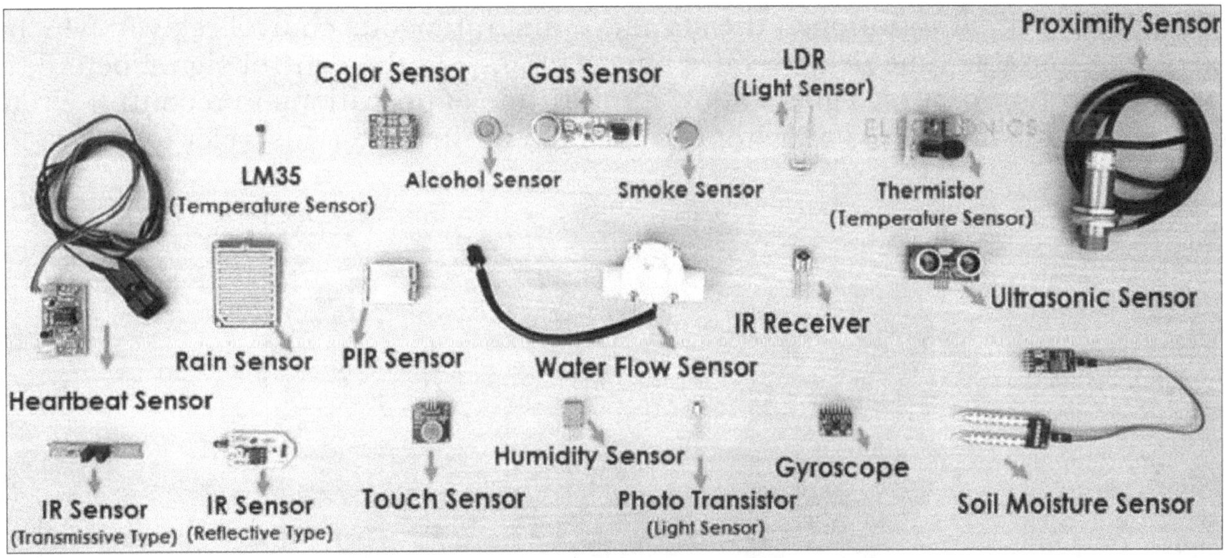

Figure-19. Types of sensors

ACTUATORS

Actuators are the electrical components that convert electrical energy to mechanical movement. There various types of actuators in electrical control design like relay, solenoids, and motors. The brief description of these actuators is given next.

Relays

A relay consists of two parts; relay coil and contact which are packed as a unit with separate contact points. When current is passed through relay coil then due to electromagnetism, it attracts iron arm connected to contact. The symbol for relay is shown in Figure-20. The working principle of relay is shown in Figure-21. So, whenever you need high power contacts controlled by low power switches then you can use relays.

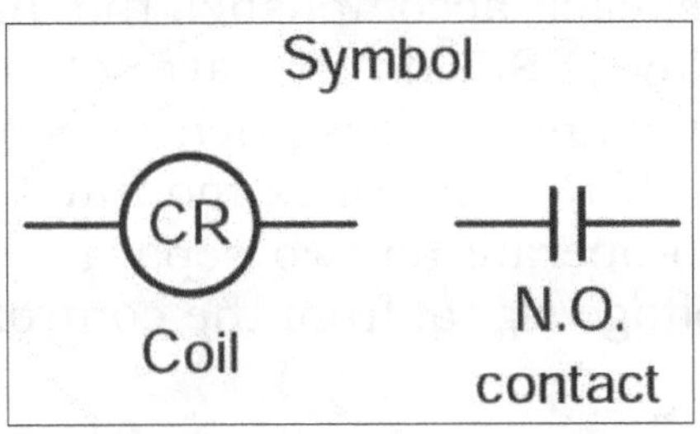

Figure-20. Relay symbol

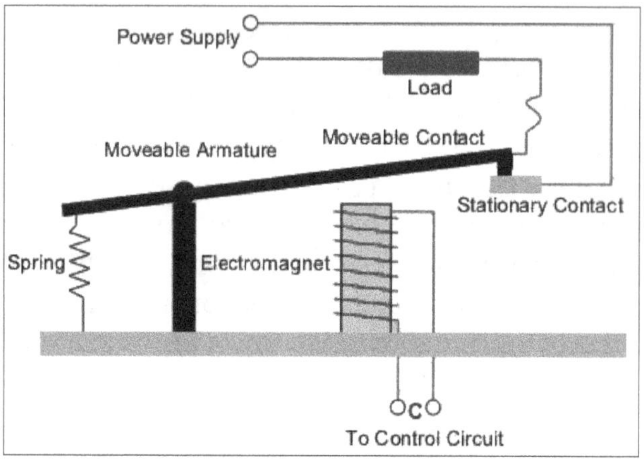

Figure-21. Relay working principle

Time Relays

Timing relays are a variation of the standard instantaneous control relay in which a fixed or adjustable time delay occurs after a change in the control signal before the switching action occurs. Timers allow a multitude of operations in a control circuit to be automatically started and stopped at different time intervals; refer to Figure-22.

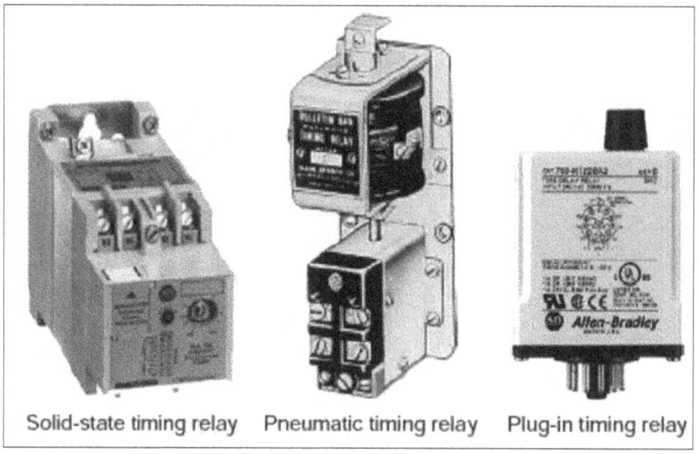

Figure-22. Time relays

Latching Relays

Latching relays typically use a mechanical latch or permanent magnet to hold the contacts in their last energized position without the need for continued application of coil power. They are especially useful in applications where power must be conserved, such as a battery-operated device, or where it is desirable to have a relay stay in one position if power is interrupted.

Solid State Relays

A solid-state relay (SSR) is an electronic switch that, unlike an electromechanical relay, contains no moving parts. Although EMRs and solid-state relays are designed to perform similar functions, each accomplishes the final results in different ways. Unlike electromechanical relays, SSRs do not have actual coils and contacts. Instead, they use semiconductor switching devices such as bipolar transistors, MOSFETs, silicon-controlled rectifiers (SCRs), or triacs mounted on a printed circuit board. All SSRs are constructed to operate as two separate sections: input and output. The input side receives a voltage signal from the control circuit and the output side switches the load.

Solenoid

Solenoid is an electrical component in which uses electrical energy to move an armature electromagnetically. There are mainly two types of solenoids, linear and rotary with linear and rotary working directions respectively for mechanical output. The symbol for solenoid coil is shown in Figure-23.

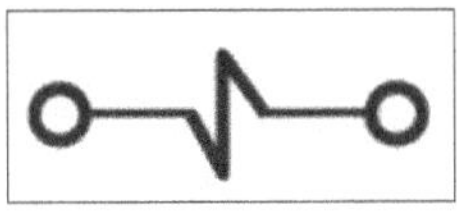

Figure-23. Symbol for solenoid coil

There are two power supply options for solenoid coils, AC and DC. The choice mainly depends on the type of supply available. If both the supplies are available then AC solenoids are more powerful in fully open position as compared to DC solenoids. AC operated solenoids are faster in response as compared to DC solenoids. The working principle of solenoid coil is shown in Figure-24.

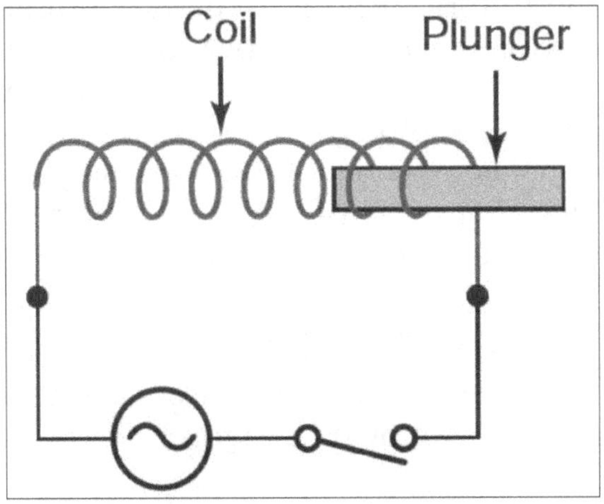

Figure-24. Solenoid coil working principle

Solenoid Valve

In solenoid valves, the opening and closing of valve is controlled by solenoid coil. These valves are mainly used to control opening and closing for hazardous fluids like nitrogen, gas, acids, and so on. The working principle of solenoid valve is shown in Figure-25. The symbols for solenoid valve are shown in Figure-26.

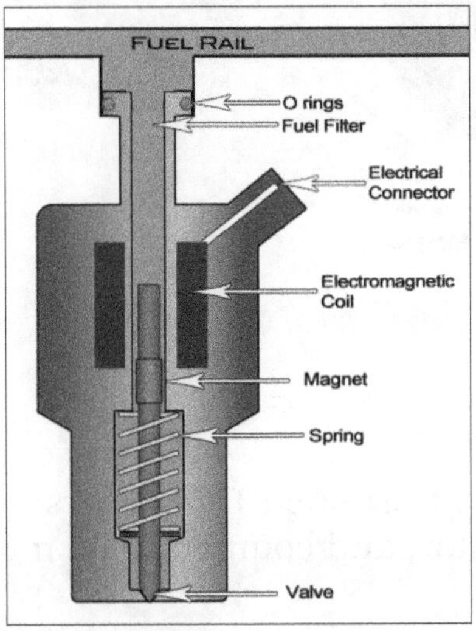

Figure-25. Solenoid valve

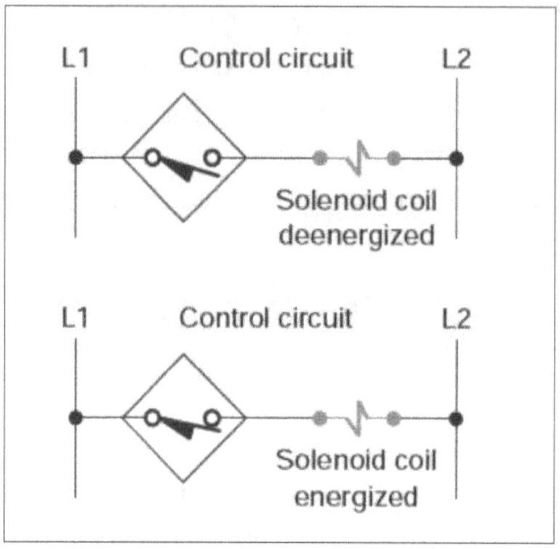

Figure-26. Solenoid valve symbols

MOTORS

Motors are also a type of actuators with rotary motion but since they are widely used in industries, it is better to discuss their types also. All motors work on the same electromagnetism principle: you pass electricity through a coil, it gets magnetized and starts to pull/push a mechanical plunger. In case of motors, there are two magnets instead of one. The same poles (N-N or S-S) of two magnets repel each-other while different poles (N-S) of two magnets attract each other. We will first discuss DC motors and then we will discuss with AC motors.

DC Motors

In DC motors, the current does not change direction and remains constant. The working principle of DC motor is shown in Figure-27.

Brushed Electrical DC Motor

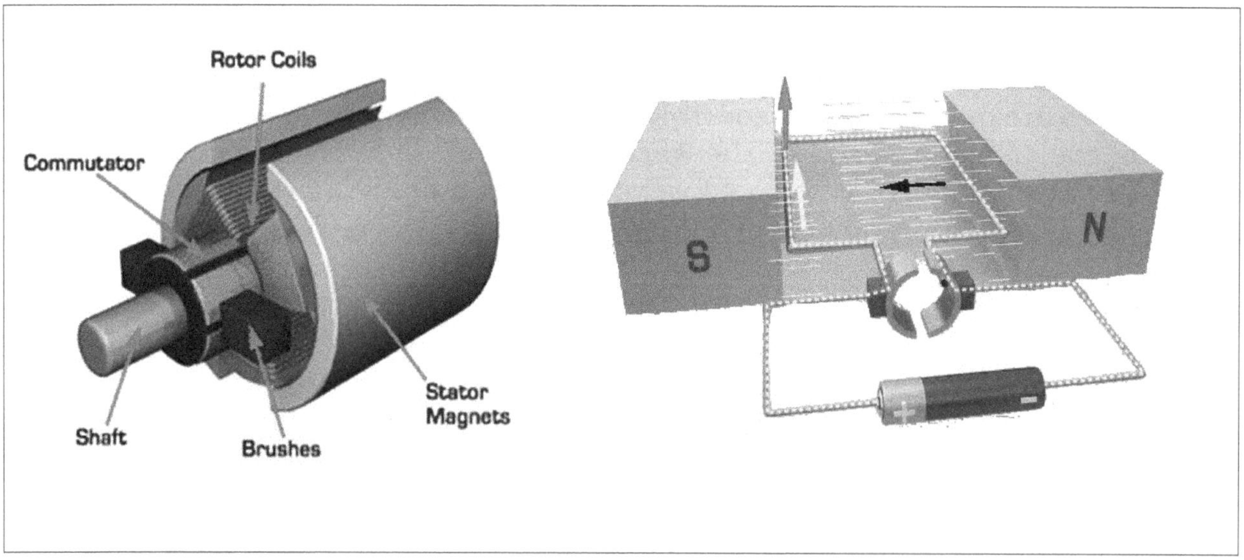

Figure-27. DC motor construction and working

There are mainly four types of brushed DC motors: permanent magnet dc motor, series dc motor, shunt dc motor, and compound dc motor. The applications of these motors are discussed next.

1. **Permanent Magnet DC Motor** : In this type of DC motor, permanent magnets of fixed capacity are placed for field magnets while rotor has coils for electromagnetism; refer to Figure-28. This type of DC motor provides great starting torque and has good speed regulation, but torque is limited so they are typically found on low horsepower applications.

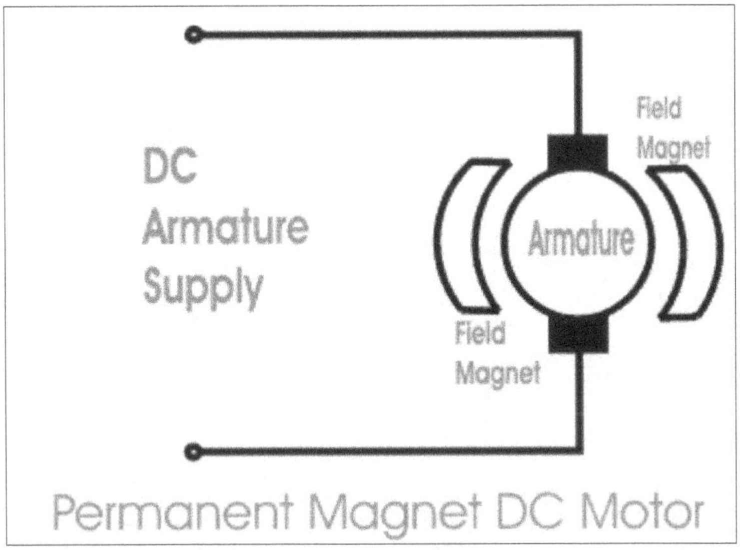

Figure-28. Permanent magnet dc motor

2. **Series DC Motor** : In this type of DC motor, the same current flows through rotor winding with is flowing through field winding; refer to Figure-29. The series DC motors create a large amount of starting torque, but cannot regulate speed and can even be damaged by running with no load. These limitations mean that they are not a good option for variable speed drive applications.

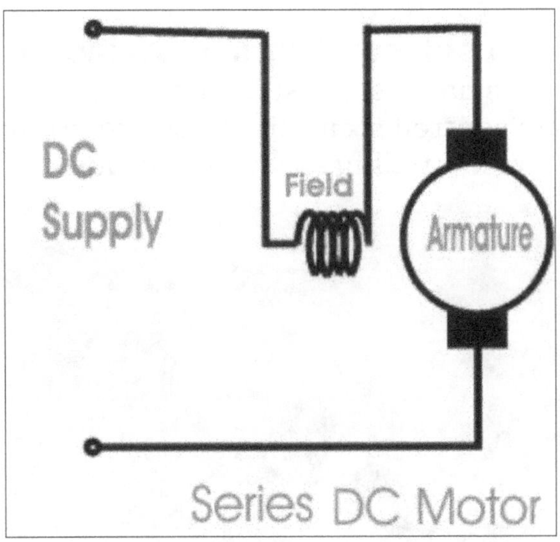

Figure-29. Series DC motor

3. **Shunt DC Motor** : In this type of DC motor, the field winding is connected in parallel to rotor winding so the voltage for both windings is same; refer to Figure-30. These motors offer great speed regulation due to the fact that the shunt field can be excited separately from the armature windings, which also offers simplified reversing controls.

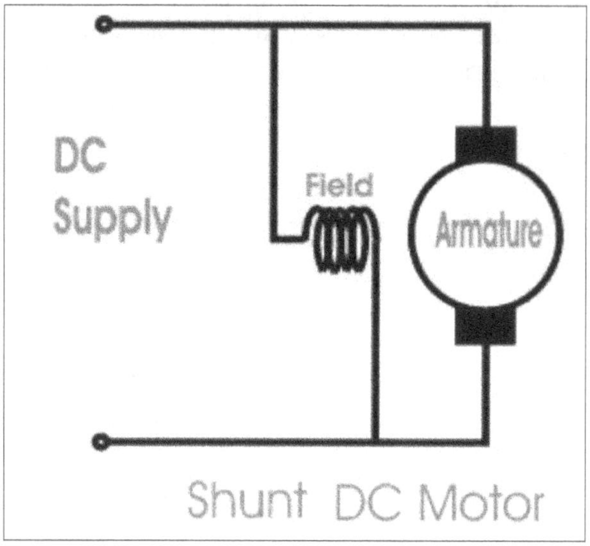

Figure-30. Shunt DC motor

4. **Compound DC Motor** : This type of DC motor uses combination of both shunt type and series type winding connections. Compound DC motors have good starting torque but may experience control problems in variable speed drive applications. There are two types of compound DC motors: cumulative and differential. When the shunt field flux assists the main field flux, produced by the main field connected in series to the armature winding then its called cumulative compound DC motor. In case of a differentially compounded self excited DC motor i.e. differential compound DC motor, the arrangement of shunt and series winding is such that the field flux produced by the shunt field winding diminishes the effect of flux by the main series field winding.

Brushless Electrical DC Motor

The brushless DC motors as name suggest do not have brushes to control rotation of motor. Instead a separate controller circuit is used to produce pulses of current for rotor or stator windings for controlling speed and torque of motor; refer to Figure-31.

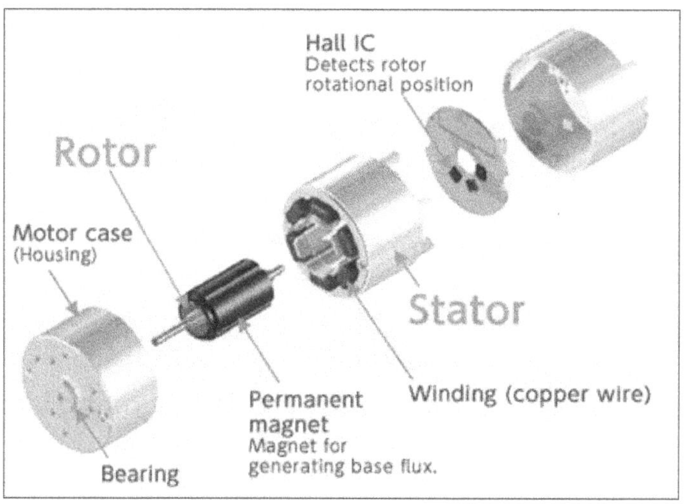

Figure-31. Brushless motor

Advantages of BLDC (BrushLess DC) motors :

- Brushless motors are more efficient as its velocity is determined by the frequency at which current is supplied, not the voltage.
- As brushes are absent, the mechanical energy loss due to friction is less which enhanced efficiency.
- BLDC motor can operate at high-speed under any condition.
- There is no sparking and much less noise during operation.
- More electromagnets could be used on the stator for more precise control.
- BLDC motors accelerate and decelerate easily as they are having low rotor inertia.
- It is high performance motor that provides large torque per cubic inch over a vast sped rang.
- BLDC motors do not have brushes which make it more reliable, high life expectancies, and maintenance free operation.
- There is no ionizing sparks from the commutator, and electromagnetic interference is also get reduced.
- Such motors cooled by conduction and no air flow are required for inside cooling.

Disadvantages of BLDC Motors:

- BLDC motor cost more than a brushed DC motor.
- The limited high power could be supplied to BLDC motor, otherwise, too much heat weakens the magnets and the insulation of winding may get damaged.

AC Motors

Alternating Current motors work on the principle of rotating magnetic field. The magnetic field of stator is made to rotate in circle and the electromagnetically charged rotor follows the rotation by attraction and repulsion due to polarity. There are mainly two types of AC motors : Induction Motor and Synchronous Motor.

Induction Motor

In an induction motor, there is no slip ring or DC excitation applied to rotor. The AC current in the stator induces a voltage across an air gap and into the rotor winding to produce rotor current and associated magnetic field. The stator and rotor magnetic fields then interact and cause the rotor to turn. The construction of induction motor is shown in Figure-32. There are two variations in induction motor: squirrel cage and wound rotor. In Squirrel cage induction motor, the rotor is constructed using a number of single bars short-circuited by end rings and arranged in a hamster-wheel or squirrel-cage configuration; refer to Figure-33. When voltage is applied to the stator winding, a rotating magnetic field is established. This rotating magnetic field causes a voltage to be induced in the rotor, which, because the rotor bars are essentially single-turn coils, causes currents to flow in the rotor bars. These rotor currents establish their own magnetic field, which interacts with the stator magnetic field to produce a torque

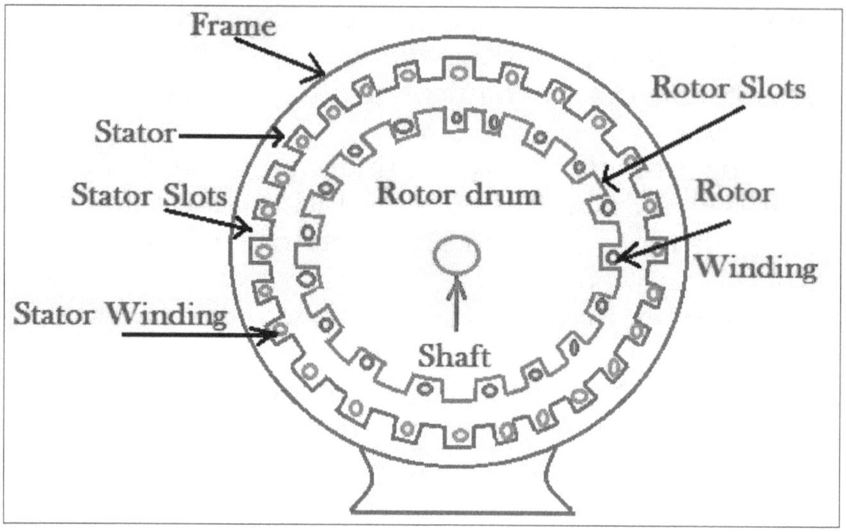

Figure-32. Construction of induction motor

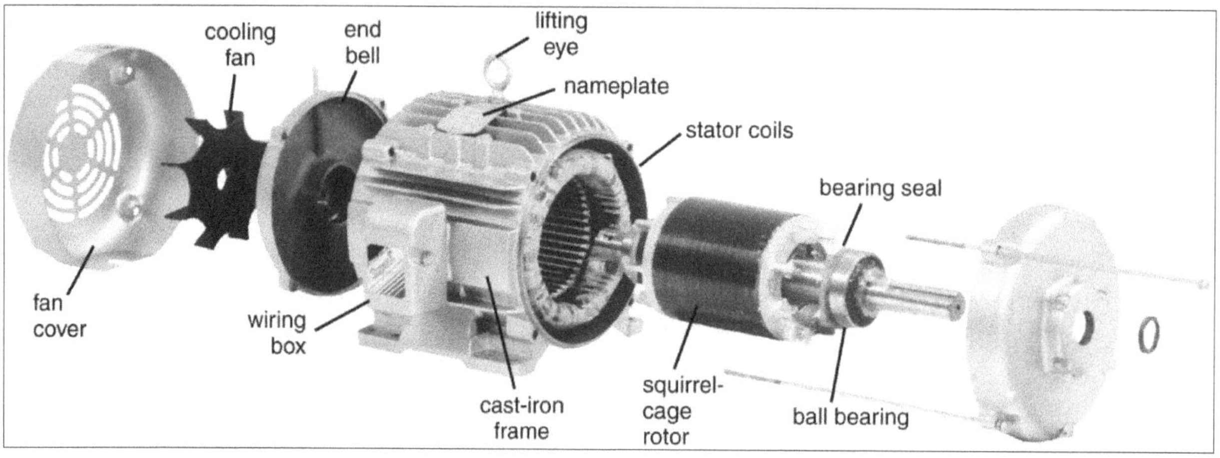

Figure-33. Squirrel cage induction motor

In wound rotor induction motor, also called slip-ring motor, wiring of rotor coils is terminated at slip rings; refer to Figure-34. A wound-rotor motor is used for constant-speed applications requiring a heavier starting torque than is obtainable with the squirrel-cage type. With a high-inertia load a standard cage induction motor may suffer rotor damage on starting due to the power dissipated by the rotor. With the wound rotor motor, the secondary resistors can be selected to provide the optimum torque curves and they can be sized to withstand the load energy without failure. Starting a high-inertia load with a standard cage motor would require between 400 and 550 percent start current for up to 60 seconds. Starting the same machine with a wound-rotor motor (slip-ring motor) would require around 200 percent current for around 20 seconds. For this reason, wound rotor types are frequently used instead of the squirrel-cage types in larger sizes.

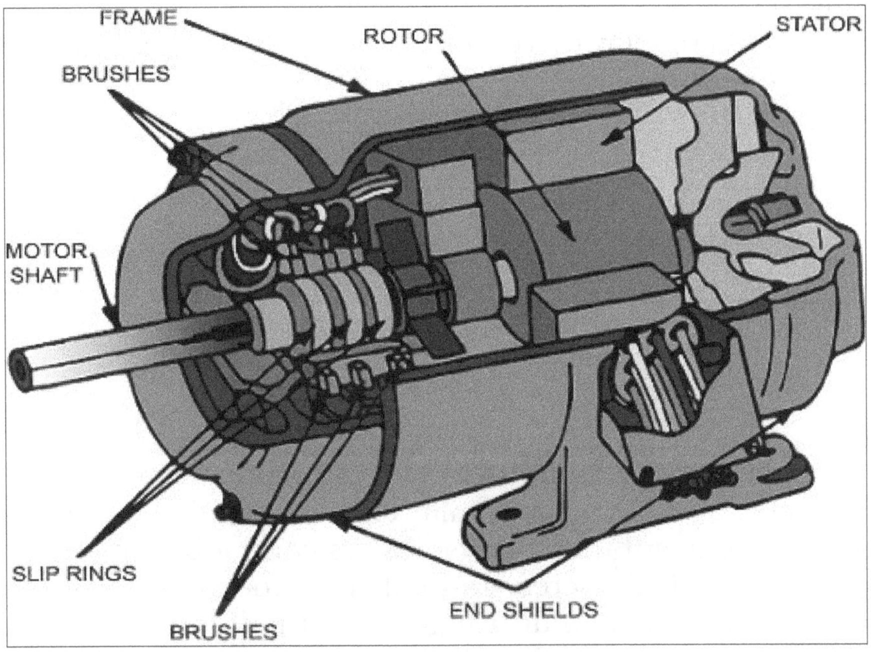

Figure-34. Wire wound induction motor

Synchronous Motor

In a synchronous ac motor, the stator magnetic field is produced by three phase power supply whereas the rotor magnetic field is produced by constant dc supply; refer to Figure-35. As you can see from the figure, the rotor behaves like permanent magnet with fixed polarity and follow the polarity force of stator. The rotor attracts towards the pole of the stator for the first half cycle of the supply and repulse for the second half cycle. Thus the rotor becomes pulsated only at one place. This type of motor are not self-starting and need an external method to bring rotor to synchronous speed.

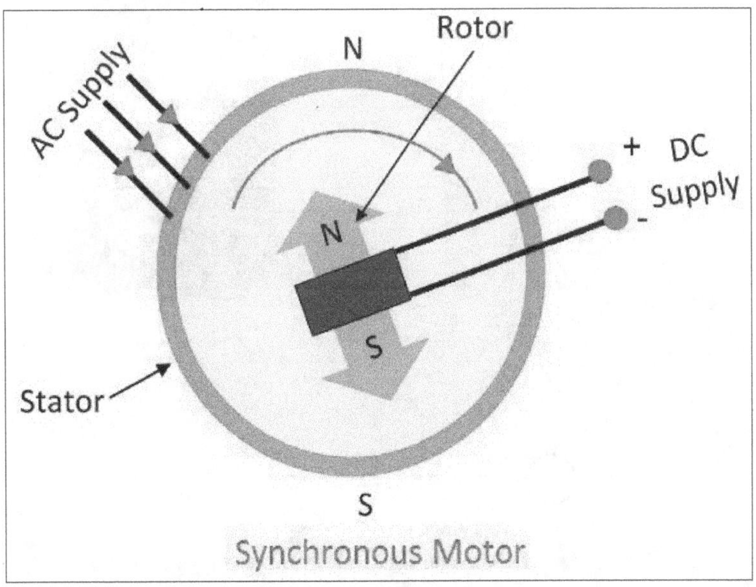

Figure-35. Synchronous motor

Selection of Motor

The electric motors are selected to satisfy the requirements of the machines on which they are to be installed. Following are some of the important factor the define the selection criteria of motor.

- Power Supply Type
- Mechanical power rating (in hp or Watt)
- Current Rating (in ampere)
- NEMA Code
- Efficiency
- Frame size
- Frequency
- Full-load RPM
- External requirements for running motor
- Temperature rating
- Duty cycle

CONTACTORS

The basic operation of a contactor is similar to that of a relay but contactor contacts can carry much more current than relays. Relays cannot be directly used in circuits where current exceeds 20 amperes. In such conditions contactors can be used. Contactors are available in a wide range of ratings and forms. Contactors are available up to the ampere rating of 12500A. Contactors cannot provide short circuit protection but can only make or break contacts when excited. When relatively lower voltage current is passed through coils of contactors, they cause high voltage supply contacts to close and when there is no current in the coils then contacts open; refer to Figure-36. There is no short-circuit protection in contactors so a separate circuit breaker or overload relay should be used with contactor in circuits.

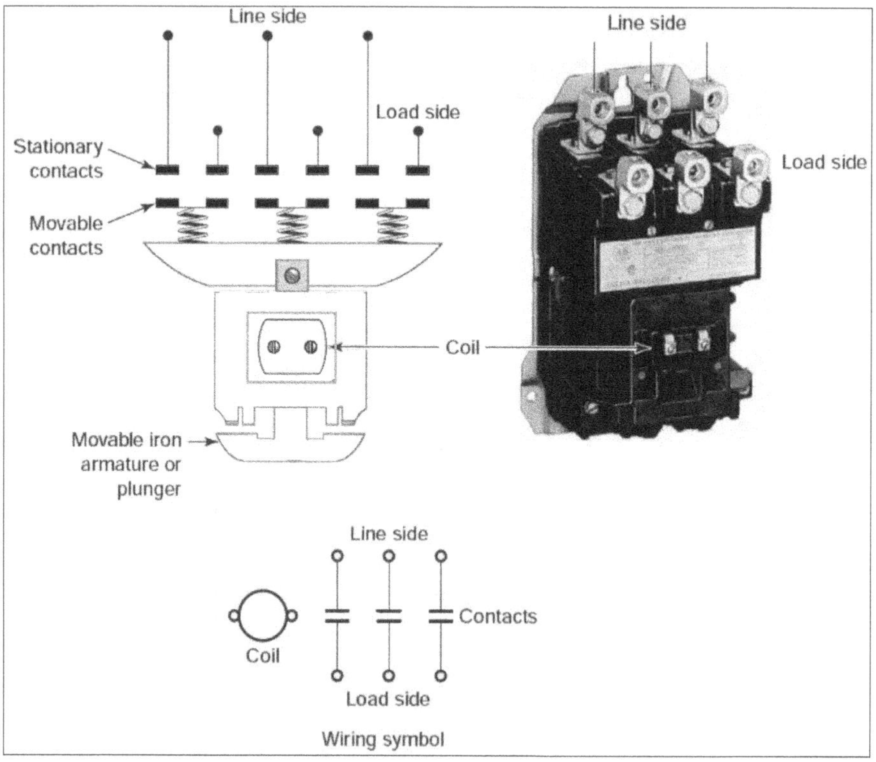

Figure-36. Contactor

Contactors are categorized based on the type of load (IEC utilisation categories - 60947) and current and power rating (NEMA size). Few important IEC utilisation

categories are below:
AC-1: Non-inductive or slightly inductive and resistive heating type of loads
AC-2: Starting of slip ring induction motor
AC-3: Starting and switching off Squirrel-cage motors during running time
AC-15: Control of AC electromagnets.
AC-56b:- Switching of capacitor banks
DC–1: Non-inductive or slightly inductive and resistive heating type of loads
DC-2: Starting, inching and dynamic breaking of DC shunt motors
DC-3: Starting, inching and dynamic breaking of DC series motors
DC-13: Control of DC electromagnets

NEMA size:
NEMA size is based on the maximum continuous current and horse power rating of the induction motor controlled by the contactor. In NEMA standard contactors are designated as size 00,0,1,2,3,4,5,6,7,8,9. These numbers represent current ratings; refer to Figure-37.

60 Hz AC contactor NEMA ratings 600 volts max	
NEMA size	Continuous amps
00	9
0	18
1	27
2	45
3	90
4	135
5	270
6	540
7	810
8	1215
9	2250

DC contactor NEMA ratings 600 volts max	
NEMA size	Continuous amps
1	25
2	50
3	100
4	150
5	300
6	600
7	900
8	1350
9	2500

Figure-37. NEMA ratings for contactors

Solid State Contactor

In solid state contactor; refer to Figure-38, electronic switching is used to open or close contacts of main line. In this type of contactor, there is no magnetic coil. Instead, semiconductors are used to perform contactor work. The most common high-power switching semiconductor used in solid-state contactors is the silicon controlled rectifier (SCR). An SCR is a three-terminal semiconductor device (anode, cathode, and gate) that acts like the power contact of a magnetic contactor. A gate signal, instead of an electromagnetic coil, is used to turn the device on, allowing current to pass from cathode to anode. Since there is not moving part so they are faster in switching as compared to magnetic contactors. The only drawback with solid state contactors is that when they fail, they fail causing short circuit instead of open circuit. So, they can become dangerous if not made in good quality.

Figure-38. Solid state contactor

MOTOR STARTERS

Motor starter is a combination of contactors, disconnectors, and overload protection circuits. When starting a motor, the circuit has to be upto 20 times more current than regular current required to run motor at normal speed. So, the system should have an overload circuit that counts for this phase. Once the motor is running on normal speed then it should protect the motor in high current or short-circuit mishaps. Figure-39 shows a typical motor starter construction.

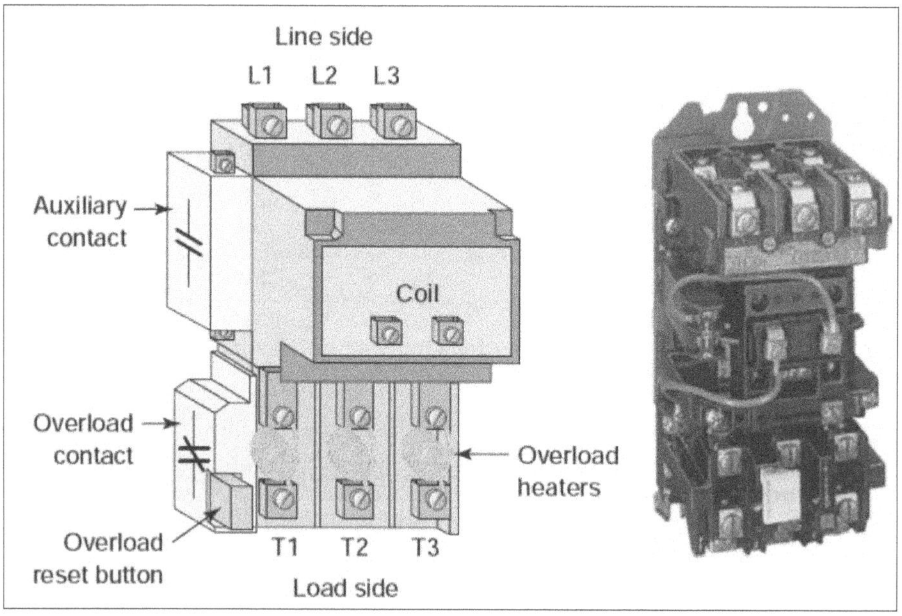

Figure-39. Motor starter

There are mainly four types of motor starters: Across-The-Line, the Reversing Starter, the Multispeed Starter, and the Reduced Voltage Starter.

Across-the-line or Full Voltage Non-Reversing (FVNR) is the most commonly used general purpose starter. This starter connects the incoming power directly to the motor. It can be used in any application where the motor runs in only one direction, at only one speed, and starting the motor directly across the line does not create any "dips" in the power supply.

The **Reversing Starter or Full Voltage Reversing (FVR)** reverses a motor by reversing any two leads to the motor. This is accomplished with two contactors and one overload relay. One contactor is for the forward direction and the other is for reverse. It has both mechanically and electrically interlocked sets of contactors.

The **Multispeed Starter** is designed to be operated at constant frequency and voltage. There are two ways to change the speed of an AC motor: Vary the frequency of the current applied to the motor or use a motor with windings that may be reconnected to form different number of poles. The multispeed starter uses the latter option to change speed.

Reduced Voltage Starter (RVS) is used in applications that typically involve large horsepower motors. The two main reasons to use a reduced voltage starter are to reduce the inrush current and to limit the torque output and mechanical stress on the load. Power companies often won't allow this sudden rise in power demand. The reduced voltage starter addresses this inrush problem by allowing the motor to get up to speed in smaller steps, drawing smaller increments of current. This starter is not a speed controller. It reduces the shock transmitted to the load only upon start-up.

FUSES

Fuses are the most simple devices used to protect circuit from current overload. Fuses are sacrificial devices which get blown when there is too much current through them. In a fuse, wire of specific size is installed between two terminals. This wire has fixed rating in amperes upto which it can pass the current. When current of higher level is passed, it gets melt and disconnects the circuit. You can find fuses in almost every electrical appliance. Fuses are divided into two categories, Low Voltage Fuse (LV Fuse) and High Voltage Fuse (HV Fuse). The low voltage fuses are good upto 230 V and beyond that you should use HV fuses. The low voltage fuses are further divided into four categories rewire-able, cartridge, striker and switch fuses; refer to Figure-40.

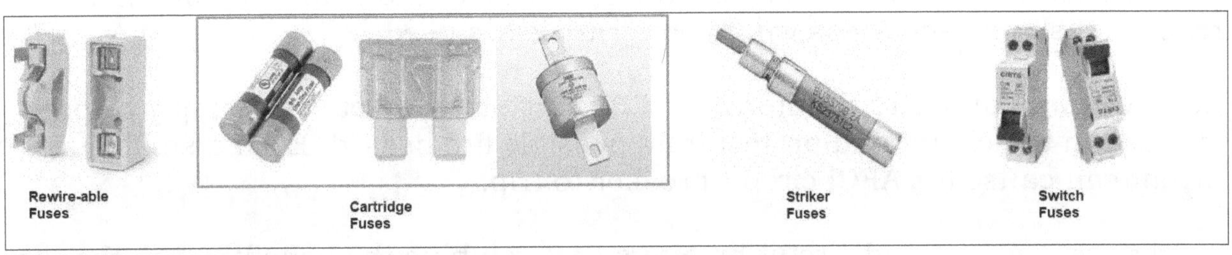

Figure-40. Low Voltage fuses

The high voltage fuses are used to protect transformers and power supplies. There are mainly three types of high voltage fuses; Cartridge Type HRC Fuse, Liquid Type HRC Fuse, and Expulsion Type HRC Fuse; refer to Figure-41.

Selection of a fuse depends on the normal current rating of connected components that you want to protect.

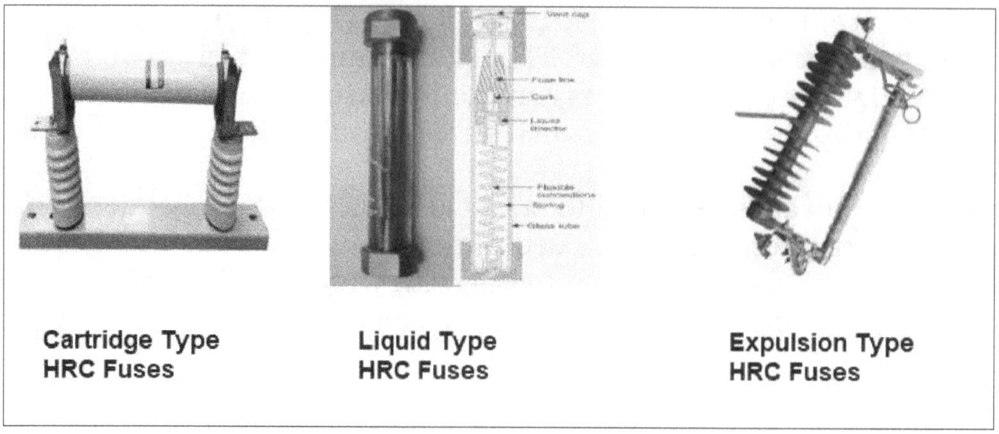

Figure-41. High voltage Fuses

CIRCUIT BREAKERS

Circuit Breaker as the name suggests is used to break the circuit when current is above the specified limits. The functioning of circuit breaker is similar to fuse as it also protects circuit from high current. The difference is that a circuit breaker changes from closed to open position when current is higher than its limits (It does not become sacrificial lamb). When fault has been identified and cured then you use the same circuit breaker again. The circuit breakers are available in three categories: Standard, GFCI and AFCI.

The standard circuit breakers are available in two configurations single pole and double-pole. The single pole circuit breaker can handle 120V supply with 15 amp to 30 amp current where as double pole circuit breaker can handle 230V supply with 15 to 200 amps of current. The standard circuit breakers are good for normal appliances.

The GFCI (Ground Fault Circuit Interrupter) Circuit Breakers are used when a line to ground fault can occur causing life threatening situations to consumer like in wet areas, workshops, and areas where power tools are used.

The AFCI (Arc Fault Circuit Interrupter) Circuit Breakers cut the supply before arc is formed in the circuit and protect from possible fire hazard. Even a small surge of current can cause the AFCI circuit breaker to trip.

Now a days, you can find circuit breakers that combine the benefits of all the three types in one unit.

Depending the applications of circuit breakers, they can be classified as shown in Figure-42.

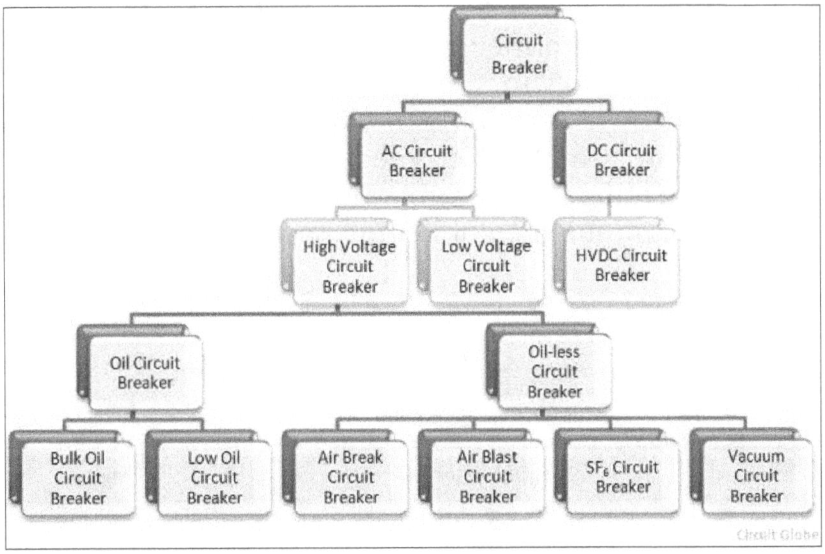

Figure-42. Circuit Breaker Types based on applications

DISCONNECTORS

Disconnectors are used to completely de-energize the circuits for maintenance work. A disconnector should also ensure that no accidental connection is formed when circuit is open. Disconnectors are generally used in high voltage transmission lines carrying current at 32 A to 2000 A with voltage ranging from 36 kV to 100kV. You can find disconnectors for low voltage circuits also but contactors and circuit breakers are good alternatives in such cases. The disconnectors used in low voltage applications are called switch disconnectors or safety switches; refer to Figure-43.

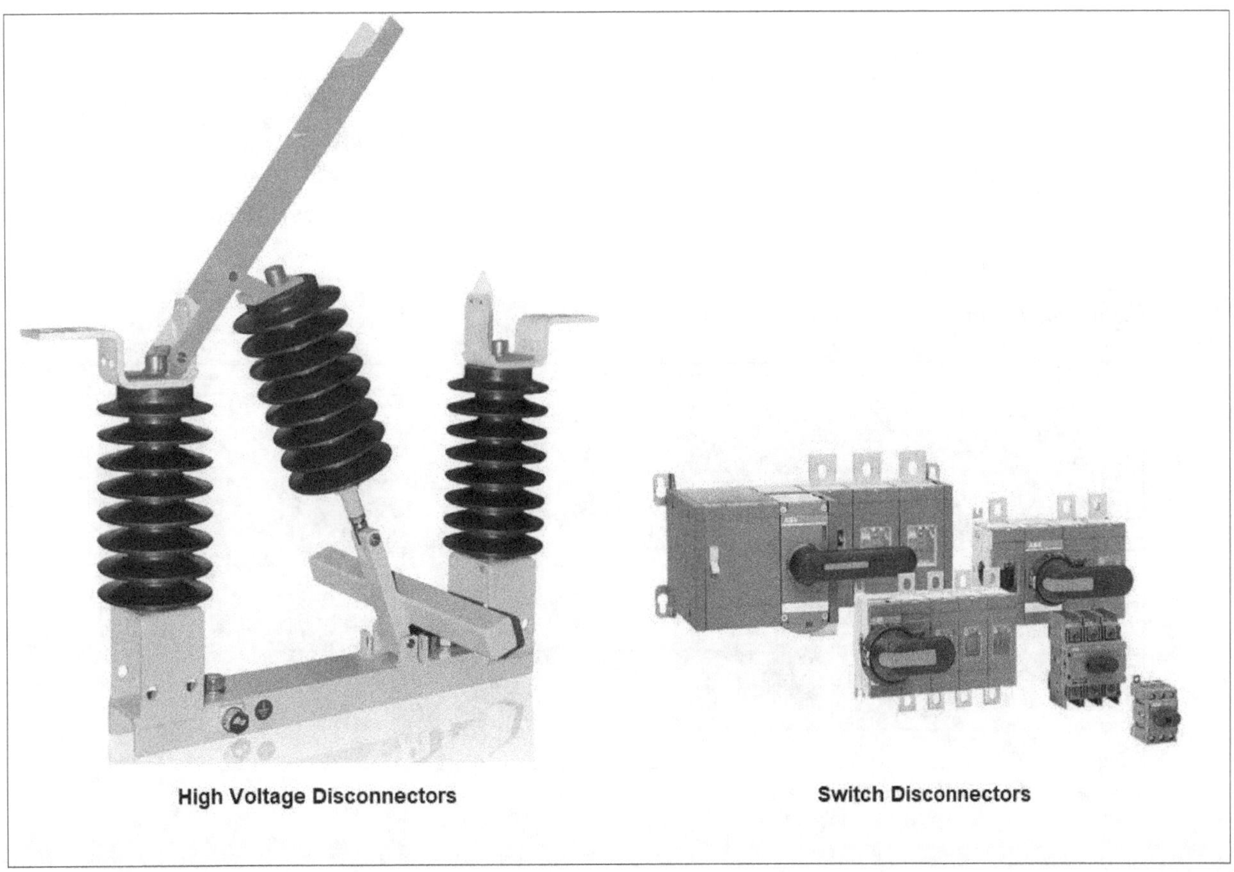

Figure-43. Disconnectors

TERMINALS AND CONNECTORS

Electric Terminal are used to connects wires and systems with low resistance. In every well designed system, the wires should end with terminals for connections. A connector has fixed number of terminals in which wires are fixed. Most of the time, you will find connectors in pair: Male connector and Female connector which fit in each other to form connection. There are various types of terminals are shown in Figure-44. A 9 pin male and female connector pair is shown in Figure-45.

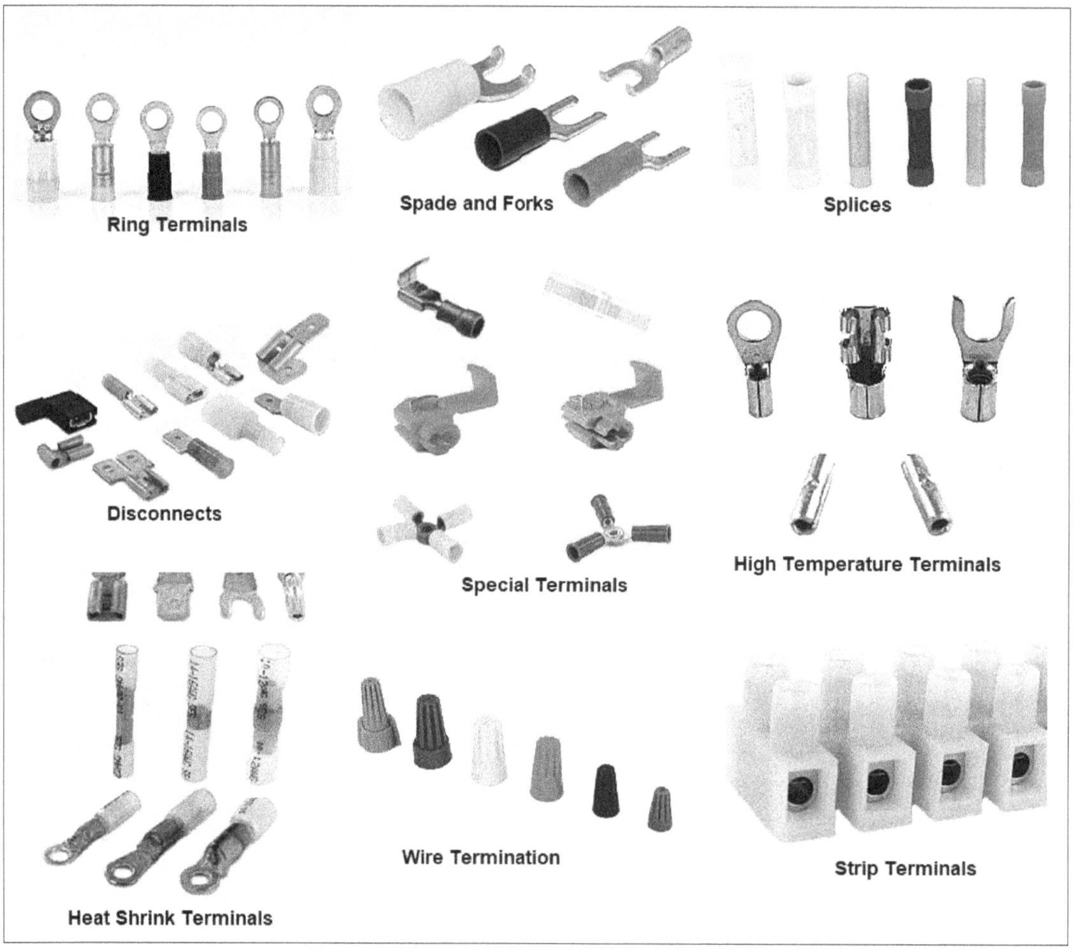

Figure-44. Terminals

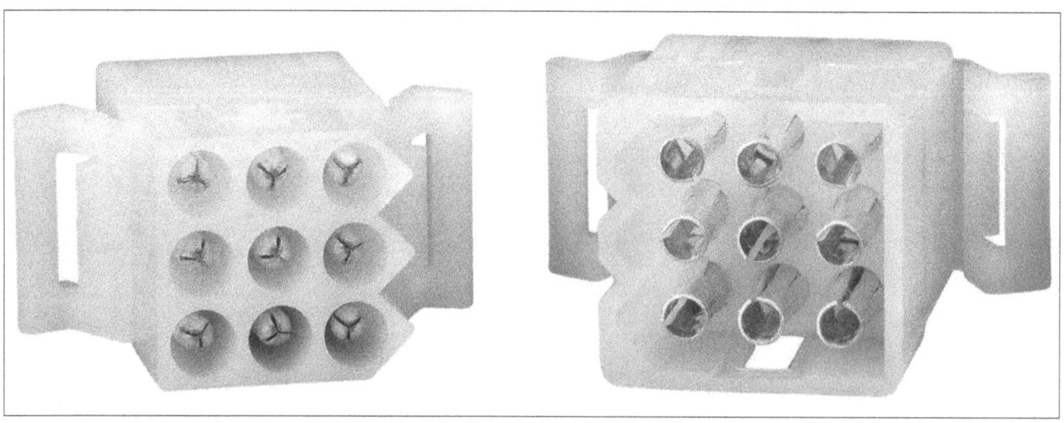

Figure-45. Electrical connectors

Review Questions

Q1. What is the primary function of push buttons in electrical control circuits?
A. To measure voltage in circuits
B. To start and stop motors or machines
C. To detect the position of actuators
D. To convert AC to DC

Q2. What type of contact material is typically used in push buttons for switching at logic levels?
A. Copper
B. Aluminum
C. Gold
D. Iron

Q3. What does a glowing pilot light indicate in an electrical control system?
A. Circuit is faulty
B. Power is disconnected
C. Circuit is connected and power is running
D. Motor has failed

Q4. What is the purpose of a legend plate?
A. To disconnect power
B. To amplify signals
C. To identify the purpose of push buttons and pilot lights
D. To provide additional contacts

Q5. What mechanism is commonly used for the operation of selector switches?
A. Electromagnetic coils
B. Bimetallic strips
C. Rotating cam
D. Push action

Q6. Which type of toggle switch is also known as a three-way switch?
A. SPST
B. DPST
C. SPDT
D. DPDT

Q7. What is one key selection criterion for drum switches?
A. Number of joystick directions
B. Color of LED
C. Number of poles
D. Size of push button

Q8. What causes a limit switch to activate?
A. Signal from a microcontroller
B. Change in power supply
C. Reaching a threshold of pressure, temperature, or position
D. Actuation by joystick

Q9. What is the main feature of a micro-limit switch?
A. Gold-plated contacts
B. Snap action switching
C. Light indicators
D. Pressure detection

Q10. Which type of limit switch is used to cut off power at specific rotational positions?
A. Micro-limit switch
B. Proximity switch
C. Rotating cam limit switch
D. Drum switch

Q11. Which mechanism is most common in a temperature switch?
A. Mercury sensor
B. Rotating shaft
C. Bimetallic strip
D. Ultrasonic sensor

Q12. What is typically used in a liquid-filled temperature switch to detect temperature change?
A. Gold wire
B. Gears and pulleys
C. Brass bulb with chemical fluid
D. Light sensors

Q13. In an air compressor, which component stops the motor after reaching a pressure threshold?
A. Toggle switch
B. Pressure switch
C. Flow switch
D. Limit switch

Q14. What is another name for a liquid level switch?
A. Drum switch
B. Joystick
C. Float switch
D. Selector switch

Q15. What is the function of a flow switch?
A. Measure voltage
B. Toggle between circuits
C. Monitor and control fluid flow
D. Activate a joystick

Q16. What is the use of joystick switches in electrical systems?
A. Toggle circuit modes
B. Control motion in multiple directions
C. Limit high pressure
D. Illuminate panels

Q17. What does the hysteresis zone in proximity switches refer to?
A. The area where voltage is zero
B. Distance between two limit switches
C. The range between operating and release points
D. The area with no electrical signal

Q18. What are typical power supply voltages for proximity switches?
A. 3V or 5V
B. 9V or 12V
C. 24V DC or 120V AC
D. 220V or 380V

Q19. What is the main purpose of sensors in control systems?
A. To add mechanical load
B. To transmit digital or analog signals based on detected parameters
C. To display system errors
D. To rotate shafts

Q20. Which actuator uses electromagnetic force to move an armature?
A. Proximity switch
B. Solenoid
C. Pressure switch
D. Joystick

Q21. What is a key difference between electromechanical and solid-state relays?
A. SSRs use magnets while EMRs use coils
B. SSRs contain no moving parts
C. EMRs are always battery-powered
D. SSRs cannot handle high voltage

Q22. Why are latching relays useful?
A. They can operate without electricity
B. They reset automatically
C. They hold their position without continuous power
D. They provide temperature control

Q23. What type of relay introduces a time delay before switching?
A. Solid state relay
B. Time relay
C. Pressure relay
D. Latching relay

Q24. What type of solenoid is typically faster in response?
A. Hydraulic solenoid
B. AC solenoid
C. DC solenoid
D. Thermal solenoid

Q25. What is the main application of solenoid valves?
A. Turning on lights
B. Switching between power supplies
C. Controlling hazardous fluid flow
D. Heating fluids

Q26. What is the basic working principle of all motors?
A. Electromagnetic induction using a capacitor
B. Magnetic repulsion between two coils
C. Passing electricity through a coil to generate magnetic interaction
D. Rotating the rotor manually to generate current

Q27. Which type of DC motor uses permanent magnets for field magnets?
A. Shunt DC Motor
B. Series DC Motor
C. Compound DC Motor
D. Permanent Magnet DC Motor

Q28. Why is a Series DC motor not suitable for variable speed drive applications?
A. It requires high voltage
B. It generates high heat
C. It has poor speed regulation and can be damaged without load
D. It consumes too much current

Q29. In a Shunt DC Motor, the field winding is:
A. In series with rotor winding
B. In parallel with rotor winding
C. Externally disconnected from rotor winding
D. Magnetized permanently

Q30. What distinguishes a cumulative compound DC motor from a differential compound DC motor?
A. Speed control type
B. Load capacity
C. Direction of shunt field flux relative to main field
D. Efficiency rating

Q31. Which of the following is NOT an advantage of brushless DC motors?
A. High-speed operation
B. No sparking and reduced noise
C. High rotor inertia
D. Maintenance-free operation

Q32. What is a key limitation of BLDC motors?
A. Require too much voltage
B. Cannot operate at high speeds
C. Cost more and have limited power handling
D. Difficult to install

Q33. Which component induces voltage in the rotor of an induction motor?
A. Slip ring
B. Wound rotor
C. DC excitation
D. Stator

Q34. What is a key benefit of wound rotor induction motors over squirrel cage motors?
A. Lower cost
B. Better noise control
C. Higher starting torque and better handling of high-inertia loads
D. Simpler construction

Q35. Which type of motor is not self-starting and requires external help to start?
A. Squirrel cage motor
B. Series DC motor
C. Synchronous motor
D. Brushless DC motor

Q36. What factor is NOT considered in the selection of a motor?
A. Frequency
B. Frame color
C. Power supply type
D. Duty cycle

Q37. What distinguishes a contactor from a relay?
A. Lower current handling
B. Higher voltage range
C. Smaller size
D. Can handle much more current

Q38. Which IEC category is used for controlling AC electromagnets?
A. AC-3
B. AC-1
C. AC-15
D. AC-2

Q39. What is a disadvantage of solid state contactors?
A. Slow switching
B. High maintenance
C. May fail as short circuit
D. Use of mechanical parts

Q40. What is the purpose of a motor starter?
A. Reduce voltage supply permanently
B. Provide overload and short-circuit protection at startup
C. Speed control of motor
D. Control circuit voltage only

Q41. What does a Full Voltage Reversing (FVR) starter do?
A. Increases speed
B. Starts motor in both directions using two contactors
C. Reduces power supply
D. Eliminates inrush current

Q42. How does a multispeed starter change the motor speed?
A. By varying frequency
B. By changing voltage
C. By using variable resistors
D. By reconnecting windings to change the number of poles

Q43. What is the function of a fuse?
A. Converts AC to DC
B. Measures circuit voltage
C. Protects circuit by blowing during overcurrent
D. Increases current supply

Q44. What is a key advantage of circuit breakers over fuses?
A. Lower price
B. Can be reused after fault is cleared
C. Higher resistance
D. Built-in light indicators

Q45. What type of circuit breaker is used in areas with high arc risk?
A. GFCI
B. Standard
C. Double pole
D. AFCI

Q46. What is the main purpose of disconnectors?
A. Amplify signal
B. Change voltage supply
C. Completely de-energize circuits for maintenance
D. Provide overload protection

Q47. What is true about terminals and connectors?
A. They reduce current flow
B. Connect wires without forming circuits
C. Ensure low resistance connections
D. Replace fuses in modern systems

Index

Ethics of an Engineer

- Engineers shall hold paramount the safety, health and welfare of the public and shall strive to comply with the principles of sustainable development in the performance of their professional duties.
- Engineers shall perform services only in areas of their competence.
- Engineers shall issue public statements only in an objective and truthful manner.
- Engineers shall act in professional manners for each employer or client as faithful agents or trustees, and shall avoid conflicts of interest.
- Engineers shall build their professional reputation on the merit of their services and shall not compete unfairly with others.
- Engineers shall act in such a manner as to uphold and enhance the honor, integrity, and dignity of the engineering profession and shall act with zero-tolerance for bribery, fraud, and corruption.
- Engineers shall continue their professional development throughout their careers, and shall provide opportunities for the professional development of those engineers under their supervision.

Page left blank intentionally

www.ingramcontent.com/pod-product-compliance
Lightning Source LLC
LaVergne TN
LVHW082014200926
848481LV00010B/2860